Modern aspects of inorganic chemistry

Modern aspects of inorganic chemistry

H. J. Emeléus and A. G. Sharpe

A Halsted Press Book

John Wiley & Sons
New York

Published in the U.S.A. by Halsted Press,
a Division of John Wiley & Sons, Inc., New York.
Printed in Great Britain

Library of Congress Cataloging in Publication Data
Emeléus, Harry Julius.
Modern aspects of inorganic chemistry.
"A Halsted Press book."
Editions for 1938–1960 by H. J. Emeléus and J. S. Anderson.
1. Chemistry, Inorganic. I. Sharpe, A. G., joint author. II. Title.
QD151.2.E43 1973 546 73-2577
ISBN 0 470 23902 6

Contents

Preface to the fourth edition

In 1938 one of us (H.J.E.) published with J. S. Anderson an earlier version of this book.

Since then, there have been important advances in many branches of inorganic chemistry. In this new book we have retained the same general approach, but have brought the treatment up to date by dealing with new subjects which have emerged. The text has been entirely rewritten, an undue increase in length being avoided by omitting or curtailing some sections of the earlier work, especially those relating to theoretical chemistry, which are now a part of normal undergraduate courses. In its new form the book is suitable for use by both undergraduate and post-graduate students. We hope that it will also serve to give the non-specialist a readable account of some of the recent developments in the subject.

We are indebted to Professor J. S. Anderson, who was unable to take part in preparing this new edition, for allowing us to do so. We also thank Dr G. M. Sheldrick for reading critically a number of the chapters. The use of S.I. units in the text has been largely restricted to the expression of energies in both kJ and kcal: other recommended symbols are not yet in general use, and their adoption in this book could, in our view, lead to confusion.

H. J. Emeléus
A. G. Sharpe

Acknowledgments

The authors gratefully acknowledge permission to reproduce the following diagrams, which first appeared in the books listed. Figs 3.3, 9.3, 9.4, and 9.7: A. F. Wells, *Structural Inorganic Chemistry*, The Clarendon Press, 3rd edition, 1962; Fig. 4.2: N. N. Greenwood, *Ionic Crystals, Lattice Defects and Nonstoichiometry*, Butterworths, 1968; Fig. 4.1: A. K. Galwey, *The Chemistry of Solids*, Chapman and Hall, 1967; Figs 8.17–8.22: R. N. Grimes, *Carboranes*, Academic Press, 1971; Figs 10.6, 10.8, and 10.29: J. R. Van Wazer, *Phosphorus and its Compounds*, Vol. 1, Wiley, 1958; Figs 10.5, 10.7, 10.16 and 14.3: F. A. Cotton and G. Wilkinson, *Advanced Inorganic Chemistry*, Wiley, 2nd edition, 1966; Figs 10.17 and 10.18: E. L. Muetterties, *The Chemistry of Boron and its Compounds*, Wiley, 1967; Figs 9.8, 9.9, 9.24 and 10.36: D. C. Corbridge, in E. J. Griffith and M. Grayson, eds, *Topics in Phosphorus Chemistry*, Vol. 3, Wiley, 1966; Figs 10.3, 10.4 and 10.9: M. Becke-Goehring and H. Hoffman, *Komplex-chemie*, Springer, 1970; Figs 8.25, 9.6, 9.10, 13.6 and 13.7: H. J. Emeléus and A. G. Sharpe, *Advances in Inorganic Chemistry*, Vols 1, 2, 7 and 10, Academic Press, 1959–67; Figs 15.1 and 15.2: S. F. A. Kettle, *Coordination Compounds*, Nelson, 1969.

1 Isotopes and their applications in chemistry

1.1 Introduction[1, 2, 3]

Modern physics recognizes the existence of a large number of fundamental particles. Only three of these, however, are of major importance in chemistry under ordinary conditions: the electron, mass 0·00055 (on the scale on which that of ^{12}C is 12·00000) and charge $-1{\cdot}6021 \times 10^{-19}$ C, or $-4{\cdot}803 \times 10^{-10}$ e.s.u.; the proton, mass 1·00728 and charge opposite to that of the electron; and the neutron, mass 1·00867 and charge zero. In addition to the electron, four other particles are emitted during radioactive decay: the α-particle or helium nucleus, mass 4·0015 and charge +2 on the electronic scale; the positron, of mass equal to and charge opposite to that of the electron; and the neutrino and antineutrino, of zero or near zero resting mass and zero charge, which are required to possess angular momentum in order that this quantity, as well as energy, may be conserved in nuclear reactions.

Nearly all the mass of an atom is concentrated in the nucleus, which, for a species of mass number (the mass to the nearest integer) A and atomic number Z, contains Z protons and $A - Z$ neutrons; electrical neutrality of the whole atom is maintained by the presence of Z extranuclear electrons. The number of positive charges on the nucleus, Z, is equal to the ordinal number of the element in the periodic classification (i.e. its atomic number), and defines the chemical identity of an atom. Different atoms of the same element may, however, contain different numbers of neutrons; such atoms, differing in mass but not in nuclear charge, are called isotopes, and their separation and applications form the main subjects of this chapter.

Many nuclei persist unchanged over an indefinite period of time; others undergo natural radioactive decay, the chemical consequences of which may be summarized as follows. Loss of an α-particle lowers the atomic number by two and the mass number by four; loss of a β-particle (an electron of nuclear origin) raises the atomic number by one and leaves the mass number unchanged; and loss of γ-radiation (X-rays) leaves both the atomic number and the mass number unchanged. The rate constant of the first-order disintegration of an initial number N_0 of nuclei, of which N remain after a time t, is the decay constant λ, defined by the equations

$$\frac{dN}{dt} = -\lambda N$$

and

$$N = N_0 \, e^{-\lambda t}$$

Instead of quoting λ, it is common practice to refer to the half-life $t_{\frac{1}{2}}$, the time taken to halve the number of radioactive nuclei. This is related to λ by the equation

$$t_{\frac{1}{2}} = \frac{\ln 2}{\lambda} = \frac{0{\cdot}693}{\lambda}$$

Alternatively, reference may be made to the specific activity, the number of disintegrations per second per gram, normally expressed in curies per gram (1 curie $= 3{\cdot}7 \times 10^{10}$ disintegrations per second). The curie, it may be noted, is the unit in which amounts of radioactive isotopes are commonly measured; the number of atoms in one curie of a particular isotope is, of course, inversely related to its half-life. As an illustration we may consider the early stages in the decay of uranium-238, in which the nature of the emissions and the half-lives of the nuclei are as follows:

$$^{238}_{92}U \xrightarrow[4{\cdot}5\times 10^{9}\text{ years}]{\alpha} {}^{234}_{90}Th \xrightarrow[24\text{ days}]{\beta} {}^{234}_{91}Pa \xrightarrow[1{\cdot}1\text{ min}]{\beta}$$

$$^{234}_{92}U \xrightarrow[2\times 10^{5}\text{ years}]{\alpha} {}^{230}_{90}Th \xrightarrow[8\times 10^{4}\text{ years}]{\alpha} {}^{226}_{88}Ra$$

To a reader familiar with the periodic table it is unnecessary to cite both the atomic number and the symbol of an atom, and in this book the atomic number will often be omitted. The very wide variation in half-lives will be noted; thus we normally think of uranium as an ordinary element which is slightly radioactive, whilst ^{234}Pa is so short-lived that it is extremely difficult to use this isotope for the study of the chemistry of protactinium.

There is no fundamental difference between natural radioactivity and nuclear changes resulting from bombardment of nuclei with charged particles or neutrons. Such changes conform to the principle of conservation of mass number and atomic number, and are expressed as, for example,

$${}^{14}_{7}\mathrm{N} + {}^{4}_{2}\mathrm{He} \rightarrow {}^{17}_{8}\mathrm{O} + {}^{1}_{1}\mathrm{H} \quad \text{or} \quad {}^{14}\mathrm{N}(\alpha,p)\,{}^{17}\mathrm{O}$$

$${}^{27}_{13}\mathrm{Al} + {}^{1}_{0}n \rightarrow {}^{28}_{13}\mathrm{Al} + \gamma \quad \text{or} \quad {}^{27}\mathrm{Al}(n,\gamma)\,{}^{28}\mathrm{Al}$$

Isotopic masses are not exactly integers, and in nuclear transformations mass is destroyed. From Einstein's equation describing the conversion of matter into energy,

$$E = mc^2$$

(where c is the velocity of light, 3×10^8 m sec^{-1}), it is seen that one gram of matter destroyed produces 9×10^{10} kJ ($2{\cdot}15 \times 10^{10}$ kcal) of energy. The energies of covalent bonds lie in the range 80–900 kJ (20–220 kcal) mole^{-1}, so it is evident that in a nuclear reaction drastic changes in the chemical environment of an atom may be expected.

In discussing nuclear changes it is convenient to express changes in atomic mass in millions of electron volts (MeV) per atom. One atomic mass unit (amu) is $1/N$ (i.e. $1/(6{\cdot}02 \times 10^{23})$ grams), and one MeV is equivalent to $1{\cdot}6 \times 10^{-13}$ J; thus,

$$1 \text{ amu} = 931 \text{ MeV}$$

The binding energy E of a nucleus may be calculated from the decrease in mass that would occur if the appropriate numbers of neutrons and hydrogen atoms were brought together to form the atom concerned:

$$\begin{aligned} E &= 931\,\Delta M \\ &= 931[Z \times \text{mass of H atom} + (A - Z) \times \text{mass of neutron} - M] \end{aligned}$$

where M is the atomic mass. For ^{12}C, for example,

$$\begin{aligned} \Delta M &= (6 \times 1{\cdot}00783) + (6 \times 1{\cdot}00867) - 12{\cdot}00000 \\ &= 0{\cdot}0990 \text{ amu} \\ E &= 92{\cdot}3 \text{ MeV} \end{aligned}$$

The binding energy per nuclear particle is this value divided by the number of protons and neutrons (i.e. $92{\cdot}3/12 = 7{\cdot}69$ MeV). This quantity varies with mass number: it is (by definition) zero for ^{1}H, and reaches a maximum at about ^{56}Fe, thereafter slowly declining. Species of mass number about 56 therefore represent the thermodynamically most stable form of nuclear matter.

It is the decrease in binding energy per nuclear particle as atomic mass increases that results in the liberation of very large amounts of energy when atoms of the heaviest elements undergo fission. Consider as a typical example the following reaction of $^{235}_{92}U$ under the action of slow neutrons (i.e. neutrons which have been slowed down to thermal velocities by passage through media such as graphite or paraffin wax in which the probability of a nuclear reaction is particularly small): only the initial products are shown.

$$^{235}_{92}U + ^{1}_{0}n \rightarrow ^{92}_{36}Kr + ^{141}_{56}Ba + 3^{1}_{0}n$$

The binding energy per nuclear particle is about 8·4 MeV in the mass number range 90–140, but only 7·5 MeV at mass number 235. The fission of a single atom of ^{235}U therefore results in the liberation of about 200 MeV; fission of 1 g-atom of this isotope would liberate about 2×10^{10} kJ ($4{\cdot}5 \times 10^{9}$ kcal), more than a million times the energy liberated by burning the same weight of coal. The fusion of light nuclei to form heavier ones is capable of producing even larger amounts of energy; such processes are believed to take place in the sun and stars.

1.2 The separation of stable isotopes[3, 4, 5]

1.2.1. The consequences of differences in isotopic masses. The difference in mass between isotopes of an element is the origin of all separation procedures. In some cases the dependence on mass difference is readily understood (e.g. in the electromagnetic method or in simple diffusion); in others (e.g. the chemical exchange method) the mass difference exerts its influence in a more subtle way through its effect on the fundamental vibration frequencies of isotopic molecules. According to quantum mechanics, a molecule has a certain amount of energy even at 0°K. For a diatomic molecule, for instance, this zero-point energy per mole E_0 is given by

$$E_0 = \tfrac{1}{2}Nh\nu_0 = \tfrac{1}{2}Nhc\bar{\nu}_0$$

where N is Avogadro's number, h Planck's constant, c the velocity of light, ν_0 the fundamental vibration frequency in sec^{-1}, and $\bar{\nu}_0$ the fundamental vibration frequency in cm^{-1} (as it is loosely but generally called—see Section 2.1), which is in turn related to the restoring force constant f and the reduced mass μ by the equation

$$2\pi\bar{\nu}_0 c = (f/\mu)^{\frac{1}{2}}$$

If m_a and m_b are the masses of the two atoms in the molecule, μ is given by

$$\mu = \frac{m_a . m_b}{m_a + m_b}$$

The effect of isotopic substitution on the restoring force constant is negligible, so the fundamental vibration frequencies of isotopic molecules are different, and the effect is greatest for isotopes of light elements; since the zero-point energy enters into expressions for free energies of molecules and activation energies of reactions, thermodynamic and kinetic factors are both involved. Many applications of isotopes depend upon these small differences in zero-point energies.

1.2.2. The electromagnetic method. This method, essentially that first used by J. J. Thomson and F. W. Aston for the identification of isotopes, depends upon the fact that particles of different charge/mass ratio move differently in electrical and magnetic fields. A volatile compound is made to form charged particles (of atomic or molecular dimensions) by the action of a beam of electrons obtained from a heated filament. The positive ions formed are accelerated by an electric field, and are then subjected to the action of a magnetic field, which causes them to follow circular paths of radii dependent on their charge/mass ratios.

A potential difference V accelerates the ions of mass m and charge e to a velocity v, where

$$\tfrac{1}{2}mv^2 = Ve$$

In a magnetic field of strength H in a direction perpendicular to that of the ions, the path of each ion becomes a circle of radius r such that the centrifugal force mv^2/r is equal to the force Hev exerted by the field. Thus,

$$\frac{mv^2}{r} = Hev$$

Elimination of v leads to

$$\frac{e}{m} = \frac{2V}{r^2H^2}$$

Passage of a beam of ions of differing charge/mass ratios through the same fields therefore leads to separation into components which can be collected individually.

The special feature of the mass spectrographic method is that it results in complete separation in a single operation; it has been employed in, for example, the separation of ^{235}U and ^{238}U, present to the extent of 0·7 and 99·3 per cent respectively in the form of their volatile hexafluorides (the only naturally occurring isotope of fluorine is ^{19}F). The method is not, however, economically sound for large-scale application under normal conditions, and its principal value is for analytical and structural work, in which it is usual to measure the current transported by the different ions, the instrument then being termed a mass spectrometer. Such an instrument provides the easiest way of investigating the distribution of a non-radioactive isotopic tracer such as ^{18}O or ^{13}C. In recent years it has been found that there is a valuable correlation between the structure of a molecule and the relative abundances of the various-sized fragments that are formed when it disintegrates under the influence of electron-bombardment. This technique has been applied extensively in the study of organic and organometallic compounds and more complex volatile inorganic compounds such as the hydrides of boron; further reference is made to it in Section 2.10.

1.2.3. Gaseous diffusion. The ratio of the rates of diffusion of two isotopic species, of molecular weights M_1 and M_2, through a porous membrane containing holes of diameter much smaller than the mean free paths of the molecule is given by

$$\frac{\text{Rate}_1}{\text{Rate}_2} = \left(\frac{M_2}{M_1}\right)^{\frac{1}{2}}$$

For such a process the separation factor α is defined as

$$\alpha = \frac{n_1'/n_2'}{n_1/n_2}$$

where n_1 and n_2 are the concentrations of the species before, and n_1' and n_2' the concentrations of the species after, the operation. Except for a few systems such as H_2/D_2 and HD/D_2, α will clearly be not far removed from unity; for the separation of gaseous $^{235}UF_6$ from $^{238}UF_6$, for instance, α should ideally be 1·0043. Such a value, however, takes no account of back-diffusion or of the less efficient separation which results if bulk flow at higher pressures accompanies molecular diffusion. Furthermore, the change in composition of the material on the starting side of the barrier makes the separation become less effective as the operation proceeds. Although, therefore, a fair degree of enrichment can be obtained by the use of a number of diffusion units in cascade at low pressures, a very large number of stages is necessary to effect a complete separation, and at low pressures only small amounts of material can be handled. Nevertheless, this method has been applied to the separation of deuterium and to the substantial enrichment of ^{13}C and ^{15}N present as methane and molecular nitrogen respectively.

When separation on a large scale is required (e.g. of $^{235}UF_6$) it is usual to work with a system involving diffusion from a region of higher to one of lower pressure. The best flow arrangement for the successive stages is one in which half the gas pumped into each stage diffuses through the barrier, the residual half being returned to the feed of the next lower stage. For the production of 99 per cent pure $^{235}UF_6$, several thousand diffusions through a suitable inert porous material are required. This method is used on an industrial scale.

1.2.4. Thermal diffusion. It can be shown from an advanced treatment of the kinetic theory of gases that if a mixture of light and heavy molecules is confined in the space between two surfaces at different temperatures, the heavy molecules tend to diffuse towards the region of lower temperature, and the light molecules towards the region of higher temperature; this process continues until its effect is counterbalanced by that of ordinary diffusion. The degree of separation is very low, but if the effect of thermal diffusion is enhanced by convection a useful method results. This may be done by mounting a very long tube vertically and heating a metal wire stretched axially along the tube to a temperature of several hundred degrees; several tubes may be arranged in series. Although the method is slow, the apparatus is readily constructed, needs little attention, and may be made to yield a regular take-off of the enriched product. Among many separations to which it has been applied are $H^{37}Cl$, $^{13}CH_4$, $^{15}N_2$ and $^{18}O_2$.

1.2.5. Other physical methods. Among other physical methods which have been shown to result in isotopic enrichment are molecular distillation, centrifugal separation, ion migration, and gas–liquid chromatography. None of these, however, is comparable in practical importance to the electromagnetic and diffusion methods and the chemical method now to be described.

1.2.6. The chemical exchange method. The differences in the zero-point energies of isotopic molecular species lead to small differences in the free energies of the reactants and the products in exchange reactions, the equilibrium constants of which are therefore slightly removed from unity. The following are typical values at 25°C:

$$^{15}NH_3(g) + {}^{14}NH_4^+(aq) \rightleftharpoons {}^{14}NH_3(g) + {}^{15}NH_4^+(aq) \quad K = 1{\cdot}031$$
$$H^{12}CN(g) + {}^{13}CN^-(aq) \rightleftharpoons H^{13}CN(g) + {}^{12}CN^-(aq) \quad K = 1{\cdot}026$$
$$HC^{14}N(g) + C^{15}N^-(aq) \rightleftharpoons HC^{15}N(g) + C^{14}N^-(aq) \quad K = 1{\cdot}003$$
$$^{13}CO_2(g) + H^{12}CO_3^-(aq) \rightleftharpoons {}^{12}CO_2(g) + H^{13}CO_3^-(aq) \quad K = 1{\cdot}012$$

Such reactions are particularly advantageous in that they can readily be used in multistage processes; those which involve only proton transfer are also very rapid.

In the production of an ammonium salt enriched in ^{15}N, for example, a solution of ammonium sulphate is fed into the top of a packed column up which ammonia is blown. Most of the liquid reaching the bottom is treated with alkali to regenerate ammonia which is blown up the column again, but some of it passes to a second column in which further enrichment takes place, and the process is then repeated several times. The carbon isotope ^{13}C is produced in the form of the $H^{13}CO_3^-$ anion by similar equilibration between gaseous carbon dioxide under pressure and potassium hydrogen carbonate solution in the presence of a fibre-glass or alumina catalyst; the carbon dioxide is enriched in ^{18}O at the same time.

Exchange reactions in purely gaseous systems such as

$$^{15}NO + {}^{14}NO_2 \rightleftharpoons {}^{14}NO + {}^{15}NO_2$$
$$^{12}CO_2 + {}^{13}CO \rightleftharpoons {}^{13}CO_2 + {}^{12}CO$$

may also be utilized for isotopic enrichment; if such equilibria are established in a thermal diffusion column the effects of the exchange and thermal diffusion may be combined.

1.2.7. The electrolytic method. This method is important only for the separation of deuterium. Its theoretical basis is complicated, like that of most electrode processes: since the separation factor varies markedly with variation in the nature of the cathode used and the pH of the electrolyte, kinetic as well as thermodynamic (i.e. standard electrode potential) factors must be involved. The starting material is usually water distilled from electrolyte which has accumulated in commercial electrolytic hydrogen cells and has thereby become appreciably enriched. A 0·5 M solution of sodium hydroxide in this water is electrolysed between nickel electrodes until the volume is about one-tenth of its original value. Nine-tenths of the liquid is now neutralized with carbon dioxide, and the water is distilled and added to the remaining one-tenth of the electrolyte. This process is repeated for as many stages as are necessary to obtain the desired concentration of deuterium; in the later stages of electrolysis it is economic to burn the gas evolved at the cathode and return it to the appropriate electrolytic stage. In order to obtain water containing 99 per cent of the heavier isotope, it is necessary to electrolyse ordinary water until about only 10^{-4} per cent of the original volume remains. Cheap electric power is therefore essential for this process to compete successfully with thermal diffusion or chemical exchange.

1.3 The separation of unstable isotopes[6, 7]

The principles underlying the production of isotopes by artificial means have been mentioned earlier. Because of its zero charge, the neutron is particularly effective as an entity for penetrating atoms to bring about nuclear reactions, and at the present time the chief sources of synthetic radioactive isotopes are (n,γ), (n,p) and (n,d) processes in nuclear reactors and the products of the fission of the nuclei of certain very heavy isotopes. Some species, however, can be obtained only by bombardment with charged particles (e.g. protons, α-particles or C^{6+} nuclei) in an accelerating machine such as a cyclotron. For irradiation in the pile it is normal to use liquids or solids; for targets in the cyclotron solids are, of course, essential.

The methods used for the separation of the isotopes produced depend on the nature of the nuclear reaction. Where the desired product is not isotopic with the target material, or where separation from fission products is involved, the operation is essentially one of chemical separation of a very small amount of one element from a large amount of one or more others; volatilization, solvent extraction, electrodeposition, ion-exchange or precipitation on a 'carrier' (sometimes in the form of a compound of an inactive isotope of the element concerned) are commonly employed.

Iodine-131, for example, is obtained by an (n,γ) reaction on elemental tellurium, the tellurium isotope which is formed by slow neutron capture rapidly decaying according to the scheme

$$\begin{array}{c} {}^{130}Te + {}^{1}n \rightarrow {}^{131}Te + \gamma \\ \downarrow \\ {}^{131}I + \beta \end{array}$$

The target is oxidized with dichromate and acid, when tellurate and iodate are formed; mild reduction with oxalic acid then converts the iodate into free iodine, which is distilled off, trapped in dilute alkali, and reduced to iodide by sulphite.

Iron-59, obtained by an (*n,p*) reaction using fast neutrons on ^{59}Co, may be separated by dissolving the target in hydrochloric and nitric acids, giving a solution containing iron (III) and cobalt (II) chlorides, and extracting with an ether; suitable choice of the hydrochloric acid concentration makes it possible to extract the iron (probably as $H_9O_4{}^+FeCl_4{}^-$, i.e. $[H_3O^+(H_2O)_3](FeCl_4{}^-)$, whilst the cobalt remains in the aqueous layer. Copper-64, made by the same type of reaction from ^{64}Zn, can be separated by dissolving the target in acid under oxidizing conditions followed by preferential electrodeposition, there being a difference of more than one volt in the standard potentials of the Cu^{2+}/Cu and Zn^{2+}/Zn electrodes. Carbon-14, best made by an (*n,p*) reaction on $Be_3{}^{14}N_2$ in the pile, is liberated as $^{14}CH_4$ by the action of water; this is oxidized to $^{14}CO_2$, absorbed in barium hydroxide solution, and sold in the form of $Ba^{14}CO_3$ containing 10–30 per cent of the carbon as ^{14}C.

Because of the small amounts of material involved, special care is necessary to avoid losses arising from hydrolysis in extremely dilute solutions, adsorption on solids present in the solution or on the walls of the containing vessel, or solid solution formation if processes involving precipitation are used; small amounts of carriers are often added to facilitate control of the distribution of the desired isotope. The isotopes ^{90}Sr and ^{140}La, for instance, are commonly separated from fission products, inactive strontium ions being added to a sulphate-containing solution of the isotopes until strontium sulphate is precipitated. Under such conditions all of the small quantity of lanthanum would be adsorbed on the precipitate if an inactive lanthanum salt were not previously added; after such an addition, however, nearly all of the adsorbed lanthanum will be inactive, most of the ^{140}La being 'held back' in the solution. When ^{32}P is made by an (*n,p*) reaction on sulphur, the target is dissolved in aqua regia, and a little iron (III) salt is added. Partial neutralization results in the formation of iron (III) hydroxide, which acts as a 'scavenger' for the phosphate present. Dissolution of the precipitate in acid, followed by removal of the iron in an acidic ion-exchange column, then gives a solution containing the radiophosphorus in the form of phosphoric acid.

The isolation of elements which do not occur naturally follows these general methods. Astatine, the homologue of iodine, is obtained by the reaction ^{209}Bi (α,2n) ^{211}At, and may be separated by taking advantage of its volatility.

Because of its technological importance, the fissile plutonium isotope ^{239}Pu, an α-emitter with a half-life of 2·4 × 10^4 years, calls for special mention. Its production in a nuclear reactor is represented by the scheme

$$^{238}_{92}\mathrm{U} + {}^{1}_{0}n \rightarrow {}^{239}_{92}\mathrm{U}$$

$$^{239}_{92}\mathrm{U} \xrightarrow[2\cdot3\ \mathrm{min}]{\beta} {}^{239}_{93}\mathrm{Np} \xrightarrow[2\cdot3\ \mathrm{days}]{\beta} {}^{239}_{94}\mathrm{Pu}$$

The reactor consists essentially of rods of uranium metal, containing the isotopes of mass 235 and 238, inserted in graphite or deuterium oxide. By the action of neutrons of thermal velocities on ^{235}U, some atoms of this isotope undergo fission, an average of 2–3 fast neutrons being produced per atom that is split. Passage of these neutrons through the graphite or deuterium oxide slows them down, and they are then captured either by the uranium-238, leading to the changes above, or by the uranium-235, leading to the supply of further neutrons in a chain process. Owing to the short half-life of neptunium-239, this element is not present in very large amounts in the irradiated rods a few weeks after their removal from the reactor, and the essential process is therefore one of separation of plutonium from uranium and fission products. Uranium and plutonium are removed from the solution obtained after dissolving the rods in nitric acid by solvent extraction of their nitrates of formula $MO_2(NO_3)_2$, usually with tributyl phosphate. In these nitrates both elements are in the oxidation state +6, but plutonium is more easily reduced than uranium, and washing the organic layer with an aqueous solution of a mild reducing agent results in reduction of the plutonium and its transfer to the aqueous solution.

Where the product is isotopic with the target, as happens in an (n,γ) reaction not followed rapidly by β-decay, the product is inevitably diluted with a large amount of inactive isotope. However, the energy liberated as γ-radiation may well lead to bond-breaking and the separation of the radioisotope in a form chemically different from that of the target. This effect (the Szilard–Chalmers effect[8]) was first observed in the irradiation of liquid ethyl iodide containing ^{127}I with slow neutrons, after which it was found that most of the radioactive ^{128}I formed could be extracted into aqueous solution. For the effect to be useful in a separation it is, of course, essential that the fragments produced do not recombine rapidly, and that there is no rapid exchange of the isotope between the species in which it is present in the target and in the product; the target should therefore be a kinetically inert molecule or ion. In the example described here, addition of a little molecular iodine before irradiation helps, by leading to the formation of $^{128}I\,^{127}I$ molecules, to avoid re-formation of ethyl iodide or the production of other organic iodine compounds.

The Szilard–Chalmers effect has been studied for many other elements: when a permanganate is irradiated with slow neutrons, for example, most (but not all) of the radioactive manganese is present in the form of lower oxidation state species, but its distribution among different oxidation states depends on the nature of the cation present and the history of the target after irradiation. The discussion of these changes lies beyond the scope of this book, however.

1.4 The applications of isotopes[1, 6]

1.4.1. Classification. Applications of isotopes fall into four main groups:

(a) those in which determination of the location and concentration of an isotope is used in analytical and mechanistic studies;
(b) those which depend upon the production of isotopes;
(c) geochemical and radiocarbon dating;
(d) those which depend upon the small differences between isotopes of a given element, principally in various branches of spectroscopy (especially vibrational spectroscopy) and in reaction kinetics.

We shall now discuss and illustrate each type of application in turn.

1.4.2. Tracer methods. Most of these, except for studies of exchange reactions, depend upon the very small quantities of radioactive materials that can be determined by modern counting equipment. For an isotope of half-life 14 days emitting β-radiation (e.g. ^{32}P), 10^{-9} curie (10^{-16} g) can readily be detected; thus the method is about a million times more sensitive than the flame test for sodium. Smaller amounts of shorter-lived isotopes can be detected, but when the half-life is of the order of only a few minutes (e.g. ^{13}N, 10 min; ^{15}O, 2 min) it becomes very difficult to isolate and use the isotope within a period brief enough for useful results to be obtained. This is why nearly all tracer studies with nitrogen or oxygen are carried out using the stable isotopes ^{15}N and ^{18}O.

Tracer methods are very widely used in biochemistry, but these applications lie outside the scope of this book.

1.4.3. Analytical studies. Among the most important applications are the study of diffusion in liquids and solids, the determination of solubilities of sparingly soluble compounds, partition coefficients and vapour pressures, and the investigation of solid solution formation and adsorption on precipitates. Thus the diffusion of atoms in solids and its dependence on crystallographic direction may be examined by producing a layer of a radioisotope on one surface of a specimen and then measuring the variation in the distribution of activity throughout the specimen after a known time. The partition coefficient of iodine between immiscible liquids is easily measured by determining the activity of labelled iodine in the same volumes of different layers. The solubility of strontium sulphate may be found by diluting $^{90}SrSO_4$ with the inactive salt, determining the activity per gram of the resulting mixture, and then measuring the activity of the residue obtained by evaporation of a solution saturated with the solid. Red phosphorus is only very slightly volatile; its latent heat of sublimation may be obtained by investigating the dependence of its vapour pressure (measured by Knudsen's method) on temperature.

1.4.4. Exchange reactions. For any isotopic exchange reaction ΔH^0 is near zero because of the very close similarity in the thermodynamic properties of the reactants and products; there must always, however, be an increase in entropy arising from the more random distribution of isotopes after exchange, and hence ΔG^0 for an exchange reaction must always be significantly negative and the reaction must be thermodynamically feasible. For the reaction

$$H_2O + D_2O \rightleftharpoons 2HOD$$

at the ordinary temperature, for example, the equilibrium constant is near the statistical value of 4; but since ΔH^0 for the reaction is only -125 J (-30 cal), $T\Delta S^0$ must provide most of the driving energy of -3500 J (-840 cal) for the exchange process. The investigation of exchange reactions is therefore concerned with kinetic and mechanistic, rather than with thermodynamic, factors.

Exchange reactions are normally followed by mixing two species AX and BX, one of which is labelled with a suitable isotope of X, taking portions at suitable times, separating the two species, and determining the distribution of the isotope used for labelling, using counting methods if this is radioactive and mass spectroscopy or infrared spectroscopy if it is not. Absence of exchange is unambiguous; but exchange is sometimes induced by the separation procedure, especially if precipitation is involved, and where exchange is found to occur it is desirable to seek confirmation using a different technique for separation.

Studies of hydrogen exchange show that compounds, whether organic or inorganic, that contain hydrogen bonded to nitrogen or oxygen always exchange this hydrogen for deuterium when they are treated with deuterium oxide. Compounds containing hydrogen bonded to carbon, on the other hand, undergo exchange very slowly indeed unless the hydrogen is known from other evidence to be acidic (e.g. in acetylene or nitromethane). The rapid exchange of hydrogen attached to oxygen is undoubtedly related to the high mobility (as inferred from conductivity data) of the proton in water and the chain mechanism of the conduction process; the polarity of the O—H bond favours the attainment of a transition state in which one such bond is being broken while another is being made. The absence of exchange when sodium hypophosphite (NaH_2PO_2) is crystallized from heavy water indicates the absence of O—H, and therefore the presence of P—H, bonds in the $H_2PO_2^-$ anion; this conclusion is confirmed by the presence of a P—H stretching frequency in the infrared spectrum of the sodium salt.

Non-equivalence of the sulphur atoms in the $S_2O_3^{2-}$ ion is shown by forming thiosulphate from ^{35}S-labelled sulphur and inactive sulphite, and then decomposing the product with acid; all of the ^{35}S reappears in the sulphur produced. Most oxy-anions do not undergo measurable exchange of their oxygen with ^{18}O in the form of water—nor does hydrogen peroxide. This makes it possible to deduce something of the nature of the hydrogen peroxide–permanganate reaction in the presence of an acid: when peroxide containing ^{18}O is oxidized by permanganate not containing this isotope, the ^{18}O is present in the gaseous oxygen which is liberated and not in the solution, showing that the permanganate acts as an oxidizing agent by deprotonating the hydrogen peroxide molecule without breaking the O—O bond.

The exchange of the ligand in a complex with labelled ligand in solution is a process which has been investigated extensively. Complexes may be divided into two groups, labile and kinetically inert, on the basis of such studies. The $[Fe(CN)_6]^{4-}$ and $[Fe(CN)_6]^{3-}$ ions, for example, do not exchange cyanide with ^{14}C-labelled aqueous cyanide, whereas the corresponding exchange for the $[Ni(CN)_4]^{2-}$ ion is very fast. The lability of the latter species arises not from a large dissociation constant (it is, in fact, thermodynamically a very stable ion), but from the formation of the $[Ni(CN)_5]^{3-}$ ion, which is also labile, when it is treated with aqueous cyanide. Among hydrated ions it is interesting to note that studies involving $^{18}OH_2$ labelling have shown that whereas $[Fe(H_2O)_6]^{3+}$ is labile, $[Cr(H_2O)_6]^{3+}$ is kinetically inert to substitution; since the ions are closely similar in size and bear the same charge, it is clear that subtle electronic factors are involved. Finally, mention may be made of the very rapid apparent exchange of radioactive iron between the hexacyanoferrate (III) and hexacyanoferrate (II) ions; both are kinetically inert to substitution, and it is in fact an electron that is being transferred, so that hexacyanoferrate (III) becomes hexacyanoferrate (II) and vice versa—a process very much more rapid than dissociation or complex formation. Many other isotopic exchange reactions are mentioned in later chapters.

1.4.5. Activation analysis.[9] In this method of analysis an element E is determined by bombarding the sample (normally with neutrons in a reactor, but occasionally with neutrons or other particles from different sources), and measuring the intensity of radioactivity induced. A sample containing a known weight of E is irradiated at the same time and under exactly the same conditions; after completion of the irradiation period (which is usually a few times the half-life of the isotope which it is desired to produce), the sample and standard are normally treated to separate the isotopes of the element being determined from other products, and the ratio of their activities then gives directly the ratio of the amounts of E. It should be noted that E must have a reasonably high capture cross-section (i.e. probability of capturing) for the bombarding particles, and the half-life of the product should be sufficiently long for subsequent chemical separation and measurement.

The method is particularly suitable for the determination of trace impurities or constituents (e.g. arsenic in transistor germanium, rubidium and caesium in rocks, or phosphorus in biological materials); in some instances it may be employed without destruction of the sample. For this to be so, however, it is necessary that one or more of certain conditions should be fulfilled: the element being determined (or, strictly, the particular isotope being determined) should have a much higher cross-section for neutron capture than other species present, or the product of its irradiation should have a half-life markedly different, or a type of emission different, from those obtained from other species present. Gold, for example, has a high cross-section for neutron capture, whilst lead has a very low cross-section and therefore does not interfere in the determination of gold; osmium in rhodium can be determined because one of the active products, ^{193}Os, has a half-life (32 hr) much longer than that of ^{104}Rh (4 min), so after a few hours the activity of the sample is virtually that of ^{193}Os.

Sometimes different products of similar half-life can be distinguished because of the different energies of their radiations. The determination of arsenic in germanium depends upon the reaction $^{75}As(n,\gamma)^{76}As$, a 27-hour β-emitter being formed. This isotope and ^{77}As of half-life 40 hours may also be produced from germanium, however:

$$^{74}Ge(n,\gamma)^{75}Ge \xrightarrow[80\ \text{min}]{\beta} {}^{75}As(n,\gamma)^{76}As$$

and

$$^{76}Ge(n,\gamma)^{77}Ge \xrightarrow[60\ \text{sec}]{\beta} {}^{77}As$$

Suitable choice of irradiation period and neutron intensity can minimize the production of ^{76}As from ^{74}Ge by the former process, whilst ^{76}As can be distinguished from ^{77}As by counting through an aluminium absorber which is impenetrable to the electrons of lower energy from ^{77}As (0·7 MeV as compared with 2·9 MeV from those from ^{77}As).

Activation by charged particles in accelerating machines frequently leads to the formation of a variety of products, but is nevertheless employed occasionally. Owing to the long half-life of ^{14}C (5570 years) it is not feasible to determine carbon by neutron irradiation, but the reaction $^{12}C(d,n)^{13}N$ in a cyclotron can be utilized. Under the same conditions ^{16}O is converted into ^{17}F, but the half-life of this isotope (1·1 min) is only one-tenth of that of ^{13}N, so fifteen minutes after the irradiation its activity is negligible.

1.4.6. Dating applications. Several methods which involve radioactive decay have been developed for the estimation of the age of mineral and vegetable deposits. These depend upon deducing the time at which a sample was isolated from free exchange with its surroundings. This is done by measurement of the activity of either a species formed at a rate indicated by the half-life of its progenitor, or a species undergoing decay at a rate indicated by its own half-life.

One such method utilizes the measurement of the very small amounts of helium present in minerals containing uranium and thorium. Such helium is almost certainly formed from α-particles, and in granites and similar rocks of close texture is likely to remain trapped; from the amount present, the time during which it has been accumulating may be calculated. One gram of uranium in equilibrium with its decay products generates roughly 10^{-7} ml of helium per year, and the rate of production from thorium is rather less than one-third of this. It is necessary to determine also the uranium and thorium contents of the rock; then if the uranium equivalent per gram of mineral is taken as (U + 0·3 Th) the age in years is given by the quantity

$$\frac{10^7 \times \text{He (ml per gram of mineral)}}{\text{U} + 0{\cdot}3\ \text{Th}}$$

In view of the likelihood that some helium may have escaped, ages determined by this method are likely to be lower limits.

Radiocarbon dating is based on the formation of ^{14}C in the upper atmosphere by the action of cosmic radiation on ^{14}N; it decays to ^{14}N by electron emission, with a half-life of 5570 years. Carbon removed from the carbon dioxide cycle and isolated in a solid such as wood or bone therefore loses its specific activity at a rate indicated by the decay constant, and measurement of the specific activity thus enables the time of removal of the carbon from the cycle to be deduced. The utility of the method is necessarily limited to times not longer than a few multiples of the half-life, and its reliability depends upon the truth of the assumption that the rate of cosmic activation has been constant over the period involved. Tritium, which has a half-life of 12·5 years, has been used in a similar way for determining the age of wine.

1.4.7. Applications to spectroscopy and other methods of structure determination.[10] Because of the inverse dependence of vibration frequencies on the square roots of reduced masses of molecules, isotopically substituted molecules have different vibrational (Raman and infrared) spectra. For $H^{35}Cl$ and $D^{35}Cl$, for example, the ratio of the square roots of the reduced masses is 1 : 1·395 and the observed vibration frequencies are 2886 and 2091 cm^{-1} respectively, a ratio of 1·380 : 1. So long as the other atom A in a diatomic hydride is significantly heavier than hydrogen or deuterium, the ratio of the A—H and A—D vibrational frequencies will approach $\sqrt{2}$ (i.e. 1·41). In the interpretation of the vibrational spectrum of a complex molecule it is usual to assign frequencies to particular groups, the assumption being made that the vibrations of one group are relatively independent of those of the rest of the molecule. (This assumption is nearly true if a vibration involves motion of a group containing light atoms in a molecule consisting otherwise of heavy atoms, but under other conditions coupling (or mixing) of group vibrational modes occurs.) In seeking to identify, say, an O—H vibration frequency, it is often valuable to compare the original spectrum with that of the deuterated compound (hydrogen attached to oxygen always exchanges rapidly); a band which moves to about $1/\sqrt{2}$ of its original frequency on deuteration is very probably associated with the O—H group.

The most important spectroscopic uses of isotopic substitution are, however, in the determination of interatomic distances and angles in molecules, for example by means of microwave (far infrared) spectroscopy. For a diatomic molecule having a permanent dipole moment in the ground state, the lowest frequency absorption band in the microwave (pure rotational) spectrum, corresponding to the difference between successive rotational energy levels, is given by

$$\nu = \frac{h}{4\pi^2 I}$$

where h is Planck's constant and I the moment of inertia, in this case μr_0^2, μ being the reduced mass and r_0 the equilibrium internuclear separation. In a larger molecule the moments of inertia will obviously be related to both lengths and angles in a more complex manner, and it will be necessary to solve a series of simultaneous equations in order to determine all the structural parameters. (This subject is discussed in more detail in Section 2.5.) In order to get enough experimental information to make possible the solution of these equations, isotopic substitution, which affects moments of inertia in a calculable way by altering the distribution of mass but not significantly changing molecular dimensions, is employed; different frequencies can then be observed, and if enough isotopically substituted molecules are investigated several moments of inertia can be obtained and the detailed structure can be elucidated.

For many elements (e.g. chlorine, silicon, germanium) the natural abundances of isotopes are such that ordinary compounds contain more than one species: in GeH_3Cl, for example, species containing ^{35}Cl and ^{37}Cl, and ^{70}Ge, ^{72}Ge, ^{74}Ge and ^{76}Ge are all present in proportions sufficient to give rise to microwave spectra for $^{70}GeH_3{}^{35}Cl$, $^{70}GeH_3{}^{37}Cl$, and so on. Frequently, however, synthesis of isotopically substituted species is necessary, just as it is for the study of exchange reactions. In such syntheses it is desirable to utilize the special isotope to the fullest extent possible, and conventional preparations often have to be modified considerably. Deuterium chloride or deuteroammonia, for example, would not be prepared by dissolving hydrogen chloride or ammonia in an excess of deuterium oxide and then recovering the deuterated species; the action of deuterium oxide on anhydrous aluminium chloride or magnesium nitride would be much more economical. Isotopic carbon in the usual form of barium carbonate may be converted into compounds particularly useful for synthetic purposes by processes such as the following:

(a) $BaCO_3 \xrightarrow{\text{acid}} CO_2 \xrightarrow{\text{C}} CO \xrightarrow{H_2} CH_3OH$

(b) $BaCO_3 \xrightarrow{\text{acid}} CO_2 \xrightarrow{\text{RMgBr}} RCOOH$

(c) $BaCO_3 \xrightarrow{NH_4Cl,K} KCN$

The devising of simple preparations for molecules isotopically substituted in different ways (e.g. of CH_2ClSiH_3, $^{13}CH_2ClSiH_3$, CD_2ClSiH_3 and CH_2ClSiD_3 for a recent microwave study) provides scope for considerable ingenuity.

Isotopically substituted species are also important in nuclear magnetic resonance spectroscopy (Section 2.7). Many common nuclei (e.g. ^{12}C, ^{16}O, ^{32}S) have nuclear spin quantum number, I, zero (i.e. are non-magnetic). For nuclei such as ^{2}D, ^{14}N, ^{35}Cl and ^{37}Cl which have $I > \frac{1}{2}$, the observation of nuclear magnetic resonance spectra is rarely possible (e.g. for ^{11}B). The most suitable nuclei for nuclear magnetic resonance spectroscopy are those with $I = \frac{1}{2}$, such as ^{1}H, ^{13}C, ^{19}F and ^{31}P. Sometimes, therefore, it is desired to incorporate such a nucleus, of low natural abundance (e.g. ^{13}C), into a molecule in order to be able to obtain a spectrum; often, however, the reverse process is carried out in order to simplify a complex spectrum. Substitution of deuterium for hydrogen is particularly important in the latter connection, and the use of deuterochloroform as a solvent (in order to avoid interaction of solvent protons with the material being investigated) is now common practice.

Finally, mention may be made of neutron diffraction (Section 2.3) as a method for the location of hydrogen atoms in crystals. Whereas the scattering factors of atoms for X-rays increase rapidly with increase in atomic number, thereby making the accurate location of light atoms very difficult in most compounds, scattering factors for neutrons show only a small and irregular variation with atomic number; furthermore, the value for deuterium is considerably larger than that for ordinary hydrogen. There is therefore a marked advantage in preparing deuterated species to be studied by neutron diffraction. By this method it has been found, for example, that titanium deuteride (TiD_2) has the fluorite (CaF_2) structure, and that the deuterium atom in the DF_2^- anion in $NaDF_2$ is in the middle of the anion.

1.4.8. The kinetic isotope effect. It can be shown from the theory of absolute reaction rates that molecules containing different isotopes should react at different rates if the bonds to the different isotopes are involved in the transition state (the origin of this effect is in the different zero-point energies of isotopic species). The activation energy for the reaction involving the heavier isotope is slightly greater and the species containing this isotope therefore reacts more slowly. The difference is maximal for bonds involving tritium and hydrogen, substantial for those involving deuterium and hydrogen, and large enough to be measured for ^{14}C, ^{15}N and ^{18}O compared with ^{12}C, ^{14}N and ^{16}O; for heavier elements it is too small to be of practical importance.

If we consider molecules AH and AT (T = tritium) reacting with a species C and the rate-determining step involves fission or substantial stretching of the bond in AH or AT, a kinetic isotope effect is found; if, however, the bond in AH or AT is not involved in the rate-determining step, no such effect appears. The most famous application of this principle is the demonstration, by showing that there is no selective displacement of hydrogen rather than tritium in the nitration of tritiated benzene by nitric acid in sulphuric acid, that the reaction has the mechanism

$$C_6H_6 + NO_2^+ \xrightarrow{\text{slow}} C_6H_6NO_2^+$$

$$C_6H_6NO_2^+ \xrightarrow{\text{fast}} C_6H_5NO_2 + H^+$$

The rate of oxidation of $(CH_3)_2CDOH$ by acidified aqueous dichromate is only one-seventh that of the oxidation of $(CD_3)_2CHOH$ or $(CH_3)_2CHOH$ under the same conditions, showing that the C—H bond in the secondary alcohol group must be attacked in the rate-determining step. This leads to a postulated mechanism

$$(CH_3)_2CHOH \xrightarrow{\text{fast}} (CH_3)_2CH{\cdot}O{\cdot}CrO_2OH$$

$$(CH_3)_2C(H \leftarrow OH_2)(O{\cdot}CrO_2OH) \xrightarrow{\text{slow}} (CH_3)_2C{=}O + H_3O^+ + CrO_3H^-$$

This mechanism is of considerable interest to the inorganic chemist, for the species CrO_3H^- is a derivative of chromium (IV), an oxidation state not existing in appreciable concentration in aqueous media, in which it disproportionates to chromium (III) and chromium (VI). The presence of some unfamiliar entity in the reaction mixture may, however, be confirmed by a completely independent method: although acidified aqueous dichromate has no action on manganous salts, addition of manganous sulphate to a mixture of acidified dichromate solution and isopropanol results in the precipitation of manganese dioxide.

In complex ion chemistry, the use of D_2O in place of H_2O has been adopted in attempts to find out whether O—H bond fission is involved in the rate-determining step. The fact that the Fe^{2+}–Fe^{3+} exchange reaction, for instance, is twice as fast in H_2O as in D_2O suggests that a hydrogen atom is transferred in a process such as

$$Fe^{2+}OH_2 + H_2O^*Fe^{3+} \rightarrow \left[Fe^{2+}O(H)\text{—}H \ldots\ldots O^*(H)(H)Fe^{3+} \right]^{5+}$$

$$\rightarrow \begin{matrix} FeOH^{2+} & + & H_3O^*Fe^{3+} \\ \rightleftharpoons & & \rightleftharpoons \\ Fe^{3+} + OH^- & & {}^*Fe^{2+} + H_3O^+ \end{matrix}$$

(An alternative view is that only coupling of the hydration shells by hydrogen bonding is involved.) The interpretation of experiments of this kind is sometimes complicated by the fact that D_2O has slightly different physical properties from H_2O, but the kinetic isotope effect seems certain to play an increasingly important part in mechanistic studies.

References for Chapter 1

1 Heslop, R. B., and Robinson, P. L., *Inorganic Chemistry*, 3rd edn. (Elsevier, 1967).
2 Phillips, C. S. G., and Williams, R. J. P., *Inorganic Chemistry*, Vol. II (Oxford University Press, 1966).
3 Barnard, A. K., *Theoretical Basis of Inorganic Chemistry* (McGraw-Hill, 1965).
4 Glueckauf, E., *Endeavour*, 1961, **20**, 42.
5 Dawton, R. H. V. M., and Smith, M. L., *Quart. Rev. Chem. Soc.*, 1955, **9**, 1.
6 Friedlander, G., and Kennedy, J., *Nuclear and Radiochemistry*, 2nd edn. (Wiley, 1964).
7 Katz, J. J., and Seaborg, G. T., *The Chemistry of the Actinide Elements* (Methuen, 1957).
8 Maddock, A. G., *Chem. Britain*, 1970, **6**, 287.
9 Atkins, D. H. F., and Smales, A. A., *Adv. Inorg. Chem. Radiochem.*, 1959, **1**, 315.
10 Wheatley, P. J., *The Determination of Molecular Structure*, 2nd edn. (Oxford University Press, 1968).

Physical methods in inorganic chemistry

2

2.1 Introduction

Most of this chapter is concerned with diffraction and spectroscopic methods, though not all such methods are considered now. These are by no means the only physical techniques used in inorganic chemistry. However, some methods will already be familiar from elementary courses in physical chemistry. Among these, for example, are potentiometric determination of equilibrium constants (to which further reference is made in Chapter 6) and thermochemical measurements, including the spectroscopic determination of bond energies (which are referred to in Chapter 5 and in many other places). Other methods, which in inorganic chemistry are used chiefly in the study of compounds of transition metals, are discussed in later chapters; these include optical rotatory dispersion (Chapter 14), magnetic susceptibility measurements and electron spin resonance (Chapter 16) and electronic (visible and ultraviolet) spectroscopy (Chapter 17).

Much of the present chapter is devoted to an outline of methods of structure determination. Relatively few inorganic molecules are big enough to require identification of functional groups by 'fingerprinting' methods, and purely qualitative spectroscopic methods, though certainly very important, do not occupy quite the dominating position that they hold in organic chemistry. We shall do no more here than provide a brief introduction to the principles, scope and limitations of the methods discussed; many excellent fuller accounts of various physical methods are available, and several references are given later.

Before the uses of individual methods are summarized, two general points should be made. The first concerns the units in which workers in different regions of the electromagnetic spectrum express results. In the X-ray, ultraviolet and visible regions it is usual to refer to the wavelength of radiation and to express it in Ångström units (1 Å = 10^{-10} m or 10^{-8} cm). The recommended S.I. unit for bond lengths is the nanometre (10^{-9} m), but since bond lengths are usually 1–3 Å it is very likely that the use of the Ångström unit will persist, and we have retained it here. In the infrared it is customary to work in wave-numbers (i.e. the number of wavelengths per centimetre); this is the frequency divided by the velocity of light, and its unit is therefore $s^{-1}/cm\ s^{-1}$ (i.e. cm^{-1}). (The micron, μ (10^{-4} cm), and millimicron, mμ (10^{-7} cm), are also used occasionally as units of wavelength.) In microwave spectroscopy true frequencies are used and results are given in cycles or megacycles (millions of cycles) per second. Wave-numbers in reciprocal centimetres are often loosely referred to as 'frequencies' and the symbol ν is commonly used for both; the context, however, usually makes the meaning quite clear.

The second point concerns the approximate time-scales of the methods described in this chapter. These are (in seconds): electron diffraction (10^{-20}), neutron diffraction and X-ray diffraction (10^{-18}), vibrational spectroscopy (10^{-13}), microwave spectroscopy (10^{-11}) and nuclear magnetic resonance spectroscopy (10^{-1} to 10^{-5}, depending upon the system being examined). If the time-scale of an atomic movement is less than that of a physical technique used to examine the species in which such movement is taking place, an erroneous conclusion will be reached. The probability of the occurrence of a particular atomic movement depends on the size of the quantum of energy required, and hence is affected by the temperature at which the measurement is made. It is, therefore, not surprising that the use of different methods at different temperatures sometimes indicates different structures; some examples are given later.

2.2 X-ray diffraction[1, 2, 3]

X-ray diffraction is the most generally applicable and most widely used of all methods of structure determination, and it is used mainly for this purpose. The X-ray powder photograph of a substance is, however, also a valuable characteristic property which can be used for identification, particularly for solids which give no useful vibrational spectra. For the later actinide elements (Chapter 23), which are usually available only in quantities of a milligram or less, compounds are sometimes identified by comparison of their powder photographs with those of analogous compounds of earlier elements.

A crystal is a periodic three-dimensional array of atoms which acts as a three-dimensional diffraction grating for X-rays of wavelength comparable to the interatomic distances and interplanar spacings within it. When a beam of X-rays passes through the crystal, the X-rays diffracted by atoms in successive planes will mutually extinguish one another unless reflections from adjacent planes are in phase. If the distance between successive planes is d and the grazing angle of incidence θ, the path difference between X-rays scattered from neighbouring planes is $2d \sin \theta$. The condition that X-rays of wavelength λ shall be reflected by any family of planes is then given by the Bragg relationship

$$n\lambda = 2d \sin \theta$$

where d is directly related to the dimensions of the unit cell and the Laue indices of the reflecting planes. (The Laue indices of a family of planes are the reciprocals of the fractional intercepts which the plane nearest the origin makes with the crystallographic axes.) In essence, the determination of the dimensions of the unit cell (the smallest repeating unit from which the lattice can be built up by packing without any particular orientation) involves measurement of the Bragg angles, allocation of the correct indices and evaluation of the interplanar spacings.

All modern methods of obtaining diffraction patterns employ monochromatic X-radiation. In the powder method the X-ray beam impinges on a fine powder in which the particles are orientated at random; some, however, are so placed as to satisfy the Bragg law for each value of d. The reflections are recorded photographically on a thin cylindrical film in a circular camera. Reflections from planes with closely similar spacing but different Laue indices are not separated adequately, however, and the powder method is really suitable only for crystals having unit cells of fairly high symmetry and, consequently, fewer possible values of d for a given value of θ.

In the rotating crystal method the X-ray beam strikes a small crystal which is slowly rotated about a vertical axis; the crystal is mounted so that the axis of rotation is also one of the principal axes of the crystal. The angle of incidence is thus continuously varied. When Bragg's law is satisfied, a diffraction pattern is obtained and is recorded photographically or, for very accurate work, by means of an ionization counter. The crystal is then remounted so that all its important axes in turn are perpendicular to the X-ray beam. For a compound with a large unit cell the closeness of the reflections causes difficulties; these may be diminished by oscillation of the crystal through a small angle instead of by rotation through 360°, or by synchronized movement of both crystal and film. The last method, due to Weissenberg, is the most powerful photographic method for the determination of crystal structures.

After the unit cell parameters have been obtained, the independently measured density of the crystal allows the number of molecules in the unit cell to be deduced (the converse process, the detection of defects by density measurements, is discussed in Chapter 4). In some simple cases the systematic absence of certain reflections corresponding to particular values of the Laue indices suffices to determine the structure completely. Usually, however, it is necessary to measure accurately the relative intensities of the reflections. The intensity of reflection I from a particular plane is proportional to the square of a quantity called the structure factor F, which is a function of the positions of the atoms present and their scattering powers. These last depend on the numbers of extranuclear electrons for low-angle scattering, but decrease by calculable amounts as θ increases. The problem is then to compare values of F calculated for a postulated structure with those obtained by experiment, and to 'refine' the postulated structure to give the best agreement between observed and calculated structure factors. The calculation of structure factors by means of Fourier series, and the discussion of the solution of the difficulties which arise from the fact that since $I \propto F^2$ the phase of F is not known, are beyond the scope of this account. We should, however, summarize the particular merits of the X-ray diffraction method. Although the successive refinements of a structure to allow for thermal vibrations of the atoms and to improve the agreement between observed and calculated structure factors beyond what Fourier refinement can achieve are lengthy and call for access to a powerful electronic computer, the result is a direct and unique solution of the structure. In the most favourable cases bond lengths can be determined to within about $\pm$ 0·01 Å and bond angles to within 1°. Since the method can be applied at low temperatures, virtually all substances are susceptible to X-ray determination of their structures. Finally, it is the only method capable of solving the structures of compounds containing large numbers of atoms.

The structures of many compounds are known as a result of partial structure determinations. If two compounds of similar empirical formula possess the same space group and have similar unit cell dimensions and reflection intensity distributions, it is very probable that they are isostructural. This is the case, for example, for many complexes of Mn^{2+}, Fe^{2+}, Co^{2+} and Ni^{2+}.

Since X-rays are scattered by electrons, it is very difficult to locate light atoms accurately in the presence of heavy atoms, and accurate bond lengths involving hydrogen atoms cannot be obtained by X-ray diffraction. Distinctions between elements of adjacent atomic number (e.g. carbon and nitrogen in complex cyanides, magnesium and aluminium in minerals) are also difficult. Thermal motion leads to uncertainties in atomic positions, and X-ray methods alone do not distinguish between random orientation and molecular or ionic rotation (see Section 4.4). In addition, although X-rays can be diffracted by gases and liquids, the information obtained from studies of these effects does not lead to the determination of structure in the sense in which it is used in this chapter; only for a few compounds of heavy elements in aqueous solution have X-ray methods been used successfully to yield structures of substances not in the solid phase.

2.3 Neutron diffraction[1, 4]

The general principles of neutron diffraction by single crystals or powders, which is used mainly for structure determinations, are similar to those of X-ray diffraction. Neutrons produced by fission in a nuclear reactor and slowed down to thermal velocities by passage through a moderator have wavelengths of about 1·5 Å, and reflection of the neutron beam from a crystal of calcite or lead gives a beam of a small band of wavelengths (not a strictly monochromatic beam). Even the strongest neutron beams available are considerably weaker than an ordinary X-ray beam used in crystallography, and larger crystals or samples are therefore needed for neutron diffraction than for X-ray diffraction. Because the beam is not monochromatic, resolution is rather poor, and relative intensities are less accurate than in X-ray work; furthermore, neutron counting by means of a boron trifluoride counter is a less convenient operation than the photographic recording of X-rays.

Despite these limitations, neutron diffraction has some very powerful advantages. For diamagnetic species all the scattering is done by nuclei; most of it is resonance scattering resulting from the formation and immediate disintegration of unstable nucleus–neutron combinations (isotopes which have high cross-sections for neutron capture must obviously be avoided in neutron diffraction). The overall scattering factors of different atoms vary over a factor of only about four, and in an erratic manner; ^{2}H, ^{12}C, ^{14}N and ^{16}O have scattering factors which are about the same as those of the heavy elements. Neutron diffraction is thus particularly useful for the location of light atoms in structures where the X-ray scattering would be dominated by heavy atoms. It is, in fact, common practice to determine the positions of heavy atoms first by X-ray diffraction, and then to determine the positions of light atoms by neutron diffraction. Partly because of its high scattering factor, and partly because its use results in less incoherent scattering (i.e. scattering with change of wavelength), deuterium is often substituted for hydrogen in compounds to be examined by neutron diffraction; ways in which this is done have been described in Chapter 1.

Among compounds whose structures have been elucidated by neutron diffraction are the hydrides, carbides and nitrides of transition metals, the oxides of the actinide elements, and xenon fluorides; many hydrogen-bonded structures (e.g. those of KDF_2, solid D_2O, $KHCO_3$ and solid ND_3) have also been examined. Compounds containing elements of adjacent atomic number can usefully be studied by neutron diffraction if their neutron scattering factors differ; thus $MgAl_2O_4$, in which the Mg^{2+} and Al^{3+} ions cannot be distinguished by X-ray diffraction, has been shown to be a normal spinel (see Section 3.4).

Since the neutron has a magnetic moment by virtue of its having a nuclear spin of $\frac{1}{2}$, additional scattering results from the interaction of a neutron beam and paramagnetic atoms or ions. The different scattering factors of high-spin Fe^{2+} and Fe^{3+}, for example, make it possible to distinguish between these ions in the inverse spinel structure of Fe_3O_4. Neutron diffraction is also used extensively to study magnetic ordering in ferromagnetic and antiferromagnetic substances; this aspect of the structure of Fe_3O_4 is mentioned later in Chapter 16, but its further discussion lies beyond the scope of this book.

2.4 Electron diffraction[1, 5]

The uses of electron diffraction are restricted to the study of gases and vapours and of thin films. Because of their negative charge, the scattering of electrons is about a thousand times as effective as that of X-rays. Scattering is by the potential field of the species under study; for heavy elements this can be considered to reside at nuclear positions. Like X-rays, electrons are scattered much more by heavy than by light atoms, and, as in X-ray diffraction, it is impossible to locate hydrogen atoms accurately in the presence of much heavier atoms. Any substance which is stable in the vapour phase and has a vapour pressure of a few millimetres of mercury at a reasonable working temperature may be studied, but obviously the higher the temperature used the greater the margin of uncertainty arising from thermal vibrations.

In a gas the molecules have all possible orientations. Because of the use of high energy electrons (a typical accelerating potential is about 40 kV, corresponding to a wavelength for the electrons of about 0·06 Å), the motion of a single molecule is negligible during the time of its interaction with the electron beam. Careful control of the accelerating potential is necessary to ensure a monochromatic beam. Only a fraction of a second is needed for exposure, which is achieved by directing a jet of vapour perpendicular to the electron beam in a highly evacuated apparatus; the vapour is immediately condensed in a cold trap on the other side of the beam, and the molecular diffraction pattern is recorded on a photographic plate. In the scattering process, electrons scattered from different atoms in any one molecule will be out of phase by an amount that depends upon the interatomic distances and the orientation of the molecule with respect to the incident beam. The image produced on the photographic plate consists of a series of diffuse concentric rings around the undeflected electron beam. Much of the scattering is, however, incoherent atomic scattering (which is caused by excitation of atoms to higher energy levels with production of electrons of lower energy) or coherent (i.e. elastic) atomic scattering, and these have to be subtracted from the total intensity to give the coherent molecular scattering from which the structure is worked out. In the case of carbon tetrachloride, for example, the intensity of the coherent scattering I is given approximately by the expression

$$I/K = Z_{\mathrm{C}}^2 + 4Z_{\mathrm{Cl}}^2 + 6Z_{\mathrm{Cl}}^2 \frac{\sin[sr_0(\mathrm{Cl{-}Cl})]}{sr_0(\mathrm{Cl{-}Cl})} + 4Z_{\mathrm{C}}Z_{\mathrm{Cl}} \frac{\sin[sr_0(\mathrm{C{-}Cl})]}{sr_0(\mathrm{C{-}Cl})}$$

where $s = (4\pi \sin \theta/2)/\lambda$, K is a constant depending upon the geometry of the apparatus, Z_{C} and Z_{Cl} are the atomic numbers of carbon and chlorine, and $r_0(\mathrm{Cl{-}Cl})$ and $r_0(\mathrm{C{-}Cl})$ are the chlorine–chlorine and carbon–chlorine distances; for a regular tetrahedral structure, $r_0(\mathrm{Cl{-}Cl}) = \sqrt{8/3}\, r_0(\mathrm{C{-}Cl})$. The first two terms represent coherent atomic scattering and are independent of the scattering angle θ, and their relative importance therefore decreases as θ increases; the third and fourth terms represent molecular scattering, the integers being the numbers of chlorine–chlorine and carbon–chlorine distances in the molecule. Since I depends upon $\sin[sr_0(\mathrm{C{-}Cl})]/sr_0(\mathrm{C{-}Cl})$, and s in turn depends upon θ, I_m must pass through a series of maxima and minima as θ increases—hence the form of the diffraction pattern. It can now be found by trial and error which values of sr_0(C–Cl) correspond to the observed maxima and minima in I, and hence if λ is known from the accelerating potential a series of values for r_0(C–Cl) can be obtained. In a more accurate treatment Z is replaced by $Z-f$, where f is a function of sin θ/λ; thermal motion has to be taken into consideration.

Two variations in this procedure are normal in modern electron diffraction studies. In order to diminish the importance of atomic scattering a 'rotating sector' of suitable shape is interposed between the beam and the photographic plate so as to diminish exposure times near the direct beam and hence bring out more detail at high θ. It is now customary to analyse the molecular scattering intensity curve with the aid of a computer so as to obtain a plot of a radial distribution function related to the probability of finding a particular internuclear distance in the molecule against internuclear distance. Bond lengths can then be read from the plot and a limited number of other molecular parameters can be calculated from them.

Where the molecule being investigated is a small one, or where it is highly symmetrical, the electron diffraction method gives results nearly as accurate as spectroscopic methods (e.g. bond lengths to within $\pm 0{\cdot}005$ Å and bond angles to within $\pm 1°$). For simple molecules such as O_2, Br_2, NH_3 and C_2H_6, structural parameters obtained by modern electron diffraction methods and spectroscopic studies are in very close agreement. It is rare, however, for more than ten diffraction rings to be observed, and for a more complex molecule possessing more than about six parameters the experimental data are usually insufficient to enable a structure to be worked out unless reasonable assumptions are made; unfortunately, assessments of the reasonableness of assumptions are both subjective and time-variable. As we have mentioned already, light atoms cannot be located accurately in molecules containing much heavier atoms, and bond lengths are mean values over all vibrational states populated at the temperature at which the diffraction pattern is obtained. Nevertheless, the value of electron diffraction in providing a non-spectroscopic method for the study of gases and vapours is very great, and since the advent of modern computing methods its application and reliability have increased considerably.

2.5 Pure rotational spectroscopy[1, 6, 7]

When a molecule absorbs energy in accordance with Planck's relationship

$$\Delta E = h\nu$$

the energy may go to increase its rotational energy only, its rotational and vibrational energy, or its rotational, vibrational and electronic energy. The difference between successive electronic energy levels is much greater than that between successive vibrational energy levels, which is in turn much greater than that between successive rotational energy levels. By studying spectra in the very low energy (low frequency, or high wavelength) region it is possible to study rotational spectra in isolation. Pure rotational spectra occur in the far infrared and microwave regions (frequency 10^4–10^6 megacycles sec^{-1}); absorption in this range corresponds to a gain in energy of a fraction of a kiloJoule per mole.

In order to show a pure rotational spectrum, a molecule must possess a permanent dipole moment. (All changes in rotational and vibrational energy levels are subject to certain restrictions which can be deduced from quantum mechanics; these restrictions are known as selection rules.) The rotational energy levels for a diatomic molecule are given by

$$E_{rot} = \frac{h^2}{8\pi^2 I} J(J+1)$$

where I is the moment of inertia about an axis through the centre of gravity and perpendicular to the internuclear axis and J is the rotational quantum number, which can have any integral value including zero. For the pure rotational spectrum of a diatomic molecule the selection rule $\Delta J = \pm 1$ applies, and the energy difference between two adjacent levels is given by

$$\Delta E_{rot} = \frac{h^2}{8\pi^2 I}[J'(J'+1) - J''(J''+1)]$$

where J' refers to the higher energy level and J'' to the lower. Since the selection rule for this particular system requires

$$J' - J'' = 1$$

then

$$\Delta E_{rot} = \frac{h^2}{8\pi^2 I} \times 2J'$$

Thus the pure rotational spectrum consists of a series of equally spaced lines, and from their separation the moment of inertia and hence, knowing the atomic masses, the internuclear distance may be derived.

Similar principles apply to the determination of the structure of a linear triatomic molecule which has a permanent dipole moment, such as carbonyl sulphide (OCS) or hydrogen cyanide. Here, however, we have only one moment of inertia but two bond lengths to be found. This is done by obtaining the spectrum of an isotopically substituted species to give a second moment of inertia, it being assumed that there is no change in bond lengths when one isotope of an element replaces another (the change is in fact very small indeed if account is taken of the different zero-point energies of isotopic species). The high sensitivity and resolution of microwave spectroscopy often make it possible to observe spectra of molecules composed of naturally occurring isotopes without recourse to isotopic syntheses.

Non-linear molecules are usually classified on the basis of their moments of inertia. Spherical-top molecules like CH_4 have three equal moments of inertia; symmetric-top molecules like CH_3Cl and NH_3 have two moments equal and the third different; and asymmetric-top molecules, such as CH_2Cl_2 and H_2O, have three unequal moments. In the case of a symmetric top, rotation round the unique threefold axis of symmetry causes no change in the dipole moment with respect to the direction of the incident radiation, and this mode of rotation is inactive. Rotation about the other axes is similar to rotation of a linear molecule, and leads to only one moment of inertia; for the determination of interatomic distances in a symmetric-top molecule isotopic substitution is therefore essential. For asymmetric tops, differences in energy levels depend upon more than one moment of inertia, and several isotopic substitutions may be necessary before the structure of even a simple asymmetric-top molecule can be worked out. For a planar molecule it may be noted that there are only two independent moments of inertia (the sum of the two smaller moments being equal to the third); thus for ClF_3, a planar T-shaped molecule with three independent geometrical parameters (e.g. the two Cl—F bond lengths and the F—Cl—F angle), it is necessary to use both chlorine isotopes to determine the structure.

Diatomic and linear triatomic molecules which have no permanent dipole moment may be studied by means of rotational Raman spectroscopy (see the following section), to which different selection rules apply. Among molecules which have been studied in this way are H_2, N_2, F_2 and CS_2.

Microwave measurements must be made on gases or vapours at low pressures (about 10^{-4} mm of mercury). The accuracy of the method results from the availability of klystron generators and is very high (for diatomic and linear triatomic molecules the limit is, in fact, set by the accuracy with which Planck's constant is known). The chief disadvantage of the method is that its use for the complete determination of a structure is restricted to small molecules; for larger molecules the number of parameters soon comes to exceed what can be obtained even by extensive isotopic substitution, and values for some bond lengths or angles must then be assumed in order to simplify the problem.

It may be mentioned that the splitting of the lines of microwave spectra in an electric field can be utilized in the determination of dipole moments, but since the nature of the effect is somewhat different for species of different symmetries we shall not go into details here. Values obtained agree well with those from the standard methods for the determination of dipole moments (i.e. from the temperature dependence of the dielectric constant, and from the determination of the dielectric constant at radio frequencies and of the refractive index at visible frequencies), and since measurements are made on gases at very low pressures, corrections for deviations from ideal gas behaviour, or (in the case of solution measurements) for solute–solvent interaction, are not involved. The measurement of nuclear quadrupole coupling constants from the hyperfine structure of the rotational lines for certain species is discussed briefly in Section 2.8.

2.6 Vibrational-rotational spectroscopy[1, 6, 8, 9]

Molecular spectra in the near infrared region correspond to changes in vibrational energy levels accompanied by (relatively very small) changes in rotational energy levels, and thus consist of bands of closely spaced lines. They may be observed either by absorption in the infrared using a conventional infrared spectrometer or in the Raman spectrum. In Raman spectroscopy a beam of monochromatic light of frequency ν_0 strikes the sample, and the light scattered at right angles to the incident beam is examined. Most of it is unchanged in frequency, but weak Raman lines at lower and higher frequencies are sometimes observed. A line at a lower frequency

ν' means that the molecule has absorbed an amount of energy $h(\nu_0 - \nu')$ where $\nu_0 - \nu' = \nu$ is the Raman frequency; a line at a higher frequency (usually a much weaker line) means that the molecule, originally in a higher energy level, has reverted to a lower energy level after collision with the incident photon. Raman lines may be caused by changes in either vibrational or rotational energy levels, and in each case may be observed at $\nu_0 \pm \nu$. For a rotational Raman spectrum to be observed, the selection rules are that the polarizability of the molecule (i.e. the proportionality constant between the dipole moment induced in an electric field and the strength of the field) perpendicular to the axis of rotation must be anisotropic (i.e. must be different in different directions in the plane perpendicular to the axis), and that $\Delta J = \pm 2$. The rotational fine structure of Raman lines corresponds in energy range to microwave spectroscopy and can be used for the determination of the structures of simple molecules having low moments of inertia, as mentioned in the preceding section. For most molecules, however, limitations imposed by intensity and resolving power result in fine structure not being observed.

In the infrared spectrum, which can be studied under much higher resolution, rotational fine structure is observed; in fact, because rotation occurs at the same time as vibration but with a lower frequency, the two forms of motion sometimes couple so that the pure vibrational frequency is not observed and its value has to be obtained by intrapolation. As for the pure rotational spectrum in the microwave region, the difference in frequency of the rotational lines can be used for the determination of moments of inertia of simple molecules; but since rotational lines associated with different vibrational transitions sometimes overlap, and since it is not safe to assume that the moment of inertia remains constant during vibration, the use of pure rotational spectra whenever possible for this purpose is to be preferred. For molecules possessing no permanent dipole moment, however, vibrational–rotational spectra must be used; this applies, for example, to CO_2, SiH_3D, BF_3 and *trans*-N_2F_2. (We should also mention here that electronic spectra also show rotational fine structure, analysis of which is used especially for the determination of bond lengths and angles in excited states of simple molecules known under ordinary conditions or in unstable species produced in photochemical reactions or in the electrical discharge.)

We can now concentrate on vibrational frequencies. For a polyatomic molecule, all vibrating motions may be described in terms of a limited number of normal modes of vibration, but combinations of these very frequently appear in spectra. Given the infrared and Raman spectra of a compound, therefore, it is often a very difficult matter to assign all vibrational bands, but some help is given by the general selection rules for infrared and Raman vibrational spectroscopy, which we shall now state.

For a vibration to be active in the infrared spectrum, the motion of the nuclei must result in a change in the dipole moment of the molecule (the intensity depends on the square of the ratio of the change in dipole moment to the change in internuclear distance). For activity in the Raman spectrum, the motion of the nuclei must result in a change in the polarizability of the molecule (in this case the intensity depends on the square of the ratio of the change in polarizability to the change in internuclear distance). Unfortunately, it is often very difficult to tell by inspection whether a particular vibration will result in a change in polarizability, and the application of these selection rules, except in very simple cases, is a matter for the expert. 'Forbidden' vibrations do, in fact, sometimes appear in spectra, but they are always very much weaker than 'allowed' vibrations.

A molecule or ion containing n atoms has $3n - 5$ normal modes of vibration if it is linear and $3n - 6$ modes if it is not. A diatomic molecule thus has one vibration, in which the internuclear distance successively increases and decreases (i.e. a stretching vibration). A linear triatomic molecule or ion XY_2 (e.g. CO_2 or NO_2^+) should have four normal modes of vibration, conventionally labelled ν_1, ν_2, etc., in order of descending symmetry and frequency. These are shown in Fig. 2.1, from which it will be seen that two of the modes (marked ν_{2a} and ν_{2b}) are identical; in spectroscopic terminology, the ν_2 mode is said to be doubly degenerate. Of these, the symmetrical stretching frequency (ν_1) is infrared-forbidden but Raman-active, and the reverse holds for ν_2 and ν_3. There is, in fact, a useful generalization that for a molecule which possesses a centre of symmetry no fundamental is active in both Raman and infrared spectra. For a linear triatomic molecule YXZ (e.g. OCS), ν_1, ν_2 and ν_3 are active in both, and the same is true for a bent triatomic molecule (e.g. H_2O or ONCl). Planar and pyramidal molecules XY_3 both have four normal modes of vibration, but the symmetrical stretching frequency is infrared-active only for the pyramidal case. Thus we see that the number of fundamental frequencies and their presence in or absence from the Raman or infrared spectrum can tell us the structure of a molecule; in the case of a compound of unknown structure, we must

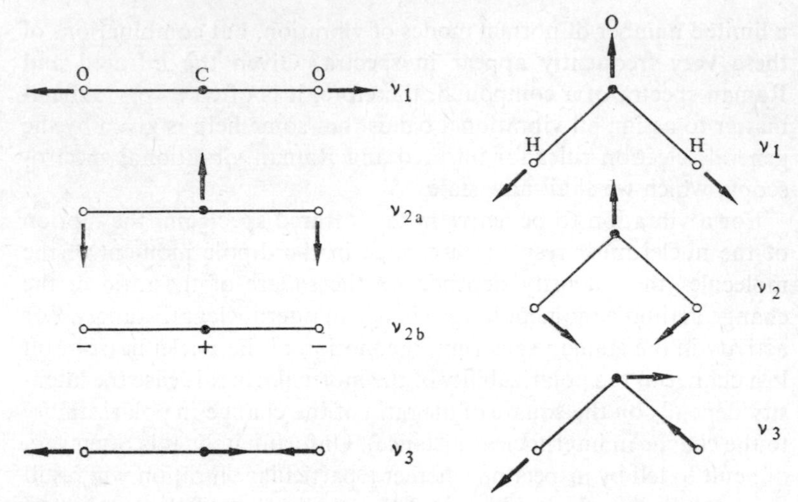

Figure 2.1 Normal modes of vibration in CO_2 and H_2O molecules (for ν_{2b} for CO_2, + and − denote upward and downward motion in the direction perpendicular to the plane of the paper)

write down possible structures and, making due allowance for degeneracy, predict the number of fundamentals and where they will occur, and compare prediction with observation.

In practice this is, unfortunately, less simple than it looks. It is not always possible to say which vibrations are fundamentals (overtones and combination bands, if unusually intense, may confuse the interpretation); some fundamentals may be too weak to be observed (or may lie outside the range of observation) and two fundamentals may by chance have nearly the same frequency. For substances which can be investigated under normal conditions, further information can be obtained by determining whether the Raman-scattered radiations are polarized, by considering infrared band shapes and by comparing observed frequencies and band shapes with those for related compounds of known structure; but for molecules containing more than about a dozen atoms or having unsymmetrical structures the reliability of the operation becomes doubtful.

It has been assumed implicitly throughout the foregoing discussion that the spectra were due to the molecule or ion whose structure was being examined and were not influenced by other species. This is true so long as spectra are measured in the gas phase or in solution in inert solvents (Raman spectra, because of their low intensities, are usually measured in the liquid phase or in concentrated solution).

Vibrational spectra of solids, and especially of ionic solids, are, however, subject to lattice effects to a marked extent, and splitting of frequencies is very common. Thus although the $[Co(CN)_6]^{3-}$ ion, for example, shows a single asymmetrical C≡N stretching frequency (ν_6) in the infrared spectrum in aqueous solution, solid $K_3[Co(CN)_6]$ shows two such frequencies when pure and three when present in solid solution in sodium chloride. Bands arising from combinations with low frequency lattice modes are also common. It would not be far from true even to say that in the solid state selection rules do not apply. The recent development of laser sources for Raman spectroscopy[10] makes it much easier to study gases and solids (including those which are coloured, i.e. have an electronic spectrum in the visible region); this will probably lead to a large number of structural assignments for solid compounds during the next few years, but some caution should be exercised in accepting conclusions which appear to be at variance with other evidence.

The foregoing discussion of vibrational spectroscopy has been concerned with the determination of molecular symmetry, but many applications of this method have as their objectives only the identification of a molecule or a functional group. In such 'fingerprinting' applications (mention of which has been made in the previous chapter) the frequency, shape and intensity of a band are all useful characteristics. The C≡N stretching band in cyano compounds, for example, not only normally appears in the range 2040–2170 cm^{-1}, but is also identifiable by its sharpness and high intensity; Si—F stretching bands are also very strong whereas S—H stretching bands are usually very weak. The uses of C≡O stretching modes in metal carbonyls (which are very strong in the infrared spectra) for the identification of different kinds of carbonyl group are discussed in detail in Chapter 20. In complex molecules such as those of these substances the establishment of empirical correlations between infrared spectra and structures determined by X-ray methods is essential. It must be said that in general more care is needed in the use of 'fingerprinting' methods in inorganic than in organic chemistry, chiefly because the range of atoms involved is so much greater; shifts on substitutions may be much larger, and it is sometimes impossible to assign observed bands to individual vibrations. 'Cage' molecules such as S_4N_4, P_4S_3 and P_4S_{10}, for example, tend to vibrate as a whole.

In a few instances really large variations in the frequency of a particular vibration denote a marked change in bond order. Complexes containing NO, for example, may have this ligand present as

NO^+ (stretching frequency about 2200 cm^{-1}), NO^- (stretching frequency about 1150 cm^{-1}) or an intermediate form; NO^+ and NO^-, it will be noticed, are isoelectronic with N_2 and O_2 respectively; and in NO^- and O_2 there are two more electrons than in NO^+ and N_2. These are necessarily accommodated in antibonding orbitals, thus weakening the bond. The quantitative measure of the resistance of a bond to a *small* deformation is its force constant (discussed earlier in Section 1.4); the correlation between bond lengths, bond energies, force constants and vibrational frequencies is discussed further in Chapter 5.

Finally, we may mention the use of comparisons of spectra of isoelectronic species and predictions of spectra (based on spectra and structures of related compounds) in the characterization of new compounds. When arsine and hydrogen iodide combine at low temperatures, for example, the formation of a species having an infrared spectrum like that of GeH_4 shows that the arsonium ion AsH_4^+ has been produced; a similar comparison of the infrared spectrum of solid 'perchloric acid monohydrate' with the vibrational spectrum of ammonia indicates that the former compound is really $H_3O^+ClO_4^-$. A particularly striking example of the use of infrared spectroscopy is the demonstration that HOF is formed by photolysis of mixtures of fluorine and water suspended in a nitrogen matrix at 15°K and the calculation of its thermodynamic functions.[11]

2.7 Nuclear magnetic resonance spectroscopy[1, 8, 12, 13]

Most of the applications of nuclear magnetic resonance in chemistry are concerned with the identification of functional groups, the determination of the relative numbers of nuclei in these groups, and the ascertaining of the relative positions of the groups in molecules in the gas phase or in solution. It is, however, possible to determine the positions of hydrogen atoms (and, in principle, those of a few other elements) in certain compounds and to obtain information about molecular symmetry by nuclear magnetic resonance spectroscopy, and we shall give a brief account of these subjects first, beginning with the location of hydrogen atoms in solids.

Many nuclei, notably those of ^{1}H, ^{13}C, ^{19}F, ^{29}Si and ^{31}P, possess a nuclear spin of $\frac{1}{2}$ and are suitable for nuclear magnetic resonance studies; spectra can never be obtained for nuclei having zero spin and rarely (e.g. ^{11}B) for nuclei having spin $\geqslant 1$. Nuclear magnetic moments are less than electronic magnetic moments by a factor of about 10^3. In a magnetic field the nuclear magnet for a nucleus of spin $\frac{1}{2}$ may be aligned parallel or antiparallel to the field, these configurations being of lower and higher energy respectively. The difference in energy between these two states is very small relative to kT, so that the states are very nearly equally populated, and the observation of transitions between them necessitates the use of intense fields and of a resonance effect. Since the energy difference between the states is proportional to the strength of the magnetic field, the resonance frequency is also proportional to the strength of the field: for a hydrogen nucleus the resonance frequency in a field of 10^4 gauss is $4{\cdot}3 \times 10^6$ cycles sec^{-1} (i.e. in the radiofrequency region). It is usual to study the effect by the use of a very carefully controlled constant frequency and a small variation of the strength of the magnetic field; a large permanent magnet or electromagnet supplies a static field and a small variable field is superimposed on this by a subsidiary electromagnet. Absorption of electromagnetic radiation is then plotted against the change in the applied field strength; since, however, for a given nucleus the field strength and frequency are linearly proportional, nuclear magnetic resonance spectra may be measured in either field or frequency units.

For isolated nuclei of spin $\frac{1}{2}$ (e.g. protons), the absorption spectrum is a single sharp line. For a crystal containing protons grouped in pairs (e.g. for a crystal of a salt containing a single molecule of water of crystallization in the unit cell), a doublet is obtained; triangular arrangements of protons give a triplet, and tetrahedral arrangements a quadruplet. In each case the lines are broadened somewhat owing to weaker interactions with more distant nuclei. For a doublet, for example, the form of the absorption curve depends upon the distance apart of the protons r and the angle θ made by the line joining them with the direction of the field; the separation of the doublet therefore varies with the orientation of the crystal in the field. If a finely powdered solid is used in place of a single crystal, the absorption is an average resulting from the superposition of the lines from all of the possible values of θ. In either case, the second moment or mean-square width of the observed absorption line (the moment of inertia of the absorption curve about its axis of symmetry divided by the area under the curve) is a complicated function of r^{-6}. The procedure is to determine the structure as far as possible by X-ray methods, and then to calculate second moments of the absorption curve for various values of r and see which agrees best with the experimental result. This has been done, for example, for ammonium chloride for which it can safely be assumed that the bond angle in the cation is regular tetrahedral, and for nitric acid monohydrate, in reality $H_3O^+NO_3^-$, for which the general shape of the absorption line shows the presence of an equilateral-triangular three proton unit and the second moment gives the hydrogen–hydrogen distance, and hence the oxygen–hydrogen distance, in the pyramidal cation.

When the compound under investigation contains nuclei of more than one kind having spin $\frac{1}{2}$, as in the case of a phosphonium salt, account must be taken of all interactions between such atoms. Measurements in solids are made at low temperatures since internal motions (such as free rotation) narrow the absorption line; at higher temperatures the spectrum of ammonium chloride, for example, becomes a single line. Paramagnetic species, which produce strong magnetic fields, broaden the absorption and make it impossible to apply the method, which is therefore useless for the determination of the positions of hydrogen atoms in many transition metal compounds. For diamagnetic substances, however, the method provides a valuable alternative to neutron diffraction, though it should be noted that for the location of a proton the nuclear magnetic resonance method gives a result which refers to a time average of about 10^{13} times that obtained by neutron diffraction.

So far only occasional use has been made of nuclei other than ^1H in low resolution nuclear magnetic resonance studies of solids, but mention may be made of one example: ^{19}F studies of $K_2[TiF_6]$, which is known from X-ray diffraction to contain a regular octahedral anion, have been used to obtain the fluorine–fluorine, and hence the titanium–fluorine, distance in this compound.

For liquids and gases, the individual molecules are moving relatively freely, and very sharp resonance lines result. Moreover, since the magnetic field experienced by a nucleus is very slightly modified by the shielding effect of extranuclear electrons, the line appears at a change in the applied field strength which depends upon the chemical environment of the nucleus and on the field strength. The characteristic positions of these lines are known as chemical shifts, and their intensities are related to the relative numbers of nuclei in each type of environment. Ethanol which has not been specially purified from traces of acid, for example, under medium resolution gives three proton resonance lines of relative intensity 1 : 2 : 3. Under high resolution, these lines are seen to be a singlet, a quartet of relative intensity 1 : 3 : 3 : 1, and a triplet of relative intensity 1 : 2 : 1 respectively. The magnitudes of these spin–spin splittings (known as spin–spin coupling constants), which arise from interactions with protons on adjacent carbon atoms, are independent of the field strength, a fact which enables spin–spin splittings to be distinguished from the presence of non-equivalent protons having different chemical shifts.

Because of the difficulty of providing a standard reference magnetic field, a chemical standard is normally employed for the measurement of chemical shifts; for protons this is customarily tetramethylsilane (Me_4Si), which gives a single strong sharp line, the position of which is very nearly independent of the solvent used and is well away from regions in which other protons absorb. Chemical shifts for protons are preferably expressed on the τ scale for protons defined by

$$\tau = [10^6(\nu - \nu_{Me_4Si})/\nu_{Me_4Si}] + 10 \text{ p.p.m.}$$

Fluorine chemical shifts are usually measured with respect to CCl_3F and phosphorus shifts with respect to 85 per cent H_3PO_4.

Let us consider further the spectrum of pure ethanol. The medium resolution resonances of relative intensity 1 : 2 : 3 correspond to protons of the OH, CH_2 and CH_3 groups respectively, the intensities being proportional to the numbers of protons of each kind. The splitting patterns under high resolution may be explained by considering the methyl group triplet. Each of the CH_3 protons is under the influence of the nuclear spins of the two protons of the CH_2 group. The total spin of the CH_2 group protons is $\frac{1}{2} + \frac{1}{2} = 1$ leading to $(2 \times 1) + 1 = 3$ different orientations of their total spin and thus three different fields and, in consequence, three lines for the methyl group. But since there is only one way for the CH_2 group protons both to add to, or both to subtract from, the applied field, whereas there are two ways in which their spins can cancel out, the central line of the CH_3 triplet is of twice the intensity of the other two. The quartet associated with the CH_2 protons may be similarly explained by considering the effect of the three CH_3 protons. In general, spin coupling to one proton leads to a doublet, the lines being of equal intensity; and coupling to two protons leads to a 1 : 2 : 1 triplet, to three protons to a 1 : 3 : 3 : 1 quartet, to four protons to a 1 : 4 : 6 : 4 : 1 quintet, and so on. Under ordinary conditions the hydroxyl group proton exchanges very rapidly with those of other molecules; its average spin state is therefore zero and the protons of the CH_2 group have no effect on it. If, however, ethanol is very thoroughly purified, the OH proton resonance becomes a 1 : 2 : 1 triplet like that of the CH_3 protons, though the total intensity remains, of course, only one-third as great. At the same time the pattern of the CH_2 protons, now under the influence of the spins of the OH and CH_3 protons, becomes a (poorly resolved) octet. Thus the detailed spectrum can tell us something not only about structure but also about rates of exchange processes. Ethanol will provide a final example of such information. In the staggered configuration one CH_3 proton of any CH_3CH_2X compound should be slightly different from the other two; however, no such difference is detectable by proton magnetic resonance, from which it follows that rotation about the C—C single bond must be a faster process than observation of the spectrum.

Few inorganic compounds contain more than one or two kinds of proton, so inorganic applications of proton resonance spectroscopy are limited. However, special mention should be made of the very high field proton resonance of hydrogen bonded to transition metals ($\tau = 10$–50); this is very useful in, for example, studies of the protonation of metal carbonyls (see Chapters 8 and 20). A distinction can sometimes be made between coordinated and solvent water by means of proton magnetic resonance, and for cations having a relatively low rate of hydrogen exchange between the first coordination sphere and the bulk solvent, hydration numbers can be calculated. Similar studies using ^{17}O-labelled water have the advantage that many hydrated cations exchange oxygen much less rapidly than hydrogen with solvent water; they are referred to in Section 6.2.

The equilibrium and kinetics of redistribution reactions, typified by

$$PCl_3 + P(OEt)_3 \rightleftharpoons PCl_2OEt + PCl(OEt)_2$$

may often be followed conveniently by nuclear magnetic resonance if an element having a nuclear spin of $\frac{1}{2}$ is involved. In this case the ^{31}P chemical shifts for all four compounds are appreciably different, and since the intensities for a given nucleus in different sites are always proportional to the number of nuclei in those sites, both the rate and the thermodynamics of the reaction can be studied.

The interpretation of high resolution nuclear magnetic resonance spectra is obviously immensely important in organic chemistry, but it also has applications in inorganic chemistry. The fluorine resonance in bromine pentafluoride, for example, consists of two peaks of intensity ratio 4 : 1, the intense line being a doublet and the weak line a quintet with the relative intensities 1 : 4 : 6 : 4 : 1. This shows that the molecule is a square pyramid (though it does not, it may be noted, tell us exactly where the bromine atom is). At 180°C the spectrum collapses to give a single line, showing that at this temperature the fluorine atoms are, for nuclear magnetic resonance purposes, identical. A similar state of affairs holds for sulphur tetrafluoride and chlorine trifluoride. For the former, the spectrum at $-98°$ consists of two triplets, indicating two non-equivalent pairs of fluorine atoms in agreement with the structure established by other methods as a trigonal bipyramid with one equatorial position occupied by a lone pair of electrons; at $-58°$ only two broad peaks are seen; and at room temperature the spectrum consists of a single peak. This again arises from the relatively long time-scale of nuclear magnetic resonance spectroscopy. For $Fe(CO)_5$ (containing ^{13}C) and PF_5, intramolecular exchange is found to occur at all temperatures in the range $-100°$ to room temperature; its probable mechanism is shown in Chapter 14. Although, therefore, high resolution spectra obtained at low temperatures can provide information about molecular symmetry, nuclear magnetic resonance is not always an adequate substitute for vibrational spectroscopy in this field, and the restricted range of nuclei which can be used imposes further limitations.

It will be noticed that we have treated nuclear magnetic resonance in an empirical way, and have said nothing about the interpretation of chemical shifts and spin–spin coupling constants or the relationship between these quantities and the nuclear spin quantum numbers of atoms in the species to which they refer. For discussions of these subjects more advanced accounts[8, 13] should be consulted.

2.8 Nuclear quadrupole resonance spectroscopy[8, 14]

If the spin quantum number I of a nucleus is greater than $\frac{1}{2}$, the positive charge distribution in it is non-spherical, and the potential energy of such a nucleus in an inhomogeneous electric field depends upon the orientation of the nuclear quadrupole moment (the measure of the deviation of the charge distribution from spherical symmetry) to the field. The allowed orientations are quantized, just as the energy of a spinning electron in the positive field of a nucleus is quantized, and the nucleus has $2I + 1$ orientations, describable by a nuclear magnetic quantum number m, which may have values $-I$, $-(I - 1)$, $\ldots, 0, \ldots, (I - 1), I$. In the lowest energy level the greatest amount of positive nuclear charge is as near as possible to the greatest density of negative charge in the environment. The difference in energy between different orientations is very small (it corresponds to electromagnetic radiation in the radiofrequency region), and at ordinary temperature different energy levels are almost equally populated. Direct measurement of the energy of transition between different orientations, which depends upon the quadrupole moment eQ and the field gradient q, must be carried out in solids, because molecular rotation in a gas or a liquid averages the field gradient and splitting of the quadrupole energy levels does not occur. There is also no splitting of energy levels if the nuclear charge distribution is symmetrical (i.e. if I is 0 or $\frac{1}{2}$) or if the electron distribution round the nucleus is symmetrical (e.g. if the outer electronic configuration of the species is s^1, s^2 or sp^3).

The value of eQq, which is called the nuclear quadrupole coupling constant, may then be obtained in much the same way as the energy required for a transition in nuclear magnetic resonance spectroscopy, except that no magnet is required, the internal crystal field of the specimen replacing the magnetic field. Alternatively, the nuclear quadrupole coupling constant may be obtained from the fine structure of the microwave spectrum, which originates in the fact that different nuclear orientations give rise to slightly different moments of inertia (see Section 2.6).

If the nuclear quadrupole moment of an atom is known from other sources (e.g. nuclear magnetic hyperfine structure), the field gradient at the nucleus can be calculated; even if the moment is not known, however, field gradients in different compounds can be compared. The method has so far been used mainly for compounds of ^{35}Cl, ^{37}Cl, ^{79}Br, ^{81}Br (all of which have $I = \frac{3}{2}$) and ^{127}I (which has $I = \frac{5}{2}$).

The simplest application of nuclear quadrupole resonance spectroscopy is to the detection of the same isotope in different environments. Thus it has been shown that all the bromine atoms in $K_2[SeBr_6]$ are equivalent above $-33°$, but that between this temperature and $-52°$ two types of crystallographically non-equivalent bromine atoms are present. Below $-52°$ there are three types of bromine atom present. This suggests a progressive drop in lattice symmetry (e.g. cubic → tetragonal → orthorhombic). Phase transitions in several other complex halides of formula type $M^I_2[M^{IV}X_6]$ ($X = Cl$, Br or I) have been found.

Of much greater chemical interest, however, is the variation in the nuclear quadrupole coupling constant in a series of compounds of the same element. Since the distribution of s-electron density about the nucleus is symmetrical and d- and f-electrons do not penetrate near the nucleus, the field gradient can be considered to be determined by the distribution of p-electron density. Some values for eQq of ^{35}Cl in chlorine compounds are (in Mc sec^{-1}): BrCl, -104;

ICl, −82; SiF_3Cl, −43; KCl, 0. Since the value for ^{35}Cl in atomic chlorine is −110, this suggests a progressive decrease in field gradient along the series corresponding to a decrease in covalent character as the chlorine atom becomes more nearly a Cl^- ion. Unfortunately the quantitative assessment of the distribution of p-electron density is fraught with difficulties, since the type of hybridization and the degree of π-bonding, as well as the extent of ionic character in the bond, may affect the field gradient. Thus although the ionic character of bonds in complex chlorides, bromides and iodides estimated from nuclear quadrupole coupling constants agrees roughly with expectations based on Pauling's electronegativity values (see Chapter 5), it is difficult at present to know how much significance to attach to these observations.

There are, however, possibilities in the use of nuclear quadrupole resonance spectra for 'fingerprinting', especially since most species which give such spectra do not give nuclear magnetic resonance spectra. Some work has already been done on ^{14}N compounds, and further developments in this field should follow.

2.9 Mössbauer spectroscopy[8, 15, 16]

Mössbauer spectroscopy involves nuclear transitions resulting from absorption of γ-radiation (i.e. radiation of very short wavelength); the conditions for absorption depend upon the electronic environment and site symmetry of the nucleus. The method has so far been used chiefly for ^{57}Fe, ^{119}Sn, ^{127}I and ^{129}I, but is capable of application to some seventy other isotopes.

Consider a gaseous system consisting of a radioactive source of γ-rays and a sample capable of absorbing them. The source nucleus, on emission of a γ-ray, decays to the ground state, and the energy of the γ-radiation E_γ is given by

$$E_\gamma = E_r + D - R$$

where E_r is the energy difference between the excited and ground states of the source nucleus, D is the Doppler shift due to the translational motion of the nucleus, and R its recoil energy. The energy of a γ-ray emitted from a nucleus moving in the same direction as the γ-ray is slightly different from that of one emitted from a nucleus moving in another direction. Since the source nuclei may be moving in many directions the result is a band of values for E_γ rather than a single value; this is referred to as Doppler broadening.

In Mössbauer spectroscopy the source and sample nuclei are the same (e.g. ^{57}Fe, produced in the source in an excited state by a decay of ^{57}Co in which the nucleus captures an extranuclear electron). The energy of the γ-ray absorbed to effect a transition in the sample is then

$$E_{\gamma} = E_{r} + D + R$$

(In this case R is positive, since the existing γ-ray has to bring about recoil of the absorbing nucleus.) Values of E_{γ} for source and absorber are thus two bands whose centres are $2R$ apart, and there is only a very low probability (corresponding to the degree of overlap of the bands) that a γ-ray from the source will be absorbed and effect a transition. If, however, source and sample nuclei are both in crystals, the probability of a recoilless transition is much increased, and a spectrum can be obtained if the nuclei are in identical environments. If they are not, resonant absorption occurs only when the energy of the emitted γ-ray is varied by moving the source towards the absorber or away from it; the greater the velocity of such movement towards the absorber, the greater the energy of the γ-ray (by the Doppler effect), and vice versa. Thus the measurement of a Mössbauer spectrum consists essentially of varying the source velocity and plotting it against the absorption of γ-rays. The shifts in the absorption maxima (relative to the value for stainless steel as an arbitrary zero) are known as chemical or isomer shifts, and are conventionally expressed in mm sec^{-1} rather than as energies.

From the chemical point of view most of the interest in the Mössbauer effect centres on the use of isomer shifts for the identification of atoms in different electronic environments. Differences in isomer shifts arise from the fact that the relative energies of the ground and excited nuclear states are linear functions of the electron density at the nucleus. Since only *s*-electrons have a finite electron density at the nucleus, *p*-, *d*- and *f*-electrons affect the isomer shift only by virtue of any effect of screening the *s*-electrons from the nuclear charge. We cannot discuss the theoretical calculation of isomer shifts here, however, and will treat the isomer shift as a quantity to be compared for different substances at the same temperature in the same way as the nuclear magnetic resonance chemical shift or the nuclear quadrupole resonance coupling constant. Before doing this, however, it should be mentioned that if the ground or excited state of the nucleus possesses a quadrupole moment, the energy levels in that state will be split to an extent depending on the quadrupole moment and the field gradient at the nucleus (just as in nuclear quadrupole resonance spectroscopy). In the case of ^{57}Fe, *I* for the ground state is $\frac{1}{2}$, but for the first excited state it is $\frac{3}{2}$. For high-spin complexes, there is a virtually spherical distribution of electron density at the nucleus in the case of iron (III) (d^5), but this is not so for iron (II), and in the latter case a much larger quadrupole splitting results. Conversely, $[Fe(CN)_6]^{4-}$ shows no quadrupole splitting, but $[Fe(CN)_6]^{3-}$ does. For octahedral iron (II) complexes not containing six equivalent ligands, however (such as $[Fe(CN)_5NH_3]^{3-}$), quadrupole splitting occurs.

A few isomer shifts for ^{57}Fe relative to stainless steel at ordinary temperatures are: high-spin iron (II), +0·93 to +1·01; high-spin iron (III), +0·10 to +0·16; $K_4[Fe(CN)_6]$, −0·33; $K_3[Fe(CN)_6]$, −0·41; K_2FeO_4, −1·20 mm sec^{-1}. Comparison of their isomer shifts with these values may be used to show that the dark blue precipitates formed from Fe^{3+} and $[Fe(CN)_6]^{4-}$ or from Fe^{2+} and $[Fe(CN)_6]^{3-}$ are both iron (III) hexacyanoferrates (II). Similar investigations using ^{123}Sb and ^{129}I have shown that there are two types of antimony atoms in SbO_2 and of iodine atoms in CsI_3; these facts are, of course, already known from X-ray data.

Information about symmetry is provided for *cis* and *trans* compounds of the type SnA_2B_4 (the *trans* compounds give simpler spectra) and for IF_6^+ in IF_6AsF_6 (which gives only a one-line spectrum and hence must have a strictly octahedral structure). Comparison of the ^{129}Xe Mössbauer spectrum of XeF_4 and that obtained from the β-decay product of $K^{129}ICl_4.H_2O$ suggests planar $XeCl_4$ is produced in the decay. Finally, information about bonding is given for a number of compounds, especially complex cyanides of iron. The closeness of isomer shifts for $[Fe(CN)_6]^{3-}$ and $[Fe(CN)_6]^{4-}$ suggests that the real oxidation states of the iron atoms in these species are less different than their formal oxidation states. The fact that the shift is more negative for $[Fe(CN)_5NO]^{2-}$ supports the idea that the 'nitrosyl' ligand in this complex is NO^+ which, by removing *d*-electron density by means of π-bonding, results in less shielding of the 4*s*-electrons and hence raises the *s*-electron density at the nucleus. (In the case of ^{57}Fe, increase in *s*-electron density has been shown to produce a more negative shift; this is not necessarily true for other isotopes.) Conversely, $[Fe(CN)_5(NH_3)]^{3-}$, in which NH_3 cannot form π-bonds, has a less negative isomer shift than $[Fe(CN)_6]^{4-}$. Other aspects of bonding in these complexes are discussed in Chapters 15 and 18.

2.10 Mass spectrometry[8, 17]

The principles underlying this technique were mentioned in connection with the separation of isotopes in Chapter 1. Since it is applicable mainly to substances having a vapour pressure of the order of 1 mm of mercury at room temperature or, at most, a few hundred degrees higher, mass spectrometry is far more widely used in organic than in inorganic chemistry, and we shall discuss it only briefly here.

Bombardment of a molecule A with electrons of the necessary energy may lead to the formation of A^+ (the molecular ion), A^{n+} and a variety of species formed by fragmentation of such ions. For most substances a peak (often intense) due to A^+ appears in the mass spectrum, and it is usually the peak of highest m/e, thus affording a simple and accurate method of determining molecular weight (occasionally combination of A with a neutral fragmentation product results in a peak of higher mass number). Consideration of how the molecular weight can be built up from atomic weights of elements present will often make it possible to arrive at a molecular formula also; this is the case, for example, with the phosphorus sulphides, which are very difficult to characterize by chemical methods, and with many organometallic compounds. Account must, of course, be taken of the fact that most elements have more than one isotope; thus a peak at m/e due to an ion RCl^+ (in which R contains only elements occurring as one isotope) is always accompanied by a peak, roughly one-third as intense, at $(m + 2)/e$, the abundances of ^{35}Cl and ^{37}Cl being 75·4 and 24·6 per cent respectively.

The fragmentation pattern of the molecular ion is invaluable in the determination of structures of organic compounds, for which extensive correlations between such patterns and known structures have now been established. As well as simple fragmentation of the molecular ion, however, a number of secondary processes may take place and complicate the interpretation of the mass spectrum; a molecule in which the atomic sequence is PQRS, for instance, may undergo the following changes:

$$PQRS + e \rightarrow PQRS^+ + 2e$$
$$PQRS^+ \rightarrow PS^+ + QR$$

The rearrangement of the ion $PQRS^+$, leading to the formation of a species PS^+ containing a bond not present in the original molecule, can be misleading, the more so since it may be suspected only if the structures and fragmentation patterns of other compounds closely related to PQRS are known. This type of rearrangement occurs extensively with cage structures such as the phosphorus sulphides. However, in areas of inorganic chemistry in which structures of several compounds are known independently, mass spectrometry is a useful tool; it has recently been used to show, for example, that the hydride B_8H_{18} contains two tetraborane fragments linked by a B—B bond. Correlations between structures and spectra of organometallic compounds are now being made, and it may be predicted with confidence that mass spectrometry will play an important part in the elucidation of the structures of new organometallic compounds.

A number of physicochemical applications of mass spectrometry also yield information which is useful in inorganic chemistry. The equilibrium between a solid and its vapour can be measured at different temperatures by determining the variation in the intensity of a particular peak in the mass spectrum, and from this heats of sublimation of, for example, metals and salts can be determined. Measurements of appearance potentials provide one of the best ways of determining bond dissociation energies; these are useful not only in mechanistic studies but can also, if enough of them are measured, be used to provide thermochemical data.

2.11 Photoelectron spectroscopy[18]

This method is briefly mentioned here more because of its potential usefulness than its past applications. It is essentially an application of the photoelectric effect to gases; high energy monochromatic radiation (produced by the action of an electric discharge on helium) is absorbed by the gas, and the energy spectrum of the electrons which are ejected is measured using a suitable analyser. Then

$$E = h\nu - I$$

where I is an ionization potential. The special interest of the technique in inorganic chemistry arises from the fact that the nature of the photoelectron spectrum depends upon the bonding character of the electron which has been removed. Excitation of an electron from a non-bonding orbital gives rise to a peak with a strong vibrationless component, followed often by a few weaker components with some vibrational energy. This is because on excitation of a non-bonding electron there are no significant changes in equilibrium bond lengths, and the molecular ion formed is therefore in its vibrational ground state. Excitation of an electron from a bonding or antibonding orbital, on the other hand, does lead to changes in equilibrium bond lengths, and so the molecular ion formed can be in various vibrational levels.

In the photoelectron spectra of HCl, HBr and HI, for example, the band at the lowest ionization potential shows a typical non-bonding structure, the vibrationless peak being very narrow and by far the strongest component; much the same is true for CH_3Cl. For SiH_3Cl and GeH_3Cl, on the other hand, the shape of the band at the lowest ionization potential suggests $p_\pi \rightarrow d_\pi$-bonding from the halogen to the Group IV element.[19]

For H_3NBH_3 and $OCBH_3$, correlation of the spectra of the donor molecules with those of the complexes leads to the conclusion that whereas in the former compound there is only a σ-bond between nitrogen and boron, in the latter compound the σ-carbon $\rightarrow$ boron bond is reinforced by a π-boron $\rightarrow$ carbon bond.

Ionization potentials for the removal of a particular electron (e.g. a chlorine 3*p*-electron from a series of organic chlorine compounds) can be correlated with structure, so this new branch of spectroscopy also has promising applications in analytical chemistry and in molecular structure determination.

References for Chapter 2

1 Wheatley, P. J., *The Determination of Molecular Structure*, 2nd edn. (Oxford University Press, 1968).
2 Stout, G. H., and Jensen, L. H., *X-ray Structure Determination* (Macmillan, 1968).
3 Woolfson, M. M., *An Introduction to X-ray Crystallography* (Cambridge University Press, 1970).
4 Bacon, G. E., *Adv. Inorg. Chem. Radiochem.*, 1966, **8**, 225.
5 Almenningen, A., Bastiansen, O., Haarland, A., and Seip, H. M., *Angew. Chem.* (*International*), 1965, **4**, 819.
6 Whiffen, D. H., *Spectroscopy* (Longmans, 1966).
7 Sugden, T. M., and Kenney, C. N., *Microwave Spectroscopy of Gases* (van Nostrand, 1968).
8 Drago, R. S., *Physical Methods in Inorganic Chemistry* (Reinhold, 1965).
9 Nakamoto, K., *Infrared Spectra of Inorganic and Coordination Compounds*, 2nd edn. (Wiley, 1970).
10 Booth, M. R., and Gillespie, R. J., *Endeavour*, 1970, **24**, 89.
11 Noble, P. N., and Pimentel, G. C., *Spectrochim. Acta*, 1968, **24A**, 797.
12 Eaton, D. R., in *Physical Methods in Advanced Inorganic Chemistry*, eds. H. A. O. Hill and P. Day (Interscience, 1968).

13 Lynden-Bell, R. M., and Harris, R. K., *Nuclear Magnetic Resonance Spectroscopy* (Nelson, 1969).
14 Kubo, M., and Nakamura, D., *Adv. Inorg. Chem. Radiochem*, 1966, **8**, 257.
15 Greenwood, N. N., and Gibb, T. C., *Mössbauer Spectroscopy* (Chapman & Hall, 1971).
16 Fluck, E., *Adv. Inorg. Chem. Radiochem.*, 1964, **6**, 433.
17 Lewis, J., and Johnson, B. F. G., *Accts. Chem. Research*, 1968, **1**, 245.
18 Turner, D. W., *Chem. Britain*, 1968, **4**, 435.
19 Cradock, S., and Ebsworth, E. A. V., *Chem. Comm.*, 1971, 57.

3 The structures and energetics of ionic crystals

3.1 Introduction[1, 2]

The structures of solid inorganic compounds are nearly always determined by X-ray diffraction. Since the scattering powers of different species depend upon the total number of electrons they contain, X-ray methods suffer from two important limitations: they are not suitable for the determination of the positions of light atoms in the presence of much heavier atoms, and they are generally incapable of identifying the state of ionization of the species present. The former difficulty can often be met by the use of neutron diffraction, since (for reasons mentioned in Chapter 2) scattering powers of different atoms for neutrons vary only narrowly and irregularly with increase in atomic number; this method has been used extensively in the study of hydrides, oxides, carbides and fluorides. The latter difficulty, however, requires a very precise determination of electron density for its solution by X-ray methods, and this has so far been achieved in only a few instances. In sodium chloride, for example, it is established that, within a small margin of error, ten and eighteen electrons are associated with the sodium and chlorine nuclei respectively, showing that the species present are Na^+ and Cl^-. More commonly, the ionic nature of a structure is inferred from the adoption of a regular three-dimensional structure and the absence of units recognizable to the chemist as molecules, and is supported by evidence from a variety of other sources (e.g. hardness, conductivity, solubility, and spectroscopic and thermochemical data), sometimes of doubtful reliability. In this chapter we shall be concerned mainly with substances for which the ionic model is generally accepted as correct, but mention will also be made of some compounds for which it is unsatisfactory.

3.2 Typical MX structures

There are four basic structures in which salts of formula type MX commonly crystallize: those of caesium chloride, sodium chloride, zinc sulphide (zinc blende or sphalerite) and zinc sulphide (wurtzite). These structures are illustrated in Fig. 3.1 (a) to (d); the first three have cubic symmetry, the last is hexagonal. Of more significance

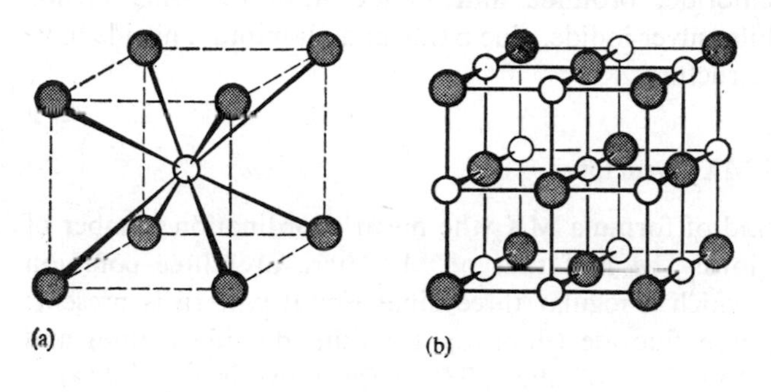

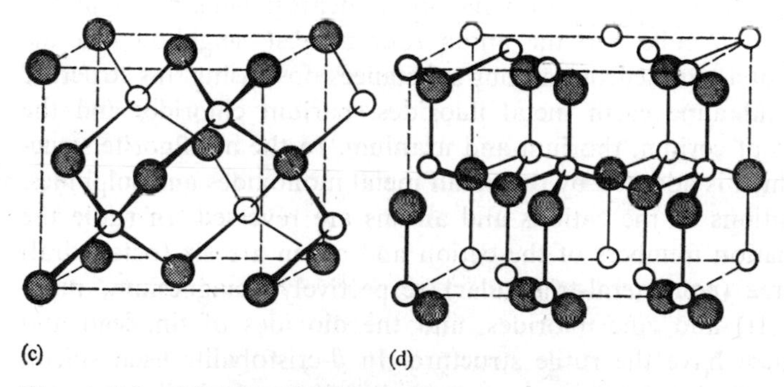

Figure 3.1 (a) The CsCl structure
(b) The NaCl structure
(c) The Zinc blende (cubic ZnS) structure
(d) The Wurtzite (hexagonal ZnS) structure

than the external symmetry, however, are the environments of the individual ions. In the caesium chloride structure each ion of either sign is surrounded by eight nearest neighbours at the corners of a cube; this structure is relatively uncommon, but is adopted by caesium chloride, bromide and iodide, thallium (I) chloride and bromide, and ammonium chloride and bromide at ordinary temperatures. Each ion in the sodium chloride structure has six nearest neighbours at the corners of an octahedron; most alkali metal halides and hydrides, alkaline earth metal oxides, silver chloride, caesium fluoride, and ammonium chloride and bromide above 184° and 138° respectively have this structure. In both the zinc blende and wurtzite structures, each ion has four nearest neighbours distributed tetrahedrally around it, and the structures differ only in the orientation of the tetrahedra; copper (I) chloride, bromide and iodide have the zinc blende structure, whilst silver iodide, zinc oxide and aluminium nitride have the wurtzite structure.

3.3 Typical MX_2 structures

In a compound of formula MX_2 the mean coordination number of X must obviously be half that of M. There are three common structures in which a regular three-dimensional pattern is present: those of calcium fluoride (fluorite), titanium dioxide (rutile) and β-cristobalite (one form of silica). These are shown in Fig. 3.2 (a) to (c). Fluorite and β-cristobalite have cubic symmetry; rutile is tetragonal. In fluorite the cation has eight nearest neighbours at the corners of a cube and the anion four nearest neighbours at the corners of a tetrahedron; among substances possessing this structure are the alkaline earth metal fluorides, barium chloride, and the dioxides of cerium, thorium and uranium. In the antifluorite structure, which is adopted by the alkali metal monoxides and sulphides, the positions of the cations and anions are reversed. In rutile the coordination numbers of the cation and anion are six (octahedral) and three (equilateral-triangular) respectively; magnesium, manganese (II) and zinc fluorides, and the dioxides of tin, lead and manganese have the rutile structure. In β-cristobalite each silicon atom has four oxygen atoms surrounding it tetrahedrally and each oxygen atom has two silicon atoms as nearest neighbours (the angle Si–O–Si is actually somewhat less than 180°); beryllium fluoride and germanium dioxide also have this structure, but it may be noted that the latter compound also exists in the rutile structure.

Many compounds of formula type MX_2 crystallize in so-called layer structures, typified by that of cadmium iodide illustrated in Fig. 3.3. In this, each cadmium atom is surrounded octahedrally by

---

Figure 3.2 (a) The CaF_2 structure
(b) The TiO_2 structure
(c) The SiO_2 (β-cristobalite) structure

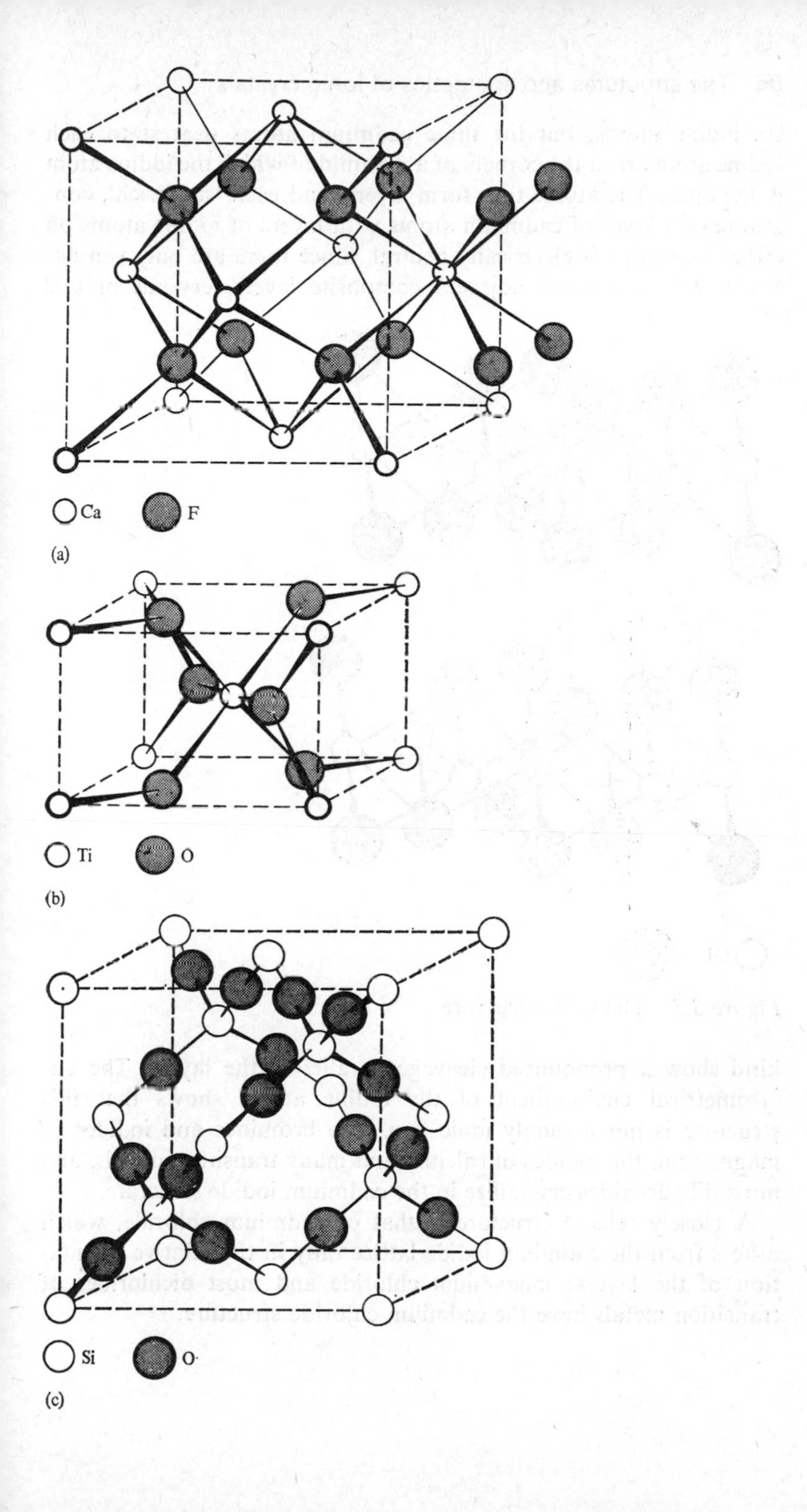
Ca
F
(a)
Ti
O
(b)
Si
O
(c)

six iodine atoms, but the three cadmium atoms nearest to each iodine atom are at the corners of a pyramid of which the iodine atom is the apex. The atoms thus form layers, and each 'sandwich', consisting of a layer of cadmium atoms with layers of iodine atoms on either side of it, is electrically neutral. Since there are only van der Waals forces between adjacent composite layers, crystals of this

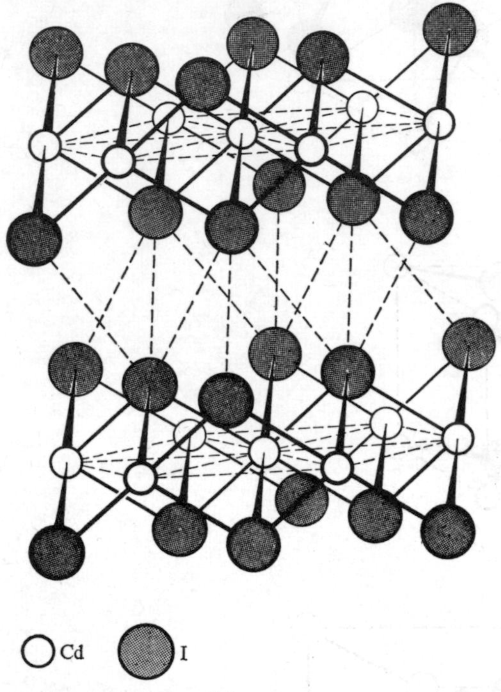

Figure 3.3 The CdI_2 structure

kind show a pronounced cleavage parallel to the layers. The unsymmetrical environment of the iodine atoms shows that this structure is not a purely ionic one. The bromides and iodides of magnesium, the iodides of calcium and many transition metals, and most dihydroxides crystallize in the cadmium iodide structure.

A closely related structure is that of cadmium chloride, which differs from the cadmium iodide lattice only in the relative orientation of the layers; magnesium chloride and most dichlorides of transition metals have the cadmium chloride structure.

3.4 The perovskite and spinel structures

To conclude this brief survey of simple structures, mention will be made of two ternary oxide structures in which no oxy-anion is present.

In the cubic unit cell of perovskite ($CaTiO_3$) there is a titanium atom at the centre of the cube, with calcium atoms placed at the corners and oxygen atoms occupying the midpoints of all faces (Fig. 3.4). Thus the calcium atom has twelve oxygen atoms as nearest neighbours, their arrangement being that of cubic close-packing; the titanium atom has six oxygen atoms distributed around it octahedrally. Among other compounds which have this structure

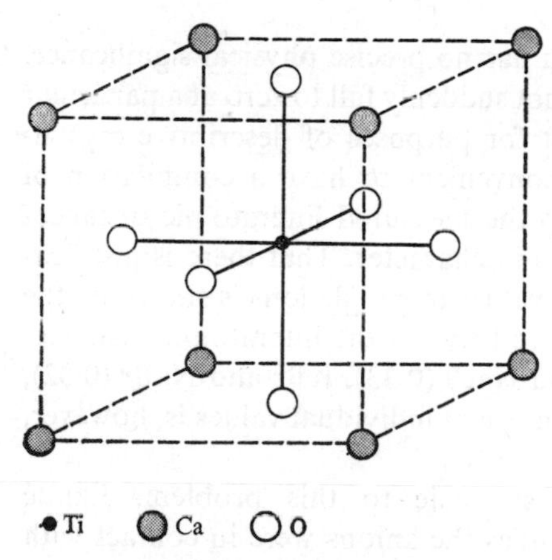

Figure 3.4 The perovskite structure

(or slightly deformed versions of it) are $NaNbO_3$, $LaGaO_3$, $KMgF_3$ and $KZnF_3$.

The compound $MgAl_2O_4$ (spinel) has a structure based on a cubic close-packed array of oxygen atoms. In such an array, each oxygen atom has twelve other oxygen atoms equidistant from it, and the 'holes' between oxygen atoms are of two types. The octahedral holes, of which there is one per oxygen atom, are surrounded by six oxygen atoms and the smaller tetrahedral holes, of which there are two per oxygen atom, are surrounded by four oxygen atoms. In spinel itself, one-eighth of the tetrahedral holes are occupied by magnesium atoms and one-half of the octahedral holes by aluminium atoms. Many other mixed oxides of formula $A^{II}B_2^{III}O_4$ have this structure, such as $Zn^{II}Fe_2^{III}O_4$ and $Fe^{II}Cr_2^{III}O_4$. In others, such as $Fe^{II}Fe_2^{III}O_4$ (for which neutron diffraction will distinguish Fe^{II} and Fe^{III} because of their different numbers of unpaired electrons) and $Co^{II}Fe_2^{III}O_4$, half of the B^{III} species are in tetrahedral holes, with the other half and the A^{II} species in octahedral holes; this structure, which may be written $B[AB]O_4$ to distinguish it from the normal $A[B_2]O_4$ (the square brackets enclosing the species in the octahedral holes), is known as the inverse spinel structure. The factors which determine whether a compound has the normal or inverse structure are discussed further in Section 18.3, and the conducting properties of Fe_3O_4 in Section 4.4.

3.5 Ionic radii

The radius of an individual ion has no precise physical significance, since the electron density does not suddenly fall to zero at a particular distance from the nucleus; but for purposes of descriptive crystallography and chemistry it is convenient to have a compilation of values obtained by dividing up the measured interatomic distances in compounds taken to be ionic in character. That there is justification for assigning approximate values to simple ions is shown by the rough constancy of the difference between the interatomic distances in KF and NaF (0·35), KCl and NaCl (0·33), KBr and NaBr (0·32), and KI and NaI (0·30 Å); the choice of individual values is, however, a somewhat arbitrary process.

Three approaches have been made to this problem. Landé assumed that in the lithium halides the anions were in contact with one another. Wasastjerna used molar refractivities of the ions as a basis for division of the internuclear distance, assuming a proportionality between molar refractivity and ionic volume. Pauling considered a series of alkali metal halides containing isoelectronic ions (NaF, KCl, RbBr and CsI) and took the radius of an ion to be inversely proportional to its effective nuclear charge (i.e. its actual nuclear charge less an estimated screening effect). In view of the approximate nature of the concept of ionic radius, great significance should not be attached to small differences in quoted values so long as self-consistency is maintained in any one compilation. Since the fact of an interatomic distance implies a balance between attractive and repulsive forces, some dependence of size on coordination number would be expected unless both types of force were affected to the same degree by change in the number of nearest neighbours, and it will be seen later that this is not so. The following values, derived from data for compounds having the sodium chloride structure, must be increased slightly for the same ions in eightfold coordination and decreased slightly for fourfold coordination: F^-, 1·36; Cl^-, 1·81; Br^-, 1·95; I^-, 2·16; O^{2-}, 1·40; Li^+, 0·60; Na^+, 0·95; K^+, 1·33; Rb^+, 1·48; Cs^+, 1·69; Mg^{2+}, 0·65; Ca^{2+}, 0·99; Sr^{2+}, 1·13; Ba^{2+}, 1·35 Å.

For species such as Si^{4+} and Cl^{7+} the radii commonly quoted (0·50 Å and 0·26 Å respectively) are highly artificial, and are obtained by subtracting the value for O^{2-} from Si–O in SiO_2 and Cl–O in ClO_4^-; the values for successive ionization potentials of silicon and chlorine make it inconceivable that ions of such high charges are really present. Similarly, the statement that, owing to the lanthanide contraction, the radii of Mo^{6+} and W^{6+} are almost identical really means that Mo–O and W–O in analogous compounds of molybdenum (VI) and tungsten (VI) are almost the same. Values for transition metal ions are sometimes uncertain owing to crystal field effects; this subject is discussed separately in Chapter 15.

The radius attributed to the hydride ion, in which the nucleus consists of only a single proton, shows how the nature of the cation present can exert a marked influence on the size of the anion. Subtraction of the appropriate cation radius from the metal–hydrogen distance in the appropriate hydride leads to radii of H^- of 1·54 in CsH, 1·42 in LiH, 1·35 in CaH_2 and 1·30 Å in MgH_2.

Finally, it should be mentioned that in a few instances in which the variation in electron density along a line joining the nuclei has been determined, the minimum in electron density does not occur at the point indicated by the ionic radii in general use. In NaCl, for example, the minimum occurs at 1·64 Å from the chlorine nucleus, and in LiF at 0·92 Å from the lithium nucleus.

These considerations make the discussion of crystal structures in terms of radius ratios at best a rough guide to the incidence of different structural types, and for this reason only a brief treatment of this topic is given in the following section. In the discussion of lattice energetics, however, radius ratios are not involved, since only the total interionic distance r_0 enters into the expression for the lattice energy; this subject is therefore treated more fully in Sections 3.7 to 10.

3.6 Radius ratio rules

To a first approximation, the structures of ionic crystals are determined by the relative numbers and relative sizes of the ions they contain. If, notwithstanding the considerations mentioned in the preceding section, we consider ions as rigid spheres, it is easy to calculate geometrically the largest number of ions which can be in contact with another ion of smaller radius. Since, for monatomic ions, cations are generally smaller than anions, it is customary to consider the radius ratio r_+/r_-; the coordination numbers possible and the arrangements of anions round a cation are shown for different values of r_+/r_- in Table 3.1.

Table 3.1 *Influence of radius ratios on structures of salts*

r_+/r_-	*Maximum coordination number of cation*	*Arrangement of anions round cation*
<0·15	2	Linear
0·15–0·22	3	Triangular
0·22–0·41	4	Tetrahedral
0·41–0·73	6	Octahedral
>0·73	8	Cubic

In discussing the structures of salts in terms of radius ratios, it is assumed that maximum stability is attained when every ion is surrounded by as many ions of opposite sign as possible. Although this assumption is valid, it is less simple than it at first appears to be. In the lattice of a salt of formula MX, for example, the two ions must have the same coordination number; increase in the coordination number of M^+ to give it more X^- ions as nearest neighbours inevitably leads to its having more M^+ ions as next nearest neighbours, and the increase in electrostatic attraction is then largely cancelled by an increase in electrostatic repulsion. Furthermore, the requirement that the structure be an infinite one precludes very high coordination numbers in salts. If M^+ and X^- have the same radius, for instance, it is possible to pack twelve X^- ions round the M^+ ion; but it is impossible to do so in such a way that each X^- ion is then surrounded by twelve M^+ ions.

The radius ratio rule is therefore a guide to what is possible on geometrical grounds, rather than a reliable guide to the actual structures of salts. Thus, amongst the alkali metal halides, only caesium chloride, bromide and iodide have the caesium chloride structure under ordinary conditions; caesium fluoride and all the halides of lithium, sodium, potassium and rubidium have the sodium chloride structure. However, when caesium chloride is sublimed onto an amorphous surface it is obtained in the sodium chloride structure, and the interionic distance decreases from 3·566 Å in the normal form to 3·474 Å. Conversely, at high pressures rubidium chloride, bromide and iodide adopt the caesium chloride structure. The change of structure along the series BeF_2 (β-cristobalite), MgF_2 (rutile), CaF_2, SrF_2 and BaF_2 (fluorite) is readily understandable in terms of radius ratios, and the same explanation is often given for the series CO_2 (linear molecule), SiO_2 (polymorphic, but always having 4:2 coordination), GeO_2 (dimorphic, β-cristobalite or rutile), SnO_2 and PbO_2 (rutile), and ThO_2 (fluorite). Some members of the last series, however, are certainly not purely ionic species. For these, and for halides having the cadmium chloride and cadmium iodide layer structures, it is often suggested that an ionic structure has been modified by polarization (the distortion of the electronic charge density round an ion, normally of the electronic cloud of a large anion by a small, highly charged cation). This approach is helpful in leading to an understanding of why ionic structures are not adopted, but it does not lead to the prediction of which structure will result, and the discussion of polarizing powers and polarizabilities lacks a firm quantitative basis. We shall therefore make only a limited use of the concept of polarization in this book.

3.7 Lattice energy[2, 3, 4, 5, 6]

Consider a salt MX having the sodium chloride structure, the ions carrying charges z_+e and z_-e, where e is the electronic charge and z_+ and z_- are positive integers (necessarily the same in this case, but it is desirable to develop the argument in a general way). Each M^{z+} ion is surrounded by six X^{z-} ions at a distance r, twelve M^{z+} ions at $\sqrt{2}\,r$, eight X^{z-} ions at $\sqrt{3}\,r$, six M^{z+} ions at $\sqrt{4}\,r$, twenty-four X^{z-} ions at $\sqrt{5}\,r$, and so on. The electrostatic energy change when a M^{z+} ion is brought from infinity to its stable position in a sodium chloride type lattice is therefore given by

$$-\frac{6e^2}{r}z_+z_- + \frac{12e^2}{\sqrt{2}\,r}(z_+)^2 - \frac{8e^2}{\sqrt{3}\,r}z_+z_- + \frac{6e^2}{\sqrt{4}\,r}(z_+)^2 - \frac{24e^2}{\sqrt{5}\,r}z_+z_- \ldots$$

$$= -\frac{z_+z_-e^2}{r}\left(6 - \frac{12z_+}{\sqrt{2}\,z_-} + \frac{8}{\sqrt{3}} - \frac{6z_+}{\sqrt{4}\,z_-} + \frac{24}{\sqrt{5}} - \ldots\right)$$

Since the ratio of the charges on the ions (z_+/z_-) is constant for a given type of structure (unity for NaCl, two for CaF_2, etc.), the series in parentheses is a function only of the crystal geometry, and is independent of the values of z_+, z_- and r; it is called the Madelung constant A after its first evaluator, and the electrostatic energy of interaction of a cation with all other ions in the crystal is then $-z_+z_-e^2A/r$. The environments of both ions in the sodium chloride structure being identical, the same expression gives the electrostatic energy of interaction of an anion with all other ions. The total electrostatic interaction energy is then given by half the sum of the interaction energies of cations and anions, since addition of the two would take each interaction into account twice. Thus for a mole of MX (N pairs of ions, where N is Avogadro's number) the total electrostatic energy change on formation of the lattice from ions at infinity is $-z_+z_-e^2NA/r$. (There is also a decrease in energy of $2RT$ representing the mechanical energy liberated when the gaseous ions form a solid, but this is very small relative to the electrostatic energy term and may be neglected here.)

For the sodium chloride structure, the series for the Madelung constant converges to a value of 1·7476. For the caesium chloride structure, the Madelung constant is 1·7627. It might seem surprising that, for a given interionic distance, the total electrostatic interaction energy is almost the same for sixfold and eightfold coordination, but the slowness with which the series for Madelung constants converge emphasizes how important it is to consider all interactions in the lattice and not just the interaction of an ion and its nearest neighbours. Values for some other Madelung constants are given in Table 3.2. Attention should be drawn to the fact that different values are sometimes given for structures containing ions having different charges (e.g. 5·038 instead of 2·519 for the Madelung constant of the fluorite structure). This practice results from including the charges on the ions in the Madelung constant (i.e. quoting z_+z_-A) and is not to be recommended.

Many expressions have been used to take into account the repulsive contribution to the interaction energy. The repulsive energy increases very steeply as interionic distance decreases after the ions have come into contact and is most simply represented by a term $+NB/r^n$, where B is a constant, the repulsion coefficient, which is approximately proportional to the number of nearest neighbours and n is another constant, called the Born exponent, which can be evaluated from compressibility data. The Born exponent generally

Table 3.2 *Madelung constants*

Structure type	*A*
CsCl	1·7627
NaCl	1·7476
ZnS (zinc blende)	1·638
ZnS (wurtzite)	1·641
CaF_2	2·519
TiO_2 (rutile)	2·408*
CdI_2	2·191*

* Values for these structures vary slightly according to the ratio of the lattice constants for the unit cell.

lies between 8 and 10, but is lower if both ions have the helium electronic configuration and higher if both have the xenon electronic configuration. The total interaction energy is then given by

$$-\frac{NAz_+z_-e^2}{r}+\frac{NB}{r^n}$$

and the lattice energy (defined as the energy required at 0°K to convert one mole of the salt into gaseous ions infinitely removed from one another) is

$$U=\frac{NAz_+z_-e^2}{r}-\frac{NB}{r^n}$$

Since $dU/dr = 0$ at the equilibrium distance $r = r_0$,

$$0=-\frac{NAz_+z_-e^2}{r_0{}^2}+\frac{nNB}{r_0{}^{n+1}}$$

whence

$$B=\frac{Az_+z_-e^2r_0{}^{n-1}}{n}$$

and

$$U=\frac{NAz_+z_-e^2}{r_0}\left(1-\frac{1}{n}\right)$$

This is the Born–Landé expression for lattice energy. More detailed calculations of lattice energies employ more sophisticated expressions for the repulsive energy, and also take into account London or van der Waals forces, which are attributed to synchronized oscillating dipoles in adjacent ions, and zero-point energies; the results are not greatly modified, however, as may be seen from the data for sodium chloride and magnesium oxide in Table 3.3. In estimating lattice energies of compounds which are unknown or whose structures are unknown (or are of low symmetry, so that the Madelung constants are unknown), chemists usually employ the simple formula and often assume $n = 9$; in a comparative treatment a small change in n is relatively unimportant. Physicists, on the other hand, being mainly concerned with the most accurate description of a very few compounds whose properties are known in great detail, naturally prefer more rigorous (and, unfortunately, more difficult to handle) formulae for lattice energies.

Table 3.3 *Contributions to lattice energies*

Energy	*NaCl**	*MgO**
Electrostatic attraction	+860 (205·6)	+4634 (1108)
Repulsion	−99 (23·6)	−699 (167)
London	+12 (2·9)	+6 (1·5)
Zero-point	−7 (1·7)	−18 (4·4)
Total	766 (183·2)	3923 (938)

* Values are in kJ and, in parentheses, in kcal.

Lattice energies derived from formal ionic charges, Madelung constants, Born exponents and interionic distances are often referred to as 'calculated' values to distinguish them from values obtained by means of thermochemical cycles. It should, however, be pointed out that interionic distances measured by X-ray diffraction are themselves experimental data and may conceal departures from ideal ionic behaviour. Furthermore, the real charges on ions may be less than their formal charges, and the evaluation of the non-electrostatic part of the lattice energy is subject to some uncertainty. Nevertheless, the concept of lattice energy is of immense importance in inorganic chemistry, and many examples of its utility will be given later.

We may conclude this section by returning briefly to the subject of interionic distances and ionic radii. The expression for the repulsion constant B may be rewritten in the form:

$$r_0 = \left(\frac{nB}{Az_+z_-e^2}\right)^{1/(n-1)}$$

If we consider a salt existing in the caesium chloride and sodium chloride structures, and assume that, since repulsive forces are significant only over short distances, B is proportional to the number of nearest neighbours, and that $n = 9$,

$$\frac{r_0(\text{CsCl form})}{r_0(\text{NaCl form})} = \left(\frac{8 \times 1{\cdot}748}{6 \times 1{\cdot}763}\right)^{\frac{1}{8}} = 1{\cdot}036$$

For the CsCl and NaCl forms of CsCl, RbCl, RbBr and RbI the ratio of the experimental values of r_0 is found to be 1·027, 1·036, 1·029 and 1·025 respectively; the agreement is at least good enough to suggest that the premise that the repulsive forces are important over much shorter distances than the electrostatic forces is correct. The effect of the change in r_0 on the lattice energy is, however, diminished by the change in the Madelung constant, and the overall conclusion is that there is probably very little difference between the lattice energies of the two forms of the alkali metal halides being discussed in this paragraph. This point could be tested experimentally by, for example, measuring the heats of dissolution of the two forms in water, but no such measurements have yet been made.

3.8 The Born-Haber cycle

The lattice energy of a salt at 298°K (which is very slightly different from that at 0°K by an amount depending on the difference in heat capacities of the salt and the gaseous ions) may be related to several other quantities by means of the Born–Haber thermochemical cycle. In the particular cases where the anion is a halide, all of the other quantities have been measured independently, and we shall therefore first discuss the cycle for a halide of formula MX_n. In the diagram below small corrections for expansion terms have been omitted for simplicity; the effect is only to make the lattice energy obtained from the cycle slightly too large by an amount $(n + 1)RT$.

In this cycle, L is the latent heat of sublimation of the metal, ΣI the sum of its first n ionization potentials, D the dissociation energy of the gaseous halogen, E the electron affinity of the gaseous halogen

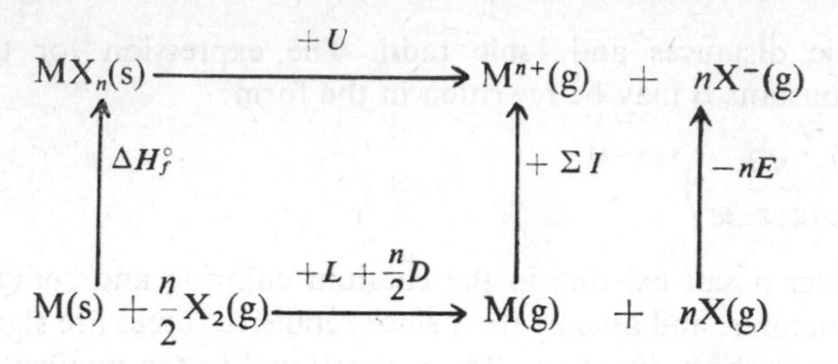

atom, and ΔH_f^0 the standard heat of formation of the halide; it is assumed here that the halogen is gaseous, but if it is bromine or iodine the necessary latent heat term can easily be included. The physicochemical properties of the halogens are discussed in Chapter 12, where the lattice energies of halides are mentioned again. For the present we may consider sodium chloride as a typical example: here $L = 109$, $I = 496$, $D = 243$, $E = 356$ and $\Delta H_f^0 = -411$ kJ (or $L = 26{\cdot}0$, $I = 118{\cdot}5$, $D = 58{\cdot}0$, $E = 85$ and $\Delta H_f^0 = -98{\cdot}2$ kcal). Substitution in the relationship

$$U = L + \frac{n}{2}D + \Sigma I - nE - \Delta H_f^0$$

gives $U = 780$ kJ (186·5 kcal), and subtraction of $2RT$ then leads to a lattice energy at 298°K of 775 kJ (185·3 kcal). The value at 0°K calculated from the Born–Landé formula, the observed interionic distance of 2·81 Å, and a Born exponent of 9 is 766 kJ (183 kcal); a more refined calculation gives 778 kJ (186 kcal).

The agreement between 'calculated' and 'cycle' lattice energies is as good as this for all the alkali metal halides and for those halides of the alkaline earth metals that have the rutile or the fluorite structure. Whilst this agreement is not a rigid proof that all these substances are completely ionic in character (as was mentioned earlier, 'calculated' lattice energies involve the use of experimental interionic distances), it does at least show that the ionic model provides a very satisfactory basis for the discussion of the thermochemistry of these compounds. For compounds having layer structures, however, the position is very different; the 'calculated' lattice energy for cadmium iodide, for instance, is 1986 kJ (475 kcal), whilst the 'cycle' value is 2435 kJ (582 kcal), showing that the ionic model is inadequate. A similar conclusion holds for copper (I) chloride, for which the 'calculated' and 'cycle' lattice energies are 904 and 979 kJ (216 and 234 kcal) respectively. For the silver halides the discrepancy between 'calculated' and 'cycle' values (the latter always being greater) increases from 29 kJ (7 kcal) for the fluoride to 113 kJ (27 kcal) for the iodide. In this case there is also an interesting variation in the interionic distance; whereas the distances in NaCl and AgCl are almost identical, that in AgF is 0·15 Å longer than the distance in NaF, whilst that in AgBr is 0·11 Å shorter than the distance in NaBr (silver iodide has the zinc blende structure). These data indicate that if AgF is taken as an essentially ionic crystal, covalent bonding increases the lattice energy to an increasing extent along the series AgCl, AgBr and AgI; this is the factor underlying the decrease in solubility in water along the series.

3.9 The Kapustinskii equation

Methods of obtaining lattice energies which have been discussed so far involve knowledge of all the other quantities in the Born–Haber cycle or of the interionic distance and the Madelung constant. It has, however, been shown that there is probably little difference in the lattice energies of a given alkali metal halide in the caesium chloride and sodium chloride structures; it seems reasonable to suppose the same would be true of a dihalide existing in the fluorite and rutile structures. Furthermore, it was pointed out by Kapustinskii in 1933 that the Madelung constants for structures of typical salts divided by the number of ions in the structure are all 0·80–0·88. This led Kapustinskii to suggest that lattice energies could all be represented approximately by means of a general equation:

$$U_0 = \frac{0{\cdot}874 N \nu z_+ z_- e^2}{r_+ + r_-}\left(1 - \frac{1}{n}\right)$$

where 0·874 is half the Madelung constant for the sodium chloride structure (two ions), ν is the number of ions in the formula of the compound whose lattice energy is being estimated, r_0 is replaced by $r_+ + r_-$, radii for sixfold coordination, and the other symbols have their usual meanings. If n is taken to be 9, conversion to kJ (kcal) mole^{-1} yields

$$U = \frac{1071 \nu z_+ z_-}{r_+ + r_-}\ \text{kJ} \quad \left(\frac{256 \nu z_+ z_-}{r_+ + r_-}\ \text{kcal}\right)$$

This equation and a more sophisticated later version have been used extensively for estimating lattice energies either by direct insertion of values of $(r_+ + r_-)$ derived from X-ray measurements or by the following procedure. If, for example, an anion X^- forms compounds

M_1X and M_2X with cations M_1^+ and M_2^+, it is easily shown that

$$U[M_1X] - U[M_2X] = \Delta H_f^0[M_1^+(g)] - \Delta H_f^0[M_2^+(g)] - (\Delta H_f^0[M_1X] - \Delta H_f^0[M_2X])$$

If the data on the right-hand side are known, the difference in lattice energies of the two salts can be calculated, and the radius of X^- may be calculated by using those of M_1^+ and M_2^+. In this way, for example, $r(NO_3^-)$ is calculated as 1·89 Å and $r(BF_4^-)$ as 2·28 Å. These values are roughly in line with expectations based on a knowledge of bond lengths and the radii of monatomic ions; since they have been obtained from thermochemical data they are commonly referred to as 'thermochemical radii'. Once obtained, they can be used to estimate lattice energies of nitrates and fluoroborates of other cations of known or estimated 'thermochemical radius'. So long as like species are being compared, much of the uncertainty in this procedure disappears. It is, however, essential in reading any material on this subject to be careful to see exactly what assumptions are being made.

There are various other empirical methods of estimating lattice energies, but we shall not discuss them here.

3.10 Applications of lattice energetics[2, 3, 4, 7, 8]

3.10.1. Estimation of electron affinities. Electron affinities can be measured directly only for the conversion of atoms into singly charged anions. If we want to know ΔH^0 for the reaction

$$O(g) + 2e = O^{2-}(g)$$

we apply the Born–Haber cycle to a metal oxide whose lattice energy can be calculated accurately if an ionic structure is assumed (e.g. MgO). This has the sodium chloride structure, and its compressibility has been measured accurately. Combination of the lattice energy with the dissociation energy of oxygen (determined spectroscopically), the heat of sublimation of magnesium, its first two ionization potentials and the standard heat of formation of magnesium oxide gives a value of about −630 kJ (150 kcal) mole^{-1} for the double electron affinity of oxygen. Since the combination of an oxygen atom with a single electron liberates 142 kJ (34 kcal) mole^{-1}, addition of the second electron must absorb about 770 kJ (185 kcal) mole^{-1}. In fact it appears that the only reason why the oxide ion O^{2-} exists at all is the high lattice energy of oxides resulting from the double charge.

Similar calculations may be used to estimate the heats of reactions leading to the formation of complex ions, for example,

$$BF_3(g) + F^-(g) = BF_4^-(g)$$

from the estimated lattice energy of KBF_4 and related quantities.

3.10.2. Estimation of proton affinities. Let us consider as an example the estimation of the proton affinity P of ammonia from the cycle:

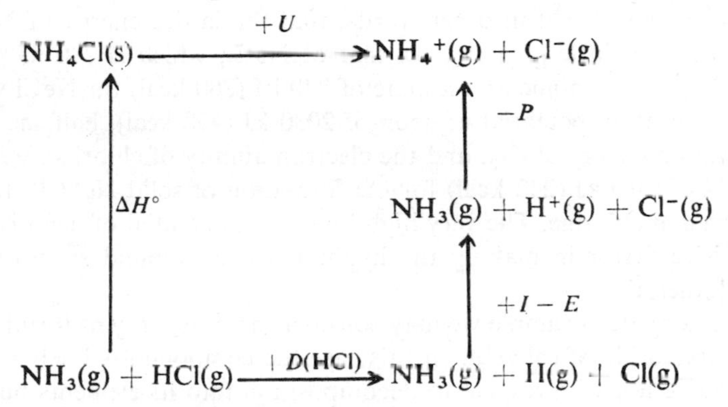

For this purpose we need to know the lattice energy of ammonium chloride, its heat of formation from ammonia and hydrogen chloride, the dissociation energy of hydrogen chloride, the ionization potential of hydrogen and the electron affinity of chlorine. The first of these can be calculated from r_0 and all the others have been determined independently. Use of the value of 653 kJ (156 kcal) for the lattice energy leads to a proton affinity of 895 kJ (214 kcal) for ammonia, and similar values are obtained from analogous cycles involving ammonium bromide and ammonium iodide (ammonium fluoride has the wurtzite structure owing to hydrogen-bonding, and a cycle based upon its estimated lattice energy leads to an anomalous result). A value for phosphine (816 kJ or 195 kcal) may be obtained similarly from the lattice energy of phosphonium iodide.

Since hydroxonium perchlorate ($H_3O^+ClO_4^-$) is isomorphous and almost isodimensional with ammonium perchlorate, the lattice energies of the two compounds may be taken to be equal; comparison of their heats of formation then leads to a value of 762 kJ (182 kcal) for the proton affinity of gaseous water. It is interesting to note that this value is lower than that for phosphine, indicating that the lower hydration energy of the PH_4^+ ion is the reason for the decomposition of phosphonium salts by water.

3.10.3. Estimation of heats of formation. It is only rarely that the lattice energy of a compound is known (or can be calculated from X-ray data) and the heat of formation is not; the chief application of the Born cycle in this connection is therefore to the estimation of the heats of formation of unknown compounds from reasonable estimates of what their lattice energies would be. If, for example, neon chloride existed as a salt, it would be reasonable to assume that it would have the NaCl structure and that Ne^+ would be a little smaller than Na^+; in other words, that the lattice energy of NeCl would be a little greater than that of NaCl, which is 770 kJ (184 kcal). If we combine an estimate of 840 kJ (200 kcal) for NeCl with the ionization potential of neon of 2080 kJ (498 kcal), half the dissociation energy of Cl_2, and the electron affinity of chlorine, we get $\Delta H_f^0 = 1010$ kJ (242 kcal) for the formation of solid Ne^+Cl^- from neon and chlorine. The very high ionization potential of neon is the decisive factor in making the hypothetical compound so strongly endothermic.

As a related example we may consider the possibility of forming a monofluoride of calcium. In this case the compound is likely to be unstable not with respect to decomposition into its elements but to *disproportionation* according to the equation

$$2CaF(s) = Ca(s) + CaF_2(s)$$

The oxidation state of one half of the calcium is increased and that of the other half decreased. In this case we know the standard heat of formation of CaF_2, and that of metallic calcium, is, by definition, zero; if we assume that the lattice energy of CaF would be about the same as that of KF (Ca^+ would be substantially larger than Ca^{2+}) we can obtain an estimate for the heat of formation of CaF from the Born–Haber cycle and hence derive ΔH^0 for the disproportionation. In a reaction involving only solids the entropy change would be very small, and ΔH^0 would therefore be almost identical with ΔG^0, which is related to the equilibrium constant of the reaction. This calculation leads to a value of ΔH^0 of -610 kJ (146 kcal) for the reaction as written, from which it appears that the likelihood of an *ionic* monofluoride of calcium being prepared, even at high temperatures, is very slight. The decisive factor in the instability of calcium monofluoride is its lattice energy, which, if it were the same as that of KF, would be only about one-third of that of CaF_2; the increased ionic charge, the increased Madelung constant for the MX_2 structure and the decrease in ionic radius on losing a second electron all play important parts. For the alkali metals, even these factors are insufficient to overcome the very high second ionization potentials (4561 kJ, or 1091 kcal, for sodium, compared with the first ionization potential of 496 kJ, or 118 kcal); but for elements such as calcium (first and second ionization potentials of 590 and 1146 kJ (141 and 274 kcal) respectively) the relatively small gap between ionization potentials favours disproportionation of a lower ionic oxidation state. (This argument does not, however, preclude the formation of a covalent molecule CaF, and such a species is, indeed, formed at very high temperatures.) It is a characteristic feature of the chemistry of the transition metals that the balance between ionization potentials and lattice energies is such that most of them form two salt-like fluorides of comparable stability (see Chapter 18 for a fuller discussion).

Thermochemical arguments of the type discussed above played an important part in the beginnings of noble-gas chemistry. The discovery that the compound O_2PtF_6 was a salt $O_2^+PtF_6^-$, coupled with the realization that the ionization potential of xenon was slightly lower than that of molecular oxygen, gave rise to the idea that $Xe^+ PtF_6^-$ might have a lattice energy sufficient to permit its production from Xe and PtF_6 (which clearly has a high electron affinity), and hence led to the work described in more detail in Chapter 13.

3.10.4. Further applications. Many other applications of lattice energetics appear later in this book, some of the most important being the following:

(a) the discussion of the special properties of fluorine in relation to the other halogens (Chapter 12);
(b) the account of the stabilities of metal hydrides (Chapter 8), polyhalides (Chapter 12) and peroxides and superoxides (Chapter 11);
(c) the derivation of crystal field stabilization energies (Chapter 18);
(d) the discussion of the characterization of high oxidation states of metals as fluorides, and of their low oxidation states as iodides (Chapters 18, 22 and 23).

Compilations of lattice energies appear in references 3, 4 and 10.

References for Chapter 3

1 Wells, A. F., *Structural Inorganic Chemistry*, 3rd edn. (Oxford University Press, 1962).
2 Greenwood, N. N., *Ionic Crystals, Lattice Defects, and Nonstoichiometry* (Butterworths, 1968).
3 Johnson, D. A., *Some Thermodynamic Aspects of Inorganic Chemistry* (Cambridge University Press, 1968).
4 Waddington, T. C., *Adv. Inorg. Chem. Radiochem.*, 1959, **1**, 157.
5 Harvey, K. B., and Porter, G. B., *Introduction to Physical Inorganic Chemistry* (Addison-Wesley, 1963).
6 Phillips, C. S. G., and Williams, R. J. P., *Inorganic Chemistry*, Vol. I (Oxford University Press, 1965).
7 Sharpe, A. G., in *Halogen Chemistry*, ed. V. Gutmann, Vol. I, p. 1 (Academic Press, 1967).
8 Dasent, W. E., *Inorganic Energetics* (Penguin Books, 1970).
9 Bartlett, N., *Proc. Chem. Soc.*, 1962, 218.
10 Ladd, M. F. C., and Lee, W. H., *Progress in Solid State Chem.*, 1963, **1**, 37; 1965, **2**, 378.

The defect solid state 4

4.1 Introduction[1, 2]

In Chapter 3 it was assumed implicitly that the compounds under discussion had ideal crystals in which each ion occupied an appropriate lattice site and each site was tenanted by the right kind of ion. This state of affairs, however, represents equilibrium only at the absolute zero of temperature. Above absolute zero, the atoms in a real crystal are always undergoing thermal vibration. In the case of a discrete molecule or complex ion, this may result in random orientation or a change of structure; the stoichiometry, however, is unaffected. In the case of infinite structures like those of ionic salts or metals, two major types of defect may be recognized: those in which stoichiometry is maintained but atoms are randomly distributed or missing from their ideal positions in the lattice, and those in which the solid phase is stable over a range of composition, sometimes narrow and sometimes wide (i.e. is non-stoichiometric).

In the following account, after a brief qualitative summary of the band theory of solids, we shall adopt the classification of Greenwood,[2] and first discuss defects which are inherent in the thermodynamics of the infinite solid state (and which must therefore occur in all macromolecular structures) and then those which are specific to particular compounds, describing in turn specific defect structures in stoichiometric and in non-stoichiometric crystals. We shall thus be dealing with defects of an essentially equilibrium nature. It must, however, be recognized that there are several other types of defect, among them dislocations (which are essentially defects concentrated along lines or surfaces instead of being distributed uniformly throughout the crystal) and defects caused by the action of charged or uncharged radiations, such as those which arise from nuclear reactions within a crystal (as in the Szilard–Chalmers effect). Among the properties of real crystals which arise from their imperfections may be mentioned electronic and photoconductivity, diffusion, catalytic activity and the ability to react chemically; some of these will be mentioned again later in this chapter.

4.2 The band theory of solids[2, 3, 4, 5]

In this approach to the study of the properties of solids, we are concerned with the energy distribution of the electrons within the crystal as a whole. The band theory was developed to account for the characteristic properties of metals (high coordination number of twelve or eight in the solid state, and thermal and electrical conductivity), and we shall begin by referring to a metallic lattice; the theory is, however, capable of application also to ionic solids and some non-metallic crystals.

The electrical properties of metals require that they contain free electrons comparable in number with the number of atoms present. On classical electromagnetic theory, the free electrons constitute an 'electron gas' which diffuses freely through the crystal lattice formed by the positive ions of the metallic element. Such a picture immediately encounters a fundamental difficulty, however, which is overcome only by quantum mechanical conceptions. The energy of the electrons in the 'electron gas' would necessarily contribute to the specific heat. This means that the specific heats of metals would be higher than those of insulators, whereas in fact the Dulong and Petit rule (that the atomic heat $= 3R$) applies equally to conductors and insulators. Thus it appears either that the electrons contribute very little to the specific heat or that there are relatively few free electrons.

The solution of this problem is associated with the work of Sommerfeld, Fermi and Bloch. As will be seen, it follows in essence directly from the application of the Pauli exclusion principle to the metallic electrons, whereby not more than two electrons of opposed spin angular momentum may occupy any one electronic energy level. In the molecule of a compound, the electrons must be referred to molecular energy levels (orbitals), and not to the levels originally occupied in the uncombined atoms. A single crystal of a metal constitutes a giant molecule; the electrons are held in common by all the atoms contained in the unit, and have to be fitted into a very large, but finite, number of energy levels. If N atoms are condensed from the gas phase and each contributes one electron and one energy level, a band of N levels of nearly equal energy results, and at absolute zero the lowest $N/2$ levels will all be doubly occupied.

Since the electron has a wave-nature, each possible electronic energy level may be described in terms of a wave-function which is a solution of the appropriate Schrödinger equation. If, then, as the simplest model, the potential within the metal is taken as uniformly zero, rising to a finite value at the boundary, the electron is represented as a stationary wave within the crystal. It emerges from the solution of the wave-equation that the wavelength of the electrons in the highest energy levels is of the same order of magnitude as the distance between the atoms, and it follows that the energy of such electrons greatly exceeds the thermal energy of an atom at the ordinary temperature. Some of the electrons at temperatures above the absolute zero may be promoted to levels higher than the $N/2$ th, and their energy distribution then roughly follows the Maxwell–Boltzmann law. It is only these electrons, few at ordinary temperatures in comparison with the total number of electrons, that contribute to the heat capacity of the metal, so that the electronic contribution to the specific heat is inevitably small. The major part of the specific heat is used to increase the thermal energy of the atoms, so that metals conform to the Dulong and Petit rule.

In an actual metallic crystal, the potential field is not uniform, but periodic; the potential rises to a maximum at each positive ion and is at a minimum between them. In such a case, the solution of the wave-equation shows that electrons may be accommodated only in bands of closely spaced energy levels between permitted limits; between such permitted energy bands are forbidden regions within which no electron may be accommodated. The positions and widths on the energy scale of the permitted bands are a property of the structure and the chemical nature of the solid. The breadth of a band depends on the amount of overlap between the wave-functions, so that for inner electrons of the constituent atoms it is extremely small; thus $2s$- and $2p$-electrons of sodium ($2s^2 2p^6 3s^1$) have essentially the same energy in metallic sodium as in isolated sodium atoms. For valence electrons, however, the overlap is large, and whereas the single valence electron in an isolated sodium atom is completely defined by its energy level $3s$, in a crystal of the metal it occupies one of the many very closely spaced energy levels within the $3s$ energy band. Each energy band corresponds, in principle, to one quantum level of an isolated atom, and the ranges of forbidden energies correspond to the jumps in energy from one quantum state to the next.

When a metallic crystal at absolute zero containing N electrons in the lowest $N/2$ energy levels of a band is heated, a few of the electrons in the uppermost filled levels (i.e. just below the middle of the band) may acquire some thermal energy and hence be promoted to the lowest unoccupied levels and singly occupy some of these. Once promoted, the electrons give rise to conductivity. The distribution of electrons in such a conducting solid is represented diagrammatically in Fig. 4.1. Electrons in filled inner orbitals are associated with particular nuclei, and to transfer them from one atom to another

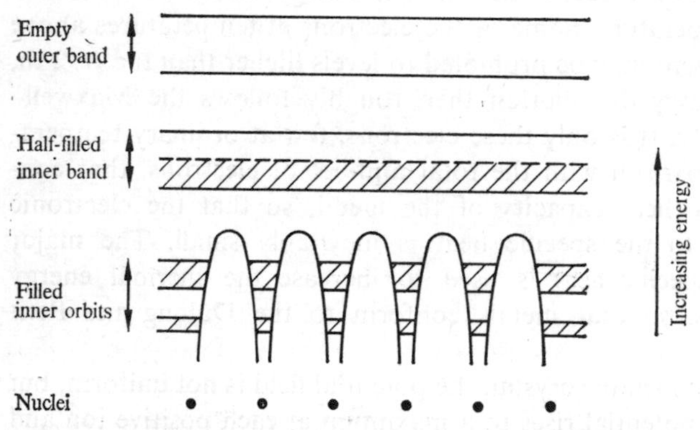

Figure 4.1 Diagrammatic representation of band structure in a conducting solid

would require a large amount of energy; such processes cannot occur where these levels are already doubly occupied. The outermost electrons are accommodated within the band, and any electron present in a singly occupied level may migrate through the crystal without having to surmount energy barriers. The dotted line represents the half-filled band, the levels below it being doubly occupied at absolute zero.

Fig. 4.1 represents the band structure of a conducting solid with a half-filled band. We may now consider dispositions of electrons and levels in general, and in Fig. 4.2 (from which filled inner orbitals are omitted) four cases are represented. In (a), each occupied band is exactly filled and there is a large energy gap ΔE_0 to the next allowed level, which is therefore completely empty; solids having this type of band structure are insulators (e.g. the alkali metal halides, for which

ΔE_0 is 670–1170 kJ (160–280 kcal) mole^{-1}). In (b), the forbidden energy gap is not much greater than thermally accessible energies ($RT = 2{\cdot}5$ kJ (0·6 kcal) mole^{-1} at room temperature and 9·1 kJ (2·15 kcal) mole^{-1} at 800°C). Substances for which diagram (b) holds are insulators at low temperatures, but with increasing temperature electrons are promoted from the filled band into the

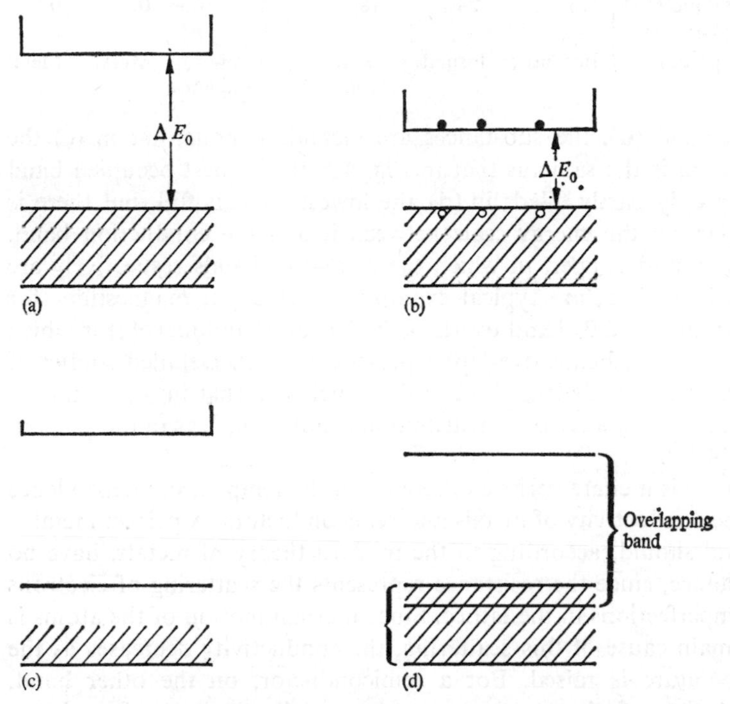

Figure 4.2 Energy band structure of various types of solid
(a) insulator
(b) intrinsic semiconductor
(c) metal; highest occupied band only partly filled
(d) metal; two overlapping bands

conduction band, and the material becomes a semiconductor with a conductivity which increases exponentially with temperature; singly filled levels ('positive holes') in the almost filled band, as well as electrons in the conduction band, contribute to the conductivity. Typical values of ΔE_0 for semiconductors vary continuously from the values for salts to a fraction of a kiloJoule (or kilocalorie); as one would expect from the pattern of variation of energy levels with atomic number, they decrease down a periodic group. For example, in Group IV the values are:

	Diamond	*Si*	*Ge*	*Sn (grey)*	*Sn (white)*	*Pb*
ΔE_0 (kJ mole^{-1})	750	117	48	9	0	0
(kcal mole^{-1})	115	28	18	2	0	0
Conducting properties	Insulator	Insulator	Semi-conductor	Semi-conductor	Metal	Metal

In (c) and (d), the substances are metallic conductors: in (c) the situation is the same as that in Fig. 4.1, the highest occupied band being only partly filled; in (d) the lowest band is full, but there is overlap on the energy scale between it and the conduction band. Sodium and copper, in which the 3*s* and 4*s* bands respectively are only half filled, are typical examples of case (c); magnesium (for which the filled 3*s* band overlaps the 3*p* band) and nickel (for which the 3*d* and 4*s* bands overlap) represent case (d). Detailed studies of the properties of nickel lead to the conclusion that in the metal the average actual electronic distribution is not $3d^8 4s^2$, as in the gas, but $3d^{9\cdot5}4s^{0\cdot5}$.

There is a characteristic difference in the temperature dependence of the conductivity of metals and semiconductors. A perfect metallic crystal should, according to the modern theory of metals, have no resistance, since the resistance represents the scattering of electrons by imperfections in the lattice. Since thermal motion of the atoms is the main cause of this scattering, the conductivity decreases as the temperature is raised. For a semiconductor, on the other hand, promotion of electrons into conduction bands is much easier at higher temperatures, and the conductivity increases exponentially with increase in temperature.

4.3 Defects inherent in the thermodynamics of the solid state[2, 3, 4]

There are two types of inherent defect, called Schottky defects and Frenkel defects after the workers who first suggested their existence. A Schottky defect is: in the case of a metal, a vacant site; in the case of a salt of formula MX, a vacant cation and a vacant anion site; in the case of a salt of formula MX_2, a vacant cation site and two vacant anion sites; the atom or ion in each case is removed to the surface of the crystal. A Frenkel defect is an interstitial atom or ion and a vacant lattice site; the ion is usually a cation because anions tend to be larger and their accommodation in interstitial sites is therefore difficult.

Any pure stoichiometric crystal at a temperature above absolute zero contains Schottky and/or Frenkel defects, since although energy is required to form them, their presence increases the entropy of the system and thus lowers its free energy. Defect formation may be considered a chemical process; for the formation of a Frenkel defect the reaction in the case of an ionic salt is

$$\begin{pmatrix}\text{ion in equilibrium}\\ \text{lattice position}\end{pmatrix} + \begin{pmatrix}\text{vacant interstitial}\\ \text{position}\end{pmatrix} \rightleftharpoons \begin{pmatrix}\text{ion in interstitial}\\ \text{position}\end{pmatrix} + \begin{pmatrix}\text{vacant lattice}\\ \text{position}\end{pmatrix}$$

If n_f is the number of ions in interstitial positions in unit volume at equilibrium, n_0 the number of lattice vacancies in unit volume at equilibrium, P the total number of ions in unit volume, and P_f the total number of possible interstitial positions in unit volume,

$$K = \frac{n_f n_0}{(P - n_0)(P_f - n_f)}$$

But for a Frenkel defect, $n_0 = n_f$ and, if the crystal is nearly perfect, $n_f \ll P, P_f$ so that

$$n_f^2 = PP_fK$$

If the energy of formation of a single Frenkel defect is w_f,

$$n_f = \sqrt{(PP_f)}\, \mathrm{e}^{-w_f/2kT}$$

or if W_f is the energy required for the formation of one mole of Frenkel defects,

$$n_f = \sqrt{(PP_f)}\, \mathrm{e}^{-W_f/2RT}$$

The value of W_f is usually between about 40 and 400 kJ (10 and 100 kcal) mole^{-1}. It will be noted that this treatment is general for a metal or salt irrespective of the formula type.

A similar expression may be derived for the number of Schottky defects, but in this case it is necessary to specify the formula type; for a salt MX with n_s vacant cation sites and n_s vacant anion sites (i.e. n_s defects), P cations and P anions in unit volume of the crystal,

$$n_s = P\, \mathrm{e}^{-W_s/2RT}$$

where W_s is the energy needed to create a mole of Schottky defects. The relative number of Schottky defects thus depends only on the absolute temperature and their energy of formation. This latter quantity is considerably less than the lattice energy of MX, since ions are being removed only to the crystal surface and not to infinity. For sodium chloride, for example, it is about 190 kJ (45 kcal) mole^{-1} compared with the lattice energy of 770 kJ (183 kcal) mole^{-1}. This value corresponds to values of n_s/N of about 3×10^{-17} at 25°, 5×10^{-8} at 400° and 3×10^{-5} at 800° (i.e. just below the melting point). At the last temperature the number of Schottky defects is about 4×10^{17} cm^{-3}. Since the expressions for the formation of both types of defect involve exponential factors, and W_s and W_f are different, it may be inferred that for a given substance there will normally be an overwhelming preponderance of one or other type of defect at a particular temperature.

Among several methods which have been used to study the occurrence of Schottky and Frenkel defects in stoichiometric crystals are measurements of density, conductivity and specific heats, and brief mention will now be made of each of these. Other methods which have been used in the investigation of these defects are dielectric loss measurements and nuclear magnetic resonance, electron spin resonance and Mössbauer spectroscopy; for an account of these, however, Greenwood's book[2] must be consulted.

Low concentrations of Schottky defects lower the pyknometric density of a crystal because of the increase in volume consequent upon removal of ions or atoms to the surface; low concentrations of Frenkel defects, on the other hand, leave the pyknometric density unaffected. For low concentrations of defects, the density determined by the X-ray method (from the number of atoms in the unit cell and the lattice constants) will be independent of the type of defect. Accurate comparisons of pyknometric and X-ray densities therefore provide useful information about defects. Silver chloride, for example, shows no evidence for the presence of Schottky defects between room temperature and its melting point, and its conductivity (discussed below) must thus be ascribed to Frenkel defects. Copper, silver and gold, however, have Schottky defects, and by studying the temperature dependence of the concentration of defects their energies of formation can be derived; for silver, for example, W_s is found to be 105 kJ (25 kcal) mole^{-1}—about one-third of the molar heat of sublimation.

Ionic conduction by salts arises from the migration of defects under the influence of an applied electric field. The transport numbers of the ions may be determined by methods analogous to those used for aqueous solutions. In one method two plates of crystal of equal thickness are placed in contact between weighed electrodes of metal to make the cell: (cathode) M|MX,MX|M (anode). After passage of a current the thicknesses of the plates are measured and the electrodes are weighed. Increase in the thickness of the plate nearer the anode denotes anion transport, the anion having combined with the anode to form more salt; cation transport causes a transfer of metal from anode to cathode. If transport measurements show that both cation and anion are carrying current, Schottky defects must be present; if only one ion is found to carry current, either only Frenkel defects are present or Schottky defects are present, but the transport numbers of the ions are very different. Only the negative ions are mobile in BaF_2, $BaCl_2$, PbF_2 and $PbCl_2$; only positive ions are mobile in AgCl, AgBr and AgI, and the alkali metal halides below about 400°; and both ions are mobile in the alkali metal halides (other than those of lithium) at temperatures between about 400° and the melting point.

Heats of formation of defects can also be determined from conductivity measurements at different temperatures. In this case a plot of log (specific conductivity) against $1/T$ gives $(U + \frac{1}{2}W)$, where U is the activation energy for ionic migration and W is the heat of formation per mole of defects; U can be determined independently by self-diffusion studies using radioactive isotopes or by other methods. For the alkali metal halides, W_s lies in the range 125–250 kJ (30–60 kcal) mole^{-1}, being greatest for the fluoride and least for the iodide for a given metal; for the silver halides, W_f is 155 kJ (37 kcal) mole^{-1} for AgCl and 117 kJ (28 kcal) mole^{-1} for AgBr.

Since the formation of lattice defects requires energy, there should be an anomalous rise in the specific heat of a crystal at temperatures high enough for the concentration of defects to become appreciable. This has been shown to be so for silver bromide near the melting point; once again, the study of the temperature dependence of the concentration of defects can lead to a value for their heat of formation. For silver bromide, W_f so determined is 121 kJ (29 kcal) mole^{-1}, in good agreement with the value obtained by the conductivity method.

For a salt of formula MX with ions of about equal size and a coordination number of six or more, Schottky defects are more likely than Frenkel defects owing to the absence of suitable interstitial sites. Low coordination numbers lead to open structures like those of zinc blende and wurtzite and hence permit Frenkel defect formation, and this is also favoured by a large difference in the sizes of the ions; normally the cation, being smaller, moves to the interstitial site. This generalization must be applied with caution, however, for in any salt of formula MX_2 the coordination number of the anion can be only half that of the cation. The lower coordination number makes movement of the anion easier, and this is presumably why CaF_2 and $PbCl_2$ exhibit anion Frenkel defect formation.

4.4 Specific defect structures in stoichiometric crystals[2, 6]

A rough division of specific defects found in stoichiometric crystals may be made into the following categories:

(a) orientational disorders;
(b) stacking faults;
(c) rotation or random orientation of molecules or complex ions;
(d) order–disorder phenomena, including the random distribution of atoms or ions of more than one kind in equivalent positions;
(e) complete disorder of cations.

Some of these categories may be dismissed with only brief comment. Orientational disorders occur in crystals of CO, N_2O, NO and ClO_3F. In the case of CO, for example, the entropy calculated by means of statistical mechanics from the vibration frequency and the interatomic distance is 4·6 J (1·1 cal) $mole^{-1}$ deg^{-1} greater than that obtained by the third-law method. This indicates that the crystal of CO, even at absolute zero, still has some degree of disorder, and suggests that some molecules are turned through 180° from the positions they should adopt in the completely ordered crystal. Since for an entirely random arrangement the entropy of the crystal would be $R \ln 2$ (5·8 J (1·38 cal) $mole^{-1}$ deg^{-1}) the degree of randomness is substantial. Similar conclusions apply to the other three molecules.

Stacking faults are departures from ideal layer and close-packed structures; graphite may be taken as an example. In the well-known normal graphite structure the carbon atoms of alternate layers of hexagons are vertically above one another, and the hexagonal unit cell has $a = 2{\cdot}456$ Å and $c = 6{\cdot}696$ Å. Another graphite structure exists in which the carbon atoms of every third layer are vertically above one another; this is also hexagonal but has the *c*-axis half as big again as in normal graphite. Some natural and artificial graphites have partly the normal (ABAB . . .) structure and partly the new

(ABC ABC...) one.[6] The structure of cobalt below 500° is a rather similar mixture of cubic close-packing and hexagonal close-packing.

Changes in structure with the adoption of a structure of higher symmetry are known for both molecular and ionic crystals: solid hydrogen chloride has an orthorhombic unit cell below 96°K, and a cubic unit cell above this temperature; and sodium and potassium cyanides change from an orthorhombic structure in which the CN^- ions are all parallel to the cubic sodium chloride structure at 283°K and 167°K respectively. In the case of hydrogen chloride, the change in structure is accompanied by a sharp increase in specific heat and a large increase in dielectric constant (this is followed by an unusually small increase in dielectric constant at the melting point). These observations are generally interpreted as indicating that in the cubic

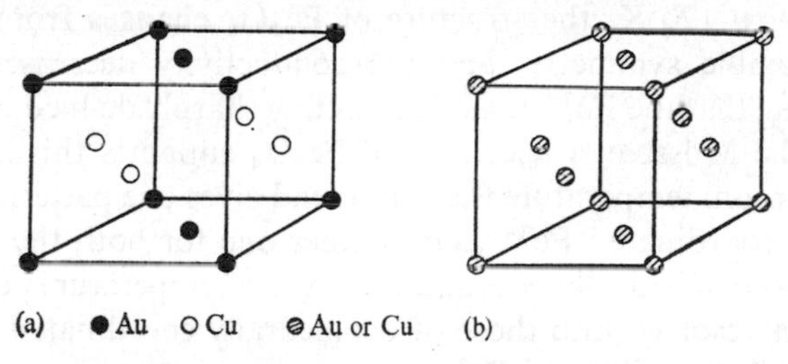

Figure 4.3 (a) ordered CuAu
(b) disordered CuAu

form of hydrogen chloride the molecules are rotating freely. For sodium and potassium cyanides only the X-ray evidence is available, and this does not serve to distinguish between random orientation of ions and continuous rotation. The change in the structures of ammonium chloride, bromide and iodide from the low-temperature CsCl forms to the high-temperature NaCl forms is also accompanied by more random orientation of the NH_4^+ tetrahedra.

Some of the simplest examples of order–disorder transformations, in which constituents in the lattice exchange places, are provided by alloys. The alloy CuAu, for example, has a tetragonal (nearly cubic) unit cell, shown in Fig. 4.3(a). When this ordered structure is heated, the unit cell becomes cubic and lattice positions are occupied at random by copper or gold atoms (Fig. 4.3b). A similar change takes place on heating the alloy Cu_3Au, though here the change is only from a face-centred cubic structure to a cubic close-packed one.

Among oxides, the existence of normal and inverted spinels provides an example of a group of substances in which the octahedral and tetrahedral holes in the cubic close-packed oxide ion lattice are occupied in a very irregular way by different cations (see Section 3.4 for an account of these structures). In this case the distribution of ions is generally well accounted for if crystal field stabilization energies (see Chapter 18) are taken into consideration. Magnetite (Fe_3O_4), for example, has the inverse spinel structure because high-spin Fe^{3+} has no crystal field stabilization energy and high-spin Fe^{2+} has a higher crystal field stabilization energy in an octahedral environment than in a tetrahedral one. There is, however, a further point of interest in this substance. In the inverse structure, best written $Fe^{III}[Fe^{II},Fe^{III}]O_4$, there is a random array of Fe^{2+} and Fe^{3+} ions at equivalent positions in the lattice, and this gives rise to a very high electronic conduction by a rapid electron-switch between these ions. Below 120°K, the structure of Fe_3O_4 changes from cubic to orthorhombic symmetry, and the conductivity decreases rapidly, suggesting that the Fe^{2+} ions then occupy sharply defined octahedral sites. (The Mössbauer spectrum of Fe_3O_4 supports this interpretation. At room temperature the compound gives one pattern for tetrahedrally coordinated Fe^{3+} and another one for both the Fe^{2+} and Fe^{3+} ions octahedrally coordinated; at low temperatures the second pattern is resolved into those of octahedrally coordinated Fe^{2+} and octahedrally coordinated Fe^{3+}.)

Many compounds having disordered fluorite structures exist. So long as electrical neutrality is preserved, replacement of Ca^{2+} or F^- by ions of similar size leaves the structure unchanged. Thus an oxide ion ($r_- = 1{\cdot}40$ Å) may replace one fluoride ion ($r_- = 1{\cdot}36$ Å) so long as Ca^{2+} ($r_+ = 0{\cdot}99$ Å) is replaced by a tripositive ion, such as Y^{3+} ($r_+ = 0{\cdot}93$ Å), and YOF is isomorphous and almost isodimensional with CaF_2, the oxide and fluoride ions being distributed statistically over the anion sites in the fluorite lattice. The ions Na^+ ($r_+ = 0{\cdot}95$ Å) and Cd^{2+} ($r_+ = 0{\cdot}97$ Å) may also replace Ca^{2+}; thus $NaYF_4$ contains a random distribution of Na^+ and Y^{3+} ions in equal numbers replacing Ca^{2+} ions in the fluorite structure, and it is even possible to obtain the compound $NaCaCdYF_8$ in which four ions of similar size are distributed randomly over the cation sites. Similar disordered systems based on the LaF_3 structure are also found.

The best example of a compound in which there is complete disorder of the cations is provided by the high-temperature α-form of silver iodide. Below 146° the β-form is stable and has the hexagonal wurtzite structure, which may be described as a system of hexagonally close-packed iodide ions with silver ions occupying half of the tetrahedral holes. At 146° the structure undergoes a sharp endothermic change to the cubic α-structure in which the iodide ions are in a body-centred cubic arrangement whilst the silver ions are distributed over a large number of sites having twofold, threefold or fourfold co-ordination by iodide ions. The $\beta \rightarrow \alpha$ structure change is accompanied by an increase in specific conductivity from $3{\cdot}4 \times 10^{-4}$ ohm^{-1} cm^2 to 1·3 ohm^{-1} cm^2; transport measurements show that all the current is carried by the silver ions, and self-diffusion studies show that the silver ions move freely between the easily deformed iodide ions. These remain as a rigid framework up to the melting point of silver iodide at 555°.

4.5 Non-stoichiometric compounds[1, 2, 5, 7, 8]

In the preceding sections of this chapter we have dealt with various types of defect in stoichiometric crystals, including some for which the basic structure has been the starting point of the description, followed later by a statement of what the structure contained. We now consider chemical species that have variable compositions and for which maxima or minima in physical properties such as melting points, conductivity or lattice order do not coincide with a simple atomic ratio of the components (i.e. non-stoichiometric compounds). It will become clear that there are no rigid distinctions between ideal stoichiometric solids, defect stoichiometric solids and non-stoichiometric compounds, and that the study of the solid state involves the examination of a continuous series of systems from one extreme type to the other.

Let us consider the general case of the incorporation of an excess of element B in a crystal phase of ideal composition AB_n. This can be done in only three ways: substitutional incorporation (B atoms replace A atoms in lattice positions normally occupied by A atoms); interstitial incorporation (additional B atoms are located interstitially); and subtractive incorporation (all B atoms occupy proper B lattice positions, but some A lattice positions are left vacant). Substitutional incorporation is likely only when A and B are both metals or both metalloids, so that ionic repulsions are not involved. Thus β-brass, nominally CuZn, is a homogeneous phase in the composition range $CuZn_{0{\cdot}55-1{\cdot}16}$ at room temperature. In the β-phase of the Na–Pb system of ideal composition $NaPb_3$ some of the lead atoms are replaced by sodium; the phase has an actual composition

$NaPb_{1.86-2.70}$ and the stoichiometric phase even lies outside the range of homogeneity. Gallium antimonide (GaSb) can take up excess of either component by substitution in the zinc blende lattice, the excess or deficit of electrons leading to electronic conductivity.

Interstitial incorporation results in a pyknometric density higher than the X-ray density. When zinc oxide is heated a little oxygen is lost, and the zinc formed occupies interstitial positions, the composition of the resulting phase at 800° being $Zn_{1.00007}O$. This type of defect is analogous to a Frenkel defect except that the interstitial atom is not compensated by a vacant lattice site.

Subtractive incorporation is analogous to a Schottky defect for one type of ion only. Iron (II) oxide and iron (II) sulphide, for example, have variable composition but are shown by density measurements to be deficient in iron rather than rich in sulphur; their composition ranges are therefore written $Fe_{0.84-0.94}O$ and $Fe_{0.88-1.00}S$. Cadmium oxide, like zinc oxide, loses oxygen when heated, but in this case the resulting phase, $CdO_{0.9995}$ at 650°, is deficient in oxygen rather than rich in metal; this may well be due to the fact that cadmium oxide has the sodium chloride structure, in which there is much less room for interstitial ions than in the four-coordinated zinc blende structure.

The deviation from stoichiometry in an ionic crystal must necessarily involve a change of oxidation state. Incorporation of excess of metal means conversion of some cations into neutral atoms or into ions of the same element in lower oxidation states. Conversely, incorporation of excess of non-metal means conversion of some cations into ions of higher oxidation state (which requires overcoming an ionization potential) or production of anions in less negative oxidation states than those originally present. A sample of iron (II) oxide of composition $Fe_{0.90}O$, for example, is really $Fe^{2+}_{0.70}\ Fe^{3+}_{0.20}O$, but (as has been pointed out in Section 4.4) the presence of cations of the same element in different oxidation states and in crystallographically equivalent positions leads to electronic conduction. Where conductivity arises from excess of negative charge, the compound is called a n-type semiconductor; where there is a deficit of negative charge (or an excess of positive charge) the compound is called a p-type semiconductor. On the band theory of solids non-stoichiometric compounds have levels ('impurity levels', so called because they are formed also by deliberate addition of suitable impurities) between the highest filled band and the conductivity band. These impurity levels can either act as a source of electrons for promotion with absorption of energy into the conduction band, or they can act as electron sinks for the acceptance of electrons promoted (again with absorption of energy) from the highest filled band; these mechanisms apply to n- and p-type non-stoichiometric compounds respectively. In the latter case, the conductivity is often known as 'positive-hole' conductivity. In both cases the conductivity, like that of other semiconductors, increases exponentially with increase in temperature.

The investigation of the variation in conductivity with pressure at constant temperature provides an alternative method to density measurements for finding whether a given compound contains an excess or a deficiency of metal. For zinc oxide, for example, increase in oxygen pressure above the solid phase lowers the conductivity of the system, and this must mean that the conductivity arises from species with more electrons than the Zn^{2+} ion; zinc oxide is therefore a n-type semiconductor. Conversely, copper (I) oxide, the conductivity of which is increased by a higher pressure of oxygen, is a p-type semiconductor.

Quantitative examinations of the dependence of the conductivity on pressure have been made in some instances. When nickel (II) oxide is heated in oxygen, some of the cations are oxidized and vacant cation sites are created according to the equation

$$4Ni^{2+}(l) + O_2(g) = 4Ni^{3+}(l) + 2\square_+ + 2O^{2-}(l)$$

where the symbol $\square_+$ denotes a vacant cation site and (l) denotes an ion in the lattice. For small deviations from stoichiometry the concentrations of Ni^{2+} and O^{2-} on normal lattice sites are nearly constant, so we may write

$$K = \frac{[Ni^{3+}]^4[\square_+]^2}{p(O_2)}$$

Since

$$[\square_+] = \tfrac{1}{2}[Ni^{3+}], \quad K = \frac{[Ni^{3+}]^6}{4p(O_2)}$$

and since the conductivity is proportional to $[Ni^{3+}]$ it is proportional to the sixth root of the oxygen pressure, a result confirmed by experiment. For the corresponding process for copper (I) oxide the equation is

$$4Cu^{+}(l) + O_2(g) = 4Cu^{2+}(l) + 4\square_+ + 2O^{2-}(l)$$

from which

$$K = [Cu^{2+}]^8/p(O_2)$$

However, a dependence of conductivity on the fifth root of the oxygen pressure is found experimentally, showing that in this instance not all copper (II) ions contribute to the acceptance of electrons from the full band and to the formation of positive holes in this band.

In addition to pressure, conductivity and density measurements, visible and ultraviolet spectra can provide useful information about non-stoichiometric systems. When sodium chloride is heated in sodium vapour it becomes yellow; potassium chloride heated in sodium vapour or potassium vapour becomes violet. In each case the metal vapour condensed at the surface forms electrons, cations which diffuse into the crystal and, consequentially, an equal number of vacant anion sites; these trap the electrons, giving rise to the so-called F-centres (German *Farbzentren*) which are the source of the absorption giving rise to the colour. The intensity of the colour produced is proportional to the stoichiometric excess of metal present (or more correctly the deficit of halogen present, since the lattice has been expanded and its density is found experimentally to have decreased). The colour is characteristic of the crystal used and independent of the metal in the vapour phase, and its wavelength of maximum absorption increases as the lattice energy of the alkali metal halide decreases. Many other types of colour centre have been obtained by suitable treatment of alkali metal halide crystals, among them V-centres produced by heating with halogens; the discussion of these, however, is beyond the scope of this book.

In conclusion, we may consider the factors that lead to the incidence of non-stoichiometry. Like the existence of defects in stoichiometric crystals, the formation of defects involving change of oxidation state increases the entropy of the system; it is, indeed, possible to regard a stoichiometric crystal of an alkali metal halide as an exceptional entity which is distinguished by having subtractive and interstitial defects present in equal concentrations under specified conditions of temperature and pressure of alkali metal or halogen. The concentration of defects, however, is determined by the usual exponential relationship between an equilibrium constant and the free energy change of a process. For a compound to be stable over an appreciable range of composition, therefore, the expenditure of energy required to produce defects must not be too large. In the case of formation of a non-metal excess (i.e. metal deficient) defect, two factors which favour defect production are a small difference in energy between the two oxidation states involved (as measured by the appropriate ionization potential) and a small difference in size, so that the lattice is not distorted to the point of collapse. Increase in the oxidation state of the cation results in increased lattice energy, both because of a slight decrease in the interionic distance and because of the increased charge. Formation of high concentrations of metal-excess defects is in general less favoured on energetic grounds, since although the production of cations in lower oxidation states is an exothermic process, their creation would distort the lattice and lower the lattice energy substantially. (The formation of anions carrying higher negative charges, such as O^{3-} from O^{2-}, would be highly endothermic and would result in a large increase in size; this has not been observed.) These considerations are compatible with the experimental generalization that marked deviations from stoichiometry have so far been found mainly in the oxides and sulphides of the transition metals. The detailed structures of markedly non-stoichiometric phases have still to be determined, however. In a crystal of composition $Fe^{2+}_{0\cdot70}\,Fe^{3+}_{0\cdot20}O$, for example, neutron diffraction studies show that the structure is not that of a random homogeneous array of defects, but that within the phase there are microdomains of different structure. However, at the present time the nature of these microdomains is not known.

References for Chapter 4

1 Anderson, J. S., *Ann. Rep. Chem. Soc.*, 1946, **43**, 104; *Proc. Chem. Soc.*, 1964, 166.
2 Greenwood, N. N., *Ionic Crystals, Lattice Defects, and Nonstoichiometry* (Butterworths, 1968).
3 Galwey, A. K., *The Chemistry of Solids* (Chapman and Hall, 1967).
4 Barnard, A. K., *The Theoretical Basis of Inorganic Chemistry* (McGraw–Hill, 1965).
5 Phillips, C. S. G., and Williams, R. J. P., *Inorganic Chemistry*, Vol. I (Oxford University Press, 1965).
6 Wells, A. F., *Structural Inorganic Chemistry*, 3rd edn. (Oxford University Press, 1962).
7 Rees, A. L. G., *Chemistry of the Defect Solid State* (Methuen, 1954).
8 Mandelcorn, L. (ed.), *Non-stoichiometric Compounds* (Academic Press, 1964).

The structures and energetics of inorganic molecules

5

5.1 Introduction

The general methods available for the determination of molecular structure were summarized in Chapter 2. In this chapter we begin with a survey of the results obtained for inorganic molecules and discrete ions which are isoelectronic with them, and then discuss briefly the interpretation of the results. We shall largely restrict attention here to compounds of the typical elements (compounds of the transition elements are discussed in Chapters 14, 15 and 18) and shall assume an elementary knowledge of valence theory, which is dealt with in several introductory[1–5] and more advanced[6–9] texts.

The strength of a chemical bond is as important a property as its direction, and Section 5.3 deals with the thermochemistry of the covalent bond. In general, for bonds between a given pair of atoms there is a qualitative relationship between the strength of a bond, its length and its order assessed theoretically or intuitively. Since the best-known attempt to assess electronegativities and bond polarities has a thermochemical basis, the discussion of bond energies leads to a summary of estimates of these quantities from bond energy data and comparison of the results with those obtained by other methods.

5.2 The shapes of molecules and ions of non-transition elements[5, 10–13]

Whereas the molecule of $BeCl_2$ in the vapour phase is linear, that of OCl_2 is bent; similarly, BF_3 is plane-triangular, NF_3 pyramidal and ClF_3 T-shaped; CCl_4 is regular-tetrahedral, $TeCl_4$ irregular-tetrahedral and ICl_4^- planar; and PF_5 is trigonal-bipyramidal, but IF_5 is square pyramidal. In a famous review of inorganic stereochemistry published in 1940, Sidgwick and Powell[10] put forward the suggestion that for a molecule or ion of formula AB_n containing only single bonds, the structure could be derived by assuming that the pairs of electrons in the valence shell of A were as far apart as possible, whether they were involved in bonding or not. Thus for two electron pairs in the valence shell of A the distribution is linear; for three, triangular; for four, tetrahedral; for five, trigonal-bipyramidal; and for six, octahedral. On this approach the general features of the structures of all the compounds mentioned in the first sentence of this paragraph (some of which have been established since 1940) can be represented as follows, a pair of crosses denoting an unshared pair of electrons.

AB_2 Cl—Be—Cl; O with Cl, Cl and x x, x x

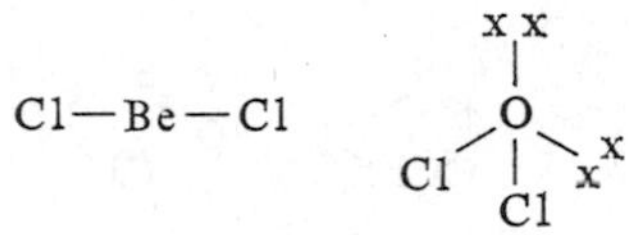

AB_3 B with F, F, F; N with F, F, F and x x; Cl with F, F, F and x x, x x

AB_4 C with Cl, Cl, Cl, Cl; Te with Cl, Cl, Cl, Cl and x x; I with Cl, Cl, Cl, Cl and x x, x x

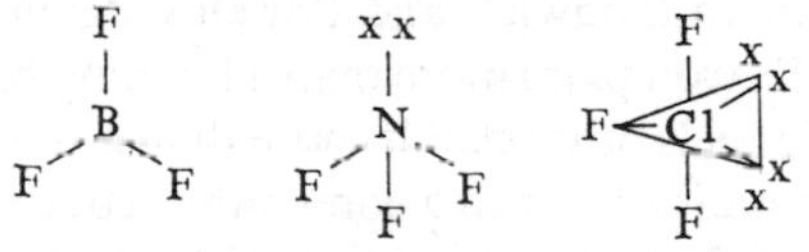

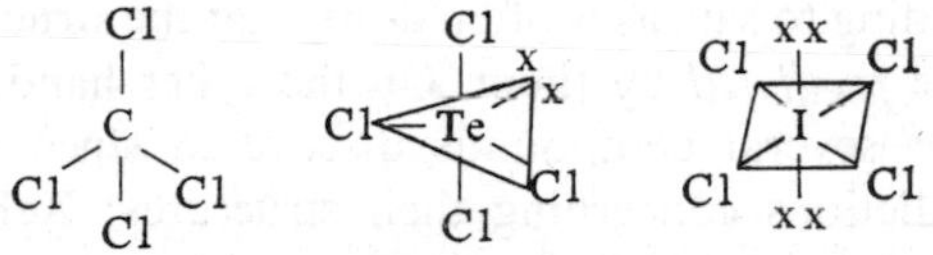

AB_5 P with F, F, F, F, F; I with F, F, F, F, F and x x

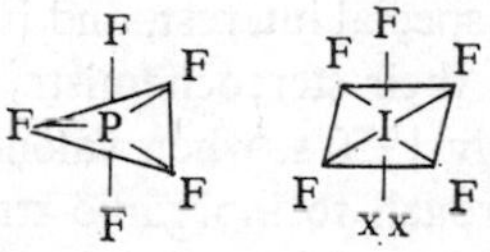

More detailed knowledge of the structures of those species which contain unshared pairs of electrons, and of related compounds, shows that bond angles are not exactly those of completely symmetrical structures. Along the series CH_4, NH_3 and OH_2, for example, the angle decreases steadily, the values being 109·5°, 107·3° and 104·5° respectively. In NF_3, the FNF angle is 102°, whilst in ClF_3 the F(equatorial)ClF(apical) angle is 87° and the F(apical) ClF(apical) angle is 186°. In a particular group of the periodic table, there is usually a decrease in bond angle as atomic number increases, as, for example, in the sequence OH_2(104·5°), SH_2(92·2°), SeH_2(91·0°) and TeH_2(89·5°). Double bonds appear to exert approximately the same stereochemical influence as single bonds: thus CO_2 and HCN are linear, whilst SO_2 and ONCl are bent, the bond angles being 119° and 117° respectively; and SO_2Cl_2 is tetrahedral (with the OSO, ClSO and ClSCl angles 120°, 106° and 111° respectively) and $SOCl_2$ pyramidal (with the ClSO and ClSCl angles 106° and 114° respectively). These four species may be represented as follows.

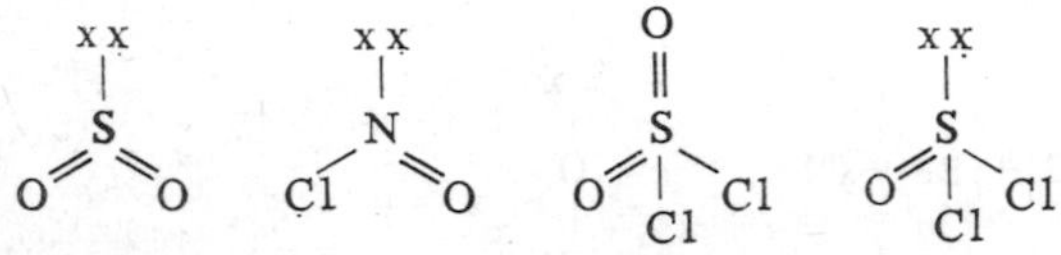

As further generalizations based on the experimental facts available at the time, Gillespie and Nyholm[11] in 1957 put forward the following additions to Sidgwick and Powell's statement: (1) a lone pair repels other electron pairs more than a bonding pair of electrons; (2) a double bond repels other bonds more than a single bond; (3) repulsion between bonding pairs in a compound AB_n depends upon the electronegativity of B, and decreases as the latter increases. These generalizations, it should be emphasized, are empirical; thus it is quite misleading to say, as is often done, that the structure of chlorine trifluoride is *predicted* by them. On the other hand, it has proved possible, for several compounds discovered since 1957, to make correct predictions concerning their structures: XeF_2 (linear) and XeF_4 (planar) are examples.

We see that the structures of molecules having unshared electrons on the central atom are thus of special interest, and it should now be mentioned that other views on their stereochemistry have been put forward. In the 1940's and early 1950's, when valence bond theory was the most fashionable approach to inorganic structures, it was customary to discuss the structures of species containing two, three, four and five pairs of electrons on the central atom of a molecule or ion AB_n in terms of sp, sp^2, sp^3 and sp^3d hybrid orbitals—in the last case, $sp^3d_{z^2}$ orbitals for a trigonal bipyramid (see Chapter 15 for a discussion of the shapes of orbitals). There is, however, no way of predicting on the basis of this approach which positions will be occupied by non-bonding pairs of electrons, and we shall, therefore, not discuss it further. Alternatively, it may be argued that there is little real evidence that justifies the invoking of d-orbitals and *d*-electrons in discussing the stereochemistry of, for example, PF_5, ICl_4^- and ICl_2^- (which is linear and could therefore be described as a trigonal bipyramid with three equatorial pairs of unshared electrons); these species are often described as involving multi-centre bonds. Thus for ICl_2^- we can consider the s, p_x and p_y orbitals on each of the atoms as containing an unshared electron pair; the three p_z orbitals can then be combined to yield one bonding, one non-bonding and one antibonding molecular orbital. With four electrons to go into these orbitals, the distribution would be full bonding and non-bonding orbitals and an empty antibonding orbital, corresponding to an average bond of one-half between the pairs of halogen atoms. Structural data for some interhalogen molecules and ions provide some support for this view; the matter is discussed further in Chapter 12. Certainly the finer details of the structures of, say, SF_4 and ClF_3 suggest that all bonds in these molecules are not equivalent: in SF_4, S–F(apical) and S–F(equatorial) are different, being 1·65 Å and 1·55 Å respectively; and in ClF_3, Cl–F(apical) and Cl–F(equatorial) are 1·70 Å and 1·60 Å respectively. In PF_5 the apical and equatorial fluorine atoms, formerly thought to be equidistant from the phosphorus atom, are known to be at 1·58 Å and 1·53 Å respectively.

So far we have said little about molecules or ions having (formally at least) six or more pairs of electrons in the valence shell of the central atom beyond noting that the structure of IF_5 can be described approximately as an octahedron with one position occupied by an unshared pair of electrons; the structures of ClF_5 and BrF_5 are similar, and so are those of the isoelectronic XeF_5^+, SbF_5^{2-} and $TeCl_5^-$ ions. The ions AlF_6^{3-}, GaF_6^{3-}, SiF_6^{2-}, GeF_6^{2-}, PF_6^- and AsF_6^- and the molecules SF_6 and SeF_6 all have regular octahedral structures, as would be expected since there are no unshared electron pairs. On valence bond theory the orbitals involved would be hybrid $sp^3d_{z^2}d_{x^2-y^2}$ orbitals. Alternatively, bonding schemes involving multicentre bonds can be constructed; in this instance there is no evidence on which to base a decision.

There are only a few species known which appear to have fourteen electrons in the valence shell of a central non-transition element. Extensive investigations of the structure of IF_7 suggest that it is approximately a pentagonal bipyramid, but that the molecule may not be rigid. The ion $[Sb(C_2O_4)_3]^{3-}$ is a pentagonal bipyramid in which the lone pair occupies an apical position. The ion IF_6^- and the molecule XeF_6 are almost certainly not regular octahedra at least at ordinary temperatures. On the other hand, the ions $SeCl_6^{2-}$, $SeBr_6^{2-}$, $TeCl_6^{2-}$ and $TeBr_6^{2-}$ (the structures of which are determined with certainty by X-ray diffraction) are regular octahedra. Possibly the stereochemically inert pair of electrons occupies an *s*-orbital, or it may be that the bond order in these ions is less than unity and that three-centre bonds are present.

Eight-coordination within a molecule or discrete ion has not yet been established for a typical element, nor is there at the present time any example of a derivative of such an element which formally involves a valence shell of sixteen electrons.

Summing up the discussion of the structures of molecules and ions of the type AB_n, we see that the principles enunciated by Sidgwick and Powell and by Gillespie and Nyholm (sometimes referred to collectively as the Valence Shell Electron Pair Repulsion Theory) are a useful systematization of the facts for species which contain central atoms formally having valence shells of not more than twelve electrons. They do not, however, predict the detailed structures of unsymmetrical molecules; and for species involving central atoms formally having valence shells of more than twelve electrons the theory gives only limited help in interpreting a complicated state of affairs.

5.3 Covalent bond energies[14, 15, 16]

It is conventional to work with covalent bond energy terms such that their summation gives heats of formation of molecules at 25°C from gaseous atoms in their ground states. Since it is possible to have bonds of different orders between many pairs of atoms, it is also necessary to specify bond order and desirable to state the identity of the compound from the standard heat of formation of which the bond energy term was derived. For a diatomic molecule, the bond energy is necessarily the same as the energy required to break the bond (usually called the bond dissociation energy, a quantity of great importance in chemical kinetics), but for larger molecules the two must be distinguished. Thus for water the heat of the reaction

$$H_2O(g) = 2H + O \tag{1}$$

which is derived from the standard heat of formation and the dissociation energies of oxygen and hydrogen, is twice the O—H bond energy in water, and it is also the sum of the heats of the reactions

$$H_2O = H + OH \tag{2}$$

and

$$OH = H + O \tag{3}$$

The heats of (2) and (3), however, are not identical, being 494 and 427 kJ (118 and 102 kcal) respectively, whilst that of (1) is 921 kJ (220 kcal); evidently breaking one O—H bond does not leave the other one unchanged, and this non-identity of successive bond dissociation energies is general.

For the halogens and hydrogen it would be generally agreed that the diatomic molecules contain single bonds, and the bond energies in these molecules, measured from the convergence limit in the ultraviolet spectrum (allowance being made for the fact that one atom is produced in an excited state) or from the temperature-dependence of log K for the reaction $X_2 = 2X$ are as follows:

	F—F	Cl—Cl	Br—Br	I—I	H—H
kJ	159	243	193	151	436
kcal	38	58	46	36	104

For the hydrogen halides, these values can be combined with standard heats of formation from gaseous halogens to yield:

	H—F	H—Cl	H—Br	H—I
kJ	566	431	366	299
kcal	135	103	87	71

The bonds in molecular nitrogen and oxygen, however, are not single bonds; the electronic structure of nitrogen is generally accepted as N≡N, but it is impossible to write a simple bond diagram for oxygen (which has two unpaired electrons in degenerate antibonding orbitals). Thus the dissociation energy of molecular nitrogen into atoms (940 kJ, or 225 kcal) is the N≡N bond energy, but it is impossible to cite a value for O=O, since there is no species known to contain this bond. The N—N and O—O single bond energies are based on values derived from hydrazine and hydrogen peroxide respectively. If, for example, the heat of formation of water from ground-state atoms is subtracted from that of hydrogen peroxide, and it is assumed that the O—H bond energies are the same in both, the O—O bond energy is derived as 155 kJ (37 kcal); a similar process based on N_2H_4 and NH_3 leads to the N—N bond energy of 121 kJ (29 kcal).

In order to obtain the C—H bond energy, the standard heat of formation of methane is found from the heats of combustion of methane, carbon and hydrogen, and this is then combined with data for the latent heat of sublimation of carbon and the bond energy in molecular hydrogen to yield a value of 416 kJ (99 kcal). It would, of course, be interesting to be able to measure also the change in energy when methane is formed from sp^3 hybridized carbon atoms and atomic hydrogen, since it is this quantity, rather than the conventional bond energy, that corresponds to the overlap of atomic orbitals; unfortunately, sp^3 hybridized carbon atoms are neither isolatable nor observable spectroscopically, and although theoretical estimates of the energy of their formation have been made, no experimental data can be available. Values for the C—F (485 kJ, or 116 kcal), C—Cl (327 kJ, or 78 kcal) and C—Br (285 kJ, or 68 kcal) bond energies may be obtained similarly from the heats of formation of the tetrahalides. For the C—C, C═C and C≡C bond energies the heats of formation of ethane, ethylene and acetylene respectively are required, and the C—H bond energies are assumed to be the same as in methane. Although this assumption is certainly not strictly true in the case of acetylene (in which the C—H distance is 0·04 Å shorter than it is in methane), this point is of less practical importance than at first seems to be the case, since the main requirement of a set of bond energy terms is self-consistency, and in predicting heats of formation a small error in the value for C—H (acetylenic) is compensated by one in the value for C≡C. The values compatible with one of 416 kJ (99 kcal) for C—H are: C—C, 347 kJ (83 kcal); C═C, 612 kJ (146 kcal); C≡C, 838 kJ (200 kcal).

Thermochemical data are often used in organic chemistry to provide evidence for electron delocalization. Their use for this purpose in inorganic chemistry is inevitably very restricted owing to the relatively small number of compounds to be considered, the considerably greater difficulty of determining heats of formation of many inorganic compounds and the difficulty of obtaining reference values. Although, for example, it would be interesting to be able to compare the heat of formation of borazine ($B_3N_3H_6$) (see Chapter 9) with that for a system containing 3 B—H, 3 N—H, 3 B═N and 3 B—N bonds, it is difficult to estimate the B—H bond energy (all boranes contain hydrogen in bridging as well as in terminal positions) and impossible to find a simple compound of known heat of formation in which a B═N bond can be taken to be present. Nevertheless, consideration of bond energies can often throw light on a problem in inorganic chemistry, and bond energy data will be referred to frequently in subsequent chapters.

It is convenient at this stage to interpolate a mention of bond stretching force constants. For a diatomic species the force constant f is given by the equation

$$\nu = \frac{1}{c}\sqrt{\frac{f}{4\pi^2\mu}}$$

where ν is the vibration frequency (in cm^{-1}), c is the velocity of light and μ the reduced mass. In polyatomic molecules there are frequently strong interactions between the various bonds, and evaluation of force constants is more difficult and subject to some uncertainty. There is a rough general relationship between bond energy and force constant, but nothing like proportionality: for example, the force constants for C—C, C=C and C≡C bonds are 4·5, 11 and 17×10^5 dynes cm^{-1} (or 4·5, 11, and 17×10^2 Newtons m^{-1}) respectively. The existence of simple quantitative relationship would, indeed, hardly be expected, since bond energies refer to complete rupture of a bond whilst stretching frequencies relate to small displacements from equilibrium positions. We shall not, therefore, make much use of force constants except in occasional discussions of closely similar molecules of the same symmetry.

5.4 Electronegativity[13, 15, 17]

If we compare the energies of single bonds between different atoms, E(A—B), with those in the appropriate reference molecules, E(A—A) and E(B—B), we notice that E(A—B) is nearly always greater, and often much greater, than the average of E(A—A) and E(B—B). For the hydrogen halides, for example, ΔE, given by

$$\Delta E = E(\text{A—B}) - \tfrac{1}{2}[E(\text{A—A}) + E(\text{B—B})]$$

has the following values:

	HF	HCl	HBr	HI
ΔE (kJ)	268	92	46	4
(kcal)	64	22	11	1
(eV)	2·8	0·96	0·48	0·04

It was suggested by Pauling in 1932 that ΔE is a measure of the extra energy of the A—B bond arising from a contribution A^+B^- or A^-B^+ to the pure covalent structure (i.e. arising from ionic–covalent resonance), and is a function of the difference in electronegativity Δx of A and B; for these purposes electronegativity is defined as the power of an atom *in a molecule* to attract electrons to itself. Pauling expressed ΔE in electron volts (to keep numbers small), and, by considering all the bond energy data then available, found empirically that an approximately self-consistent set of values for Δx could be obtained by taking

$$\Delta x = (\Delta E)^{\frac{1}{2}}$$

In order to avoid giving any element a negative value for x, x_H was taken as 2·1, leading to the electronegativities of the halogens as: F, 4·0; Cl, 3·0; Br, 2·8; and I, 2·4. Similar calculations were made for other elements.

In the forty years since Pauling introduced this approach, many of the thermochemical data he used have been revised, and several minor modifications have been suggested, the most important being to relate ΔE to the difference between the bond energy E(A—B) and the geometric mean of E(A—A) and E(B—B); this is found empirically to give a slightly better degree of self-consistency. The most recent values[15] for some of the elements, calculated in this way, are given in Table 5.1. It may be noted that these values are strictly those in the compounds which were the source of the thermochemical data; some slight dependence on valence state is to be expected.

Table 5.1 *Electronegativities of some elements*

		H	2·2						
B	1·9	C	2·5	N	3·1	O	3·6	F	4·0
Al	1·6	Si	1·9	P	2·2	S	2·7	Cl	3·2
		Ge	2·2	As	2·2	Se	2·6	Br	3·0
		Sn	2·1	Sb	2·1	Te	2·0	I	2·8

Many other methods for the assessment of electronegativity have been suggested. Mulliken proposed the use of the mean value of the first ionization potential and the electron affinity; this scale has the advantages of simplicity and rigour, but unfortunately relatively few electron affinities are known. Also, to be strictly relevant to molecules, valence state ionization potentials and electron affinities should be used, and these are not directly measurable. Allred and Rochow calculated the force f acting on the electrons at a distance from the nucleus equal to the covalent radius r by means of the equation

$$f = \frac{e^2 Z^*}{r^2}$$

where Z^* is the effective nuclear charge estimated theoretically using a set of shielding parameters. Values for f were plotted against Pauling electronegativities and the best straight line drawn through the points. From the slope and intercept of this line the following equation to give Allred–Rochow electronegativities x_{AR} on the Pauling scale was suggested:

$$x_{AR} = 0{\cdot}359\frac{Z^*}{r^2} + 0{\cdot}744$$

Correlations of electronegativity values with dipole moments, nuclear quadrupole coupling constants and nuclear magnetic resonance spin-spin coupling constants have also been attempted; since these quantities all involve factors other than electronegativity, however, we shall not discuss them here.

Electronegativities on the Mulliken and Allred–Rochow scales are in roughly the same order as on the Pauling scale using modern data, and we shall simply refer to 'electronegativity' in a general sense on the rare occasions when we use this concept. Its chief application is in the qualitative discussion of properties which appear to depend, in part at least, on unsymmetrical electron distributions in bonds—notably the inductive effect in organic chemistry. In simpler systems it is preferable to base the discussion on measurable thermochemical properties (such as ionization potentials, electron affinities, bond energies and lattice energies) whenever possible.

5.5 Covalent bond lengths[17, 18]

The methods described in Chapter 2 have made it possible to accumulate a large body of data on bond lengths. Problems of reference standards for single, double and triple bonds apply in the same way as they do to the choice of source compounds for the determination of bond energies. The molecules H_2, F_2, Cl_2, Br_2, I_2, H_2O_2, N_2H_4 and C_2H_6 and other alkanes are universally accepted as reference species; to them may be added elemental silicon, germanium and (grey) tin, orthorhombic sulphur and selenium (S_8 and Se_8), and (if the possible effect of an unusually small bond angle is discounted) white phosphorus and arsenic (P_4 and As_4). Halving the bond length in these species gives the values (in Å) for single covalent radii assembled in Table 5.2.

The N≡N bond length in N_2 (1·10 Å) is much less than that of N—N in N_2H_4 (1·48 Å), corresponding to a much stronger bond. In

Table 5.2 *Single covalent radii*

H	C	N	O	F
0·37	0·77	0·74	0·74	0·72
	Si	P	S	Cl
	1·17	1·10	1·04	0·99
	Ge	As	Se	Br
	1·22	1·22	1·17	1·14
	Sn			I
	1·40			1·33

organic chemistry, isolated C—C and C=C bonds have well-defined lengths of 1·54 Å and 1·33 Å respectively; bonds intermediate between these values (such as 1·42 Å in graphite and 1·39 Å in benzene) are explained in terms of fractional bond orders (in these two cases of $1\frac{1}{3}$ and $1\frac{1}{2}$ respectively, arising from electron delocalization).

Most elements, however, form no compounds in which there can be assumed to be single covalent bonds between two atoms of the element. If, for example, we want to ascribe a covalent radius to mercury, it is not satisfactory to take half of the distance in the metal, and the Hg—Hg distance in mercury (I) compounds is somewhat variable; the best source would appear to be CH_3HgCl, for which a microwave determination of the structure gives C—Hg = 2·06 Å and Hg—Cl = 2·28 Å. If we can assume additivity of covalent radii, subtraction of the radii of carbon and chlorine gives 1·29 Å in each case for the covalent radius of mercury. For nickel, however, the only simple molecular compound of known structure is $Ni(CO)_4$; for the past twenty years there has been vigorous argument concerning the degree of π-bonding between nickel and carbon in this compound, and therefore no reference compound for the determination of the single covalent radius of Ni^0 is available.

Even for mercury, however, the procedure described above is open to question. A survey of bond lengths in simple heteronuclear compounds discloses that covalent radii are not always additive. For example, although Si—C in $Si(CH_3)_4$ (1·89 Å) is almost equal to the sum of the covalent radii (1·94 Å), Si—F in SiF_4 (1·56 Å) is much less than the mean of Si—Si in Si and F—F in F_2 (i.e. the sum of the covalent radii). One reason for this is probably d_π–p_π Si—F bonding in SiF_4; but a similar reason cannot be invoked to account for the bond shortening in CF_4 (in which C—F is 1·32 Å compared with a sum of the covalent radii of 1·49 Å). There is a similar (though smaller) shortening of the C—O bond in methanol or of the N—F bond in NF_3. It has therefore been suggested that, since bond shortening is most marked between atoms of substantially different electronegativities, partial ionic character is also a possible cause of the shortening, which parallels an increase in bond strength mentioned in the last section.

When departures from an additivity relationship appear to arise from both π-bonding and partial ionic character, it is clear that bond lengths do not lend themselves to any rigorous classification, and their chief use is therefore in comparisons of similar systems. The increase in the B—F bond length from 1·30 Å in BF_3 to 1·42 in BF_4^-, for instance, supports the idea of p_π–p_π bonding in BF_3 only; and the decrease in the B—N bond length on going from $[(H_3N)_2BH_2]^+$ (1·59 Å) to borazine (1·42 Å) establishes the B—N bond order in the latter compound as substantially greater than unity.

We may conclude this chapter by again drawing attention to one of the inherent limitations in the comparison of bond properties in inorganic chemistry—there are so few compounds that it is usually difficult, and sometimes impossible, to change one variable whilst keeping others constant. It is, therefore, essential in all comparisons to state explicitly the sources of data, and to exercise great caution in reaching conclusions based on small differences in a bond property between compounds of somewhat different structure.

References for Chapter 5

1 Coulson, C. A., *Valence*, 2nd edn. (Oxford University Press, 1961).
2 Cartmell, E., and Fowles, G. W. A., *Valency and Molecular Structure*, 3rd edn. (Butterworths, 1966).
3 Douglas, B. E., and McDaniel, D. H., *Concepts and Models of Inorganic Chemistry* (Blaisdell, 1965).
4 Harvey, K. B., and Porter, G. B., *Introduction to Physical Inorganic Chemistry* (Addison–Wesley, 1963).
5 Mackay, K. M., and Mackay, R. A., *Introduction to Modern Inorganic Chemistry* (Intertext, 1968).
6 Drago, R. S., *Physical Methods in Inorganic Chemistry* (Reinhold, 1965).
7 Murrell, J. N., Kettle, S. F. A., and Tedder, J. M., *Valence Theory* (Wiley, 1965).
8 Linnett, J. W., *Wave Mechanics and Valency* (Wiley, 1960).
9 Ballhausen, C. J., and Gray, H. B., *Molecular Orbital Theory* (Benjamin, 1964).
10 Sidgwick, N. V., and Powell, H. M., *Proc. Roy. Soc. A*, 1940, **176**, 153.

11 Gillespie, R. J., and Nyholm, R. S., *Quart. Rev. Chem. Soc.*, 1957, **11,** 339.
12 Gillespie, R. J., *J. Chem. Education*, 1963, **40,** 295.
13 Day, M. C., and Selbin, J., *Theoretical Inorganic Chemistry*, 2nd edn. (Reinhold, 1969).
14 Cottrell, T. L., *The Strengths of Chemical Bonds*, 2nd edn. (Butterworths, 1958).
15 Johnson, D. A., *Some Thermodynamic Aspects of Inorganic Chemistry* (Cambridge University Press, 1968).
16 Dasent, W. E., *Inorganic Energetics* (Penguin, 1970).
17 Cotton, F. A., and Wilkinson, G., *Advanced Inorganic Chemistry,* 2nd edn. (Interscience, 1966).
18 Wells, A. F., *Structural Inorganic Chemistry*, 3rd edn. (Oxford University Press, 1962).

Inorganic chemistry in aqueous media

6

6.1 Introduction

The properties of aqueous solutions depend so much upon the nature of the solvent that we begin this chapter with a résumé of some of the physical properties of water. These, together with corresponding values for methanol and dimethyl ether, which may be regarded as mono- and di-methyl derivatives, are assembled in Table 6.1. Ice is known from a neutron diffraction study of D_2O to have the wurtzite structure with unsymmetrical hydrogen bonds between oxygen atoms, the O—H distance being 1·01 Å compared with 0·96 Å in water vapour. The wurtzite structure is a very open one (hence the low density of ice), and on melting a partial breakdown of the hydrogen-bonded system results in individual molecules being accumulated in the cavities in the network and hence in an increase in density; with rise in temperature this effect outweighs that of thermal expansion up to 4°C (the temperature of maximum density), but above this temperature the effect of thermal expansion is greater; the high entropy of vaporization at the boiling point shows that much of the hydrogen bonding remains even at 100°C. The strength of the O—H ..... O hydrogen bond in ice has been estimated as about 20 kJ (5 kcal) $mole^{-1}$; because of uncertainty about how many bonds are broken in the liquid, quantitative data for this phase are hardly meaningful.

Hydrogen bonding in water is the principal reason for the insolubility of non-polar substances in this solvent. Two substances will always tend to mix because of the increase in entropy on mixing. In the case of two liquids forming an ideal solution (i.e. one which obeys Raoult's law exactly), there is no heat change on mixing, and for gases or solids forming ideal solutions the only heat change is that associated with conversion of the solute into the liquid phase. It is easily shown by a thermodynamic treatment that the ideal

Table 6.1 *Physical properties of water, methanol and dimethyl ether*

	H_2O	CH_3OH	$(CH_3)_2O$
Critical temperature (°C)	374·2	240·6	126·7
Boiling point (°C)	100	64·7	−23·7
Melting point (°C)	0	−97·8	−138·5
Molar heat of vaporization (kJ)	40·67	35·23	21·50
(kcal)	9·72	8·42	5·14
Molar heat of fusion (kJ)	6·02	2·17	5·05
(kcal)	1·44	0·52	1·18
Molar entropy of vaporization			
($J\ deg^{-1}$)	109	102	86
($cal\ deg^{-1}$)	26·1	24·3	20·6
Dielectric constant	78·5	32·6	5·0
Dipole moment (Debye units)	1·84	1·68	1·30

solubility (expressed as a mole fraction) of a solute should be independent of the solvent chosen; for solutions of, for example, nitrogen and naphthalene in organic solvents like hydrocarbons, chloroform and carbon tetrachloride, this is nearly true. For water to act as a solvent, however, hydrogen bonds have to be broken and the partially ordered structure of the liquid has to be changed. The second consideration is one of the entropy of the system and will be discussed further below; the first one, however, is a simple heat effect, and its consequence is that unless the units of the potential solute become bonded more strongly to water molecules than to one another, and also more strongly than water molecules are bonded to one another, dissolution will not take place. Thus the only molecular species that dissolve unchanged in water to an appreciable extent are those which, by virtue of their containing a high proportion of polar H—F, H—O or H—N bonds, can form large numbers of hydrogen bonds with the solvent.

Water undergoes self-ionization to a slight extent according to the equation

$$2H_2O \rightleftharpoons H_3O^+ + OH^-$$

On the classical Arrhenius theory of acids and bases, these terms denote substances which produce (hydrated) protons and hydroxyl ions respectively in aqueous solution. Wider definitions are provided by the Lowry–Brønsted approach, in which an acid is a substance that tends to give up a proton and a base a substance that tends to accept one. Thus the anions of weak acids are strong bases, and so are pyridine and ammonia. In the ionization of hydrogen chloride in water according to the equation

$$HCl + H_2O \rightleftharpoons H_3O^+ + Cl^-$$

water is acting as a base, and the fact that ionization is almost complete shows it is a much stronger base than the chloride ion. When ammonia is dissolved in water, however, the solvent acts as an acid:

$$NH_3 + H_2O \rightleftharpoons NH_4^+ + OH^-$$

Addition of a stronger base containing the hydroxyl ion deprotonates the ammonium ion with formation of ammonia and water. The Lowry–Brønsted concept has been of great value in physical chemistry since it lends itself to quantitative treatment; still wider definitions are, however, useful in the case of non-aqueous solvent systems and will be mentioned in Chapter 7.

We conclude this section with a few points about conventions and units before going on to discuss hydration energies, the solubility of ionic salts, the strengths of acids, complex formation and redox processes.

(a) Concentrations of aqueous solutions are almost invariably expressed in the units moles (including g-ions) per litre or per kilogram, a convention which tends to obscure the fact that, owing to the very low molecular weight of water, mole fractions of solutes are lower by a factor of about 55 than molar concentrations. Thus the value of $K_W = 10^{-14}$ for the equilibrium constant of the reaction

$$2H_2O \rightleftharpoons H_3O^+ + OH^-$$

is based on the convention that the activity of water is unity, and the true autoprotolysis constant is considerably lower. Choice of different standard states for solute and solvent sometimes leads to difficulty in thermodynamic treatments of aqueous solutions.

(b) Because it is impossible to measure the potential of a single electrode, it is conventional to express all potentials relative to that of the standard hydrogen electrode (i.e. hydrogen at one atmosphere pressure in equilibrium with hydrogen ions at unit activity). This is equivalent to the convention that the standard free energy change of the reaction

$$H^+(a = 1, aq) + e \rightleftharpoons \tfrac{1}{2}H_2(g, 1\ atm)$$

is zero. (It is actually about -430 kJ, or -103 kcal.) Thus great care must be exercised in combining thermochemical data based on electrochemical measurements with values obtained calorimetrically and therefore based on the convention that only heats and free energies of formation of elements in their standard states at 25°C are zero. The need for such care will be apparent when the factors which influence redox potentials are discussed in Section 6.5. It may be mentioned that heats of hydration of gaseous ions and standard entropies of ions in aqueous solution are sometimes given relative to those of H^+(g) and H^+(aq), but where this is the case it is normally clear from tabulated data.

(c) The standard free energy of solution of a substance ΔG^0_s is related to the equilibrium constant K by means of the equation

$$\Delta G^0_s = -RT \ln K$$

The nature and units of K, however, depend upon the nature of the solute when dissolved (activities of solids are, by convention, taken as unity). For a non-electrolyte, K is the product of the solubility (expressed in moles kg^{-1}) and the activity coefficient; for an electrolyte, K is the thermodynamic solubility product K_{sp}, the expression for which depends on the formula of the electrolyte. For sodium chloride, for example,

$$K_{sp} = m^2\gamma^2$$

where m is the solubility in moles kg^{-1} and γ is the mean ionic activity coefficient in the saturated solution. For calcium fluoride,

$$K_{sp} = m^3\gamma^3$$

If we think of several salts of the same molal solubility and ignore any variations in γ, the magnitude of the standard free energy change on dissolution is therefore proportional to the number of species formed in aqueous solution from one 'molecule' of the solid.

(d) Solubility products, dissociation constants of acids and electrode potentials are measured in the absence of added electrolytes, and corrections are made when possible to allow for activity coefficients not being unity. Formation constants of complexes, on the other hand, are often measured in a constant ionic medium such as 2M or 4M sodium perchlorate solution without consideration of activity coefficients. Heats of reaction are determined at all ionic strengths. Some uncertainty inevitably arises when data obtained under one set of conditions are compared with those obtained in a medium of quite different ionic strength, and very detailed discussions of thermodynamic properties, for example, are pointless unless all refer to the same conditions.

6.2 Hydration energies and the solubilities of ionic salts[1–5]

The equilibrium between a solid salt M^+X^- and its ions in saturated aqueous solution may be considered in terms of the following thermodynamic cycle.

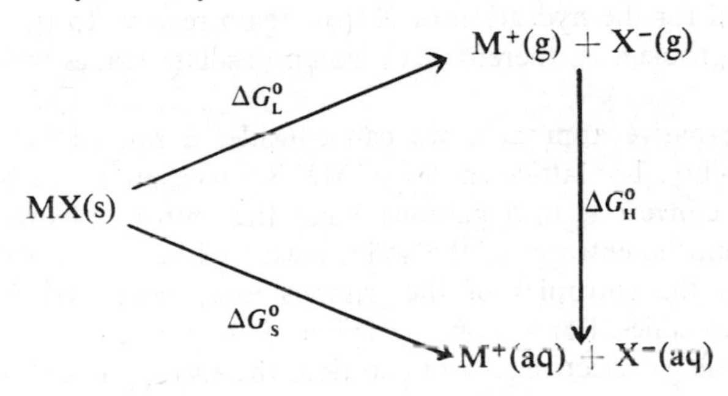

In this cycle ΔG_L^0 is the free energy change for conversion of the solid into gaseous ions at infinity (this is very nearly the same as the lattice energy, but it includes a small term to take into account the entropy of vaporization of the ions); ΔG_H^0 is the sum of the free energies of hydration of the ions; and ΔG_S^0 is the free energy change for the dissolution process, related to the solubility product by the equation

$$\Delta G_S^0 = -RT \ln K_{sp}$$

Since ΔG_S^0 is the small difference between two large quantities, accurate quantitative treatment of the solubility of salts is very difficult, and we shall concentrate on a comparative treatment, confining attention almost entirely to simple salts of formula M^+X^-. Even so, entropy effects cannot be neglected, and we shall therefore have to discuss both heats and free energies of hydration.

We can obtain the free energy of hydration of $M^+(g)$ relative to that of $H^+(g)$ from measurement of the standard potential E^0 of the M^+/M electrode. This gives us the standard free energy change ΔG^0 for the reaction

$$M(s) + H^+(aq) \rightleftharpoons M^+(aq) + \tfrac{1}{2}H_2$$

From E^0, ΔG^0 can be obtained by the relationship

$$\Delta G^0 = -nFE^0$$

where n is the number of electrons transferred (one in this case) and F is the Faraday. To complete the calculation we also need to know ΔG^0 for the sublimation and ionization of the metal and for the dissociation and ionization of hydrogen. The temperature dependence of E^0 enables us to obtain the entropy of hydration of M^+(g) relative to that of H^+(g), and from ΔG^0 and ΔS^0, ΔH^0 can easily be calculated. Similar measurements can be used to obtain the thermodynamic data for the hydration of X^-(g), again relative to that of H^+(g). The problem is, therefore, to assign absolute values to one ion.

In an alternative approach, we can consider a salt M^+X^- of known solubility. The lattice energy of MX is a measure of the heat required to convert it into gaseous ions; the entropy change is obtained from the entropy of the solid, measured by the third-law method, and the entropies of the gaseous ions, calculated from statistical mechanics. For a monatomic ion of atomic weight M and having a noble gas electronic configuration, the entropy is given by the equation

$$S^0 = \tfrac{3}{2}R \ln M + 109 \quad (\text{J mol}^{-1} \text{ deg}^{-1})$$

(A further small term has to be added if the ion does not have a noble gas configuration.) The free energy of dissolution of MX can be obtained from the solubility product and the heat of dissolution from calorimetric measurements; hence the entropy of dissolution can be calculated. In this case, however, we can only derive values for the sum of the hydration heats, free energies or entropies for both ions, and again the problem is to assign values to one ion. Whichever approach is used, it is essential that the set of values produced shall be as nearly self-consistent as possible. Since, for example, the lattice energy of sodium fluoride is 134 kJ (32 kcal) greater than that of sodium chloride and both salts dissolve in water without appreciable liberation or absorption of heat, the heat of hydration of F^- must be 134 kJ (32 kcal) greater than that of Cl^-. In like manner the differences in heats of hydration between Cl^- and Br^-, and between Br^- and I^-, are 34 and 44 kJ (8 and 11 kcal) respectively.

Many attempts have been made to get over the problem of subdividing heats, free energies or entropies of hydration, either by manipulation of the data or by calculations based on a particular electrostatic model. Recent attempts by both methods, by Halliwell and Nyburg[4] and by Noyes[5] respectively, have yielded results which are in good agreement and which satisfy the requirement of self-consistency; a set of values[1, 3] based on Noyes's work are given in Table 6.2. Data for some transition metal ions are given later in Chapter 18.

Table 6.2 *Absolute thermodynamic data for some aqueous ions at 25°*

Ion	ΔH^0 *Hydration* (kJ mole^{-1}*)	ΔS^0 *Hydration* (J mole^{-1} deg^{-1}†)	ΔG^0 *Hydration* (kJ mole^{-1}*)
H^+	−1120 (−268)	−121 (−29)	−1084 (−259)
Li^+	−544 (−130)	−134 (−32)	−506 (−121)
Na^+	−435 (−104)	−100 (−24)	−406 (−97)
K^+	−352 (−84)	−67 (−16)	−330 (−79)
Rb^+	−326 (−78)	−54 (−13)	−310 (−74)
Cs^+	−293 (−70)	−50 (−12)	−276 (−66)
Mg^{2+}	−1980 (−474)	−293 (−70)	−1895 (−453)
Ca^{2+}	−1650 (−395)	−238 (−57)	−1582 (−378)
Sr^{2+}	−1480 (−354)	−222 (−53)	−1415 (−338)
Ba^{2+}	−1365 (−326)	−188 (−45)	−1310 (−313)
Al^{3+}	−4750 (−1135)	−506 (−121)	−4600 (−1099)
La^{3+}	−3370 (−806)	−406 (−97)	−3250 (−777)
F^-	−473 (−113)	−142 (−34)	−432 (−103)
Cl^-	−339 (−81)	−84 (−20)	−314 (−75)
Br^-	−306 (−73)	−67 (−16)	−284 (−68)
I^-	−260 (−62)	−46 (−11)	−247 (−59)

* Values in parentheses are in kcal mole^{-1}.
† Values in parentheses are in cal mole^{-1} deg^{-1}.

Examination of the data in Table 6.2 discloses certain features of general interest. Ions of high charge have much greater heats of hydration and much more negative entropies of hydration than ions of low charge. This variation in heat of hydration is understandable on an electrostatic model; the negative entropy term may be thought of as arising from the tendency of ions of high charge to introduce more order into the system by fixing the orientations of a large number of water molecules. Some dependence of heat and entropy of hydration on ionic size is also to be expected. It is, in fact, found that free energies of hydration are roughly reproduced if the Born equation (derived from electrostatics)

$$\Delta G^0_{\text{hydration}} = -\frac{Nz^2e^2}{2r}\left(1 - \frac{1}{\varepsilon}\right)$$

is modified empirically. In this equation ε is the dielectric constant of the solvent and r is the radius of the ion; the other symbols have their usual meanings. In the modified equation r is altered to $r_+ + 0{\cdot}7$ Å for cations, where r_+ is the solid state ionic radius; for anions, r is taken to be the solid state radius r_-. (The fair degree of success of this modified equation in representing experimental data does not, of course, prove that the particular model on which it is based is correct. In particular, it is most unlikely that the dielectric constant in the immediate neighbourhood of an ion is as high as that of the pure solvent; in Noyes's alternative treatment of the problem,[5] all crystal radii are assumed to apply in solution, and attention is concentrated on calculating the change in the dielectric constant.)

Since entropies of monatomic ions in solids or in the gaseous state vary only slightly from one element to another, it is permissible to discuss the variation in solubility for salts having nearly the same heat of solution in terms of the entropies of hydration of the ions. Thus, for example, the entropy factor alone would suggest that most salts containing two ions of high charge would be very sparingly soluble. Conversely, for salts containing a given anion and a series of cations which have nearly the same entropies of hydration (e.g. the alkali metal cations, for which there is a slight increase in the entropy of hydration along the series), the variation in the solubility product can be accounted for largely in terms of the difference between the variation in lattice energy and that in heat of hydration of the cation. Since the lattice energy of a series of salts of similar structure is proportional to $1/(r_+ + r_-)$, there will be only a relatively small variation in this quantity if $r_- \gg r_+$. This does not affect the substantial decrease in heat of hydration along the series Li^+ to Cs^+, however, and hence the solid is increasingly stabilized with respect to the ions in solution. Thus for salts of large anions (e.g. perchlorate, fluorosilicate, chloroplatinate, cobaltinitrite) there is a steady decrease in solubility as the size of the alkali metal cation increases. It may be noted that for a very small anion, such as fluoride, this trend is reversed, whilst for chlorides there is a minimum solubility at potassium. It is, in fact, possible to treat the solubility of salts from a mathematical point of view with fair success. For details of this operation the original sources[3, 6] must be consulted, but it is worth noting one experimental generalization related to such a treatment: the best precipitant for a large complex ion is an ion of equal and opposite charge but of the same size. This lattice stabilization of species such as $[Ni(CN)_5]^{3-}$, $[FeCl_6]^{3-}$ and $[W(CO)_4(C_5H_5)]^+$ by precipitation as salts containing the ions $[Cr\ en_3]^{3+}$, $[Co(NH_3)_6]^{3+}$ and $[PF_6]^-$ respectively is very useful in preparative chemistry.[7]

The discussion in the last paragraph relates to the solubility of what are taken to be ideal ionic salts. Many marked variations in the pattern of solubilities are, of course, due to completely different factors. Thus the sharp decrease in solubility between silver fluoride and silver chloride (whereas there is a small increase between sodium fluoride and sodium chloride) can be traced to a substantial non-Coulombic contribution to the lattice energy of silver chloride (see Section 3.8). Although, therefore, the solubility of a salt in water is one of the first properties encountered in the study of chemistry, the quantitative discussion of this phenomenon remains a very difficult problem.

Nothing has yet been said about the interpretation of hydration energy on a molecular model. It is generally accepted that the interaction between a simple cation like Na^+ and water is mainly of an ion–dipole nature, the water molecules nearest to the ion having their oxygen atoms adjacent to it. Conversely, for an anion the hydrogen atoms of the solvent are adjacent to the ion. Limited evidence for the correctness of these models is provided by X-ray diffraction studies of concentrated aqueous solutions; in such solutions of sodium and potassium hydroxides, for example, each cation appears to be surrounded tetrahedrally by four oxygen atoms.[8]

Many indirect methods have been used to try to determine hydration numbers of ions, but most of them suffer from the limitation that they really examine only the total interaction between the ion and the environment. Thus the well-known fact that the lithium ion has the lowest ionic mobility of all the alkali metal ions does not prove it has the highest coordination number, but only that the total resistance to motion is a maximum. It would seem reasonable that for an ion as small as Li^+ there should be more interaction with next nearest neighbours than there is for the larger cations. Recently, however, the nature of some hydrated cations in solution has been investigated successfully by isotopic tracer studies (see Section 1.4), visible spectroscopy and nuclear magnetic resonance methods.

The basis of the spectroscopic method is comparison of the spectra of ions in solution with those of species in solids whose structures are known; its application is restricted to transition metal ions. By this method it can be shown, for example, that the predominant ion in the pink dilute solutions of cobalt (II) salts is $[Co(H_2O)_6]^{2+}$ rather than $[Co(H_2O)_4]^{2+}$. For cations with a relatively slow rate of exchange of water molecules between the first coordination sphere and the bulk solvent, separate ^{1}H or ^{17}O magnetic resonance signals can be observed for the two different kinds of water molecule, and hydration numbers can be calculated from the areas of peaks for solvated ions. Separation of signals can be achieved by lowering the temperature and/or dilution of aqueous solutions with organic solvents. In this way the existence of $[Be(H_2O)_4]^{2+}$, $[Mg(H_2O)_6]^{2+}$ and $[Al(H_2O)_6]^{3+}$ in solution has been established.[9,10] These results have been confirmed by another nuclear magnetic resonance method in which addition of a suitable paramagnetic cation (e.g. Co^{2+}) is used to shift the signal of non-coordinated water, that of coordinated water not being affected; for this work it is necessary to use water enriched in ^{17}O, since ^{16}O and ^{18}O have zero nuclear spin.[11] For the alkali and alkaline earth metal cations, however, the rate of exchange between coordinated and bulk solvent water (which can be measured by modern fast reaction techniques) is too high for these methods to be applicable. Nor has it yet proved possible to apply any of these modern methods to the study of anions, about the hydration of which very little is known.

6.3 The strengths of acids in aqueous solution[1,2,12,13]

The strengths of different acids in aqueous solution have been the subject of much comment; for an analytical discussion, however, it is necessary to consider two separate types of acid, hydrides and oxy-acids.

The factors which influence the degree of dissociation of a hydrogen halide in aqueous solution are apparent from consideration of the following cycle:

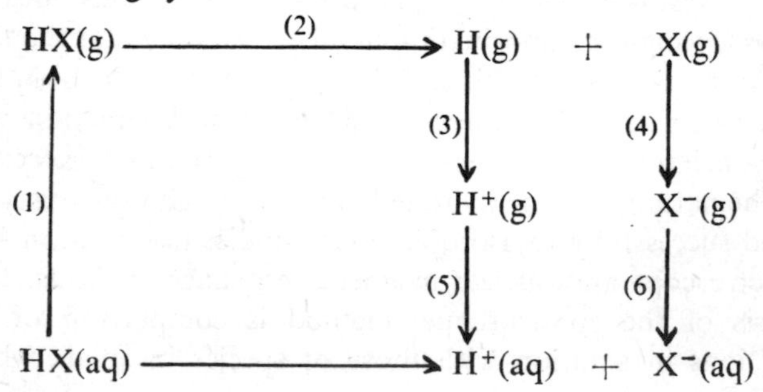

In this, stages (3) and (5) are common to all of the hydrogen halides. If we consider heat changes first, stage (2) represents the dissociation energy of the hydrogen halide, stage (4) the electron affinity of the halogen atom and stage (6) the heat of hydration of the halide ion; the sources of values for all of these have already been indicated. Stage (1), the heat of vaporization of the undissociated hydrogen halide from aqueous solution, can be measured (in reverse) for hydrogen fluoride (since this is a very weak acid), but it has to be estimated for the other hydrogen halides from comparisons with values for the noble gases and the methyl halides. We can thus compute ΔH^0 for the cycles and combine these with values for ΔS^0 (also, in part, estimated) to get values for ΔG^0.

In view of the fact that some of the data are estimates we shall not tabulate them all, but merely summarize the conclusions concerning the weakness of hydrogen fluoride ($pK_a = 3{\cdot}2$) (the other three are all very strong in aqueous solution with pK_a in the range -7 to -10). Of the difference in ΔG^0 obtained from the cycles between HF and HCl (63·6 kJ, or 15·2 kcal), nearly half (30·5 kJ, or 7·3 kcal) is due to stage (1): gaseous hydrogen fluoride has a high heat of hydration because of hydrogen bonding to water. Stages (2), (4) and (6) together contribute 23·8 kJ (5·7 kcal); although the fluoride ion has a higher heat of hydration, the greater bond strength in hydrogen fluoride, and the lower electron affinity of fluorine, outweigh the hydration term; and the remaining 9·2 kJ (2·2 kcal), the difference in $T\Delta S^0$, arises almost entirely from the more negative entropy of hydration of the smaller fluoride ion. It will be noticed that in this analysis we have not mentioned electronegativity or the inductive effect of fluorine, but it will also be noticed how many different factors contribute to the overall effect. Similar treatments may be devised for the series H_2O, H_2S, H_2Se and H_2Te, and NH_3, PH_3, AsH_3 and SbH_3, but not all the necessary data are available. Nevertheless, it can be seen that the increase in the strengths of the hydrides as acids as the elements in a given group become more metallic in character has a rational thermodynamic basis.

The largest group of oxy-acids are the carboxylic acids in organic chemistry, many of which have pK_a values in the range 1 to 5 (e.g. HCOOH, 3·8; CH_3COOH, 4·8; $CH_2ClCOOH$, 2·9; $CHCl_2COOH$, 1·3; CCl_3COOH, 0·7). These variations are commonly attributed to the inductive effects of CH_3 and Cl, which are believed to be electron-repelling and electron-attracting groups respectively; it is, however, interesting to see just how these inductive effects operate. Studies of the temperature-dependence of pK_a values show that ΔH^0 for the dissociation of HCOOH, CH_3COOH and CCl_3COOH is in each case nearly zero, and the effect is really one of $T\Delta S^0$; apparently chlorine atoms, by withdrawing electrons from the —COO^- end of the anion, cause it to bring about less ordering of the orientations of water molecules, and CH_3 groups have the opposite effect. Changes in heat terms (bond energy, electron affinity, heat of hydration and heat of solution of gaseous molecules of the acids) appear to cancel one another.

If we now consider the generality of oxy-acids, in which many different elements in different oxidation states are present, and for most of which adequate thermodynamic data are lacking, it is obvious that there is not going to be a simple explanation of the wide variation found. An empirical rule of Bell[12] states that for an acid of formula $XO_n(OH)_m$ the pK_a is given approximately by the relationship

$$pK_a = 8 - 5n$$

Some typical values are Cl(OH), 7·2, and $B(OH)_3$, 9·2; ClO(OH), 2·0, and NO(OH), 3·3; $ClO_2(OH)$, −1, and $NO_2(OH)$, −1·4; and $ClO_3(OH)$, −10. (It is interesting to note that H_2CO_3, commonly considered a very weak acid, has a pK_a of 3·9 if account is taken of the fact that carbon dioxide in aqueous solution is nearly all present in molecular form.) It is clear that for the oxy-acids of chlorine there must be a considerable variation in ΔH^0 for dissociation, but with drastic change in the nature of the ion we cannot sort out the contributions of possible variations in the O—H bond energy, the electron affinity of the radical from which the anion is derived, the extra delocalization energy and the hydration energy of the anion, and the heat of solution of the undissociated acid.

6.4 Complex formation[1, 14–18]

6.4.1 Formation constants. Complex formation in aqueous solution may lead to the production of positively charged species (e.g. $[Cu(NH_3)_4]^{2+}$ in ammoniacal solutions of copper (II) salts and $[Mg_2(OH)_3]^+$ in partially hydrolysed solutions of magnesium salts), neutral species (e.g. $[Fe(CH_3COCHCOCH_3)_3]$, iron (III) acetylacetonate, and $[Ni(CH_3C({=}NO)C({=}NOH)CH_3)_2]$, nickel dimethylglyoximate) or negatively charged species (e.g. the tetracyanonickelate and stannate ions, $[Ni(CN)_4]^{2-}$ and $[Sn(OH)_6]^{2-}$

respectively). The formation of a complex may be recognized by measurements of colligative properties, spectra (electronic or vibrational), solubility, conductivity or electrode potentials, or by chemical methods. Some caution is necessary in drawing negative conclusions from chemical evidence, since all reactions have equilibrium constants, and if a test reaction for a cation has an equilibrium constant greater than one of complex formation for the same ion, the test reaction of the ion will be given despite complex formation. Thus in a solution of a silver salt containing a large excess of ammonia, virtually all the silver is present as $[Ag(NH_3)_2]^+$, and when excess of ammonia is present addition of a solution of a chloride produces no precipitate; silver iodide, however, is much less soluble than silver chloride (the values for K_{sp} are 10^{-16} and 10^{-10} respectively), and addition of iodide precipitates silver iodide even in the presence of the highest concentration of ammonia that can be obtained in aqueous solution.

Neutral complexes (or inner complex salts as they are sometimes called) are often very sparingly soluble in water but soluble in organic solvents; iron (III) acetylacetonate, for example, commonly written [Fe acac$_3$], is a red solid (m.p. 179°) which is readily extracted from aqueous media by benzene or chloroform. Since the anion in this substance is obtained by removal of a proton from the weak acid acetylacetone, the formation of the compound in aqueous solution involves both the equilibria

$$CH_3COCH_2COCH_3 \rightleftharpoons CH_3COCHCOCH_3^- + H^+ \qquad K = 10^{-10}$$

and

$$Fe^{3+} + 3CH_3COCHCOCH_3^- \rightleftharpoons [Fe(CH_3COCHCOCH_3)_3] \qquad K = 10^{26}$$

The amount of complex formed therefore depends upon the hydrogen ion concentration; if the medium is too acidic, hydrogen ions compete successfully with iron (III) ions for the acetylacetonate anion; if, on the other hand, the medium is too alkaline, hydroxyl ions precipitate the very sparingly soluble iron (III) hydroxide (for which K_{sp} is about 10^{-37}). In the extraction of iron (III) from aqueous media by chloroform and acetylacetone there is thus an optimum pH for the process. Clearly, a study of the dependence of the extent of complex formation by the anion of a weak acid on the pH of the solution is capable of yielding interesting information, and we shall now turn to a discussion of quantitative aspects of complex formation. As in the preceding section, we shall consider the subject in general terms in this chapter and deal with complex formation by transition metal ions in more detail in Chapter 18.

In order to simplify formulae, we shall in the remainder of this subsection denote any metal ion in aqueous solution by M and any ligand by L, charges being omitted; furthermore, square brackets will be used to denote concentrations (or, strictly, activities) rather than, as before, the complexes themselves. We shall also assume for the sake of simplicity that no insoluble product or complex containing more than one metal ion is formed, and we shall for the present ignore the fact that complex formation involves changes in the hydration of the ions. Finally, it should perhaps be emphasized that we are discussing systems in thermodynamic equilibrium; mention will be made later of a few systems for which equilibria are reached only very slowly. If the conditions mentioned above are fulfilled, the interaction of M and L may be written as a series of equilibria:

$$\mathrm{M} + \mathrm{L} \rightleftharpoons \mathrm{ML} \qquad K_1 = \frac{[\mathrm{ML}]}{[\mathrm{M}][\mathrm{L}]}$$

$$\mathrm{ML} + \mathrm{L} \rightleftharpoons \mathrm{ML_2} \qquad K_2 = \frac{[\mathrm{ML_2}]}{[\mathrm{ML}][\mathrm{L}]}$$

$$\mathrm{ML}_{n-1} + \mathrm{L} \rightleftharpoons \mathrm{ML}_n \qquad K_n = \frac{[\mathrm{ML}_n]}{[\mathrm{ML}_{n-1}][\mathrm{L}]}$$

The equilibrium constants K_1, K_2, K_3, ..., K_n are known as stepwise formation constants. Alternatively, we may consider a series of overall equilibrium constants, for which the general symbol β is used. Thus

$$\mathrm{M} + \mathrm{L} \rightleftharpoons \mathrm{ML} \qquad \beta_1 = \frac{[\mathrm{ML}]}{[\mathrm{M}][\mathrm{L}]}$$

$$\mathrm{M} + 2\mathrm{L} \rightleftharpoons \mathrm{ML_2} \qquad \beta_2 = \frac{[\mathrm{ML_2}]}{[\mathrm{M}][\mathrm{L}]^2}$$

$$\mathrm{M} + n\mathrm{L} \rightleftharpoons \mathrm{ML}_n \qquad \beta_n = \frac{[\mathrm{ML}_n]}{[\mathrm{M}][\mathrm{L}]^n}$$

Obviously

$$\beta_n = K_1, K_2 \; . \; . \; . \; . \; . \; K_n$$

From such equilibrium constants, standard free energies of complex formation may be evaluated. By making measurements at different temperatures, heats and entropies of complexing may also be obtained, though since ΔH is small for complexing in aqueous solution and is not quite independent of temperature, it is better in accurate work to combine a calorimetric determination of ΔH with a potentiometric one of ΔG.

If it is established that only one complex, of known empirical formula, is present in a solution of known total concentrations of M and L, a determination of the concentration of uncomplexed M or L will suffice to give the formation constant. This determination may be made in many ways: for example, by polarographic or e.m.f. measurements if a suitable reversible electrode exists, by potentiometric determination of pH if the ligand is the anion of a weak acid, or by ion-exchange, spectrophotometric or distribution methods (for details the full account by Rossotti and Rossotti[17] should be consulted).

Generally, however, more than one complex will be present in a given solution, and successive formation constants are then obtained by determining the equilibrium concentrations of one (or, better, two) species present by physical methods which do not disturb the equilibrium, and combining these values with those for the total concentrations in the system of metal C_M and ligand C_L, which are known from analysis of the initial solutions. Let us consider the relationships between these total concentrations and the equilibrium concentrations of the free metal ion [M], free ligand [L], or of the cth complex $[ML_c]$. The degree of formation of the cth mononuclear complex α_c is given by

$$\alpha_c = \frac{[ML_c]}{C_M} = \frac{[ML_c]}{\sum_0^N [ML_n]} = \frac{\beta_c[L]^c}{\sum_0^N \beta_n[L]^n}$$

where

$$\sum_0^N \alpha_c = 1$$

Similarly, the average number of ligands bonded to each metal $\bar{n}$ is given by

$$\bar{n} = \frac{(C_L - [L])}{C_M} = \frac{\sum_0^N n[ML_n]}{\sum_0^N [ML_n]} = \frac{\sum_0^N n\beta_n[L]^n}{\sum_0^N \beta_n[L]^n}$$

Study of the dependence of α_c or $\bar{n}$ on [L] will therefore provide the information from which values of $\beta_1, \beta_2, \ldots\ldots, \beta_n$ (and hence of $K_1, K_2, \ldots\ldots, K_n$) may be obtained by mathematical analysis. If mixed complexes containing two ligands (e.g. OH^- and Cl^- in the system Sn^{2+}—Cl^-—H_2O) or polynuclear complexes (e.g. $[Fe_2(OH)_2]^{4+}$ in solutions of iron (III) salts) are formed, correspondingly more experimental work and treatment of data are required. When n is large, the determination of a full set of formation constants is a considerable operation, and for most systems only the overall formation constant β_n has been measured. It has been stated already, but should perhaps be repeated, that determinations of successive formation constants are very often made in media of high ionic strength; small differences in reported values often arise from differences in the media employed.

6.4.2. Thermodynamics of complex formation. It is usually found that there is a progressive decrease in the stepwise formation constant as n increases; the following are values for the Ni^{2+}—NH_3 system, obtained from a pH study (using a glass electrode) at 30°C in 2M ammonium nitrate solution.

$$Ni^{2+} + NH_3 \rightleftharpoons [Ni(NH_3)]^{2+} \quad K_1 = 10^{2\cdot79}$$
$$[Ni(NH_3)]^{2+} + NH_3 \rightleftharpoons [Ni(NH_3)_2]^{2+} \quad K_2 = 10^{2\cdot26}$$
$$[Ni(NH_3)_2]^{2+} + NH_3 \rightleftharpoons [Ni(NH_3)_3]^{2+} \quad K_3 = 10^{1\cdot69}$$
$$[Ni(NH_3)_3]^{2+} + NH_3 \rightleftharpoons [Ni(NH_3)_4]^{2+} \quad K_4 = 10^{1\cdot25}$$
$$[Ni(NH_3)_4]^{2+} + NH_3 \rightleftharpoons [Ni(NH_3)_5]^{2+} \quad K_5 = 10^{0\cdot74}$$
$$[Ni(NH_3)_5]^{2+} + NH_3 \rightleftharpoons [Ni(NH_3)_6]^{2+} \quad K_6 = 10^{0\cdot03}$$

The overall formation constant β_6 in this case is $10^{8\cdot74}$. From these data it will be seen that the nature of the main complex or complexes present in an ammoniacal solution of a nickel salt depends upon the $[NH_3]/[Ni^{2+}]$ ratio; only at high concentrations of ammonia, for example, is the pentammine complex the dominant species present, and even in very high concentrations of ammonia conversion of the pentammine into the hexammine is incomplete.

In this case it has been found by two independent calorimetric investigations, at 26·8° and 25° respectively, that ΔH for each substitution by ammonia is approximately −17 kJ (4 kcal), and is roughly constant (though the degree of constancy is different in the two sets of data[16]). If we disregard the possible variation in ΔH and consider the equilibria in general terms as

$$ML'_{N-n}L_n + L \rightleftharpoons ML'_{N-n-1}L_{n+1} + L'$$

(in the particular case discussed above, $N = 6$, $L' = H_2O$ and $L = NH_3$), the rate of formation of $ML'_{N-n-1}L_{n+1}$ is given by $k_f[ML'_{N-n}L_n][L]$, where k_f is the forward rate constant. If the statistical effect is the only variable that changes with change in n, the forward rate constant will be proportional to the number of sites for replacement of L' by L, so that the rate of the forward reaction is given by $k'_f(N-n)[ML'_{N-n}L_n][L]$. Similarly, for the back reaction the rate constant is proportional to $(n+1)$ and the rate is given by $k'_b(n+1)[ML'_{N-n-1}L_{n+1}][L]$. Thus the equilibrium constant $K_{n+1} = k_f/k_b$ is given by

$$K_{n+1} = \frac{k'_f(N-n)}{k'_b(n+1)}$$

In a similar way it can be shown that for the preceding substitution

$$K_n = \frac{k'_f(N-n+1)}{k'_b n}$$

and so the ratio between successive formation constants is given by

$$\frac{K_{n+1}}{K_n} = \frac{n(N-n)}{(n+1)(N-n+1)}$$

Ratios calculated from this formula and those derived from the data for K_1, K_2, etc., already given are compared in Table 6.3.

Table 6.3 *Statistical and experimental values for* K_{n+1}/K_n *for the* Ni^{2+}—NH_3 *system*

	Statistical ratio	*Experimental ratio*
K_2/K_1	0·42	0·29
K_3/K_2	0·53	0·43
K_4/K_3	0·56	0·36
K_5/K_4	0·53	0·31
K_6/K_5	0·42	0·20

The rough agreement shows that the statistical aspect is important but that other factors also play some part. These may include steric hindrance (where ligands bulkier than water are involved) and Coulombic effects (where charged ligands are involved). In the system Ni^{2+}—NH_3, ΔG for each stage is less negative than ΔH, and each substitution therefore has a negative entropy change; ΔS becomes more negative as n increases. In systems where the ligand is charged, on the other hand, the removal of charge when cation and anion come together is accompanied by displacement of water from the hydration spheres of the reactants. Thus, for example, in the Al^{3+}—F^- system, data for which are given in Table 6.4, the increase in entropy is the factor mainly responsible for complex formation.

Table 6.4 *Thermodynamic data for the* Al^{3+}—F^- *system*

n	1	2	3	4	5	6
$\ln K_n$	6·13	5·02	3·85	2·74	1·63	0·47
ΔH_n (kJ)	4·81	3·26	0·80	1·17	−3·14	−6·48
(kcal)	1·15	0·78	0·19	0·28	−0·75	−1·55
ΔS_n (J deg^{-1})	134	109	75	54	21	−13
(cal deg^{-1})	32	26	18	13	5	−3

An abrupt change in the ratio of successive formation constants is an indication of a change in coordination number, the intervention of special steric effects or a change in the electronic structure of the metal ion. For the Ag^+—NH_3 system, K_2 is greater than K_1, suggesting that coordination of one molecule of ammonia converts the hydration shell round the Ag^+ ion into a monosubstituted tetrahedron or octahedron, but that the introduction of the second ligand effects a change in structure to give the hydrated linear $Ag(NH_3)_2^+$ ion. With very bulky ligands such as 6,6′-dimethyl-2,2′-dipyridyl, many metals which form tris-2,2′-dipyridyl complexes can, for steric reasons, form only mono- or bis- complexes. And for the 2,2′-dipyridyl complexes of iron (II), a change from being paramagnetic to being diamagnetic takes place when the third ligand is coordinated; this leads, for reasons which will be indicated in Chapter 18, to K_3 being larger than K_2.

The stabilities of complexes are much increased when ligands which coordinate at two or more positions (polydentate ligands) replace monodentate ligands. Coordination by such ligands produces ring structures, the metal atom forming part of the ring; this process is known as *chelation* (derived from a Greek word meaning a crab's claw). Typical examples are the oxalate and ethylenediamine (en) complexes of many metals. Equilibrium constants for the Ni^{2+}—NH_3 and Ni^{2+}—en systems are given below (data for the Ni^{2+}—CH_3NH_2 system are, unfortunately, not available).

$$Ni^{2+} + 2NH_3 \rightleftharpoons [Ni(NH_3)_2]^{2+} \quad \ln \beta_2 = 5{\cdot}0$$
$$Ni^{2+} + 4NH_3 \rightleftharpoons [Ni(NH_3)_4]^{2+} \quad \ln \beta_4 = 8{\cdot}0$$
$$Ni^{2+} + 6NH_3 \rightleftharpoons [Ni(NH_3)_6]^{2+} \quad \ln \beta_6 = 8{\cdot}7$$
$$Ni^{2+} + en \rightleftharpoons [Ni\,en]^{2+} \quad \ln \beta_1 = 7{\cdot}5$$
$$Ni^{2+} + 2en \rightleftharpoons [Ni\,en_2]^{2+} \quad \ln \beta_2 = 13{\cdot}9$$
$$Ni^{2+} + 3en \rightleftharpoons [Ni\,en_3]^{2+} \quad \ln \beta_3 = 18{\cdot}3$$

In each case it is found that, for the same number of the same ligand atoms, the formation constant of the chelate complex is substantially greater. (This effect, known as the *chelate effect*, is even greater if complexes of a tridentate or tetradentate amine are compared with those of ammonia.) Calorimetric studies show that in the case of nickel and other transition metal ions having incomplete *d*-shells of electrons, the preference for complexing by ethylenediamine rather than ammonia is partly a heat effect and partly an entropy effect; for d^0 or d^{10} ions such as Zn^{2+}, however, it is almost entirely an entropy effect; the contribution of $T\Delta S^0$ to the more negative ΔG^0 of complexing by the chelate is almost independent of the nature of the metal ion.

The usual interpretation of this increase in entropy when, say, ethylenediamine displaces two molecules of ammonia is that the freeing of an extra molecule results in an increase in the disorder of the system, just as the formation of a gas contributes to the entropy increase during a reaction; this increase in entropy is offset to some extent by the loss of rotational entropy when the ring system is formed. Another way of looking at preferential complex formation by chelating agents is to consider the attachment of a second ligand after one molecule of ammonia or one end of ethylenediamine has coordinated: the probability of coordination by the other end of the ethylenediamine molecule is greater because it is forced to stay in the vicinity of the metal ion. At the same time it should, however, be mentioned that the chelate effect is, from a thermodynamic point of view, quite small; it is made to look larger by the convention of writing the activity of water as unity instead of as 55M. If we re-write the overall equilibrium constant of the general complex formation reaction

$$M + nL \rightleftharpoons ML_n$$

in terms of mole fractions we get, for dilute solutions,

$$\beta'_n = \frac{\left[\dfrac{ML_n}{55}\right]}{\left[\dfrac{M}{55}\right]\left[\dfrac{L}{55}\right]^n}$$

or

$$\beta'_n = \beta_n . 55^n$$

i.e. $\ln \beta'_n = \ln \beta_n + n \ln 55 = \ln \beta_n + 1{\cdot}74n.$

For Ni^{2+} and Zn^{2+}, $\ln \beta'_2$ is 8·5 and 8·3 respectively for $L = NH_3$, and $\ln \beta'_1$ is 9·6 and 7·7 respectively for L = en. Thus by working in mole fractions the chelate effect almost disappears for zinc and becomes very small for nickel; it does, indeed, approach zero generally for the case in which monodentate uncharged ligands are compared with chemically similar polydentate uncharged ligands in which the same atoms act as donors.

The effect on the ordinary equilibrium constant for complex formation of replacing a monodentate ligand by a chelating one is usually a maximum for formation of five-membered rings, slightly smaller for six-membered rings, and much smaller for larger rings. (This generalization may not hold, however, if electron delocalization through a conjugated system is possible in the ligand.) Probably the best known chelating agent is the anion of ethylenediamine tetracetic acid (EDTA), $(HOOCCH_2)_2NCH_2CH_2N(CH_2COOH)_2$, which can act as a sexadentate ligand with formation of six five-membered rings; this forms very stable complexes with most metal ions, and is extensively used in titrimetric analysis. Formation of a complex from a cation and the anion of EDTA, which carries four negative charges, is accompanied by a large increase in entropy, most of which is probably the consequence of partial neutralization of charge, just as in the Al^{3+}—F^- system mentioned earlier.

For all the systems so far discussed in this section, measurement of equilibrium constants by standard physicochemical methods has been possible. There are, however, many systems for which this is not so. If a complex is not formed rapidly and stoichiometrically, if it is not in thermodynamic equilibrium with its dissociation products, or if no satisfactory method for the analysis of the system can be found, a formation constant must be estimated from thermal data.

These considerations apply, for example, to such well-known systems as

$$Fe^{2+} + 6CN^- \rightleftharpoons [Fe(CN)_6]^{4-}$$

and

$$Fe^{3+} + 6CN^- \rightleftharpoons [Fe(CN)_6]^{3-}$$

The ratio of the equilibrium constants of these reactions can be obtained from values of E^0 for the electrodes Fe^{3+}/Fe^{2+} and $[Fe(CN)_6]^{3-}/[Fe(CN)_6]^{4-}$, but an absolute value for one of them has to be obtained from a calorimetrically determined heat of reaction and an estimated entropy change. In view of the exponential relationship of K to ΔG^0, it is not surprising that equilibrium constants obtained by such methods are often not very reliable.

6.4.3. Factors affecting the stability of complexes. There are no universally valid principles on the basis of which the relative values of formation constants of complexes of different metal ions with the same ligand, or of the same metal ion with different ligands, can be predicted. A number of interesting correlations have, however, been found, and we shall now summarize the most important of these.

For non-transition metal ions the stability of complexes normally increases with decreasing size (in the crystallographic sense) of the cation (e.g. among the alkaline earth metal ions the order is $Ba^{2+} < Sr^{2+} < Ca^{2+}$). The same behaviour is found for the series of lanthanide tripositive cations and (for most ligands) along the series of first-row transition metal dipositive and tripositive ions; for the transition metal ions, however, an additional effect (ligand field stabilization) is operative, and the data for these ions are therefore discussed further in Chapter 18. Increase in charge, for a given size of ion, nearly always results in a substantial increase in the stability of complexes (e.g. along the series Li^+, Mg^{2+}, Al^{3+}). Where a metal exists in two different oxidation states, the ion of the higher oxidation state is always smaller than that of the lower oxidation state, and the effects of change of charge and size reinforce one another. Thus metals in their higher oxidation states nearly always form more stable complexes than in their lower oxidation states (this is true, for example, for nearly all complexes of Fe^{3+} and Fe^{2+}, those with *o*-phenanthroline and dipyridyl being the only important exceptions; special factors are involved in these two cases).

So far as acceptor behaviour towards different ligands is concerned, metal ions fall into two groups, though the division is not always clear-cut. If we consider halide ions as ligands, it is found that for the lighter typical elements, metals at the beginning of transition series, lanthanides and actinides the usual order of stability of complexes is fluoride > chloride > bromide > iodide. For tellurium, polonium, the platinum metals, copper, silver, gold, cadmium, mercury and thallium this order is reversed. Elements of the former group also form more stable complexes with ligands containing oxygen and nitrogen as the donor atom than with analogous ligands containing sulphur and phosphorus respectively as donor; they were classified by Ahrland, Chatt and Davies[19] as class 'a' acceptors, and this term has passed into general use. Very few quantitative data concerning the stabilities of complexes of the second group of elements (class 'b' acceptors) with ligands other than the halide ions are available, but qualitative evidence (often, however, relating to non-aqueous conditions) supports the generalization that they form more stable complexes with sulphur donors than with oxygen donors, and with phosphorus donors than with nitrogen donors.

Although there is considerable variation because of the effect of change of oxidation state, a rough qualitative generalization about the stability of complexes formed by class 'b' elements with ligands containing different atoms as donors is that the order is

$$S \sim C > I > Br > Cl > N > O > F$$

For class 'a' elements this order is reversed and for some of these elements fluoride is the only ligand which will displace water.

At this point we should comment further on the distinction between class 'a' and class 'b' acceptors. If, for example, we consider a general equilibrium in aqueous solution such as

$$[MF_6]^{n-} + 6Cl^- \rightleftharpoons [MCl_6]^{n-} + 6F^-$$

we see that the factors determining the position of equilibrium are the metal–halide ion bond energies, and the heats and entropies of hydration of the simple and complex halide ions. If the equilibrium lies on the right-hand side this does not necessarily mean that the M—Cl bond is stronger than the M—F bond, but merely that the M—F bond is not stronger by a large enough margin to override hydration effects.

A recent survey of experimental data shows that for class 'a' acceptors the differences in formation constants of complex halide ions are in fact due mainly to entropy factors; for class 'b' acceptors, on the other hand, the entropy factors are overridden by a change in ΔH^o, which becomes more negative as we go from formation of the complex fluoride to that of the complex iodide.

Another general division of donors (Lewis bases) and acceptors (Lewis acids) which has gained some popularity recently is Pearson's classification into 'hard' and 'soft' acids and bases.[20] 'Hard' acids are ions which parallel the proton in their attachment to ligands; they correspond to class 'a' species; 'soft' acids bond more strongly to highly polarizable ligands and correspond to class 'b' species. Ligands are similarly divided into bases that are non-polarizable ('hard') and those that are polarizable ('soft'), examples being H_2O, NH_3, OH^- and SO_4^{2-} in the former category and R_2S, R_3P, I^- and CN^- in the latter. It is found empirically that the most stable complexes are those of hard acids with hard bases or of soft acids with soft bases. At the present time, however, this approach lacks a quantitative basis, and we shall not employ it in this book.

6.5 Redox processes[1, 3, 21, 22]

6.5.1. Electrode potentials. The electromotive force E^o of the half-reaction

$$Zn^{2+}(a = 1) + 2e \rightleftharpoons Zn(s)$$

is a measure of the standard free energy change of this reaction relative to that for the half-reaction

$$H^+(a = 1) + e \rightleftharpoons \tfrac{1}{2}H_2 \text{ (1 atm)}$$

If n electrons are involved in the reduction process,

$$-\Delta G^o = nFE^o$$

and the equilibrium constant is given by

$$\log K = \frac{nFE^o}{2{\cdot}303RT} = \frac{nE^o}{0{\cdot}059} \quad \text{at } 25°$$

Although it is customary to express the free energies of half-reactions in terms of volts (because most are obtained from measurements of electrode potentials relative to that of the standard hydrogen electrode), they can also be expressed in terms of equilibrium constants; for example,

$$Zn^{2+}(a = 1) + 2e \rightleftharpoons Zn(s) \quad K = 10^{-26}$$

This practice has the advantage of drawing attention to the value of combining them with solubility products and formation constants of complexes, but it does perhaps encourage the user to forget that,

Table 6.5 *Selected standard electrode potentials in aqueous solution at 25°*

Half-reaction	E^0 (V)
$Li^+ + e \rightleftharpoons Li$	−3·05
$K^+ + e \rightleftharpoons K$	−2·93
$Ca^{2+} + 2e \rightleftharpoons Ca$	−2·87
$Na^+ + e \rightleftharpoons Na$	−2·71
$Mg^{2+} + 2e \rightleftharpoons Mg$	−2·37
$Al^{3+} + 3e \rightleftharpoons Al$	−1·66
$Mn^{2+} + 2e \rightleftharpoons Mn$	−1·18
$Zn^{2+} + 2e \rightleftharpoons Zn$	−0·76
$Fe^{2+} + 2e \rightleftharpoons Fe$	−0·44
$Cr^{3+} + e \rightleftharpoons Cr^{2+}$	−0·41
$H^+ + e \rightleftharpoons \frac{1}{2}H_2$	0 (by convention)
$Cu^{2+} + e \rightleftharpoons Cu^+$	+0·15
$Cu^{2+} + 2e \rightleftharpoons Cu$	+0·34
$[Fe(CN)_6]^{3-} + e \rightleftharpoons [Fe(CN)_6]^{4-}$	+0·36
$\frac{1}{2}I_2 + e \rightleftharpoons I^-$	+0·54
$Fe^{3+} + e \rightleftharpoons Fe^{2+}$	+0·77
$Ag^+ + e \rightleftharpoons Ag$	+0·80
$\frac{1}{2}Br_2 + e \rightleftharpoons Br^-$	+1·07
$[Fe\ phen_3]^{3+} + e \rightleftharpoons [Fe\ phen_3]^{2+}$	+1·12
$\frac{1}{2}O_2 + 2H^+ + 2e \rightleftharpoons H_2O$	+1·23
$\frac{1}{2}Cr_2O_7^{2-} + 7H^+ + 6e \rightleftharpoons Cr^{3+} + \frac{7}{2}H_2O$	+1·33
$\frac{1}{2}Cl_2 + e \rightleftharpoons Cl^-$	+1·36
$MnO_4^- + 8H^+ + 5e \rightleftharpoons Mn^{2+} + 4H_2O$	+1·51
$Co^{3+} + e \rightleftharpoons Co^{2+}$	+1·95
$Ag^{2+} + e \rightleftharpoons Ag^+$	+1·98
$\frac{1}{2}S_2O_8^{2-} + e \rightleftharpoons SO_4^{2-}$	+2·01
$\frac{1}{2}F_2 + e \rightleftharpoons F^-$	+2·87

unlike other equilibrium constants, those based on standard potentials are relative to that for the reduction of hydrogen ions.

Some representative reduction potentials are given in Table 6.5.

It should be noted that the half-reactions are sometimes written in reverse; for example,

$$Zn \rightleftharpoons Zn^{2+} + 2e$$

The standard free energy change of this reaction is necessarily that of

$$Zn^{2+} + 2e \rightleftharpoons Zn$$

with the sign changed, and hence the sign of E^0 changes, too, when the half-reaction represents an oxidation instead of a reduction. The practice adopted in this book is the one recommended by the International Union of Pure and Applied Chemistry (IUPAC) and has the advantage that it gives the potential of the actual electrode (e.g. of the Zn metal in a solution of Zn^{2+} at $a = 1$, or of the Pt electrode in a solution containing Fe^{III} and Fe^{II}, each at unit activity) relative to that of the standard hydrogen electrode.

Most of the values given in Table 6.5 have been obtained directly by electrical measurements, but a few have been calculated from thermal data; these are those for which measurements cannot be made in aqueous media because of decomposition of the solvent (e.g. $\frac{1}{2}F_2/F^-$) and those for which the equilibrium is established only very slowly (i.e. for which the electrode is not thermodynamically reversible, such as $\frac{1}{2}O_2$, $2H^+/H_2O$).

Any half-reaction involving a reduction may be written in the form:

$$aA + bB + \ldots\ldots + ne \rightleftharpoons pP + qQ + \ldots\ldots$$

The value of E^0 then refers to all species at unit activity, and under non-standard conditions, if we assume activity coefficients to be unity, E relative to the standard hydrogen electrode is given by

$$E = E^0 - \frac{RT}{nF} \ln \frac{[P]^p[Q]^q \ldots\ldots}{[A]^a[B]^b \ldots\ldots}$$

$$= E^0 - \frac{0{\cdot}059}{n} \log \frac{[P]^p[Q]^q \ldots\ldots}{[A]^a[B]^b \ldots\ldots} \quad \text{at } 25^\circ\text{C}$$

Thus for the reduction of Zn^{2+}, since activities of solids are by convention taken as unity,

$$E = E^0 - \frac{0{\cdot}059}{2} \log \frac{1}{[Zn^{2+}]}$$

This means that as the solution of Zn^{2+} becomes more dilute, E becomes more negative and ΔG for the reduction more positive (i.e.

it is harder to reduce Zn^{2+} in dilute than in concentrated solution). Similarly, for the reduction of permanganate in acidic solution, the activity of water being taken as unity,

$$E = E^0 - \frac{0{\cdot}059}{5} \log \frac{[Mn^{2+}]}{[MnO_4^-][H^+]^8}$$

In this instance E^0 is positive relative to that for the standard hydrogen electrode, denoting that permanganate under standard conditions (including $[H^+] = 1$) is reduced with a more negative ΔG^0 than hydrogen ions (i.e. is a better oxidizing agent). Increase in $[H^+]$ above $[H^+] = 1$, keeping $[MnO_4^-]$ and $[Mn^{2+}]$ constant, thus makes E more positive than E^0, and because there is an eighth power dependence upon $[H^+]$ the effect is considerable.

The potentials for the reduction of water to hydrogen and for the reduction of oxygen to water (the reverse of the oxidation of water) are of great importance, since they determine (within the general limitation of thermodynamics) the range of chemical species that exist in aqueous media. For the reduction:

$$H^+ + e \rightleftharpoons \tfrac{1}{2}H_2$$

$$E = E^0 - 0{\cdot}059 \log \frac{[P_{H_2}]^{\frac{1}{2}}}{[H^+]}$$

Thus any system with E more negative than $-0{\cdot}059$ pH is thermodynamically unstable with respect to liberation of hydrogen, even under unit pressure of hydrogen; for neutral water E is $-0{\cdot}41$ V and for molar alkali (pH ~ 14) E is $-0{\cdot}83$ V. Therefore it is much harder to reduce water in alkaline than in acidic media.

For reduction of oxygen the relevant standard potential is

$$\tfrac{1}{2}O_2 + 2H^+ + 2e \rightleftharpoons H_2O \quad E^0 = +1{\cdot}23 \text{ V}$$

Then

$$E = E^0 - \frac{0{\cdot}059}{2} \log \frac{1}{[P_{O_2}]^{\frac{1}{2}}[H^+]^2}$$

At $P_{O_2} = 1$, $E = 1{\cdot}23 - 0{\cdot}059$ pH. For neutral water, E is thus $+0{\cdot}83$ V and for molar alkali it is $+0{\cdot}41$ V. Thus oxygen should, from the thermodynamic point of view, oxidize any system of standard potential less positive than $+1{\cdot}23$ V at pH 0, $+0{\cdot}83$ V at pH 7 or $+0{\cdot}41$ V at pH 14. Conversely, any system with E^0 more positive than $+1{\cdot}23$ V should oxidize water at pH 0, and so on. Any 'system' refers, of course, to one in which all species are present at unit activity (e.g. one of Fe^{3+} and Fe^{2+} each at $a = 1$ has $E^0 = +0{\cdot}77$ V). Such a system would be oxidized or reduced only until its own value of E (which may also depend upon $[H^+]$—see later) was the same as that of the $H^+/\frac{1}{2}H_2$ or $\frac{1}{2}O_2,2H^+/H_2O$ system. Since in experimental chemistry we are usually concerned not with standard systems but with reactants as pure as possible (e.g. iron (II) salts containing as little iron (III) as possible), it will be seen that corrections for non-standard conditions are very important when we apply standard electrode potentials to the interpretation of experimental observations.

We have referred in passing to the general limitation of thermodynamics. This is, of course, that thermodynamics tells us nothing about the rates of reactions, and many processes which are thermodynamically feasible have high activation energies; this is especially true of those which involve the breaking of covalent bonds. Thus many species which should oxidize water do not do so at an appreciable rate at ordinary temperatures; examples include solutions of permanganate at unit hydrogen ion concentration and persulphate under all conditions. Surface effects may also hinder reactions; thus although magnesium does not normally liberate hydrogen from water owing to the formation of a thin film of oxide, if the film is broken up by amalgamation a rapid reaction occurs.

6.5.2. The stabilization of oxidation states. Neglect of activity coefficients seldom introduces any serious inaccuracy into discussions of redox equilibria, but removal of species by precipitation, complex formation or neutralization (which can be regarded as complexing of H^+) may have very large effects. These, however, may easily be calculated if the necessary equilibrium constants are known.

Silver, for example, forms a complex iodide (silver iodide is readily soluble in a concentrated solution of potassium iodide), and for the reaction

$$Ag^+ + 3I^- \rightleftharpoons AgI_3{}^{2-}$$

β_3 is 10^{14}. Combination of this value with $E^0 = +0{\cdot}8$ V for

$$Ag^+ + e \rightleftharpoons Ag$$

leads to a value of $E^0 = -0{\cdot}02$ V for

$$AgI_3{}^{2-} + e \rightleftharpoons Ag + 3I^-$$

Thus at unit hydrogen ion and iodide ion concentrations metallic silver is a very slightly better reducing agent than gaseous hydrogen, and at high hydrogen ion and iodide ion concentrations the gap between the two is increased. Although, therefore, silver does not displace hydrogen from most acids, it will do so readily from concentrated aqueous hydriodic acid.

Manganese (III) is a powerful oxidizing agent:

$$Mn^{3+} + e \rightleftharpoons Mn^{2+} \qquad E^0 = +1{\cdot}5\ V$$

In alkaline media, however, manganese (III) and manganese (II) are present as $Mn(OH)_3$ and $Mn(OH)_2$, having solubility products of 10^{-36} and 10^{-13} respectively. At pH 14 ($[OH^-] = 1$), E^0 for the system

$$Mn(OH)_3(s) + e \rightleftharpoons Mn(OH)_2(s) + OH^-$$

is therefore only $+0{\cdot}15$ V; at the same pH, E^0 for the system

$$\tfrac{1}{2}O_2 + 2H^+ + 2e \rightleftharpoons H_2O$$

is, as we have shown previously, $+0{\cdot}41$ V; under these conditions oxygen can thus oxidize manganese (II) to manganese (III), the consequence essentially of the very much lower solubility of the hydroxide of the metal in the higher oxidation state. This state of affairs is quite general, and similar observations on the greater ease of oxidation in alkaline medium might be made for many other metals (e.g. Tl, Pb, Ti, V, Cr, Fe, Co and Ni).

When both oxidation states of an element are subject to complex formation, it is usually found that the overall formation constant is greater for the higher oxidation state, and thus complexing makes reduction more difficult (i.e. E^0 less positive and ΔG^0 less negative). (The example above, in which silver (I) is complexed by iodide, whilst silver (O) is not, merely represents a special case of this general effect.) Often modification of E^0 by complexing can be used for the estimation of formation constants in non-labile systems. Consider, for example, the following experimental data:

$$Co^{3+} + e \rightleftharpoons Co^{2+} \qquad E^0 = +1{\cdot}95\ V$$

$$[Co(NH_3)_6]^{3+} + e \rightleftharpoons [Co(NH_3)_6]^{2+} \qquad E^0 = +0{\cdot}1\ V$$

These two equations are linked by the formation of the hexammine complexes; that of cobalt (III) is kinetically inert with respect to dissociation, and addition of ammonia to a solution of a cobalt (III) salt in aqueous acid (itself metastable with respect to the liberation of oxygen) results in precipitation of cobalt (III) hydroxide. However, β_6 for the reaction

$$Co^{2+} + 6NH_3 \rightleftharpoons [Co(NH_3)_6]^{2+}$$

can be determined potentiometrically and is 10^5. Combination of this value with the difference of the two standard potentials gives β_6 for the reaction

$$Co^{3+} + 6NH_3 \rightleftharpoons [Co(NH_3)_6]^{3+}$$

as 10^{36}. Where neither formation constant can be measured, as for $[Fe(CN)_6]^{3-}$ and $[Fe(CN)_6]^{4-}$, study of the modification of E^0 by complexing will, of course, lead only to the ratio of the overall formation constants. Thus the relative values for the standard potentials

$$[Fe(CN)_6]^{3-} + e \rightleftharpoons [Fe(CN)_6]^{4-} \quad E^0 = +0{\cdot}36 \text{ V}$$
$$Fe^{3+} + e \rightleftharpoons Fe^{2+} \quad E^0 = +0{\cdot}77 \text{ V}$$
$$[Fe\ phen_3]^{3+} + e \rightleftharpoons [Fe\ phen_3]^{2+} \quad E^0 = +1{\cdot}12 \text{ V}$$

show that, unlike cyanide and nearly all other ligands, *o*-phenanthroline (phen) complexes more strongly with iron (II) than with iron (III). They also show why *o*-phenanthroline iron (II) sulphate can be used as an indicator in the titration of iron (II) with powerful oxidizing agents: virtually all the $Fe^{2+}(aq)$ is oxidized before appreciable oxidation of $[Fe\ phen_3]^{2+}$ begins. *Why* *o*-phenanthroline (and also 2,2-dipyridyl) have this effect is discussed in Chapters 15 and 18; we are concerned here with the evidence for preferential complexing of the lower oxidation state, and must defer its interpretation.

There are some chemical changes in which an element is converted from one into two different oxidation states; such *disproportionations* may be exemplified by the reactions

$$2Cu^{+} \rightleftharpoons Cu + Cu^{2+}$$

and

$$3MnO_4^{2-} + 4H^{+} \rightleftharpoons 2MnO_4^{-} + MnO_2 + 2H_2O$$

which take place when copper (I) sulphate (prepared from copper (I) oxide and dimethyl sulphate) is added to water and when acid is added to an alkaline solution of a manganate (made by fusing manganese dioxide with potassium nitrate and excess of potassium hydroxide, and dissolving the product in water). Equilibrium constants of such reactions may be calculated from potential data (most simply *via* conversion into standard free energies) or the operation may be conducted in reverse. Once again, removal of reactants or products by neutralization, precipitation or complex formation may upset equilibria. Thus for the above disproportionation of copper (I) the equilibrium constant is only 10^6, and in the presence of chloride ion or cyanide, copper (I) is stable in the presence of water because of the equilibria:

$$Cu^+ + Cl^- \rightleftharpoons CuCl(s) \qquad K = 10^7$$
$$Cu^+ + 4CN^- \rightleftharpoons [Cu(CN)_4]^{3-} \qquad K = 10^{30}$$

Since hydrogen ions are not involved, the extent of disproportionation of copper (I) is independent of the pH of the solution, whilst the opposite is true of the disproportionation of manganese (VI), which exists in appreciable concentrations only in strongly alkaline media.

The interplay of redox reactions, precipitation and complex formation is the key to the understanding of a large part of solution inorganic chemistry, and in later chapters we shall often refer to this subject.

6.5.3. Potential diagrams[21] and oxidation state diagrams.[2, 23–25] The presentation of all the potential data for an element which exists in several oxidation states in aqueous media is a lengthy operation, and two ways of summarizing this information have been used, the so-called potential diagrams and oxidation state diagrams. These may be illustrated for manganese, for which the relevant experimental data for the principal oxidation states (O, II, III, IV, VI and VII) are

$$Mn^{2+} + 2e \rightleftharpoons Mn \qquad E^0 = -1{\cdot}18\ V$$
$$Mn^{3+} + e \rightleftharpoons Mn^{2+} \qquad E^0 = +1{\cdot}51\ V$$
$$MnO_2 + 4H^+ + 2e \rightleftharpoons Mn^{2+} + 2H_2O \qquad E^0 = +1{\cdot}23\ V$$
$$MnO_4^- + 8H^+ + 5e \rightleftharpoons Mn^{2+} + 4H_2O \qquad E^0 = +1{\cdot}51\ V$$
$$MnO_4^- + e \rightleftharpoons MnO_4^{2-} \qquad E^0 = +0{\cdot}56\ V$$

These data may be used (care being taken to take due account of numbers of electrons involved) to calculate E^0 for such 'missing' changes as

$$MnO_4^- + 4H^+ + 3e \rightleftharpoons MnO_2 + 2H_2O \qquad E^0 = +1{\cdot}69\ V$$

and to derive the potential diagram at pH 0:

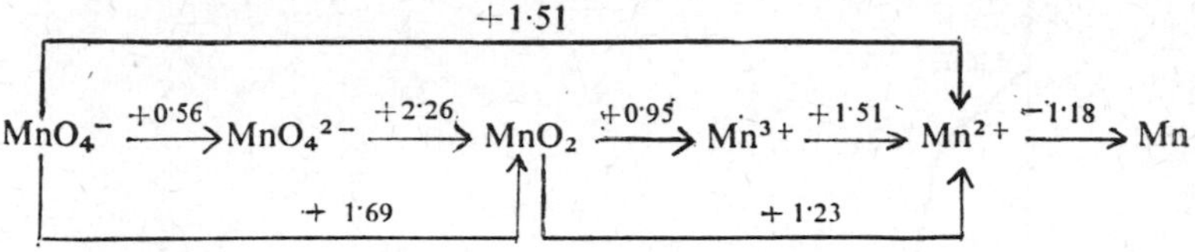

$$\begin{array}{c} \overbrace{\hspace{20em}}^{+1{\cdot}51} \\ MnO_4^- \xrightarrow{+0{\cdot}56} MnO_4^{2-} \xrightarrow{+2{\cdot}26} MnO_2 \xrightarrow{+0{\cdot}95} Mn^{3+} \xrightarrow{+1{\cdot}51} Mn^{2+} \xrightarrow{-1{\cdot}18} Mn \\ \underbrace{MnO_4^- \rightarrow MnO_2}_{+1{\cdot}69} \quad \underbrace{MnO_2 \rightarrow Mn^{2+}}_{+1{\cdot}23} \end{array}$$

Such diagrams quickly reveal oxidation states which are unstable with respect to disproportionation (e.g. MnO_4^{2-}, which is seen to be a more powerful oxidizing agent than MnO_4^- at pH = 0, and will therefore not accumulate during the reduction of permanganate), and they enable the reader to get a general view of the solution chemistry of an element more rapidly than is possible from a set of equations. However, there is a tendency to forget that the potentials sometimes depend (to extents which are not so clear as from the equations) on pH, and if complexing and precipitation have to be considered (as in real chemistry they so often must be), the diagrams become very elaborate.

In the usual graphical method of illustrating redox relationships, the free energy change (normally in eV though other energy units may be used) for the half-reaction

$$M(0) \rightleftharpoons M(Z) + Ze$$

is plotted against the oxidation state Z. The lowest points on the graph then represent the most stable oxidation states; half-reactions with negative ΔG^0 are represented by downward slopes, whilst half-reactions with positive ΔG^0 have upward slopes, the slope of the line joining any two points giving E^0. Any state represented by a 'convex' point is thermodynamically unstable with respect to disproportionation into states adjacent to it on the graph, whilst the opposite is true for any state represented by a 'concave' point. The reader may verify that the graph in Fig. 6.1 corresponds to the data already given for standard potentials of manganese systems. Such oxidation state diagrams are very useful for impressing on the mind a general picture of the solution chemistry under particular conditions, but suffer from the same limitations as potential diagrams if pH-dependence of the relative stabilities of oxidation states, complexing or precipitation is involved, and a three-dimensional model constructed of the oxidation state diagrams at different values for another variable is then needed. (In the case of manganese,

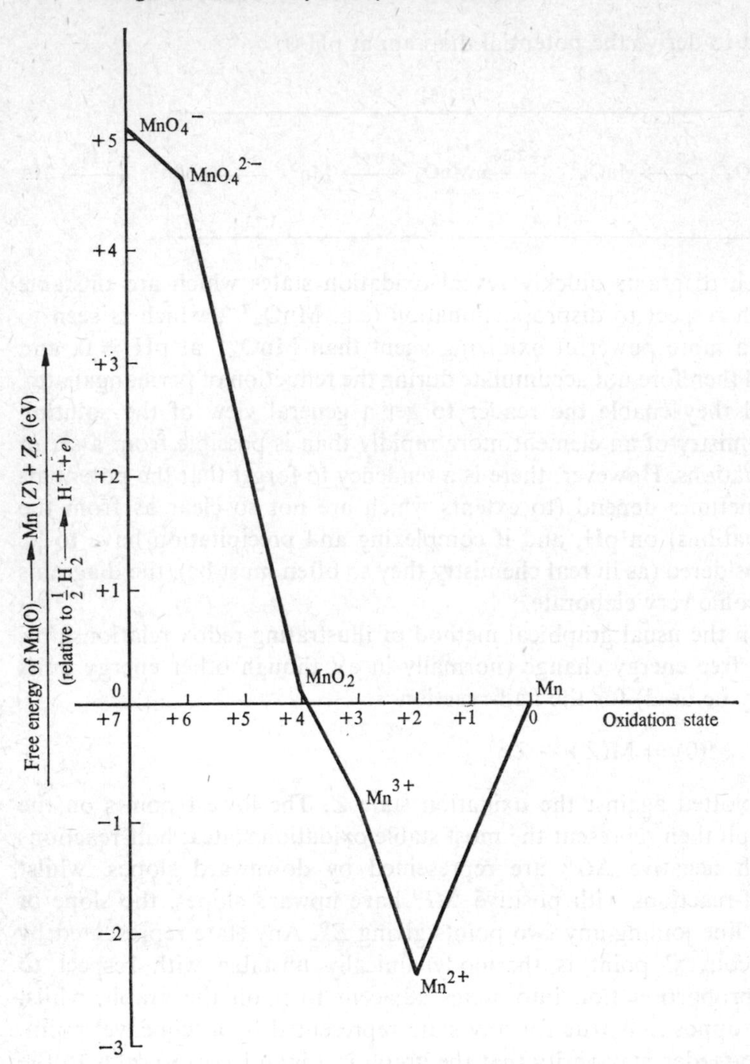

Figure 6.1 Free energies of formation of oxidation states of Mn at pH 0

for example, the diagram for pH 14 is quite different from that for pH 0.)

6.5.4. Factors determining the magnitude of redox potentials.[3, 21, 22]
We conclude this chapter with a brief discussion of the way in which

standard potentials are related to independently measurable thermodynamic quantities, choosing as examples the systems

$$Na^+ + e \rightleftharpoons Na \quad E^0 = +2{\cdot}7 \text{ V}$$

and

$$Ag^+ + e \rightleftharpoons Ag \quad E^0 = -0{\cdot}8 \text{ V}$$

In each case E^0 is an experimental value (in the case of sodium it is derived from measuring E for a series of amalgams of very low sodium concentration and then measuring the potential of the amalgam against that of the pure metal using a non-aqueous solvent). If we disregard entropy changes in this context (taking them into account does, in fact, make little difference here), we can split the half-reaction

$$M^+(aq) + e \rightleftharpoons M(s)$$

into the stages

$$M^+(aq) \xrightarrow{(1)} M^+(g) \xrightarrow{(2)} M(g) \xrightarrow{(3)} M(s)$$

in which (1) is absorption of the heat of hydration of $M^+(g)$, (2) is liberation of heat equivalent to the ionization potential of M, and (3) is liberation of the latent heat of sublimation of M. Since every E^0 is relative to that for the half-reaction

$$H^+(aq) + e \rightleftharpoons \tfrac{1}{2}H_2(g)$$

we must also consider the sum of the stages

$$H^+(aq) \xrightarrow{(1)} H^+(g) \xrightarrow{(2)} H(g) \xrightarrow{(3)} \tfrac{1}{2}H_2(g)$$

where (1) and (2) are again hydration and ionization terms, and (3) is half the bond energy in molecular hydrogen. Values for these stages are given in Table 6.6.

Table 6.6 *Factors determining the magnitude of redox potentials*

ΔH^0 (kJ; values in kcal in parentheses)

Stage	(1)	(2)	(3)	*Sum*
$Na^+ \rightarrow Na$	435(104)	−494(118)	−109(26)	−168 (−40)
$H^+ \rightarrow \tfrac{1}{2}H_2$	1120(268)	−1306(312)	−217(52)	−403 (−96)
$Ag^+ \rightarrow Ag$	502(120)	−728(174)	−280(67)	−506(−121)

Conversion to potentials yields

$$Na^+ + e \rightleftharpoons Na \quad E^0 = +2{\cdot}5\ V$$
$$Ag^+ + e \rightleftharpoons Ag \quad E^0 = -1{\cdot}0\ V$$

It is worth noting that the difference in the latent heats of sublimation of sodium and silver is nearly as large as the difference in their ionization potentials. If we carry out a similar analysis of the factors determining E^0 for the systems Cu^{2+}/Cu and Zn^{2+}/Zn we find that the difference in E^0 arises almost entirely from the difference in heats of sublimation. These examples show how important purely 'physical' quantities may be in influencing chemical behaviour. We can now state the general conditions for a metal ion–metal system to have a strongly negative redox potential: the metal must have a low latent heat of sublimation and a low ionization potential, and the ion must have a high heat of hydration. Since for the alkali metals the latent heats of sublimation and the ionization potentials decrease from lithium to caesium, and the heats of hydration decrease by nearly the same amounts as the sums of the other two quantities, E^0's for the M^+/M systems for all the alkali metals are in the range $-3{\cdot}0$ to $-2{\cdot}7$ V, and the great difference in reactivity towards water between lithium and caesium is therefore kinetic and not thermodynamic in origin (it probably originates in the low melting point of caesium, which allows the metal to melt easily and thus present a larger surface for reaction).

A corresponding analysis of the factors that determine the magnitudes of the $\frac{1}{2}X_2/X^-$ standard potentials for the halogens shows the overriding importance of the hydration energy term. For more complex systems such as

$$MnO_4^- + 8H^+ + 5e \rightleftharpoons Mn^{2+} + 4H_2O$$

however, no profitable treatment can be given at the present time.

References for Chapter 6

1 Harvey, K. B., and Porter, G. B., *Physical Inorganic Chemistry* (Addison-Wesley, 1963).
2 Phillips, C. S. G., and Williams, R. J. P., *Inorganic Chemistry*, Vol. I (Oxford University Press, 1965).
3 Johnson, D. A., *Some Thermodynamic Aspects of Inorganic Chemistry* (Cambridge University Press, 1968).
4 Halliwell, H. F., and Nyburg, S. C., *Trans. Faraday Soc.*, 1963, **59**, 1126.

5 Noyes, R. M., *J. Amer. Chem. Soc.*, 1962, **84,** 513.
6 Johnson, D. A., *J. Chem. Education*, 1968, **45,** 236.
7 Basolo, F., *Coordination Chemistry Reviews*, 1968, **3,** 213.
8 Wells, A. F., *Structural Inorganic Chemistry*, 3rd. edn. (Oxford University Press, 1962).
9 Matwiyoff, N. A., and Taube, H., *J. Amer. Chem. Soc.*, 1968, **90,** 2796.
10 Hinton, J. F., and Amis, E. S., *Chem. Rev.*, 1971, **71,** 627.
11 Connick, R. E., and Fiat, D. N., *J. Chem. Phys.*, 1963, **39,** 1349.
12 Bell, R. P., *The Proton in Chemistry* (Cornell University Press, 1959).
13 Sharpe, A. G., in *Halogen Chemistry*, ed. V. Gutmann, Vol. I (Academic Press, 1967).
14 Rossotti, F. J. C., in *Modern Coordination Chemistry*, eds J. Lewis and R. G. Wilkins (Interscience, 1960).
15 Kettle, S. F. A., *Coordination Compounds* (Nelson, 1969).
16 Chemical Society Special Publications Nos. 17 and 25 (compilations of data with an introduction).
17 Rossotti, F. J. C., and Rossotti, H., *The Determination of Stability Constants* (McGraw-Hill, 1961).
18 Nancollas, G. H., *Interaction in Electrolyte Solutions* (Elsevier, 1966).
19 Ahrland, S., Chatt, J., and Davies, N. R., *Quart. Rev. Chem. Soc.*, 1958, **12,** 265.
20 Pearson, R. G., *J. Amer. Chem. Soc.*, 1963, **85,** 3533.
21 Latimer, W. L., *Oxidation Potentials*, 2nd edn. (Prentice-Hall, 1952).
22 Sharpe, A. G., *Principles of Oxidation and Reduction*, 2nd edn. (Royal Institute of Chemistry, 1968).
23 Frost, A. A., *J. Amer. Chem. Soc.*, 1951, **73,** 2680.
24 Ebsworth, E. A. V., *Education in Chemistry*, 1964, **1,** 123.
25 Heslop, R. B., and Robinson, P. L., *Inorganic Chemistry*, 3rd edn. (Elsevier, 1967).

Reactions in non-aqueous solvents 7

7.1 Introduction[1, 2]

The special properties of water mentioned at the beginning of the last chapter are shared to some extent by a variety of protonic solvents. The dielectric constants and specific conductivities of some of these are given in Table 7.1; values for specific conductivity are subject to some uncertainty because of the difficulty of removing the last traces of impurities (e.g. of water in ammonia). All the molecules shown are polar and in some there is clearly a donor atom.

Table 7.1 *Dielectric constants and specific conductivities of some protonic solvents*

Solvent	*Dielectric constant*	*Specific conductivity* K (ohm^{-1} cm^{-1})
H_2O	78·5 (25°)	5×10^{-8} (25°)
NH_3	23 (−33°)	5×10^{-11} (−33°)
HF	84 (0°)	1×10^{-6} (0°)
HCl	9·3 (−73°)	3×10^{-8} (−80°)
H_2SO_4	100 (25°)	1×10^{-2} (25°)
C_2H_5OH	24·3 (25°)	1×10^{-9} (20°)

Acids and bases in water were originally defined by Arrhenius as substances which, when in solution, increased the concentration of hydrogen ions or hydroxide ions respectively. It is now known that the hydrogen ion is solvated to H_3O^+ and an acid in any other protonic solvent may likewise be defined as a solute which yields solvated protons. In liquid ammonia, for example, self-ionization occurs according to the equation

$$2NH_3 \rightleftharpoons NH_4^+ + NH_2^-$$

and a solution of an ammonium salt is therefore acidic. These protonic solvents ionize to give different anions, and a base is defined as a solute which increases the concentration of the anion formed in the self-ionization of the solvent. Thus in ammonia a soluble amide such as KNH_2 gives a basic solution and in anhydrous hydrogen fluoride, the self-ionization scheme for which is

$$3HF \rightleftharpoons H_2F^+ + HF_2^-$$

the corresponding species is a soluble fluoride.

An alternative definition of acids and bases, suggested independently by Brønsted and Lowry in 1923, regards acids as proton donors and bases as proton acceptors. This is consistent with the above, but self-ionization will involve a proton transfer; for example,

$$H_2O + H_2O \rightleftharpoons OH^- + H_3O^+$$
$$NH_3 + NH_3 \rightleftharpoons NH_2^- + NH_4^+$$

One important result of this definition is that it becomes possible to classify non-aqueous protonic solvents as acidic or basic, according to whether the solvent molecule has a low or a high affinity for protons. This may be illustrated by considering two extreme cases, hydrogen fluoride and ammonia. The former is a very acidic solvent and, as is shown later (Section 7.3), will protonate several different types of solute. Most substances that are acids in water also accept rather than yield protons when dissolved in hydrogen fluoride. Ammonia, on the other hand, is a very basic solvent and the ammonium ion is formed very readily from protonic solutes. It is, in fact, a 'levelling' solvent for acids, and any acid having a dissociation constant greater than about 10^{-5} in water is almost completely ionized to NH_4^+ and its own anion in liquid ammonia solution. Water is also sufficiently basic (in the Brønsted–Lowry sense) to be a levelling solvent for a range of mineral acids. Nitric, hydrochloric, hydrobromic, hydriodic and sulphuric acids appear equally strong in aqueous solution, though they can be differentiated in a more acidic solvent such as anhydrous acetic acid. It should be noted that, in terms of the Brønsted–Lowry definition, a given substance may well be classed as an acid in one solvent and as a base in another. This is so, for example, in the case of urea in liquid ammonia and in water:

$$H_2NCONH_2 + NH_3 \rightleftharpoons H_2NCONH^- + NH_4^+ \text{ (acid)}$$
$$H_2NCONH_2 + H_2O \rightleftharpoons H_2NCONH_3^+ + OH^- \text{ (base)}$$

Lewis defined an acid as an electron pair acceptor and a base as an electron pair donor. This is again consistent with the other uses of the terms in this chapter, though, to avoid confusion, it is simpler, in considering non-aqueous solvents, to refer to the acceptor and donor functions of a molecule. The Lewis definition is, of course, used in a much wider context.

The terms acid and base are also used in considering aprotic solvents (i.e. solvents the self-ionization of which does not yield a solvated proton). Examples of what is believed to be the self-ionization of typical substances of this class are:

$$NOCl \rightleftharpoons NO^+ + Cl^-$$
$$COCl_2 \rightleftharpoons COCl^+ + Cl^-$$
$$2BrF_3 \rightleftharpoons BrF_2^+ + BrF_4^-$$

An acid or a base is then defined as a solute which increases the concentration of the cation or anion associated with the ionization of the parent solvent. The term solvent system is also in common use in discussing phenomena in a particular liquid.

7.2 Liquid ammonia[1–4]

7.2.1. General properties. Liquid ammonia boils at $-33{\cdot}4°$, freezes at $-77{\cdot}8°$ and has a specific conductivity at its boiling point of about 5×10^{-11} ohm^{-1} cm^{-1}. The dissociation constant at $-50°$ is 10^{-33} g-ion^2 l^{-2}. The dielectric constant at the boiling point is 23 and the liquid is a moderate or good solvent for a variety of organic and inorganic compounds. The former include alcohols, amines, ethers, esters, halocarbons and also aromatics. Soluble inorganic compounds which do not undergo solvolysis (i.e. react with the solvent) give conducting solutions. Most ammonium salts are freely soluble, as are many nitrites, nitrates, cyanides and thiocyanates. Solubilities are less for salts of multivalent cations; as in water, lattice energy and entropy effects outweigh solvation energies in this case. Iodides of univalent and bivalent cations dissolve and, for a given cation, the halides become less soluble in going from the iodide to the fluoride. Other insoluble types are metal hydroxides, oxides, sulphides and the salts of certain oxy-acids, including sulphates, carbonates and phosphates. Various ammoniates of inorganic salts are known which are analogous to hydrates in the water system. For example, the hexammoniate of nickel iodide, which has the structure $[Ni(NH_3)_6]I_2$ analogous to $[Ni(H_2O)_6]I_2$, loses ammonia only above 200°C, whereas $[Mg(NH_3)_6]I_2$ has an appreciable dissociation pressure at 20°C.

The simplest ionic reaction in liquid ammonia is the acid–base neutralization reaction between an ammonium salt (an acid) and a soluble metal amide (a base); for example,

$$NH_4NO_3 + KNH_2 = KNO_3 + 2NH_3$$

This may be followed by either a conductimetric or a potentiometric titration. In either case a sharp break is observed at the end-point though, in the former case, the change in gradient of the conductivity/concentration plot after neutralization with acid is less marked than in water. This is because the ammoniated proton in liquid ammonia, unlike the hydrated proton in water, does not travel by a chain mechanism, probably owing to the fact that hydrogen bonding is weaker in ammonia than in water. In fact, the NH_4^+ ion is only about as mobile as other cations in liquid ammonia. Visual indicators may also be used for titrations in liquid ammonia; phenolphthalein, for example, is present as the colourless undissociated molecule in acid solution, and yields its red anion in basic solution (i.e. with a slight excess of KNH_2 in the above acid–base titration).

Other types of ionic reaction are shown below:

$$Ba(NO_3)_2 + 2AgCl = BaCl_2(s) + 2AgNO_3$$
$$AgNO_3 + KNH_2 = AgNH_2(s) + KNO_3$$
$$3HgI_2 + 6KNH_2 = Hg_3N_2(s) + 6KI + 4NH_3$$

In the first reaction silver chloride dissolves as the complex $[Ag(NH_3)_2]Cl$. The second is analogous to the precipitation of silver hydroxide, while the third corresponds to the precipitation of an insoluble oxide.

Amphoteric behaviour has also been observed. Zinc amide, for example, is insoluble, just as zinc hydroxide is insoluble in water, but dissolves on addition of potassium amide to yield the ammonozincate $K_2[Zn(NH_2)_4]$. This is decomposed by addition of an ammonium salt with reprecipitation of the amide, just as zinc hydroxide is precipitated by decomposing the soluble aquozincate with an aqueous acid.

$$K_2[Zn(NH_2)_4] + 2NH_4NO_3 = Zn(NH_2)_2(s) + 2KNO_3 + 4NH_3$$

There are a number of other examples of this type of reaction.

A final point of resemblance between the solvent properties of ammonia and water is the frequent occurrence of solvolysis. Alkali metal hydrides and oxides, for example, are decomposed:

$$M^{I}H + NH_3 \rightarrow M^{I}NH_2 + H_2$$
$$M^{I}_2O + NH_3 \rightarrow M^{I}NH_2 + M^{I}OH$$

Many halides of non-metals or weakly basic metals (e.g. those of Ti^{IV} or V^{V}) also react rapidly with liquid ammonia to form an amide, which loses ammonia progressively when heated, forming in turn an imide and a nitride. This parallels the thermal degradation of an hydroxide to an oxide; for example,

$$BCl_3 \xrightarrow[<0^\circ]{NH_3} B(NH_2)_3 \xrightarrow{700^\circ} BN$$

$$SiCl_4 \longrightarrow Si(NH_2)_4 \xrightarrow{0^\circ} Si(NH)(NH_2)_2 \xrightarrow{1200^\circ} Si_3N_4$$

Much work has also been done on purely physicochemical aspects of liquid ammonia solutions, and this has provided thermodynamic and electrochemical data. These, however, will not be considered here.

7.2.2. Solutions of metals in liquid ammonia.[1,4,5] All of the alkali and alkaline earth metals with the exception of beryllium have the remarkable property of dissolving in liquid ammonia. Europium and ytterbium, both of which have a +2 oxidation state, also dissolve. All are characterized by relatively low latent heats of sublimation and low ionization potentials, and their ions have high solvation energies; all have strongly negative aqueous reduction potentials. For example, the E°_{aq} values for the couples Eu^{2+}/Eu (calculated from Eu^{3+}/Eu and Eu^{3+}/Eu^{2+}), K^+/K and Na^+/Na are −3.4, −2·9 and −2·7 V respectively. The values for the two alkali metals are known in liquid ammonia; these (relative to the standard hydrogen electrode in the same solvent) are −2·0 and −1·9 V for K^+/K and Na^+/Na respectively. Samarium, which, like europium and ytterbium, has a well-defined +2 oxidation state, is anomalous in not dissolving in liquid ammonia; possibly a kinetic factor is involved in this case.

The solubilities of these metals in ammonia are moderate to high, the values for Li, Na and K being 10·9, 24·8 and 46·4 g per 100 g of NH_3 at −33°, while the saturation concentration for caesium at −50° is 334 g per 100 g. These values vary little with temperature. The alkali metals may be recovered unchanged by evaporating the solvent, while the remainder form ammoniates, the composition of which approximates to $M.6NH_3$ for the alkaline earth metals. All of the solutions are, however, metastable with respect to decomposition to the metal amide and hydrogen (e.g. $Na + NH_3 \rightarrow NaNH_2 + \frac{1}{2}H_2$). This reaction is promoted by ultraviolet light and by certain metallic catalysts (e.g. iron or finely divided platinum). The alkali metals also dissolve in a number of amines (e.g. methylamine) and in some ethers (e.g. diethyl ether and tetrahydrofuran). Such solutions have, in general, been less fully studied than those in ammonia. Their physical characteristics are similar, but solubilities are often much lower.

The metal solutions in ammonia are deep blue in colour when dilute and bronze when concentrated. The blue colour of the dilute solutions is due to the short wavelength tail of a broad absorption band which extends into the infrared with a maximum at about 15 000 Å. This absorption is almost the same for all the metals, including europium and ytterbium, though an additional band centred in the visible spectrum is found for amine solutions.

All of the solutions also have a high electrical conductivity which varies in an anomalous manner with concentration. This is shown in Fig. 7.1 for the alkali metals Li, Na and K, for which the curves are almost identical.

At high dilution the conductivity of these solutions is roughly an order of magnitude greater than that of a fully ionized salt in water, while in the most concentrated solutions the value approaches that for a pure metal. The minimum (Fig. 7.1) occurs at a concentration of about 0·05M. It has been shown for sodium that the temperature coefficient of specific conductivity is 1·9 per cent per degree for dilute solutions. It increases to a maximum of 4 per cent per degree at about one molar and then decreases to almost zero at higher concentrations. This is a clear indication that the mechanism of conduction must also vary with concentration. For dilute solutions there is good evidence, which is discussed more fully later, that the species present are solvated metal cations and solvated electrons. The conducting species are believed to be the cation and the electron (rather than the solvated electron), the latter making much the greater contribution.

The dilute solutions are paramagnetic, the value of the susceptibility at very low concentrations corresponding to the formation of one free electron from each metal atom. The susceptibility falls with increasing concentration and the solution becomes diamagnetic in

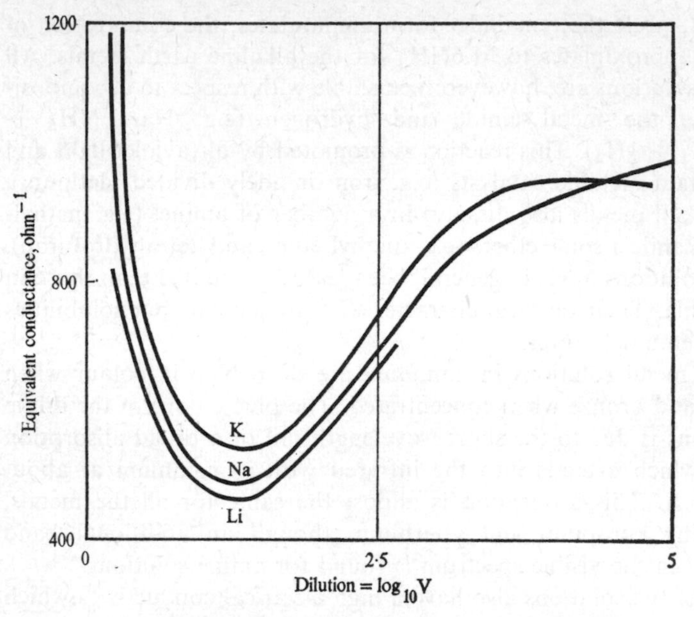

Figure 7.1 Equivalent conductance of metal–ammonia solutions at $-33{\cdot}5°$ containing 1 g-atom of the metal in *V* litres

the region of the minimum conductance, becoming weakly paramagnetic again at higher concentrations.

Solution of a metal in liquid ammonia causes a large increase in volume (i.e. the density decreases). This observation has been linked directly with the idea that, in dilute solutions, metal ions and electrons are present, the latter occupying what is in effect a 'hole'. The radius of this cavity has been estimated from the partial molar volumes of the alkali metal ions to be about 3 Å, the electron displacing 2–3 ammonia molecules. It interacts with protons of the surrounding ammonia molecules and the characteristic absorption band of metal–ammonia solutions is associated with a transition of the electron from a 1*s* to a 2*p* state. The light absorption is clearly not dependent on the metal present in solution.

In order to account for the variations in conductivity and magnetic susceptibility with concentration, it is now believed that three equilibria are involved:

$$M \rightleftharpoons M^+ + e \qquad K = 9{\cdot}9 \times 10^{-3} \qquad (1)$$
$$M^- \rightleftharpoons M + e \qquad K = 9{\cdot}7 \times 10^{-4} \qquad (2)$$
$$M_2 \rightleftharpoons 2M \qquad K = 1{\cdot}9 \times 10^{-4} \qquad (3)$$

At low concentrations only (1) is important, but as [M] increases (2) and (3) must be taken into consideration. The M species can be envisaged as a solvated metal ion and a solvated electron held together electrostatically, and the M^- and M_2 species as ionic assemblies of one and two solvated cations respectively and two solvated electrons; in the latter species the interaction between the electrons must be strong enough to lead to spin-pairing. In highly concentrated solutions the properties are like those of a liquid metal in which the cations are ammoniated.

7.2.3. Chemical properties of metal-ammonia solutions.[6, 7] Solutions of metals in liquid ammonia have, as might be expected, very strong reducing properties. They are extensively used in organic chemistry, as are solutions of the alkali metals in methylamine. Ammonium salts, as strong acids, decolorize the blue solutions readily:

$$NH_4^+ + e \rightarrow NH_3 + \tfrac{1}{2}H_2$$

Many other substances which are more weakly acidic in ammonia react similarly; for example,

$$\begin{aligned} GeH_4 + e &\rightarrow GeH_3^- + \tfrac{1}{2}H_2 \\ AsH_3 + e &\rightarrow AsH_2^- + \tfrac{1}{2}H_2 \\ RC{\equiv}CH + e &\rightarrow RC{\equiv}C^- + \tfrac{1}{2}H_2 \end{aligned}$$

Anions such as those shown may sometimes be isolated as the alkali metal salt or used directly in further synthesis. Reaction of an alkali metal in ammonia with a non-metallic halide will yield a similar product; for example,

$$(C_6H_5)_3GeCl + 2Na \rightarrow (C_6H_5)_3GeNa + NaCl$$

Electron addition may occur without bond cleavage, as in the formation of alkali metal peroxides and superoxides.

$$\begin{aligned} O_2 + e &\rightarrow O_2^- \rightarrow M^IO_2 \\ + 2e &\rightarrow O_2^{2-} \rightarrow M^I_2O_2 \end{aligned}$$

Sulphur reacts similarly to form sulphides and polysulphides, while carbon monoxide gives products which were formerly thought to be alkali metal carbonyls M^ICO. X-ray structural work has now shown that the correct formulation is $M^I_2[O{-}C{\equiv}C{-}O]$.

These solutions are also able to react with certain transition metal complexes to yield products in which the metal is in an unusually low oxidation state; for example,

$$K_2Ni(CN)_4 \xrightarrow[NH_3 \text{ at } -33°]{\text{excess K in}} K_4Ni(CN)_4$$

$$Pt(NH_3)_4Br_2 \xrightarrow[\text{at } -33°]{\text{K in } NH_3} Pt(NH_3)_4 + 2KBr$$

The product of the second reaction is a yellow solid which decomposes above 25°C to platinum and ammonia. The course of the reduction may be followed by a conductimetric or potentiometric titration, and the presence of excess of alkali metal is also evident from the persistence of the blue colour. A product which is stable to 90°C, $Ir(NH_3)_5$, is obtained by reduction of $[Ir(NH_3)_5Br]Br_2$ under similar conditions. It also gives only metal and ammonia when decomposed thermally.

Another interesting application of solutions of the alkali metals in ammonia is in the reduction of carbonyls to carbonyl metallates (see Section 20.5); for example,

$$Mn_2(CO)_{10} + 2K^+ + 2e \rightarrow 2K^+ + 2[Mn(CO)_5]^-$$
$$Fe(CO)_5 + 2Na^+ + 2e \rightarrow 2Na^+ + [Fe(CO)_4]^{2-} + CO$$

They are also able to reduce salts of a variety of main-group elements to form polyanions. Examples of the resulting compounds are Na_4Sn_9, Na_3Sb_3 and Na_3Sb_7. The majority are soluble in liquid ammonia, in which they form conducting solutions.

7.3 The hydrogen halides[1, 2]

These are all highly acidic in the sense that they yield protons readily. The acid strengths increase in the order HF < HCl < HBr < HI. Relevant physical properties are shown in Table 7.2.

Hydrogen fluoride is highly associated (Trouton constant 103·4 J (24·7 cal) mole^{-1} deg^{-1}) as the result of hydrogen bonding. In the solid there are linear zig-zag chains while in the gas phase the average number of molecules in aggregates is 3·5; both short chains and rings are present. It is probable that the same is true of the liquid. In contrast, the three other hydrogen halides have Trouton constants in the normal range (ca. 84 J (20 cal) mole^{-1} deg^{-1}) and hydrogen bonding must therefore be much less developed.

Table 7.2 *Physical properties of the hydrogen halides*

	HF	HCl	HBr	HI
Melting point	−83°	−114·6°	−88·5°	−50·9°
Boiling point	19·5°	−84·1°	−67·0°	−35·0°
Dielectric constant	84 (0°)	9·28 (−95°)	7·0 (−85°)	3·39 (−50°)
Specific conductivity (ohm^{-1} cm^{-1})	10^{-6} (0°)	3×10^{-8} (−80°)	1×10^{-8} (−84°)	8×10^{-9} (−45°)

Self-ionization for all four liquids is of the type

$$3HX \rightleftharpoons H_2X^+ + HX_2^-$$

There is direct spectroscopic evidence for the existence of the H_2F^+ ion in association with SbF_6^-, but no comparable evidence for the other cations. The anions HF_2^-, HCl_2^-, HBr_2^- and HI_2^- have all been identified in solid salts.

Hydrogen fluoride, because of its high dielectric constant, is potentially a good solvent for ionic compounds, but most undergo solvolysis. Chlorides, bromides and iodides of the alkali and alkaline earth elements, for example, are converted into fluorides, as are some oxides and hydroxides. Group I and Group II metal fluorides are soluble, as are the fluorides of silver (I) and thallium (I). A few nitrates and sulphates also dissolve unchanged. Solvolysis occurs much less readily with the other hydrogen halides, but the dielectric constants are much lower and they are poor solvents for ionic compounds apart from halides with a large cation such as NR_4^+ or Cs^+. Many covalent halides are soluble. These and a wide range of organic compounds which are freely soluble will be referred to later. A few simple ionic reactions have been observed in hydrogen fluoride; for example,

$$Ba(ClO_4)_2 + 2AgF = 2AgClO_4 + BaF_2$$

Applying the familiar definition of an acid to these solvents, the concentration of the solvated proton may be increased in either of two ways, the first involving addition of a protonic acid. It is found that, because of the very high acidity of hydrogen fluoride, only perchloric and fluorosulphonic acids are sufficiently strong to give up protons and function as acids. Other substances which are strong acids in water accept protons in hydrogen fluoride and therefore either function as bases by increasing the HF_2^- concentration or undergo solvolysis; for example,

$$HNO_3 + 4HF \rightleftharpoons H_3O^+ + NO_2^+ + 2HF_2^-$$
$$H_2SO_4 + 3HF \rightleftharpoons HOSO_2F + H_3O^+ + HF_2^-$$

The remaining hydrogen halides are even more acidic and no strong oxy-acids are capable of yielding protons. There is, however, some evidence that HBr and HI are acids in HCl, as would be expected from their relative acidities.

There are many examples of the formation of an acidic solution by adding a covalent halide to act as a halide ion acceptor; for example,

$$2HF + SbF_5 \rightleftharpoons H_2F^+ + SbF_6^-$$
$$2HF + BF_3 \rightleftharpoons H_2F^+ + BF_4^-$$
$$2HCl + BCl_3 \rightleftharpoons H_2Cl^+ + BCl_4^-$$
$$4HCl + B_2Cl_4 \rightleftharpoons 2H_2Cl^+ + B_2Cl_6^{2-}$$
$$4HCl + SnCl_4 \rightleftharpoons 2H_2Cl^+ + SnCl_6^{2-}$$

Such solutions are good electrolytes and the presence of the complex halide anions has been shown spectroscopically in a number of instances. Anions containing two different halides may also be formed. Thus BF_3 in HCl gives $[BF_3Cl]^-$. It should be noted that a number of halides do not act as halide ion acceptors (e.g. $SiCl_4$ and PCl_3 in HCl).

Basic solutions may be formed by adding either the appropriate halide ion or a proton acceptor. For hydrogen fluoride a salt such as potassium fluoride is readily soluble, and electrolysis of the resulting solution forms the basis for methods for producing fluorine. The other hydrogen halides require the salt of a large cation such as $[NR_4]^+$ to secure a satisfactory solubility. Other halides can, however, provide halide ions. Thus chlorine and bromine trifluorides are miscible with hydrogen fluoride and there is appreciable ionization:

$$ClF_3 + HF \rightleftharpoons ClF_2^+ + HF_2^-$$
$$BrF_3 + HF \rightleftharpoons BrF_2^+ + HF_2^-$$

Phosphorus pentachloride and pentabromide also give basic solutions in HCl and HBr respectively:

$$PX_5 + HX \rightleftharpoons PX_4^+ + HX_2^-$$

Several different types of proton acceptor may be used. In hydrogen fluoride a number of organic compounds containing oxygen, nitrogen or sulphur atoms capable of exercising a donor function can take up protons and yield conducting solutions. These include alcohols, ethers, ketones, acids and amines, and electrolysis of such solutions has been used in the electrochemical fluorination process for preparing perfluoro compounds (see Section 11.6):

$$(C_2H_5)_2O + 2HF \rightleftharpoons (C_2H_5)_2OH^+ + HF_2^-$$
$$CH_3COOH + 2HF \rightleftharpoons CH_3COOH_2^+ + HF_2^-$$

The same range of organic compounds will also dissolve in the other hydrogen halides, though the resulting conducting solutions have been less fully studied.

The phosphorus atom in phosphine or triphenylphosphine is also protonated in HCl, HBr or HI, and a similar reaction has been observed for triphenylarsine in hydrogen chloride. When the phosphine solution is allowed to react with an equivalent quantity of a solvo-acid such as $H_2Cl^+BCl_4^-$ in HCl solution a range of new phosphonium salts may be isolated; for example,

$$PH_4^+HCl_2^- + H_2Cl^+BCl_4^- \rightarrow PH_4BCl_4 + 3HCl$$

Protonation of a few π-bonded systems to form a carbonium ion has also been studied in liquid hydrogen chloride; for example,

$$Ph_2C{=}CH_2 + 2HCl \rightleftharpoons Ph_2CCH_3^+ + HCl_2^-$$

Such a solution gives a sharp 1 : 1 end-point when titrated conductimetrically with an acid solution such as $H_2Cl^+BCl_4^-$, though no salt can be isolated. Indeed, this technique has been used quite generally in investigating the solution chemistry of hydrogen halides; with hydrogen fluoride it requires the use of apparatus that is resistant to chemical attack and the most useful material for the construction of cells, etc., is Kel F, a polymer of $CF_2{=}CFCl$ which can be machined. Apart from the obvious use of acid–base titrations in providing direct evidence for the reality of any ionization scheme that may be postulated, they are of value in making new halo-complexes. Some examples have been given above.

7.4 Anhydrous sulphuric acid[8, 9]

Anhydrous sulphuric acid, like hydrogen fluoride, is a highly acidic solvent, and there are many points of resemblance between the two systems. It has a specific conductivity of $1{\cdot}04 \times 10^{-2}$ ohm^{-1} cm^{-1} at 25°C, at which temperature the dielectric constant is 100. Like hydrogen fluoride it is highly associated as a result of hydrogen bonding. This is also evident in the solid, which has a layer structure with each molecule hydrogen-bonded to four others.

The chief mode of self-ionization is shown in the first of the equations below. The ionic product for this equilibrium is $2{\cdot}7 \times 10^{-4}$ at 25°C. The pure acid is, however, appreciably dissociated at 25° into SO_3 and H_2O, and the other equilibria shown are a consequence of this.

$$\begin{aligned}
2H_2SO_4 &\rightleftharpoons H_3SO_4^+ + HSO_4^- \\
H_2SO_4 &\rightleftharpoons H_2O + SO_3 \\
H_2O + H_2SO_4 &\rightleftharpoons H_3O^+ + HSO_4^- \\
SO_3 + H_2SO_4 &\rightleftharpoons H_2S_2O_7 \\
H_2S_2O_7 + H_2SO_4 &\rightleftharpoons H_3SO_4^+ + HS_2O_7^- \\
2H_2SO_4 &\rightleftharpoons H_3O^+ + HS_2O_7^-
\end{aligned}$$

The equilibrium constants for these reactions have been evaluated on the basis of freezing point studies, and it has been found that the concentrations of $H_3SO_4^+$ and HSO_4^- are greater by a factor of more than ten than those of any other species. In considering the solvent system chemistry, therefore, only the first equilibrium need be considered.

Because of its very high viscosity, it would be expected that ionic mobilities would be very low. This is generally the case, but the $H_3SO_4^+$ and HSO_4^- ions have mobilities of the same order of magnitude as H_3O^+ and OH^- in water, showing that proton-transfer mechanisms are operative. Thus in solutions of alkali metal hydrogen sulphates in sulphuric acid the transport numbers of the cations and anions are 0·02 and 0·98 respectively.

As in hydrogen fluoride, the majority of strong aquo-acids accept rather than yield protons in sulphuric acid. Perchloric, fluorosulphuric and disulphuric acids, however, behave as weak acids. Phosphoric acid is typical of a number of strong aquo-acids which behave as bases:

$$H_3PO_4 + H_2SO_4 \rightleftharpoons H_4PO_4^+ + HSO_4^-$$

A comparatively strong acid in H_2SO_4 is formed when B_2O_3 is dissolved in oleum:

$$B_2O_3 + 3H_2S_2O_7 + 4H_2SO_4 \longrightarrow 2H_3SO_4^+ + 2[B(HSO_4)_4]^-$$

There is spectroscopic evidence that the formulation of the anion is more complex than is shown above. The anion may be thought of as analogous to BF_4^- in hydrogen fluoride and the acid may be titrated conductimetrically with a soluble bisulphate, which represents the simplest type of base. The two acids $H_2Sn(HSO_4)_6$ and $H_2Pb(HSO_4)_6$ have also been described.

There is a close analogy between sulphuric acid and the hydrogen halides in that both readily yield basic solutions on addition of a proton acceptor, assuming that solvolysis does not occur; for example,

$$R_2CO + H_2SO_4 \rightleftharpoons R_2COH^+ + HSO_4^- \text{ (R = alkyl or aryl)}$$
$$RNH_2 + H_2SO_4 \rightleftharpoons RNH_3^+ + HSO_4^-$$
$$R_3P + H_2SO_4 \rightleftharpoons R_3PH^+ + HSO_4^-$$

Water itself is a strong base ($H_2O + H_2SO_4 \rightleftharpoons H_3O^+ + HSO_4^-$). There are also examples of bases which result from the reaction of an inorganic solute with sulphuric acid. Of outstanding importance is the behaviour of nitric acid, where there is cryoscopic and spectroscopic evidence for the formation of the nitronium ion (NO_2^+), which is responsible for aromatic nitration:

$$HNO_3 + 2H_2SO_4 \rightleftharpoons NO_2^+ + H_3O^+ + 2HSO_4^-$$

Certain organic compounds yield carbonium ions. Examples of this behaviour are provided by the following equations:

$$Ph_3COH + 2H_2SO_4 \rightleftharpoons Ph_3C^+ + H_3O^+ + 2HSO_4^-$$
$$Ph_2C{=}CH_2 + H_2SO_4 \rightleftharpoons Ph_2CCH_3^+ + HSO_4^-$$

As a final example of the formation of bases, mention may be made of the behaviour of certain transition metal complexes in sulphuric acid. It is found that the triphenylphosphine and triphenylarsine iron carbonyls ($Ph_3PFe(CO)_4$, $Ph_3AsFe(CO)_4$,$(Ph_3P)_2Fe(CO)_3$ and $(Ph_3As)_2Fe(CO)_3$) all form yellow solutions in sulphuric acid, the proton resonance spectrum of which shows a high-field resonance characteristic of a metal–hydrogen bond:

$$Ph_3PFe(CO)_4 + H_2SO_4 \rightleftharpoons HFe(CO)_4(PPh_3)^+ + HSO_4^-$$

A similar high-field resonance is observed with complexes such as $[\pi\text{-}C_5H_5Mo(CO)_3]_2$ and $[\pi\text{-}C_5H_5W(CO)_3]_2$. In some types of carbonyl–olefin–metal complex, however, protonation no longer occurs on the metal atom. Thus tricarbonyl cyclo-octatetrene iron gives a red solution in sulphuric acid, but the proton resonance provides no evidence for a metal–hydrogen bond. In this case the ring is protonated and rearranges to form a bicyclic system.

Several other highly acidic protonic solvents have been investigated in considerable detail—notably fluorosulphonic acid—but they will not be discussed here since the behaviour closely resembles that for the two examples taken.

7.5 Other protonic solvents[7, 10]

Liquid ammonia is the best known of a large number of protonic solvents, some of which, like ammonia, are highly basic. Primary and secondary amines belong to this class, as do anhydrous hydrazine, formamide and acetamide. Hydrazine (m.p. 2°; b.p. 113·5°C) has a specific conductivity of 1×10^{-6} ohm^{-1} cm^{-1} at 0° and ionizes according to the scheme

$$2N_2H_4 \rightleftharpoons N_2H_5^+ + N_2H_3^-$$

The dielectric constant at 25°C is 52 and a number of ionic compounds dissolve to give conducting solutions (e.g. alkyl ammonium halides, alkali metal iodides and perchlorates). As would be expected from the analogy with liquid ammonia, hydrazinium salts (e.g. $N_2H_5ClO_4$) function as acids and alkali metal salts of the $N_2H_3^-$ ion as bases. The latter may be made by adding hydrazine to the finely divided alkali metal suspended in ether. Acid–base neutralization reactions have been studied conductimetrically.

Both formamide and acetamide, the ionization schemes for which are set out below, have also been extensively studied:

$$2HCONH_2 \rightleftharpoons HCONH_3^+ + HCONH^-$$
$$2CH_3CONH_2 \rightleftharpoons CH_3CONH_3^+ + CH_3CONH^-$$

Both are associated by hydrogen bonding and the latter, which melts at 82°, must be studied at an appropriate higher temperature. The general pattern of the acid–base behaviour follows the same lines as for ammonia.

Two solvents which call for special mention because of their relationship to water are ethanol (or alcohols in general) and hydrogen sulphide. Self-ionization in the former is of the type

$$2ROH \rightleftharpoons ROH_2^+ + RO^-$$

Specific conductivities of the pure liquids are of the order of 10^{-9} $ohm^{-1}\ cm^{-1}$ for the lower aliphatic alcohols, with dielectric constants in the approximate range 15–30.

Liquid hydrogen sulphide (b.p. −60·4°) has a dielectric constant of only 10·2 at this temperature. It is a poor solvent for most ionic inorganic compounds, though salts where the large cation size reduces the lattice energy tend to be more soluble (e.g. $[N(C_2H_5)_4]Cl$). The self-ionization scheme postulated is

$$2H_2S \rightleftharpoons H_3S^+ + SH^-$$

the specific conductivity at −78° being $3 \times 10^{-11}\ ohm^{-1}\ cm^{-1}$. Aqueous mineral acids such as hydrogen chloride or sulphuric acid are also acids in this solvent, though their conductivities are low, and they may be titrated conductimetrically with soluble hydrosulphides (e.g. NaSH). A proton acceptor such as a tertiary amine also yields a basic solution:

$$Et_3N + H_2S \rightleftharpoons Et_3NH^+ + SH^-$$

There is evidence for amphoteric behaviour as freshly precipitated arsenic (III) sulphide will dissolve in either acids or bases:

$$As_2S_3 + 3H^+ \rightleftharpoons 2As^{3+} + 3SH^-$$
$$As_2S_3 + 6R_3N + 3H_2S \rightleftharpoons 2[R_3NH]_3[AsS_3]$$

The complex formed in the second reaction decomposes with reprecipitation of As_2S_3 when hydrogen chloride is added.

There is, too, a variety of interesting solvolytic reactions; for example,

$$SnCl_4 + 2H_2S \rightarrow SnS_2(s) + 4HCl$$
$$PCl_5 + H_2S \rightarrow PSCl_3 + 2HCl$$
$$CH_3COCl + H_2S \rightarrow CH_3COSH + HCl$$
$$CaC_2 + H_2S \rightarrow CaS + C_2H_2$$

Various other basic or weakly acidic solvents have been examined in considerable detail (e.g. HCN, CH_3COOH), but they will not be considered further here as their behaviour is to a large extent predictable.

7.6 Aprotic solvent systems[1, 10]

Protonic solvent systems are very numerous but, in contrast, aprotic solvent systems are encountered only among a limited range of halides and oxy-halides, and in the case of the two oxides N_2O_4 and SO_2. As will be shown later, much of the evidence relating to the second of these oxides can be interpreted without assuming its self-ionization. The discussion of this class of solvent will begin with consideration of the liquid halogens and interhalogen compounds, which serve to illustrate most of the principles involved.

7.7 The halogens

The measured specific conductivities of the liquid halogens are: Cl_2, 7×10^{-8} ohm^{-1} cm^{-1} (−35°); Br_2, $1{\cdot}1 \times 10^{-9}$ ohm^{-1} cm^{-1} (25°); and I_2, $5{\cdot}2 \times 10^{-5}$ ohm^{-1} cm^{-1} (114°). Only for iodine does the conductivity give a clear indication of ionization, though values comparable with those for the other two halogens are encountered in many other instances.

The most probable mode of self-ionization is

$$3X_2 \rightleftharpoons X_3^+ + X_3^-$$

A number of covalent chlorides (e.g. $SiCl_4$, $TiCl_4$ and $AsCl_3$) dissolve in liquid chlorine. It is not known, however, if the conductivity is increased as would be the case if halide ion transfer occurred (e.g. $2Cl_2 + AsCl_3 \rightleftharpoons Cl_3^+ + AsCl_4^-$). The Cl_3^+ ion has been detected spectroscopically in the salt Cl_3AsF_6, which is formed when ClF_2AsF_6 is treated with chlorine. Sodium chloride and potassium chloride do not dissolve in liquid chlorine, but it is not known if salts of larger cations (e.g. NR_4^+) dissolve and give conducting solutions.

Many halides of non-metals are also soluble in liquid bromine (e.g. CCl_4, BBr_3, $TiBr_4$ and $AsBr_3$), but there is as yet no evidence that they behave as solvo-acids. Conducting solutions in liquid bromine are formed by bromides of large cations such as Cs^+ or Bu_4N^+. Bromine also forms addition compounds with organic bases such as pyridine and acetamide. Addition of these bases to liquid bromine enhances its conductivity, which is probably due to formation of the ions $(Base)_n\ Br^+$ and Br_3^-.

The evidence for the existence of both solvo-acids and solvo-bases is much clearer for liquid iodine. Iodine monohalides are weak electrolytes, as are iodides of the alkali metals. Furthermore, conductimetric titration of, for example, IBr with KI in liquid iodine solution shows a sharp break at a 1 : 1 molar ratio:

$$I^+ + Br^- + K^+ + I_3^- \rightarrow KBr + 2I_2$$

Such reactions have also been followed potentiometrically. The I^+ cation is probably solvated to I_3^+, though the I_2^+ species has been identified in the blue solutions formed by iodine in oleum or iodine pentafluoride. In each case it has been shown to be paramagnetic, the moment being 1·7–2·0 Bohr magnetons.

7.8 The interhalogen compounds[11]

The structures and general reactions of the interhalogen compounds are considered elsewhere (Section 11.3), this section being concerned only with their behaviour as ionizing solvents. The available evidence, which varies in completeness from case to case, suggests that self-ionization involves halide ion transfer, as shown below, for the four types of interhalogen compound:

$$3AB \rightleftharpoons A_2B^+ + AB_2^-$$
$$2AB_3 \rightleftharpoons AB_2^+ + AB_4^-$$
$$2AB_5 \rightleftharpoons AB_4^+ + AB_6^-$$
$$2AB_7 \rightleftharpoons AB_6^+ + AB_8^-$$

The two iodine monohalides, ICl and IBr, yield conducting solutions in polar solvents. The monochloride, the two solid forms of which have melting points of 27·2° and 13·9°C, boils at 100° and is appreciably dissociated into its elements throughout the liquid range (1·1 per cent at 100°). The specific conductivity at 35° is $4 \cdot 6 \times 10^{-3}$ ohm^{-1} cm^{-1}, and various ionic halides (e.g. KCl and NH_4Cl) dissolve and enhance the conductivity. There is no doubt that they act as solvo-bases, forming the ICl_2^- ion, which is known as a linear ion in its solid alkali metal salts and in $PCl_4^+ICl_2^-$. Organic bases such as pyridine also form 1 : 1 and 1 : 2 adducts which are electrolytes in ICl and can be formulated as (Base)$I^+ICl_2^-$.

A number of covalent halides will dissolve in iodine monochloride, some (such as $AlCl_3$ and $SbCl_5$) with an increase in the conductivity but others (such as $SiCl_4$ and $TiCl_4$) with little effect. The former are formulated as solvo-acids ($ICl_2^+AlCl_4^-$ and $ICl_2^+SbCl_6^-$) and, when titrated conductimetrically with a solvo-base such as NH_4Cl, show a break at 1 : 1 molar ratio. With $SnCl_4$ the break occurs at a 1 : 2 acid–base ratio, in keeping with the formation in solution of the dibasic acid $(ICl_2)_2SnCl_6$. The behaviour of iodine monobromide is very similar though its thermal dissociation is greater (13·4 per cent at 100°C).

Bromine monofluoride and monochloride and iodine monofluoride are all unstable and have not been investigated as solvents. The same is true of chlorine monofluoride, though compounds are now known which correspond in type with acids and bases, assuming the liquid halogen fluoride to ionize as follows:

$$3ClF \rightleftharpoons Cl_2F^+ + ClF_2^-$$

Boron trifluoride and arsenic pentafluoride yield 1 : 1 adducts, which may be formulated as $Cl_2F^+BF_4^-$ and $Cl_2F^+AsF_6^-$. Caesium and nitrosyl fluorides also form 1 : 1 adducts and the former has been shown by X-ray analysis to contain the linear ClF_2^- ion. Neither type of adduct has, however, been studied in chlorine monofluoride solution.

Three interhalogen compounds of the type AB_3, namely ICl_3, ClF_3 and BrF_3, lend themselves to investigation in the present context. Iodine trichloride is a solid melting at 101° and is appreciably dissociated to the monochloride and chlorine when liquid. The specific conductivity at 101° is $8 \cdot 5 \times 10^{-3}$ ohm^{-1} cm^{-1}. It is reasonable to assume ionization in accordance with the scheme

$$2ICl_3 \rightleftharpoons ICl_2^+ + ICl_4^-$$

X-ray structural evidence on the 1 : 1 adducts with $AlCl_3$ and $SbCl_5$ is consistent with their formulation as $ICl_2^+AlCl_4^-$ and $ICl_2^+SbCl_6^-$ (i.e. as acids), while the planar ICl_4^- ion is known in the salt $KICl_4.H_2O$. This would be a base in the parent solvent, though acid–base behaviour has not been investigated.

Chlorine and bromine trifluorides, the specific conductivities of which are $6{\cdot}5 \times 10^{-9}$ and $8{\cdot}0 \times 10^{-3}$ ohm^{-1} cm^{-1} at 0° and 25°C respectively, have been much more fully studied in spite of experimental difficulties arising from their high chemical reactivity. The self-ionization equations should be

$$2ClF_3 \rightleftharpoons ClF_2^+ + ClF_4^-$$
$$2BrF_3 \rightleftharpoons BrF_2^+ + BrF_4^-$$

In keeping with this, a series of adducts of chlorine trifluoride with fluorides capable of acting as fluoride ion acceptors has been isolated. These may be formulated as solvo-acids (e.g. ClF_2AsF_6, ClF_2BF_4, $(ClF_2)_2TiF_6$ and ClF_2PtF_6). It is also found that caesium fluoride reacts with chlorine trifluoride to form the solvo-base $CsClF_4$, and analogous salts of Li^+, Na^+, Ag^+, Tl^+, Ba^{2+} and NO^+ have also been described. The structures of these compounds have not been established, nor has it been confirmed that they are electrolytes in chlorine trifluoride solution.

Bromine trifluoride forms similar solvo-acids and solvo-bases, examples of which are shown below:

Solvo-acids	*Solvo-bases*
$BrF_2^+SbF_6^-$	$K^+BrF_4^-$
$(BrF_2^+)_2SnF_6^{2-}$	$Ag^+BrF_4^-$
$BrF_2^+AuF_4^-$	$Ba^{2+}(BrF_4^-)_2$
$BrF_2^+TaF_6^-$	$NO^+BrF_4^-$

Members of both groups form conducting solutions in bromine trifluoride and the structures of solid $KBrF_4$, $NOBrF_4$ and BrF_2SbF_6 have been confirmed by neutron and X-ray diffraction. In BrF_2SbF_6 there is evidence for cation–anion fluorine bridging, and the conductivity of BrF_3 may be due to a fluorine chain mechanism analogous to that of the proton in water or sulphuric acid. Furthermore, conductivity measurements on solutions containing a series of different molar ratios of several acids and bases have shown the breaks expected for neutralization reactions; for example,

$$BrF_2^+SbF_6^- + Ag^+BrF_4^- \longrightarrow AgSbF_6 + 2BrF_3$$
$$(BrF_2^+)_2SnF_6^{2-} + 2K^+BrF_4^- \longrightarrow K_2SnF_6 + 4BrF_3$$

Bromine trifluoride is a sufficiently powerful fluorinating agent to convert most compounds into fluorides, and this fact has been used in preparing a series of fluoro-complexes. Examples are shown below:

$$Au + Ag \rightarrow AgAuF_4$$
$$SnF_4 + NOCl \rightarrow (NO)_2SnF_6$$
$$VCl_3 + KF \rightarrow KVF_6$$
$$Sb_2O_3 + N_2O_4 \rightarrow NO_2SbF_6$$

The correct molar proportions of reactants (e.g. Au + Ag) are taken and dissolved in excess of bromine trifluoride. This first fluorinates the metals to AuF_3 and AgF, which react as solvo-acid and solvo-base, producing the salt ($AgAuF_4$) and solvent. After the reaction bromine and the free bromine trifluoride are removed in vacuum. A similar explanation is applicable in the other instances, even if, as in the case of BrF_2VF_6, the acid cannot be isolated in the pure state. Many oxides (e.g. MoO_3 and WO_3) are converted to fluorides by BrF_3, with liberation of oxygen, though replacement of oxygen in oxy-anions may be incomplete (e.g. $K_2Cr_2O_7$ gives $KCrOF_4$) or complete (e.g. $KMnO_4$ gives $KMnF_5$ and KPO_3 gives KPF_6).

Although iodine trifluoride is so unstable that it cannot be used as a solvent, the solvo-bases KIF_4, $RbIF_4$ and $CsIF_4$ have been prepared from the alkali metal iodides and iodine pentafluoride in cold solution. This reaction must occur through the intermediate formation of IF_3, and the same compounds are obtained from the metal fluorides and IF_3 in acetonitrile solution at −45°. Arsenic pentafluoride and boron trifluoride also form unstable adducts with IF_3, which may be formulated as the solvo-acids $IF_2^+AsF_6^-$ and $IF_2^+BF_4^-$, though proof of these structures is lacking.

Of the three pentafluorides ClF_5, BrF_5 and IF_5, only the last two have been investigated in any detail. The specific conductivity of IF_5 at 25° is $5{\cdot}4 \times 10^{-6}$ ohm^{-1} cm^{-1}, that of bromine pentafluoride at the same temperature being 9×10^{-8} ohm^{-1} cm^{-1}. The self-ionization scheme

$$2IF_5 \rightleftharpoons IF_4^+ + IF_6^-$$

is supported by the formation of a compound which may be formulated as $IF_4^+SbF_6^-$ by reaction with SbF_5. Reaction with KF gives KIF_6. Both compounds dissolve in iodine pentafluoride and enhance its conductivity, and their reaction may be followed conductimetrically:

$$IF_4^+SbF_6^- + K^+IF_6^- \rightarrow KSbF_6 + 2IF_5$$

An increase in conductivity is also observed when BF_3 is passed into IF_5 and, if the latter contains KF in solution (as KIF_6), potassium fluoroborate may be isolated. The blue solution formed when iodine is dissolved in the pentafluoride is believed to contain the uncoordinated paramagnetic I_2^+ cation.

There is little doubt that the behaviour of bromine pentafluoride will resemble that of the iodine analogue. Compounds which may be formulated as $BrF_4^+SO_3F^-$, $KBrF_6$ and $CsBrF_6$ have indeed been isolated. There should be a parallel between the acid–base chemistry of IF_5 and BrF_5 and that of the structurally similar $XeOF_4$, which is known to form $Cs^+[XeOF_5]^-$ and apparently also $XeF_3O^+[Sb_2F_{11}]^-$.

Iodine heptafluoride, which melts at 5–6°C at 2 atm pressure, would be expected to ionize according to the equation

$$2IF_7 \rightleftharpoons IF_6^+ + IF_8^-$$

The conductivity of the pure liquid has not been measured and no compound is formed with NaF, KF or RbF, all of which are insoluble. A 1 : 1 adduct is, however, formed with AsF_5 and there is a much less stable adduct with BF_3. Either could be formulated as a solvo-acid, but there is no structural evidence; nor have conductivities in IF_7 been measured.

7.9 Other covalent halides

The treatment of other covalent halides as the basis for solvent systems follows very much the same lines as for the interhalogen compounds. Self-ionization is believed to involve halide ion transfer and solvo-acids and solvo-bases can then be formulated. Most attention has been directed to Group V halides, and the following have been studied in fair detail (values of the dielectric constant are given in brackets):

AsF_3 (5·7 at −6°)	$SbCl_3$ (33 at 75°)	$BiCl_3$ (13·9 at 175°)
$AsCl_3$ (12·6 at 17°)	$SbBr_3$ (20·9 at 100°)	
$AsBr_3$ (8·8 at 35°)		

All of these trihalides are believed to ionize when in the pure liquid state according to the scheme

$$2AB_3 \rightleftharpoons AB_2^+ + AB_4^-$$

Arsenic trichloride is typical and will suffice for detailed consideration. As would be expected from the relatively low dielectric constant, the only ionic chlorides that are readily soluble are those with a large univalent cation (e.g. NR_4^+ and PCl_4^+), while covalent chlorides such as $AlCl_3$, $FeCl_3$ and $SnCl_4$, which are capable of accepting chloride ions, also dissolve. These two types of halide both yield conducting solutions and the usual solvo-acid–solvo-base conductimetric titration may be done; for example,

$$(AsCl_2^+)_2SnCl_6^{2-} + 2Me_4N^+AsCl_4^- \rightarrow (NMe_4)_2SnCl_6 + 4AsCl_3$$

Although solid adducts formed by $AsCl_3$ with halide ion acceptors and donors can often be isolated, structural information is meagre. There is, for example, no direct proof that the $AsCl_2^+$ ion can exist in the solid state.

Arsenic pentachloride is unknown, but when either PCl_5 or $SbCl_5$ is dissolved in $AsCl_3$ and the solution is treated with chlorine, the adducts PCl_5AsCl_5 and $SbCl_5AsCl_5$ can be isolated. Aluminium chloride similarly gives $AlCl_3AsCl_5$. These could be written as $AsCl_4^+$ salts—solvo-acids in the hypothetical $AsCl_5$ solvent system ($2AsCl_5 \rightleftharpoons AsCl_4^+ + AsCl_6^-$).

Mercuric bromide (m.p. 238°; b.p. 320°C) merits special mention here as, contrary to what might have been expected, it has a specific conductivity of the order of 10^{-4} ohm^{-1} cm^{-1} just above the melting point, and is therefore sharply differentiated from ionic melts such as NaCl or CaF_2, in which ionization is often complete.

Here the following self-ionization scheme has been postulated:

$$2HgBr_2 \rightleftharpoons HgBr^+ + HgBr_3^-$$

It is a good solvent for various mercury (II) salts (e.g. $Hg(ClO_4)_2$ and HgI_2) and also for many halides (e.g. CsBr, $CuBr_2$ and $AlBr_3$). In each case the resulting solution has a greatly enhanced electrical conductivity and the solutes have been classified as solvo-acids and solvo-bases; for example,

$$\left.\begin{array}{l} Hg(ClO_4)_2 + HgBr_2 \rightleftharpoons 2HgBr^+ClO_4^- \\ Hg(NO_3)_2 + HgBr_2 \rightleftharpoons 2HgBr^+NO_3^- \end{array}\right\} \text{acids}$$

$$HgBr_2 + KBr \rightleftharpoons K^+HgBr_3^- \quad \text{base}$$

Here again, many neutralization reactions have been studied conductimetrically; for example,

$$HgBr^+ClO_4^- + K^+HgBr_3^- \rightarrow KClO_4 + 2HgBr_2$$

It seems likely that other metallic halides with bonding which is largely covalent in character would behave similarly.

7.10 Oxy-halides[10, 12, 13]

At first sight it would be expected that the discussion of oxy-halides as solvents would follow the same lines as for halides (i.e. the conductivity of the pure liquid would be explained by postulating self-ionization involving halide ion transfer). Recently, however, it has been realized that the donor function of oxygen in the solvent molecule may complicate the issue. The nature of the complication may be illustrated by considering phosphorus oxychloride, the self-ionization of which was taken to be

$$POCl_3 \rightleftharpoons POCl_2^+ + Cl^-$$

This was used to account for the small observed specific conductivity of 2×10^{-8} ohm^{-1} cm^{-1} at 20°C, a value which may well be too high because of the extreme difficulty of purifying this solvent.

Various halides (e.g. $SbCl_5$, $FeCl_3$, $AlCl_3$ and $TiCl_4$) form solid adducts with $POCl_3$ and also dissolve to give conducting solutions. They were accordingly formulated as solvo-acids (e.g. $POCl_3 + FeCl_3 \rightleftharpoons POCl_2^+ + FeCl_4^-$). Conducting solutions were also given by quaternary ammonium halides which, acting as solvo-bases, could be titrated conductimetrically with acids; for example,

$$POCl_2^+SbCl_6^- + Et_4N^+Cl^- = Et_4NSbCl_6 + POCl_3$$

This treatment was first questioned when it was shown by X-ray diffraction that oxygen coordination occurs in the solid adducts $SbCl_5POCl_3$ and $TiCl_4POCl_3$, the latter being a dimer with a double halogen bridge.

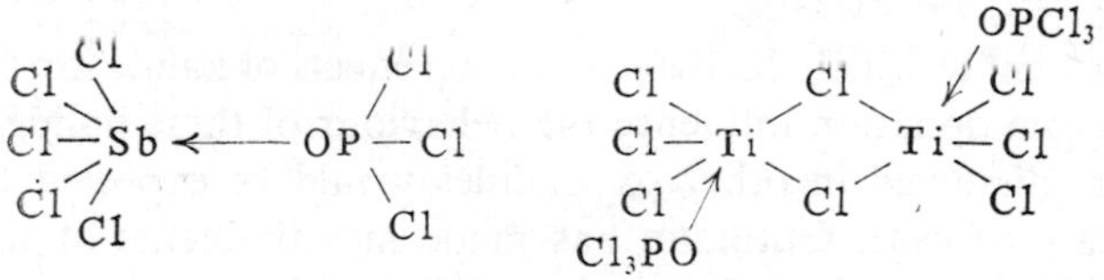

```
                                                         OPCl3
     Cl          Cl                                     /
 Cl   \         /              Cl        Cl           ↙ Cl
   \   \       /                 \      /  \         /
 Cl—Sb ←—— OP—Cl              Cl—Ti          Ti—Cl
   /   /       \                 / ↗ \      /  \
 Cl   Cl        Cl             Cl      Cl        Cl
                                  /
                              Cl3PO
```

There was no evidence for the existence of the $POCl_2^+$ ion in the solid. There was also spectroscopic evidence for oxygen coordination in the adducts $AlCl_3POCl_3$ and $GaCl_3POCl_3$.

The nature of the ions present in solutions of these halides in $POCl_3$ is still rather uncertain. Iron (III) chloride solutions have been studied in detail and their colour varies with concentration. Only at concentrations as low as 10^{-4} M is the characteristic ultraviolet absorption of the $FeCl_4^-$ ion observed. Meek and Drago found that an identical ultraviolet absorption spectrum was observed for dilute solutions of $FeCl_3$ in $PO(OEt)_3$, the oxygen in which should have a donor function similar to that in $POCl_3$. In this solvent all of the chloride to form the $FeCl_4^-$ ion must have come from $FeCl_3$, and an ionization scheme based on oxygen coordination was postulated, the inference being that a similar scheme was operative for $POCl_3$:

$$FeCl_3 + PO(OEt)_3 \rightleftharpoons Cl_3Fe \leftarrow OP(OEt)_3 \rightleftharpoons$$
$$Cl_2Fe \leftarrow OP(OEt)_3^+ + Cl^-(\rightarrow FeCl_4^-)$$
$$FeCl_3 + POCl_3 \rightleftharpoons (Cl_3Fe \leftarrow OPCl_3) \rightleftharpoons$$
$$Cl_2Fe \leftarrow OPCl_3^+ + Cl^-(\rightarrow FeCl_4^-)$$

Conductimetric titration of a solution of $FeCl_3$ in $PO(OEt)_3$ with a chloride ion also gave a sharp break at a 1 : 1 molar ratio, and a similar 1 : 1 end-point was observed on titrating an $SbCl_3$ solution in $PO(OEt)_3$, the essential ionic reaction being

$$Cl_2M \leftarrow OPX_3^+ + Cl^- \rightarrow Cl_3M + POX_3$$

Transport experiments on solutions in nitrobenzene of $FeCl_3POCl_3$ labelled with ^{59}Fe showed migration of iron to both the anode and the cathode compartments.

The behaviour of $FeCl_3$ in $POCl_3$ can clearly be explained without invoking the self-ionization of $POCl_3$ and the concept of solvo-acids. The same is probably true of some other halides (e.g. $AlCl_3$ and $SbCl_5$), though the evidence is less complete. It seems, however, that in the presence of a strong Lewis base such as pyridine, halide ion transfer takes place:

$$C_5H_5N + POCl_3 + FeCl_3 \rightleftarrows C_5H_5NPOCl_2^+ + FeCl_4^-$$

The extent to which the two competing effects of halide ion transfer and oxygen donation influence the behaviour of these halides when they are dissolved in other oxy-halides would be expected to vary from case to case. Gutmann has made measurements of the heat evolved in the reaction of a number of O- and N-containing solvents with $SbCl_5$ as a reference acceptor, using very dilute solutions in dichloroethane, setting up in this way a scale of 'donor numbers'. For $POCl_3$, $C_6H_5POCl_2$ and $SeOCl_2$ the donor number is high, indicating that there is a strong bond from oxygen to antimony. For other oxy-halides the heat effect and consequently the donor numbers are low (e.g. for NOCl, $COCl_2$ and CH_3COCl), and halide ion transfer is believed to be the major effect. The self-ionization equations then follow the usual pattern:

$$NOCl \rightleftharpoons NO^+ + Cl^-$$
$$COCl_2 \rightleftharpoons COCl^+ + Cl^-$$
$$CH_3COCl \rightleftharpoons CH_3CO^+ + Cl^-$$

Only nitrosyl chloride will be considered in any detail. It is a good solvent for nitrosonium salts, the NO^+ ion being solvated to $[O{=}N{-}Cl{-}N{=}O]^+$, and the numerous NOCl adducts with metal halides (e.g. $NOCl.AlCl_3$, $NOCl.AsCl_3$, $NOCl.AuCl_3$ and $2NOCl.SnCl_4$) may be formulated as solvo-acids. Many are good conductors in NOCl solution and the Raman spectrum of $NOCl.AlCl_3$ confirms the presence of $AlCl_4^-$ in the solid adduct. There is also rapid exchange of the radioactive isotope ^{36}Cl between NOCl and the halides of Al^{III}, Ga^{III}, In^{III}, Tl^{III}, Fe^{III} and Sb^{V}, which is also indicative of halo-complex formation.

Conducting solutions are also formed by the tetraalkyl ammonium chlorides, solutions of which give the expected break when titrated conductimetrically with the solvo-acids.

It must be emphasized in concluding this discussion of oxy-halides that the latter are numerous and that some (e.g. NOF, NO_2F and ClO_2F) have been little studied. Thio-halides are also likely to be similar in behaviour. The competing effects of halide ion transfer and donation from oxygen or sulphur are, however, likely to arise in each case, and much more complete structural evidence is needed in many instances before their roles in adduct formation and in the behaviour of solutions can be assessed.

7.11 Dinitrogen tetroxide[14, 15]

Dinitrogen tetroxide (b.p. 21·3°; m.p. −12·3°) has a low dielectric constant (2·42 at 0°) and, as would be expected, is a poor solvent for ionic compounds. However, many types of organic compound are soluble and a number form adducts. The specific conductivity is very low (2×10^{-13} ohm^{-1} cm^{-1} at 17°), the mode of ionization, which is postulated entirely on indirect evidence, being

$$N_2O_4 \rightleftharpoons NO^+ + NO_3^-$$

On this basis nitrosonium salts would be solvo-acids and nitrates solvo-bases. Conductimetric titrations have not been reported, though the reaction between nitrosyl chloride and solid silver nitrate appears to involve a neutralization reaction:

$$NOCl + AgNO_3 = AgCl + N_2O_4$$

It has, however, been suggested that it would occur in media other than N_2O_4.

Tetramethylammonium nitrate in which the nitrate ion is labelled with ^{15}N is found to undergo quantitative exchange of NO_3^- with N_2O_4, and this is best interpreted by assuming that the latter ionizes as shown above. There is also evidence from the infrared spectrum for the presence of NO^+ and NO_3^- in solutions of N_2O_4 in anhydrous HNO_3.

Unquestionably the main interest in this solvent stems from its solvolytic reactions, especially those leading to anhydrous nitrates. The latter, when prepared in aqueous solution, usually give hydrates from which in most cases water cannot be removed without forming either an oxide nitrate or an oxide. A few chlorides give nitrates directly; for example,

$$KCl + N_2O_4 \rightarrow KNO_3 + NOCl$$

More often an N_2O_4 adduct of the nitrate is formed from which N_2O_4 can usually be driven off by heating in a vacuum to about 100°. The nature of these adducts is referred to later.

Some carbonyls are decomposed by liquid N_2O_4 to form N_2O_4 adducts of the metal nitrate; for example,

$$Mn_2(CO)_{10} + 6N_2O_4 = 2Mn(NO_3)_2.N_2O_4 + 4NO + 10CO$$

Only a few metals (e.g. the alkali metals, Zn and Hg) are completely attacked. Others are unaffected, or only very slowly attacked, because of the formation of an insoluble film of nitrate—N_2O_4 adduct.

Solutions of tetraethylammonium nitrate in liquid N_2O_4, which would be basic on the solvent system concept, are more reactive than the oxide itself. They will, for example, dissolve copper. Ionization of N_2O_4 is also promoted by its solution in a Lewis base such as ethyl acetate or acetonitrile (Base + $N_2O_4 \rightleftharpoons$ Base NO^+ + NO_3^-), and solid adducts of N_2O_4 with the base may often be isolated. These solutions are conducting and also show enhanced reactivity towards metals, in addition to solvolysing various chlorides; for example,

$$NiCl_2 \rightarrow Ni(NO_3)_2.N_2O_4 \xrightarrow{\text{heat}} Ni(NO_3)_2$$

It is of interest that liquid dinitrogen tetroxide reacts at $-78°$ with $Me_4N[BCl_4]$, converting it to $Me_4N[B(NO_3)_4]$.

An alternative route to the preparation of anhydrous nitrates is by the use of dinitrogen pentoxide, which is normally handled as a liquid as a result of some decomposition to N_2O_4 and oxygen. It is believed to ionize to NO_2^+ and NO_3^-. Typical reactions of this reagent are the conversion of V_2O_5 to $VO(NO_3)_3$ and of $ZrCl_4$ to $Zr(NO_3)_4$. It also provides a means of dehydrating the hydrated nitrates of elements such as Ti, Zr and Hf without decomposition.

The adducts formed by metal nitrates with N_2O_4 are usually formulated as salts of the NO^+ cation (e.g. $Zn(NO_3)_2.2N_2O_4$ as $(NO^+)_2[Zn(NO_3)_4]^{2-}$). Many such adducts show infrared absorption at about 2300 cm^{-1} which is characteristic of NO^+. It has also been shown that when $Zn(NO_3)_2.2N_2O_4$ is electrolysed in nitromethane solution, zinc migrates to the anode. The behaviour of $Fe(NO_3)_3$ on electrolysis is similar. Conductimetric titration in nitromethane solution of $Zn(NO_3)_2$ with NEt_4NO_3 also gives a sharp break at a 1 : 2 molar ratio, in keeping with the formation of the $[Zn(NO_3)_4]^{2-}$ anion.

The investigations outlined above have made available for the first time many anhydrous metallic nitrates. There is clear structural evidence that those of the most electropositive elements are ionic, but the volatility of the nitrates of a number of the less basic metals suggests that they are covalent. Thus $Cu(NO_3)_2$, $Ti(NO_3)_4$ and $Au(NO_3)_3$ can all be sublimed in a vacuum without decomposition at temperatures below 150°. Nitrates such as $Cu(NO_3)_2$ and $Zn(NO_3)_2$ are also highly soluble in polar organic solvents. Copper (II) nitrate is monomeric in the vapour phase, in which it has the structure

$$\bar{O}-\overset{+}{N}\langle{}^{O}_{O}\rangle Cu \langle{}^{O}_{O}\rangle\overset{+}{N}-\bar{O}$$

(Whether the molecule is planar is not yet established.) In the solid the structure consists of infinite cationic chains of composition $[CuNO_3]^+$ (in which each nitrate ion is bonded to two copper atoms) and NO_3^- anions; these units are so packed that the copper (II) atom has its usual distorted octahedral environment (see Section 15.3) of oxygen atoms.

7.12 Sulphur dioxide[1]

Sulphur dioxide (b.p. −10·0°; m.p. −75·46°) has a dielectric constant of 15·4 at 0°. Iodides of the alkali metals are fairly soluble, with bromides, chlorides and fluorides increasingly less so. The solubility increases with the cation size. Other salts of univalent metals which show small to moderate solubility are cyanides, thiocyanates, acetates and sulphites. Ionic halides of bivalent metals show low solubilities but many covalent halides dissolve readily (e.g. BCl_3, $AlCl_3$ and $AsCl_3$). The same is true of many types of organic compound (e.g. amines, ethers, alcohols and esters).

The electrical conductivity of a range of univalent electrolytes in liquid sulphur dioxide has been extensively studied and there have also been measurements of electrode potentials, a typical cell being Ag,AgBr/Br^- in SO_2/Hg_2Br_2,Hg. The pure solvent has a specific conductivity of $3\text{–}4 \times 10^{-8}$ ohm^{-1} cm^{-1} at −10°, to account for which Jander suggested the following ionization scheme:

$$2SO_2 \rightleftharpoons SO^{2+} + SO_3^{2-}$$

The sulphite ion is solvated to $S_2O_5^{2-}$. This is the only solvent for which dissociation into two doubly charged ions has been postulated, and it seems unlikely on electrostatic grounds. Nevertheless, there appeared to be support. Thionyl derivatives (e.g. $SOCl_2$) gave weakly conducting solutions, attributed to ionization to $SO^{2+} + 2Cl^-$, and were thought of as solvo-acids. They could be titrated conductimetrically with solutions of sulphites (solvo-bases), well-defined breaks at the correct end-point being obtained; for example,

$$SOCl_2 + Cs_2SO_3 = 2CsCl + 2SO_2$$

Various solvolytic reactions were also observed; for example,

$$PCl_5 + SO_2 = POCl_3 + SOCl_2$$
$$WCl_6 + SO_2 = WOCl_4 + SOCl_2$$

Several reactions which were interpreted as involving amphoteric behaviour were reported. The equations below, for example, show the precipitation of aluminium sulphite, its solution in excess of solvo-base and its reprecipitation by a solvo-acid. These stages were followed by conductimetric titrations:

$$2AlCl_3 + 3(Me_4N)_2SO_3 = Al_2(SO_3)_3 + 6Me_4NCl$$
$$Al_2(SO_3)_3 + 3(Me_4N)_2SO_3 = 2(Me_4N)_3[Al(SO_3)_3]$$
$$2(Me_4N)_3[Al(SO_3)_3] + 3SOCl_2 = Al_2(SO_3)_3(s) + 6Me_4NCl + 6SO_2$$

The first indication that the self-ionization postulated for liquid sulphur dioxide was incorrect came when it was found by isotope exchange studies that $SOCl_2$ and $SOBr_2$ do not exchange ^{35}S or ^{18}O with liquid SO_2, as they would inevitably do if both gave the SO^{2+} ion. This suggests that the thionyl halides ionize to SOX^+ and X^-, a more probable mode, which is also consistent with the rapid exchange of radioactive chlorine between thionyl chloride and an ionic chloride in SO_2. This could, however, also be explained by an associative equilibrium

$$SOCl_2 + Cl^- \rightleftharpoons SOCl_3{}^-$$

There is also rapid exchange of radioactive sulphur between thionyl chloride and thionyl bromide in liquid SO_2, which is accounted for either by dissociation to $SOX^+ + X^-$ or by formation of a transition complex

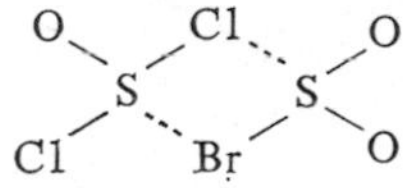

O   Cl   O
  S   S
Cl   Br   O

A further relevant observation is that exchange of radioactive sulphur occurs between liquid SO_2 and the sulphite ion. This is more readily understood as $SO_3{}^{2-}$ is known to be in equilibrium with its solvated form $S_2O_5{}^{2-}$.

Although there is no exchange of ^{35}S between $SOCl_2$ or $SOBr_2$ and SO_2, addition of chloride or bromide ion, as the case may be, in the form of an ammonium salt catalyses this reaction. In each case a kinetic study has been made. Taking the chloride-catalysed reaction as an example, the kinetic results can be explained by postulating the following equilibria:

$$^*SO_2 + Cl^- \rightleftharpoons {}^*SO_2Cl^- \quad \text{(fast)}$$
$$^*SO_2Cl^- + SOCl_2 \rightleftharpoons {}^*SOCl_2 + SO_2Cl^- \quad \text{(slow)}$$

A different rate law is found for the bromide catalysed reaction, though the mechanism proposed again involves the intermediate SO_2Br^-, which appears to be less stable than SO_2Cl^-. This in turn is less stable than the fluorosulphinate ion SO_2F^-, which is well known as a fluorinating agent (see Section 9.4), KSO_2F being formed from KF and SO_2.

There has been considerable further work on the kinetics of catalysis of the $SOCl_2$–SO_2 exchange by various other species (e.g. tertiary amines, $SbCl_5$). Nowhere, however, is there any evidence from the exchange studies to support Jander's original postulate for the self-ionization of sulphur dioxide.

The key observation which appears to support the self-ionization scheme postulated for liquid sulphur dioxide is the occurrence of 'neutralization' reactions between thionyl derivatives and sulphites. These, however, can also be explained in terms of nucleophilic attack of SO_3^{2-} (or $S_2O_5^{2-}$) on $SOCl_2$ with expulsion of two chloride ions:

$$SO_3^{2-} + SOCl_2 \rightleftharpoons (SO_3SOCl_2^{2-}) \rightarrow (SO_3SOCl^-) + Cl^- \rightarrow (SO_3SO) + 2Cl^- \rightarrow 2SO_2 + 2Cl^-$$

If this reaction is fast and the equilibrium $SO_3^{2-} + SOCl_2 \rightleftharpoons 2SO_2 + 2Cl^-$ is sufficiently over to the right, conductimetric titration would break at a 1 : 1 molar ratio. The only open question then is whether liquid sulphur dioxide, when pure, has any residual conductivity and, if so, what causes it?

References for Chapter 7

1 Waddington, T. C. (ed.), *Non-aqueous Solvent Systems* (Academic Press, 1965).
2 Lagowski, J. J. (ed.), *The Chemistry of Non-aqueous Solvents*, Vols I & II (Academic Press, 1966 and 1967).
3 Jander, J., *Anorganische und allgemeine Chemie in flüssigem Ammoniak* (Interscience, 1966).
4 Jolly, W. L., *Progress in Inorg. Chem.*, 1959, **1**, 235.
5 Symons, M. C. R., *Quart. Rev. Chem. Soc.*, 1959, **13**, 99; Catterall, R., and Symons, M. C. R., *J. Chem. Soc.*, 1965, 3763.
6 Birch, A. J., *Quart. Rev. Chem. Soc.*, 1950, **3**, 68.
7 Audrieth, L. F., and Kleinberg, J., *Non-aqueous Solvents* (Wiley, 1953).
8 Gillespie, R. J., and Robinson, E. A., *Adv. Inorg. Chem. Radiochem.*, 1959, **1**, 386.
9 Gillespie, R. J., and Robinson, E. A., in *Ref. 1*, p. 117.
10 Gutmann, V., *Coordination Chemistry in Non-aqueous Solvents* (Springer-Verlag, 1968).
11 Sharpe, A. G., in *Ref. 1*, p. 285.
12 Drago, R. S., and Purcell, K. F., in *Ref. 1*, p. 211.
13 Payne, D. S., in *Ref. 1*, p. 301.
14 Addison, C. C., and Logan, N., *Adv. Inorg. Chem. Radiochem.*, 1964, **6**, 72.
15 Addison, C. C., *Chemistry in Liquid Dinitrogen Tetroxide* (Pergamon Press, 1967) (Vol. 3, Part I of *Chemistry in Non-aqueous Ionising Solvents*).

8 Hydrogen and the hydrides

8.1 Introduction

This chapter opens with brief accounts of the two hydrogen isotopes, deuterium and tritium, some of their applications in chemistry and of *ortho*- and *para*-hydrogen. In reviewing the hydrides themselves we deal first with those formed by the alkali and alkaline earth elements and a few other electropositive elements. These are best described as salt-like, since they have ionic lattices in which the negatively charged hydrogen ion can be identified. Some of the more recent work on the chemistry of the large group of covalently-bonded hydrides is then considered, though no attempt is made to treat the subject systematically, and much that is reasonably familiar is omitted. Anionic hydride complexes, the best-known examples of which are $[BH_4]^-$ and $[AlH_4]^-$, are treated in this section since the bond between hydrogen and the central atom of the complex is also covalent.

The concluding sections of the chapter cover hydrides of the transition metals, the lanthanides and the actinides. Some of the transition metals form complexes in which one or more atoms of hydrogen are linked to the metal atom by a σ-bond, which is usually stabilized by the presence of certain high-field ligands. Carbonyl hydrides are excluded as they fit better into the systematic treatment of carbonyl chemistry. The final section is devoted to the metallic hydrides formed by a number of *d*- and *f*-block elements, in which metallic characteristics are retained. The nature of the bonding in such compounds is still uncertain, but differences from other hydrides are sufficiently great to justify a separate category.

8.2 *Ortho*- and *para*-hydrogen

The nucleus of the hydrogen atom has spin angular momentum defined by the nuclear spin quantum number $I = \frac{1}{2}$, measured in units of $h/2\pi$. In the hydrogen molecule the spins may be either parallel, with a resultant of unity, or antiparallel, with a resultant of zero. This gives rise to two spin isomers of molecular hydrogen, *ortho* and *para*, the first of which has the spins parallel. They are identical chemically, but differ in thermodynamic properties and in their rotational spectra: indeed it was the observation made in 1924 of an alternation in the intensities of successive rotational lines in the electronic band spectrum of hydrogen that led to the discovery of the two forms. The ratio of the intensities of successive strong and weak lines in the rotational spectrum is 3 : 1, and this is the ratio of the abundance of the *ortho* and *para* isomers at equilibrium at room temperature.

The molecules of *ortho*- and *para*-hydrogen differ in internal energy, and a temperature-dependent equilibrium exists between them. The *para* form has the lower internal energy and its formation is therefore favoured by reducing the temperature. Interconversion is not, however, spontaneous but requires the presence of a catalyst or a source of hydrogen atoms. Paramagnetic oxides such as Fe_2O_3, active charcoal, palladium and platinum are good catalysts, as are paramagnetic species such as nitric oxide or the triphenylmethyl radical. When ordinary hydrogen, which contains the 3 : 1 equilibrium mixture, is cooled to liquid air temperature in the presence of active charcoal, for example, it is converted into almost pure *para*-hydrogen, which will persist unchanged at room temperature. If, however, a catalyst is introduced, or the gas is submitted to conditions which produce hydrogen atoms, such as passage through a discharge or heating to 1000°C, the 3 : 1 equilibrium mixture is reformed.

The two forms of hydrogen differ significantly in their physical properties, and differences in thermal conductivity and heat capacity form the basis of a method for assaying mixtures, which involves measuring the heat loss from a wire heated to 160–180°K and surrounded by the mixture under study. The vessel in which the wire is mounted is immersed in liquid air or liquid hydrogen.

Ortho- and *para*-hydrogen have recently been separated by gas chromatography, using helium as carrier gas on an Al_2O_3 column at 77°K. There is a difference between the heats of adsorption of about 0·4 kJ (0·1 kcal). Pure or enriched o-H_2 will of course revert to the 3 : 1 equilibrium mixture in the presence of a catalyst.

The full theoretical explanation of why there is a limiting ratio of the two forms lies outside the scope of this book; it is based on statistical mechanics. If there are two states, one with spins in one direction and the other with opposed spins, their relative probabilities are 3 : 1 because, with the spins in the same direction, the resultant spin is able to orient itself in an external magnetic field in three directions. When, however, the spins are opposed, the orientation of the resultant vector is indeterminate, a condition which is assigned the probability of one.

As shown below, there are significant differences in the triple points and vapour pressures of the two forms of molecular hydrogen. The values for the *ortho* isomer have been extrapolated from data for mixtures of known composition.

	o-H_2	p-H_2	*Ordinary* H_2
Triple point (°K)	13·99°	13·83°	13·95°
Vapour pressure (20·4°K)	751 mm	787 mm	760 mm

One of the applications of *para*-hydrogen is in determining the stationary concentration of hydrogen atoms in a reacting mixture. If pure *para*-hydrogen is used, then its conversion to the equilibrium mixture will proceed at a rate proportional to the hydrogen atom concentration. This has been applied in studying both the photochemical union of hydrogen and chlorine and also the photodecomposition of ammonia. A further point of interest is that, in dealing with liquid hydrogen, the heat effect accompanying the transformation of the equilibrium mixture to *para*-hydrogen is sufficient to cause substantial losses by evaporation. It is usual, therefore, to pass hydrogen over a catalyst at a low temperature prior to liquefaction, in order to convert it largely to the *para* form.

Deuterium also exists in an *ortho* and *para* form, but the nuclear spin of the atom is 1 and not $\frac{1}{2}$. This results in the even rotational levels (including $J = 0$) occurring in the *ortho* state, which is the stable form at very low temperatures, with a limiting ratio at higher temperatures of o-D_2 : p-D_2 = 2 : 1. All other molecules made up of like atoms with nuclear spin also exist in *ortho* and *para* modifications, with parallel and antiparallel spins respectively. Examples of such molecules and the nuclear spins are $^{15}N_2$, $\frac{1}{2}$; $^{19}F_2$, $\frac{1}{2}$; $^{14}N_2$, 1; $^{35}Cl_2$, 1; and $^{17}O_2$, $\frac{5}{2}$. Evidence for spin isomerism in these cases comes only from the intensities of rotational lines in the spectra, and there is no detectable difference in other physical properties. It should be noted in relation to this topic as a whole that the magnetic moment due to nuclear spin is very much smaller than that due to an electron, which is why molecules such as p-H_2 with a resultant *nuclear* spin are not paramagnetic.

8.3 Deuterium and tritium[1]

Deuterium, the hydrogen isotope of mass 2, was not detected in early applications of the mass spectrograph because of its confusion with ionized hydrogen molecules. Its existence was first suspected when a discrepancy was observed between the chemical atomic weight of hydrogen and the mass spectrographic value. The former, based on O = 16·00000, had a value of 1·00780, in good agreement with Aston's value of 1·00778, also based on O = 16·00000. The chemical value had to be changed, however, when it was found that oxygen had three isotopes, ^{16}O, ^{17}O and ^{18}O, with abundances of 99·757, 0·039 and 0·204 per cent respectively, leading to a mean isotopic mass of 16·00447. Correction of the chemical atomic weight of hydrogen by a factor of 16·00447/16·00000 (=1·00028) gives a new chemical atomic weight for hydrogen of 1·008 on the ^{16}O scale, which is greater than the physical value by an amount which exceeds the errors of experiment. The discrepancy can be accounted for only on the assumption that ordinary hydrogen contains one or more isotopes of mass greater than unity.

This point was rapidly put to the test, and the optical spectrum of the residues left, after evaporating a considerable volume of liquid hydrogen, was found to contain lines in the positions calculated for the Balmer series of a hydrogen isotope of mass 2. The discovery of this new isotope differing in mass from hydrogen itself by 100 per cent led to many attempts at its separation. Success came rapidly when Washburn and Urey found that if water is electrolysed there is a preferential release of hydrogen and the residue becomes slightly enriched in deuterium. In one of the earliest full-scale applications of this discovery 2310 gallons of water were reduced in a seven-stage electrolytic process to a volume of 82 ml, which proved to be 99 per cent pure deuterium oxide. The composition was readily monitored by a density determination ($d_{D_2O} = 1{\cdot}1059$, $d_{H_2O} = 0{\cdot}9982$ at 20°). The electrolyte was a caustic soda solution, the electrodes being nickel sheets, and there were periodic distillations of the residues to free them from accumulated alkali.

Various alternative methods for preparing deuterium, or one of its compounds, in a pure or enriched form are available. They include the fractional distillation of water to separate the less volatile D_2O

from HDO and H_2O. This can also be applied to other deuterium compounds (e.g. ND_3 and NH_3), since there is always a small difference in volatility. Molecular hydrogen and deuterium may also be separated by fractional diffusion (see page 6), gas chromatography and by chemical exchange processes (e.g. H_2O(liq)/HSD(g)). Most deuterium compounds are, however, now made either directly or indirectly from deuterium oxide, which is prepared on a large scale by the electrolytic process for use as a moderator in nuclear reactors.

Deuterium and its compounds differ appreciably in physical properties from their hydrogen analogues, as can be seen from the examples given in Table 8.1 below.

Table 8.1 *Physical properties of deuterium and its compounds*

	m.p. (°C)	*b.p.* (°C)		*m.p.* (°C)	*b.p.* (°C)
H_2	−259·2	−252·7	HCl	−114·6	−84·1
D_2	−254·5	−249·5	DCl	−114·9	−81·5
H_2O	0	100	C_6H_6	5·5	80·1
D_2O	3·82	101·4	C_6D_6	6·6	79·2
NH_3	−77·8	−33·3	HCN	−14	25·3
ND_3	−73·5	−31·0	DCN	−12	25·9

Deuterium compounds can be prepared as a rule by the same methods as are used for their hydrogen analogues, provided they are suitable for small-scale work and give reasonably high yields. In some instances exchange reactions may also be employed: an ammonium salt, for example, becomes fully deuterated on treatment with successive quantities of deuterium oxide. Fully or partially deuterated benzene is also produced when benzene vapour and deuterium are passed over a palladium catalyst. Examples of the use of deuterated compounds in structural and mechanistic studies are given elsewhere in this book.

Ionic mobilities are less in D_2O than in H_2O, and the viscosity of D_2O is also smaller. In most cases salts are also somewhat less soluble and heats of solvation of ions are smaller in deuterium oxide. Salt hydrates and deuterates also show small differences in dissociation pressures and transition temperatures.

The zero-point energy for H_2 (i.e. the vibrational energy at the absolute zero) is 26·0 kJ (6·2 kcal) mole^{-1}, whereas for D_2 it is lower by a factor of $\sqrt{2}$ and equal to 18·4 kJ (4·4 kcal) mole^{-1}. The dissociation energy of the molecule is the difference between the electronic binding energy (458 kJ (109·4 kcal) mole^{-1}) and the zero-point energy, and the value for H_2 at the absolute zero is 432 kJ (103·2 kcal) mole^{-1}, whereas for D_2 it is 439 kJ (105·0 kcal) mole^{-1}. There are similar differences in zero-point energies for other bonds in which H and D are involved (e.g. C—H and C—D), and this is the underlying cause of the appreciable differences in reaction rates and in positions of equilibria involving the two isotopes and their compounds.

The mass difference between hydrogen and deuterium influences not only zero-point energies but also vibrational frequencies in general. If we consider, for example, M—H and M—D, the ratio of the vibrational frequencies is given by

$$\frac{\nu_H}{\nu_D} = \left(\frac{\mu_D}{\mu_H}\right)^{\frac{1}{2}}$$

where μ, the reduced mass, is $M_M M_H/M_M + M_H$. Since in this case $M_M \gg M_H$ the frequency ratio is approximately $(M_D/M_H)^{\frac{1}{2}} = \sqrt{2}$. There is thus a predictable shift in the position of a vibrational frequency on deuteration, which is of great value in identifying molecular vibrations in which hydrogen is involved. The following are examples from the very large literature:

O—H in H_2O 3657 cm^{-1}
O—D in D_2O 2727 cm^{-1}
S—H in H_2S 2615 cm^{-1}
S—D in D_2S 1892 cm^{-1}
P—H in PH_3 2327 cm^{-1}
P—D in PD_3 1694 cm^{-1}

An example of the use of deuterium substitution in vibrational spectroscopy is in the identification of metal–hydrogen frequencies in transition metal carbonyl hydrides and related compounds. The M—H frequency falls in roughly the same spectral region (at about 1900 cm^{-1}) as the C—O stretching frequency, and is therefore masked. There is a shift on deuteration to about 1400 cm^{-1}, which is free of interference and readily observed.

Applications of deuterium in n.m.r. spectroscopy are indirect because the ^{2}D spectrum is much weaker than the ^{1}H spectrum, and is broadened by quadrupole effects (^{2}D has spin $I = 1$). The ^{2}D resonance occurs at 15·4 MHz for a magnetic field which gives a ^{1}H resonance at 100 MHz. One of the common applications is the use of $CDCl_3$ and other deuterated compounds as solvents in observing ^1H resonances in organic compounds, thus eliminating any interference by the solvent. Partial deuteration is sometimes useful in elucidating ^1H coupling constants in hydrides. Deuterides are also used in place of hydrides in neutron diffraction structural studies so as to take advantage of the greater scattering by the deuterium nucleus, introduction of which is assumed to produce no appreciable change in structure. There is also substantial incoherent scattering of neutrons by protons, leading to high background scattering if hydrides are used instead of deuterides for neutron diffraction.

Tritium occurs in natural hydrogen (*ca* 1 part in 10^{-7}). This minute concentration stems from the nuclear reaction $^{14}N + n = {}^{12}C + {}^3H$, caused by neutrons arriving in the upper atmosphere from outer space. The isotope was discovered in 1934 by Oliphant, Harteck and Rutherford as a product of the bombardment of deuterium compounds such as D_3PO_4 or $(ND_4)_2SO_4$ with fast deuterons ($D + d = T + p$). Several other methods are now available, the most convenient being the irradiation of lithium or one of its compounds with slow neutrons in a nuclear reactor:

$$^6Li + n = {}^4He + {}^3T$$

The tritium formed is separated from helium by absorption on uranium metal as UT_3, and is then regenerated by heating to 500°. Alternatively, it may be oxidized to T_2O.

Tritium is a radioactive isotope which decays by emission of low-energy β-particles (max. energy 18·3 keV; half-life 12·4 years) to form ^3He. The boiling point is −248·1°. It serves as an alternative to deuterium in tracer experiments and is monitored by measurement of the radioactivity.

8.4 The saline hydrides[1a, 2]

The alkali and alkaline earth metals combine directly with hydrogen when heated, forming solid hydrides of the types MH and MH_2 respectively. Reaction occurs in the temperature range 250–600°, depending on the metal and its state of purity. Sodium hydride, the most important of these compounds, is made at 250–400° and it is usual to hydrogenate the metal in the presence of a surface-active agent in kerosene, or to use some form of agitation, in order to overcome difficulties due to the surface film of hydride, which can lead to incomplete conversion.

All of these hydrides dissociate reversibly at higher temperatures, the thermal stability decreasing from Li to Cs and from Ca to Ba. Lithium hydride, for example, has a dissociation pressure of 27 mm at its melting point (688°) whereas sodium hydride and potassium hydride have dissociation pressures at 400° of 14·5 and 178 mm respectively. All are white crystalline solids when pure. The alkali metal hydrides have the face-centred cubic structure of NaCl. Bonding appears to be largely ionic with M^+ and H^- as the ions. The latter is formed by uptake of an electron by the hydrogen atom from the electropositive metal and the ionic radius, obtained by deducting the cation radius from observed M—H distances, is about 1·4 Å. The position of the hydrogen atoms in lithium hydride has been established by neutron diffraction.

Lattice energies, calculated from the structures, decrease from 904 kJ (216 kcal) for LiH to 653 kJ (156 kcal) for CsH. These values, taken in conjunction with other quantities in the Born cycle for alkali metal hydrides, lead to a value of $\Delta H = -67$ kJ (-16 kcal) for the electron affinity of the hydrogen atom. It is interesting to ask why LiH is the hydride most stable with respect to decomposition to its elements. Though lithium has the highest ionization potential of the alkali metals, the difference in ionization potential between Li and Cs (145 kJ, or 35 kcal) is smaller than the difference in the lattice energies of the hydrides. This, it may be noted, is a consequence of the relatively small size of the hydride ion, which has roughly the same radius as F^-. For a larger anion like I^- ($r = 2{\cdot}16$ Å) the variation in lattice energy (138 kJ, or 33 kcal) is less than the difference in ionization potential, and CsI is therefore more stable than LiI with respect to decomposition into the elements.

The hydrides of the alkaline earth metals have complicated orthorhombic structures based on approximately hexagonal close-packing of the metal atoms: there are two sets of M—H distances (e.g. in SrH_2 each Sr has three H atoms at 2·35 Å and four H atoms at 2·71 Å).

It would clearly be desirable to establish the anionic nature of hydrogen in these compounds by showing that it migrates to the anode on electrolysis of the melt. This has been done for LiH (m.p. 688°), though the experiments cannot be interpreted quantitatively because thermal dissociation of the hydride is appreciable at the melting point. For the other hydrides, thermal dissociation even below the melting point is so high that migration studies are difficult. Electrolysis of CaH_2 dissolved in a LiCl–KCl eutectic at 360° does, however, give hydrogen at the anode in an amount that accords with Faraday's laws.

The standard heats of formation of the hydrides (except RbH and CsH, which have not been measured) are shown below.

	LiH	NaH	KH	CaH_2	SrH_2	BaH_2
$-\Delta H_f^0$ (kJ)	90·8	56·5	57·8	188·7	176·7	171·5
(kcal)	21·7	13·5	13·8	45·1	42·2	41·0

There are small differences in the values of these and other properties for the deuterides. It is also of interest in comparing these compounds with other metal hydrides referred to later to note that their densities are greater than those of the parent metals, by 25–45 per cent for the alkali metal compounds and by 5–10 per cent for those of the alkaline earth metals. For the alkali metal hydrides this increase in density is associated with the more efficient packing of atoms in going from the body-centred cubic structure of the metal to the face-centred cubic structure of the hydride.

The chemical reactivity of the hydrides increases from LiH to CsH and from CaH_2 to BaH_2. The alkali metal compounds react with water at room temperature with increasing vigour, as the series is ascended, to form the hydroxide and hydrogen, which may ignite. This reaction is less violent for the alkaline earth metal hydrides. Oxidation in air or oxygen shows the same trend: hydrides of the lighter elements oxidize only when heated, but those of the heavier elements may burn spontaneously, especially if they are finely divided and a trace of moisture is present.

Lithium hydride stands alone in being soluble to a small extent in ether and some other polar solvents; this may be indicative of some covalent character in the bonding but the same is probably also true of the other hydrides. It is a useful reagent in ether solution for preparing hydrides from halides; for example,

$$4LiH + SiCl_4 \rightarrow SiH_4 + 4LiCl$$

The reaction with $AlCl_3$ in ether gives $LiAlH_4$ and that with B_2H_6 the borohydride:

$$4LiH + AlCl_3 \rightarrow LiAlH_4 + 3LiCl$$
$$2LiH + B_2H_6 \rightarrow 2LiBH_4$$

Various other reactions of lithium hydride are known: that with nitrogen at elevated temperatures leads to $LiNH_2$, Li_2NH or Li_3N, according to the conditions, and, with ammonia, $LiNH_2$ and hydrogen are formed.

Sodium hydride and the other alkali metal hydrides resemble the lithium compound in being strong reducing agents. Only the sodium compound is made on a technical scale and finds some applications in organic chemistry. It is also used in the preparation of sodium borohydride by the following reaction:

$$4NaH + B(OMe)_3 \longrightarrow NaBH_4 + 3NaOMe$$

Reactions of the alkaline earth metal hydrides are exemplified by those of calcium hydride, which forms the nitride Ca_3N_2 with nitrogen above about 500° and, at elevated temperatures, is converted by carbon dioxide to calcium formate and by ammonia to calcium amide. A number of metal oxides and halides are reduced to metal, while sulphates are reduced to sulphides. The alkaline earth metal hydrides also form mixed hydride halides of the type MHX (M = Ca, Sr, Ba; X = Cl, Br).

As has been mentioned earlier, there are only minor differences in physical and chemical properties between the hydrides and the deuterides or tritides of the elements in these two groups. Preparative methods are the same, though few tritides are known. Dissociation pressures of deuterides are greater than those of the corresponding hydrides.

Magnesium hydride (MgH_2) should almost certainly be included among the saline hydrides. It may be made in good yield by the interaction of solid magnesium and hydrogen in the presence of magnesium iodide at 570° at 200 atm pressure, though the original preparation was by pyrolysis of diethylmagnesium in vacuum at 175°. Ethylmagnesium bromide decomposes similarly:

$$Et_2Mg \xrightarrow{175^\circ} MgH_2 + 2C_2H_4$$

$$2EtMgBr \xrightarrow{220^\circ} MgH_2 + MgBr_2 + 2C_2H_4$$

The crystal has a body-centred tetragonal lattice of the rutile type. This has been confirmed by a neutron diffraction study on MgD_2. The heat of formation is $\Delta H_f^0 = -75$ kJ (-18 kcal) mole^{-1} (i.e. less than half that for the alkaline earth metal hydrides). The density (1·45 g cm^{-3}) is appreciably less than that of the parent metal (1·74 g cm^{-3}). The thermal dissociation to metal and hydrogen is greater at a given temperature than that of calcium hydride. Very little is known about the chemistry of this hydride. It is decomposed by water with evolution of hydrogen, is a reducing agent and also reacts with B_2H_6 and $(AlH_3)_n$ to give $Mg(BH_4)_2$ and $Mg(AlH_4)_2$.

The classification of several other hydrides which will now be considered briefly is very much less well established. A beryllium hydride which is formulated as BeH_2 is formed in the reaction of alkyl beryllium compounds with $LiAlH_4$ in ether.

$$LiAlH_4 + 2R_2Be \xrightarrow{\text{ether}} 2BeH_2 + LiAlR_4$$

This and other preparative methods have not, however, given a pure product, and the synthesis from beryllium metal and hydrogen has not been achieved. The structure has not been determined, but it is possibly a polymer with hydrogen bridges.

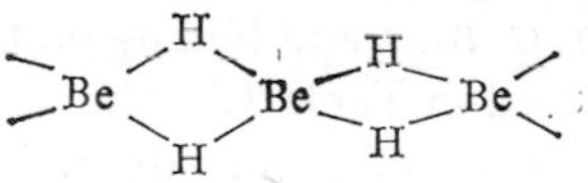

Zinc forms a hydride, ZnH_2, in the reactions shown below, and analogous reactions yield cadmium hydride (CdH_2).

$$LiAlH_4 + 2R_2Zn \xrightarrow{\text{ether}} 2ZnH_2 + LiAlR_4$$

$$2LiAlH_4 + ZnI_2 \xrightarrow[-40^\circ]{\text{ether}} ZnH_2 + 2AlH_3 + 2LiI$$

In neither case is the structure known: the zinc compound decomposes to its elements at below 100°, and cadmium hydride is unstable at room temperature. An exceedingly unstable mercury hydride (HgH_2) is also formed from HgI_2 and $LiAlH_4$ in an ether solvent at −135°, but it decomposes at slightly higher temperatures.

There are a few other hydrides, including those of copper (I), europium (II) and ytterbium (II), which are described in a later section, and which emphasize the inadequacy of any rigid classification into ionic and covalent types.

8.5 Boranes

8.5.1. Boranes and their derivatives.[3–10] The classical researches of Alfred Stock in the 1920's laid the foundation of present-day borane chemistry. Decomposition of magnesium boride (Mg_3B_2) by aqueous hydrochloric or phosphoric acid was found by him to give a mixture of volatile products from which, by using special techniques for the vacuum manipulation of volatile air-sensitive compounds developed earlier in investigating the silanes, he was able to separate and characterize the four hydrides, B_4H_{10}, B_5H_9, B_6H_{10} and $B_{10}H_{14}$. Diborane (B_2H_6) is not formed in this reaction as it is rapidly decomposed by water. It was, however, made subsequently in Stock's laboratory, together with B_5H_{11}, by heating B_4H_{10} at 100°.

Since this early work there have been many important advances. A number of new boranes have been isolated, as well as new types of derivative, notably the polyhedral ions and the carboranes. X-ray analysis and spectroscopic techniques have also been used extensively in elucidating structural problems, and new theories of electron-deficient bonding have been developed. It will be possible here to describe this new work only in outline, but a fuller treatment is available in a number of review articles and monographs. For reference purposes the formulae and, where known, melting and boiling points of the more important boranes are collected in Table 8.2. The nomenclature in current use denotes the number of boron atoms in the molecule by a Greek prefix, with the number of hydrogen atoms in brackets at the end. Thus B_5H_9 is pentaborane (9) and $B_{10}H_{14}$ is decaborane (14).

Table 8.2 *Melting and boiling points of some boranes*

Borane	*m.p.*	*b.p.*	*Borane*	*m.p.*	*b.p.*
B_2H_6	−165°	−90°	B_8H_{12}	*ca.*−20°	—
B_4H_{10}	−120°	16·1°	B_8H_{18}	—	—
B_5H_9	−46·1°	58·4°	B_9H_{15}	2·6°	—
B_5H_{11}	−123·3°	63°	$B_{10}H_{14}$	98·8°	213°
B_6H_{10}	−63·2°	108°	$B_{10}H_{16}$	—	—
B_6H_{12}	−82°	80–90°	$B_{18}H_{22}$	—	—

8.5.2. Diborane (6). Stock's original method for preparing B_2H_6 by heating B_4H_{10} is no longer used, as it may be made more conveniently and in high yield by the reaction of BCl_3 with $LiAlH_4$ in ether solution. Several other methods are also available. For example, sodium borohydride ($NaBH_4$), which is produced by the reaction of NaH with $B(OCH_3)_3$ at 250° (see page 178), may be decomposed by BF_3 in diglyme ($CH_3O(CH_2)_4OCH_3$):

$$3NaBH_4 + 4BF_3 \rightarrow 2B_2H_6 + 3NaBF_4$$

This route has the advantage of being based on readily available starting materials, and the same is true of a direct synthesis, in which B_2O_3 is heated with hydrogen at 150° at 750 atm in the presence of $AlCl_3$ and aluminium metal, and which is reported to give conversions of 40–50 per cent.

The structure of diborane (6) has been determined by electron diffraction. The two boron atoms are joined by two bridging hydrogen atoms (Fig. 8.1), with four terminal hydrogen atoms in a plane at right angles to that containing the bridging groups. This is in keeping with observations on the infrared and Raman spectra.

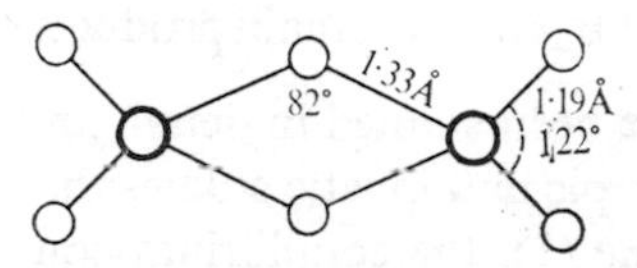

Figure 8.1

The molecule is electron deficient in the sense that eight bonds are formed but only twelve electrons are available, three from each boron and one from each hydrogen. The four terminal B—H bonds are of the same length as in BH_3 adducts, where there is no electron deficiency, and each is therefore a normal two-electron bond. This leaves only two electrons for each B—H—B bridge bond where, it will be noted, the B—H distance is greater. In the bridge bond two sp^3 boron orbitals combine with a 1*s* hydrogen orbital to form one bonding molecular orbital, which covers the three atoms and contains two electrons (Fig. 8.2), and two empty antibonding orbitals.

This type of bridge bond occurs in the higher boranes. The latter also contain open and closed three-centre B—B—B bonds shown diagramatically below (Figs. 8.3 and 8.4), where two electrons occupy

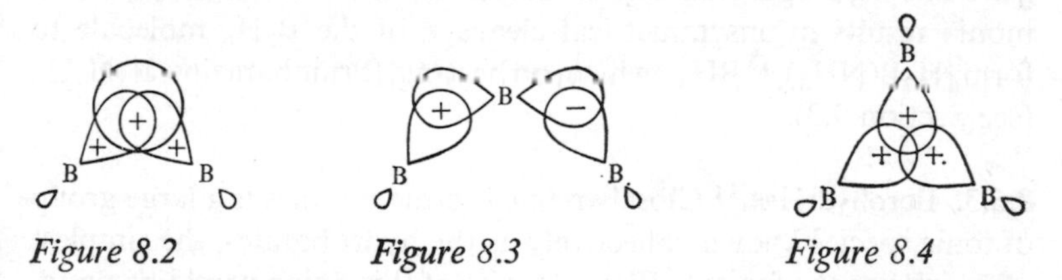

Figure 8.2 *Figure 8.3* *Figure 8.4*

a molecular orbital derived from three sp^3 boron orbitals. The only direct chemical evidence for the bridge structure of diborane (6) is the fact that in a substitution reaction (e.g. methylation) only the terminal hydrogen atoms can be replaced. Any attempt to effect further substitution results in cleavage of the molecule into two fragments, each containing one boron atom.

Diborane (6), in common with the higher boranes, is broken down completely to boron and hydrogen at temperatures above about 400°C. In the range 100–250° there is a complex series of 'cracking' reactions which, by control of the conditions, may be used in the synthesis of higher boranes; for example,

(a) B_5H_9 is formed as a major product by circulating B_2H_6 with H_2 though a glass tube at 200–250°;
(b) B_2H_6 at 115° gives B_5H_{11} in moderate yield;
(c) B_5H_{11} and H_2 at 100° give B_4H_{10} and B_2H_6 as the main products.

The complicated equilibria involved have been studied in detail, and it has been established by mass spectroscopic and kinetic studies that B_2H_6 is in equilibrium with BH_3, borane (3), the equilibrium constant for the reaction $B_2H_6 \rightleftharpoons 2BH_3$ being $1{\cdot}63 \times 10^{-5}$ at 155°.

Various adducts of borane (3) with Lewis bases are known. When, for example, B_2H_6 is heated with CO at 100° at 60 atm the equilibrium shown below is set up:

$$B_2H_6 + 2CO \rightleftharpoons 2H_3BCO$$

If the mixture is cooled rapidly to freeze the equilibrium, H_3BCO (b.p. −64°) may be isolated. Other BH_3 adducts are mentioned in discussing boron polymers (Chapter 9).

Diborane (6) is very rapidly decomposed by water to give H_3BO_3 and H_2. Mixtures with air or oxygen explode when heated to temperatures over about 100°. Reaction with chlorine may also be explosive, but terminally halogenated diborane (6) derivatives have been isolated from the products of its controlled reaction with halogens and hydrogen halides or boron halides. Reaction with ammonia results in unsymmetrical cleavage of the B_2H_6 molecule to form $[H_2B(NH_3)_2]^+BH_4^-$ which, on heating, forms borazine, $B_3N_3H_6$ (see Section 9.3).

8.5.3. Borohydrides.[11] Closely related to the boranes is a large group of ionic species known collectively as the hydroborates, the simplest of which has the formula $[BH_4]^-$: salts of this anion would be designated as tetrahydromonoborates (−1). The name borohydride is, however, in common use and will be adopted here to avoid confusion. Other hydroboron ions containing more than one boron atom will be described later.

Borohydrides are formed by a number of the metallic elements, the most important being the alkali metal compounds, which can be made by reaction of the alkali metal hydride with diborane (6) in one of the higher ethers (e.g. diglyme, $CH_3O(CH_2)_4OCH_3$). Formally this can be regarded as a Lewis acid–base interaction between BH_3 and H^- to give $[H_3B{\leftarrow}H]^-$. There are, however, a number of better preparative methods that avoid the isolation of B_2H_6, one being the reaction of sodium hydride with methyl borate in tetrahydrofuran (THF) where the sodium salt is precipitated:

$$4NaH + 4B(OCH_3)_3 \xrightarrow{THF} NaBH_4 + 3Na[B(OCH_3)_4]$$

Aluminium borohydride, the first compound of this class to be characterized, is formed when trimethylaluminium and diborane (6) react at 80°C:

$$Al(CH_3)_3 + 2B_2H_6 \rightarrow Al(BH_4)_3 + B(CH_3)_3$$

Alkyls of a number of other elements (e.g. Li, Be, Mg, Zn and Ga) undergo a similar reaction to give the metal borohydrides. Metathetical reactions are also widely used, as is shown by the following examples:

$$NaBH_4 + LiCl \xrightarrow{\text{iso-PrNH}_2} LiBH_4 + NaCl$$

$$2NaBH_4 + BeCl_2 \xrightarrow{125°} Be(BH_4)_2 + 2NaCl$$

$$3NaBH_4 + AlCl_3 \xrightarrow{\text{ether}} Al(BH_4)_3 + 3NaCl$$

$$2Al(BH_4)_3 + MF_4 \xrightarrow{20°} M(BH_4)_4 + 2AlF_2(BH_4) \quad (M = U,Th)$$

$$8LiBH_4 + 2TiCl_4 \xrightarrow{20°} 2Ti(BH_4)_3 + 8LiCl + B_2H_6 + H_2$$

A number of borohydrides can be formed only at low temperatures and decompose at or below 20°. The indium compound, for example, is made by the reaction shown below, but decomposes at −10° to indium, hydrogen and diborane:

$$In(CH_3)_3 + 2B_2H_6 \xrightarrow[THF]{-40°} In(BH_4)_3 + B(CH_3)_3$$

The thallium (I) compound, on the other hand, which can be prepared in aqueous solution from $TlNO_3$ and KBH_4, is stable to about 40°. Other unstable borohydrides are those of Sn^{II}, Fe^{II}, Cu^{I}, Ag^{I} and

NH_4^+. The quaternary ammonium borohydrides are much more stable, as are the tetraphenylphosphonium and triphenylsulphonium salts. Most can be prepared from sodium borohydride and the halide. It should be noted that many derivatives of substituted borohydride ions (e.g. $[BH_3CN]^-$ and $[BHF_3]^-$) are also known, as well as compounds where $[BH_4]^-$ and other groups are bonded to a metal atom (e.g. $Al(CH_3)_2BH_4$).

Structural information on borohydrides is meagre. The melting and boiling points are consistent with a transition from an ionic structure in the alkali metal compounds to purely covalent bonding in derivatives of some of the less basic metals. This is illustrated by the following examples:

	m.p.	*b.p.*
$LiBH_4$	275°	—
$Be(BH_4)_2$	123°	Subl. 91·3° at 1 atm
$Al(BH_4)_3$	−64·5°	44·5°

The alkali metal compounds have a sodium chloride structure containing the tetrahedral BH_4^- ion, apart from $LiBH_4$, which has this structure only above 110°; at lower temperatures there is evidence for hydrogen bridging between lithium and boron. Recent electron diffraction studies have shown the beryllium compound to have the structure shown in Fig. 8.5, in which the three heavier atoms form

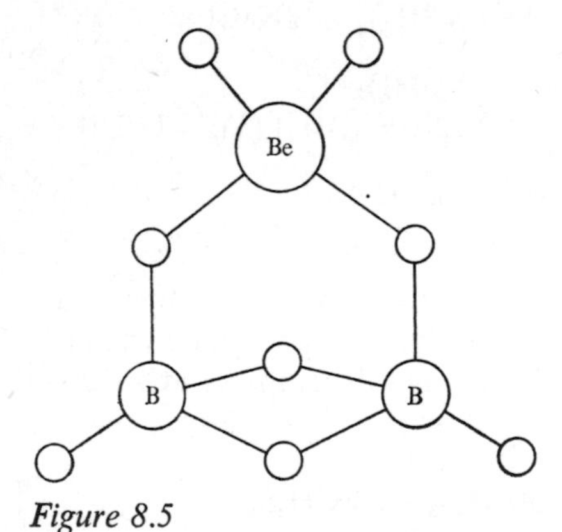

Figure 8.5

an approximately equilateral triangle with hydrogen bridges as shown. From its physical properties the aluminium compound is clearly covalent, and electron diffraction results are in keeping with the hydrogen-bridged structure (Fig. 8.6). Raman and infrared absorptions in the frequency range 1500–2000 cm^{-1} are associated with the B—H—B bond. These borohydrides with hydrogen bridges are electron deficient, whereas the $[BH_4]^-$ ion is not.

Lithium borohydride decomposes in the absence of air at about 275°, but the sodium and potassium compounds are stable under these

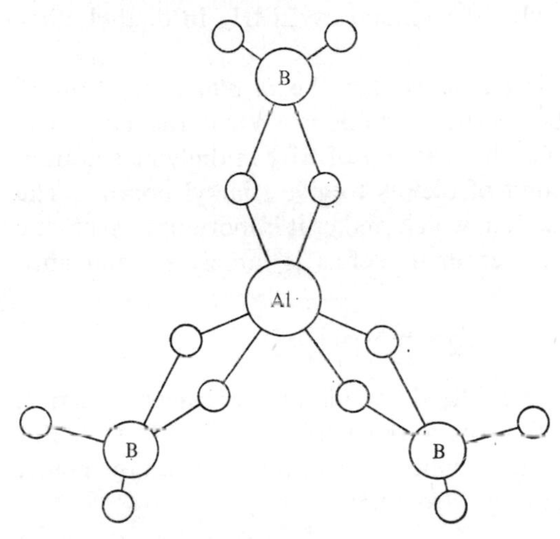

Figure 8.6

conditions to at least 400°. The alkali metal compounds dissolve in a variety of polar solvents whereas the covalent compounds (e.g. $Al(BH_4)_3$) dissolve in non-polar solvents. Sodium and potassium borohydrides also dissolve in water and undergo slow hydrolysis at pH <9, whereas the aluminium and beryllium compounds undergo violent decomposition; the lithium compound is also decomposed by water.

The alkali metal borohydrides are a convenient source of diborane (6); for example,

$$3NaBH_4 + 4BF_3 \xrightarrow{\text{diglyme}} 3NaBF_4 + 2B_2H_6$$

Diborane (6) is also formed in protonation reactions, the essential step of which is $BH_4^- + H^+ \rightarrow H_2 + BH_3$; for example,

$$NaBH_4 + HCl \rightarrow NaCl + H_2 + \tfrac{1}{2}B_2H_6$$

They are powerful reducing agents and, in an ethereal solvent, will convert non-metallic halides to hydrides (e.g. $SiCl_4$ to SiH_4). The reducing action is also apparent in attempts to prepare borohydrides of readily reducible metals from their halides; for example,

$$FeCl_3 + LiBH_4 \xrightarrow[-80^\circ]{\text{ether}} FeCl_2 + \tfrac{1}{2}B_2H_6 + \tfrac{1}{2}H_2 + LiCl$$

$$FeCl_2 + 2LiBH_4 \xrightarrow{0^\circ} Fe + B_2H_6 + H_2 + 2LiCl$$

Copper (I) chloride is reduced to metal by $LiBH_4$ in diethyl ether at 0°.

Sodium borohydride is employed for the *in situ* generation of B_2H_6 in so-called hydroboration reactions. When the borane is generated by, for example, the reaction of BF_3 in diglyme solution, it adds to the double bond of olefins to give trialkyl boranes. The special feature of this reaction, which makes it important in synthetic work, is that addition to terminal olefins is largely *cis* and anti-Markownikoff:

$$6RCH{=}CH_2 + B_2H_6 \rightarrow 2B(CH_2CH_2R)_3$$

The product may then be oxidized by alkaline hydrogen peroxide, without change of configuration, to the alcohol RCH_2CH_2OH. For a more detailed review of the applications of this reaction the reader is referred to a monograph by Brown.[12]

8.5.4. Higher boranes. No binary triborane is known, though salts of the octahydrotriborate (−1) ion ($B_3H_8^-$) have been isolated. Thus NaB_3H_8 is one of the products of the reaction of B_2H_6 with sodium amalgam:

$$2B_2H_6 + 2NaHg \rightarrow NaBH_4 + NaB_3H_8\ (+2Hg)$$

Several other preparative routes are available, one of the simplest being the reaction

$$NaBH_4 + B_2H_6 \xrightarrow[100^\circ]{\text{diglyme}} NaB_3H_8 + H_2$$

The structure of this ion (Fig. 8.7) has been established by X-ray analysis; comparison with that of B_4H_{10} (Fig. 8.8) shows that it can be regarded as a segment of the latter, formed by loss of BH_2.

This method of representing the structures of boranes and their derivatives is in common use. It has the advantages of simplicity and of showing very clearly the position of bridging bonds. In the case of B_4H_{10} the two segments of the molecule are actually folded upwards about the B—B axis.

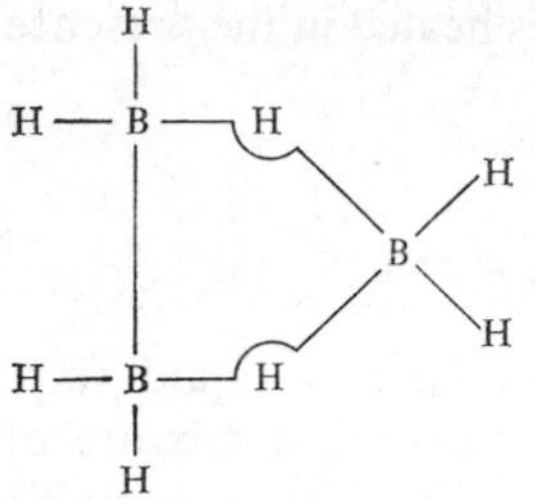

Figure 8.7

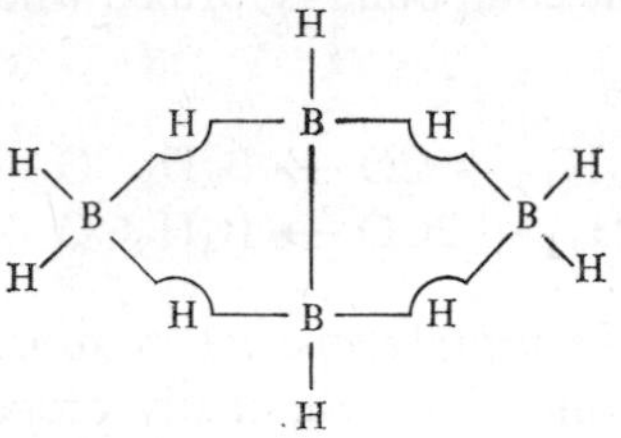

Figure 8.8

(a) *Tetraborane (10)*. This was first isolated by Stock from the products of the acid hydrolysis of magnesium boride, and he also prepared it from B_2H_5I by a Wurtz-type synthesis with sodium amalgam:

$$2B_2H_5I + 2Na \xrightarrow{-45^\circ} B_4H_{10} + 2NaI$$

A more convenient method involves the decomposition of B_2H_6 in a 'hot-cold' tube apparatus, in which the reaction zone through which the borane is passed is the space between two coaxial cylindrical glass tubes, the inner of which is heated internally to 120° while the outer is cooled externally to −78°. Tetraborane(10) (b.p. −16·1°) condenses on the latter as it is produced in the reaction

$$2B_2H_6 \rightarrow B_4H_{10} + H_2$$

and is thus largely removed from the reaction zone and prevented from undergoing secondary reactions; conversion is 80–90 per cent. With some modification of the temperatures, the same apparatus may be used in preparing B_5H_{11} from B_2H_6.

Tetraborane (10) decomposes to other boranes at about 60° and consequently its chemistry has not been fully investigated. The reaction with ammonia is of interest as it involves an unsymmetrical cleavage similar to that observed for B_2H_6. A BH_2 fragment is split off and forms the $[H_2B(NH_3)_2]^+$ cation, leaving the $[B_3H_8]^-$ anion. Ethers and amines cleave the molecule differently, leaving an adduct of the B_3H_7 moiety; for example,

$$B_4H_{10} + 2N(CH_3)_3 \rightarrow (B_3H_7)N(CH_3)_3 + H_3B.N(CH_3)_3$$

There is also a reaction of B_4H_{10} with carbon monoxide at 120° in a 'hot–cold' flow reactor, which gives the carbonyl adduct B_4H_8CO.

The same compound is formed when B_5H_{11} is heated in the presence of CO:

$$B_4H_{10} + CO \rightarrow B_4H_8CO + H_2$$
$$B_5H_{11} + 2CO \rightarrow B_4H_8CO + BH_3CO$$

(b) *The pentaboranes.* Pentaborane (9) (B_5H_9) is a liquid (b.p. 58·4°) which is conveniently prepared by circulating a mixture of B_2H_6 with excess hydrogen through a tube at 200–250°. It is stable with respect to decomposition to other boranes up to approximately 200° and is also resistant to hydrolysis by water, with which it is immiscible. In contrast, pentaborane (11) (b.p. 63°) is much less stable to heat and decomposes at 60° giving mainly B_2H_6 and B_5H_9. It is also rapidly hydrolysed by water. The usual method of synthesis of B_5H_{11} is to heat B_2H_6 to 100°, though other boranes are also formed in this reaction.

The structures of these two compounds are shown below; in both, the boron atoms are situated at the apexes of a square pyramid. In B_5H_9 (Fig. 8.9), there is one terminal B—H bond on each boron and

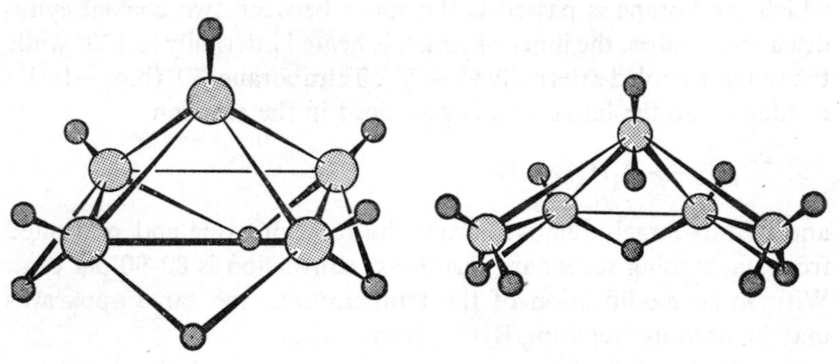

Figure 8.9 *Figure 8.10*

four B—H—B bridge bonds spanning the four boron atoms in the base. In B_5H_{11} (Fig. 8.10), one of these bridge bonds is missing in the base and the boron atoms flanking the open side each carry two terminal B—H bonds, as does the apical boron atom.

The chemistry of pentaborane (9) has been studied in considerable detail. One of its most interesting reactions is that which occurs when its vapour is passed through a silent electrical discharge. Hydrogen is formed together with a new borane, $B_{10}H_{16}$, the structure of which is made up of two pyramidal B_5H_8 units, each with a structure like that of B_5H_9, joined in the apical position by a two-electron B—B bond:

$$2B_5H_9 \rightarrow B_{10}H_{16} + H_2$$

The conversion of B_5H_9 to small *closo*-carboranes by reaction with acetylene is referred to later (Section 8.7). Direct halogenation of B_5H_9 with bromine leads to monosubstitution of hydrogen in the apical position, though substitution in other positions can also occur. Reaction with an alkyl halide in the presence of $AlCl_3$ also leads to monosubstitution in the apical position: the preferred attack by electrophiles at this position has been related to the high electron density at this point in the ground state of the B_5H_9 molecule. Various other alkylated derivatives are, however, also known, and preparative vapour-phase chromatography provides a convenient method for their separation. Several ionic species may also be derived from B_5H_9: lithium metal in liquid ammonia, for example, gives $Li_2B_5H_9$ and reaction with lithium hydride in diethyl ether gives $Li_2B_5H_{11}$.

As already mentioned, few of the reactions of B_5H_{11} have been investigated because of its instability. That with carbon monoxide has been referred to already, giving BH_3CO and B_4H_8CO (page 188). The reaction with ethylene is also consistent with a tendency for B_5H_{11} to break down into B_4H_8 and BH_3. One of the products is the adduct $(C_2H_4)B_4H_8$, for which the bridged structure shown below (Fig. 8.11) has been suggested. The other is ethylpentaborane (11).

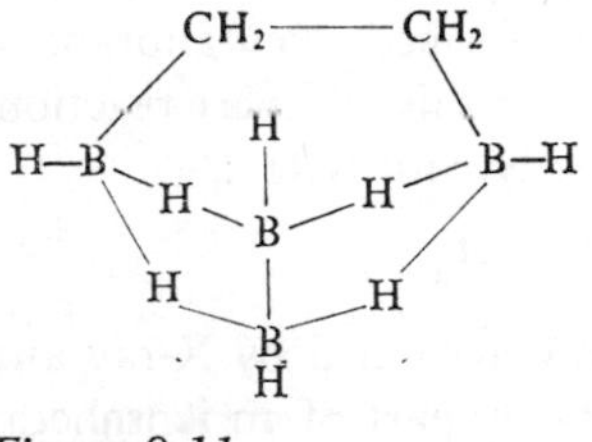

Figure 8.11

Very brief mention may be made of B_6H_{10} and B_6H_{12}, neither of which has been prepared in quantity. Hexaborane (10) (b.p. 108°) was first isolated by Stock from the mixture of boranes resulting from the decomposition of Mg_3B_2 by aqueous acids. It is also produced when B_5H_{11} decomposes in the presence of a weak Lewis base, such as dimethyl ether:

$$2B_5H_{11} \rightarrow B_6H_{10} + 2B_2H_6$$

It is stable at room temperature. The structure has been determined and the boron atoms are found to form a pentagonal pyramid with one hydrogen atom bonded to each by a two-electron 'terminal' bond. In addition there are four B—H—B bridge bonds in the base. The second of these two boranes, B_6H_{12}, is formed in very small yield when a salt of the $B_3H_8^-$ ion reacts with phosphoric acid; the structure is unknown.

The richness of this field can be gauged from the fact that two octaboranes, B_8H_{12} and B_8H_{18}, have also been isolated, though only in small amounts and with little investigation of their chemistry. Nonoborane (15), B_9H_{15}, is also known and may be made by several methods, for example, by reaction of B_2H_6 with B_5H_{11} at 30° at 25 atm:

$$2B_2H_6 + B_5H_{11} \rightarrow B_9H_{15} + 4H_2$$

The structure is known, but will not be described here; there is virtually no information on its reactions.

(c) *Decaborane* (*14*). This borane, a white crystalline solid (m.p. 98·8°; b.p. 213°), has, like B_5H_9, been studied in great detail because of its possible interest in connection with high-energy fuels. Stock's early characterization was completed with quantities of less than one gram, but recent semitechnical preparations have been on a ton scale and have involved heating B_2H_6 in a flow system at 100–150°. As Stock found, this gives a mixture of products and much attention has been directed to optimizing the conditions for $B_{10}H_{14}$ formation. Lewis bases such as dimethyl ether catalyse the conversion, and it is usual to recycle the lower boranes formed. Pentaborane (9) is believed to be a precursor of $B_{10}H_{14}$ in this complex reaction and, indeed, has been shown to react directly with B_2H_6:

$$2B_5H_9 + 5B_2H_6 \rightarrow 2B_{10}H_{14} + 10H_2$$

The structure (Fig. 8.12) has been determined by X-ray analysis, the boron framework being recognizably part of an icosahedron of twelve boron atoms. Fig. 8.13 is a topographical representation of the same molecule, and also illustrates the numbering convention in current use. Only the four B—H—B bridge bonds spanning positions 5,6, 6,7, 8,9 and 9,10 in the open face are shown in Fig. 8.13, the ten terminal B—H bonds, one on each boron atom, being omitted for the sake of clarity. This structure is fully supported by investigations of the ^{11}B and ^{1}H n.m.r. spectra, and characteristic infrared vibrational frequencies at 2640 and 1900 cm^{-1} are associated with the terminal and bridging B—H bonds respectively.

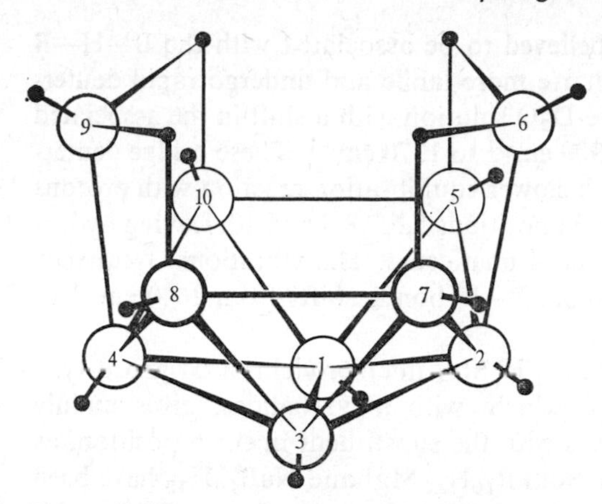

Figure 8.12

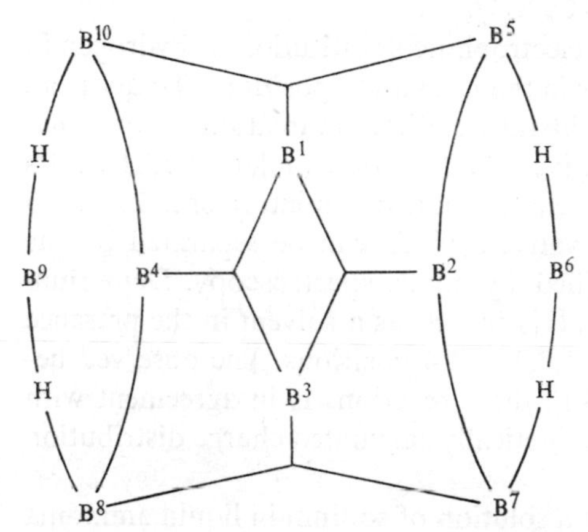

Figure 8.13

In the decaborane (14) molecule there are fifty-four bonding orbitals, four for each boron and one for each hydrogen, and forty-four bonding electrons. Twenty of the latter are used in the ten terminal B—H bonds and eight in the B—H—B bridge bonds, leaving sixteen for bonding in the boron framework.

Decaborane (14) behaves as a strong monoprotic acid and may be titrated with bases such as amines in various solvents. The sodium salt, $NaB_{10}H_{13}$, is formed in the reaction with NaH in diethyl ether. Such salts are found to react with strong acids with regeneration of $B_{10}H_{14}$:

$$B_{10}H_{14} + NaH \longrightarrow NaB_{10}H_{13} + H_2$$

The acidic function is believed to be associated with the B—H—B bridge hydrogens, which are more labile and undergo rapid deuterium exchange in dioxane-D_2O solution with a shift in the associated B—H frequency from 1900 cm^{-1} to 1370 cm^{-1}. These bridge deuterons then undergo a much slower equilibration reaction with protons on terminal B—H bonds in positions 5,6,7,8,9 and 10, leaving hydrogens in positions 1,2,3 and 4 unaffected. The vibrational frequency associated with the terminal B—H bonds shifts from 2640 cm^{-1} to 1900 cm^{-1} on deuteration.

Reaction of $B_{10}H_{14}$ with CH_3MgI in ether yields a Grignard-type compound, $B_{10}H_{13}MgI$, which, with alkyl halides, gives mainly monoalkylated derivatives with the substituent in the 6 position, as shown by n.m.r. studies. Both $B_{10}H_{13}MgI$ and $NaB_{10}H_{13}$ have been used to effect various syntheses. Thus for example the former with $(C_6H_5)_2PCl$ gives $B_{10}H_{13}P(C_6H_5)_2$.

In contrast to the above, electrophilic substitution of hydrogen in $B_{10}H_{14}$ occurs preferentially in the 1,2,3 and 4 positions. To quote an example from the extensive literature, direct iodination or bromination gives 2,4-dihalo derivatives. Methylation under Friedel Crafts conditions with CH_3Br and $AlBr_3$ results in a mixture of 2, 2,4, 1,2,3 and 1,2,3,4 methylated derivatives, which can be separated by gas chromatography and identified by n.m.r. spectroscopy. Deuterium exchange using DCl and $B_{10}H_{14}$ in CS_2 as a solvent in the presence of $AlCl_3$ also occurs in the 1,2,3 and 4 positions. The observed behaviour of decaborane (14) in these reactions is in agreement with what is expected from the theoretically calculated charge distribution in the molecule.

When $B_{10}H_{14}$ reacts with a solution of sodium in liquid ammonia or ether it takes up two electrons to form the $B_{10}H_{14}^{2-}$ ion, which is isolated as the salt $Na_2B_{10}H_{14}$. The same ion is formed when $B_{10}H_{14}$ is added to aqueous KBH_4, and its caesium and rubidium salts are sufficiently stable to be recrystallized from hot water. When an aqueous solution containing the $B_{10}H_{14}^{2-}$ ion is acidified, protonation to $B_{10}H_{15}^-$ takes place and the tetramethylammonium salt of this ion is also isolable.

A number of weakly basic ligands, including acetonitrile, tertiary amines, phosphines and dialkyl sulphides, can displace two atoms of hydrogen from $B_{10}H_{14}$:

$$B_{10}H_{14} + 2L \rightarrow B_{10}H_{12}L_2 + H_2$$

Structural studies on some of the resulting compounds have shown the ligands to be bonded to boron atoms in the 6 and 9 positions. They are thus related to the $B_{10}H_{14}^{2-}$ ion, which has hydrogen atoms bridging the 7,8 and 5,10 positions and BH_2 groups in positions 6 and 9, with one hydrogen in each BH_2 group carrying a negative charge. The chemistry of these various species has been studied in depth, but will not be further considered here.

(d) *Other boranes.* Several other more complex boranes have been isolated, but only two need be referred to in this necessarily brief treatment of the subject. The first, $B_{10}H_{16}$, was mentioned earlier as a derivative of B_5H_9 and has two square pyramidal B_5H_8 units joined, apex to apex, by a B—B bond. The second, $B_{18}H_{22}$, exists in two isomeric forms, which result from the union of two $B_{10}H_{14}$ cages so that they have a B—B edge in common, the boron atoms involved being those in the 5 and 6 positions.

8.6 Polyhedral $(BH)_n^{2-}$ ions[13]

In the preceding section we have described the boranes and various ions derived from them, and we now turn to a further group of ions of higher symmetry in which there is a closed polyhedral framework of linked boron atoms. The general formula is $(BH)_n^{2-}$, where n can have values from 6 to 12. These species have considerable chemical stability which is associated with extensive electron delocalization, and indeed they show some aromatic character in their reactions. The most accessible and best known are $B_{12}H_{12}^{2-}$ and $B_{10}H_{10}^{2-}$, but before discussing them in detail we will give some indication of how the smaller ions are prepared.

Pyrolysis of alkali metal salts of the $B_3H_8^-$ ion in vacuum at 200–230° leads to loss of hydrogen and formation of a mixture of salts of BH_4^-, $B_9H_9^{2-}$, $B_{10}H_{10}^{2-}$ and $B_{12}H_{12}^{2-}$. From this, salts of $B_9H_9^{2-}$ may be separated and air oxidation of the sodium salt in 1,2-dimethoxyethane solution gives $Na_2B_8H_8$ in moderate yield. The B atoms in $B_9H_9^{2-}$ are believed to be arranged as a symmetrical tricapped trigonal prism, while an X-ray study of $[Zn(NH_3)_4]B_8H_8$ shows the B atoms to form a distorted dodecahedron.

Air oxidation of hydrated $Na_2B_8H_8$ in 1,2-dimethoxyethane gives salts of $B_7H_7^{2-}$ and $B_6H_6^{2-}$ as major products. They can be separated and the ^{11}B n.m.r. spectrum of $Cs_2B_7H_7$ shows two doublets with relative areas in the ratio of 5 : 2. A pentagonal bipyramidal structure has been assigned on this basis. X-ray studies have shown the $B_6H_6^{2-}$ ion to be octahedral.

All of these structures are electron-deficient, as may be seen readily for the octahedral $B_6H_6^{2-}$ ion, for example, which has twenty-six electrons, twelve of which form normal two-electron B—H bonds. This leaves fourteen electrons for the eight B—B bonds in the polyhedral framework.

The existence and stability of the $B_{12}H_{12}^{2-}$ ion was predicted on theoretical grounds some five years before a salt was first isolated. There are now several high-yield syntheses from readily available materials; for example,

$$2NaBH_4 + 5B_2H_6 \xrightarrow[100\text{–}180°]{(C_2H_5)_3N} Na_2B_{12}H_{12} + 13H_2$$

The reaction in this case is almost quantitative.

The $B_{11}H_{11}^{2-}$ ion is relatively inaccessible and has been little studied, but $B_{10}H_{10}^{2-}$ salts are formed in good yield in the reaction of $B_{10}H_{14}$ with triethylamine at high temperatures:

$$B_{10}H_{14} + 2R_3N \rightarrow B_{10}H_{12}(R_3N)_2 + H_2$$

$$B_{10}H_{12}(R_3N)_2 \xrightarrow{R_3N} 2R_3NH^+ + B_{10}H_{10}^{2-}$$

The structure of $B_{12}H_{12}^{2-}$ has been established by X-ray analysis as icosahedral (Fig. 8.14). The diagram also shows the numbering

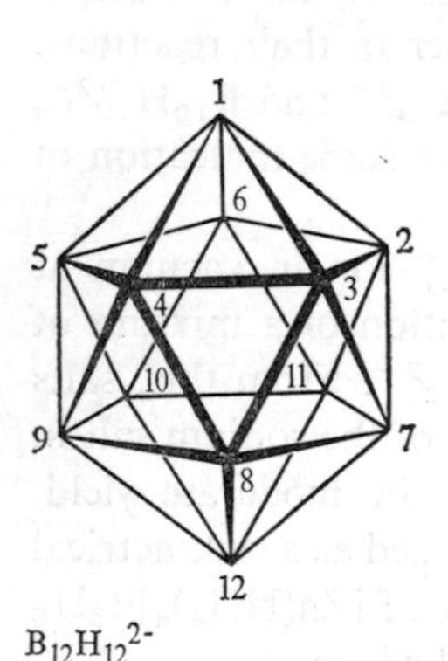

Figure 8.14

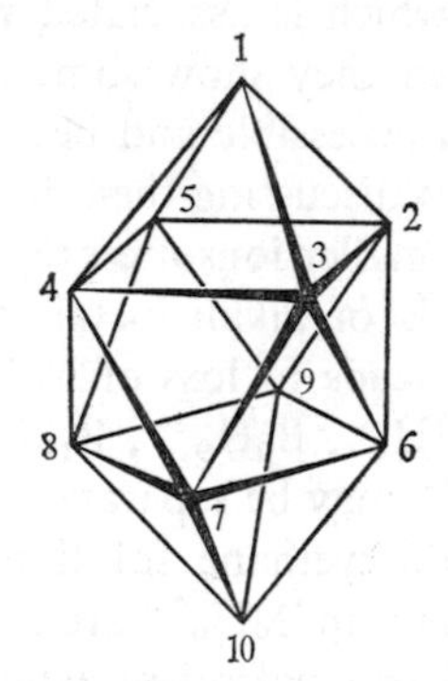

Figure 8.15

system used. In keeping with this structure, the ^{11}B n.m.r. spectrum consists of a single doublet. The $B_{10}H_{10}^{2-}$ ion has a structure (Fig. 8.15) made up of two square pyramids placed base to base with the upper one rotated through 45° about an axis joining the two apical

B atoms (1,10). Here the ^{11}B n.m.r. spectrum is found to consist of two doublets of relative areas 1 : 4.

The chemistry of these two ions has been investigated in great detail, but only the salient points will be mentioned. The alkali metal salts are stable in aqueous sodium hydroxide at 95°, but undergo slow attack by 3M aqueous hydrochloric acid at this temperature with hydrogen evolution. Attack on $B_{12}H_{12}^{2-}$ is the more rapid. Complete exchange of H for D also occurs in acidified D_2O. Passage of salts through a cation exchange resin gives the free acids $(H_3O^+)_2B_{12}H_{12}^{2-}$ and $(H_3O^+)_2B_{10}H_{10}^{2-}$, which are comparable in strength with sulphuric acid. The alkali metal salts are stable in vacuum at temperatures up to at least 500°, and the Ag^+, Cu^+ and Hg^{2+} salts are precipitated from aqueous solution without any reduction to the metal. This may be contrasted with the strong reducing properties of some of the boranes.

Both ions may be partially or, ultimately, completely halogenated by the direct reaction of chlorine, bromine or iodine in aqueous or alcoholic solution, and acids can be prepared from the halogenated ions. The halogenated ions are also unattacked by strong acids or bases. Halogenation will also occur with hydrogen fluoride or elementary fluorine, though the latter produces some degradation. The final product of the reaction of $Na_2B_{10}H_{10}$ with fluorine in aqueous solution is $Na_2B_{10}F_9(OH)$.

Many other reactions of these two ions have been described, most of them involving electrophilic replacement of hydrogen in the polyhedral framework, and a number of positional isomers have been identified. In the main, detailed consideration of this work is outside the scope of this book, though one group of reactions leading to the formation of diazonium and carbonyl derivatives is of special interest and will be mentioned briefly. The $B_{10}H_{10}^{2-}$ ion reacts with nitrous acid to form an unidentified explosive intermediate and this, with $NaBH_4$ in methanol, gives 1,10-$B_{10}H_8(N_2)_2$, which is sufficiently stable to be sublimed in vacuum at 90–100°. The nitrogen may be replaced by ligands such as amines or nitriles, and the diazonium derivative also reacts with carbon monoxide at 120–140° giving 1,10-$B_{10}H_8(CO)_2$ as the main product.

This neutral carbonyl derivative undergoes a number of interesting reactions, some of which are shown overleaf.

Reaction with chlorine in water first gives a chlorinated carboxylic acid which on thermal dehydration forms the product shown. Salts of the acid may be isolated. The neutral carbonyl $B_{12}H_{10}(CO)_2$ is

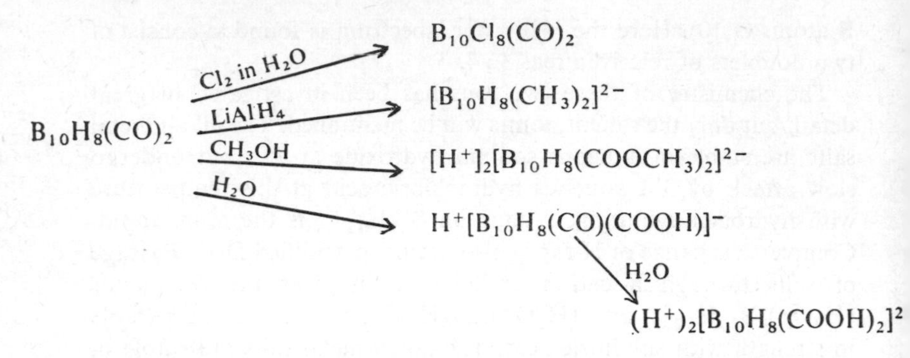

obtained in the reaction of $B_{12}H_{12}{}^{2-}$ with CO under acidic conditions, and has similar reactions.

8.7 Carboranes and metallocarboranes[14–19]

The carboranes are mixed hydrides of carbon and boron having both carbon and boron atoms in an electron-deficient skeletal framework. They are of two main types, the *closo*-carboranes with the general formula $C_2B_{n-2}H_n$ ($n = 5$–12) and the *nido*-carboranes with open-cage structures derived formally from one or other of several boranes, and containing from one to four carbon atoms in the skeleton. In addition, there are a number of carboranes with an additional hetero-atom such as phosphorus built into the basic structure and a fairly numerous family of metallocarboranes, some of which are similar to ferrocene. This field has developed very rapidly in the last few years and only its salient features can be described here; for a more detailed treatment the reader is referred to recent monographs.

The nomenclature most commonly used gives first the position and number of carbon atoms, then the type of carborane (*closo* or *nido*) and finally the name of the borane from which the compound is formally derived and the number of hydrogen atoms. Thus the three isomers of $C_2B_{10}H_{12}$ are 1,2-, 1,7- and 1,12-dicarba-*closo*-dodecaborane (12), and CB_5H_9 is monocarba-*nido*-hexaborane (9). Atoms in the cage structure are numbered consecutively, starting from that in the apical position and proceeding through successive rings in a clockwise direction (see, for example, Fig. 8.17). This is important in naming isomers arising either from different positions of the carbon atoms in the framework or different positions of substituents on carbon and/or boron.

8.7.1. The *nido*-carboranes. The small *nido*-carboranes are usually prepared by reaction of a borane with an acetylene under mild conditions. With more drastic conditions (e.g. at a higher temperature or in an electric discharge) the same reactants yield *closo*-carboranes. The borane–alkyne reactions usually give mixtures of products, which include methylated carboranes when acetylene itself is used. The boranes most commonly used in making the smaller carboranes are B_4H_{10}, B_5H_9 and B_5H_{11}. Thus, for example, B_5H_9 and C_2H_2 react in the gas phase at 215° to give mainly the *nido*-carborane 2,3-$C_2B_4H_8$, together with methyl derivatives of CB_5H_9. The same reactants at 450°, or in an electrical discharge, give the *closo*-carboranes 1,5-$C_2B_3H_5$, 1,6-$C_2B_4H_6$ and 2,4-$C_2B_5H_7$. The *nido*-carborane 2,3-$C_2B_4H_8$ is converted to the *closo*-carboranes $C_2B_3H_5$, $C_2B_4H_6$ and $C_2B_5H_7$ on pyrolysis or ultraviolet irradiation. Another example of a preparative reaction is that between C_2H_2 and B_4H_{10} at 25–50° in the gas phase, from which 2,3-$C_2B_4H_8$, 1,2-$C_2B_3H_7$ and methyl derivatives of 2-CB_5H_9 and 2,3,4-$C_3B_3H_7$ result.

These reactions are clearly very complex and difficult to systematize, but the structures of four of the *nido*-carboranes are closely related to that of B_6H_{10}, as shown below (Fig. 8.16). The structures

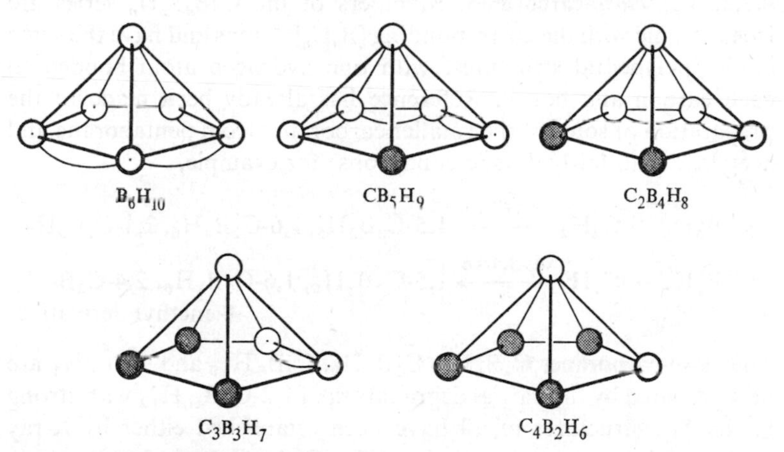

Figure 8.16 The *nido*-carboranes formally related to B_6H_{10}. All have eight pairs of electrons bonding the six cage atoms together. Hydrogen bridges are represented by curved lines

of $C_2B_4H_8$ and B_6H_{10} are based on X-ray analysis and those of the remainder mainly on ^{11}B n.m.r. spectroscopy. In the diagrams, hydrogen bridges are shown by curved lines, but terminal B—H and C—H bonds are omitted. It can be seen that the introduction of successive carbon atoms to the framework involves the elimination of one bridge hydrogen atom and one B—H (i.e. the replacement of BH_2 by an isoelectronic CH unit). Like all the carboranes these compounds are electron-deficient, with multicentre bonds and delocalization extending over the entire framework. In much the same way, $C_2B_3H_7$ has a square pyramidal structure that is formally derived from that of B_5H_9, with two BH_2 replaced by 2CH.

Largely because of preparative difficulties, relatively little is known about the reactions of the smaller *nido*-carboranes. They are only moderately stable to heat and are less resistant to hydrolysis and oxidation in air than the *closo* species. Halogen substitutions have been observed, as has the formation of anions; for example,

$$C_2B_4H_8 + NaH \xrightarrow{\text{diglyme}} Na^+C_2B_4H_7^- + H_2$$

The larger *nido*-carboranes (e.g. $C_2B_9H_{13}$) are produced by the degradation of large *closo*-carboranes by bases and will be referred to later in this connection.

8.7.2. The *closo*-carboranes. Members of the $C_2B_{n-2}H_n$ series are isoelectronic with the corresponding $[B_nH_n]^{2-}$ ions and have the same closed polyhedral structures, with one hydrogen atom bonded to each carbon and boron. Reference has already been made to the preparation of some of the smaller carboranes from pentaborane and acetylene with fairly drastic conditions; for example,

$$B_5H_9 + C_2H_2 \xrightarrow{490^\circ} 1,5\text{-}C_2B_3H_5,\ 1,6\text{-}C_2B_4H_6,\ 2,4\text{-}C_2B_5H_7$$

$$B_2H_6 + C_2H_2 \xrightarrow{\text{discharge}} 1,5\text{-}C_2B_3H_5,\ 1,6\text{-}C_2B_4H_6,\ 2,4\text{-}C_2B_5H_7 + \text{methyl derivatives}$$

The *closo*-carboranes $C_2B_6H_8$, $C_2B_7H_9$, $C_2B_8H_{10}$ and $C_2B_9H_{11}$ are best prepared by the partial degradation of 1,2-$C_2B_{10}H_{12}$ with strong bases. The structures of all have been established either by X-ray studies or from spectroscopic data. The C_2B_3, C_2B_4 and C_2B_5 *closo*-carboranes, for example, have trigonal bipyramidal, octahedral and pentagonal bipyramidal skeletal structures respectively, and positional isomers have been identified. The chemical reactions, in so far as they are known, are very similar to those of $C_2B_{10}H_{12}$, which are described below. Various substitution reactions have been studied and the hydrogen atoms bonded to carbon are weakly acidic.

The best known of all the carboranes, 1,2-dicarba-*closo*-dodecaborane (12), $C_2B_{10}H_{12}$, is prepared in good yield by the reaction of $B_{10}H_{14}$ with acetylene in the presence of a Lewis base such as acetonitrile, a dialkyl sulphide or an alkylamine. Decaborane (14) and the base (L) alone react with elimination of hydrogen, forming $B_{10}H_{12}L_2$ (see Section 8.5). This may be preformed and then treated with acetylene, but this step is not necessary. The essential reaction is

$$B_{10}H_{12}(CH_3CN)_2 + HC{\equiv}CH \longrightarrow C_2B_{10}H_{12} + 2CH_3CN + H_2$$

In effect, the acetylene molecule is inserted across the open face of the decaborane cage, which is where the CH_3CN is also attached, the product having an icosahedral molecular framework with one atom of hydrogen bonded to each boron and each carbon. The reaction goes equally well with a substituted acetylene and also with a diacetylene:

$$2B_{10}H_{14} + HC{\equiv}C{-}C{\equiv}CH \xrightarrow{CH_3CN} \underset{B_{10}H_{10}}{HC{-}C}{-}\underset{B_{10}H_{10}}{C{-}CH}$$

The icosahedral structure (Fig. 8.17) has been established for brominated derivatives by X-ray analysis and ^{11}B n.m.r. spectroscopy, and shows the carbon atoms to be in the 1 and 2 positions. The other two isomers are produced by heating, as shown below:

$$1{,}2\text{-}C_2B_{10}H_{12} \xrightarrow{480\text{–}500°} 1{,}7\text{-}C_2B_{10}H_{12} \xrightarrow{600°} 1{,}12\text{-}C_2B_{10}H_{12}$$

The three isomers are sometimes referred to as the *ortho*, *meta* and *para* forms.

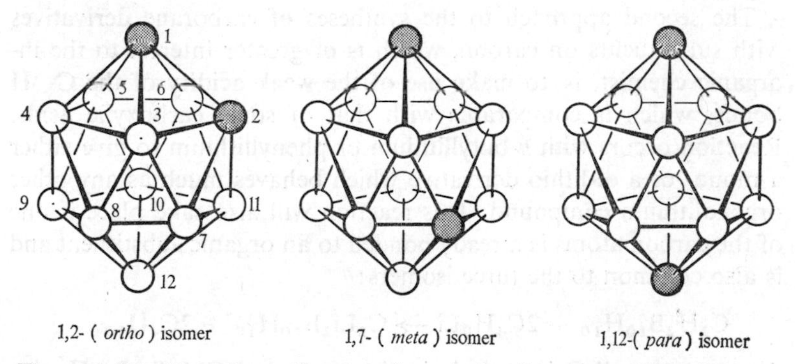

Figure 8.17 The three dicarba-*closo*-dodecaboranes, $C_2B_{10}H_{12}$

Conversion of the 1,2 to the 1,7 form is almost quantitative, but formation of the 1,12-isomer occurs with a yield of only some 20 per cent, and is accompanied by considerable degradation. The icosahedral structure is similar to that of $B_{12}H_{12}^{2-}$ (Fig. 8.14) and is electron-deficient, with electron delocalization extending over the whole framework. It is thus in effect a three-dimensional aromatic molecule, with marked electron withdrawing character, the most important result of which is to render the two hydrogen atoms bonded to carbon acidic. All the C—H and B—H bonds are of the normal two-electron type and the electron deficiency is associated with the framework, in which there are multicentre bonds.

All three of the icosahedral isomers are stable both to heat and to chemical attack, and much more so than decaborane (14). They are white crystalline solids which resist both strong oxidizing agents and strong reducing agents and are also stable to hydrolysis. This is important because it allows reactions to be carried out on substituents, often under quite drastic conditions, without destroying the cage structure, rather as the chemistry of derivatives of an aromatic ring such as benzene can be developed without destroying the ring.

Most chemical studies have been concerned with substituents on the two carbon atoms. These may be introduced in the first place by employing a substituted acetylene in the carborane syntheses. Such groups as C-alkyl, -haloalkyl, -aryl, -alkenyl and -alkynyl may be introduced into the structure in this way. Further reactions on the substituent groups may then be carried out by the usual synthetic methods of organic chemistry to give, for example, carboxylic acid, ester, alcohol, ketone, amine or unsaturated groups in the side chain. This aspect of carborane chemistry is primarily concerned with organic derivatives and will not be discussed here.

The second approach to the syntheses of carborane derivatives with substituents on carbon, which is of greater interest to the inorganic chemist, is to make use of the weak acidity of the C—H bonds, which is comparable with that of some carboxylic acids. Reaction occurs with *n*-butyllithium or phenyllithium to give either a mono- or a di-lithio derivative which behaves much as any other organolithium compound. This reaction will also take place if one of the carbon atoms is already bonded to an organic substituent and is also common to the three isomers:

$$C_2H_2B_{10}H_{10} + 2C_4H_9Li \rightarrow C_2Li_2B_{10}H_{10} + 2C_4H_{10}$$

A mono- or di-Grignard derivative (e.g. $CH(CMgBr)B_{10}H_{10}$ or

$(CMgBr)_2B_{10}H_{10}$) may also be made in the usual way from an alkyl magnesium halide and the carborane. The lithium compounds are useful in preparing carboranes with organic substituents, as shown by the examples below.

$$LiC{-}CLi\ (B_{10}H_{10})$$

$$\xrightarrow{NOCl} ONC{-}CNO\ (B_{10}H_{10})$$

$$\xrightarrow{(CH_2)_2O} HO(CH_2)_2C{-}C(CH_2)_2OH\ (B_{10}H_{10})$$

$$\xrightarrow{CH_2{=}CHCH_2Br} H_2C{=}CHCH_2C{-}CCH_2CH{=}CH_2\ (B_{10}H_{10})$$

$$\xrightarrow{(a)\ CO_2\ (b)\ H^+} HOOCC{-}CCOOH\ (B_{10}H_{10})$$

Reactions of the normal type involving lithium halide elimination have also been observed with a number of non-metallic halides, though not all of the halogen may be eliminated because of steric factors; for example,

$$PCl_3 + C_2PhLiB_{10}H_{10} \rightarrow (C_2PhB_{10}H_{10})_2PCl$$

$$2(C_6H_5)_2PCl + C_2Li_2B_{10}H_{10} \rightarrow (C_6H_5)_2P{-}C{-}C{-}P(C_6H_5)_2\ (B_{10}H_{10})$$

$$\xrightarrow{Ni(CO)_4} (OC)_2Ni \leftarrow (C_6H_5)_2PC{-}CP(C_6H_5)_2\ (B_{10}H_{10})$$

It will be noted in the second of these reaction sequences that the phosphorus atom retains its donor function. Very similar lithium halide elimination reactions occur with the halides of Si, Ge, Sn, As and Sb. Special attention has been given to silicon halides because of the possibility of incorporating the stable carborane units into silicone polymers either as pendant groups or as part of the silicone chain.

Some elastomers with very high thermal stability have been produced, but their applications are likely to be of a highly specialized nature. Dimethylsilicon dichloride reacts with the 1,2-di-lithio carborane as follows:

$$\underset{B_{10}H_{10}}{LiC\!-\!CLi} \xrightarrow{(CH_3)_2SiCl_2} \underset{B_{10}H_{10}}{Cl(CH_3)_2SiC\!-\!CSi(CH_3)_2Cl} \xrightarrow{H_2O} (CH_3)_2Si\overset{O}{\frown}Si \text{ bridged by } \underset{B_{10}H_{10}}{C\!-\!C}$$

It will be seen that hydrolysis of the initial product gives an exocyclic ring rather than a chain structure with Si—O—Si bonds. There are small differences in the reactivity of C—H bonds on the three isomers which arise from differences in the electron density distribution in the cage structure. Exocyclic rings are also formed less readily, if at all, from the 1,7 and 1,12 positions.

Various mercurials have been prepared in the normal way; for example,

$$\underset{B_{10}H_{10}}{RC\!-\!CLi} \xrightarrow{HgCl_2} \underset{B_{10}H_{10}}{RC\!-\!C}\!-\!Hg\!-\!\underset{B_{10}H_{10}}{C\!-\!CR}$$

A few σ-bonded derivatives of other metals are also known; for example,

$$Ph_3PAuCl + C_2RLiB_{10}H_{10} \rightarrow Ph_3PAuC(CR)B_{10}H_{10} + LiCl$$

Very much less interest attaches to B-alkyl and B-aryl derivatives of $C_2B_{10}H_{12}$, though a number are known. The B-halo derivatives do, however, have some very unusual properties. The icosahedral skeleton is sufficiently stable to resist breakdown during halogenation. When the 1,2-carborane is treated with halogen in the presence of an aluminium halide, the first B—H bonds to be attached are those furthest from carbon (i.e. in the 9 and 12 positions for 1,2-$C_2B_{10}H_{12}$). This is followed by attack on positions 8 and 10, but reaction does not go further. Only by photochemical chlorination is it possible to convert all ten B—H bonds to B—Cl. The same degree of halogenation is obtained with elementary fluorine in liquid hydrogen fluoride:

$$C_2H_2B_{10}H_{10} \xrightarrow[0^\circ]{HF} C_2H_2B_{10}F_{10}$$

The C—H bonds in these products retain their acidity, giving lithium derivatives, direct halogenation of which gives the fully halogenated species $C_2B_{10}X_{12}$. The fully fluorinated compounds are readily hydrolysed by water, but the other perhalo derivatives are not, and require concentrated alkali. Otherwise all of these compounds are thermally and chemically very stable, resembling the $B_{12}X_{12}^{2-}$ ions in this respect.

8.7.3. Degradation of the icosahedral carborane cage. Both 1,2- and 1,7-$C_2B_{10}H_{12}$, though not the 1,12-isomer, are degraded by strong bases such as the methoxide ion:

$$\text{1,2- or 1,7-}C_2B_{10}H_{12} + CH_3O^- + 2CH_3OH \xrightarrow[40°]{KOH,\ CH_3OH} \text{1,2- or 1,7-}C_2B_9H_{12}^- + B(OCH_3)_3 + H_2$$

Attack occurs at one or other of the equivalent boron atoms 3 and 6 in the 1,2-isomer and at 2 or 3 in the 1,7-isomer, where greatest positive charge is concentrated. The resulting ion if B(3) is removed has an open pentagonal face (Fig. 8.18) with adjacent carbons (1,2) in the first instance and non-adjacent carbons (1,7) in the second. These isomers may be reversibly protonated to yield the *nido*-carboranes, 1,2- and 1,7-$C_2B_9H_{13}$. At 100° both lose hydrogen and give the *closo*-carborane 1,8-$C_2B_9H_{11}$. The most important reaction of the $C_2B_9H_{12}^-$ ion is, however, that with NaH:

$$C_2B_9H_{12}^- + NaH \rightarrow Na^+C_2B_9H_{11}^{2-} + H_2$$

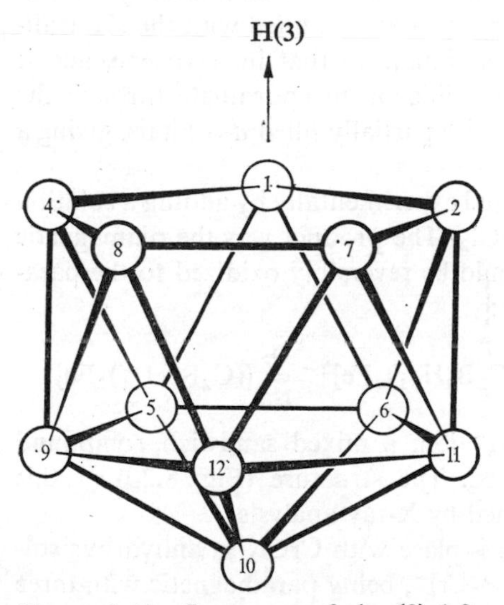

Figure 8.18 Structure of the (3)-1,2- and (3)-1,7-$C_2B_9H_{12}^-$ ions, showing numbering of cage atoms

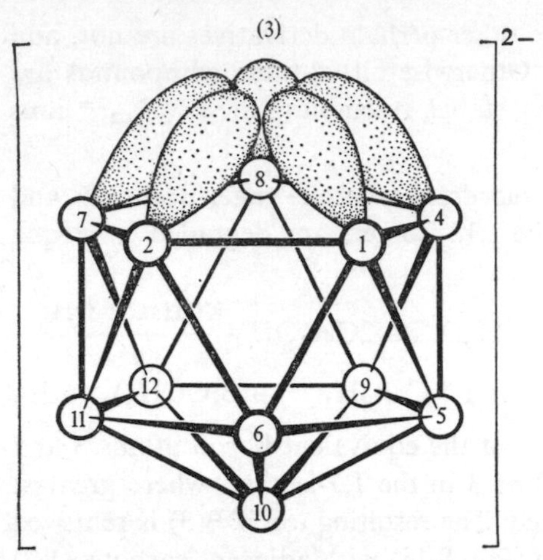

Figure 8.19 Structure of the (3)-1,2- and 1,7-$C_2B_9H_{11}^{2-}$ ions, showing approximately sp^3-orbitals directed at the vacant (3) vertex of the icosahedron

The $C_2B_9H_{11}^{2-}$ ion produced retains the open pentagonal face (Fig. 8.19) and, in it, there are five approximately sp^3-orbitals directed towards the vacant vertex of the icosahedron. These five atomic orbitals combine to give three bonding and two antibonding molecular orbitals. When the bonding orbitals are filled with the six available electrons the situation is similar to that in ferrocene, and it becomes possible for the dicarbollide ion to coordinate through the open face to transition metals with partially filled *d*-orbitals, giving a sandwich compound.

This possibility was first tested experimentally by adding a solution containing $C_2B_9H_{11}^{2-}$ to $FeCl_2$. The product was the diamagnetic ferrocene analogue, which could be reversibly oxidized to the paramagnetic iron (III) derivative:

$$2C_2B_9H_{11}^{2-} + Fe^{2+} \rightarrow [(C_2B_9H_{11})_2Fe]^{2-} \underset{Na}{\overset{O_2}{\rightleftharpoons}} [(C_2B_9H_{11})_2Fe]^-$$

In the presence of the $C_5H_5^-$ ion, a mixed sandwich compound $(C_5H_5)Fe(C_2B_9H_{11})$ is formed. The structure (Fig. 8.20) of this compound has been established by X-ray analysis.

A very similar reaction takes place with $CrCl_3$ in anhydrous solvents, the product, $[(C_2B_9H_{11})_2Cr]^-$, being paramagnetic with three unpaired electrons. Reaction with $BrMn(CO)_5$ or $BrRe(CO)_5$ is believed to form a σ-bonded complex initially, but this loses CO when

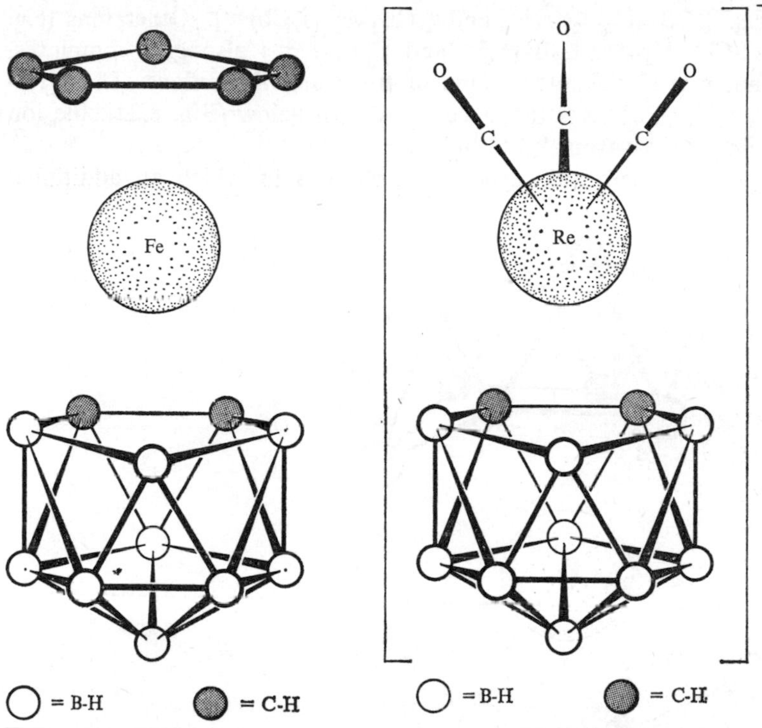

Figure 8.20 Structure of $(\pi\text{-}(3)\text{-}1,2C_2B_9H_{11})Fe(\pi\text{-}C_5H_5)$

Figure 8.21 Structure of the $[\pi\text{-}(3)\text{-}1,2\text{-}C_2B_9H_{11}]Re(CO)_3^-$ ion

refluxed in tetrahydrofuran and leaves $(C_2B_9H_{11})M(CO)_3^-$ (M = Mn or Re). The structure of the rhenium complex (Fig. 8.21) has been confirmed. There is a similar reaction with the hexacarbonyls of Cr, Mo and W under the influence of ultraviolet light, and the air-sensitive products are of the type $(C_2B_9H_{11})M(CO)_3^{2-}$ (M = Cr, Mo, W). Closely related complexes of other transition metals (Co, Ni, Pd, Cu and Au) have also been made, including some with substituents on the ion.

This subject has now become exceedingly complex and the reader is referred to the literature for a full treatment. It must suffice here to mention briefly the main additional types that are known, without any attempt at a detailed description. In the first place formation of π-bonded complexes based on carborane structures is not restricted to the $C_2B_9H_{11}^{2-}$ ion; there are a number formed on the same principle by $CB_{10}H_{11}^{3-}$ and some of its amine-substituted derivatives (e.g. $[(CB_{10}H_{11})_2Cr]^{3-}$ and $[(CB_{10}H_{11})_2Co]^{3-}$). Other ions (e.g. 1,2-$C_2B_8H_{10}^{4-}$, $C_2B_7H_9^{2-}$ and $C_2B_4H_6^{2-}$) also give complexes, and, it may be noted, some of these are *nido*-anions. Thus [1,6-$C_2B_7H_9)_2Co]^-$ has the structure shown below (Fig. 8.22), the ion being derived from 1,3-$C_2B_7H_{13}$.

Finally there are carborane structures in which an additional

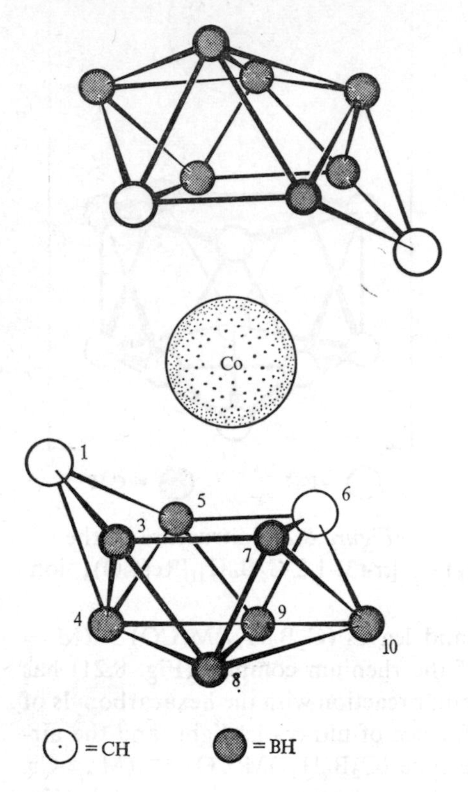

Figure 8.22 Structure of the (π-1,6-$C_2B_7H_9)_2Co^-$ ion

heteroatom is incorporated in the skeletal framework, and a few of these are also known to give π-bonded complexes. The heteroatoms capable of incorporation in the cage structure are Be, Al, Ga, Ge, Sn, Pb, P, As and Sb. As a first example, the reaction of $C_2B_9H_{11}^{2-}$ with $SnCl_2$ in refluxing benzene may be taken. In the product, $SnC_2B_9H_{11}$, the tin atom occupies a position above the open pentagonal face of the $C_2B_9H_{11}$ structure, thereby completing the icosahedron. Lead and germanium analogues are known.

Trimethylgallium reacts with *nido*-2,3-$C_2B_4H_8$ in the gas phase at 180–215° to give the volatile galladicarba-*closo*-heptaborane (7), $(CH_3)GaC_2B_4H_6$. An atom of phosphorus has also been inserted into the icosahedral structure by reaction of $Na_3CB_{10}H_{11}$ with PCl_3 to give $CPB_{10}H_{11}$, in which phosphorus is *ortho* to carbon (i.e. 1,2). This rearranges at 485° to what is believed to be the 1,7-isomer, corresponding with 1,7-$C_2B_{10}H_{12}$. These products may be degraded to the 1,2- and 1,7-$CPB_9H_{10}^{2-}$ (carbaphosphollide) ions, which have an open pentagonal face like $C_2B_9H_{11}^{2-}$, in which there is one atom of carbon and of phosphorus; they too form π-complexes with several transition metals (e.g. the $[(CPB_9H_{10})_2Fe]^{2-}$ ion).

8.8 Hydrides of aluminium, gallium, indium and thallium[2, 20]

Boron stands alone among the Group III elements in forming a large number of hydrides and hydride derivatives. Transient monohydrides (AlH, GaH, InH and TlH) have been observed spectroscopically, and the ions AlH_3^+ and $Al_2H_6^+$ can be detected with the time-of-flight mass spectrometer in the gas obtained when aluminium is evaporated from a tungsten filament at 1090–1170° into hydrogen at 0·3 mm pressure. Otherwise the only binary hydrides known are polymeric forms of AlH_3 and GaH_3. The existence of indium and thallium hydrides is not established, and they would be expected to be unstable in view of the ready decomposition of gallane.

Polymeric alane, $(AlH_3)_n$, is most conveniently prepared by treating $AlCl_3$ with LiH in ether, or when $AlCl_3$ is added to $LiAlH_4$ in ether:

$$AlCl_3 + 3LiH \rightarrow AlH_3 + 3LiCl$$
$$AlCl_3 + 3LiAlH_4 \rightarrow 4AlH_3 + 3LiCl$$

The solution remains clear for a short time and the monomer may be stabilized as an ether adduct, but the white solid polymer deposits rapidly and is difficult to free from solvent without decomposition. The existence of several crystalline forms of the alane polymer has been reported recently, though without details of the method of preparation. One form contains alternate planes of aluminium and hydrogen in the lattice, which are held together by three-centre Al—H—Al bonds. Each metal atom is surrounded by a slightly distorted octahedral array of hydrogens, three in the plane above and three in that below. The fact that aluminium has a coordination number of six, compared with four for boron, is probably at the root of the differences between borane and alane chemistry.

Decomposition of $(AlH_3)_n$ to metal and hydrogen occurs at 160–200°. It is a reducing agent which is very similar in reactivity to $LiAlH_4$. Otherwise the known chemistry relates mainly to adduct formation by AlH_3 with Lewis bases. Its reaction with LiH to form $LiAlH_4$ is referred to later. Amines, which are strong donors, give 1 : 1 and 2 : 1 adducts (e.g. $(CH_3)_3NAlH_3$ and $[(CH_3)_3N]_2AlH_3$), the second of which melts at 95° and is sufficiently stable to be sublimed in vacuum; the first is a dimer. The compounds $AlH_2Cl.O(CH_3)_2$ and $AlHCl_2.O(CH_3)_2$ are formed on adding $AlCl_3$ to a solution of AlH_3 in $(CH_3)_2O$. When activated aluminium powder suspended in tetrahydrofuran is heated with triethylenediamine under a hydrogen pressure of 350 atm, there is a direct synthesis of the adduct as shown.

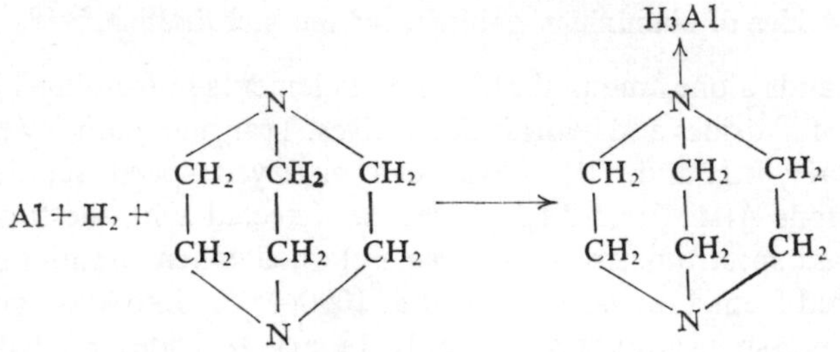

$$\begin{array}{ccc} & & H_3Al \\ & & \uparrow \\ & N & N \\ & CH_2 \quad CH_2 \quad CH_2 & CH_2 \quad CH_2 \quad CH_2 \\ Al + H_2 + & | \qquad | \qquad | \quad \longrightarrow & | \qquad | \qquad | \\ & CH_2 \quad CH_2 \quad CH_2 & CH_2 \quad CH_2 \quad CH_2 \\ & N & N \end{array}$$

Early reports of the preparation of digallane (Ga_2H_6), by the reaction of trimethylgallium with hydrogen in the electrical discharge and subsequent disproportionation of the product in the presence of triethylamine, have not been confirmed. Lithium gallium hydride is, however, well known and results from the reaction of gallium trichloride with lithium hydride in ether:

$$GaCl_3 + 4LiH \xrightarrow{-80^\circ} LiGaH_4 + 3LiCl$$

It decomposes slowly at room temperature to LiH, Ga and H_2 and bears a general resemblance to $LiAlH_4$, but is less reactive as a reducing agent. Very unstable tetrahydridogallates of other elements can also be prepared; for example,

$$AgClO_4 + LiGaH_4 \xrightarrow{\text{ether}} AgGaH_4 + LiClO_4$$

The adducts of GaH_3 with Lewis acids are very similar to those formed by AlH_3. The stable volatile 1 : 1 trimethylamine adduct is formed in the following reaction, and undergoes several ligand replacement reactions:

$$Me_3NHCl + LiGaH_4 \longrightarrow Me_3NGaH_3 + LiCl + H_2$$

Adducts of GaH_3 with phosphines, arsines, ethers and dialkyl sulphides are also known, as well as adducts of the chlorogallanes. The trimethylaminegallane adduct reacts with boron trifluoride to give gallane itself, a viscous liquid with a melting point of −15° which is believed to be polymeric, though it is insoluble in non-coordinating solvents so that the molecular weight cannot be determined.

$$Me_3NGaH_3 + BF_3 \xrightarrow{-15^\circ} GaH_3 + Me_3NBF_3$$

The compound decomposes to Ga and H_2 at room temperature but, interestingly, $GaHCl_2$, made by the reaction

$$Me_3SiH + GaCl_3 \rightarrow Me_3SiCl + GaHCl_2$$

is stable to 150°, when it decomposes to the dichloride and hydrogen.

There are reports of the preparation of $LiInH_4$, $LiTlH_4$ and an InH_3 polymer by methods exactly parallel to those used for $LiAlH_4$ and $(AlH_3)_n$; InH_3 is also reported to give InH. All of these substances, except the last, decompose at or below room temperature, and the observations on InH need confirmation.

By far the best known of the aluminium–hydrogen compounds are the tetrahydroaluminates,[21] which contain the AlH_4^- moiety and are the analogues of the tetrahydroborates. The lithium compound, commonly known as lithium aluminium hydride, was made originally by the reaction of LiH with $AlCl_3$ in ether:

$$4LiH + AlCl_3 \rightarrow LiAlH_4 + 3LiCl$$

It is fairly soluble in several ethers and the diethyl ether solution has a specific conductivity of 5×10^{-5} ohm^{-1} cm^{-1}, consistent with some dissociation to Li^+ and AlH_4^-.

The lithium and sodium compounds may also be synthesized directly in high yield from the alkali metal, aluminium and hydrogen in an ether solvent (e.g. tetrahydrofuran or monoglyme):

$$M + Al + 2H_2 \xrightarrow[\text{60–300 atm at 120–150}^\circ]{\text{ether}} MAlH_4$$

This method has the advantage that all of the lithium is converted to the desired product. The compound Na_3AlH_6, which is isomorphous with Na_3AlF_6, can also be made in a direct high-pressure synthesis from Na, Al and H_2. The lithium compound decomposes above about 150° to LiH, Al and H_2. Its reaction with water is violent and quantitative:

$$LiAlH_4 + 4H_2O \rightarrow LiOH + Al(OH)_3 + 4H_2$$

In higher ethers under controlled conditions, NH_3, PH_3 and AsH_3 all liberate hydrogen from $LiAlH_4$:

$$LiAlH_4 + 4MH_3 \rightarrow Li[Al(MH_2)_4] + 4H_2 \quad (M = N, P, As)$$

These products give NH_3, PH_3 or AsH_3 with water, but the phosphine derivative also gives ethylphosphine in high yield in its reaction with C_2H_5I. Reaction of $LiAlH_4$ with alcohols results in replacement of H by alkoxy groups to form $Li[Al(OR)_4]$, and, with certain protonic acids, hydrogen is again eliminated; for example,

$$LiAlH_4 + 4HCN \rightarrow LiAl(CN)_4 + 4H_2$$
$$LiAlH_4 + 4HN_3 \rightarrow LiAl(N_3)_4 + 4H_2$$

Reaction of $LiAlH_4$ with non-metallic hydrides in ether represents the standard method for preparing hydrides (e.g. $SiCl_4 \rightarrow SiH_4$; $BCl_3 \rightarrow B_2H_6$). Some metallic halides are converted to the corresponding tetrahydroaluminates, though some of these decompose very readily (e.g. $SnCl_4 \rightarrow Sn(AlH_4)_4$, decomposition at $-40°$; $MgCl_2 \rightarrow Mg(AlH_4)_2$, decomposition at 140°). The well-known applications of lithium aluminium hydride as a reducing agent in organic chemistry will not be described here.

8.9 Hydrides of silicon, germanium, tin and lead[2, 22–25]

These four elements form hydrides of the type M_nH_{2n+2}, which correspond with the paraffins. They decrease in thermal stability in going from silicon to lead, so that PbH_4 in fact cannot be isolated. Silanes up to Si_6H_{14} are, however, known and some of the isomeric forms have been identified. Similarly germanes up to Ge_5H_{12} have been characterized, and mass spectrometry and vapour-phase chromatography show that in each series a number of the higher members are formed, though they have yet to be separated. The boiling points of some of the simpler hydrides are shown below.

SiH_4	−111·9°	GeH_4	−88·4°	SnH_4	−52°
Si_2H_6	−14°	Ge_2H_6	29°		
Si_3H_8	53°	Ge_3H_8	110·5°		
Si_4H_{10}	108°	Ge_4H_{10}	176·9°		

No hydrides with multiple bonds, corresponding with olefinic, acetylenic or aromatic hydrocarbons, are formed, and the solid hydrides with ill-defined empirical formulae such as SiH_2 or SiH are involatile and certainly polymeric. It has been suggested that the failure of silicon to form multiple bonds of the $p_\pi - p_\pi$ type is associated with its larger size ($r = 1{\cdot}17$ Å), compared with that of carbon ($r = 0{\cdot}77$ Å), which is too great for effective sideways overlap of the orbitals needed to give a strong π-bond between two silicon atoms.

There is a progressive decrease in the M—H and M—M bond energy in going from carbon to silicon and heavier elements, as is shown below.

C—H	414 kJ (99 kcal) mole^{-1}
Si—H	318 kJ (76 kcal) mole^{-1}
Ge—H	289 kJ (69 kcal) mole^{-1}
Sn—H	251 kJ (60 kcal) mole^{-1}
Pb—H	205 kJ (49 kcal) mole^{-1}
C—C	347 kJ (83 kcal) mole^{-1}
Si—Si	226 kJ (54 kcal) mole^{-1}
Ge—Ge	188 kJ (45 kcal) mole^{-1}
Sn—Sn	151 kJ (36 kcal) mole^{-1}
Pb—Pb	— —

This is reflected in the increased ease of thermal dissociation of the hydrides and also in the inability of the heavier elements in the group to form higher hydrides. The relationship between bond energy and observed stability is not, however, a simple one and kinetic factors are also involved.

The classical method for the preparation of silanes and germanes is by decomposing magnesium silicide or germanide with aqueous hydrochloric acid in an oxygen-free atmosphere. Stock, in his pioneer work on the silanes, found this reaction to yield a mixture of products from which he isolated approximately 40 per cent SiH_4, 30 per cent Si_2H_6, 15 per cent Si_3H_8 and 10 per cent Si_4H_{10}; a small residue of higher hydrides was not separated. Dennis prepared GeH_4, Ge_2H_6 and Ge_3H_8 similarly and also noted the formation of less volatile higher germanes.

The three lowest silanes are now usually made on a laboratory scale by reaction of $SiCl_4$, Si_2Cl_6 or Si_3Cl_8 with $LiAlH_4$ in ether solution. There is currently an interest in monosilane as a source of high-purity silicon, which is produced by its pyrolysis and used in transistors. Consequently, there have been attempts to develop new methods for its cheap large-scale production. One, which gives yields up to 80 per cent, involves hydrogenation of a mixture of silica, aluminium and aluminium chloride at 175° at 400 atm in an NaCl–$AlCl_3$ eutectic (m.p. 120°). All of the alkyl and aryl silanes and germanes are produced readily by reaction of the corresponding alkyl or aryl halide with $LiAlH_4$ in ether.

The silanes decompose at about 500° to the elements, but at intermediate temperatures they undergo a reaction that is similar to the cracking of hydrocarbons, though it occurs at lower temperatures. Monosilane gives a mixture of higher silanes and solid polymeric material of variable composition in the range SiH_{1-2}. Germanes behave similarly, and the same result is obtained when SiH_4 is passed through an ozonizer-type discharge. Silanes up to Si_5H_{12}, including the *n*- and iso-isomers of Si_4H_{10} and Si_5H_{12}, have been made in this way. Vapour-phase chromatography also shows the existence of a number of higher hydrides up to at least Si_8H_{18} and of numerous isomers. Some separations of straight and branched-chain silanes have also been made by passing mixtures through a 5 Å molecular sieve. Work with the higher germanes has followed similar lines and clathrate compounds have also been found to form between some higher silanes and urea or thiourea. Interestingly, the Si_5H_{12} urea complex is stable in air although Si_5H_{12} is spontaneously inflammable.

Mixed hydrides of silicon and germanium are produced when a mixture of SiH_4 and GeH_4 is passed through a discharge, one of the products isolated being SiH_3GeH_3. Silylgermanes are also obtained when magnesium silicide-germanide, prepared by heating magnesium with powdered silicon and germanium, is hydrolysed by acid, and a number have been separated.

Monostannane (SnH_4) can be prepared, though in very low yield, by dissolving a tin–magnesium alloy in dilute acid, or by cathodic reduction of a stannous sulphate solution. Higher yields are obtained by reducing $SnCl_4$ with $LiAlH_4$ in ether, or from $SnCl_2$ and $NaBH_4$ in acid solution. Alkyl and aryl stannanes, which decompose less readily, can also be made from alkyl or aryl tin halides with the aid of $LiAlH_4$.

Reaction of $PbCl_4$ with $LiAlH_4$ in ether yields only metallic lead, but Me_3PbCl, Et_3PbCl and the corresponding dialkyl lead dihalides gave moderate yields of the corresponding plumbanes; for example,

$$4(CH_3)_3PbCl \xrightarrow[Me_2O,\,-78^\circ]{LiAlH_4} 4(CH_3)_3PbH + AlCl_3 + LiCl$$

These compounds have been analysed, though they decompose at room temperature to lead, hydrogen and the tetraalkyl lead.

The main evidence for the existence of plumbane (PbH_4) comes from the classical radioactive tracer studies of Paneth. He found that, when a lead–magnesium alloy labelled with the radioactive lead isotope ThB(^{212}Pb, $t_{\frac{1}{2}} = 10{\cdot}6$ hours) was dissolved in acid the hydrogen evolved carried off a minute amount of a volatile lead compound, identifiable only because of the great sensitivity of methods for detecting radioactivity. The active material could be condensed and re-evaporated, and was decomposed and rendered involatile by heating. A similar radioactive material was obtained by electrolysing aqueous sulphuric acid with a lead cathode containing the radioactive tracer. This is the only evidence for the formation of PbH_4; in view of the ready decomposition of SnH_4 at room temperature it would, in any case, be expected to be very unstable.

8.9.1. Properties of the hydrides. All of the silanes are spontaneously inflammable, though SiH_4 can be mixed with air or oxygen outside the critical explosion limits associated with its oxidation by a branching-chain mechanism. The germanes oxidize less readily, GeH_4 reacting with oxygen only at about 170° and Ge_2H_6 at about 100°. The silanes are also rapidly hydrolysed by dilute alkali with evolution of one H_2 for each Si—H or Si—Si bond. Monogermane, on the other hand, is unaffected by 20 per cent aqueous alkali, while SnH_4 is decomposed only by moderately strong acids or alkalis. The silanes are also stronger reducing agents than the germanes.

Both SiH_4 and Si_2H_6 react with hydrogen halides at 100–125° in the presence of an aluminium halide as catalyst to give halosilanes:

$$SiH_4 + 3HX \longrightarrow SiH_3X + SiH_2X_2 + H_2 \quad (X = Cl, Br, I)$$

This method also yields GeH_3Cl, GeH_2Cl_2, GeH_3Br and GeH_2Br_2, but when it is applied to the synthesis of GeH_3I only GeI_2 and GeI_4 result. Direct reaction of iodine with GeH_4 or Ge_2H_6 at room temperature, however, gives GeH_3I or Ge_2H_5I as the main product. Preparation of SiH_4 as an intermediate in the synthesis of silyl halides is avoided by cleaving a phenyl silane, which is readily prepared by $LiAlH_4$ reduction of the appropriate halide and is safer to handle, with a liquid hydrogen halide; for example,

$$C_6H_5SiH_3 + HX \longrightarrow SiH_3X + C_6H_6 \quad (X = F, Cl, Br, I)$$

This method is also useful in the syntheses of Si_2H_5Cl and Si_2H_5Br.

Although no Grignard-type compounds such as SiH_3MgI are yet known, alkali metal derivatives have been prepared (e.g. from SiH_4 and potassium in glyme (1,2-dimethoxyethane) at −78°):

$$2SiH_4 + 2K \longrightarrow 2KSiH_3 + H_2$$

The product, though moisture- and air-sensitive, is stable in vacuum to 240°. Monosilane is solvolysed by liquid ammonia, but both GeH_4 and SnH_4 react at low temperatures with sodium in this solvent with replacement of one or two atoms of hydrogen, forming, for example, $NaGeH_3$ and Na_2GeH_2. This type of reaction also occurs with organosilanes and germanes.

Much of the recent chemistry of the silicon and germanium hydrides centres round the preparation of silyl and germanyl derivatives. Some of the reactions of SiH_3Cl and SiH_3I are shown below.

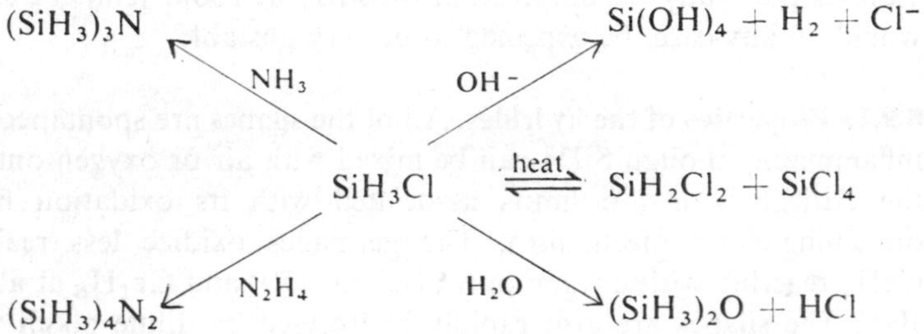

$$(SiH_3)_3N \xleftarrow{NH_3} SiH_3Cl \xrightarrow{OH^-} Si(OH)_4 + H_2 + Cl^-$$

$$SiH_3Cl \xrightleftharpoons{heat} SiH_2Cl_2 + SiCl_4$$

$$(SiH_3)_4N_2 \xleftarrow{N_2H_4} SiH_3Cl \xrightarrow{H_2O} (SiH_3)_2O + HCl$$

$$(SiH_3)_2S \xleftarrow{HgS} SiH_3I \xrightarrow{AgNCS} SiH_3NCS$$

Some of these products are similar in type to well-known methyl compounds, but they are more reactive, and trisilylamine has very weak donor properties compared with trimethylamine. Thus it does not form quaternary salts or an adduct with trimethylboron. This is attributed to N → Si p_π–d_π bonding, which reduces the availability of the electron pair on nitrogen for donation. The molecule is also planar (sp^2 hybridization) rather than pyramidal, like $(CH_3)_3N$, where the lone electron pair occupies the fourth orbital arising from sp^3 hybridization.

$(SiH_3)_2O$, $(SiH_3)_2S$ and $(SiH_3)_2Se$ do not form oxonium, sulphonium or selenonium compounds. The donor properties are clearly very weak. In disilyl oxide the bond angle (144°) is considerably larger than that in $(CH_3)_2O$ (111°), but the angles in disilyl sulphide and selenide and dimethyl sulphide and selenide all fall in the range 96–99°.

Trisilyl phosphine, formed in the reaction shown below,

$$3SiH_3Br + 3KPH_2 \xrightarrow[-100°]{(CH_3)_2O} (SiH_3)_3P + 2PH_3 + 3KBr$$

differs from trisilylamine in forming adducts, and electron diffraction shows that the molecule, like trimethylphosphine, is pyramidal. The arsenic and antimony analogues are also pyramidal. Several germanyl analogues have also been prepared, and the structures of $(GeH_3)_3N$ and $(GeH_3)_3P$ resemble those of the corresponding silyl derivatives.

8.10 Sigma-bonded hydrides of the transition metals[26]

The carbonyl hydrides of iron and cobalt, which were discovered by Hieber and his co-workers in 1931, provide the earliest examples of compounds with a σ-bond between hydrogen and a transition metal. The nature of the bond was not established for some time, but many carbonyl hydrides have since been isolated and they now constitute an important facet of carbonyl chemistry. It is for this reason that they are described in Chapters 20 and 21. Other ligands, notably tertiary arsines and phosphines, the cyclopentadienyl radical and the cyanide ion, have, however, also been found to stabilize the M—H bond in transition metal complexes. The chief interest in these compounds centres on the metal–hydrogen bond, and they are therefore considered here as hydrides. The reader will note many points of resemblance to the carbonyl hydrides. Only in two compounds, K_2ReH_9 and its technetium analogue, is hydrogen known with certainty to be σ-bonded to a transition metal in the absence of other ligands. These remarkable compounds will be described at the end of this section, together with a few which may be similarly constituted.

The characteristic common to ligands such as tertiary phosphines and the cyanide ion is that they produce strong ligand fields. They also exert a similar influence in stabilizing the metal–carbon σ-bond in transition metal alkyls and aryls. For the latter, Chatt and Shaw have suggested that one important factor in determining M—C bond stability is the energy gap Δ between the highest occupied electron level and the lowest unoccupied level; the greater Δ the higher is the stability. Since the ligands in question all increase Δ, this effect may be related to their influence on both the M—C and the M—H bonds.

The structures of complexes containing an M—H bond are established by the standard methods, though location of the hydrogen atom by X-ray diffraction is difficult and neutron diffraction is preferred (see Chapter 2). The main diagnostic method is the use of the ^1H n.m.r. spectrum, where high-field shifts in the range $\tau = 12$–45 are observed, as for the carbonyl hydrides. In the infrared spectrum metal–hydrogen stretching frequencies occur in the range 1700–2200 cm^{-1}, which are reduced by the factor of $\sqrt{2}$ on deuteration. In most of the complexes to be discussed hydrogen behaves as a strong-field ligand.

8.10.1. Complexes with tertiary phosphines and arsines. Transition metal hydrides with these two types of ligand are the most numerous, and also among the most stable, of the compounds under discussion. The first such compound to be prepared was *trans*-$[PtHCl(PEt_3)_2]$, which was obtained by reducing *cis*-$[PtCl_2(PEt_3)_2]$ at 90° with a dilute solution of hydrazine hydrate. Reduction of a complex halide is a fairly general method and various other reducing agents may be employed, such as $LiAlH_4$ in tetrahydrofuran, $NaBH_4$ in water, or hydrogen, as in the reaction

$$cis\text{-}[PtCl_2(PEt_3)_2] \xrightarrow[\text{ethanol}]{H_2,\ 95°\text{ at }50\text{ atm}} trans\text{-}[PtHCl(PEt_3)_2]$$

Reduction may be brought about in some cases by an alcohol in the presence of a base; for example,

$$IrCl_3(PEt_2Ph)_2 + OH^- + C_2H_5OH \rightarrow IrHCl_2(PEt_2Ph)_2 + CH_3CHO + H_2O + Cl^-$$

In this particular case all three chlorine atoms may be replaced by hydrogen if excess of reagent is used. By using the deuterated alcohol CH_3CD_2OH in the above reaction, it is found that the hydrogen introduced into the metal complex is that bonded originally to the α-carbon in the alcohol. Reaction is believed to involve replacement of chlorine by the ethoxide ion with subsequent transfer of hydrogen from the ethyl group to the metal. The intermediate step can thus be formulated as

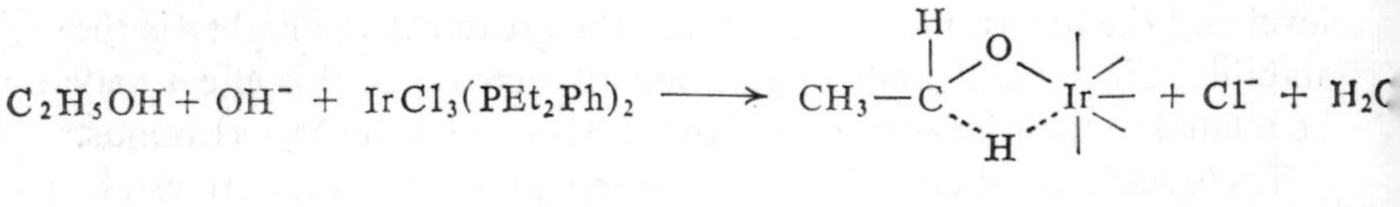

$$C_2H_5OH + OH^- + IrCl_3(PEt_2Ph)_2 \longrightarrow CH_3\text{—}C(H)(\text{—O—})\cdots H\cdots Ir + Cl^- + H_2O$$

Thermal decomposition of alkyl derivatives of some transition metals can also yield the hydride, *trans*-$[PtCl(C_2H_5)(PEt_3)_2]$, for example, giving *trans*-$[PtHCl(PEt_3)_2]$, which reverts to the starting compound when treated with ethylene at 50 atm. Mention may also be made of the formation of a volatile iron hydride complex in the following reaction, in which iron is dissolved in the diphosphine in a hydrogen atmosphere:

$$Fe + H_2 + 2o\text{-}C_6H_4(PEt_2)_2 \rightarrow FeH_2[o\text{-}C_6H_4(PEt_2)_2]_2$$

The examples cited above give some indication of the types of compound formed and of methods of synthesis. Various phosphines and arsines have been employed, and a large number of such derivatives of metals in Groups VII and VIII isolated. Reference to the section on carbonyl hydrides (Section 20.7) will show that many of the latter contain both carbon monoxide and a phosphine as ligands, the phosphine giving enhanced thermal stability.

Hydrogen functions as a normal ligand in these complexes and often exercises a strong *trans* effect. A number of isomers have been identified on the basis of measurements of dipole moment and proton n.m.r. spectra, in which coupling with ^{31}P and with other protons in the molecule yields valuable structural information.

Hydrogen may sometimes be replaced by other ligands as, for example, in the reaction shown below:

$$IrH_2Cl(PEt_3)_3 \xrightarrow{\text{dry HCl}} IrHCl_2(PEt_3)_3$$

If the complex is treated with a solution of chlorine in chloroform the product is $IrCl_4(PEt_3)_2$. In other cases, addition reactions may occur, as in the conversion of $PtHCl(PEt_3)_2$ to $PtH_2Cl_2(PEt_3)_2$ by hydrogen chloride.

8.10.2. π-Cyclopentadienyl hydrides. Like the tertiary phosphines, the cyclopentadienyl radical may occur as a ligand in association with carbon monoxide in the carbonyl hydrides. Both the preparative methods for the cyclopentadienyl carbonyl hydrides and their reactions are very similar to those of the carbonyl hydrides. They can, for example, be prepared by reduction of a halide (e.g. $\pi\text{-}C_5H_5Fe(CO)_2Cl$), or by acidification of the sodium salt of a π-cyclopentadienyl carbonylate anion; for example,

$$Mo(CO)_6 + Na^+C_5H_5^- \rightarrow Na^+[\pi\text{-}C_5H_5Mo(CO)_3]^- \rightarrow \pi\text{-}C_5H_5Mo(CO)_3H$$

The reactions that are of interest are those of the M—H bond, such as its conversion to an M—Cl bond by mild chlorinating agents (e.g. $CHCl_3$), and the loss of hydrogen on heating, with formation of a binuclear complex. Addition of olefins across the M—H bond also occurs (e.g. $\pi\text{-}C_5H_5Fe(CO)_2H$ with C_2F_4

gives π-$C_5H_5Fe(CO)_2CF_2CF_2H$), while diazomethane converts π-$C_5H_5Mo(CO)_3H$ to π-$C_5H_5Mo(CO)_3(CH_3)$.

A number of hydrides are also known in which the cyclopentadienyl radical and hydrogen are the sole ligands bonded to the transition metal. The first to be prepared was $[ReH(\pi\text{-}C_5H_5)_2]$, made by the reaction

$$\begin{aligned} ReCl_5 + Na(C_5H_5) &\xrightarrow{\text{THF}} ReH(C_5H_5)_2 \\ + NaBH_4 & \end{aligned}$$

The related compounds $MoH_2(\pi\text{-}C_5H_5)_2$, $WH_2(\pi\text{-}C_5H_5)_2$ and $TaH_3(\pi\text{-}C_5H_5)_2$ are all formed similarly.

The structure of $MoH_2(\pi\text{-}C_5H_5)_2$ (apart from the position of the H atoms) has been determined by X-ray analysis and shows the two rings to be situated as shown below (Fig. 8.23). There is some evidence for a similar oblique ring structure in the other compounds.

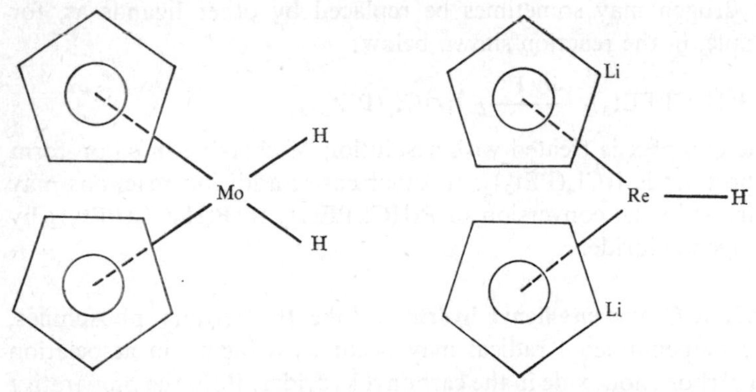

Figure 8.23 *Figure 8.24*

All are crystalline solids which sublime in high vacuum and are soluble in organic solvents. They are unaffected by water or alkali but oxidize in the air. Reaction with chloroform converts the M—H bond to M—Cl. The rhenium and tungsten compounds react with butyllithium in ether to give 1 : 1 dilithiated complexes (e.g. Fig. 8.24), from which a number of interesting derivatives may be prepared, including one in which both lithium atoms are replaced by HgCl groups in a reaction with $HgCl_2$.

8.10.3. Cyanide hydrides. Hydrogen is absorbed by an aqueous solution of a cobalt salt to which cyanide ion has been added to form

$[HCo(CN)_5]^{3-}$ and the proton n.m.r. spectrum of the product shows a chemical shift of 17·5 p.p.m. relative to water. This high value makes it certain that a Co—H bond is present in the complex. Slow decomposition of the solution occurs in the absence of oxygen at room temperature and it is very sensitive to atmospheric oxidation. The oxidation product is the anion $[Co(CN)_5OOH]^{3-}$, the potassium salt of which has been prepared by hydrolysis of $K_6[(CN)_5Co—O—OCo(CN)_5]$, and which has a characteristic electronic absorption spectrum. Solutions containing the cobalt cyanide hydride can act as homogeneous catalysts in various hydrogenation reactions (e.g. the conversion of styrene to ethylbenzene). Reaction also occurs with tetrafluoroethylene to give the ion $[Co(CN)_5CF_2CF_2H]^{3-}$, which is again strong support for the presence of the Co—H bond in the ion. A solid salt $NaCs_2[Co(CN)_5H]$ has recently been isolated.

8.10.4. Anions containing only a transition metal and hydrogen. Rhenium and technetium both form complexes containing the $[MH_9]^{2-}$ ion. Treatment of potassium perrhenate with potassium in ethylenediamine solution leads to precipitation of a solid which, after purification, is white and analyses as K_2ReH_9. The isomorphous technetium compound is prepared similarly, and various salts of these two anions are known. They may be heated to over 200° without decomposition and show strong reducing properties in aqueous solution, in which slow hydrolytic decomposition occurs. Hydrogen is evolved quantitatively on treating the salts with an aqueous acid, and the metal is deposited.

The structure of K_2ReH_9 has been established by X-ray analysis and neutron diffraction. The latter is particularly important as it enables the positions of the hydrogen atoms to be determined. The crystal is hexagonal with three molecules per unit cell. In the anion (Fig. 8.25) six of the hydrogen atoms are at the corners of a trigonal prism with the metal atom inside it, and there is one hydrogen atom outside the centre of each of the prism faces in an equatorial position.

The Re—H bonds are all approximately of the same length (1·68 Å). The proton n.m.r. spectrum shows a single high-field peak at 18·4 p.p.m. for the rhenium compound and at 19·1 p.p.m. for the technetium compound, the latter being broadened because of unresolved coupling with the ^{99}Tc nucleus (spin $\frac{9}{2}$). All of the hydrogen atoms thus appear to be equivalent on an n.m.r. time-scale, and all undergo slow deuterium exchange in alkaline D_2O. The infrared

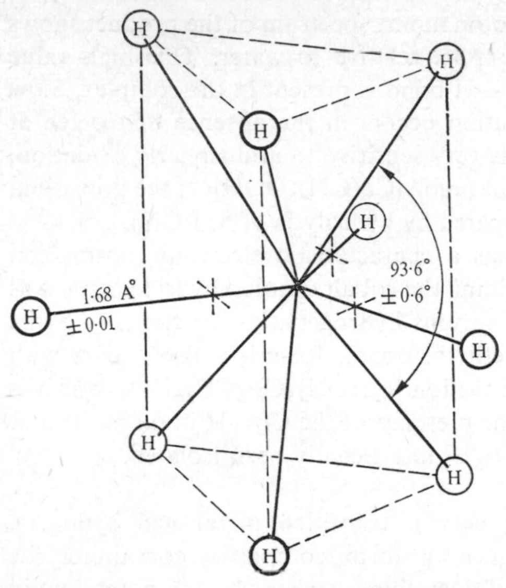

Figure 8.25 H environment of Re atom in ReH_9^{2-}

spectrum shows the bending and stretching modes which are characteristic of the M—H bonds.

Each hydrogen bonded to the metal atom may be regarded as a ligand with a charge of −1 (i.e. as contributing two electrons to the metal atom). If the latter is taken as being in the +7 state with sixty-eight electrons, addition of eighteen electrons from the nine ligands gives an effective atomic number of 86, equal to that of radon. The nine electron pairs fully occupy all of the available orbitals of the metal atom ($5d^5 6s 6p^3$) and all of the nine hydrogen orbitals.

These two compounds appear at the moment to be unique among transition metal derivatives, though rhodium forms two compounds, Li_4RhH_4 and Li_4RhH_5, which may be similar. The former is made by heating rhodium powder with $LiAlH_4$ at 600° and reacts on further heating with hydrogen to give the latter.

8.11 Metallic hydrides[1a, 2]

Many of the transition metals, as well as the lanthanides and such actinides as it has been possible to study, absorb hydrogen when heated in the gas. This phenomenon is distinct from adsorption or chemisorption, which is primarily a surface phenomenon, and involves penetration into the metal lattice. The amount of gas taken up at equilibrium is a function of pressure and temperature, and can be represented by a series of dissociation pressure isotherms, as shown diagrammatically in Fig. 8.26 for the Pd–H_2 system. The rising section of the curves on the left corresponds with solution of hydrogen in the metal, the lattice of which is expanded. The maximum hydrogen content is typically in the range $MH_{0\cdot1-0\cdot2}$. In the

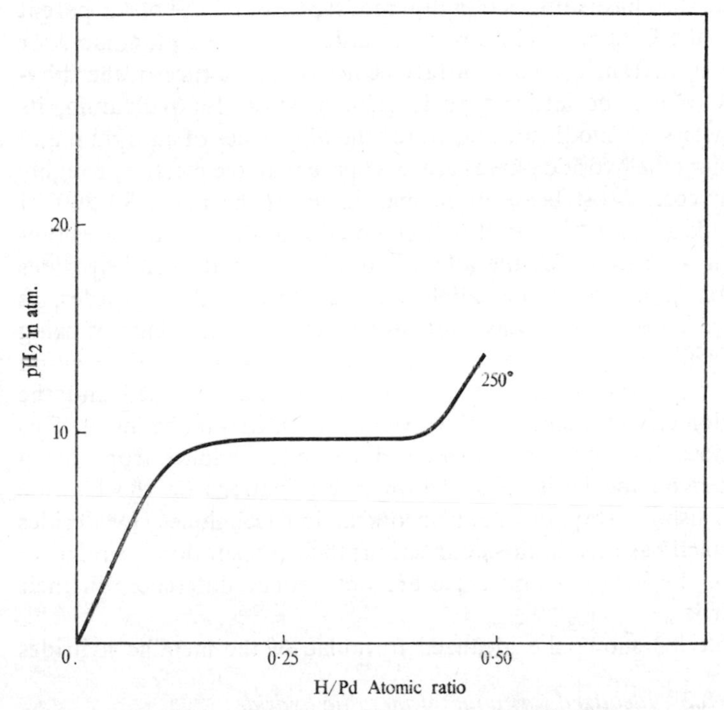

Figure 8.26 Pressure-composition curve for H_2–Pd system

plateau region a hydride phase is in equilibrium with solid solution. The steeply rising part of the curve on the right corresponds with the equilibrium between the hydride phase and hydrogen. These processes are reversible and all of the hydrogen can be removed by heating the sample in vacuum. A number of metals give a second hydride phase, for which there would be a further plateau in the equilibrium diagram.

Formation of the solid solution is endothermic, whereas hydride formation is an exothermic process. As would be expected, the rate of reaction is temperature-dependent, and is also influenced by the purity of the metal and by whether a massive sample or powdered material is used. Hydriding is commonly carried out in the temperature range 150°–400°, though some metals, including palladium, react at lower temperatures and the equilibrium studies have often been extended to 700° or higher.

There are a number of general points about metallic hydrides that may first be considered. Except for actinium, nickel, palladium and cerium, they have lattices of a different type from those of the parent metal; the hexagonal close-packed lattice of some lanthanides, for example, is transformed to a face-centred cubic lattice in the dihydrides. When the lattice type is unchanged, as for palladium, its parameters are modified. Except for the dihydrides of europium and ytterbium the hydride phases are less dense than the metal by roughly 15 per cent. Most heats of formation lie in the range 80–200 kJ (20–50 kcal) mole^{-1}, and it will be noted that these values are comparable with those for the alkali and alkaline earth metal hydrides (Section 8.4). Unlike the saline hydrides, the metallic hydrides, as prepared, are almost always non-stoichiometric in the sense of being hydrogen deficient; they have appreciable hydrogen dissociation pressures at the temperatures at which they are formed and the stoichiometry may also be influenced by impurities in the metal. This fact does not, however, prevent idealized formulae from being assigned on the basis of structure determinations on the hydride phase, using X-ray and neutron diffraction techniques. Deuterides are sometimes used in this connection; their preparation is similar to that of the hydrides and there are only minor differences in their properties.

Table 8.3 shows the idealized formulae of the metallic hydrides

Table 8.3 *Idealized formulae of metallic hydrides*

TiH_2, ZrH_2, HfH_2	Lanthanide dihydrides
VH, VH_2 NbH, NbH_2	MH_3 (M = Sm, Gd, Tb, Dy, Ho, Er, Tm, Lu, Y)
TaH	AcH_2
CrH, CrH_2	ThH_2, Th_4H_{15}
NiH	PaH_3
PdH	UH_3
CuH	PuH_3

known at present. Molybdenum and tungsten, together with the transition elements with seven or more *d* electrons (apart from nickel, palladium and copper), do not form hydrides of the type under discussion. The lanthanide hydrides have been grouped together, though there are some structural differences, which will be described later, between hydrides of the same type.

Before typical structures are described, some general features of these compounds may be noted. Firstly, they all conduct heat and electricity, though less well than the parent metal. Some are also metallic in appearance. This is in sharp contrast to the stoichiometric hydrides of the alkali and alkaline earth metals, which conduct only when in the molten state.

8.11.1. Hydrides of the transition elements. Titanium, zirconium and hafnium all react with hydrogen at a convenient rate in the temperature range 300–400°. At such temperatures the hydride phase has a considerable equilibrium dissociation pressure and the amount of hydrogen taken up increases with the hydrogen pressure applied to the system; compositions with hydrogen in excess of $TiH_{1.9}$ may be obtained, but they are more often in the range $TiH_{1.5-1.8}$. Similar considerations apply to all hydrides made directly from the elements.

Titanium crystallizes with an hexagonal close-packed structure and, in a different modification, with the body-centred cubic structure. The dihydride has a face-centred cubic fluorite structure. This is equivalent to a cubic close-packed metal sub-lattice with hydrogen atoms in the tetrahedral holes, of which there are two per metal atom, together with one octahedral hole per metal atom, which is vacant. Below 310°K this structure changes in the case of TiH_2 to a distorted tetragonal structure; ZrH_2 and HfH_2 have higher transition temperatures for the interconversion of the two corresponding forms. The structures of the dihydrides of vanadium and niobium are similarly based on a face-centred cubic metal sub-lattice. The monohydrides VH and TaH both have body-centred tetragonal metal sub-lattices, all three of the Group V metals having body-centred cubic lattices. In addition NbH and TaH each have a body-centred orthorhombic modification.

The only Group VI metal known to form a hydride is chromium. This is formed during the electrodeposition of metallic chromium from a sulphuric acid solution of CrO_3 under special conditions. One of the two phases has compositions in the range $CrH_{0.5-1}$ and has a hexagonal structure while the other, which is much less stable, has a cubic structure with the upper limit of the hydrogen content corresponding with the formula $CrH_{1\cdot7}$. This phase appears to be similar in structure to the dihydrides referred to above. A nickel hydride, $NiH_{0\cdot6-0\cdot7}$, has been made by a similar electrolytic method. It decomposes slowly at room temperature to metal and hydrogen, but neutron diffraction studies have shown the hydrogen atoms to occupy the octahedral holes of the face-centred cubic lattice of the metal.

Palladium metal has long been known to absorb hydrogen readily, even at room temperature, and the phase relations have been studied in detail. It is of interest that hydrogen uptake also occurs if the metal is made the cathode at which hydrogen ions are discharged. Limiting compositions in the range $PdH_{0\cdot6-0\cdot8}$ are observed. The metal lattice of the hydride phase is face-centred cubic, as in the metal itself, but the lattice parameter is increased by roughly 3 per cent. Neutron diffraction studies have shown that the hydrogen atoms occupy the octahedral holes in the metal structure and are randomly distributed (i.e. it may be regarded as a hydrogen-deficient sodium chloride structure). Copper hydride, which differs greatly in its mode of preparation, properties and structure from these metallic hydrides, will be described later.

The dihydrides of europium and ytterbium show some points of difference from other lanthanide dihydrides (see Section 8.11), which may be associated with the fact that these two elements also have a stable +2 oxidation state. Thus both compounds have the same orthorhombic structure as calcium hydride. They are denser than the parent metals and exhibit no metallic properties, which would be in keeping with the transfer of both valency electrons associated with the +2 state to form $2H^-$. These two hydrides also dissolve in molten lithium hydride, which is a characteristic of the saline hydrides but not of the metallic hydrides under discussion. It seems, therefore, that they would be more correctly classed with the alkaline earth hydrides.

8.11.2. Actinide hydrides. All of the actinide metals up to americium are known to form hydrides, though only those of thorium, uranium and plutonium have been studied in detail. Thorium takes up hydrogen at 200–300°C. The dihydride, which is first produced, has a body-centred tetragonal structure, which is a distorted form of the fluorite structure encountered in the lanthanide dihydrides. Further hydrogen uptake leads to Th_4H_{15}, with a complex cubic structure. Uranium gives only the trihydride, UH_3, which can be obtained at 250° using a moderate hydrogen pressure; above this temperature dissociation to metal and hydrogen becomes substantial, but the compound is unusual among the metallic hydrides in that it can be obtained with almost the correct stoichiometry. It has a complex body-centred cubic structure. Finely divided metal reacts at a much lower temperature, giving a second complex cubic form of the trihydride. In both, each hydrogen atom is surrounded tetrahedrally (or approximately so) by four atoms of uranium, but the packing of the uranium atoms is more complex than in the structures of the lanthanide trihydrides. Plutonium at 100–200° gives successively a dihydride and a trihydride, the structures of which closely resemble those of the corresponding lanthanide compounds MH_2 and MH_3; the behaviour of neptunium is similar to that of plutonium.

8.11.3. Bonding in the metallic hydrides. Formation of the metallic hydrides is exothermic, in spite of the fact that dissociation of molecular hydrogen to atoms requires 432 kJ (103 kcal) mole^{-1}. To compensate for this, the bonds formed must be strong. Any theory as to their exact nature must explain why, in contrast to the saline hydrides, these compounds exhibit metallic properties, though to a decreasing extent as the hydrogen content is increased.

Formerly these hydrides were classed with carbides, nitrides and borides as interstitial compounds, and it was thought that hydrogen atoms, like other small atoms, could be accommodated in interstices in the metal lattice producing distortion but no change in type. In fact, the metal atoms in most of these hydrides occupy a sub-lattice which is different from that of the parent metal, and the interstitial compound concept also fails to describe the bonding and changes in the physical properties of the metal which occur if it is hydrided.

Currently there are two quite distinct theories of bonding in metallic hydrides. The first considers hydrogen to lose electrons to the conduction band of the metal. Complete transfer of an electron to leave H^+ is very improbable on energetic grounds; it is more likely that there is a transfer of electron density, leaving hydrogen with a fractional positive charge. This theory is supported by studies of the migration of hydrogen under an applied potential in such materials as hydrided palladium and titanium; hydrogen migrates to the cathode. The magnetic susceptibility of palladium also falls as hydrogen is taken up, until at $PdH_{0.6}$ the material is diamagnetic. This may be interpreted as due to progressive filling of the conduction band of the metal.

The alternative theory, which postulates that H^- is formed by loss of electrons from the metal to hydrogen, also leads to the conclusion that there will be a falling away in metallic properties as the hydrogen to metal ratio increases. In LaH_2, for example, the electrical conductivity is about one-hundredth of that for the pure metal, and in LaH_3 is even less. The second theory also accounts satisfactorily for observed metal–hydrogen distances, if an effective radius of 1·3 Å is assigned to H^-, and leads to agreement between observed and calculated values for the lattice energy. For a fuller discussion of this problem the reader is referred to a review by Gibb.[1a]

8.11.4. Properties of the metallic hydrides. A great impetus to the study of the metallic hydrides discussed in this section has come from the potential use of some in nuclear reactor technology. A material such as zirconium hydride or deuteride, for example, where the metal and hydrogen both have low thermal neutron capture cross-sections, is of interest as a moderator material. Titanium and some of the other metals which form hydrides have high neutron capture cross-sections and could find specialized applications as shielding materials, while the hydrides of thorium, uranium and plutonium are all important in relation to fuel elements. It is easy to see why these materials have been so thoroughly investigated; indeed, even the fabrication of a number by the methods of powder metallurgy has been developed.

Probably the most important of the purely chemical properties is the susceptibility to atmospheric oxidation. This varies widely: finely divided UH_3, which is always formed when massive uranium is hydrided, ignites spontaneously in air, whereas hydrides of metals in the titanium group oxidize rapidly at temperatures in roughly the range 400–600°, depending on the state of division. Most of the hydrides are stable to water up to at least 100°. Most, including those of the lanthanide elements, are quantitatively decomposed by acids and show reducing properties. Their reactivity to common reagents is as a rule greater than that of the parent metals. This point may be illustrated by reference to UH_3, which has been fully studied. It reacts with HCl and HBr at 250–300° to give the trihalides, with H_2S at 400° to give US_2 and with ammonia at 250° to give nitrides.

8.11.5. Other metallic hydrides. Copper is the only transition metal, other than those included in Table 8.3, to form a well-characterized hydride. Hydrogen has a low solubility in metallic copper and no hydride phase is formed. A compound with the empirical formula CuH has, however, been known since 1844, when Wurtz prepared it by the reaction shown below:

$$2Cu^{2+} + 3H_2PO_2^- + 3H_2O \rightarrow 2CuH + 3H_2PO_3^- + 4H^+$$

It is a red-brown crystalline material and is also produced by reducing CuI with $LiAlH_4$ in ether/pyridine, or by reducing aqueous $CuSO_4$ with $NaBH_4$.

The compound has a hexagonal wurtzite structure; this has been established both by X ray and by neutron diffraction. Thermal decomposition occurs above about 50°. Decomposition by aqueous hydrochloric acid gives CuCl and H_2, and hydrogen is also evolved rapidly in alkaline solution. The hydride is a reducing agent in aqueous media, converting, for example, chlorate to chloride.

The electrical properties of copper hydride appear not to have been recorded. It seems likely, however, from lattice energy considerations, that it is not a metallic hydride and that bonding is covalent. No analogous silver and gold compounds are known. Earlier reports of the formation of hydrides of iron, cobalt, nickel and other elements by reaction of phenyl magnesium bromide with salts of the metals have not been substantiated by later investigations.

References for Chapter 8

1 Pascal, P. (ed.), *Nouveau Traité de chimie minérale*, Vol. I (Masson, 1956).
1a Gibb, T. R. P., Jr., *Prog. Inorg. Chem.*, 1962, **3**, 315.
2 Mackay, K. M., *Hydrogen Compounds of the Metallic Elements* (E. and F. N. Spon Ltd., 1966).
3 Stock, A., *The Hydrides of Boron and Silicon* (Cornell University Press, 1933).
4 Adams, R. M. (ed.), *Boron, Metallo-Boron Compounds and Boranes* (Interscience, 1964).
5 Muetterties, E. L. (ed.), *The Chemistry of Boron and its Compounds* (Wiley, 1967).
6 Lipscomb, W. N., *Adv. Inorg. Chem. Radiochem.*, 1959, **1**, 117.
7 Stone, F. G. A., *Adv. Inorg. Chem. Radiochem.*, 1960, **2**, 279.
8 Lipscomb, W. N., *Boron Hydrides* (W. A. Benjamin, 1963).
9 Eaton, G. R., and Lipscomb, W. N., *N.M.R. Studies on Boron Hydrides and Related Compounds* (W. A. Benjamin, 1969).
10 Parry, R. W., and Walter, M. K., *Preparative Inorganic Reactions*, 1968, **5**, 45.
11 James, B. D., and Wallbridge, M. G. H., *Prog. Inorg. Chem.*, 1970, **11**, 99.

12 Brown, H. C., *Hydroboronation* (W. A. Benjamin, 1962).
13 Muetterties, E. L., and Knoth, W. H., *Polyhedral Boranes* (Marcel Dekker and Edward Arnold, 1968).
14 Williams, R. E., *Prog. Boron Chem.*, 1970, **2**, 37.
15 Todd, L. J., *Prog. Boron Chem.*, 1970, **2**, 1.
16 Grimes, R. N., *Carboranes* (Academic Press, 1970).
17 Wade, K., *Electron Deficient Compounds* (Nelson, 1971).
18 Haworth, D. T., *Endeavour*, 1972, **31**, 16.
19 Todd, L. J., *Adv. Organometallic Chem.*, 1970, **8**, 87.
20 Greenwood, N. N., in *New Pathways in Inorganic Chemistry*, ed. E. A. V. Ebsworth, A. G. Maddock and A. G. Sharpe (Cambridge University Press, 1968).
21 Ashby, E. C., *Adv. Inorg. Chem. Radiochem.*, 1966, **8**, 283.
22 Stone, F. G. A., *Hydrogen Compounds of the Group IV Elements* (Prentice-Hall, 1962).
23 Ebsworth, E. A. V., *Volatile Silicon Compounds* (Pergamon, 1963).
24 Macdiarmid, A. G., *Adv. Inorg. Chem. Radiochem.*, 1961, **3**, 207.
25 Aylett, B. J., *Adv. Inorg. Chem. Radiochem.*, 1968, **11**, 249.
26 Green, M. L. H., and Jones, D. J., *Adv. Inorg. Chem. Radiochem.*, 1965, **7**, 115.

9 Polymeric inorganic compounds I: molecular species

9.1 Introduction

All polymeric substances are characterized by the presence in the molecule of a repeating structural unit. This is readily recognized in natural and synthetic organic polymers. When such a criterion is applied to inorganic compounds, however, it is more difficult to decide what to include. All ionic crystals could be thought of as polymeric, and the same is true of giant molecules such as those of silica or graphite, and of many other covalently bonded crystals (e.g. oxides and halides). It is usual to exclude ionic structures as such, but to include certain types of condensed anion, of which the condensed phosphates are typical; these are dealt with in the following chapter. The main requirement, in fact, is that bonding shall be covalent. This allows inorganic polymers to be classified under two headings, homoatomic and heteroatomic, according to whether the backbone structure is made up of atoms of the same or of different kinds.

The chief elements forming homoatomic structures are the non-metals, a familiar example being sulphur, which forms chains or rings when in the elemental form and also in several types of compound. The more important homoatomic structures will be described in the early sections of this chapter. Heteroatomic polymers are much more numerous, most being based on backbone structures in which there are various combinations of non-metallic atoms (e.g. P,O, B,N and Si,O). Since bonds between unlike atoms are in general stronger than those between like, many of these polymers are relatively stable. It will be possible here to describe only a few of the more important groups, and these only in outline, but the references given at the end of the chapter will enable the interested reader to gain a more complete picture. Heteroatomic polymers have been intensively studied in recent years as part of a search for new materials with higher thermal and chemical stability than technically important organic polymers.

9.2 Homoatomic inorganic polymers

9.2.1. Boron.[1] The preparation of pure elementary boron is difficult, largely because of the high reactivity of the element at the temperatures needed to reduce its compounds. It can, however, be made by several methods: for example, by reaction of the trichloride with hydrogen on a tungsten filament at 1300–1800°, or by thermal decomposition of the vapour of the triiodide on tantalum at 800–1000°. Pyrolysis of boranes has also been used and new purification procedures have been introduced.

Boron is a very hard material which melts at about 2200° and exists in a number of crystalline forms, for three of which, the α-rhombohedral, tetragonal and β-rhombohedral modifications, structures have been investigated. The electrical conductivity is low and when pure it is a p-type semiconductor. Interpretation of the electrical properties is, however, difficult because of the effect of small amounts of impurities in many samples. The structures of all of the forms of boron are electron deficient (i.e. there are too few electrons to allocate two per bond). This may be illustrated by considering in some detail the structure of α-boron, which is the least complicated of the three that are known. All are built up from icosahedral B_{12} units which are bonded together with different packing in the three modifications.

The icosahedral B_{12} unit is shown diagrammatically below (Fig. 9.1). There are twelve vertices which lie on a circumscribed sphere,

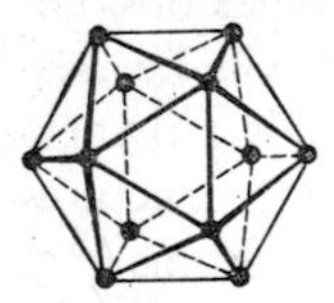

Figure 9.1

with five edges and five triangular faces meeting at each vertex. The thirty edges define twenty equilateral triangular faces. In the α-rhombohedral modification the B_{12} icosahedra are stacked with a slightly deformed cubic close-packing arrangement. The structure is thus made up of a series of superposed layers, one of which is

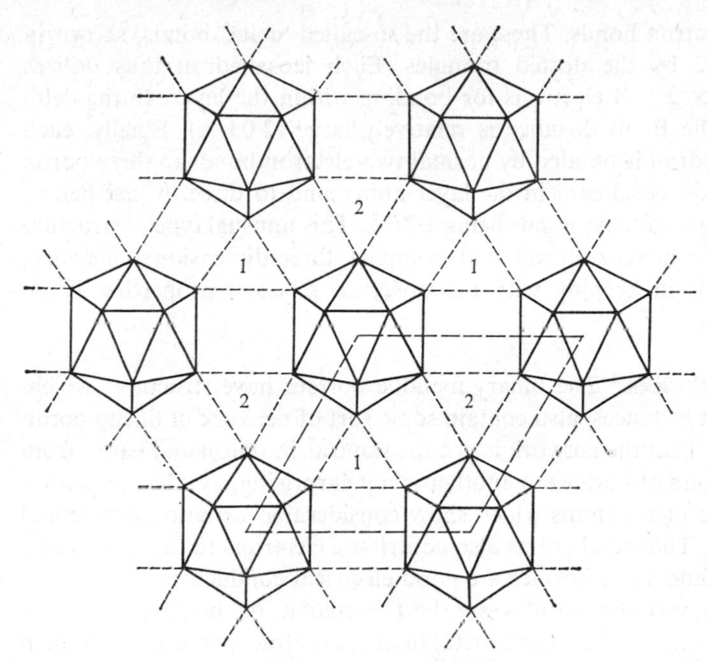

Figure 9.2 α-Rhombohedral boron. Basal plane viewed from above. Short dashes show the delta bonds, length 2·03 Å. Long dashes show the base of the unit cell. Delta bonds are parallel to the basal plane, but those marked *1* are lower than those marked *2*. Icosahedra in the layer above are centred over *1*, those in the layer below are centred below *2*

shown diagrammatically above (Fig. 9.2). The icosahedra in the layer above are centred over the position 1 and those in the layer below under the position 2. The shortest distance between the centres of icosahedra in the same layer is 4·91 Å and that between the centres of icosahedra in different layers is 5·01 Å.

There are B—B bonds of length 1·73–1·79 Å along each icosahedral edge and each boron atom is almost equidistant from five others in the same icosahedron. It has been shown by means of molecular orbital theory that the bonding within an icosahedron can be rationalized on the basis of thirteen multicentre bonds, each involving six atoms and containing two electrons. There are thirty-six electrons associated with the twelve boron atoms and the twelve remaining are used for intericosahedral bonding. Within a layer each icosahedron is bonded to six others by six three-centre two-electron bonds. These are the so-called 'delta' bonds, shown in Fig. 9.2 by the dotted triangles. Each icosahedron thus utilizes $6 \times \frac{1}{3} \times 2 = 4$ electrons for bonding within the layer. In the delta bond the B—B distance is relatively large (2·03 Å). Finally, each icosahedron is bonded by normal two-electron bonds to three boron atoms of icosahedra in the layer above and to three in that below, the length of these bonds being 1·71 Å. This unusual type of structure leads to a very strong and compact three-dimensional network, which is in keeping with the observed physical properties of the element.

9.2.2. Borides.[2] The binary metallic borides have structures which, in most instances, also contain some sort of network of linked boron atoms. That the network is not the same as in boron and varies from one group of borides to another is not surprising, as it has to accommodate metal atoms which show considerable variation in size and charge. The metal atoms also contribute electrons to the system as a whole and some borides are good electrical conductors.

Pure metallic borides are best prepared by heating powdered boron with the powdered metal in an inert atmosphere at an elevated temperature. Technically they are often made by reduction of a metal oxide in the presence of a source of boron (e.g. by a mixture of boron carbide and boron oxide). Much of the current interest in these compounds stems from their refractory nature and electrical properties. The boride TiB_2, for example, melts at 2920°, the melting point of titanium being 1700°, and the electrical conductivity is roughly an order of magnitude greater than that of the metal.

It will not be possible here to describe boride structures in detail. A brief statement of structural principles will be given, however, and it may first be noted that the formulae of boride phases usually do not conform to normal valency rules and are of little use in deciding structures. This point is well illustrated by the following formulae selected at random: Mn_2B, TiB, V_3B_4, TaB_2, CaB_6 and UB_{12}. A given element also usually forms more than one boride.

A large number of boride structures have been elucidated by X-ray analysis and some are found to have isolated boron atoms in the lattice with the nearest B—B distances equal to at least 2·1 Å, compared with a value near 1·75 Å for a true B—B bond. Borides of this type are boron-poor (e.g. Mn_4B and Ta_2B). When the proportion of boron is increased, there is a tendency for the boron atoms to be present in pairs (e.g. in V_3B_2) or in chains (e.g. in TiB and other MB phases). Here it must be pointed out that the location of the light boron atoms by X-ray analysis is less precise than that of the heavy metal atoms, and has to be based in part on steric and symmetry considerations.

In the important group of diborides, MB_2, among which are found some of the most refractory phases with the best conducting properties, boron atoms are bonded together into two-dimensional sheets like those in graphite, which alternate in the structure with layers of metal atoms. This is shown diagrammatically below (Fig. 9.3); the observed B—B distances in a layer are in the range 1·75–*ca.* 1·90 Å, and vary with the size of the metal atom. Borides of this type are formed by many transition elements and lanthanides, and also by magnesium, beryllium and aluminium.

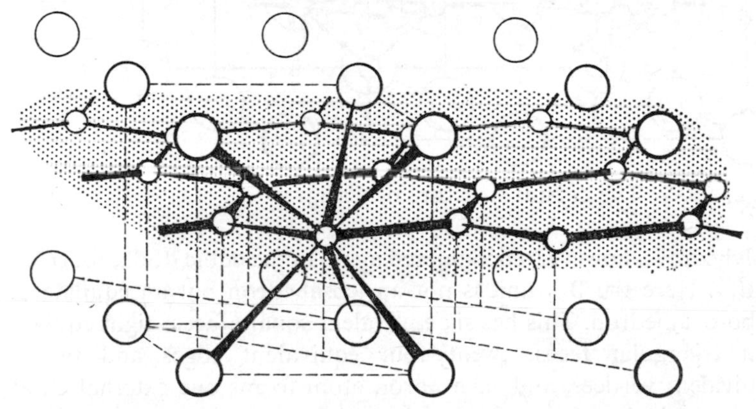

Figure 9.3 The crystal structures of AlB_2. The smaller circles represent B atoms

We will next consider the cubic MB_6 structures where, using the same criterion, it is found that the lattice contains identifiable B_6 octahedra bonded together into a cage structure with metal atoms in the cavities. Such an arrangement is shown below (Fig. 9.4). Hexaborides are formed by the rare earth metals, as well as by calcium, strontium and barium, all of which have large atomic radii. The boron framework is rigid and, within it, B—B bonds have a length of 1·75–1·80 Å.

The tetraborides have a more complex structure which can be regarded as a compromise between that of the MB_2 layer structure and the MB_6 type network. This phase is formed by a number of the metals which also give hexaborides. A framework of linked

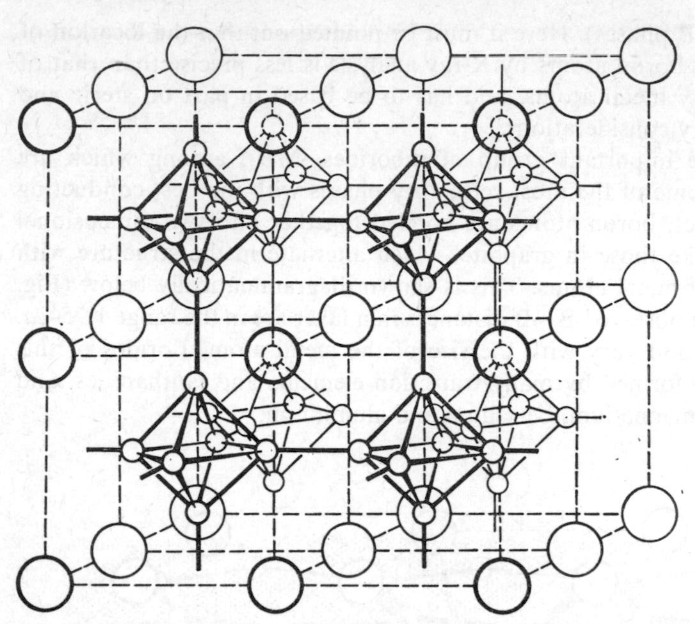

Figure 9.4 The crystal structure of CaB_6

polyhedra is, however, encountered again in the cubic dodecaborides MB_{12}. Here the B_{12} unit is not an icosahedron but an equilateral cubo-octahedron. This has six equivalent square faces, eight equivalent triangular faces, twenty-four equivalent edges and twelve equivalent vertices, and each boron atom forms one external bond to an equivalent boron atom of a neighbouring cubo-octahedron. This leads to a network in which the larger metal atoms are accommodated. Borides of this type are known for zirconium, uranium and a number of the rare earth metals.

No attempt will be made here to discuss in detail the nature of bonding in the metallic borides. One of the most successful approaches to this difficult problem is based on the assumption that electrons can be transferred from the relatively electropositive metal atoms to the electron-deficient boron framework until its bonding orbitals are filled. Excess metal valence electrons are then responsible for the metallic properties of the borides. This entails calculating from molecular orbital theory the electron deficiency of the framework, which has been done for the hexaboride and dodecaboride structures. In the network formed by the latter, for example, it is found that two electrons per metal atom are needed to stabilize the framework; these zirconium is able to provide in ZrB_{12} without ceasing to be a metallic conductor. Calculation shows that the boron framework in the hexaborides also requires two electrons per metal atom, and it is significant that in CaB_6, where calcium has only two electrons available, the electrical conductivity is low.

9.2.3. Polyboron halides.[3] The simplest of the halides with B—B bonds are of the type B_2X_4 (X = F, Cl, Br, I). The first to be isolated was diboron tetrachloride, which is formed when BCl_3 vapour is streamed at low pressure through a mercury arc:

$$2BCl_3 + 2Hg \rightarrow B_2Cl_4 + Hg_2Cl_2$$

The tetrafluoro compound is best obtained by the reaction of SF_4 with boron monoxide, which is made by dehydrating the acid $B_2(OH)_4$ resulting from the hydrolysis of $B_2(NMe_2)_4$ (*vide infra*):

$$2(BO)_n + 2nSF_4 \rightarrow nB_2F_4 + 2nSOF_2$$

There is a gas-phase reaction between B_2F_4 and BCl_3 which gives B_2Cl_4, and B_2Br_4 is obtained in a similar exchange reaction between B_2Cl_4 and BBr_3. The iodo compound has been made only by decomposing the vapour of BI_3 in a radio frequency electrical discharge.

It has been shown recently that, when boron trifluoride is passed at 1 mm pressure over boron at 2000°, a short-lived monofluoride BF is formed. If the streaming gas is condensed on a surface cooled to −190° a green solid is obtained which, when allowed to warm to room temperature, evolves a mixture of boron fluorides which, from mass spectroscopic evidence, contain up to fourteen boron atoms in the molecule.[4] Diboron tetrafluoride has been identified among the products and, when B_2F_4 is cocondensed with BF, an unstable fluoride B_3F_5 may be obtained which, from spectroscopic evidence, may be formulated as $FB(BF_2)_2$ (see Section 13.4.3).

The structures of solid B_2F_4 and B_2Cl_4 have been determined by X-ray analysis, and in both the six atoms lie in one plane. The B—B distances are 1·67 and 1·75 Å respectively and the B–halogen distances are almost identical with those in the trihalides. The Raman and infrared spectra of B_2Cl_4 in the liquid and gaseous states, however, show the two BCl_2 moieties to be in planes that are at right angles. The fluoro compound is a gas at room temperature, B_2Cl_4 and B_2Br_4 are colourless liquids, and B_2I_4 is a yellow solid.

Chemical studies have been restricted largely to B_2Cl_4, which behaves as an acid halide. Water converts it to $B_2(OH)_4$, which loses water at 100° and forms boron monoxide. Reaction with dimethylamine gives $B_2(NMe_2)_4$. This compound may be made much more conveniently by reaction of $(Me_2N)_2BCl$ with finely dispersed sodium in toluene at 125°. Methanol in the presence of a base to absorb hydrogen chloride gives the methoxide:

$$B_2Cl_4 + 4MeOH \rightarrow B_2(OMe)_4 + 4HCl$$

Diboron tetrachloride behaves as a dibasic Lewis acid, forming, for example, the adduct $B_2Cl_4.2NMe_3$ with trimethylamine. The tetrafluoride behaves similarly. Adducts with such donor species as ethers, phosphines and diphosphines are also formed normally. The chloride ion can act as a donor and with NMe_4Cl in liquid hydrogen chloride the salt $(NMe_4)_2{}^+(B_2Cl_6)^{2-}$ is formed. Cleavage of the B—B bond takes place on alkaline hydrolysis and also in the reaction with various types of olefin. A simple example is the reaction with ethylene shown below:

$$B_2Cl_4 + C_2H_4 \rightarrow Cl_2BCH_2CH_2BCl_2$$

9.2.4. Disproportionation of diboron tetrahalides. All of the diboron tetrahalides disproportionate to give the trihalide and a subhalide. Diboron tetrafluoride does so slowly at 100°, but B_2Cl_4, which has been most fully studied, decomposes slowly when in the liquid state at room temperature, and for B_2Br_4 reaction occurs even more readily. The products are highly coloured solids. The simplest isolated from B_2Cl_4 is tetraboron tetrachloride, a yellow solid the molecule of which is made up of a tetrahedron of four boron atoms, to each of which one chlorine atom is bonded (Fig. 9.5).

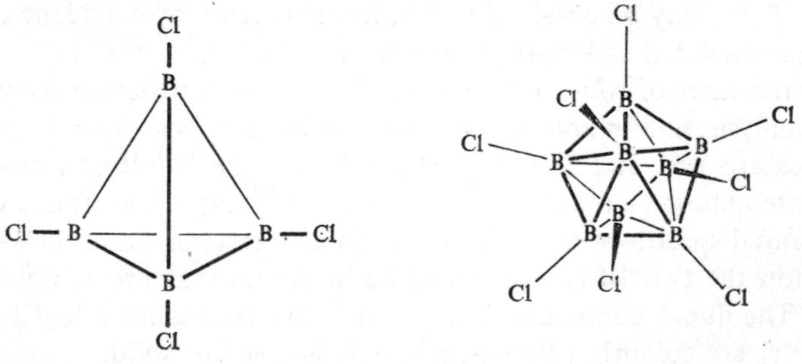

Figure 9.5

Figure 9.6

This system is electron-deficient, having ten bonds and only sixteen bonding electrons. The B—B distance is 1·70 Å, compared with 1·75 Å in B_2Cl_4, which is not electron-deficient. Few reactions have been studied because of the difficulty in preparing the compound in useful quantities, but it is known to react with dimethylamine to give $B_4(NMe_2)_4$ and partial methylation to B_4MeCl_3 occurs with dimethyl zinc. It is stable *in vacuo* to 70° and inflames spontaneously in dry air.

Octaboron octachloride (B_8Cl_8) has also been isolated from the products of the disproportionation reaction. Nothing is known of its chemistry, but the structure has been determined. It is made up of a dodecahedron of boron atoms with triangular faces, the shortest B—B distance being 1·78 Å (Fig. 9.6). This structure is also electron-deficient.

The major product from the spontaneous decomposition of B_2Cl_4 is a dark red paramagnetic compound which is usually formulated as $B_{12}Cl_{11}$, though mass spectral evidence is more consistent with $B_{11}Cl_{11}$.

The structure has been determined and consists of a cage of ten boron atoms identical with that in the $B_{10}H_{10}{}^{2-}$ ion, halogenation of which yields $B_{10}Cl_{10}{}^{2-}$ (see Section 8.6). There are eight chlorine atoms, each bonded to one of eight equatorial boron atoms, while a BCl_2 group is bonded to one of the apical borons and a BCl group to the other. Little is known about the chemistry of this interesting compound, but it gives a colourless adduct $B_{12}Cl_{11}.2NMe_3$ with trimethylamine.

The disproportionation of B_2Cl_4 is known to give other products which remain to be separated and characterized, and it can be expected that the other diboron tetrahalides will likewise yield new and interesting subhalides. The experimental work is, however, especially difficult as it involves working with small quantities of involatile solids which are moisture- and, sometimes, air-sensitive. It is possible that these halides are related structurally to a group of organoboron compounds of the type $(BR)_x$, the best known of which is $(BC_6H_5)_x$, where $x = 9$–12. This compound, which oxidizes rapidly in air, can be made by heating $(C_6H_5)BCl_2$ with dispersed sodium in boiling toluene.

9.2.5. Silicon. Elementary silicon crystallizes in only one form, which has a structure very similar to that of diamond. Germanium also has this structure; it is interesting that neither element has a modification resembling graphite. Simple catenation of silicon occurs in the hydrides, higher halides and in mixed silicon–germanium hydrides, which are described elsewhere (Section 8.9). There are also a few alkyl and aryl silicon derivatives with rings or chains of silicon atoms. Typical of these is the ring compound $((C_6H_5)_2Si)_4$ which is formed when $(C_6H_5)_2SiCl_2$ is treated with finely divided sodium in benzene. Dimethylsilicon dichloride treated similarly gives a hexameric ring compound $(Me_2Si)_6$ and also what appears to be a linear polymer $(Me_2Si)_n$, where n has an average value of about 50.

9.2.6. Silicides.[5] The most extensive range of compounds with Si—Si bonds is found among the metal silicides, the structures of many of which have been elucidated by X-ray analysis. These will not be described in detail, but as in the case of the borides, there is a wide range of structural types in which isolated silicon atoms, Si_2 groups, chains of linked silicon atoms, planar puckered networks and three-dimensional networks have all been identified. An example of a lattice with isolated silicon atoms is that of Ca_2Si, which has the $PbCl_2$ structure. There are chains of silicon atoms in USi and CaSi, planar hexagonal networks in one form of USi_2, while $CaSi_2$ has a lattice with puckered layers of silicon atoms alternating with metal atom layers. Many disilicides have layer structures, but in some (e.g. in $ThSi_2$ and $CeSi_2$) there is a honeycomb network of silicon atoms with metal atoms in the holes. The Si—Si distances are in all cases in the range 2·30–2·50 Å, compared with 2·34 Å in elementary silicon, and it is therefore reasonable to assume that Si—Si bonds are present.

9.2.7. Phosphorus.[6] Recent structural studies have shown the allotropic modifications of phosphorus to be more numerous than was at one time believed. White phosphorus, the common form, has a cubic lattice composed of discrete P_4 tetrahedra. Electron diffraction studies show that the vapour is also composed of this species, which dissociates appreciably to P_2 only above about 900°. The P—P bond distance in P_4 is 2·21 Å and the interbond angle 60°. The Raman spectra of solid and liquid phosphorus and of its solution in carbon disulphide are identical, so that the tetrahedral species must be present in each case. White phosphorus undergoes a transition at −77° to a hexagonal form, the lower symmetry of which is attributed to the fact that the P_4 units are no longer free to rotate at the lower temperature.

The second main modification, black phosphorus, exists in four forms, one of which is amorphous. The orthorhombic crystalline modification is made by heating white or red phosphorus at 220–270° in the presence of mercury, which acts as a catalyst, and with the addition of a seed of black phosphorus. It is a soft flaky material with a layer structure shown diagrammatically below (Fig. 9.7). It is a heavily puckered continuous network in which each phosphorus atom is bonded covalently to three neighbours with P—P distances of 2·22 and 2·24 Å and bond angles of 96° and 102°. This modification is a semiconductor.

Orthorhombic black phosphorus is transformed reversibly at high pressures to rhombohedral and cubic forms, the first of which also

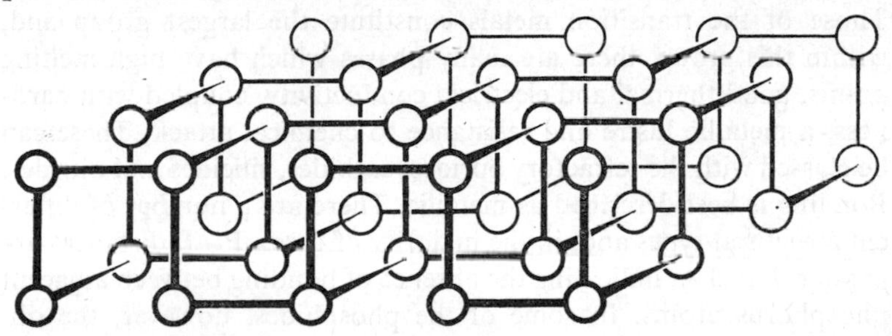

Figure 9.7 The crystal structure of black phosphorus—portion of one layer (idealized)

has a layer structure. The hexagonal network of phosphorus atoms is, however, less puckered than in the orthorhombic form. In the cubic modification each phosphorus atom is octahedrally co-ordinated to six like atoms at a distance of 2·38 Å.

Red phosphorus is assumed to be highly polymerized because of its insolubility, and is generally described as amorphous. It is, however, likely that it is not a single modification of phosphorus and it is known that crystalline materials can be produced when it is heated.

Tetrahedral As_4 molecules are present in arsenic vapour, and this may also be the structural unit in the very unstable yellow modification of the element. The metallic forms of arsenic, antimony and bismuth, on the other hand, all have layer structures, those of arsenic closely resembling the rhombohedral and orthorhombic forms of black phosphorus.

9.2.8. Phosphides. Most of the metallic elements react with phosphorus on heating to form phosphides, the structures of over one hundred of which have been established by X-ray analysis.[6] Detailed consideration of these lies outside the scope of this book, but a brief outline of the structural principles is of interest for comparative purposes.

Phosphides of Groups I and II metals have formulae which satisfy normal valency requirements (e.g. Na_3P, Ca_3P_2 and Mg_3P_2) and, since they hydrolyse readily with liberation of phosphine, are generally thought of as ionic. It is probably more correct to speak of the M—P bond as having considerable ionic character; P—P distances are at least 3·8 Å, so that there is no evidence of bonding between the phosphorus atoms.

The formulae of most phosphides lie in the range M_3P to MP_3. Those of the transition metals constitute the largest group and, within this group, there are many phases which have high melting points, good thermal and electrical conductivity, coupled with hardness, a metallic lustre and resistance to chemical attack. These can be classed with the refractory borides, carbides, silicides and nitrides. Bonding is best described as metallic. There are a number of different structural types and, in the majority of cases, P—P distances are greater than 3 Å, indicating the absence of bonding between adjacent phosphorus atoms. In some of the phosphides, however, the observed P—P distances are in the range 2·2–2·8 Å, which shows that there is covalent bonding, with chains or sheets of phosphorus atoms extending through the structure.

9.2.9. Cyclic phosphorus compounds. The first of these to be discovered was the so-called phosphobenzene, to which the formula $(PhP)_4$ was given on the basis of analysis and molecular weight determinations. Subsequent research has shown the position to be more complicated. A pentamer $(PhP)_5$ and an hexamer $(PhP)_6$ have been isolated and their structures determined, all of these species being formed in the condensation reaction between $PhPCl_2$ and $PhPH_2$. The hexameric arsenic analogue is also known and a structural investigation has shown the As—As distance in the ring to be close to the single bond value, as is the P—P distance in the phosphorus rings.

Several cyclic alkyl analogues of the above compounds are also known. They include molecules of the type $(RP)_4$, where R = Et, *n*-Pr, *n*-Bu and *n*-Oct and also a pentameric methyl compound $(MeP)_5$. Their structures have not been studied, but the n.m.r. spectra indicate ring structures. Perhaps the most interesting of these cyclic compounds are the trifluoromethyl derivatives $(CF_3P)_4$ and $(CF_3P)_5$, which are prepared by the reaction of CF_3PCl_2 with CF_3PH_2. Both are remarkably stable to heat. The structures of both have been determined, the ring in the tetramer being of the anti-pyramidal type (Fig. 9.8). The P—P distances in the ring are equal and correspond with a single bond. In the pentamer (Fig. 9.9) the P—P bond lengths in the ring vary significantly and the ring configuration, which is similar to that in $(CH_3As)_5$, is non-planar.

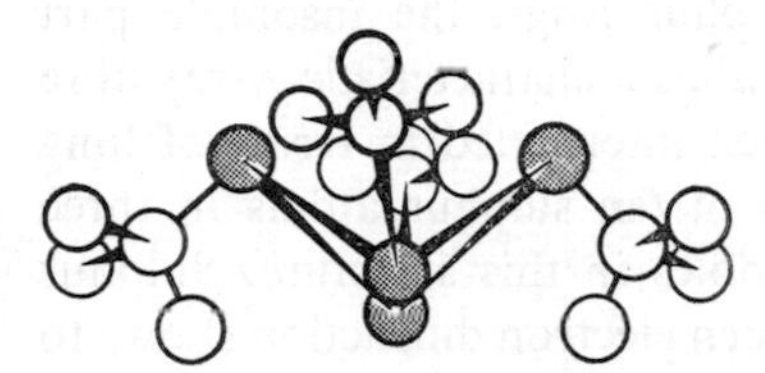

Figure 9.8 $[P(CF_3)]_4$

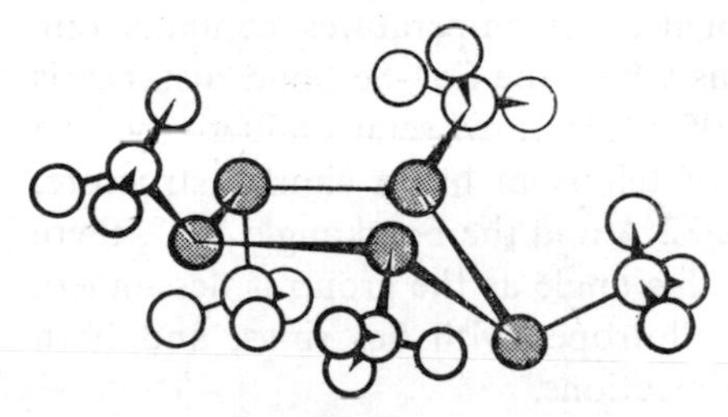

Figure 9.9 $[PCF_3]_5$

9.2.10. Group VI elements.[7] There is a very strong tendency for catenation in elementary sulphur and some of its compounds. In the stable orthorhombic form of the element (α-sulphur) X-ray analysis has shown there to be puckered crown-shaped S_8 rings with an S—S distance of 2·04 Å and an interbond angle of 108°. There is very weak bonding between these units in the solid structure, in keeping with its low melting point. This ring has been identified in another modification, γ-sulphur, though the structure of the common monoclinic form has not been determined. This is because it is rapidly transformed to the more stable α-form during the X-ray exposure. Recent work has shown that crystalline sulphur containing S_6, S_7, S_9, S_{10}, S_{12} and S_{18} molecules as the structural units can be prepared.

The S_8 ring persists as a major component of molten sulphur up to about 160°, but some chains are present in the liquid below this temperature. Between 160° and 180° the viscosity of the melt increases greatly, the simplest explanation being that long chains are formed. Paramagnetic resonance measurement indicates that these are biradicals with an average chain length of 10^6 at 170°. With further increase in temperature the melt becomes more fluid owing, it is believed, to a decrease in chain length and to the normal decrease in viscosity with rising temperature.

Plastic sulphur contains a fraction that is soluble in carbon disulphide, believed to contain sulphur rings, the insoluble part being the chain component. This exhibits a characteristic X-ray fibre pattern on stretching which has been interpreted in terms of long close-packed helices, with a period of ten sulphur atoms in three turns, there being S_8 rings in the holes in this structure. Sulphur vapour at 300° contains S_8 rings, which electron diffraction shows to be identical with those in solid α-sulphur.

Selenium, like sulphur, is polymorphic. Its grey hexagonal form, which is the stable modification at room temperatures, contains long spiral chains of selenium atoms where the Se—Se bond distance is 2·34 Å and the bond angle is 105°; this shows semiconductivity. The only known crystalline form of tellurium has a similar structure. The Te—Te bond distance is 2·82 Å and the bond angle 102°, there being a significant decrease in this angle as the group is descended. There are also two selenium allotropes with Se_8 rings, and both elements have amorphous modifications.

Sulphur, selenium and tellurium can also be oxidized to yield polyatomic cations. These are responsible for the intense colours produced when the elements are dissolved in strong acids such as HSO_3F and $H_2S_2O_7$, in which their presence can be established by cryoscopic, conductimetric and spectroscopic measurements. Oxidation of sulphur with AsF_5 or SbF_5 in anhydrous HF gives the compounds $S_{16}(AsF_6)_2$, $S_8(AsF_6)_2$, $S_{16}(SbF_6)_2$ and $S_8(Sb_2F_{11})_2$ as crystalline solids, while, with peroxydisulphuryl fluoride in HSO_3F, $S_4(SO_3F)_2$ is formed. Crystalline derivatives of Se_8^{2+}, Se_4^{2+} and Te_4^{2+} cations may be obtained by similar methods. An X-ray structural study of $Se_4(HS_2O_7)_2$ and $Se_8(AlCl_4)_2$ has shown the Se_4^{2+} cation to be square planar and the Se_8^{2+} cation to be a non-planar eight-membered ring.

9.2.11. The sulphanes and related compounds. The sulphanes, or hydrogen polysulphides, are compounds of the general formula H_2S_n. They are of interest in the present context because they show that the tendency of sulphur to form unbranched chains persists in some of its compounds. Several preparative methods are available, the first depending on the decomposition of solutions of alkali or alkaline earth metal polysulphides with acid, preferably below 0°, when an oil separates. The polysulphide solutions are formed by dissolving sulphur in a solution of a sulphide; this in itself is a remarkable phenomenon. They may also be made directly from the metal and sulphur in a melt or, preferably, in liquid ammonia solution. Distillation of the oil in high vacuum yields sulphanes from H_2S_2 to H_2S_6.

Sulphanes are also formed when a dithionite solution is decomposed by acid and in the reaction of sulphanes with chlorosulphanes at a low temperature (e.g. −50°), moisture being rigorously excluded; for example,

$$H{-}S{-}S{-}H + Cl{-}S{-}Cl + H{-}S{-}S{-}H \rightarrow H{-}S_5{-}H + 2HCl$$

This type of reaction has been used, starting from higher sulphanes, to yield members of the series with up to at least ten sulphur atoms in the chain. It is less difficult to purify the product than when the compounds are prepared from polysulphides.

Molecular weights of sulphanes can be determined cryoscopically. The Raman spectra of the liquids show characteristic S—S stretching frequencies in the range 400–500 cm^{-1} and S—S—S bending frequencies at 150–200 cm^{-1}, but there are no features in the spectrum that can be associated with chain-branching. There is also good evidence for the absence of chain-branching in organic sulphanes of the type R_2S_n. The chemistry of the sulphanes has not been extensively studied. They are thermodynamically unstable with respect to decomposition to sulphur and hydrogen sulphide, and it is not surprising, therefore, that comparable selenium and tellurium compounds are unknown, though short-chain organic analogues with two or three selenium or two tellurium atoms have been isolated.

Compounds of the type S_nX_2 (where X = Cl, Br or CN) have also been prepared. The chlorine compounds were first made by passing the vapour of S_2Cl_2 mixed with hydrogen through a narrow water-cooled tube. Extending axially within this was a second tube which could be heated to as high as 900° (the so-called 'hot–cold' tube apparatus). With this arrangement to secure shock cooling of the high-temperature reaction products, oils and solids were formed with compositions in the range $S_{20}Cl_2$ to $S_{100}Cl_2$. The limiting solubility of sulphur in S_2Cl_2 corresponds with the formula $S_{3\cdot 2}Cl_2$. There is, therefore, an *a priori* case for assuming the presence of a mixture of chlorosulphanes which is supported by Raman spectral evidence.

A more satisfactory preparative method involves the elimination of hydrogen chloride in the reaction between a sulphur chloride and a sulphane at −80°; for example,

$$\text{Cl—S—Cl} + \text{H—S}_4\text{—H} + \text{Cl—S—Cl} \rightarrow S_6Cl_2 + 2HCl$$

Only S_3Cl_2 can be distilled in vacuum without decomposition. There are also similar series of bromosulphanes (up to S_8Br_2) and cyano-sulphanes, the latter made from chlorosulphanes by reaction with mercuric thiocyanate:

$$S_xCl_2 + Hg(SCN)_2 \rightarrow S_{x+2}(CN)_2 + HgCl_2$$

Related to the above, though hardly to be regarded as polymeric except in the sense that a sulphur chain is present, are the sulphane monosulphonic and disulphonic acids. Until recently, only one member of the first series was known, and then in the form of its salts, the thiosulphates. The free acid has, however, now been prepared by reaction of chlorosulphonic acid with hydrogen sulphide in ether solution:

$$HSO_3Cl + H_2S \rightarrow HSO_3SH + HCl$$

Using higher sulphanes, this method has given members of the series $HSO_3(S_n)SH$, with $n = 2\text{–}6$. Both the acids and their salts decompose slowly at room temperature.

Salts of the sulphane dithionic acids (the polythionates) have been known for over a hundred years, though it was not until recently that the presence of an unbranched chain of sulphur atoms was established by X-ray analysis for the ions $S_4O_6^{2-}$, $S_5O_6^{2-}$ and $S_6O_6^{2-}$. The free acids are all unstable in water but have now been prepared by reaction of sulphur trioxide with sulphanes in ether solution:

$$H_2S_n + 2SO_3 \rightarrow HSO_3(S_n)SO_3H$$

Alternatively, sulphane monosulphonic acids may be condensed either with chlorosulphanes or with chlorosulphonic acid:

$$2HSO_3(S_x)H + Cl(S_y)Cl \rightarrow HSO_3(S_{2x+y})SO_3H + 2HCl$$
$$HSO_3(S_x)H + ClSO_3H \rightarrow HSO_3(S_x)SO_3H + HCl$$

Acids with up to fourteen sulphur atoms in the molecule have been prepared by these methods. They are stable for some time in ether solution at room temperature but are decomposed by water.

Although the selenium and tellurium analogues of these acids are unknown, salts of mixed anions, with a selenium or tellurium atom occupying the central position in the chain, have been prepared. The figure below (Fig. 9.10) shows the configuration of the *cis* form of the telluropentathionate ion as it occurs in barium telluropentathionate dihydrate. The *trans* form of this ion is present in the ammonium salt.

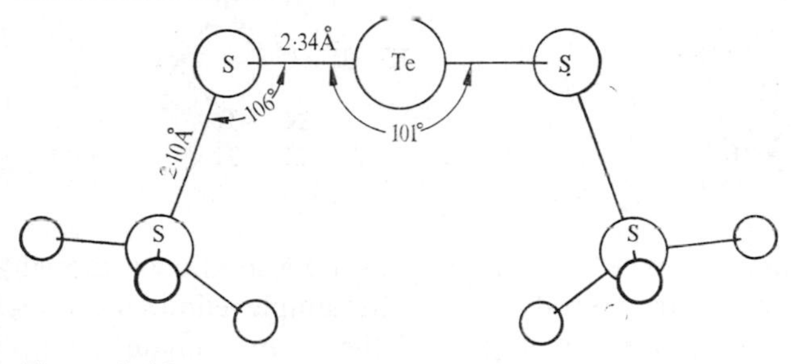

Figure 9.10

9.3 Borazines, silicones and related compounds

9.3.1. Boron–nitrogen polymers.[8] In this section we will consider three types of compound with a B—N bond. In the first, bonding is due to donation of an electron pair on nitrogen to boron as, for example, in the amine boranes, $R_3N.BR'_3$, where R may be hydrogen or an alkyl or aryl radical and R′ hydrogen, an alkyl or aryl group, an alkoxy group or a halogen. Such compounds do not polymerize and will not be discussed here, though they are referred to elsewhere (Section 8.5) in connection with the chemistry of boranes.

In the second type of compound, the aminoboranes, there is a σ-bond between boron and nitrogen and, in addition, π-bonding between the two atoms involving the free electron pair on nitrogen. The actual bond order depends on the substituents on the >B—N< group. There is evidence of restricted rotation, but the bond order is lower than for a C═C bond. Many compounds of this sort are known, the usual method for their preparation being by an elimination reaction under the action of heat, as shown below:

$$R_2NH \rightarrow BX_3 \xrightarrow{\text{heat}} R_2NBX_2 + HX$$

$$(X = H, Cl, Br;\ R = \text{alkyl})$$

A similar reaction occurs with various other substituents on boron and nitrogen.

Aminoboranes which have hydrogen and/or a halogen as one of the substituents can undergo reversible dimerization (i.e. the nitrogen atom has retained some of its donor function). Examples of compounds which behave in this way are $(CH_3)_2NBCl_2$ (the dimer of which is shown in Fig. 9.11), $(CH_3)_2NBH_2$ and $H_2NB(C_6H_5)_2$.

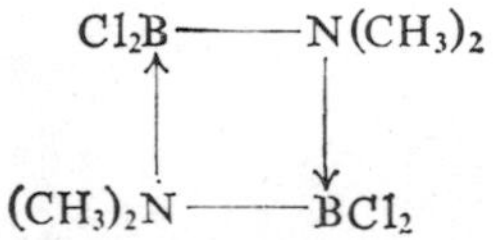

Cl_2B — $N(CH_3)_2$ (arrow N→B)
$(CH_3)_2N$ — BCl_2 (arrow N→B)

Figure 9.11

H H
N—B
H—B N—H
N—B
H H

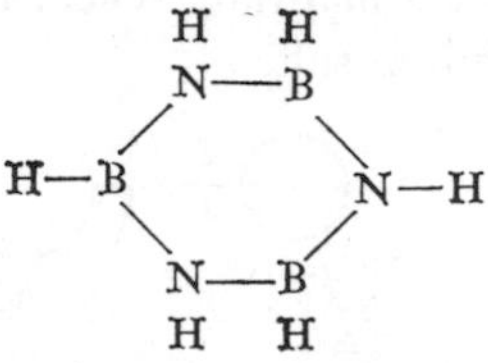

Figure 9.12

Aminoborane polymers with larger rings are also known, though they are not necessarily produced by the simple elimination reactions described above. Treatment of the 'diammoniate' of diborane with sodium amide in liquid ammonia, for example, gives $(H_2NBH_2)_x$, where $x = 2, 3, 4$ or 5. A number of compounds of this type also result from the addition reactions of borazine and its derivatives, which will be described later.

Borazine and its derivatives make up the third class of compound with B—N bonds. The parent borazine, $B_3N_3H_6$, was first made by Stock and Pohland in 1926 by heating diborane with ammonia at 125°. It is a colourless liquid (b.p. 55°) and is isoelectronic with benzene, which it resembles closely in several physical properties (e.g. b.p., m.p., liquid density and surface tension). The molecule is planar (Fig. 9.12) with one B—N bond distance of 1·44 Å throughout the ring compared with a C—C distance of 1·42 Å in benzene. The N—H and B—H distances are 1·02 and 1·20 Å respectively. The B—N distance is less than that expected for a single bond (1·60 Å in $H_3N \rightarrow BH_3$), and this is attributed to delocalized p_π—p_π bonding. Delocalization of the electrons in π-orbitals is also in keeping with the fact that the molecule shows magnetic anisotropy, though the effect is smaller than in benzene. The polarity of the B—N bond is B^+—N^-, and is the resultant of an electron drift from boron to nitrogen associated with the σ-bond and from nitrogen to boron associated with the π-bond. It is modified considerably when hydrogen is replaced by various substituents.

Borazine has often been referred to as inorganic benzene. While there is unquestionably some resemblance between the two compounds as far as bonding is concerned, there is little in the chemistry of the compound to suggest aromatic character as understood by organic chemists. The ring is broken by alkaline hydrolysis, nitrogen being eliminated as ammonia. A slow reaction occurs with water at room temperature and hydrogen is evolved. Reaction with ethanol is rapid and yields chiefly ammonia and triethoxyborane. Thermal decomposition in the absence of oxygen occurs at roughly 300°; the products are described later. All of the B-halogenated borazines are known, though only $Br_3B_3N_3H_3$ has been made by direct bromination.

Borazine is now prepared by reduction of B-trichloroborazine ($Cl_3B_3N_3H_3$) with $LiAlH_4$ in ether solution, the chloro compound being made by the reaction:

$$3BCl_3 + 3NH_4Cl \xrightarrow[\text{chlorobenzene}]{\text{refluxing}} Cl_3B_3N_3H_3 + 9HCl$$

Use of an alkylammonium halide in this reaction gives a B-trichloro-N-trialkyl borazine, which may be similarly reduced. The B-chloro compounds are converted to B-alkyl or B-aryl derivatives by reaction with Grignard reagents, and numerous B-dialkylamino derivatives are formed in reactions with secondary amines. Sodium alkoxides give B-alkoxy derivatives and the chloroborazines also react normally with silver salts such as AgF, AgCN or AgOCN. A number of derivatives with partial substitution or with mixed substituents are also known. Some borazine derivatives may be prepared directly by elimination reactions; for example,

$$3B(CH_3)_3 + 3C_6H_5NH_2 \xrightarrow{\text{heat}} (CH_3)_3B_3N_3(C_6H_5)_3 + 6CH_4$$

$$3(SiH_3)_2NBF_2 \xrightarrow{\text{heat}} F_3B_3N_3(SiH_3)_3 + 3SiH_3F$$

In contrast to the B—Cl bond, the N—H bond in borazines undergoes few reactions that leave the ring intact. Reaction of B-methyl borazines with methyl lithium has, however, given mono-, di- and tri-N-lithio-B-methyl borazines, the use of which in further synthesis is referred to later. Hexachloroborazine ($Cl_3B_3N_3Cl_3$) has also been made by reaction of BCl_3 with NCl_3 in CCl_4 solution at temperatures up to 45°. The compound melts at 176° without decomposition and reacts violently with water to form chloramine:

$$Cl_3B_3N_3Cl_3 + 9H_2O \longrightarrow 3H_3BO_3 + 3NH_2Cl + 3HCl$$

Borazine is not known to form π-bonded 'sandwich' compounds with transition metals corresponding with those which benzene and other aromatic molecules give. Recently, however, hexaalkyl borazines have been shown to do so, a typical reaction being that of hexamethyl borazine with either $(CO)_3Cr(CH_3CN)_3$ or $(CO)_3Co(NH_3)_3$ in dioxane; for example,

$$(CO)_3Cr(CH_3CN)_3 + B_3N_3(CH_3)_6 \rightarrow [B_3N_3(CH_3)_6]Cr(CO)_3 + 3CH_3CN$$

The product is stable in air, soluble in organic solvents and sublimable in vacuum. The same compound is formed when the borazine derivative and $Cr(CO)_6$ are irradiated at low pressure. The structures of these compounds have not yet been determined, but it seems highly likely that they are π-bonded complexes and, if so, would reflect the aromatic character of the borazine ring. It seems certain that related complexes of other transition metals will also be found to exist.

Reference has already been made to cyclic aminoborane polymers and to the fact that such compounds can also result from addition

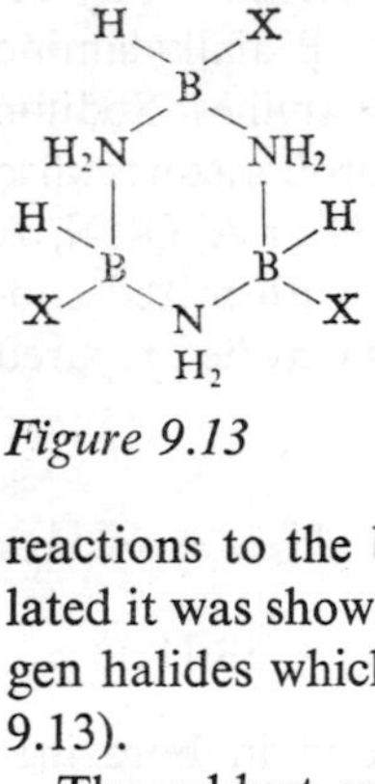

Figure 9.13

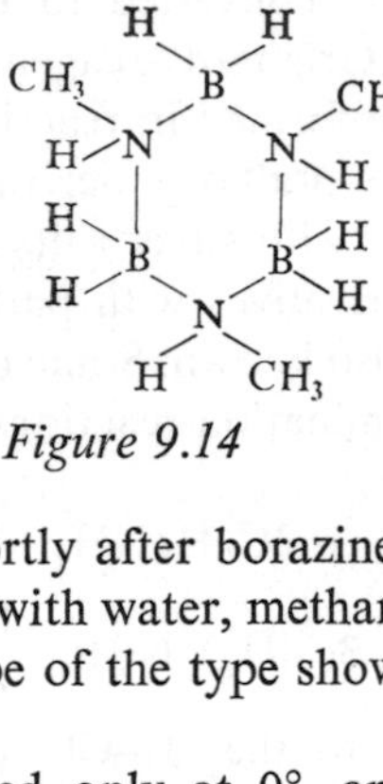

Figure 9.14

reactions to the borazine ring. Shortly after borazine was first isolated it was shown to form adducts with water, methanol and hydrogen halides which are believed to be of the type shown above (Fig. 9.13).

The adduct with water is formed only at 0°, and ring fission occurs at higher temperatures. The methoxide derivative also breaks down readily to form products with the alkoxy group bonded to boron, while the hydrogen chloride adduct loses hydrogen when heated, giving a B-chloroborazine. Addition of hydrogen chloride also occurs with B-methyl borazines, and the adducts again lose hydrogen on heating. It will be noted that the additions are in accord with the polarity of the B—N bond ($B^{\delta+}$—$N^{\delta-}$).

Formally these adducts are of the cyclohexane type and stereo isomers would be expected to exist. This has been confirmed by the isolation of two forms of $B_3H_6N_3H_3(CH_3)_3$ (Fig. 9.14), which was first prepared from methylamine and diborane. On the basis of the proton n.m.r. spectra and from analogies with the conformational analysis of the cyclohexanes, it is believed that one modification has the three methyl groups in equatorial positions and the other has two methyl groups equatorial and one axial, both isomers being probably in the chair form.

The six-membered borazine ring appears to be the preferred polymeric form. Linear polymers are formed, as far as is known, only when steric factors are unfavourable for the ring structure. This is so, for example, in the elimination reaction between phenyl boron dichloride and *n*- or *iso*butylamine, which gives a linear polymer $(C_6H_5B{-}NC_4H_9)_n$, where $n \sim 20$–40. Reaction of *tert*-butylamine with BCl_3 also gives an eight-membered ring rather than the borazine structure.

Two polynuclear borazines have been isolated from the mixture of products formed when borazine vapour is heated to 340–440° in a flow system. Hydrogen is eliminated and the analogues of naphthalene (Fig. 9.15) and biphenyl (Fig. 9.16) can be separated. A partially methylated derivative of the first has also been described.

```
    H      H
    N      N
  /   \  /   \
HB     B      BH
|      |      |
HN     N      NH
  \   /  \   /
    B      B
    H      H
```

Figure 9.15

```
      H  H           H  H
      N—B            N—B
     /    \         /    \
  HB       N—B          NH
     \    /         \    /
      N—B            N—B
      H  H           H  H
```

Figure 9.16

Polyborazines may also be synthesized with the aid of N-lithioborazines. A simple example is afforded by the reaction of N-lithiopentamethyl borazine with B-chloropentamethyl borazine in which a B—N bond is formed (see Fig. 9.17). Polynuclear borazines with a B—B bond between the rings are formed in the coupling reaction between B-chloropentaalkyl borazines and finely divided potassium in an inert solvent: $ClMe_2BNMe_3$, for example, gives the compound shown in Fig. 9.18.

Linear polymers may also be produced by using various aliphatic and aromatic diamines to bridge borazine rings. For example, ethylenediamine reacts with a borazine containing two B—Cl groups

```
     Me                  Me
     B                   N
MeN     N────────────B       BMe
 |      |            |       |
MeB     BMe       MeN        NMe
     N                   B
     Me                  Me
```

Figure 9.17

```
     Me                  Me
     N                   N
MeB     B────────────B       BMe
 |      |            |       |
MeN     NMe       MeN        NMe
     B                   B
     Me                  Me
```

Figure 9.18

as shown below (R = 2,6-dimethylphenyl). Such products are often stable to temperatures up to at least 200–250°, but their hydrolytic stabilities are usually rather low.

```
      H                                       H
      B                                       B
  RN     NR                               RN     NR
n  |      |   + nC2H4(NH2)2  ──→          |      |
  ClB    BCl                          ───B        B─NH(CH2)2NH───
      N                                       N
      R                                       R                 n

                                    + 2n HCl
```

Before leaving this topic it must be stressed that borazine chemistry is only part of a very much wider field. To mention only one facet, there are other cyclic B—N systems of the borazine type with the following sequences of atom in the ring:

B—N—B—N, N—N—B—N—B and N—N—B—N—N—B.

In addition, other rings contain boron, nitrogen and oxygen. The borazine ring is, however, of special interest not only because the basic synthesis is relatively easy but also because of the wide range of reactions that it undergoes.

Boron nitride. There is an interesting structural relationship between the borazine ring and the hexagonal form of boron nitride, which has a layer structure. Each layer is a planar network of fused borazine rings, as shown below (Fig. 9.19), the length of the B—N bond

```
     |      |      |
     B      B      B
  N     N      N      N
  |     |      |      |
  B     B      B      B
     N      N      N
     |      |      |
```

Figure 9.19

being 1·45 Å compared with 1·44 Å in $B_3N_3H_6$. The interplanar spacing is 3·30 Å and the compound resembles graphite in being soft and flaky, though it is white. The most important difference is that graphite is an electrical conductor whereas boron nitride is an insulator. In the hexagonal form the boron and nitrogen atoms of successive sheets are directly superposed, but modifications of this structure have been described with different stacking of the layers.

A modification of boron nitride with the cubic zinc blende structure is produced when the hexagonal form is heated to 1500–2000° at pressures of 50,000–90,000 atm. The transformation is catalysed by small amounts of various substances including the alkali and alkaline earth metals and their nitrides, but the mode of their action is unknown. A rather similar transformation of graphite to diamond occurs under comparable conditions; the structure of this new modification of boron nitride is very similar to that of diamond; it also has about the same hardness as diamond.

9.3.2. Boron–phosphorus compounds.[9, 10] Phosphines form a large number of Lewis acid–base complexes with boranes, BR_3, where R may be hydrogen, a halogen or an organic group. These have no tendency to polymerize but they vary greatly in stability and for some it is possible to carry out an elimination reaction and produce a phosphinoborane. Dimethylphosphine, for example, reacts with diborane to form $Me_2PH.BH_3$, which loses hydrogen at 150°:

$$Me_2PH.BH_3 \xrightarrow{-H_2} (Me_2PBH_2)_3 + (Me_2PBH_2)_4$$

This particular reaction also gives a small amount of material of higher molecular weight. The cyclic structures of the trimer and tetramer have been proved by X-ray analysis. It is not necessary to isolate the phosphine–borane adduct in every case, an example being the reaction between $(CF_3)_2PH$ and B_2H_6 at 50–75°, which gives $[(CF_3)_2PBH_2]_3$ with a little of the tetramer.

The most striking feature of these cyclic materials is their high chemical and thermal stability. The compound $[(CH_3)_2PBH_2]_3$ is, for example, stable in vacuum to about 300°. The B—H bonds are also very resistant to hydrolysis. Hydrogen atoms bonded to boron may be wholly or partially replaced by halogen by reaction with chlorine, bromine or milder halogenating agents. Reaction also occurs with alkyl halides in the presence of Lewis acid catalysts to give B-halogenated derivatives and the corresponding alkanes. A variety of other groups (alkyl, CN and SCN) have also been introduced by metathetical reactions with B-iodo derivatives.

A number of arsenic analogues of the above type are known, but will not be described here. No phosphorus analogue of borazine is known at present; thermal decomposition of the adduct $PH_3.BH_3$ results in loss of hydrogen, but the product appears to be more highly polymeric.

9.3.3. The silicones.[11] The silicones are a group of neutral inorganic polymers based on rings, chains or networks of alternating silicon and oxygen atoms. The same is true of ionic silicates, the essential point of difference being illustrated by comparing a typical silicone, hexamethylcyclotrisiloxane (Fig. 9.20) with its naturally occurring ionic counterpart, $Si_3O_9^{6-}$ (Fig. 9.21).

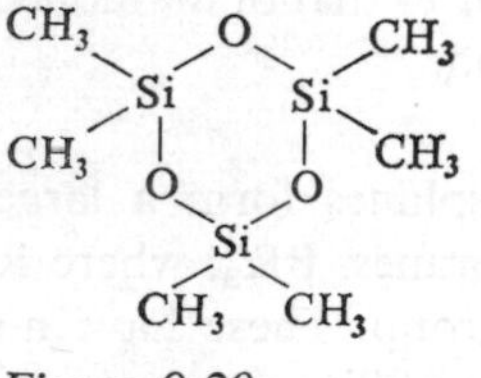

Figure 9.20

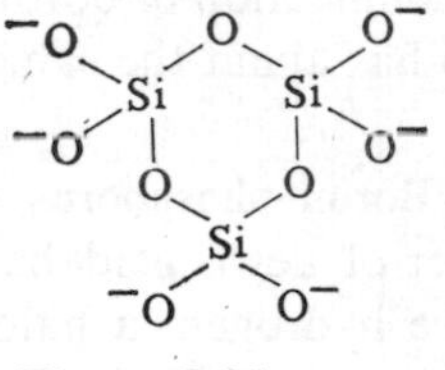

Figure 9.21

In Fig. 9.20 the residual valencies of silicon are satisfied by neutral organic groups, whereas in Fig. 9.21 they are satisfied by σ-bonded oxygen atoms, each of which carries a single negative charge.

The foundations of the subject were laid in the period 1901–40 by Kipping[12] in a far-ranging investigation of organosilicon compounds, the preparation of many of which was based on the use of the Grignard reagent. He found that hydrolysis of alkyl or aryl silicon halides yielded oils, or in some cases crystalline solids, and concluded that a condensation reaction had occurred involving intermolecular elimination of water from molecules containing Si—OH bonds; for example,

$$2\,{>}\mathrm{Si{-}Cl} \rightarrow 2\,{>}\mathrm{Si{-}OH} \rightarrow \mathrm{H_2O} + {>}\mathrm{Si{-}O{-}Si}{<}$$

The hydroxy intermediate is now known as a silanol, and may sometimes be isolated (*vide infra*). The condensation product is a polysiloxane, the trivial name silicone being used for this class of compound as a whole.

Hydrolysis of halides of the type R_3SiX can give only one possible type of condensed product, $R_3Si—O—SiR_3$, a disiloxane, many of which have been characterized. Diols formed by hydrolysis of R_2SiX_2 can condense in either of two ways, giving chains with terminal R_2SiOH groups or rings:

$$R_2SiX_2 \rightarrow R_2Si(OH)_2 \rightarrow HO(R_2SiO)_nOH + (R_2SiO)_m$$

When three halogen atoms are bonded to silicon a cross-linked structure must result on hydrolysis and condensation. These various possibilities will be considered more fully when the methods for synthesizing the organosilicon halide intermediates have been described.

Preparation of silicone intermediates. The chief intermediates are the halides, for which the main methods for preparation are the following.

(a) *Use of the Grignard reagent.* Silicon tetrachloride may be converted readily to alkyl or aryl derivatives by this means. A mixture of products always results, though one may be made to predominate by controlling the proportions of reactants. Alkoxysilanes also react with Grignard reagents and subsequent hydrolysis of the residual Si—(OR) groups follows the same course as for Si—Cl:

$$Si(OEt)_4 + RMgX \rightarrow R_xSi(OEt)_{4-x} + Mg(OEt)X$$

Use of the Grignard reagent or of similar reagents such as the alkyls of aluminium or zinc is not favoured on a large scale because of the need to handle large amounts of inflammable solvents. It also implies a batch process.

(b) *Direct silicon process.* This is particularly valuable for preparing methyl chlorosilanes, which are of great commercial importance. Methyl chloride vapour is passed at about 300° over a sinter of silicon powder with about 10 per cent of copper powder. The latter acts as a catalyst for the direct addition of the elements of the alkyl halide to silicon:

$$2MeCl + Si \xrightarrow[300°]{Cu} Me_2SiCl_2$$
$$(+ MeSiCl_3 + Me_3SiCl + \text{other compounds})$$

The yield of Me_2SiCl_2 (b.p. 69·6°) is over 50 per cent. Careful fractionation is required to separate it from $MeSiCl_3$ (b.p. 66·4°) and Me_3SiCl (b.p. 57·7°).

The overall reaction is exothermic. A mechanism involving the intermediate formation of methyl copper has been proposed:

$$MeCl + 2Cu^{o} \rightarrow MeCu + CuCl$$
$$MeCu \rightarrow Cu^{o} + Me$$
$$CuCl + Si^{o} \rightarrow Cu^{o} + SiCl$$
$$Me + SiCl \rightarrow MeSiCl$$

The above reaction cycle, if continued, will build up the observed products. Methyl radicals are known not to react directly with silicon. By-products containing hydrogen arise from breakdown of methyl radicals. The process operates less satisfactorily with higher alkyl halides, one possible reason being that the alkyl radicals become less stable as the series is ascended. A mixture of MeCl and HCl may also be used. This yields mainly $MeSiCl_3$, but other hydrogen-containing silanes are also formed:

$$MeCl + 2HCl + Si \xrightarrow[300^\circ]{Cu} MeSiCl_3 + H_2$$

Copper also catalyses the reaction of chlorobenzene with silicon at temperatures around 500°, but if it is replaced by silver reaction occurs at 400°. The role of the two metals is presumably similar, the products being phenylchlorosilanes.

(c) *Aromatic silylation.* A useful range of aryl silicon halides may be prepared by the reaction of benzene and other aromatic hydrocarbons at 230–300° with compounds containing an Si—H bond. Friedel–Crafts catalysts (BF_3, BCl_3 or $AlCl_3$) are used. A simple example is

$$C_6H_6 + HSiCl_3 \rightarrow C_6H_5SiCl_3 + H_2$$

This particular reaction also gives $(C_6H_5)_2SiCl_2$, $C_6H_5SiHCl_2$ and other products.

(d) *Silane–olefin additions.* A wide variety of olefins and acetylenes are able to react under suitable conditions with halosilanes or organohalosilanes containing one or more Si—H bonds in the molecule; for example,

$$Cl_3SiH + H_2C{=}CHR \rightarrow Cl_3SiCH_2CH_2R$$

This type of insertion reaction across the Si—H bond will often occur without a catalyst at 200–300°, but various initiators are used to secure a lower reaction temperature, which may be important if the desired products are thermally unstable. They include organic peroxides, or the mixture may also be irradiated with ultraviolet light. In such cases a free radical reaction mechanism is believed to operate.

Some unsaturated compounds (e.g. styrene, acrylonitrile) are readily polymerized by peroxides. It is found, however, that very small amounts of platinum and other Group VIII metals on a support such as carbon also catalyse olefin additions, often enabling reaction temperatures below 100° to be used. Examples are shown below:

$$Cl_3SiH + C_6H_5CH{=}CH_2 \xrightarrow{Pt} Cl_3SiCH_2CH_2C_6H_5$$

$$Cl_2SiH(CH_3) + CH_3C({=}O)OCH{=}CH_2 \xrightarrow{Pt} CH_3Cl_2SiCH_2CH_2OC({=}O)CH_3$$

Per- and polyfluoroalkyl silicon compounds are also best prepared by the above type of reaction, though a Grignard reagent or perfluoroalkyl lithium compound may also be used; for example,

$$n\text{-}C_3F_7MgI + Me_3SiCl \rightarrow n\text{-}C_3F_7SiMe_3$$

$$n\text{-}C_3F_7Li + SiCl_4 \rightarrow (n\text{-}C_3F_7)_nSiCl_{4-n}$$

Yields are, however, then low, as they also are if a perfluoroalkyl halide such as CF_3Br is substituted for an alkyl halide in the direct silicon process.

Reaction of a fluorinated olefin with the Si—H bond may be initiated in the same ways as with an olefin; for example,

$$C_2F_4 + Cl_3SiH \xrightarrow[\text{or } (PhCO)_2O_2]{\text{ultraviolet heat}} Cl_3SiCF_2CF_2H$$

$$CF_3CH{=}CH_2 + SiH_4 \xrightarrow{h\nu} CF_3CH_2CH_2SiH_3$$
$$(+ (CF_3CH_2CH_2)_2SiH_2 + (CF_3CH_2CH_2)_3SiH)$$

These reactions also give telomers of the type $H\left[\begin{array}{c}|\quad|\\ C{-}C\\ |\quad|\end{array}\right]_n Si\!\lessdot$ which multiple insertion has occurred.

Interest in such fluorinated products was stimulated by the possibility of making silicones with a thermal and oxidative stability greater than that of the alkyl and aryl analogues. The stability was found to be enhanced in some cases and to vary widely with the nature of the fluorinated groups, but aqueous bases cleaved the C—Si bond in all cases where there was fluorine on a carbon which was α or β to silicon. The desired hydrolytic stability was, however, obtained with fluorine in the γ-position (i.e. separated by at least two CH_2 groups from silicon). Some commercially important materials have been developed on this basis, among them a silicone rubber based on the polysiloxane $[CF_3CH_2CH_2SiMeO]_n$.[13]

Redistribution reactions. Some of the preparative methods outlined above give among their products materials of limited commercial use, which it is therefore desirable to convert to some more valuable material. This is often done by a so-called redistribution reaction, an example being the synthesis of Me_2SiCl_2 from unwanted Me_3SiCl and $MeSiCl_3$ formed in the direct silicon process. These materials are heated with aluminium chloride:

$$Me_3SiCl + MeSiCl_3 \underset{350°}{\overset{AlCl_3}{\rightleftharpoons}} 2Me_2SiCl_2$$

Aryl derivatives may be made similarly as, for example, in the reaction of $(C_6H_5)_4Si$ with $SiCl_4$ to give $(C_6H_5)_2SiCl_2$ as the main product.

Hydrolysis of organohalosilanes. Reference has been made already to the fact that compounds of the type R_3SiCl on hydrolysis give disiloxanes $(R_3Si)_2O$, many of which have been characterized. When there are two halogen atoms in the molecule both cyclic and linear polymers result. Thus Me_2SiCl_2 gives cyclic products of the type $[(CH_3)_2SiO]_n$, where $n = 3$–8. These are fairly volatile and may be separated from one another by fractional distillation (e.g. $(Me_2SiO)_4$ has b.p. 175° at 760 mm and $(Me_2SiO)_8$ has b.p. 168° at 20 mm). This also enables them to be separated from the less volatile linear polymers, where the chain commonly contains several hundred (Me_2SiO) units, with a considerable spread in molecular weights. The chains are usually terminated by —SiR_2OH groups, and the length is greatly influenced by the conditions of hydrolysis and the presence of impurities which may provide end groups. Cohydrolysis of a mixture of $(CH_3)_3SiCl$ and $(CH_3)_2SiCl_2$ is one of the standard ways of controlling chain growth since the former yields $(CH_3)_3SiOH$, condensation of which with the terminal OH—$Si(CH_3)_2$ group of a chain prevents its further development.

Equilibriation reactions. The Si—O—Si bond in a polysiloxane may be broken by reaction with either acids or bases, and this permits siloxane structures to be modified. A simple example is the preparation of unsymmetrical disiloxanes by heating a mixture of two symmetrical species with a strong mineral acid; for example,

$$(Me_3Si)_2O + (Et_3Si)_2O \rightleftharpoons 2Me_3Si—O—SiEt_3$$

In much the same way a cyclic polysiloxane when mixed with a disiloxane and heated with an acid or a base will give short-chain linear polymers. The rings are opened and there may be some chain growth. End groups are then provided by monofunctional groups from the disiloxane.

Properties and uses of silicones. There are certain properties common to all silicones. They have high thermal stability and, in the absence of air, will withstand temperatures up to roughly 250–300°. Oxidation of organic groups in the structure first occurs at 200 250° and leads to cross-linking. They are also unaffected by the majority of mild chemical reagents such as weak acids and alkalis or salt solutions. Many of the low molecular weight species will dissolve in solvents such as benzene, ether or carbon tetrachloride and, either in the form of films produced from solution or in bulk, all have good electrical insulating properties.

The linear polymers of low molecular weight have freezing points in the range —50° to —80°, with a marked tendency to supercool. Viscosities of low molecular weight cyclic or linear siloxanes are less than those of hydrocarbons of comparable molecular weight, and temperature coefficients of viscosity are small. The surface properties of the silicones are also distinctive. Their surface tensions are low and when applied to surfaces they form films which are highly water-repellent. It appears that in such films the molecules are oriented with the hydrocarbon groups facing outwards.

The uses of silicones may be grouped on the basis of their physical characteristics. The oils are usually chain polymers, though blending with other types of silicone is usual to secure modifications in properties. The low temperature coefficient of viscosity makes them valuable as hydraulic fluids. Their good electrical insulating properties and high thermal stability also lead to their use as dielectric fluids. Other applications are as mould-release agents, as lubricants and as antifoam agents.

Silicones with a controlled amount of cross-linking are used for resin films. They are applied either in solution or as emulsions, and the film is then cured by heating to give further cross-linking by condensation and the oxidation of pendant groups. The latter process is catalysed by incorporating an oxidation catalyst such as a heavy metal naphthenate into the silicone mix. Such films are hard and insoluble, and are used for electrical insulation and in the preparation of laminates. There is also controlled cross-linking in silicone rubbers, which are especially valuable because they retain their elasticity at lower temperatures than other rubbers and are resistant to oils. One of the major uses of silicones is in the treatment of fabrics and other surfaces to render them water-repellent. Application may be as a solution or an emulsion. There are also a number of other applications of special products which, though involving relatively small tonnages, are of great importance.

9.3.4. Other heteroatomic Group IV polymers.[14] The silicones have been investigated in great detail because of their commercial importance, but there are a number of other types of polymer formed by Group IV elements which are of great interest though, at present, of little practical use. Some of the more important of these will be referred to briefly in this section.

Hydrolysis of the organogermanium halides follows much the same course as for the analogous silicon compounds, self-condensation of the initial products giving small ring and chain structures based on a —Ge—O—Ge— framework. This structure is, however, more readily hydrolysed. Formally, hydrolysis of the corresponding tin compounds follows the same course, but the organotin oxides have entirely different structures.

A great deal of attention has been paid to the so-called polymetallosiloxanes, in which silicon and at least one other element (e.g. Al, Ti and Sn) are bonded through oxygen to provide a backbone for the polymer (—M—O—M′—O—). Such polymers may be formed in a cocondensation reaction; for example,

$$R_2SiCl_2 + R'_2SnCl_2 \xrightarrow{H_2O} \text{—O—}[SiR_2]_n\text{—O—}[SnR'_2O]_m\text{—}$$

There will then be a random distribution of the second element in the siloxane chain, governed by the proportions taken.

It is also possible to prepare polymers in which there is a main chain with oxygen and an element other than silicon (O—Al—O—Al—, for example) together with pendant $OSiR_3$ groups. In making such compounds use is made of the fact that the hydrogen atom in R_3SiOH is acidic and will form an alkali metal derivative; for example,

$$2(CH_3)_3SiOH + 2Na \xrightarrow{\text{ether}} 2(CH_3)_3SiONa + H_2$$

A few of these silonalates have been isolated, but they react readily in solution with a wide variety of halides forming products such as $Al(OSiR_3)_3$ and $Ti(OSiR_3)_4$. The latter then give the desired products on controlled hydrolysis; for example,

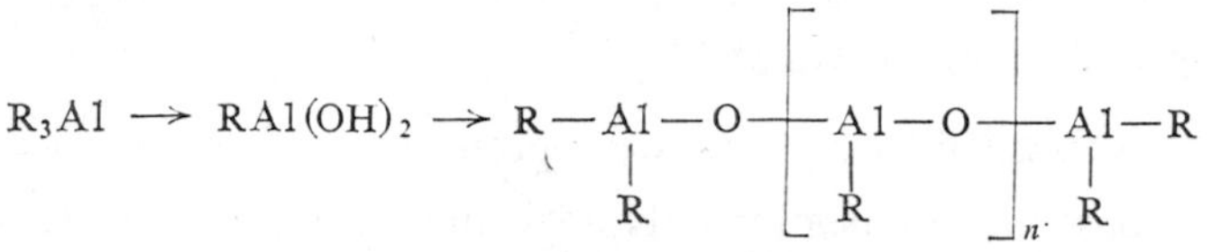

$$R_3Al \rightarrow RAl(OH)_2 \rightarrow R-\underset{\displaystyle R}{\underset{|}{Al}}-O-\left[\underset{\displaystyle R}{\underset{|}{Al}}-O\right]_n-\underset{\displaystyle R}{\underset{|}{Al}}-R$$

$$(R{=}OSiR_3)$$

Some of these species have relatively low molecular weights (<100 AlOR units) and are soluble in solvents such as benzene and acetone. Their use in making heat-resistant films has been described.

When silicon atoms are bridged by other atoms or groups instead of by oxygen it is again possible to build up polymeric structures. An obvious alternative is sulphur. Silicon sulphide (SiS_2) is itself polymeric, but its structure, which is shown below, is entirely different from that of silica.

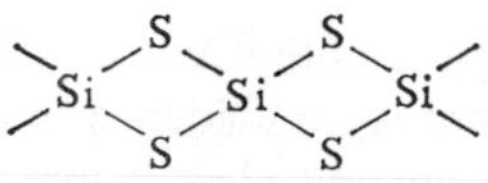

$$>\mathrm{Si}\begin{matrix} S \\ S \end{matrix}\mathrm{Si}\begin{matrix} S \\ S \end{matrix}\mathrm{Si}<$$

When hydrogen sulphide is passed into a solution of an organosilicon halide in the presence of a base such as pyridine, thiohydrolysis occurs and compounds that are formally analogous to silicones may be isolated (e.g. Me_2SiCl_2 gives $Me_2Si(SH)_2$ which condenses, *inter alia*, to the cyclic trimer $(Me_2SiS)_3$). Such compounds are, however, readily decomposed by water, as is SiS_2.

Ammonolysis of organosilicon halides likewise gives polymers in which, following intermolecular elimination of ammonia between $\geq Si-NH_2$ groups, silicon atoms are bridged by the imino group $>NH$. Here again silicone analogues are known (e.g. cyclic $(Me_2SiNH)_4$), but they hydrolyse very readily. The hydrogen atom in trialkyl imino silanes $(R_3Si)_2NH$ and other silazanes is acidic and alkali metal derivatives are readily prepared; for example, $[(CH_3)_3Si]_2NH$ reacts quantitatively with sodium amide in benzene solution to give $[(CH_3)_3Si]_2NNa$. The latter will react with a wide range of halides in an anhydrous solvent forming products of which

$[\{(CH_3)_3Si\}_2N]_3Al$ and $[\{(CH_3)_2Si\}_2N]_3Fe$ are examples. Chlorination in ether at $-50°$ also gives $[(CH_3)_3Si]_2NCl$. This interesting field lies outside the scope of this chapter, but has been reviewed recently.[15]

9.4 Phosphonitrilic polymers[16–18]

The phosphonitrilic polymers are a group of cyclic or linear compounds based on a repeating $-N{=}P\lessgtr$ unit. The earliest systematic work on this subject dates from 1895, when Stokes commenced his pioneering research on the phosphonitrilic chlorides, which are still the chief starting materials for preparative work in this field. They are now made by heating phosphorus pentachloride with a small excess of ammonium chloride in refluxing *s*-tetrachloroethane (b.p. 146°):

$$PCl_5 + NH_4Cl \rightarrow \frac{1}{n}(NPCl_2)_n + 4HCl$$

The ammonium salt dissociates appreciably at the boiling point of the solvent and the reaction involves ammonolysis of the pentahalide.

The mechanism of the reaction, which gives both cyclic and linear polymers, is not yet fully established. There is no evidence that the salt NH_4PCl_6 is an intermediate, and the current view is that PCl_5 reacts in its ionic form $[PCl_4]^+[PCl_6]^-$ in the polar solvent, the cation being attacked by ammonia with elimination of HCl and formation of $HN{=}PCl_3$. This in turn is believed to act as a nucleophile towards $[PCl_4]^+$:

$$HN{=}PCl_3 + [PCl_4]^+[PCl_6]^- \rightarrow [Cl_3P{=}N{-}PCl_3]^+[PCl_6]^- + HCl$$

The cation produced is then attacked by ammonia to form $Cl_3P{=}N{-}PCl_2{=}NH$ and HCl. This continues with increase in chain length as long as PCl_5 is available, cyclization occurring by intramolecular elimination of HCl to give rings of various sizes; for example,

$$Cl_2P\begin{matrix} \diagup N{=}PCl_3 \\ \diagdown\!\!\!\diagdown N{-}PCl_2 \end{matrix}\!\!{=}NH \longrightarrow Cl_2P\begin{matrix} \diagup N{-}PCl_2 \\ \diagdown\!\!\!\diagdown N{-}PCl_2 \end{matrix}\!\!{=}N$$

Separation of the reaction products is effected by distilling off the solvent and treating the residue, which is a mixture of oil and crystalline material, with petroleum ether. The soluble part, varying from 60–90 per cent of the whole, is a mixture of cyclic polymers $(NPCl_2)_n$, and the insoluble part consists of a mixture of linear polymers $(NPCl_2)_x.PCl_5$, with end groups provided by PCl_5. These will be considered later.

The cyclic products are separated by distilling off the trimer (m.p. 115°; b.p. 124° at 10 mm) and the tetramer (m.p. 123°; b.p. 185° at 10 mm). Other members of the series up to $(NPCl_2)_8$ may be separated by solvent extraction, sublimation and crystallization, leaving a small inseparable residue containing higher members of the series with a mean molecular composition of about $(NPCl_2)_{12}$. Proof that only ring polymers are present in this material comes from the ^{31}P n.m.r. spectrum which has a single peak, as the lower members of the series also do, showing that all the phosphorus atoms are in the same environment.

Reaction of PBr_5 with NH_4Br also yields both cyclic and linear polymers of the above types, and Me_2PCl_3 reacts with NH_4Cl to give a mixture of products from which $(NPMe_2)_3$ and $(NPMe_2)_4$ have been isolated. The phosphonitrilic fluorides are not formed from PCl_5 and NH_4F: the ammonium salt acts as a fluorinating agent and the sole product is the stable hexafluorophosphate NH_4PF_6. Presumably the same product would be obtained from PF_5 and NH_4F. The fluorides must therefore be made from the cyclic chloro compounds either before or after their separation. The best reagent is potassium fluorosulphinate (KSO_2F) made directly from KF and liquid SO_2. Other fluorinating reagents often give only partial replacement of chlorine by fluorine. The fluorine compounds are much more volatile than the corresponding chlorides and therefore easier to separate: $(NPF_2)_3$ and $(NPF_2)_{11}$, for example, have b.p. 51° and 246·7° respectively. Separation in this range is possible by fractional distillation and, using preparative vapour-phase chromatography, members of the series up to $(NPF_2)_{17}$ have been identified and characterized. Unquestionably, larger rings also exist.

The halogen atoms in the cyclic phosphonitrilic halides undergo many of the reactions normally associated with the P—Cl bond, though sometimes less readily, so that complete replacement of halogen by other groups may be difficult to effect. Most work has been done with $(NPCl_2)_3$ and $(NPCl_2)_4$ which may be partially or, under more drastic conditions, fully phenylated by reaction with benzene in the presence of a Friedel–Crafts catalyst such as $AlCl_3$.

Phenyl magnesium bromide also gives partial or complete phenylation. The trimeric and tetrameric chlorides are converted by liquid ammonia to $[NP(NH_2)_2]_3$ and $[NP(NH_2)_2]_4$, and reaction with primary and secondary amines gives a wide range of partially and fully substituted derivatives. Alkoxy derivatives are made by reaction of the chloro compounds with metal alkoxides and the chlorine atoms may be replaced by other groups by the usual types of reaction with silver or potassium salts: potassium thiocyanate in acetone, for example, converts $(NPCl_2)_3$ to $[NP(SCN)_2]_3$. The trifluoroalkyl derivatives $[NP(CF_3)_2]_3$ and $[NP(CF_3)_2]_4$ are made by first converting $(CF_3)_2PCl$ to $(CF_3)_2PN_3$ by reaction with LiN_3 at 0°. The phosphinic azide loses nitrogen at 60° and the cyclic trimer and tetramer are isolated from the product.

The phosphonitrilic chlorides react with water, the trimer being attacked more slowly than the tetramer. Well-defined acids result from which a number of salts have been prepared. It appears that the acids react in the imino form (Fig. 9.22) rather than in the normal form (Fig. 9.23) and the infrared spectra of salts show frequencies

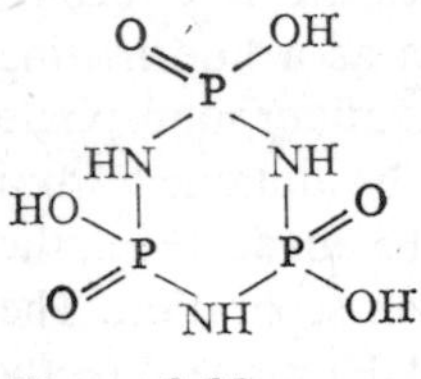

Figure 9.22

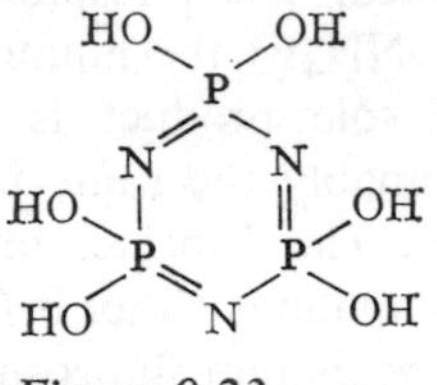

Figure 9.23

associated with P=O and N—H bonds. In the presence of silver ions, however, the acids from the trimer and tetramer give hexa- and octa-silver salts which exist in several forms, in one of which it is highly likely that silver is bonded to nitrogen.

Structure and bonding.[6,17] Most of the cyclic phosphonitrilic derivatives are crystalline solids and X-ray analysis has revealed interesting variations in the ring shape and dimensions. The trimeric halides all have almost planar rings as shown below (Fig. 9.24). There is an approximately tetrahedral distribution of valencies round phosphorus corresponding with sp^3 hybridization, with the result that the chlorine atoms in each pair lie on opposite sides of the ring. All of the P—N bonds in the ring are also of equal length. There are small differences in this value for the fluoride, chloride and bromide, but in each case the distance is less than that for a single bond. In the

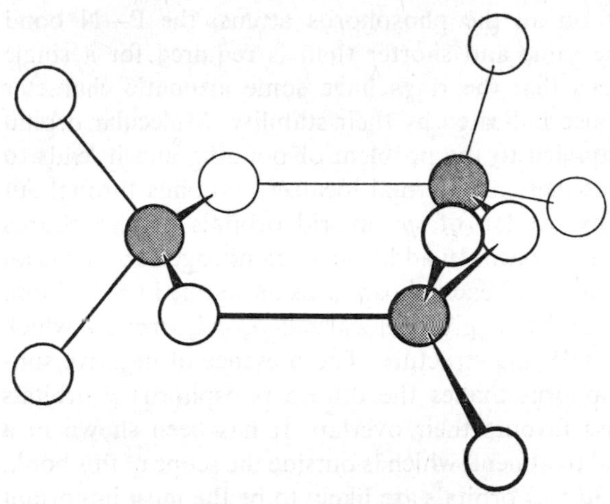

Figure 9.24 Structure of the molecule $(PNCl_2)_3$

trimeric chloride, for example, it is 1·59 ± 0·02 Å compared with 1·65 Å in $(NH_2)_3P.BH_3$ and 1·68 Å for the exocyclic P—N bonds in the tetrameric dimethylamide $[NP(NMe_2)_2]_4$.

The rings are also planar in $(NPF_2)_4$ and $(NPCl_2)_5$, but chair and boat forms have been identified for $(NPCl_2)_4$ while boat forms have been found in $(NPMe_2)_4$ and $[NP(NMe_2)_2]_4$. The structure of $P_6N_7Cl_9$ (Fig. 9.25), the only fused ring halide known at present, has

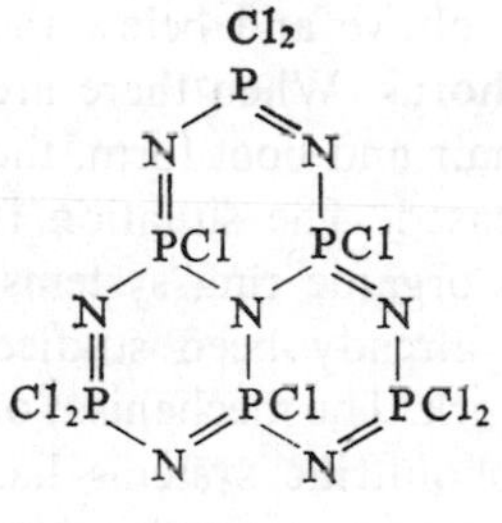

Cl₂
P
N N
PCl PCl
N N N
Cl₂P PCl PCl₂
N N

Figure 9.25

also been determined recently:[19] it is a minor product of the synthesis of phosphonitrilic chlorides from PCl_5 and NH_4Cl. The structure, which will not be considered in detail, is non-planar. An extended fused ring structure of this type may occur in phosphorus nitrides, though the structure of the only crystalline compound of this group, P_3N_5, has not been determined.

In cyclic phosphonitrilic derivatives of the type $(NPX_2)_n$ with a single substituent on all the phosphorus atoms, the P—N bond lengths are all the same and shorter than is required for a single bond. This suggests that the rings have some aromatic character which, indeed, is also indicated by their stability. Molecular orbital theory has been applied to the problem of bonding and it leads to the conclusion that there are normal localized σ-bonds throughout the ring formed by overlap of sp^3 hybrid orbitals of phosphorus with sp^2 orbitals of nitrogen. In addition, each nitrogen atom has an electron in a p_z orbital and each phosphorus atom one in a d orbital. These orbitals combine to give delocalized p_π—d_π orbitals which extend over the whole ring structure. The presence of negative substituents on phosphorus makes the diffuse phosphorus d orbitals more compact and favours their overlap. It has been shown in a detailed theoretical treatment, which is outside the scope of this book, that the $3d_{xz}$ and $3d_{x^2-y^2}$ orbitals are likely to be the most important in the ring π-bonding. Very substantial progress has been made in elucidating the relationship between the configuration of cyclic phosphonitrilic molecules and the nature of their bonds. It should be noted that the problems are much more complicated than those for benzenoid structures where p_π—p_π bonding occurs and the rings are planar.

There are many possibilities for the occurrence of isomers in these ring systems, with groups of different kinds bonded to phosphorus. In a compound such as $N_4P_4Cl_4(C_6H_5)_4$, for example, the phenyl groups may be distributed in different ways above and below the ring, or bonded singly or in pairs to phosphorus. When there are alternative shapes for the ring, such as the chair and boat form, the number of possible isomers is further increased. The situation is more complicated than that encountered in organic ring systems, but a number of groups of isomers have already been studied though there is as yet very little structural work. The mechanism of nucleophilic substitution reactions in phosphonitrilic systems has also received considerable attention. The problems are again more complex than those encountered in organic chemistry, but much the same principles are applied. Both of these subjects are closely related to the more general considerations of bonding and structure.

Basic properties of phosphonitrilic derivatives. The lone pairs of electrons on the nitrogen atoms of phosphonitrilic derivatives are potential sites for attack by Lewis acids. The basic properties of the halides are, however, very weak. The trimeric chloride forms 1 : 1 and

1 : 2 adducts with $HClO_4$, in which it is likely that protonation of nitrogen occurs. Attack is also probably on nitrogen in $N_3P_3Cl_6.3SO_3$. Substitution of chlorine by electron-releasing groups increases the basic strength, and both $N_3P_3Me_6$ and $N_4P_4Me_8$ form adducts of the type $(NPMe_2)_nRI$, where R = Me or Et. The trimer also gives a 1 : 1 adduct with $SnCl_4$ and $TiCl_4$ in which electron donation from nitrogen is likely. Substitution by dialkylamino groups increases the basicity still further and hydrochlorides are formed, though with the ambiguity that either the exocyclic nitrogen atoms or those in the ring could act as donors; structural and spectroscopic studies have shown that protonation of a ring nitrogen atom takes place.

In the adduct $N_3P_3Cl_6.2AlCl_3$ it is believed that there is donation of chloride ions to form $(AlCl_4)^-$. Crystalline complexes of the type $(NPF_2)_n.2SbF_5$ ($n = 3$–6) have also been isolated, though here fluorine-bridged structures have been suggested. Molybdenum carbonyl reacts with $N_4P_4Me_8$ to form $N_4P_4Me_8Mo(CO)_4$, in which donation from nitrogen to the metal atom almost certainly occurs and, to quote a final example, an adduct with the constitution $[N_6P_6(NMe_2)_{12}CuCl]^+CuCl_2^-$ is formed from $N_6P_6(NMe_2)_{12}$ and $CuCl_2$. The structure of this compound has been determined, and it is found that there is a five-coordinated copper atom inside the large twelve-membered ring bonded to four atoms of nitrogen and one of chlorine.[20]

Linear phosphonitrilic polymers. When phosphorus pentachloride and ammonium chloride are heated together in refluxing *S*-tetrachloroethane, that portion of the product which is insoluble in petroleum ether has the approximate composition $(NPCl_2)_n.PCl_5$, where n may be as small as 10–15. The proportion of this product is increased by using excess of PCl_5 in the preparation, and it is believed that the phosphorus halide provides the end groups for the phosphonitrilic chain as shown below.

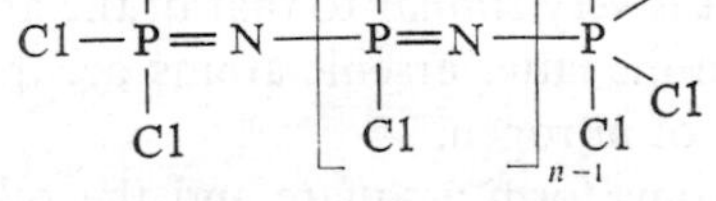

$$Cl-\underset{Cl}{\overset{Cl}{P}}=N-\left[\underset{Cl}{\overset{Cl}{P}}=N\right]_{n-1}-\underset{Cl}{\overset{Cl}{P}}\begin{matrix}Cl\\Cl\end{matrix}$$

This is consistent with the ^{31}P n.m.r. spectrum which shows two peaks, neither of which is due to free PCl_5. A very similar material is formed when $(NPCl_2)_3$ is heated in a sealed tube at 350° with PCl_5;

presumably the ring is broken, some chain growth occurs and PCl_5 functions as before. Other halides, such as $SbCl_3$, $AlCl_3$ and $TiCl_4$, are also able to provide end groups, the usual method of preparation of the polymer being to heat PCl_5 with NH_4Cl and the halide in a solvent. These linear polymers are more polar than the cyclic phosphonitrilics and also more sensitive to hydrolysis.

When a cyclic halide such as $(NPCl_2)_3$ is heated to 250–350° in a solvent such as carbon tetrachloride with access of oxygen, linear polymers with molecular weights, based on osmotic pressure measurements, which are in excess of 20,000 result. Oxygen is necessary for the reaction to occur and it may play a part both in ring-cleavage and in chain-termination. These materials are rubbers and, when stretched, give an X-ray diffraction pattern of the same type as natural rubber. They are believed to be made up of long spiral chains. Unfortunately they change gradually to horny solids on exposure to air. They are also depolymerized at temperatures above 350°, the products being mainly cyclic phosphonitrilic halides.

9.5 Tetrasulphur tetranitride and related compounds[21–23]

Tetrasulphur tetranitride (S_4N_4) is the starting point for the preparation of many compounds containing the S—N bond. Only those which are polymeric will be discussed here, the chemistry of simpler molecules such as NSF and NSF_3 being considered later (p. 352).

The tetranitride is prepared by passing ammonia into a solution of SCl_2 in benzene at 20–50°. After separation from other products it is obtained as orange-yellow needle-shaped crystals (m.p. 178°) which are sparingly soluble in benzene and other solvents. It is also formed when the vapour of SCl_2 is passed over heated ammonium chloride and, less conveniently and with a lower yield, in several other reactions.

Though kinetically stable in air, the compound is endothermic ($\Delta H_f^0 = +428$ kJ (129 kcal) mole^{-1}) and may detonate when heated to its melting point or on shock. The structure (Fig. 9.26) has been determined by X-ray analysis. It is very similar to that of the arsenic compound As_4S_4, except that in the latter arsenic atoms occupy the sulphur sites and sulphur those of nitrogen.

In S_4N_4 the four nitrogen atoms form a square and the sulphur atoms a slightly distorted tetrahedron. All of the S—N distances are equal (1·60 Å), this distance being much shorter than the sum of the covalent radii, 1·78 Å (in S_8 and N_2H_4 respectively). This has been

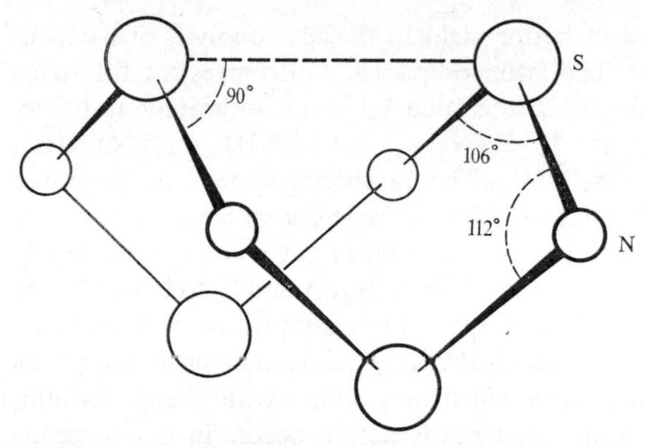

Figure 9.26 Structure of tetrasulphur tetranitride

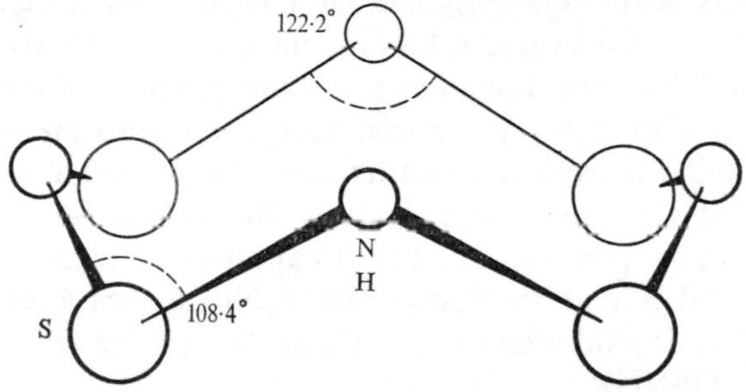

Figure 9.27 Structure of $S_4(NH)_4$

attributed to cyclic delocalization involving p_π orbitals of the nitrogen and d_π orbitals of sulphur.[24] The S—S distance is 2·58 Å, but it is not known if this implies chemical bonding between these atoms.

Tetrasulphur tetranitride is not wetted by water. It is, however, readily hydrolysed by bases, all of the nitrogen forming ammonia:

$$2S_4N_4 + 6OH^- + 9H_2O \rightarrow 2S_3O_6{}^{2-} + S_2O_3{}^{2-} + 8NH_3$$

It is reduced by hydrogen iodide to H_2S and NH_3, but with ethanolic tin (II) chloride the white crystalline tetraimide, $S_4N_4H_4$ (m.p. 124°), is formed. This may be oxidized to S_4N_4 with chlorine.

The structure of the tetraimide ring (Fig. 9.27) is very similar to that of S_8, the NSN angle being 108·4° and the SSS angle in the latter 107·8°. The N—S bond lengths are all the same (1·67 Å) and, though longer than those in S_4N_4, are still shorter than is required for a single bond. There is a family of imides which may be regarded as derived from S_8 by replacing S by NH.[25] The simplest, S_7NH (m.p. 113°), is a minor product in the preparation of S_4N_4 by ammonolysis of

SCl_2, but is formed in better yield in the ammonolysis of S_2Cl_2 in dimethylformamide. The latter preparation also gives the following products which have been separated by chromatography and fractional crystallization: 1,3-$S_6(NH)_2$, 1,4-$S_6(NH)_2$, 1,5-$S_6(NH)_2$, 1,3,5-$S_5(NH)_3$, 1,3,6-$S_5(NH)_3$. The numbering system, based on the S_8 ring, is similar to that used for benzene derivatives.

Chemical investigations on these imides have so far been largely restricted to the two most accessible, $S_4N_4H_4$ and S_7NH. The N—H bond behaves normally and may, for example, be acetylated or benzoylated. The proton is also weakly acidic and the sodium salts $S_4N_4Na_4$ and S_7NNa are formed by reaction with phenyl sodium, the other product being benzene. Reaction occurs in non-aqueous polar solvents between S_7NH and salts of mercury (II) and mercury (I), the products being $Hg(NS_7)_2$ and $Hg_2(NS_7)_2$ respectively. A few other metallic derivatives are known, including $Cu_4(NS)_4$, $Ag_4(NS)_4$ and $LiAlS_4N_4$. The last of these is the product of the reaction of $LiAlH_4$ with S_4N_4 in tetrahydrofuran. Hydrogen is evolved. The compound is explosive and its structure unknown, as indeed are those of all these metallic derivatives. Reaction of S_7NH with SCl_2 or S_2Cl_2 in the presence of a base to absorb HCl leads to the compounds $(S_7N)_2S$ and $(S_7N)_2S_2$, and S_7NBCl_2 has been obtained from BCl_3. Air oxidation of $S_4N_4H_4$ at 100° gives a tetrameric thionylimide $(OSNH)_4$.

The reaction products from the halogenation of S_4N_4 are shown below; attack is always on sulphur.

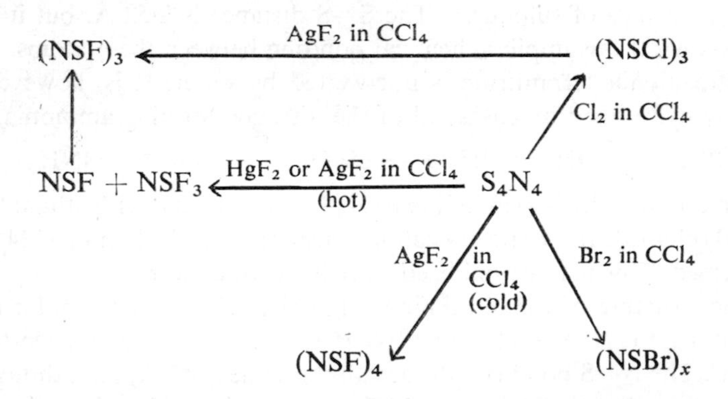

Reaction with elemental fluorine has not been investigated but it will be seen that fluorination under fairly mild conditions leaves the ring intact. The product, $S_4N_4F_4$, is a water-sensitive white solid, the

^{19}F n.m.r. spectrum of which gives a single resonance, showing all of the fluorine atoms to be equivalent. The molecule has a puckered ring structure with two alternating S—N distances of 1·66 and 1·54 Å in the ring. This indicates that the delocalization in the S_4N_4 ring has been removed, or greatly reduced, by the strongly electronegative exocyclic groups bonded to sulphur. These are believed to lead to a contraction of the sulphur *d* orbitals and consequently a better overlap. The sulphur electrons are also more strongly held and therefore less available for delocalization.

Tetrathiazyl tetrafluoride dissociates at 200 250° to NSF, which polymerizes to the trimer $(NSF)_3$. The latter is also made by fluorinating trithiazyl trichloride $(NSCl)_3$, the only product from the chlorination of S_4N_4 with chlorine at low temperatures or in solution. All attempts to make $(NSCl)_4$ have so far failed and it is assumed that chlorination proceeds via the monomer NSCl, which can be isolated by heating the trimer *in vacuo*. It polymerizes readily to $(NSCl)_3$.

The structure of $(NSF)_3$ is unknown, though its ^{19}F n.m.r. spectrum shows all the fluorine atoms to be equivalent. It is likely to resemble that of $(NSCl)_3$, the ring of which is almost planar with all of the chlorine atoms on one side of the ring in axial positions. All of its N—S distances are the same (1·60 Å), though there may well prove to be two distances in $(NSF)_3$, as there are in $(NSF)_4$, because of the more highly negative group bonded to sulphur. It is of interest that X-ray analysis has shown there to be two S—N bond distances (1·56 and 1·62 Å) in the compound $[NS(ON(CF_3)_2)]_4$, which is formed directly from the stable free radical $\cdot ON(CF_3)_2$ and S_4N_4. The electronegativity of the radical is intermediate between that of fluorine and chlorine.

The chemistry of NSF and $(NSF)_3$ is described elsewhere (Section 11.9). Bromination of S_4N_4 has not been fully investigated; the bronze-coloured crystals of the empirical formula NSBr formed on bromination in a solvent are probably polymeric.

The derivative chemistry of the trithiazyl trihalides has not been fully explored, though the ring is less stable than that in S_4N_4. One remarkable reaction shown in the equation below takes place in the presence of pyridine, which absorbs HCl. The nature of the products suggests an initial depolymerization to NSCl and NSH.

$$4N_3S_3Cl_3 + 3S_4N_4H_4 \xrightarrow[\text{heat}]{CCl_4/\text{pyridine}} 6S_4N_4 + 12HCl$$

Tetrasulphur tetranitride is itself depolymerized to S_2N_2 when its vapour is streamed at low pressure through a tube packed with silver wool and heated to 300°. The infrared spectrum shows S_2N_2 to form a flat four-membered ring of alternating sulphur and nitrogen atoms. The molecular weight has been determined in benzene solution, in which polymerization to S_4N_4 occurs rapidly on adding a trace of alkali. Spontaneous polymerization in the gas phase gives a dark blue fibrous crystalline solid, which is believed to be a linear polymer, though it is insoluble and the molecular weight is unknown. In the compressed solid form this material is a semiconductor.

The thiotrithiazyl cation ($S_4N_3^+$) is another interesting species which forms several stable salts. The chloride is made in excellent yield by boiling S_4N_4 with S_2Cl_2 in CCl_4 solution:

$$3S_4N_4 + 2S_2Cl_2 \rightarrow 4[S_4N_3]^+Cl^-$$

It is insoluble in common solvents but behaves as a 1 : 1 electrolyte in anhydrous formic acid. Nitric acid converts the chloride to the nitrate. The free base S_4N_3OH is also known. An X-ray structural study on the nitrate showed the cation to form a planar seven-membered ring (Fig. 9.28) with an average for the slightly variable S—N bond lengths of 1·56 Å. This is less than the value for a single bond and indicates some delocalization. The S—S bond length (2·06 Å) is close to that for a single bond (2·08 Å).

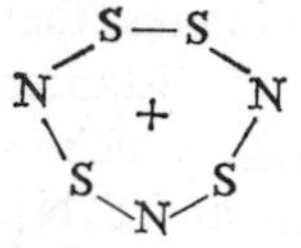

```
    S—S
   /    \
  N  +   N
   \    /
    S  S
     \/
     N
```

Figure 9.28

Tetrasulphur tetranitride is able to act as a Lewis base through the lone pairs of electrons on its nitrogen atoms. It forms adducts with several metal halides (e.g. $S_4N_4.SbCl_5$, $S_4N_4.2SnCl_4$ and $S_4N_4.MoCl_5$) and also with BF_3 and SO_3. In the antimony compound an X-ray structural study has shown nitrogen to be bonded to antimony, the ring configuration being changed to approximately that in $N_4S_4F_4$. The structure of the adduct $S_4N_4.BF_3$ is very similar. It is possible that donation from two nitrogen atoms may occur in $S_4N_4.WCl_4$ and $S_4N_4.VCl_4$, while in $S_4N_4.4SO_3$ all four basic sites may be involved. Disulphur dinitride also forms an adduct $S_2N_2.2SbCl_5$, the structure of which shows both nitrogen atoms in the planar and nearly square S_2N_2 ring to be bonded to antimony.

Thionitrosyl complexes of Fe, Co, Ni, Pd and Pt are also known.

When S_4N_4 is treated in alcoholic solution with the dichlorides of the first four of these elements or with H_2PtCl_6, three types of crystalline derivative can apparently be formed in which the S_2N_2H and S_3N groups act as bidentate ligands in the complexes $M(S_2N_2H)_2$, $M(S_2N_2H)(S_3N)$ and $M(S_3N)_2$. The hydrogen comes from the solvent and the presence of an N—H bond is proved by the infrared spectrum and the shift in frequency on deuteration. Structural evidence is not complete, but an X-ray investigation shows that in $Pt(S_2N_2H)_2$ the rings are planar with the N—H bonds in the *cis* position (Fig. 9.29).

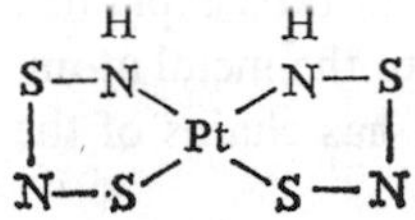

Figure 9.29

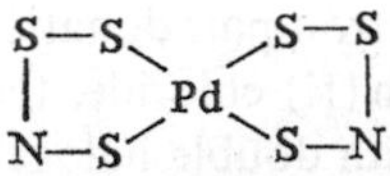

Figure 9.30

An unstable silver derivative in which the acidic hydrogen atom has been replaced by silver has been prepared from $Ni(S_2N_2H)_2$ and the stable dimethyl derivative prepared from it has the NMe groups in the *trans* position. The structure of the palladium complex $Pd(S_3N)_2$ has likewise been investigated and shown to have two planar (S_3N) rings (Fig. 9.30). In this case NH has been replaced by sulphur. Derivatives of other metals have also been described (e.g. $Hg_5(NS)_8$ and $Pb(NS)_2.NH_3$); $Pb(NS)_2.NH_3$ also has the $(NS)_2$ ring as a bidentate ligand.

9.5.1. Other sulphur-nitrogen compounds. It is important to realize that the sulphur–nitrogen compounds described above represent only one part of a very large field. A number of compounds are known containing oxygen, sulphur and nitrogen, an example being dinitrogen trisulphur dioxide (Fig. 9.31), formed in the reaction of S_4N_4

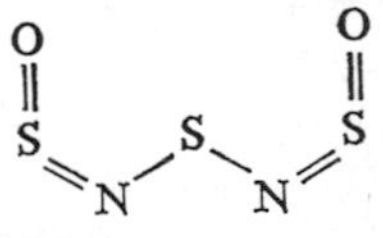

Figure 9.31

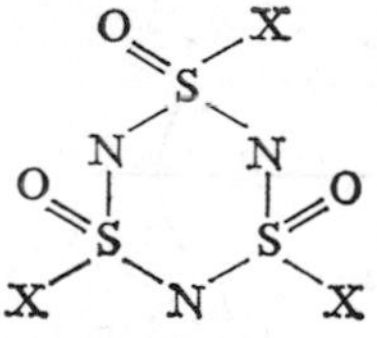

Figure 9.32

with boiling $SOCl_2$ in a stream of sulphur dioxide. The structure has been established by X-ray analysis. Finally, the existence of sulphanuryl halides should be noted (Fig. 9.32). Both the chloride and the fluoride are known and both exist in two forms in which the halogen atoms are differently oriented with respect to the ring. These and many other nitrogen–sulphur compounds are described fully in the references cited at the end of this chapter.

9.6 Coordination polymers[26, 27]

Coordination is a frequent cause of polymerization, very simple examples being the dimerization of aluminium trichloride, shown below, and the polymerization of amino boranes (see Section 9.3). The structures of a number of metallic halides can also be explained in terms of electron-pair donation from a halogen to the metal atom, as in palladium (II) chloride, the crystal of which has chains of the type shown, with double halogen bridges:

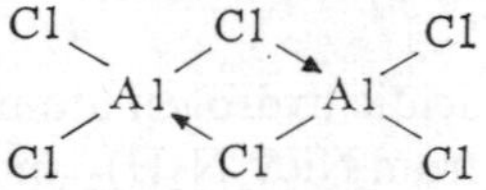

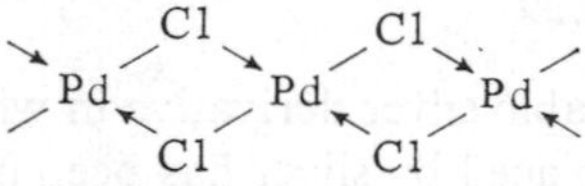

Cl Cl Cl Al Al Cl Cl Cl   Cl Cl Pd Pd Pd Cl Cl

The structures of silver (I) and gold (I) cyanides can also be understood in terms of donation from nitrogen to the metal atom [—C≡N→M—C≡N→M—]. The resulting chains are arranged parallel to one another in the lattice.

Antimony pentafluoride, a highly viscous liquid, is a further example of a halide which polymerizes as the result of electron donation by the halogen. The exact mode in which the monomer units are bonded together can be deduced from the ^{19}F n.m.r. spectrum, which shows three broad peaks with relative intensities 2 : 2 : 1. This means that there are three distinct fluorine environments in the polymer. If association were through *trans*-fluorine bridge bonding (Fig. 9.33) only two fluorine environments would occur (F_a, F_b).

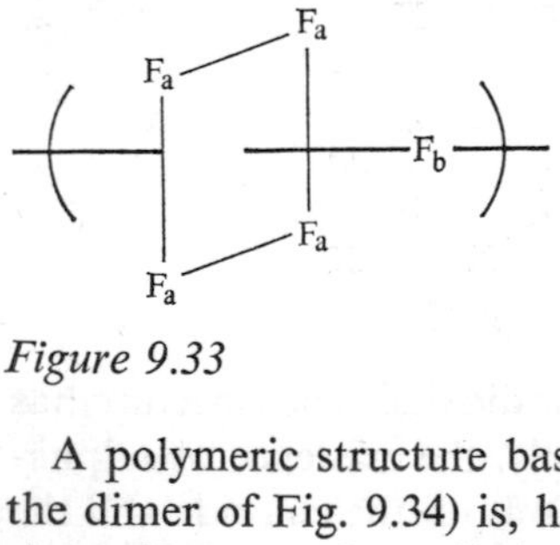

F_a F_a F_b F_a F_a

Figure 9.33

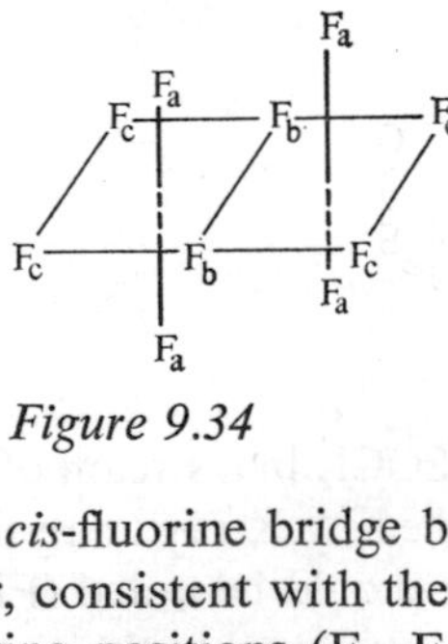

F_a F_a F_c F_b F_c F_c F_b F_c F_a F_a

Figure 9.34

A polymeric structure based on *cis*-fluorine bridge bonding (e.g. the dimer of Fig. 9.34) is, however, consistent with the n.m.r. data since there are three distinct fluorine positions (F_a, F_b, F_c) in the right abundance. This is the simplest possible structure, but a cyclic polymer or an infinite polymer would also satisfy the n.m.r. requirement if it had the same type of bridge structure.

9.6.1. Metal alkoxides. A large proportion of the metallic elements form alkoxides of the type $M(OR)_x$. The highly electropositive metals react directly with alcohols. Alkoxides of the less positive metals, including the transition metals, are commonly made from the metal halide by reaction with sodium alkoxide or an alcohol in the presence of a base to take up hydrogen chloride. The alkoxides also undergo alcohol interchange very readily.

Many of the metal alkoxides are soluble in organic solvents so that their molecular weights in solution can be determined cryoscopically. Many are found to be polymeric and, though few structures have been determined, it seems certain that polymerization occurs as a result of coordination from oxygen in an alkoxy group to the metal atom of another molecule. Thallium (I) ethoxide is tetrameric in solution and it is thought that the molecule has a cubic arrangement of four thallium and four oxygen atoms, with donation from oxygen as shown below (Fig. 9.35).

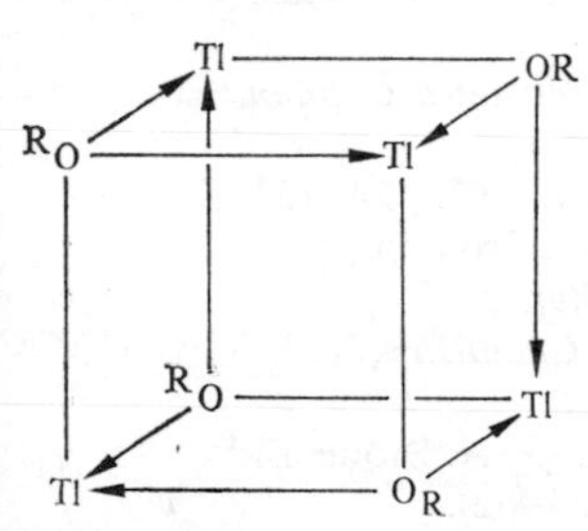

Figure 9.35

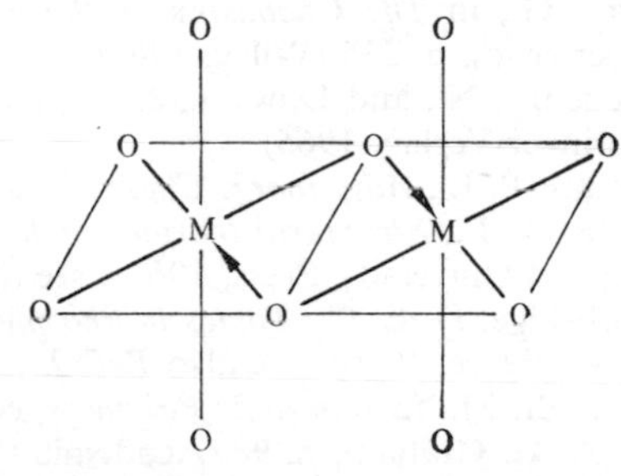

Figure 9.36

In much the same way the alkoxides of niobium, tantalum and uranium, which are dimeric, may be formulated as above (Fig. 9.36), with octahedral coordination round each metal atom; only the sites of the oxygen atoms are shown. Solution molecular weights of these alkoxides depend on temperature, concentration and the solvent used, and it is likely, therefore, that the bonding leading to polymerization is relatively weak.

9.6.2. Chelate polymers. It is also possible for a multidentate ligand to serve as a bridging group between metal atoms and thus give rise to a polymeric structure. Many such ligands are available and all act in essentially the same way. The point may be illustrated by referring to the bis-diketones, which form complexes with a variety of metals which are of the type shown below.

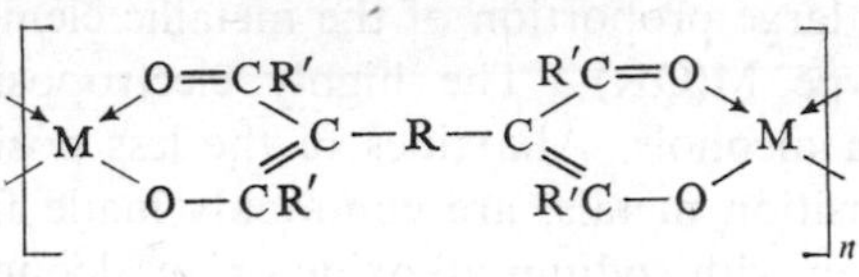

```
 ┌    O=CR'          R'C=O     ┐
 │  ↙      \        /      ↘   │
 │ M        C—R—C           M  │
 │  \      //        \\    /   │
 └    O—CR'          R'C—O     ┘n
```

The main donor atoms involved are oxygen, nitrogen and sulphur.

References for Chapter 9

1 Newkirk, A. E., in *Boron, Metallo-Boron Compounds and Boranes*, ed. R. M. Adams, p. 233 (Interscience, 1964).
Hoard, J. L., and Hughes, R. E., in *The Chemistry of Boron and its Compounds*, ed. E. L. Muetterties, p. 25 (Wiley, 1967).

2 Post, B., in *Boron, Metallo-Boron Compounds and Boranes*, ed. R. M. Adams, p. 301 (Interscience, 1964).
Arronson, B., Lundstrom, T., and Rundquist, S., *Borides, Silicides and Phosphides* (Methuen, 1965).

3 Massey, A. G., *Adv. Inorg. Chem. Radiochem.*, 1967, **10,** 1.
Urry, G., in *The Chemistry of Boron and its Compounds*, ed. E. L. Muetterties, p. 325 (Wiley, 1967).
Niedenzu, N., and Dawson, J. W., *Boron–Nitrogen Compounds* (Springer-Verlag, 1965).

4 Timms, P. L., *Adv. Inorg. Chem. Radiochem.*, 1972, **14,** 121.

5 Wells, A. F., *Structural Inorganic Chemistry*, 3rd edn., p. 770 (Oxford University Press, 1962); see also *Ref. 2.*

6 Corbridge, D. E. C., *Topics in Phosphorus Chemistry*, Vol. 3, p. 57 (Interscience, 1966); see also *Ref. 2.*

7 Schmidt, M., in *Inorganic Polymers*, eds F. G. A. Stone and W. A. G. Graham, p. 98 (Academic Press, 1962).

8 Mellon, E. K., Jr., and Lagowski, J. J., *Adv. Inorg. Chem. Radiochem.*, 1963, **5,** 259.

9 Niedenzu, K., and Dawson, J. W., in *The Chemistry of Boron and its Compounds*, ed. E. L. Muetterties, p. 377 (Wiley, 1967).

10 Parshall, G. W., in *Ref. 9*, p. 617.

11 Barry, A. J., and Beck, H. N., in *Ref. 7*, p. 189.

12 Kipping, F. S., *Proc. Roy. Soc.*, 1937, **159,** 139.

13 Haszeldine, R. N., in *New Pathways in Inorganic Chemistry*, eds. E. A. V. Ebsworth, A. G. Maddock and A. G. Sharpe (Cambridge University Press, 1968).

14 Ingham, R. K., and Gilman, H., in *Ref. 7*, p. 321.

15 Wannagat, U., *Adv. Inorg. Chem. Radiochem.*, 1964, **6,** 225.

16 Paddock, N. L., and Searle, H. T., *Adv. Inorg. Chem. Radiochem.*, 1959, **1,** 348.

17 Paddock, N. L., *Quart. Rev.*, 1964, **18**, 168.
18 Schmulbach, C. D., *Prog. Inorg. Chem.* 1962, **4**, 275.
19 Harrison, W., Oakley, R. T., Paddock, N. L., and Trotter, J., *Chem. Comm.*, 1971, 357.
20 Marsh, W. C., Paddock, N. L., Stewart, C. J., and Trotter, J., *Chem. Comm.*, 1970, 1190.
21 Allen, C. W., *J. Chem. Education*, 1967, **44**, 38.
22 Becke-Goehring, M., *Prog. Inorg. Chem.*, 1959, **1**, 207.
23 Glemser, O., and Fild, M., in *Halogen Chemistry*, ed. V. Gutmann, Vol. II, p. 1 (Academic Press, 1967).
24 Craig, D. P., and Paddock, N. L., *J. Chem. Soc.*, 1962, 4118.
25 Heal, H. G., in *Inorganic Sulphur Chemistry*, ed. G. Nickless, p. 459 (Elsevier, 1968).
26 Bradley, D., *Prog. Inorg. Chem.*, 1960, **2**, 303.
27 Block, B. P., in *Ref.* 7, p. 447.

10 Polymeric inorganic compounds II: polycations and polyanions

10.1 Introduction

The various types of condensed cations, acids and anions that will be described in this chapter can all be regarded as polymeric in the sense that they contain a repeating structural unit based on the formation of hydroxy $\left(M \genfrac{}{}{0pt}{}{\langle OH \rangle}{\langle OH \rangle} M \right)$ or oxo (M—O—M) bridges. They are, however, formed under widely different conditions. The equilibria between polycations derived from hydrated metal ions, and between polyanions, derived from oxy-anions of transition metals, are usually readily reversible in aqueous solution; they involve four- or six-coordinated metal atoms. Among the silicates, condensation reactions of $[SiO_4]^{4-}$ ions are irreversible, and discussion centres on the structures of crystalline silicates, most of which occur in nature. The borates have been selected for brief discussion because they differ from the silicates in that boron can have both threefold and fourfold oxygen coordination; again, however, most of our knowledge of the structural principles of these substances is derived from crystallographic work. The phosphates have an extensive solution chemistry to reinforce information obtained from solid state studies, but equilibrium among different species in solution is not always rapidly attained; in some respects, therefore, their chemistry presents the greatest challenge to rationalization, and so we have ended our discussion with an account of these species.

10.2 Polycations[1–3]

Discussion of the condensation processes that can occur in solution, leading to the formation of polycations and polyanions, can best be introduced by first considering briefly the behaviour of simple aquocations. Their hydration in aqueous solution is well established, though different methods of measurement may give discordant results for the number of water molecules associated with a particular species. Equally, there is abundant evidence that many cations are hydrated in crystalline compounds isolated from aqueous solutions. Thus, for example, the bivalent metals form a very large number of salts containing metal atoms and water molecules in a ratio of 1 : 6, from which the presence of a common cation of the type $[M(H_2O)_6]^{2+}$ may be inferred. A well-known example is the series of cobalt salts $[Co(H_2O)_6]X_2$, where X = Cl, Br, I, ClO_3, etc., or where X_2 is replaced by a bivalent anion such as $[SO_4]^{2-}$. Similarly, tetraquo-cations, $[M(H_2O)_4]^{n+}$, are found in many other salts. It should, however, be noted that the hydrated species in the crystal are not necessarily the same as those in solution.

Some of the most important properties of metal ions in aqueous solution, such as the relationship between complex salts and oxy-acids, the nature of hydrolytic reactions and the formation of highly condensed basic salts, find a common interpretation in the enhanced acidity of coordinated water, which may be expressed by the equilibrium:

$$[M(H_2O)_n]^{x+} \rightleftharpoons [M(H_2O)_{n-1}(OH)]^{(x-1)+} + H^+$$

The acidity constants of aquo-cations diminish at each stage of acid dissociation, but a sufficient reduction in the hydrogen ion activity must lead to an uncharged complex, and finally to anionic species: for example,

$$[Cr(H_2O)_6]^{3+} \rightarrow [Cr(H_2O)_5(OH)]^{2+} \rightarrow [Cr(H_2O)_3(OH)_3]^0 \rightarrow [Cr(H_2O)_2(OH)_4]^-, \text{ etc.}$$

The oxygen atom of a coordinated hydroxyl group is still nucleophilic, and capable of attacking and displacing labile groups, such as the water molecules of aquo complexes. The mutual interaction of two hydroxy aquo complexes can by this means lead to a diol complex:

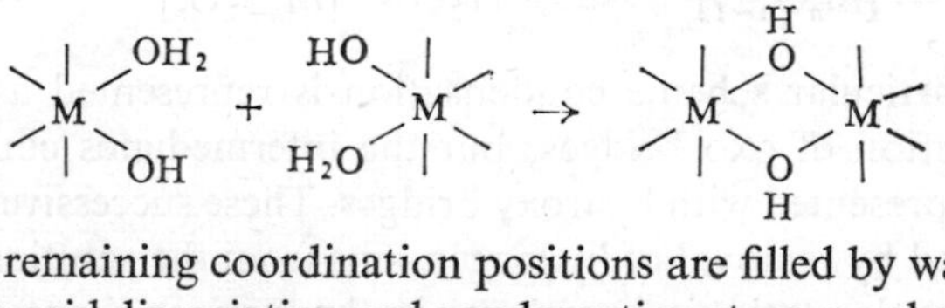

$$\text{M}(\text{OH}_2)(\text{OH}) + (\text{HO})(\text{H}_2\text{O})\text{M} \rightarrow \text{M}\begin{matrix}\text{H} \\ \text{O} \\ \\ \text{O} \\ \text{H}\end{matrix}\text{M}$$

When the remaining coordination positions are filled by water molecules, both acid dissociation and condensation steps may be repeated, so that condensation proceeds beyond the binuclear state. The average complexity of the system may then vary continuously as a function of the hydrogen ion concentration from the mononuclear aquo-cation, in the most acid solutions, through polynuclear complexes with increasing numbers of repeating links up to the stage of infinite aggregation, represented by the deposition of an hydroxide or basic salt.

If several coordination positions are filled by firmly bound ligands such as ammonia or the oxalato group, condensation will produce finite binuclear complexes. It has been shown that, for one process at least, the critical step is the dissociative loss of a water molecule from the hydroxy aquo complex:

$$[Cr(C_2O_4)_2(H_2O)_2]^- \rightleftharpoons \left[Cr(C_2O_4)_2\begin{matrix}H_2O\\OH\end{matrix}\right]^{2-} \xrightarrow{\text{slow}} [Cr(C_2O_4)_2OH]^{2-}$$

$$[Cr(C_2O_4)_2OH]^{2-} + \left[\begin{matrix}H_2O\\HO\end{matrix}Cr(C_2O_4)_2\right]^{2-} \rightarrow \left[(C_2O_4)_2Cr\begin{matrix}OH\\OH\end{matrix}Cr(C_2O_4)_2\right]^{4-}$$

A structural relationship may be expected between the solid products of reaction and their precursors, the polynuclear species in solution. This may be illustrated by the reactions leading to the deposition of bismuth oxy-salts from bismuth salt solutions, where the hydrogen ion equilibria can be analysed to give a quantitative account of the successive reactions involved. The results accord well with the idea that, as the hydrogen ion concentration is reduced, polynuclear cations are built up by the hydrolysis reaction, through the formation of oxo bridges between Bi^{3+} ions:

$$2Bi^{3+} + H_2O \rightleftharpoons [Bi_2O]^{4+} + 2H^+$$

$$Bi^{3+} + [Bi_2O]^{4+} + H_2O \rightleftharpoons [Bi_3O_2]^{3+} + 2H^+$$

$$\vdots$$

$$Bi^{3+} + [Bi_nO_{n-1}]^{(n+2)+} + H_2O \rightleftharpoons [Bi_{n+1}O_n]^{(n+3)+} + 2H^+$$

In this particular scheme condensation is represented as involving the formation of oxo bridges, but the intermediates could equally well be represented with hydroxy bridges. These successive equilibria, if displaced by removal of hydrogen ions from the solution, build up high molecular cationic complexes in the manner shown. The magnitude of the equilibrium constant at each stage determines the importance of the steps that are intermediate between the first hydrolysis product and the ultimate infinite complex, which approximates increasingly closely to the formula $[BiO^+]$. The oxygen bridging binds the atoms of the polymeric cation into a sheet, which occurs as an actual structural unit in a variety of basic bismuth salts, including BiOCl.

To take another example, there are infinite cationic chains of the type $[(M(OH)_2)_n]^{2n+}$ in the crystalline basic sulphates and chromates of thorium and uranium (IV). The initial states for the formation of these condensed cations are already involved in the solution equilibria. They are polycations which can be formulated as $[(M^{O}_{OH})_nM]^{n+4}$ ($n = 1, 2, 3, \ldots$), and these are the conjugate bases of the species finally incorporated in the crystalline basic salts. Correlation of the solution equilibria with crystal structures similarly suggests that in uranyl salt solutions at pH >4, the polycations $[(UO_2O)_nUO_2]^{2+}$ represent the initial stage of formation of the two-dimensional sheets that are characteristic of the structure of the crystalline uranyl halides, the metal uranates and uranium trioxide.

Condensation may alternatively take place so as to form polynuclear complexes of definite formulae and favoured stability. Thus, some crystalline basic salts of thorium, uranium (IV) and cerium (IV) contain polycations $[M_6O_4(OH)_4]^{12+}$ with a cage structure (Fig. 10.1). If these finite complexes also exist in the solution, each metal atom can attain the coordination number of eight by association with four molecules of water. Similarly, crystalline zirconyl chloride

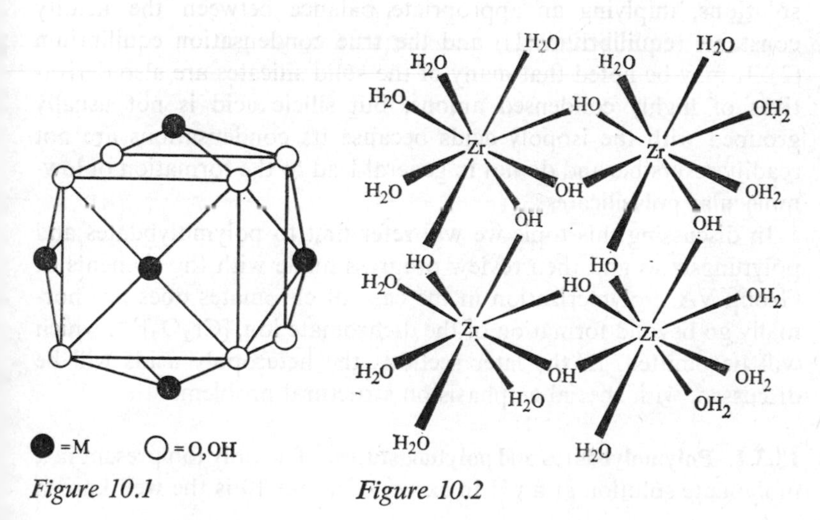

Figure 10.1 *Figure 10.2*

($ZrOCl_2.8H_2O$) actually contains the cyclic four-nuclear cation $[Zr_4(OH)_8(H_2O)_{16}]^{8+}$ (Fig. 10.2).

10.3 Isopoly and heteropoly acids of transition metals[4]

The anions of the weak acids derived from the amphoteric metals of Groups VA and VIA, and particularly vanadium, molybdenum and tungsten, have the interesting property of condensing reversibly as the pH of their solutions is lowered, to give a series of larger anions derived from what are termed *isopoly acids*. If this condensation process takes place in the presence of certain other anions, among which PO_4^{3-} and SiO_4^{4-} are the best known, the latter are also built into the product, which is then known as a *heteropoly anion*.

Salts of both types of polyanion have been known for well over a hundred years, and a great deal of research has been done to identify the various ionic species formed in solution and relate them to solid phases. In principle, the condensation process is the same in every case: it involves the formation of an oxo bridge by elimination of water from two molecules of the weak acid which is conjugate to the anions of the system; for example,

$$[XO_4]^{-n} + H^+ \rightleftharpoons [XO_3(OH)]^{-n+1} \quad (1)$$

$$2[XO_3(OH)]^{-n+1} \rightleftharpoons [O_3XOXO_3]^{-2n+2} + H_2O \quad (2)$$

These condensation reactions take place reversibly in dilute aqueous solutions, implying an appropriate balance between the acidity constants (equilibrium (1)) and the true condensation equilibrium (2). It may be noted that many of the solid silicates are also derivatives of highly condensed anions, but silicic acid is not usually grouped with the isopoly acids because its condensations are not readily reversible and do not in general lead to the formation of low-molecular polysilicates.

In discussing this topic we will refer first to polymolybdates and polytungstates and then review progress made with the elements of Group VA. Polymerization in the case of chromates does not normally go beyond formation of the dichromate ion, $[Cr_2O_7]^{2-}$, which will be omitted. In the later sections the heteropoly acids will be discussed, with special emphasis on structural problems.

10.3.1. Polymolybdates and polytungstates.[5] The only ion present in a molybdate solution at a pH in excess of about 10 is the well-known ion $[MoO_4]^{2-}$, but if the pH is lowered to about 6–8, polyanion formation commences, the first stable species to be formed being $[Mo_7O_{24}]^{6-}$:

$$7[MoO_4]^{2-} + 8H^+ \rightleftharpoons [Mo_7O_{24}]^{6-} + 4H_2O$$

The existence of this ion in crystalline salts has been proved by structural studies. Such an ion is likely to be protonated or associated with water molecules in aqueous solution, but the extent to which this occurs is unknown, both in this and in similar cases. The existence of intermediates with a lower degree of aggregation is very probable, but evidence as to their nature is conflicting.

When the solution containing $[Mo_7O_{24}]^{6-}$ is acidified further to about pH 1·5–2·9, the octamolybdate ion $[Mo_8O_{26}]^{4-}$ is formed, and it too has been identified in crystalline salts. Larger anionic complexes are probably formed at higher acidity, but none has been identified up to a pH of roughly 1, where the trioxide is precipitated.

Many different techniques have been employed in the study of these and similar polyion systems, but the results are often indecisive. In this particular case, for example, conductimetric and pH titrations, diffusion and dialysis measurements, cryoscopy, light-scattering studies, ion exchange, ultraviolet spectrophotometry and equilibrium ultracentrifugation have all been used. That the results are so often conflicting and ambiguous is due, in the main, to inherent limitations of the methods when applied to this particular type of problem. To take an example, the precision with which breaks in potentiometric or conductimetric titration curves can be interpreted is insufficient to distinguish between the formation of $[Mo_7O_{24}]^{6-}$ and $[Mo_8O_{26}]^{4-}$. Assumptions also have to be made in diffusion and dialysis studies, which in principle should give ion weights and thus provide clear answers, which tend to vitiate the results or make them only qualitative.

Another source of difficulty in all studies on isopoly and heteropoly acids is the relationship between ions in solution and solid phases that separate. The latter can be identified by analysis, but they do not necessarily contain the same ions as are in the solution. There are also many instances where a solid obtained from solution has proved to be a mixture. Salts of condensed acids can also be made in other ways (e.g. by heating the acidic oxide with a base), and these too may contain anions that are not encountered in solution studies.

10.3.2. Polytungstates. The isopolytungstates differ in type from the molybdenum compounds, as may be seen from the formula below, which summarizes the position as at present understood.

$$[WO_4]^{2-} \xrightarrow{pH\ 6\text{–}7} [HW_6O_{21}]^{5-} \rightleftharpoons [W_{12}O_{41}]^{10-}$$

para-tungstate A *para*-tungstate B

$$[WO_4]^{2-} \xrightarrow{pH\ 3{\cdot}3} [H_3W_6O_{21}]^{3-} \underset{pH > 3}{\overset{pH < 3}{\rightleftharpoons}} [H_2W_{12}O_{40}]^{6-}$$

$$[H_3W_6O_{21}]^{3-} \xrightarrow{pH < 1} WO_3$$

ψ-meta-tungstate *meta*-tungstate

When an alkaline solution containing the $[WO_4]^{2-}$ ion is acidified to pH 6–7, there is a fast reaction which can be followed by a potentiometric or conductimetric titration:

$$6[WO_4]^{2-} + 7H^+ \longrightarrow [HW_6O_{21}]^{5-} + 3H_2O$$

The ion first formed (*para*-tungstate A) ages in solution in a matter of hours or days to a second ion, $[W_{12}O_{41}]^{10-}$, known as *para*-tungstate B. This is stable, and salts may be prepared which hydrolyse slowly to a mixture of $[HW_6O_{21}]^{5-}$ and $[W_{12}O_{41}]^{10-}$. The X-ray structure of the sodium salt $Na_{10}[W_{12}O_{41}].28H_2O$ has been determined and the results disclose an interesting discrepancy since the elementary cell contains the ion $[W_{12}O_{46}]^{20-}$ rather than $[W_{12}O_{41}]^{10-}$. This difficulty can be resolved by formulating the sodium salt as $Na_{10}[W_{12}O_{36}(OH)_{10}].23H_2O$, in keeping with which its broad-line proton n.m.r. spectrum shows the presence of hydroxyl groups in addition to water molecules. This, however, does not necessarily represent the degree of hydration of the ion in solution.

At pH 3–4 the *ψ-meta*-tungstate ion, $[H_3W_6O_{21}]^{3-}$, is formed, and goes over to *meta*-tungstate, $[W_{12}O_{38}(OH)_2]^{6-}$, at a slightly lower pH. The formulation of the ion is again based on X-ray structural studies of the sodium salt. Tungsten oxide is precipitated at pH 1.

Relaxation measurements with the temperature-jump method and rapid neutralization studies in a flow system have been used to investigate the intermediate species involved in the condensation to *para*-tungstate A. They have resulted in the identification of $[HWO_4]^-$, $[W_2O_7(OH)]^{3-}$, $[W_4O_{12}(OH)_4]^{4-}$, $[HW_4O_{12}(OH)_4]^{3-}$ and $[W_6O_{20}(OH)_2]^{6-}$, and this opens the way for the eventual clarification of the mechanism by which these large ions are built up.

Relatively little work has been done on the structures of isopoly molybdates and tungstates. It is, however, clear that the underlying principle is the union of MoO_6 or WO_6 octahedra by sharing of corners or edges, but not of faces. In ammonium *para*-molybdate, for example, the $[Mo_7O_{24}]^{6-}$ anion has the octahedra arranged as shown diagramatically in Fig. 10.3(a). The way in which the octahedra are assembled can best be understood by noting that those labelled 1, 2, 3 and 4 share edges while 1 and 3 and 2 and 4 share corners, so that the centres of 1, 2, 3 and 4 form a rectangle. The remaining three

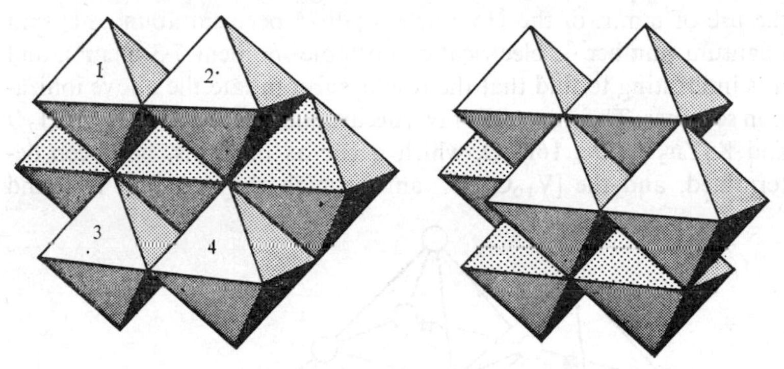

Figure 10.3 (a) and (b)

octahedra are then so placed that they share edges with their nearest neighbours. The octamolybdate ion in $(NH_4)_4Mo_8O_{26}.5H_2O$ has a similar though somewhat more complicated structure, shown in Fig. 10.3(b). The same type of structure is found in the *para*-tungstate and *meta*-tungstate ions, the former being isostructural with 12-tungsto and 12-molybdo hetero anions referred to later.

10.3.3. Vanadates, niobates and tantalates.[6] Much the same range of techniques has been used in studying ion condensation reactions in this series as for the polyanions discussed earlier. In strongly alkaline solutions the ion present is $[VO_4]^{3-}$, or possibly $[VO_3(OH)]^{2-}$. As the pH is reduced from 13 to about 8 condensation to $[V_2O_7]^{4-}$ takes place, and this occurs more readily in concentrated solutions. Further reduction of the pH to 7·2 gives so-called *meta*-vanadate. The name arises from the fact that the same ion is produced if a crystalline *meta*-vanadate with an $M_2O : V_2O_5$ ratio of unity is dissolved in water. In the solid *meta*-vanadate, however, the anion consists of chains of VO_4 tetrahedra linked by sharing corners, and the species in solution is almost certainly a trimer, $[V_3O_9]^{3-}$, or $[V_4O_{12}]^{4-}$.

As the pH is reduced still further, decavanadate is produced at about pH 6:

$$5[V_4O_{12}]^{4-} + 8H^+ \rightleftharpoons 2[V_{10}O_{28}]^{6-} + 4H_2O$$

This ion is protonated in turn to $[HV_{10}O_{28}]^{5-}$ and $[H_2V_{10}O_{28}]^{4-}$ as the pH is lowered to about 3·5. In very acid solution, at about pH 2, $[VO_2]^+$ ions are formed and finally hydrated V_2O_5 separates:

$$V_{10}O_{28}{}^{6-} + 16H^+ \rightleftharpoons 10VO_2{}^+ + 8H_2O$$

A novel approach to the study of this aggregation process has been the use of n.m.r. of the ^{51}V nucleus (99·75 per cent abundant; spin quantum number $\frac{7}{2}$; electrical quadrupole moment 7·3 e cm^2), and it is interesting to find that the results substantiate the above ionization schemes. The structure of two decavanadates, $Ca_3V_{10}O_{28}.16H_2O$ and $K_2Zn_2V_{10}O_{28}.16H_2O$, which occur as minerals, has been determined, and the $[V_{10}O_{28}]^{6-}$ anion which they contain is found

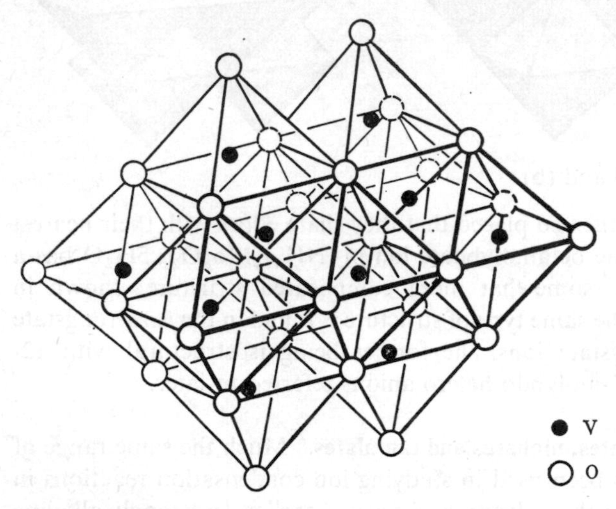

Figure 10.4

to be made up of ten VO_6 octahedra joined by sharing edges (Fig. 10.4).

Though both niobium and tantalum form condensed oxo anions in aqueous solution, these differ from the polyvanadates. Both elements are substantially more basic than the first member of the group, and their hydrated pentoxides are precipitated very readily. Hexaniobates and hexatantalates of the type $M^I_8(Nb,Ta)_6O_{19}.16H_2O$ may be prepared by fusing the pentoxides with KOH and precipitating the salt from the aqueous extract by means of ethanol. The structure of the tantalum compound, which is freely soluble in water, has been determined, and it is found to contain the highly symmetrical anion shown in Fig. 10.5. There is an octahedral arrangement of the six metal atoms. The scattering power of the oxygen atoms is too weak for their positions to be determined directly, but they can be inferred from the Ta–O distances expected and the postulated requirement that there shall be sixfold-coordination of oxygen round each tantalum atom. In the highly symmetrical structure six octahedra are packed, with some distortion, to form a larger octahedron. Of the nineteen oxygen atoms, one is shared by all six octahedra, twelve by two, and six are unshared.

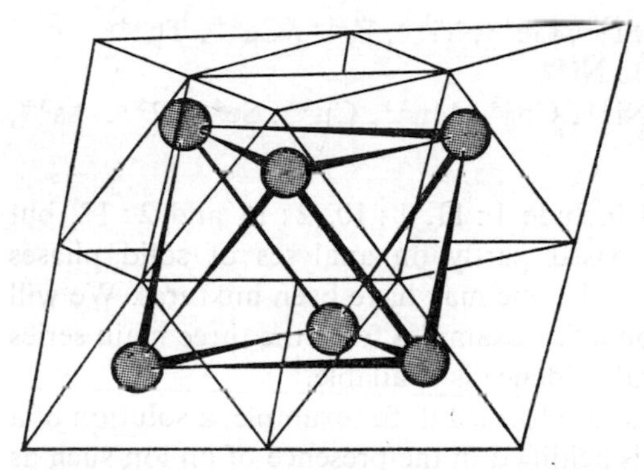

Figure 10.5

The $[Ta_6O_{19}]^{8-}$ ion is unchanged in solution in the range pH 10–13, and there is no evidence of protonation. Under more acid conditions precipitation of the hydrated pentoxide occurs. In an aqueous solution of the hexaniobate the anions $[H_2Nb_6O_{19}]^{6-}$, $[HNb_6O_{19}]^{7-}$ and $[Nb_6O_{19}]^{8-}$ have been shown to exist at pH 13·5–9·0, but there is no information to show whether other species are formed between the lower of these values and the isoelectric point (pH 4·5).

10.3.4. Heteropoly acids and their salts.[3, 7] The principle underlying the formation of the anions of heteropoly acids is believed to be that the anion of the second acid provides a central group round which octahedra such as MoO_6 and WO_6 condense by a process of oxygen-sharing, similar to that encountered in the isopoly acids. It is important to note that the central group (e.g. $[PO_4]^{3-}$) also shares its oxygen atoms with surrounding octahedra.

Heteropoly acids have been described with a wide variety of elements acting as the central atom. These include not only those capable of oxy-acid formation (e.g. B, Al, Si, Ge, Sn, P, As, Sb, Se, Te and I), but also metals of the transition series (e.g. Ti, Zr, Ce, Th,

V, Nb, Ta, Cr, Mo, W, V, Mn, Fe, Co, Ni, Rh, Os, Ir and Pt). Presumably thc latter are coordinated with surrounding oxygen atoms, but there is virtually no structural evidence available. The anions are usually classified in terms of the ratio of the number of central atoms to the number of metal atoms associated with the surrounding octahedra—most commonly molybdenum, tungsten or vanadium, the 1 : 12, 1 : 9 and 1 : 6 types occurring most frequently. For the heteropoly molybdates, for example, the central atoms that are associated with these types are shown below:

1 : 12	P^{5+}, As^{5+}, Si^{4+}, Ge^{4+}, Ti^{4+}, Zr^{4+}, Ce^{4+}, Th^{4+}
1 : 9	Mn^{4+}, Co^{4+}, Ni^{4+}
1 : 6	Te^{6+}, I^{7+}, Ni^{2+}, Co^{2+}, Mn^{2+}, Cu^{2+}, Se^{4+}, P^{3+}, As^{3+}, P^{5+}

Other ratios reported include 1 : 11, 1 : 10, 2 : 18 and 2 : 17, but these formulations are based partly on analyses of solid phases deposited from solution and some may have been mixtures. We will concentrate, therefore, on a few examples from the three main series for which some structural evidence is available.

Ions of heteropoly acids are formed if, for example, a solution of a molybdate or tungstate is acidified in the presence of an ion such as $[PO_4]^{3-}$ or a metal ion. A very familiar example is the phosphomolybdate test for phosphate, in which a solution containing a phosphate is treated with excess of ammonium molybdate solution, made strongly acid with nitric acid and warmed; yellow ammonium phosphomolybdate is precipitated. When this reaction is used in the gravimetric determination of phosphate, the material weighed is formulated as $(NH_4)_3[PMo_{12}O_{40}]$. Many of the heteropoly acids and their salts are soluble in water, and also in solvents such as ethers, alcohols and ketones. The acids are often strong and their anions are stable under more acid conditions than can be tolerated by the anions of many isopoly acids.

The same range of techniques can be used in solution studies as for the isopoly anions, and it appears that discrete heteropoly anions exist in definite pH ranges. Thus, for example, on progressive acidification of a sodium silicate–ammonium molybdate solution, 1-, 2-, 6- and 12-silicomolybdate ions are formed in turn. There is likewise evidence for the progressive degradation of the more highly condensed anions as the pH is raised. These changes can be followed by pH titration, and also by paper chromatography. Some uncertainty may exist as to the acidity range over which particular ions are stable, but for a number of the 12-heteropoly anions values between pH 1–4 are observed.

10.3.5. Structures of the heteropoly acids. The first structural study of a heteropoly acid was made by Keggin on a crystalline hydrate of

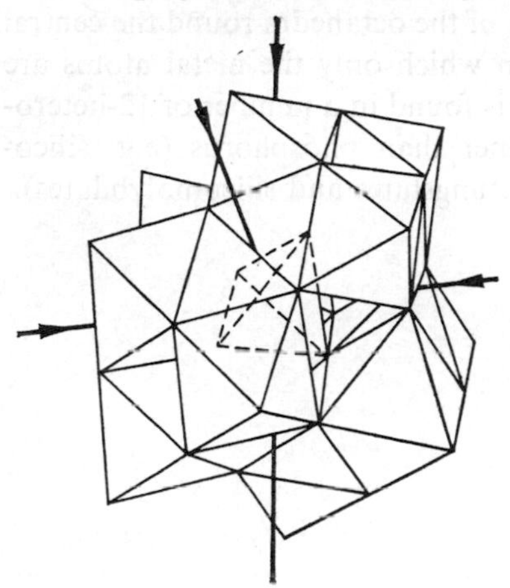

Figure 10.6

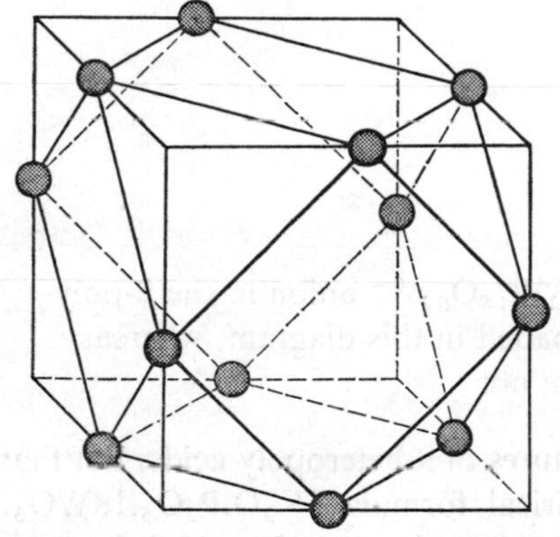

Figure 10.7 Arrangement of W atoms in the $[SiW_{12}O_{40}]^{4-}$ ion in 0·3M aqueous solution (H. A. Levy, P. A. Agron and M. D. Danford, *J. Chem. Phys.*, **30**, 1486 (1959))

12-phosphotungstic acid with the empirical formula $P_2O_5.24WO_3.9H_2O$. The anion, shown diagrammatically in Fig. 10.6, was found to have a central PO_4 group surrounded by WO_6 octahedra. Each corner oxygen of the PO_4 group is shared with three octahedra (i.e.

makes up part of the three octahedra), each of which also shares one oxygen atom with each of its two neighbours. The octahedra are then further linked together by the sharing of corners. The packing together of such anions in the crystal leaves large cavities, which permits the existence of higher hydrates (e.g. $H_3[PW_{12}O_{40}].29H_2O$). A clearer picture of the arrangement of the octahedra round the central atom is provided by Fig. 10.7, in which only the metal atoms are shown. The same anion structure is found in a number of 12-heteropoly acids based on elements other than phosphorus (e.g. silicotungstates, borotungstates, arsenotungstates and silicomolybdates).

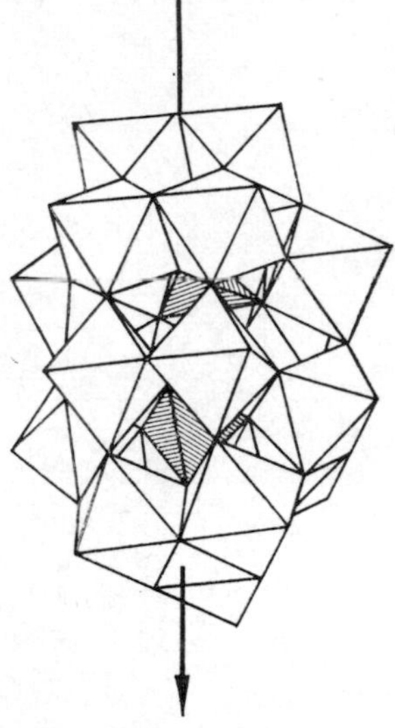

Figure 10.8 Structure of the $(P_2W_{18}O_{62})^{6-}$ anion in the 9-poly-acids. The PO_4 tetrahedra are shaded in this diagram, whereas the WO_6 or MoO_6 octahedra are not

Less is known about the structures of 9-heteropoly acids, but that of a compound with the empirical formula $3K_2O.P_2O_5.18WO_3.14H_2O$ is of interest because the anion, shown in Fig. 10.8, has two PO_4 tetrahedra enclosed within WO_6 octahedra bonded together by sharing oxygen atoms. There are other dimeric ions of this sort, but their structures are unknown.

In the 6-heteropoly acids the central atom (I or Te) is larger and capable of coordinating with six atoms of oxygen. A geometrical arrangement of octahedra which fulfils this condition for the $[XM_6O_{24}]$ group was first proposed by Anderson and is shown

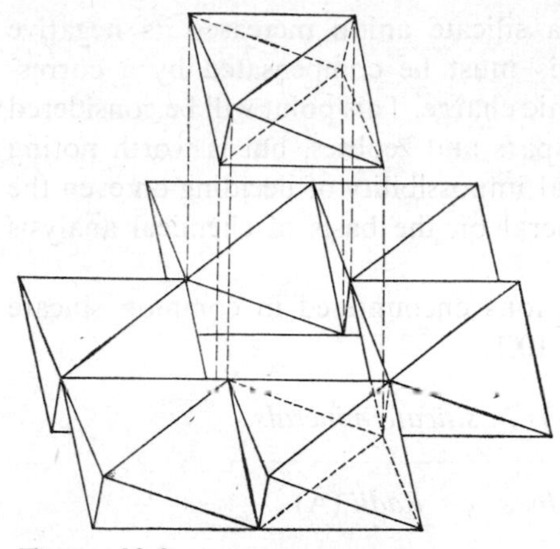

Figure 10.9

diagrammatically in Fig. 10.9. Six MO_6 octahedra are arranged in a hexagonal annulus so as to share two corners with each of two neighbouring octahedra. The central cavity of the resulting $[M_6O_{24}]^{12-}$ structure is found to be just large enough to accommodate an octahedron, corresponding with that of the hetero atom. This structure was subsequently identified in potassium and ammonium molybdotellurates which contain the $[TeMo_6O_{24}]^{6-}$ ion.

10.4 Silicates[8, 9]

Most of our knowledge of silicate structures comes from the study of naturally occurring silicate minerals, which are crystalline and lend themselves to investigation by X-ray crystallographic methods. In the processes involved in the formation of such minerals in nature isomorphous replacement is of frequent occurrence. Thus, magnesium, calcium, iron (II) and other bivalent metal ions of similar size may replace one another mutually to a greater or lesser extent, as may also Al^{3+} and Fe^{3+} or OH^- and F^-. The criterion for such replacements to be possible is that the ions in question shall have the same charge and about the same size. There is, however, a second kind of isomorphous replacement which is of frequent occurrence, and which may be far reaching in extent, thereby altering the whole apparent formulation of the compound. This is the replacement of an ion in the lattice by one of very similar size but of different charge. An instance of this sort, which is particularly important in silicate structures, is the replacement of Si^{4+} ($r = 0{\cdot}50$ Å) by Al^{3+} ($r = 0{\cdot}55$ Å).

Each such replacement in a silicate anion increases its negative charge by one unit and this must be compensated by a corresponding increase in the cationic charge. This point will be considered further in discussing the felspars and zeolites, but is worth noting here as illustrating the virtual impossibility of deciding on even the molecular formula of a mineral on the basis of chemical analysis alone.

The radii of some of the ions encountered in common silicate minerals are shown in Table 10.1.

Table 10.1 *Radii of some ions in silicate minerals*

Ions	*Radii* (Å)	*Ions*	*Radii* (Å)
Be^{2+}	0·39	Ca^{2+}	0·98
Si^{4+}	0·50	Na^{+}	0·98
Al^{3+}	0·55	K^{+}	1·33
Fe^{3+}	0·67	F^{-}	1·33
Mg^{2+}	0·71	O^{2-}	1·40
Fe^{2+}	0·83	OH^{-}	1·40

The Goldschmidt coordination number of anions of radius R_x round cations of radius R_A is determined by the ratio $R_A : R_x$, which is equal to 0·36 in the case of Si^{4+} and O^{2-}. As will be seen from Table 10.2, which gives the limiting radius ratios for different coordination numbers, and the associated geometrical arrangements, this falls in the range for a tetrahedral arrangement of silicon round oxygen. This is observed in all of the crystalline silicates, though one

Table 10.2 *Limiting radius ratios*

$R_A : R_x$	*Coordination number*	*Geometrical arrangement*
1	12	Close-packing
1–0·732	8	Corners of cube
0·732–0·414	6	Corners of octahedron
0·414–0·22	4	Vertices of tetrahedron
0·732–0·414	4	Square planar
0·22–0·15	3	Corners of triangle
<0·15	2	Linear

modification of silica with a structure of the rutile type in which there is sixfold coordination of oxygen has been reported. It is formed in several reactions which give silica if these are carried out at 500–800° under a pressure of 35,000 atm. The product is denser than the three normal forms of silica and is also more resistant to chemical attack (e.g. by HF).

It is important to realize that the reference to Si^{4+} ions in the context of silicate structures is no more than a formal representation, and does not imply that the bonding of silicon to oxygen is electrostatic. This is almost certainly not so, as is shown by the fact that the Si—O—Si bond angle is usually about 140°. The convention is, however, extremely helpful in describing the structures, and is almost invariably employed for this purpose.

10.4.1. Orthosilicates. In the orthosilicates there are discrete $[SiO_4]^{4-}$ tetrahedra so arranged in the lattice that the oxygen atoms of each are also coordinated (in the Goldschmidt sense) round the metal cations so as to form a neutral structure. The several distinct orthosilicate structures correspond to various ways in which this may be effected.

In olivine, $(Mg,Fe)_2SiO_4$, the metal ions, which can replace one another isomorphously, are so packed between SiO_4 tetrahedra that each magnesium ion is between six oxygens. Each oxygen is then linked directly to one silicon atom in its tetrahedron and coordinated jointly to three magnesium atoms. For stability in such coordination structures, according to Pauling, the charge on each negative ion must be equal and opposite to the sum of the electrostatic valency bonds reaching it from the cations to which it is common. Thus, the magnesium ion (charge +2), being coordinated in an octahedron, directs an electrostatic valency bond of strength $\frac{1}{3}$ to each oxygen ion. In the olivine structure, therefore, the net electrovalencies reaching each oxygen are:

$$\left.\begin{array}{ll}\text{from three } Mg^{2+} \text{ ions, } 3 \times \frac{1}{3} = +1 \text{ unit} & \\ \text{from the central } Si^{4+} \text{ ion} & \\ \quad \text{of its tetrahedron} & +1 \text{ unit}\end{array}\right\} = +2 \text{ units in all}$$

This balances its intrinsic charge of −2 units.

In phenacite (Be_2SiO_4) and willemite (Zn_2SiO_4) a different structure obtains, with metal ions tetrahedrally coordinated. Each oxygen is then common to one SiO_4 tetrahedron and two MO_4 tetrahedra. As may readily be seen, the Pauling rule is again exemplified.

Various other silicate minerals are based on the orthosilicate structure, but have additional negative ions—O^{2-}, OH^- or F^-—which are not coordinated in SiO_4 tetrahedra, Thus, additional oxygen ions are present in cyanite, $Al_2O(SiO_4)$; OH^- and F^- in topaz, $Al_2(OH,F)_2(SiO_4)$, and the minerals of the chondrodite group, $Mg(OH,F)_2$, $nMg_2(SiO_4)$, where $n = 1$, 2, 3 or 4, consist of sheets of the olivine structure interleaved with layers of OH^- and F^-, which are nearly identical in size. The latter are not part of SiO_4 tetrahedra,

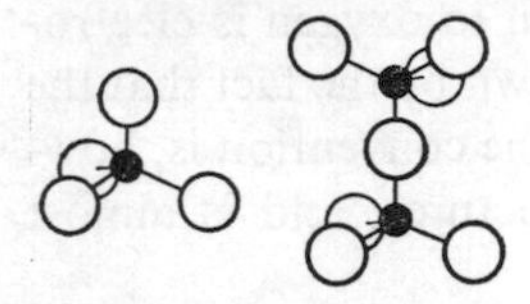

$[SiO_4]^{4-}$ $[Si_2O_7]^{6-}$

Figure 10.10 Discrete silicate anions

but are so situated that, together with the oxygens of the SiO_4 groups, they go to make up octahedra surrounding each magnesium ion. The structures of the various minerals of this group are all of this type, being derived from one another by variation in the relative positions of the olivine and OH sheets.

Relatively few larger discrete anions derived from the orthosilicate

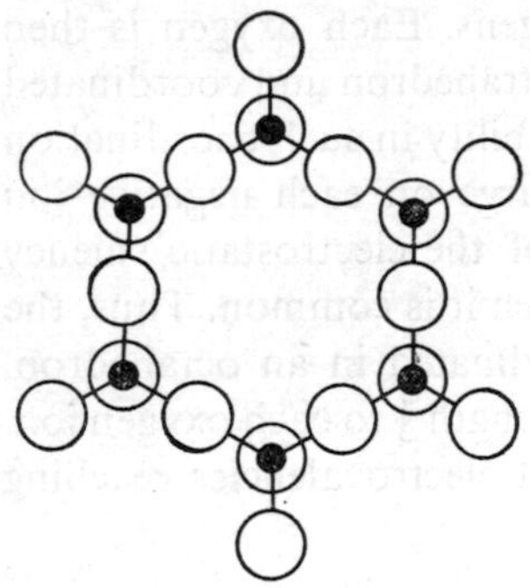

Figure 10.11 The $(Si_6O_{18})^{12-}$ group structure found in beryl

unit have been identified in minerals. The $[Si_2O_7]^{6-}$ anion, shown diagrammatically in Fig. 10.10, occurs in thortveitite ($Sc_2Si_2O_7$), in which Sc^{3+} ions are octahedrally coordinated. The bridging oxygen atom of the anion uses both of its electrovalencies in joining the two tetrahedra, and the charge of -6 arises from six free electrovalencies of the remaining oxygens.

Of the discrete cyclic anions where the tetrahedron shares two vertices, the $[Si_6O_{18}]^{6-}$ unit is important (Fig. 10.11). It occurs in beryl ($Be_3Al_2Si_6O_{18}$). The Si_6O_{18} groups are so disposed in different layers that the oxygens satisfy the coordination requirements of the cations and, incidentally, produce an open structure with wide channels. It is interesting to correlate this with the well-known occlusion of helium by beryl. In the mineral benitoite ($BaTiSi_3O_9$) the $[Si_3O_9]^{6-}$ anion contains a ring of three silicon atoms bonded together by Si—O—Si bridges.

10.4.2. Linear polyanion: metasilicates. The metasilicate anion, $[SiO_3]_n^{2n-}$, is formed by the linking up of SiO_4 tetrahedra into infinite chains, extending throughout the crystal (Fig. 10.12). Depending on the coordination requirements (i.e. the ionic radii) of the cations, the orientation of the tetrahedra about the chain axis and the periodicity

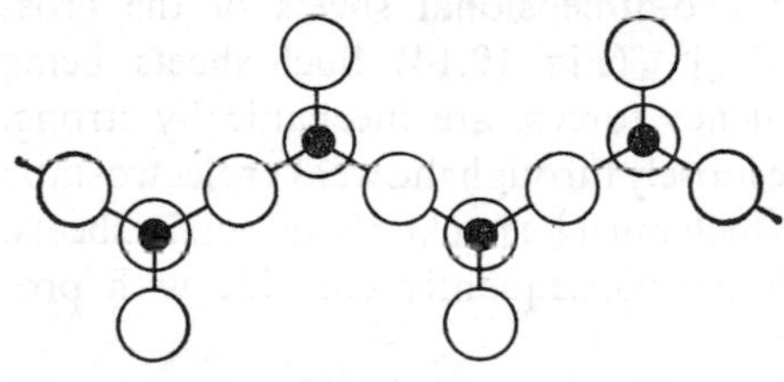

Figure 10.12 Chain-like metasilicate $(SiO_3)^{2-}$ group

in the chain, several types of structure (e.g. the pyroxene metasilicates and the wollastonite group minerals) can occur, all containing the same basic structural unit. Diopside, which has the formula $CaMg(SiO_3)_2$, is a pyroxene mineral of this type.

The amphibole mineral tremolite, $Ca_2Mg_5(Si_4O_{11})_2(OH)_2$, is one of a number of minerals in which the anion is formed by cross-linking two diopside chains by a further sharing of tetrahedron corners to give a band structure (Fig. 10.13).

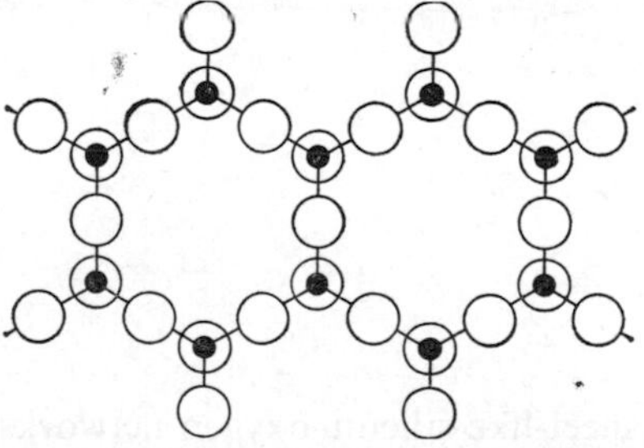

Figure 10.13 Cross-linked $[(Si_4O_{11})]^{6-}$ groups

In these metasilicates the important relationship between crystal structure and physical properties can be clearly seen. The chains or bands are in all cases arranged parallel to the *c*-axis of the crystal, and are bound laterally to one another by coordination of their oxygen atoms round metal ions. The resulting structure is mechanically strongest in the direction in which the anions are oriented, where fracture would involve breakage of the chain itself with severance of strong Si—O bonds. The structure is weaker in directions at right angles to the chains, where the bonds between metal ions and oxygen can be thought of as having a fractional order. This very crude picture is in accordance with the fact that cleavage of the crystals is highly developed parallel to the *c*-axis and the minerals frequently exhibit a fibrous structure—notably in asbestos, a form of amphibole.

The process of cross-linking of strings of SiO_4 tetrahedra, if carried to completion, will produce two-dimensional sheets of the gross composition and charge $[Si_4O_{10}]^{4-}$ (Fig. 10.14). Such sheets, being bound together by strong valency forces, are mechanically strong. Parallel sheets are bound more loosely through the weaker electrostatic bonds involving the cations, which must be packed between the sheets. The silicon–oxygen sheets should consequently coincide with pro-

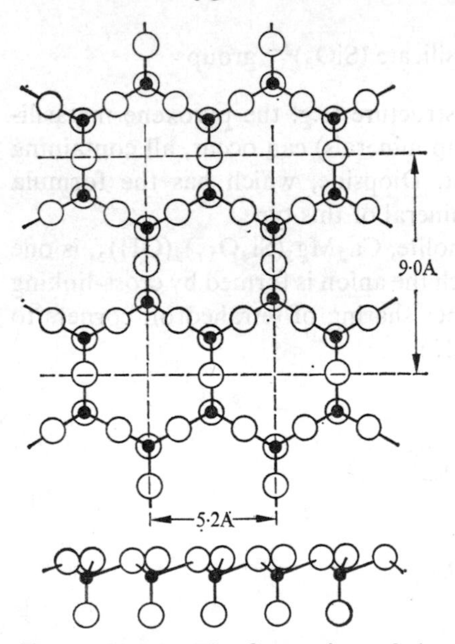

Figure 10.14 The formation of sheet-like silicon–oxygen networks. The lower figure shows a cross-section of the sheet

nounced cleavage planes of the crystal, and silicates with this type of structure should exhibit well-developed laminar cleavage.

In fact, the situation is more complicated than this simple presentation implies, as may be seen by considering a few examples. Talc, which is notable for its extreme softness, has the composition $3MgO.4SiO_2.H_2O$ and the structure proposed by Pauling is shown diagrammatically in Fig. 10.15(a). Layers of brucite ($Mg(OH)_2$) are

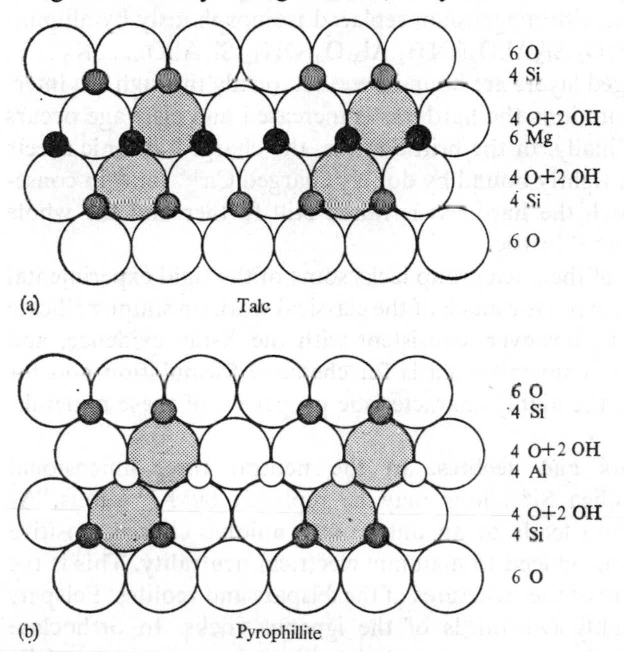

Figure 10.15

sandwiched, by the sharing of oxygen and hydroxyl, between two sheets of $Si_2O_5^{2-}$ (considered to be equivalent to single layers of the structure of β cristobalite), giving a composite layer which may be represented as . . . $O_6.Si_4.O_4(OH)_2.Mg_6.O_4(OH)_2.Si_4.O_6$ . . . The sheet is also electrically neutral.

In pyrophillite (Fig. 10.15(b)), which has the molecular formula $Al_2O_3.4SiO_2.H_2O$, four aluminiums replace the six magnesiums of talc and the equivalence of the positive and negative ions in the composite sheet is maintained, though one-third of the positions in the central layer are unfilled. In each case the softness and ready cleavage of the mineral follows from the very weak attractive forces between adjacent sheets.

If we now take phlogopite ($K_2O.6MgO.Al_2O_3.6SiO_2.H_2O$) we find that part of the silicon is replaced isomorphously by aluminium, with the result that each composite layer bears a net anionic charge. Alkali metal ions accordingly enter between the sheets in corresponding numbers to restore neutrality, and the composition may be represented by the sequence $...K_2...O_6.Si_3Al.O_4(OH)_2.Mg_6.O_4(OH)_2.Si_3Al.O_6...K_2...$. Muscovite (ordinary mica) is structurally analogous, but with magnesium replaced isomorphously by aluminium: $...K_2...O_6.Si_3Al.O_4(OH)_2.Al_4.O_4(OH)_2.Si_3Al.O_6...K_2....$

When charged layers are bound electrostatically through the interposed potassium ions the hardness is increased and cleavage occurs less readily. Finally, in the brittle micas, the charged anionic sheets are even more tightly bound by doubly charged Ca^{2+} ions, in consequence of which the hardness is raised still further and the whole structure becomes brittle.

This theory of the mica group lacks some of the rigid experimental backing which exists for much of the classical work on simpler silicate structures. It is, however, consistent with the X-ray evidence, and provides a very convenient basis for chemical formulation and for understanding the highly characteristic properties of these minerals.

10.4.3. Felspars and zeolites. In the neutral three-dimensional structure of silica Si^{4+} ions may be replaced by Al^{3+} ions. As such substitution leads to an unbalanced anionic charge, positive ions must be introduced to maintain electrical neutrality. This is the essential feature of the structures of the felspars and zeolites. Felspars make up roughly two-thirds of the igneous rocks. In orthoclase felspar ($KAlSi_3O_8$) one-quarter of the silicon ions are replaced by aluminium and a single univalent ion must therefore be introduced to restore neutrality. In celsian ($BaAl_2Si_2O_8$), to quote a further example, a bivalent ion is present and achieves the same end. Cavities in the framework are relatively large and this is why, in general, the larger cations (e.g. Na, K, Ca and Ba) occur in this type of mineral.

The structural principle of the zeolites is the same as that of the felspars (i.e. an anion is created by the replacement of silicon with aluminium). The honeycomb structure is, however, more open, with channels which have free diameters in the range 4–7 Å. This is shown diagrammatically in Fig. 10.16.

The two most striking general properties of the zeolites are their capacity for base exchange and the ability of the hydrated materials to lose water or to rehydrate without any change in optical or

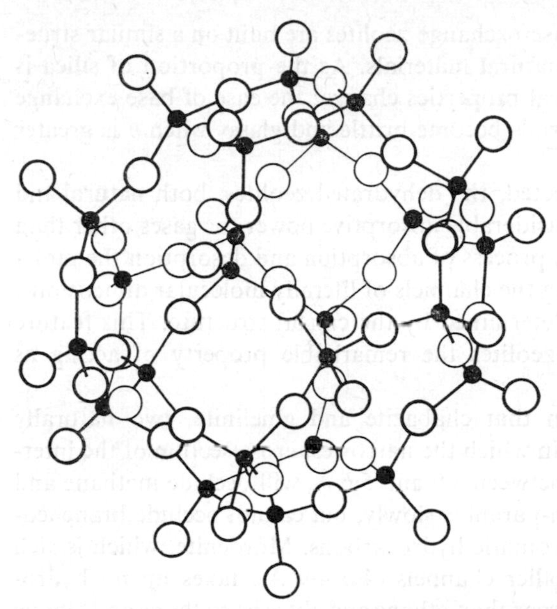

Figure 10.16 The arrangement of AlO_4 and SiO_4 tetrahedra which gives the cubo-octahedral cavity in some zeolites and feldspathoids. • represents Si or Al

crystallographic properties. By 'base exchange' is meant the replacement of one cationic constituent by another on simple treatment with a salt of the corresponding metal. Thus if a sodium zeolite such as analcite ($NaAlSi_2O_6.H_2O$) is treated with a solution of a salt of some other metal (e.g. silver nitrate) the sodium is replaced by an equivalent amount of the second cation, giving a silver zeolite in the case quoted. The reactions concerned are reversible and lead to a state of equilibrium.

The same phenomenon of base exchange is exhibited also by certain synthetic silicate masses, notably by those materials applied to the softening of water. In this case the calcium salts dissolved in the water undergo base exchange with a sodium zeolite, with the result that calcium is removed and sodium ions go into solution. At an appropriate stage the base exchange material can be regenerated by treating it with a large excess of sodium chloride solution:

$$Na_2O.Al_2O_3.nSiO_2.mH_2O + CaSO_4 \rightarrow CaO.Al_2O_3.nSiO_2.mH_2O + Na_2SO_4 \text{ (water softening)}$$

$$CaO.Al_2O_3.nSiO_2.mH_2O + 2NaCl \rightarrow Na_2O.Al_2O_3.nSiO_2.mH_2O + CaCl_2 \text{ (regeneration)}$$

These artificial base exchange zeolites are built on a similar structural model to the natural materials. As the proportion of silica is increased, the physical properties change; the ease of base exchange is lost and the materials become brittle and glassy when *n* is greater than 3.

As might be expected, the dehydrated zeolites, both natural and synthetic, have a considerable absorptive power for gases other than water vapour. In the process of absorption and desorption the molecules must pass along the channels of literally molecular dimensions, having a diameter determined by the crystal structure. This feature confers on certain zeolites the remarkable property of acting as molecular sieves.

Barrer has shown that chabazite and gmelinite, two naturally occurring minerals, in which the narrowest cross-section of the interstitial channels lies between 4·9 and 5·6 Å, will occlude methane and ethane rapidly and n-paraffins slowly, but cannot occlude branched-chain paraffins or aromatic hydrocarbons. Mordenite, which is rich in sodium, with smaller channels (4·0–4·9 Å), takes up no hydrocarbon molecules larger than ethane and absorbs methane and ethane only slowly, whereas nitrogen, oxygen and smaller molecules are occluded rapidly. The replacement of sodium in mordenite by calcium or barium (by a cation exchange process) decreases the cross-section of the channel to 3·8–4·0 Å and such mordenites will absorb nitrogen, argon and smaller molecules, but not methane and ethane. The process of absorption involves diffusion of the gas molecules through the solid, and it is the activation energy of this process which, varying from one molecule to another, confers selectivity. By operating at appropriate temperatures the selective occlusion can be utilized to effect difficult separations of gaseous mixtures—thus n-heptane and isooctane can be separated quantitatively. These materials now have a variety of technical applications and are also very useful in the laboratory (e.g. for the drying of solvents).

10.5 Borates[10]

There are several important differences in the structural chemistry of borates and silicates. Unlike silicic acids, the free boric acids are well-defined crystalline substances in which hydrogen bonding is a prominent structural feature. The radius of B^{3+} is 0·24 Å, compared with 0·50 Å for Si^{4+}, and, taking the radius of O^{2-} as 1·40 Å, this gives a radius ratio of 0 : 17. From the values given in Table 10.2, this would imply an oxygen coordination number of three for boron, whereas three and four are observed. The transition to tetrahedral sp^3 hybridization is favoured by the ready acceptance of an electron pair into the low energy fourth orbital of the boron valence shell. The measured B—O distances in trigonal borates are in the range 1·28–1·44 Å, with a mean of 1·37 Å, whereas the tetrahedral bond length is in the range 1·42–1·54 Å, with a mean of 1·48 Å.

The analysis of ^{11}B nuclear quadrupole resonance effects has been used in this field to distinguish between tetrahedrally and trigonally coordinated boron in the borates. The element has two naturally occurring isotopes, ^{10}B (18·83 per cent) with a nuclear spin of 3 and ^{11}B (81·17 per cent) with a nuclear spin of $\frac{3}{2}$, the latter being detected with greater sensitivity in low-resolution n.m.r. analysis. The tetrahedral borates, which have a nearly spherical charge distribution round boron, show sharp first-order splitting of the ^{11}B resonance, while the asymmetric field gradients in the triangular borates give broad resonance peaks with larger coupling constants. These measurements are, of course, used in conjunction with those from X-ray analysis.

Crystalline orthoboric acid (H_3BO_3) has a lattice made up of superposed planar sheets of $B(OH)_3$ molecules linked together by hydrogen bonds (Fig. 10.17), the interplanar distance being 3·18 Å.

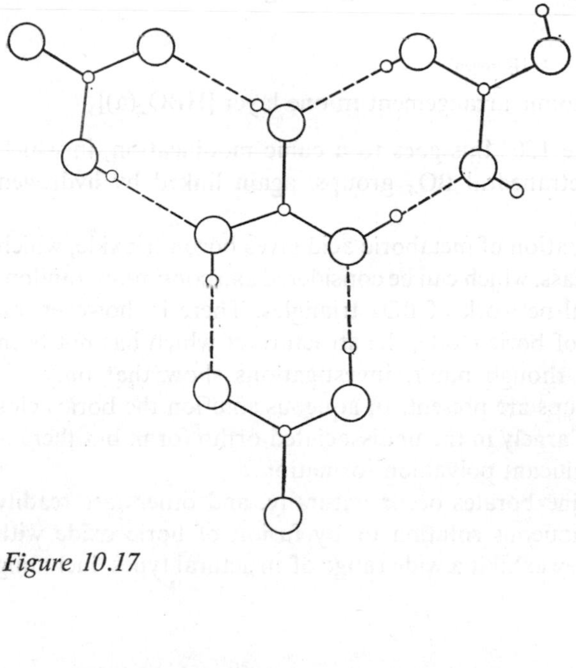

Figure 10.17

The hydrogen bond is not symmetrical, the two OH distances being 2·7 and 0·8 Å.

Dehydration of the ortho acid can give three crystalline forms of metaboric acid (HBO_2). Below 130° an orthorhombic modification is formed in which threefold oxygen coordination is retained (Fig. 10.18), with trimeric B_3O_6 rings joined by hydrogen bonds.

A monoclinic polymorph is then formed at 130–150° which has BO_4 tetrahedra and B_2O_5 groups in chains, linked by hydrogen

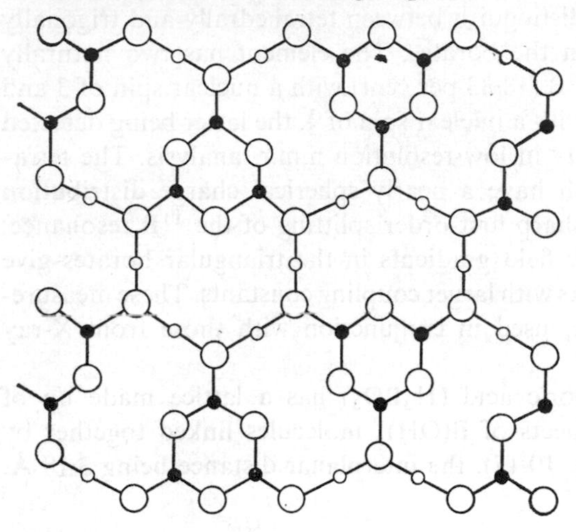

Figure 10.18 Atomic arrangement in one layer [HBO_2(a)]

bonds, and above 150° this goes to a cubic modification, in which there are only tetrahedral BO_4 groups, again linked by hydrogen bonds.

Further dehydration of metaboric acid gives boron trioxide, which readily forms a glass, which can be considered as a continuous random three-dimensional network of BO_3 triangles. There is, however, an hexagonal form of boric oxide, the structure of which has not been fully elucidated, though n.m.r. investigations show that only triangular BO_3 groups are present. In aqueous solution the boric acids are present very largely in the undissociated ortho form, but there is evidence for significant polyanion formation.

Many crystalline borates occur naturally, and others are readily prepared from aqueous solution or by fusion of boric oxide with metal oxides. They exhibit a wide range of structural types, including rings, chains, sheets and three-dimensional networks. The monomeric $(BO_3)^{3-}$ ion is encountered in a few instances (e.g. in the rare earth orthoborates, $M^{III}BO_3$) and the simple pyroborate ion, $(B_2O_5)^{4-}$, also occurs (e.g. in $Co_2B_2O_5$, where there are two triangles of oxygen atoms with one in common). It is, however, also possible to have two tetrahedral groups with an oxygen in common, as in the mineral pinnoite ($MgB_2O(OH)_6$), where the anion has the structure shown in Fig. 10.19. The trimeric ring $(B_3O_6)^{3-}$ (Fig. 10.20) is found

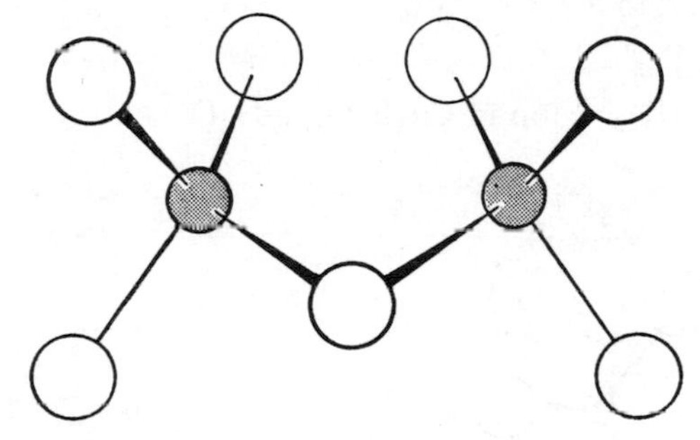

Figure 10.19

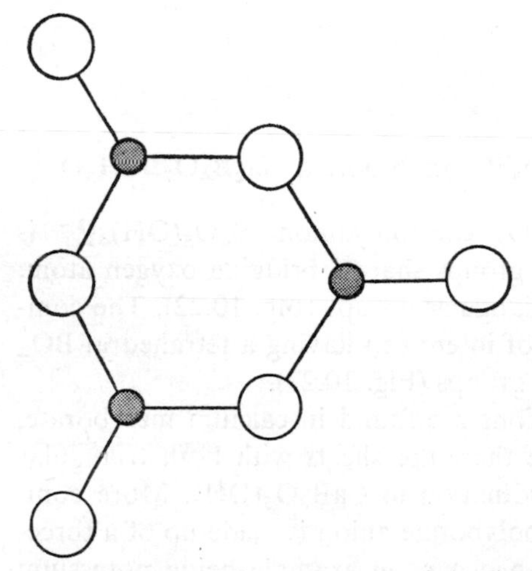

Figure 10.20 The structure of the $(B_3O_6)^{3-}$ ion

in sodium and potassium metaborates, and it is of interest that this six-membered ring also occurs in cyclic trimeric borate esters, $(ROBO)_3$. It is possible, however, for the six-membered ring to contain both trigonal and tetrahedral boron atoms, as, for example, in the ion of $Ca_2B_6O_{11}.nH_2O$ (Fig. 10.21).

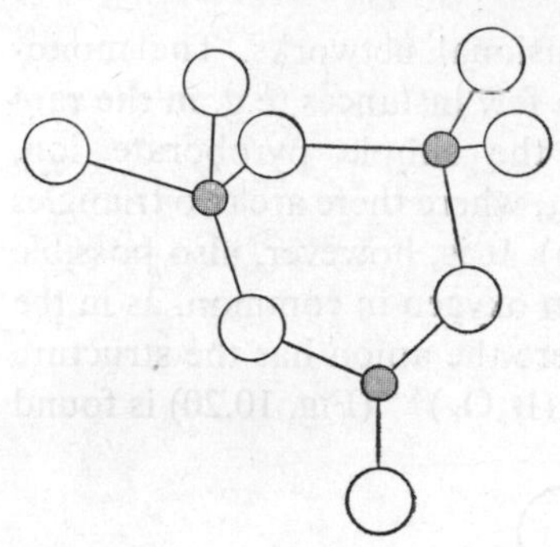

Figure 10.21 The $B_3O_3(OH)_5^{2-}$ ion in $Ca_2B_6O_{11}.nH_2O$

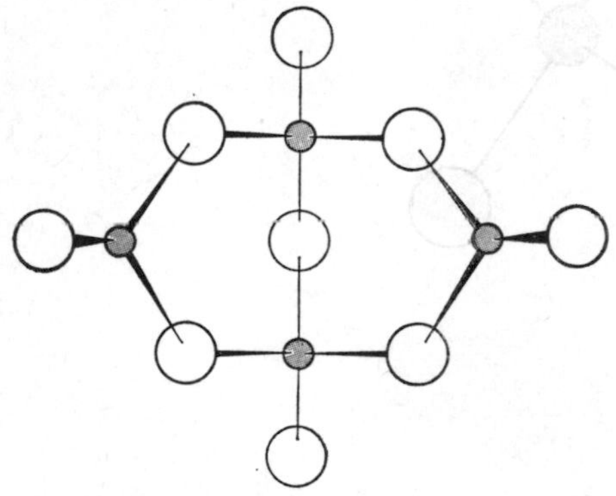

Figure 10.22 The $B_4O_5(OH)_4^{2-}$ ion in borax, $Na_2B_4O_7.10H_2O$

In borax ($Na_2B_4O_7.10H_2O$) the polyanion $[B_4O_5(OH)_4]^{2-}$ is present. The two tetrahedral groups share a bridging oxygen atom, and are also linked to two triangular groups (Fig. 10.22). The compound $KB_5O_8.4H_2O$ is also of interest in having a tetrahedral BO_4 group linking two triangular groups (Fig. 10.23).

Chains of linked $(BO_2)^-$ ions are found in calcium metaborate, $Ca(BO_2)_2$ (Fig. 10.24), while there are sheets with both triangular and tetrahedral oxygen coordination in $CaB_3O_5(OH)$. More complex structures in which the polyborate anion is made up of a three-dimensional network can also occur, an example being potassium pentaborate (KB_5O_8), in which B_3O_3 rings are built into helical chains by oxygen-sharing and these chains are similarly connected to one another.

In all of the above crystalline materials, X-ray analysis enables the structures to be determined with little ambiguity, apart from the position of hydrogen atoms. The nature of polyborate anions in

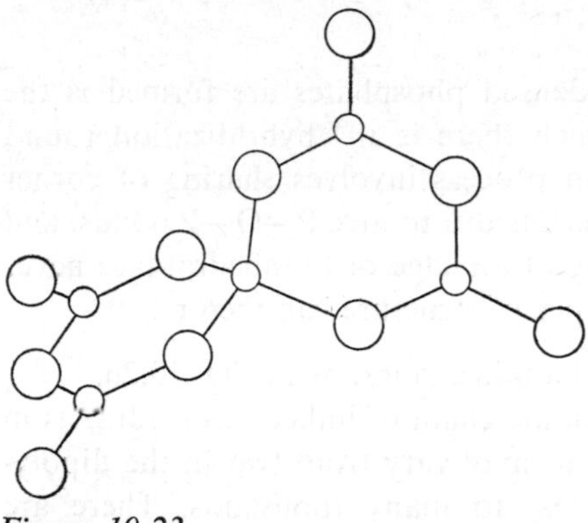

Figure 10.23

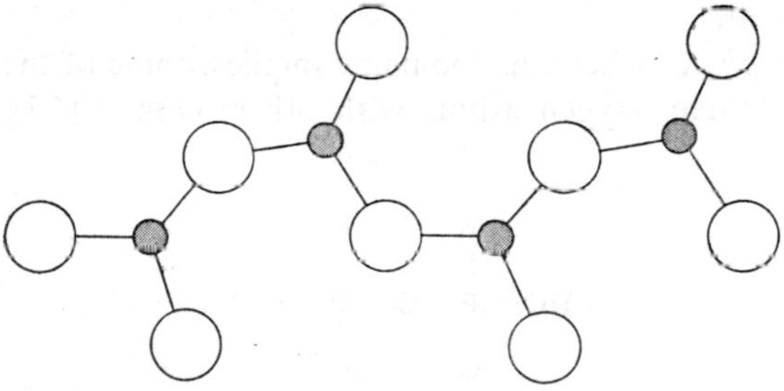

Figure 10.24

solution, however, is much less clear. Under highly acid and highly basic conditions it is generally accepted that the species present are $B(OH)_3$ and $[B(OH)_4]^-$ respectively. Physicochemical and spectroscopic techniques leave little doubt that polyanionic species are present between these extremes of pH, but there appears to be little clear evidence as to their nature.

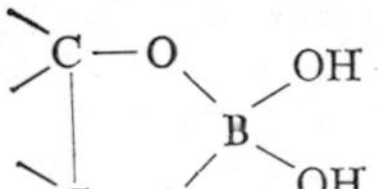

Figure 10.25

Although boric acid is very weak, its strength can be increased, so that it can be titrated with strong bases, if certain polyhydroxy compounds are added to its solution. This is applied in the analytical determination of boric acid, usually by the addition of glycerol or mannitol, which give a four-coordinated chelate complex (Fig. 10.25).

10.6 Condensed phosphates[11, 12]

The unit from which all condensed phosphates are formed is the tetrahedral PO_4^{3-} ion, in which there is sp^3 hybridization round phosphorus. The condensation process involves sharing of corner oxygen atoms between these tetrahedra to give P—O—P bonds, and it should be noted that sharing of an edge or tetrahedral face never occurs. The three following types of structure can then result:

(a) *Metaphosphates* with small anionic rings, as in Fig. 10.26.
(b) *Polyphosphates* with an anionic chain of linked tetrahedra, as in Fig. 10.27. The chain length may vary from two in the dipolyphosphates (pyrophosphates) to many thousands. There are terminal P—OH groups, which are not usually included when formulating the salts.
(c) *Cross-linked phosphates* where, as the name implies, some of the tetrahedra share three oxygen atoms with others (Fig. 10.28).

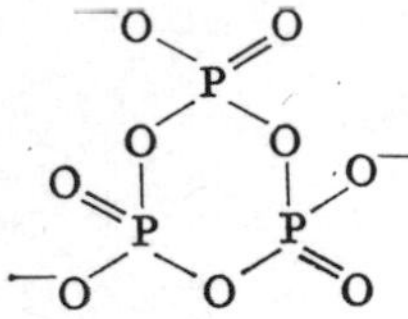

Figure 10.26

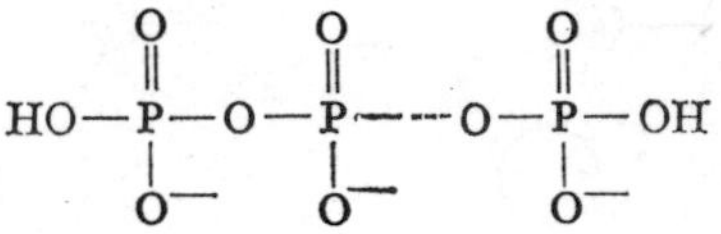

Figure 10.27

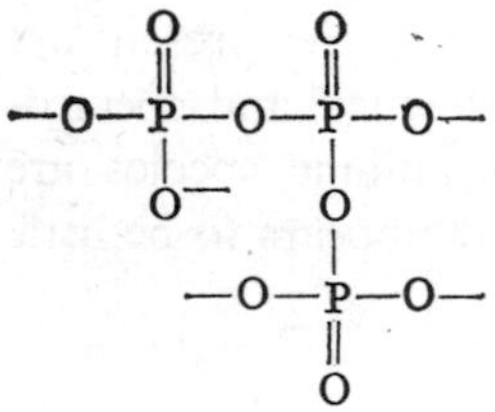

Figure 10.28

When all the tetrahedra share three corners the anionic character is lost, as *each* of three oxygen atoms forms two σ-bonds, and the condensed product is then one or other of the crystalline forms of phosphorus pentoxide which are described later.

All of the condensed phosphates are formed in reactions at elevated temperatures, in which water is eliminated intermolecularly from the anions of simpler phosphate groups with residual P—OH bonds. In aqueous solution they tend to undergo hydrolytic cleavage of the P—O—P bonds more or less readily, and the simple $[PO_4]^{3-}$ ion is the most stable form at all pH values. Hydrolysis may, however, be very slow so that in effect many condensed anions can exist in solution. The important point is that there is not the rapidly attained pH-dependent equilibrium between them that is characteristic of the iso- and heteropoly acids.

It is well known that for orthophosphoric acid the three acid dissociation constants differ widely ($pK_1 = 2{\cdot}15$, $pK_2 = 7{\cdot}1$, $pK_3 = 12{\cdot}4$). This is also true of the condensed phosphates where there are two P—OH groups on the same phosphorus atom. Thus, in the polyphosphates (Fig. 10.27) the terminal phosphorus atoms carry one weak and one strong acidic group, whereas those in the body of the chain carry only a single strongly acidic group. In the cyclic metaphosphates (Fig. 10.26), on the other hand, all the acid groups are equally strong. Acid titration thus provides a method of distinguishing between the two types and, more importantly, enables the chain length of polyphosphates of short or moderate length to be determined directly. All that is required is an accurate potentiometric titration to give the ratio of strong to weak acid groups.

10.6.1. Metaphosphates. There is no evidence for the existence of mono- or dimetaphosphate ions; formation of the latter would involve the sharing of an edge between two $[PO_4]$ tetrahedra. Trimetaphosphates of many cations are, however, well known. The sodium salt $Na_3P_3O_9$ is produced when NaH_2PO_4 is heated to 600–640°, and the melt is then held at 500° in a free atmosphere (i.e. under conditions favouring the removal of water vapour and the completion of the condensation process). The ring structure of the anion has been established by X-ray analysis of several salts, and the ^{31}P n.m.r. spectrum of trimetaphosphates shows a single peak split into a triplet by the two neighbouring phosphorus atoms in the ring, showing the three phosphorus atoms to be equivalent, as expected. In alkaline solution trimetaphosphates are hydrolysed almost quantitatively to tripolyphosphates.

Sodium tetrametaphosphate ($Na_4P_4O_{12}.4H_2O$) may be obtained by treating the volatile molecular form of P_4O_{10} with cold NaOH in $NaHCO_3$ solution. The intermediate in the reaction scheme shown overleaf has not been isolated.

There are several other ways of preparing salts of this acid (e.g. by heating NaH_2PO_4 with phosphoric acid at 400° and slowly cooling the melt). The presence of an eight-membered ring in the anion is proved by X-ray analysis of the structure of the sodium salt and by the ^{31}P n.m.r. spectrum. There is some evidence that the anion can exist in both boat and chair forms. Hydrolysis of tetrametaphosphates in alkaline solutions occurs more slowly than in the case of trimetaphosphates.

A few phosphates of this type with larger rings have been isolated,

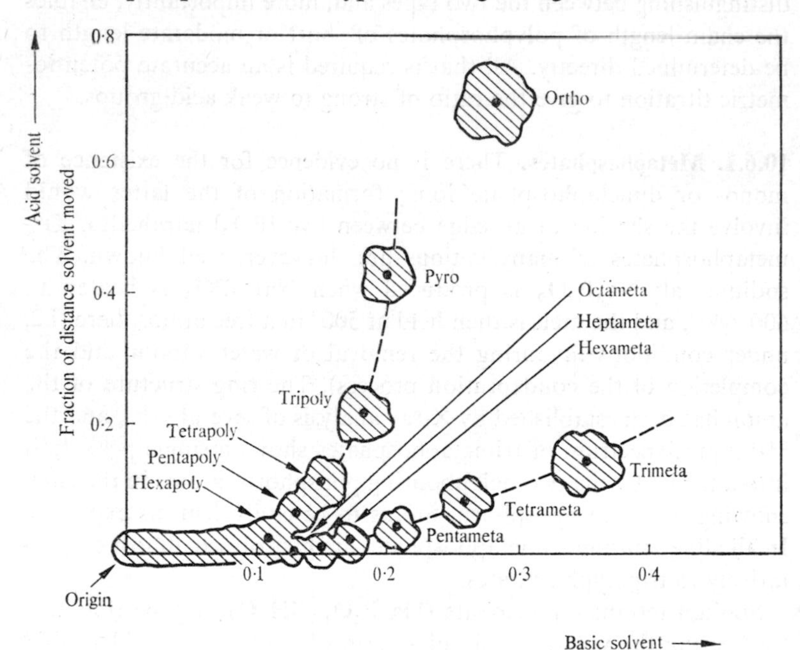

Figure 10.29 Corner of a two-dimensional paper chromatogram showing the positions of the pentameta- through octametaphosphate rings in relation to the positions of the well-known ring and chain phosphates. The basic solvent went 9 in. in 24 hr, whereas the acid solvent went 4·5 in. in 5·5 hr

including the salt $Na_6P_6O_{18}$, which is hydrolysed in alkaline solution to hexapolyphosphate. Paper chromatography is particularly useful in this field in detecting the presence of higher members of a series, even when they cannot be isolated, and also in checking the purity of a particular product. This can be seen from Fig. 10.29, which shows the type of chromatogram that would be obtained for a mixture containing both metaphosphates and short-chain polyphosphates. The two types are clearly distinguishable and it is also seen that separations become virtually impossible when there are more than six to eight PO_4 units in the chain or ring. There is no evidence that metaphosphate anions larger than $[P_8O_{24}]^{8-}$ can exist.

10.6.2. Polyphosphates. In contrast to the metaphosphates, there appears to be a complete series of polyphosphates with chain lengths ranging from two PO_4 groups in the dipolyphosphate up to several thousand. Di-, tri-, tetra- and pentapolyphosphates have been isolated, but higher members of the series as prepared are always mixtures, and their separation is impossible at present. It is usual therefore to express their compositions in terms of an average chain length, which is controlled by the conditions of preparation. In the salts, the chains have terminal P—OH groups, so that further condensation with elimination of water can occur on heating. Mean molecular weights of the long-chain species can be determined by the usual techniques applicable to polymers, including dialysis and diffusion studies, electrophoresis, the measurement of viscosities and the use of the ultracentrifuge.

The sodium salt of the dipolyphosphate ion, which is familiar as sodium pyrophosphate, is formed when NaH_2PO_4 is heated to 170° under conditions allowing the escape of water vapour. Some diphosphate is also produced when a concentrated aqueous solution is heated at 80° or above. The anion in $Na_4P_2O_7.10H_2O$ has the form shown in Fig. 10.30.

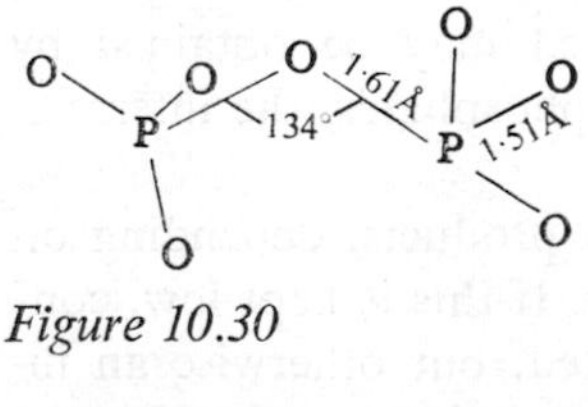

Figure 10.30

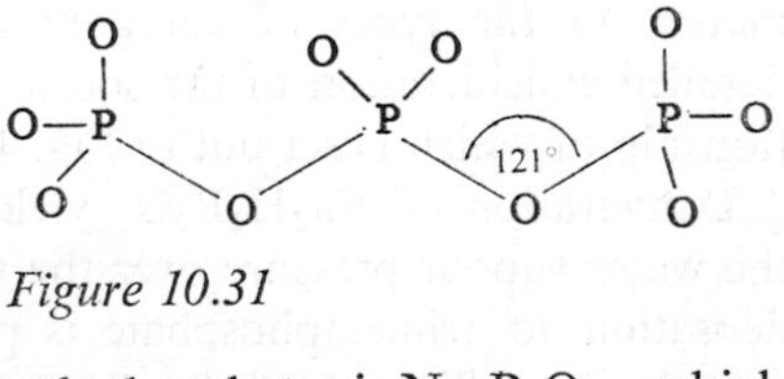

Figure 10.31

The most important of the tripolyphosphates is $Na_5P_3O_{10}$, which exists in two anhydrous forms in which the ion has almost identical dimensions (Fig. 10.31), and as a hexahydrate. This salt is prepared on a technical scale for water treatment, one process involving evaporation of solutions of Na_2HPO_4 and NaH_2PO_4 with an overall $Na_2O:P_2O_5$ ratio of 5 : 3, followed by calcination at 300–400°:

$$2Na_2HPO_4 + NaH_2PO_4 = Na_5P_3O_{10} + 2H_2O$$

End-group titration gives the expected ratio of weakly to strongly acidic hydrogens of 2 : 3 and the ^{31}P n.m.r. spectrum shows two peaks, the areas under which indicate a 2 : 1 abundance ratio for phosphorus atoms in the two environments. The peak corresponding to the end phosphorus atom is split into a doublet and the other into a triplet.

The use of tripolyphosphates in water treatment depends on the fact that stable soluble complexes are formed on the addition of excess of the sodium salt to a solution containing ions such as Ca^{2+} and Mg^{2+}, the triphosphates of which are otherwise insoluble. This effect is called sequestration: it results in the effective removal of deleterious cations from hard waters. Other polyphosphates, including Graham's salt, which is described below, act in the same way.

Relatively few tetrapolyphosphates are known, though alkaline cleavage of sodium tetrametaphosphate gives almost pure tetrapolyphosphate. The calcium salt $Ca_7(P_5O_{16})_2$, which is a pentaphosphate, has also been described, but beyond this one is always dealing with mixtures.

10.6.3. Long-chain polyphosphates. These present a complicated picture. Both glasses and crystalline materials are obtained and the mean chain length can be determined by end-group titration as well as by the normal techniques of polymer chemistry. Rough separation of a mixture on the basis of chain length can also be effected by fractional precipitation on the addition of an organic liquid, such as acetone, to long-chain polyphosphate solutions. A good general picture of the types of compound formed may be obtained by detailed consideration of the sodium polyphosphates, the interrelationship of which is set out in Fig. 10.32.

Dehydration of $Na_2H_2P_2O_7$ yields two products, depending on the water vapour pressure over the system. If this is kept low, condensation to trimetaphosphate is promoted, but otherwise an insoluble crystalline substance known as Maddrell's salt after its discoverer is obtained. This, as shown, can exist in a high and a low

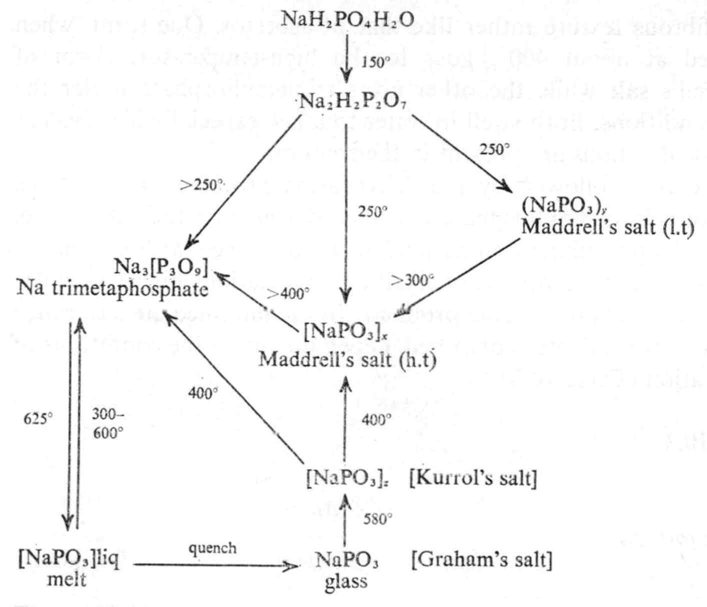

Figure 10.32

temperature form. Both are unstable with respect to the trimetaphosphate, which is the stable end-product of annealing of any of the long-chain polyphosphates at 500–600°.

Sodium trimetaphosphate melts at 526° and long-chain polyphosphate anions are produced. If the melt is quenched these persist in the glassy product, which is known as Graham's salt. It does, however, also contain up to roughly 10 per cent of metaphosphates and a little cross-linked material. Graham's salt shows no sign of crystalline structure. The average chain length, which can be determined by end-group titration and by the physical methods referred to earlier, is roughly in the range 20–500 PO_4 units, depending on the temperature and time of heating of the melt. This material swells in water and dissolves. At pH > 7 it hydrolyses only very slowly, but the rate increases progressively as the acidity is increased, trimetaphosphate predominating in the hydrolysis product. The material is used extensively in water treatment, its mode of action being similar to that of sodium tripolyphosphate mentioned earlier.

When Graham's salt is annealed above 550°, and especially if it is inoculated with some of the final product, it devitrifies, forming a substance known as Kurrol's salt, which can be obtained in two forms which differ in density. One is of platy habit and the other has a fibrous texture rather like that of asbestos. One form, when annealed at about 400°, goes to the high-temperature form of Maddrell's salt while the other gives trimetaphosphate under the same conditions. Both swell in water to a gel, especially if univalent or bivalent cations are present in the solution.

The course followed by the dehydration processes taking place when an acid orthophosphate is heated depends on the cation size. Cyclic metaphosphate anions tend to be produced with cations of medium size, while with very small or large cations high molecular polyphosphates are the sole products. In the intermediate size range a mixture of products is obtained, depending upon the conditions of dehydration (Table 10.3).

Table 10.3

Anions formed	*Cations*		
	Univalent	*Bivalent*	*Trivalent*
Polyphosphate chains	H^+ Li^+(0·78 Å)	Be^{2+}(0·39 Å)	
Cyclic anions*		Mg^{2+}, Mn^{2+}, Fe^{2+} Ni^{2+}, Co^{2+}, Cu^{2+} (0·65–0·91 Å)	Al^{3+}(0·55Å)
Cyclic anions,† polyphosphate chains	Na^+(0·98 Å)	Zn^{2+}, Cd^{2+} (0·83, 1·03 Å)	
Polyphosphate chains	K^+, Rb^+, Cs^+ NH_4^+ (>1·3 Å)	Ca^{2+}, Sr^{2+}, Ba^{2+} Pb^{2+}, Hg^{2+} (>1·06 Å)	Fe^{3+}, Cr^{3+} Bi^{3+} (>0·64 Å)

* Tetrametaphosphates.
† Trimetaphosphates (Na): tetrametaphosphates (Zn, Cd).

Polyphosphates of cations other than sodium are found to be capable of existing in forms corresponding with the sodium Maddrell and Kurrol salts. These differences are not due to chain length alone, and X-ray structural studies have now established that the anions differ in the way in which the PO_4 tetrahedra are oriented relative to one another along the chain. This point is illustrated in Fig. 10.33, which shows some of the chain types that have been found by X-ray analysis.

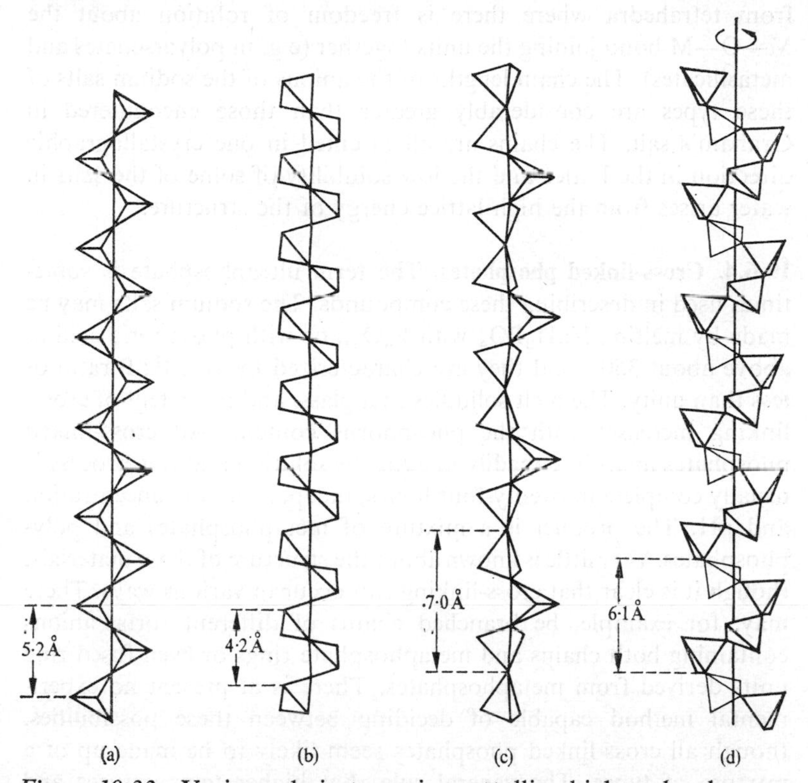

Figure 10.33 Various chain types in crystalline high-molecular phosphates: (a) $(LiPO_3)_x$ low temperature form; (b) $Rb(PO_3)_x$; (c) $(NaPO_3)_x$ Maddrell's salt (high temperature form); (d) $(NaPO_3)_x$ Kurrol's salt

Chains of type (a) occur in one of the forms of lithium polyphosphate, and it is of interest that the tetrahedra in diopside, $[CaMg(SiO_3)_2]_x$, have this arrangement of SiO_4 units. An example of type (b) is provided by rubidium polyphosphate: the tetrahedra are somewhat rotated with respect to form (a). Type (c) is found in the high-temperature form of Maddrell's salt, while the fourth type (d) has been identified in both the A and B forms of Kurrol's sodium polyphosphate, as well as in silver polyphosphate. The units of this anion are arranged in a screw pattern, each fourth tetrahedron having the same orientation as the first. These examples suffice to show the structural principle involved. These and other related types are encountered in other chain-like anions built up from tetrahedra where there is freedom of rotation about the M—O—M bond joining the units together (e.g. in polyarsenates and metasilicates). The chain lengths of the anions of the sodium salts of these types are considerably greater than those encountered in Graham's salt. The chains are all oriented in one crystallographic direction in the lattice and the low solubility of some of the salts in water arises from the high lattice energy of the structure.

10.6.4. Cross-linked phosphates. The term ultraphosphate is sometimes used in describing these compounds. The sodium salts may be made by melting NaH_2PO_4 with P_4O_{10} or with phosphoric acid at above about 350°, and they are characterized by an $M^I : P$ ratio of less than unity. The melt solidifies to a glass, and the extent of cross-linking increases with the phosphorus content. All cross-linked phosphates hydrolyse readily in aqueous solution and this process is usually complete in twenty-four hours, independent of concentration and pH. The product is a mixture of metaphosphates and polyphosphates. Very little is known about the structure of these materials, though it is clear that cross-linking can occur in various ways. There may, for example, be branched chains of different sorts, anions containing both chains and metaphosphate rings or even fused ring units derived from metaphosphates. There is at present no experimental method capable of deciding between these possibilities, though all cross-linked phosphates seem likely to be made up of a mixture of types. The general rule that higher temperatures and prolonged heating will favour more complete condensation through residual P—OH groups is true in this case also.

10.6.5. Phosphorus pentoxide. This can be regarded as the end-product of the cross-linking of PO_4 tetrahedra, each of which shares three of its four corners. As would be expected, it has an entirely different structure from silica, where all four corners of SiO_4 tetrahedra are shared. There are three crystalline modifications of known structure and also a glassy form. The vapour of the pentoxide is made up of discrete P_4O_{10} molecules, and these also form the structural unit in the common form of the solid, which is familiar as a laboratory reagent. The structure is rhombohedral and P_4O_{10} has the configuration shown in Fig. 10.34: electron diffraction measurements show that this is the same in the vapour.

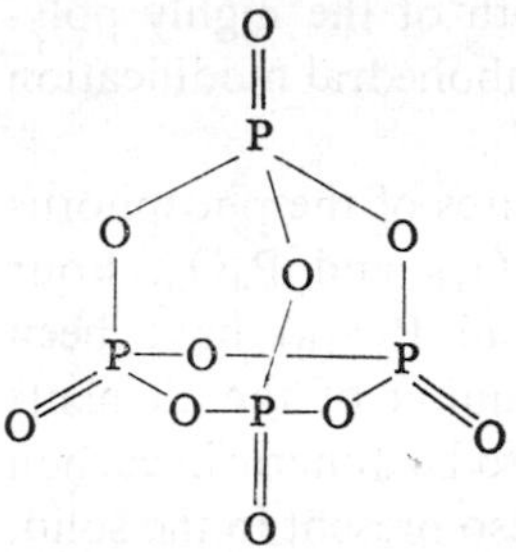

Figure 10.34

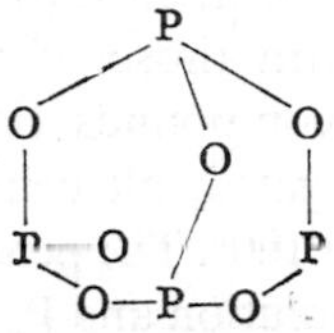

Figure 10.35

The P—O—P distance is that expected for a single P—O bond (1·60 Å), whereas the terminal P=O bonds are considerably shorter (1·40 Å). The infrared spectrum is also consistent with the presence of P=O linkages in the molecule. The structure can be seen to be based on a P_4 tetrahedron with oxygen atoms along each edge and bonded at the corners. Phosphorus trioxide has a very similar structure (Fig. 10.35), but lacks the terminal P=O bonds.

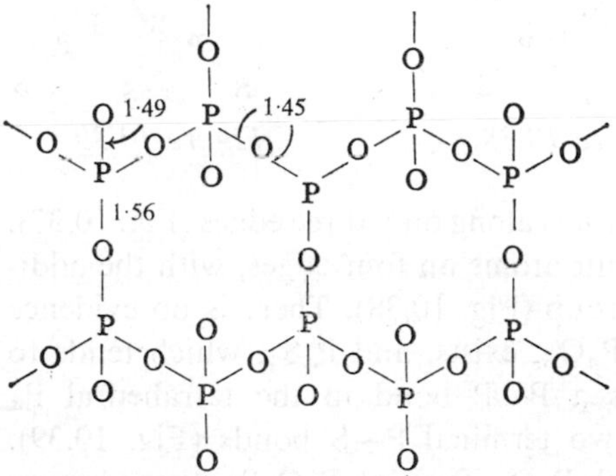

Figure 10.36

The two other forms, both of which are orthorhombic, are more highly polymerized. In one there is a continuous three-dimensional structure built up from rings of ten PO_4 tetrahedra, which are linked together. In the second there are sheets made up of rings of six PO_4 tetrahedra linked together (Fig. 10.36), the layer structure being rather similar to that found in mica. These three modifications differ considerably in the rate at which they react with water. For the first, reaction is rapid and violent. Dissolution of the second and third forms is much slower, almost certainly because the breakdown of a more complex structure is involved. Both of the highly polymeric forms are produced by heating the rhombohedral modification in a sealed tube above its melting point (420°).

It is of some interest to compare the structures of the phosphorus sulphides with those of the two oxides, P_4O_{10} and P_4O_6. Four crystalline compounds (P_4S_3, P_4S_5, P_4S_7 and P_4S_{10}) have been fully investigated, all being made by direct union of the elements above about 100°. The pentasulphide is found to be dimeric in carbon disulphide solution and P_4S_{10} molecules are also present in the solid, their configuration being very similar to that of P_4O_{10}, with sulphur replacing oxygen (Fig. 10.34). There is no evidence for the existence of more highly condensed modifications of the pentasulphide, which is considerably less stable to heat than the pentoxide. The other three sulphides also have molecular lattices, all of which are based on a P_4 tetrahedron.

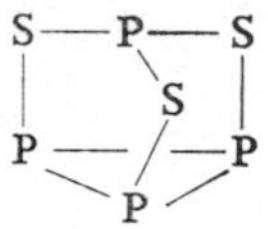

Figure 10.37

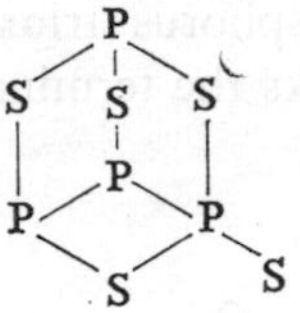

Figure 10.38

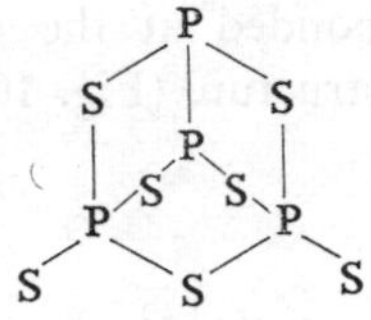

Figure 10.39

In P_4S_3 there are sulphur atoms along only three edges (Fig. 10.37), while in P_4S_5 there are sulphur atoms on four edges, with the addition of one terminal P—S group (Fig. 10.38). There is no evidence that P_4S_6, the analogue of P_4O_6, exists, and P_4S_7, which tends to be sulphur-deficient, retains a P—P bond in the tetrahedral P_4 framework, though it has two terminal P—S bonds (Fig. 10.39). Sulphur combines readily with P_4O_6, forming $P_4O_6S_4$, the structure of which is similar to that of P_4O_{10} and P_4S_{10}, with the terminal oxygens of the former replaced by sulphur.

10.6.6. Arsenatophosphates. The condensed phosphates considered above constitute only a part of a much wider field of phosphorus oxy-acid chemistry. Thus, for example, condensed acids can be obtained from the lower oxy-acids including those with P—P bonds. One of the most interesting compounds of the latter type is the salt $(CsPO_2)_6.xH_2O$, the anion of which has been shown by X-ray analysis to be a puckered ring of six PO_2 groups with P—P bonds. Detailed consideration of this field lies outside the scope of this book, however, and we will concentrate in this section on some compounds where arsenate anions are built into condensed phosphates.

When the salt $NaH_2AsO_4.H_2O$ is heated strongly, it forms a long-chain polyarsenate which is structurally similar to Maddrell's salt. On quenching the melt it is also possible to obtain a glass resembling Graham's salt, though there is no evidence for the formation of cyclic metaarsenates. One notable difference between these compounds and the polyphosphates is that they all hydrolyse very readily, with conversion to orthoarsenate. If mixtures of NaH_2PO_4 and NaH_2AsO_4 are melted and the melt is then quenched, polyarsenatophosphates result, which have anions with a statistical distribution of PO_4 and AsO_4 tetrahedra in the chain. Here again the AsO_4 tetrahedra are the points of hydrolytic instability, and breakdown occurs rapidly in aqueous solution to orthoarsenate and polyphosphates, or, with a high proportion of phosphorus in the chain, to polyphosphates and trimetaphosphate. Trimetaarsenatophosphates have also been prepared.

References for Chapter 10

1 Sillén, L., *Quart. Rev.*, 1959, **13**, 146.

2 Emeléus, H. J., and Anderson, J. S., *Modern Aspects of Inorganic Chemistry*, 3rd edn., p. 238 *et seq.* (Routledge, 1960).

3 Gimblett, F. G. R., *Inorganic Polymer Chemistry*, Ch. 3 (Butterworths, 1963).

4 See *Ref.* 2, p. 315 *et seq.*, for an account of the earlier development of this subject.

5 Kepert, D. L., *Prog. Inorg. Chem.*, 1962, **4**, 199.

6 Pope, M. T., and Dale, B. W., *Quart. Rev.*, 1968, **22**, 523.

7 Van Wazer, J. R., *Phosphorus and its Compounds*, p. 559 (Interscience, 1958).

8 Bragg, W. L., *Atomic Structure of Minerals* (Oxford, 1937).

9 Wells, A. F., *Structural Inorganic Chemistry*, 3rd edn. (Oxford, 1962).

10 Ross, V. F., and Edwards, J. O., *The Chemistry of Boron and its Compounds*, ed. E. L. Muetterties, Ch. 3 (Wiley, 1967).

11 Thilo, E., *Adv. Inorg. Chem. Radiochem.*, 1962, **4**, 1.

12 Corbridge, D. E. C., *Topics in Phosphorus Chemistry*, Vol. 3 (Interscience, 1966).

11 Some non-metallic halides

11.1 Introduction

This chapter is prefaced by a short section in which some of the physical properties of the halogens are described. In the subsequent sections we have reviewed a series of topics selected from the wealth of new material on the chemistry of non-metallic halides. This, it is hoped, will serve to give a general picture of recent developments though, of necessity, much has been omitted, and the choice of material is somewhat arbitrary. That special emphasis is placed on non-metallic fluorides is due to the preponderance of new work in this field, much of the chemistry of the other halides being now well known.

11.2 General properties of the halogens[1–3]

The bond dissociation energies of chlorine, bromine and iodine may be determined spectroscopically by observing the wavelength in the absorption spectrum at which bands, associated with molecular vibrations, converge to give continuous absorption. Absorption of light in the continuum leads to dissociation into atoms, and the energy corresponding with the convergence limit represents the minimum energy needed to bring this about. The values quoted below are obtained in this way, except that for fluorine.

Bond dissociation energy	F_2	Cl_2	Br_2	I_2
kJ mole^{-1}	158·2	243·5	193·0	151·0
kcal mole^{-1}	37·8	58·2	46·1	36·1

The main absorption spectrum of fluorine is a continuum and it is not possible, therefore, to deduce the bond dissociation energy directly from a convergence limit. Extrapolation from the values for the other halogens suggests that $D(F_2)$ should be of the order of 300 kJ (70 kcal) mole^{-1}, but it is now known to be anomalously low, and to have a value close to 158 kJ (37·8 kcal) mole^{-1}. This has been established by several methods, including a study of the ultraviolet absorption spectra of the alkali metal fluorides and also measurements, in the temperature range 230–380°C, of the rate of effusion of fluorine at low pressure through an orifice in a nickel vessel. The second of these methods measures the molecular weight (i.e. the extent to which the gas is dissociated into atoms) as a function of temperature. The dissociation has also been measured by heating fluorine in a pretreated nickel vessel at temperatures up to 850°, and observing the difference in pressure at a series of temperatures on a differential manometer between it and an identical vessel containing nitrogen at the same initial pressure, and heated to the same temperature. Since nitrogen does not dissociate to atoms in this temperature range, the observations directly furnish data on the equilibrium $F_2 \rightleftharpoons 2F$. The reason for the relatively low bond dissociation energy of fluorine has not been finally settled. One of the factors believed to be important is the repulsion of non-bonding electrons, which will be greatest in fluorine, the halogen with the smallest internuclear distance.

The low energy needed to convert molecular fluorine into atoms, coupled with the fact that it forms strong bonds with other atoms, lies at the root of its high reactivity and the exothermic character of many of its reactions. This point may be illustrated by comparing NF_3 with NCl_3. The first is exothermic ($\Delta H_f^0 = -109$ kJ (-26 kcal) mole^{-1}), whereas NCl_3 is endothermic ($\Delta H_f^0 = +230$ kJ ($+55$ kcal) mole^{-1}). If these two molecules are considered as formed from N_2 and the halogen, the same energy is absorbed in converting N_2 to 2N but less is used in producing fluorine atoms than chlorine atoms, and more is recovered in forming N—F bonds than in forming N—Cl bonds (N—F = 272 kJ (65 kcal) mole^{-1}; N—Cl = 193 kJ (46 kcal) mole^{-1}). The bond dissociation energy of the iodine molecule is close to that of fluorine, but it forms considerably weaker bonds.

The average bond energy in a fluoride is always greater than that in the other halides, as can be seen from the following typical values (in kJ or, in parentheses, kcal mole^{-1}): HF, 566 (135); HCl, 431 (103); HBr, 366 (87); HI, 299 (71); Si—F in SiF_4, 566 (135); Si—Cl in $SiCl_4$, 384 (91); P—F in PF_3, 490 (117); and P—Cl in PCl_3, 325 (78). Electronegativity, described by Pauling as 'the power of an atom in a molecule to attract electrons to itself', is also at a maximum in fluorine. The values given below form part of an electronegativity scale set up by Pauling (see Section 5.4).

B	1·9	N	3·1	F	4·0
Al	1·6	P	2·2	Cl	3·2
C	2·5	O	3·6	Br	3·0
Si	1·9	S	2·7	I	2·8

The value for oxygen, it will be noted, lies between those for fluorine and chlorine; in keeping with this, oxygen, like fluorine, forms strong covalent bonds, and there are a number of similarities between covalent oxides and fluorides.

The electron affinities of the halogen atoms, which measure the tendency to form univalent negative ions in the gas phase, show only small variations. This, however, has little influence on the thermodynamics of compound formation by the elements, as factors such as lattice energies and solvation energies, which depend on ion size, introduce much larger effects. Variations in the first ionization potential, which determine the relative ease with which univalent positive ions are formed are, on the other hand, much more significant. The value for fluorine is so high that its appearance as a positive ion in chemical compounds is virtually excluded, whereas the value for iodine is little greater than that for some metals. Cationic iodine is in fact well known, though it is often stabilized by a ligand, and the tendency to form positive ions in chemical compounds decreases in going from bromine to chlorine. The only real evidence for such ions comes from the structural analysis of solids and the results of physical measurements, including ion migration studies.

		F	Cl	Br	I
Electron affinity	(kJ)	338	355	331	302
	(kcal)	81·0	84·8	79·0	72·1
First ionization potential	(kJ)	1680	1253	1142	1008
	(kcal)	402	300	273	241

The crystal structure of all of the halogens in the solid state has been investigated by X-ray analysis. Fluorine has two solid modifications, the lattices of which are respectively cubic and monoclinic. The structures are similar to those of two forms of solid oxygen. The remaining halogens crystallize in layer structures with interlayer distances approximately those expected from the van der Waals radii. The covalent and ionic radii of the halogens are tabulated below.

	F	Cl	Br	I
Covalent radius (Å)	0·72	0·99	1·14	1·33
Ionic radius (Å)	1·36	1·81	1·95	2·16

The covalent radius of fluorine and also the radius of its ion are close to the corresponding values for oxygen (0·74 Å and 1·40 Å), and there is also a rough correspondence between the values for the other Group VI elements and the halogens. Thus the ionic radii of S^{2-}, Se^{2-} and Te^{2-} are respectively 1·84, 1·98 and 2·21 Å. This leads to a similarity in the structures of many ionic fluorides and oxides on the one hand and also between chlorides, bromides or iodides and sulphides, selenides or tellurides. The latter frequently form layer structures, rather than the true ionic lattices which characterize most metallic fluorides and oxides. The significance of ionic radii in relation to lattice and solvation energies is discussed elsewhere (Chapters 3 and 6) and it remains only to refer to the standard potentials of systems involving the halogens and their ions in aqueous solution, values for which are given below.

$$\tfrac{1}{2}F_2(g) + e \rightleftharpoons F^- \qquad E^0 = +2{\cdot}9\ V$$
$$\tfrac{1}{2}Cl_2(g) + e \rightleftharpoons Cl^- \qquad E^0 = +1{\cdot}365\ V$$
$$\tfrac{1}{2}Br_2(l) + e \rightleftharpoons Br^- \qquad E^0 = +1{\cdot}065\ V$$
$$\tfrac{1}{2}I_2(s) + e \rightleftharpoons I^- \qquad E^0 = +0{\cdot}535\ V$$

These data emphasize the unique position of fluorine as an oxidizing agent.

11.3 Interhalogen compounds[4]

The halogens combine exothermally with one another to form interhalogen compounds of four types (AB, AB_3, AB_5 and AB_7, where A is the heavier halogen), two ternary compounds ($IFCl_2$ and IF_2Cl) also being known. Table 11.1 shows the formulae and, where known, the melting and boiling points of the interhalogens that have so far been described.

The formation of these compounds is not surprising in view of differences in electronegativity within the group and the ability of the

Table 11.1 *Melting and boiling points of the interhalogens*

Formula	*m.p.*	*b.p.*
IF	—	—
ICl α	27·2°	—
ICl β	13·9°	—
IBr	40–41°	119°
BrF	*ca.* −33°	*ca.* 20°
BrCl	−54°	—
ClF	−155°	−100·1°
IF_3	—	—
IF_2Cl	—	—
$IFCl_2$	—	—
ICl_3	101°	(decomposition)
BrF_3	8·77°	125·7°
ClF_3	−76·3°	11·7°
IF_5	9·4°	100·5°
BrF_5	−60·5°	41·3°
ClF_5	−93°	−14°
IF_7	4·77°	(sublimes)
BrF_7	—	—

heavier halogens to exhibit valencies in excess of unity. It is also favoured by the increase in entropy when compounds containing two different halogens are formed from the separate elements. They are closely related to oxy-halides such as ClO_2F, ClO_3F and IOF_3, in which fluorine is replaced by oxygen, the electronegativity of which is likewise high.

Values for enthalpies of formation and mean bond energies have been determined for the majority of the interhalogen compounds and are tabulated in Table 11.2.

The iodine chlorides, iodine bromide and bromine chloride have all been known for a very long time, and are formed by mixing stoichiometric quantities of the two halogens. Apart from their self-ionization and the formation of polyhalide ions, which are referred to elsewhere (Chapter 7), their main chemical interest centres on their

Table 11.2 *Enthalpies of formation ΔH_f^0 and mean bond energies* E for interhalogen compounds (in kJ with values in kcal in parentheses)*

Compound	ΔH_f^0	*E*
ClF	−56·5 (−13·5)	257 (61)
ClF_3	−163·0 (−38·9)	172 (41)
ClF_5	−241·5 (−57·7)	153 (36)
BrF	−74·0 (−17·7)	249 (60)
BrF_3(g)	−256·0 (−61·1)	201 (48)
BrF_5(g)	−429·0 (−102·5)	187 (45)
IF	−94·5 (−22·6)	280 (67)
IF_5(g)	−837·0 (−200·0)	268 (64)
IF_7	−958·5 (−229·1)	231 (55)
$BrCl$	+14·5 (+3·5)	218 (52)
ICl(g)	+17·5 (+4·2)	207 (50)
IBr(g)	+40·5 (+9·7)	175 (42)

* It should be noted that apical and equatorial bonds in tri-, penta- and hepta-fluorides are not equivalent.

behaviour as mild halogenating agents. Apparent irregularities in their action on aromatic compounds result from the possibility of two types of reaction in polar solvents. ICl acts as an iodinating agent owing to its ionization to I^+ and ICl_2^- or related solvated species. Iodine bromide, on the other hand, is dissociated to some extent to its elements even at ordinary temperatures and, since bromine is much more reactive than iodine, always functions as a brominating agent. Iodine trichloride, the structure of which is described later, is also appreciably dissociated at room temperature to the monochloride and chlorine.

Both ClF and ClF_3 can be made by passing appropriate proportions of the two elements through a reactor constructed of copper, nickel or Monel metal and heated to 200–300°, and the monofluoride is readily converted to the trifluoride by excess of fluorine under similar conditions. The pentafluoride requires more drastic conditions for its preparation; it is produced at 350° with a $Cl_2 : F_2$ ratio of 1 : 14 with an operating pressure of 250 atm, but the reaction temperature is lowered if a fluoride of one of the heavier alkali metals is added, when intermediate formation of a compound of the type $M^{I}ClF_4$ can be assumed. It is also formed at 30° from a 2:1 mixture of F_2 and ClF_3 on irradiating with light of wavelength > 3100 Å.

Bromine monofluoride results when bromine and fluorine react at 25°, but it has never been prepared in a pure state because of its ready disproportionation ($3BrF \rightleftharpoons Br_2 + BrF_3$). Excess of fluorine gives the trifluoride at <100° and the pentafluoride is best made from the elements at 200°. Iodine burns in fluorine with a pale green flame to the pentafluoride, which is converted to the heptafluoride in a gas-phase reaction at 250–270°. The very high reactivity of all of these fluorides necessitates the use of metal apparatus in their preparation and manipulation, though up to moderate temperatures they can also be handled in vessels made of KelF (polymerized CF_2CFCl) or Teflon (polymerized C_2F_4). The preparation of BrF_7 by the reaction below has been reported, but no properties have so far been described.

$$BrF_5 + F_2 \xrightarrow[CsBrF_6]{110-340^\circ} BrF_7$$

Bands due to IF are observed in the emission spectrum of the flame of iodine burning in fluorine. The monofluoride is formed as a brown solid when a fluorine–nitrogen mixture is passed into a solution of iodine in CCl_3F at −78°, but it disproportionates readily to I_2 and IF_5 below 0°. Further passage of fluorine under the same conditions gives IF_3 as a yellow solid, but this also decomposes below 0°, though both of these compounds form more stable pyridine adducts, $I(pyr)_2F$ and $I(pyr)_2F_3$. The compound CF_3IF_2, formed by fluorinating CF_3I in CCl_3F at −78°, also decomposes below 0° to CF_4, CF_3I, I_2 and IF_5. Iodine monochloride reacts with elementary fluorine at −78° in CCl_3F to give IF_2Cl and, under the same conditions, chlorine reacts with IF to give $IFCl_2$. Both compounds are, however, unstable.

11.3.1. Reactions of the halogen fluorides. An important group of reactions of the halogen fluorides can be explained in terms of their self-ionization according to the schemes set out below:

$$3AB \rightleftharpoons A_2B^+ + AB_2^-$$
$$2AB_3 \rightleftharpoons AB_2^+ + AB_4^-$$
$$2AB_5 \rightleftharpoons AB_4^+ + AB_6^-$$

These are discussed in Chapter 7, to which the reader is referred, though some reference to further points related to the solvent system behaviour of the halogen fluorides is included in this section.

Chlorine monofluoride is a powerful fluorinating agent and will react with many metals and non-metals, either at room temperature or on warming, converting them to fluorides, often with inflammation. Hydrogen also inflames, as do many organic compounds, though reaction with the latter can often be controlled by using an inert solvent to yield fluoro or chlorofluoro derivatives. Readily dissociated adducts are formed with a number of fluorides (e.g. $2ClF.SbF_5$, $2ClF.BF_3$). These compounds were originally assigned the structures $FCl_2^+SbF_6^-$ and $FCl_2^+BF_4^-$ with the unusual bent $(Cl{\diagup}^{F}{\diagdown}Cl)^+$ cation on the basis of spectroscopic evidence, but the more probable asymmetrical structure $(Cl—Cl—F)^+$ has recently been proposed. The monofluoride also reacts with alkali metal fluorides (K, Rb, Cs) to form relatively stable salts of the type M^IClF_2.

Chlorine trifluoride is easier to handle than the more volatile monofluoride and can be stored as liquid in steel cylinders, since the metal is protected from continuing attack by an insoluble fluoride film. Reaction with water is usually explosive, but can be controlled to give initially HF, ClO_2F and ClF. Many elements (e.g. P, S, As, Sb and powdered Mo, W, Rh or Ir) inflame in the gas, which converts SO_2 to SOF_4 and a number of metal oxides to fluoride and free oxygen. These examples show the reactivity to approach that of free fluorine. Both ClF_3 and BrF_3 may be used in place of the element in the preparation of volatile fluorides such as VF_5, TaF_5 and MoF_6 from the metals or their oxides, and the ready formation of UF_6 in this way has resulted in applications of the two compounds in nuclear technology. The reactions with organic compounds resemble those of ClF.

The behaviour of ClF_3 as a fluoride ion donor or acceptor, which results in the formation of salt-like derivatives containing ClF_2^+ and ClF_4^-, is discussed with similar observations on BrF_3 in Chapter 7. Chlorine pentafluoride is also a vigorous fluorinating agent, though little of its detailed chemistry has so far been published.

Apart from the behaviour of BrF_3 as a fluoride ion donor or acceptor, its chemistry is confined largely to reports of fluorination reactions in which it takes part. Examples of these are:

$$6ClO_2 + 2BrF_3 \rightarrow 6ClO_2F + Br_2$$
$$3N_2O_5 + BrF_3 \rightarrow Br(NO_3)_3 + 3NO_2F$$

Tracer studies with ^{18}F have shown that rapid fluorine exchange occurs between either ClF_3 or BrF_3 and HF in the gas or the liquid phase. There is evidence from the infrared spectrum for the formation of a 1 : 1 adduct $ClF_3.HF$, or the corresponding BrF_3 compound, and exchange in the liquid phase can be explained if these ionize as follows:

$$ClF_3.HF \rightleftharpoons ClF_2^+ + HF_2^-$$

Bromine pentafluoride closely resembles the trifluoride in its reactions.

Since both IF and IF_3 disproportionate very readily, it is difficult to investigate their reactions. The former has, however, been shown to add readily across the double bond of olefins in low-temperature reactions. The trifluoride forms adducts with pyridine and other nitrogen bases and, though its self-ionization to IF_2^+ and IF_4^- has never been demonstrated, unstable 1 : 1 adducts are formed at low temperatures with AsF_5, SbF_5 and BF_3 which can be formulated as ionic (e.g. $IF_2^+SbF_6^-$). Alkali metal salts of the type M^IIF_4 are also known (see Section 7.6.2).

The pentafluoride IF_5 is a considerably milder fluorinating agent than the other halogen fluorides. Reaction with water gives HIO_3 and HF, but there is no reaction with hydrogen up to 100° and sulphur is converted mainly to SF_4 rather than to SF_6. Some elements (e.g. P, As, Sb) inflame in contact with liquid IF_5, but the reactions with many organic compounds are less violent than those of the other halogen fluorides and are readily controlled. Iodine heptafluoride is known to be much more reactive than the pentafluoride, though few reactions have been examined in detail. Its 1 : 1 adduct with AsF_5 has been shown to have a face-centred cubic lattice with IF_6^+ and AsF_6^- ions, which is the first clear evidence in line with the hypothetical self-ionization of liquid IF_7 to IF_6^+ and IF_8^-; there is no indication that the heptafluoride will react with fluorides of the heavier alkali metals to give M^IIF_8.

11.3.2. Structures of the interhalogen compounds. The structures of BrF, ClF, BrCl, IBr and ICl are known from spectroscopic and electron diffraction studies on their vapours.[7] Iodine monochloride exists in two crystalline forms, in both of which the molecules are arranged in zig-zag chains. There is no information on the structure of the very unstable trifluoride of iodine, but both ClF_3 and BrF_3 have been studied by X-ray diffraction as solids, as well as by microwave spectroscopy for the vapours; the results by the different methods are in good agreement. The dimensions of the planar T-shaped ClF_3 molecule are given in Fig. 11.1; the structure of BrF_3 is similar. These structures are often described as trigonal bipyramids,

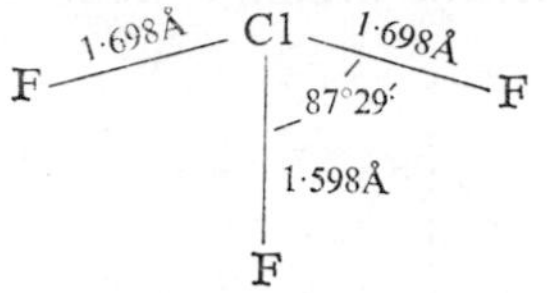

Figure 11.1

in which two equatorial positions are occupied by non-bonding pairs, but the marked difference between the Cl—F equatorial and axial bond lengths shows that such a description may be an oversimplification (see Chapter 5).

From the temperature-dependence of the ^{19}F n.m.r. spectrum of ClF_3 it has been concluded that there is rapid fluorine exchange between the apical and equatorial positions. A suggested mechanism shown below involves formation of a dimer, for the existence of which the entropy of vaporization (Trouton's constant) of 96 J (23 cal) mole^{-1} deg^{-1} provides support. The ^{19}F n.m.r. spectrum of BrF_3 shows a similar temperature-dependence.

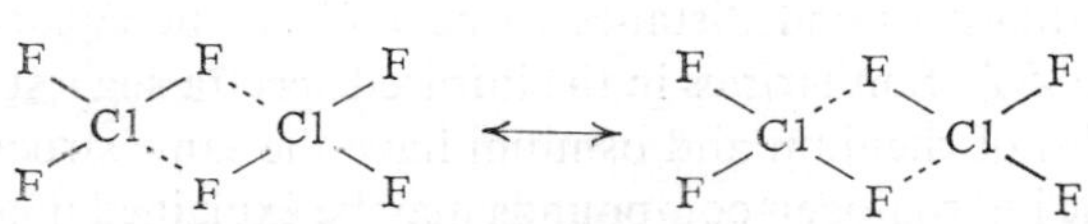

Iodine trichloride (which is largely dissociated into ICl and Cl_2 when vaporized) has a structure in the solid state made up of discrete planar I_2Cl_6 molecules (Fig. 11.2). The greater length of the bridge bonds recalls the structures of B_2H_6 and Al_2Cl_6, though in both of the latter the distribution of bonds around the Group III element is tetrahedral.

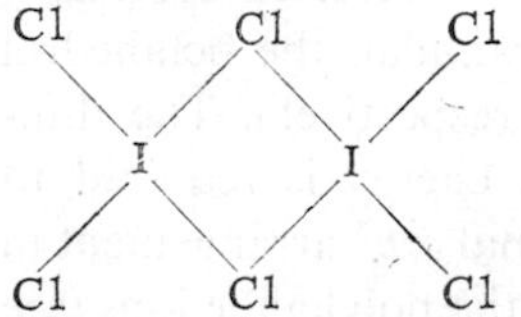

Figure 11.2

All three pentafluorides have a square pyramidal structure with a pair of electrons occupying the vacant octahedral site. Dimensions for the BrF_5 molecule, for which X-ray data are available, are shown in Fig. 11.3. It will be noted that the bond from bromine to the unique fluorine atom is again shorter. The F(apical)—Br—F(equatorial) angles are all in the range 80–87° (i.e. the bromine atom is slightly below the equatorial plane).

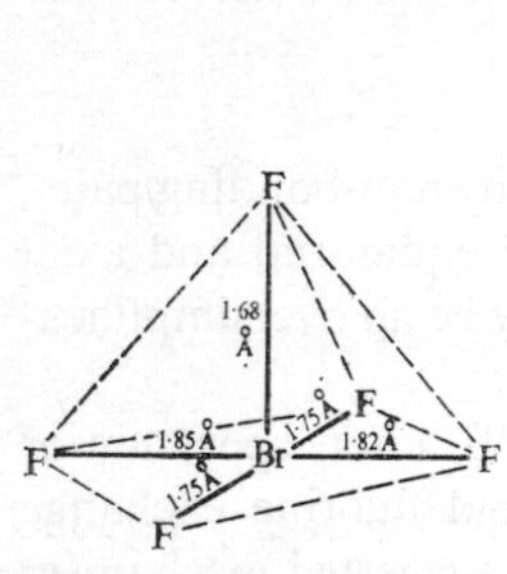

Figure 11.3

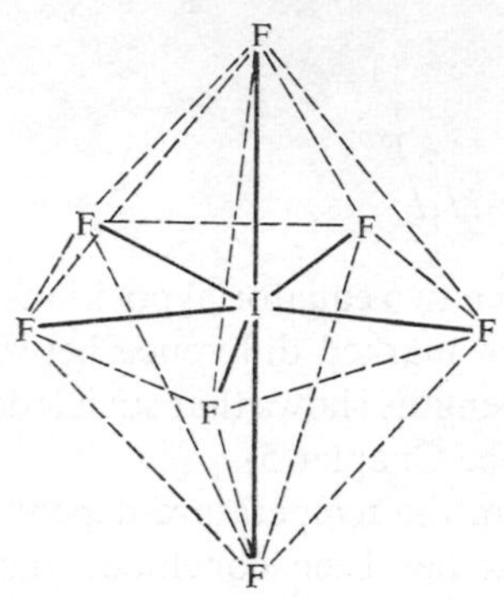

Figure 11.4

The structure of the IF_7 molecule is shown in Fig. 11.4. Fluorine atoms are grouped round the centre iodine atom at the corners of a pentagonal bipyramid, a recent electron diffraction study having shown the I—F(apical) bond distance to be 1·77 Å, the equatorial distance being 1·85 Å. Similarities in the infrared spectra suggest that the heptafluorides of rhenium and osmium have the same structure.

Bonding in the interhalogen compounds may be explained qualitatively in terms of electron-pair bonds with, as has been pointed out, non-bonding pairs occupying vacant stereochemical sites. Deviations from symmetry in the structures, which are evident from differences in the length of seemingly identical bonds, then follow naturally from the repulsive forces arising from the electron pairs, or, in the heptafluoride, the particular arrangement of bonded fluorine atoms. This approach may be linked with the use of hybridized orbitals—sp^3d, sp^3d^2 and sp^3d^3 for the trigonal bipyramidal, the octahedral and the pentagonal bipyramidal structures respectively. The difficulty in this treatment is that considerable energy is required to promote the s and p electrons from their normal s^2p^5 arrangement in the halogens to d orbitals. This also arises in the polyhalide ions (see the next section).

11.4 Polyhalide complex ions[7, 8]

It has been known for a very long time that iodine is more soluble in water or ethanol if a soluble iodide, such as potassium iodide, is added to the solution. This is due to the formation of the I_3^- ion, which is the simplest of a series of polyiodide species of the type I_n^-, n being an odd number up to at least 9. These ions can form either anhydrous or hydrated salts (e.g. KI_3,H_2O, $(NPr_4)I_5$ and $(NPr_4)I_7$). A few polybromides are also known, and there is a single example of a polychloride $(NEt_4)Cl_3$, though the stability decreases in the sequence $I_n^- > Br_n^- > Cl_n^-$. A considerable number of univalent ions of the same general type, but containing two or more different halogens, have also been identified. Some examples of these follow, and it will be seen that both anionic and cationic species are represented.

ICl_2^-, IBr_2^-	ICl_4^-	IF_6^-	IF_6^+, IF_4^+
$IBrCl^-$	ICl_3F^-	BrF_6^-	ICl_2^+
$BrCl_2^-$	IF_4^-		ClF_2^+, BrF_2^+
ClF_2^-	BrF_4^-, ClF_4^-		

The list includes a number of ions derived from the halogen fluorides (e.g. BrF_4^-, IF_6^- and ClF_2^+) which have been described elsewhere (Chapter 7). They are highly reactive and can be produced only in a solution of the appropriate halogen fluoride. The general method for preparing the remainder is by addition of a halogen or an interhalogen compound to a solution containing a halide ion, or, in a few cases, to a solid halide. The solvent must be resistant to chemical attack by the halogen or interhalogen, must dissolve the halide and must not solvolyse the desired product. The last factor frequently precludes the use of water, and many preparations are consequently done in solvents such as methanol or ethanol. Some examples are given below:

$$Me_4NI + I_2 \xrightarrow{MeOH} Me_4NI_n \quad (n = 3, 5, 7, 9, \text{ according to proportions})$$

$$RbCl + I_2 + Cl_2 \xrightarrow{H_2O} RbICl_2$$

$$Me_4NBr + ICl \xrightarrow{CH_3COOH} Me_4NIBrCl$$

$$CsI(g) + Br_2 \xrightarrow{EtOH} CsIBr_2$$

Polyhalide ions vary widely in stability and some exist only in solution, where they can be studied by such physicochemical methods as spectrophotometry, conductance measurements, potentiometry, polarography, solubility determinations and the study of distribution between liquid phases. Such methods also provide equilibrium data and formation constants, as, for example, in the system $I^- + nI_2 \rightleftharpoons (I_{2n+1})^-$ in aqueous solution, where successive equilibria involving species up to I_9^- may be identified.

Crystalline phases are formed with the larger univalent cations, such as the heavier alkali metals and the alkyl ammonium cations. Polyvalent cations, which are smaller, do not yield salts except when the size is increased by coordination as in $[Ni(NH_3)_4]^{2+}[I_7^-]_2$ and $[Co(NH_3)_6]^{3+}[I_3^-]_3$. This effect will also explain why, in the series $MICl_4$ (M = Li, Na, K, Rb, Cs), the lithium salt is known only as the tetrahydrate, whereas the remainder can be obtained in the anhydrous form. Some of the polyhalides also form complexes with certain organic donor molecules (e.g. $LiI_3.4C_6H_5CN$, $NaI_3.2C_6H_5CN$ and $KI_3.2C_6H_5CN$); here, too, the ligand has a stabilizing effect.

Polyhalide acids have been shown in a few cases to be formed when a halogen or interhalogen compound is added to an acid. The most familiar example is $HICl_4$, which results when ICl_3 is added to concentrated hydrochloric acid. It may be separated as the tetrahydrate $HICl_4.4H_2O$, which cannot be dehydrated without decomposition. Other acids such as $HICl_2$, $HIBr_2$ and HIBrCl can be formed similarly, but exist only in solution.

All of the solid polyhalides dissociate more or less readily when heated, though a number have observable melting points roughly in the range 70–200°. For a given anion the ease of dissociation decreases as the size of the cation is increased. The solid product is always the monohalide containing the lightest halogen, if more than one is present in the anion. Thus $CsICl_2$ gives CsCl and ICl rather than CsI and Cl_2. The greater lattice energy of the halide containing the smaller anion is probably the most important factor determining the mode of decomposition, but the difference in bond energies of the other products is also important.

11.4.1. Cationic species. Some of the interhalogen compounds exhibit electrical conductivity when in the liquid state. This is true both of ICl and ICl_3. The liquid monochloride has a specific conductance of $4{\cdot}5 \times 10^{-3}$ ohm^{-1} cm^{-1} at 30·6°, and for ICl_3, which undergoes some dissociation to ICl and Cl_2, the value is $9{\cdot}8 \times 10^{-3}$ ohm^{-1} cm^{-1} at 109°. Both of these substances also conduct in polar solvents. The ionization schemes shown below have been postulated to account for the conductivity:

$$3ICl \rightleftharpoons I_2Cl^+ + ICl_2^-$$
$$2ICl_3 \rightleftharpoons ICl_2^+ + ICl_4^-$$

The ICl_2^+ cation has been identified by X-ray analysis in the salts $ICl_2^+SbCl_6^-$ and $ICl_2^+AlCl_4^-$, which are formed in the sealed tube reaction of the trichloride with $SbCl_5$ or $AlCl_3$. Recently the compound $AsCl_2F_7$ has been shown to contain the FCl_2^+ ion. Other cations such as BrF_2^+ and IF_4^+ are discussed in Chapter 7, where the interhalogen compounds are treated as aprotic solvent systems.

11.4.2. Chemical properties of polyhalides. The chemical reactions of the polyhalides are essentially those of the dissociation products. All are hydrolysed to some extent in aqueous solution, and the ease of hydrolysis increases with the introduction into the ion of the more electronegative halogens, the triiodide being the most stable. In non-aqueous solutions they behave as mild halogenating agents: polybromides, for example, will brominate aniline or phenol. Fully fluorinated ions such as BrF_4^- are much more reactive, and indeed are rather similar to the parent halogen fluoride.

11.4.3. Structure and bonding in the polyhalide ions. The structures of a number of these ions have been determined by X-ray diffraction and show some interesting features. Thus the triiodide ion I_3^- may have equal or unequal I—I bond distances, depending on the cation with which it is associated. In the caesium salt these distances are 3·03 and 2·83 Å with a bond angle of 176°. They are also unequal in NH_4I_3 (3·10 and 2·82 Å), but equal in $[(C_6H_5)_4As]I_3$ (2·90 ± 0·02 Å). Tetraethylammonium iodide exists in two crystalline modifications, in one of which the anion is symmetrical and in the other unsymmetrical, while the middle bromine atom in the Br_3^- ion is central in the trimethylammonium salt, but not in the caesium salt.

It has been suggested that the symmetry of the trihalide ion depends on the size of the cation with which it is associated, the smaller cations tending to produce distortion if unevenly distributed in the solid structure. It may be noted that the I_3^- ion has been found by spectroscopic studies to be symmetrical in solution. There are also examples of both symmetrical and unsymmetrical configurations of the ICl_2^- ion.

Structural studies have been made on only three pentahalide anions. The ions ICl_4^- (in $KICl_4.H_2O$) and BrF_4^- (in $KBrF_4$) both have square planar configurations, with the heavier halogen at the centre. Deviations from a bond angle of 90° in ICl_4^- are only slight, but the four bond lengths are unequal (Fig. 11.5), and also significantly less than the sum of the covalent radii or the terminal I—Cl

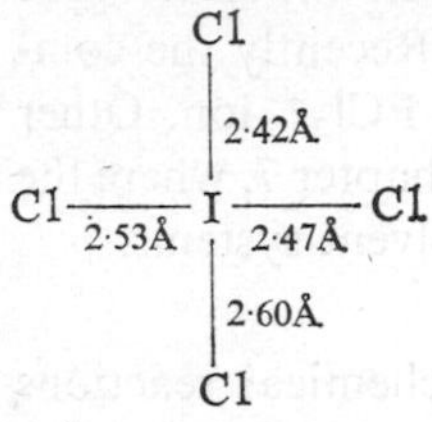

Figure 11.5

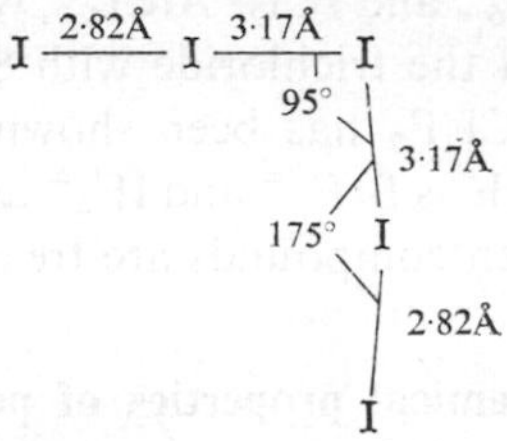

Figure 11.6

distances in I_2Cl_6, an effect that may be associated with the influence of the cation and also perhaps with the presence of water molecules in the structure.

The structure of the anion in $[Me_4N]I_5$ has also been determined. It is an almost planar L-shaped species with the dimensions shown in Fig. 11.6. In the so-called heptaiodide $[Et_4N]I_7$, structural evidence does not confirm the presence of discrete I_7^- ions, but rather of a three-dimensional array of I_3^- ions and I_2 molecules. The short distances found between iodine atoms of separate units in these two species does, however, suggest that there is strong interaction between them. In Me_4NI_9 there is an assembly of I_5^- units with loosely associated I_2 molecules.

The structures of the ions BrF_6^- and IF_6^- are of great interest, since they might be expected to be derived from the pentagonal bipyramidal structure of IF_7 by removal of one fluorine atom. The vibrational spectra of hexafluoroiodates suggest that the salts cannot contain a regular octahedral IF_6^- ion, but no X-ray structure determination has yet been reported.

Although structural information on the polyhalides is very incomplete, a great deal of attention has been paid to the nature of the bonding. The simplest of the three approaches to the problem, developed to account for the existence of the trihalide ion, considers the electrostatic interaction between the halide ion and the polarizable halogen molecule. This has proved to be of limited use in predicting a value for the heat of formation for I_3^-, and does not account for the centrosymmetrical trihalide ions such as ICl_2^-.

The postulate of the use of localized covalent bonds is complicated by the fact that it is necessary in most cases to increase the valency of the central atom by promotion of *s* and *p* electrons to *d* orbitals, so as to create hybridized sp^3d and sp^3d^2 orbitals. This implies the availability of the rather large amount of energy for promoting the electrons. Otherwise the treatment is similar to that of bonding in the neutral interhalogen compounds, and an acceptable, though not necessarily true, explanation can be given of bonding in some of the simpler ions such as ICl_4^-.

The third approach to the problem avoids the promotion of electrons to *d* orbitals and considers only the use of delocalized *p* orbitals of iodine for bonding. Thus for symmetrical I_3^- it is suggested that the ion is a resonance hybrid of the structures

$$I—I\ I^-$$

and

$$^-I\ I—I$$

Studies of nuclear quadrupole coupling constants in salts containing ICl_2^- and ICl_4^- ions support the idea that bonds in these ions have a large degree of ionic character. This observation, taken in conjunction with the long I—Cl bonds, their low stretching frequencies and evidence from Mössbauer spectroscopy that the *s* orbitals of the iodine atoms are not involved in bonding, suggests that the best description of these ions is that they contain iodine atoms carrying substantial positive charges and chlorine atoms carrying substantial negative charges, linked by delocalized *p* orbitals. But no one treatment is adequate for the discussion of all interhalogens and polyhalides, and the interpretation of the structures of these substances is one of the most unsatisfactory areas of valency theory.

11.5 Boron halides[9]

Boron forms halides of three types, the simple trihalides of the type BX_3 (X = F, Cl, Br, I), polyboron halides which are described in Chapter 7, and a monofluoride, BF, the formation and reaction of which is considered in Chapter 13, since it is closely associated with the short-lived difluoride of silicon.

Boron has an electronegativity of 1·9 on the Pauling scale, so that the polarity of the boron to halogen bond is $B^+—X^-$ in each instance. On this basis it would be expected that the acceptor properties of boron in the trihalides would decrease in the order $BF_3 > BCl_3 > BBr_3 > BI_3$. Fluorine should exercise the greatest electron-withdrawing effect and so lead to the greatest positive charge on boron, which would favour the acceptance of electrons from a donor. There is, however, much evidence that the stability of many comparable adducts of the trihalides is in the order $L{\rightarrow}BI_3 > L{\rightarrow}BBr_3 > L{\rightarrow}BCl_3 > L{\rightarrow}BF_3$. Thus, for example, values for ΔH^0 for 1 : 1 adduct formation between pyridine and a boron halide in nitrobenzene solution are −143, −189 and −217 kJ (−34·2, −45·2 and −51·8 kcal) mole^{-1} for BF_3, BCl_3 and BBr_3 respectively.

A suggested explanation for this anomaly is that p_π—p_π back-bonding occurs, involving the transfer of electron density from the halogen to a vacant *p* orbital on the boron atom, and that this is at a maximum for BF_3 and decreases along the series. This occurs in the absence of a ligand and results in a decrease in B—X bond length below the value expected for a single bond. Its effect on the interaction with a ligand is to reduce the positive charge on boron and so reduce its acceptor function. If the B—X bonds have some double bond character this will also increase the energy required to change the hybridization on boron from sp^2, which it has in the planar trihalide molecules, to sp^3, which characterizes the 1 : 1 adduct. This, in turn, would be directly reflected in the energetics of adduct formation.

Formation of the tetrahaloborates represents a special case of donor–acceptor interaction, with the halide ion acting as an electron donor. The most familiar example is the tetrafluoroborate ion BF_4^-, which is formed readily from BF_3 and a metallic fluoride. The B—F bond in BF_4^- is longer than in BF_3, in keeping with the fact that the $2p_z$ orbital of boron is no longer vacant, but forms part of the sp^3 hybrid, so that back-bonding cannot occur. It has been suggested that electrostatic repulsion between the fluorine atoms may also contribute to bond lengthening.

The remaining tetrahaloborates are more difficult to prepare and require a large cation for stability in the solid state. Thus BCl_3 will react with an alkyl ammonium chloride in chloroform solution forming the BCl_4^- salt. Similarly BCl_3 when heated in a sealed tube with KCl, RbCl or CsCl, using nitrobenzene as solvent, also yields tetrachloroborates. The most general reaction, however, is that between the boron trihalide and an alkyl ammonium chloride in anhydrous liquid hydrogen chloride (see Section 7.3). This can also give mixed haloborates; for example,

$$Me_4NCl + BF_3 \xrightarrow{HCl} Me_4NBF_3Cl$$

Anhydrous HF does not yield HBF_4 with BF_3, but an aqueous solution of the acid is obtained when BF_3 dissolves in aqueous HF, just as SiF_4 gives H_2SiF_6; neither acid can be isolated pure and either BF_3 or SiF_4 is evolved if concentration beyond a certain point is attempted. All of the tetrahaloborates other than salts of BF_4^- are rapidly decomposed by water.

Some boron trifluoride adducts exist in an ionized form, the dihydrate $BF_3.2H_2O$, for example, having a substantial electrical conductivity when in the liquid state, which is attributed to the ions H_3O^+ and $[BF_3OH]^-$. In other adducts with fluorides, fluoride ion transfer can be shown to occur. Thus X-ray analysis of $SF_4.BF_3$ shows it to be $SF_3{}^+BF_4{}^-$, while in the case of $NOF.BF_3$ the presence of the tetrafluoroborate ion may be detected by infrared spectroscopy.

Exchange reactions take place in mixtures of boron halides and these, too, can be detected spectroscopically; for example,

$$BCl_3 + BBr_3 \rightleftharpoons BBrCl_2 + BBr_2Cl$$

Such a reaction is best explained by postulating an intermediate in which the two boron atoms are joined by two B-(Hal)-B bridges, with alternative modes of breakdown. It has also been shown by ^{11}B n.m.r. spectroscopy that BClBrI is present in a mixture of the three halides. Exchange between alkyl boranes and boron halides can occur in the presence of a Lewis base catalyst; for example,

$$R_3B + 2BF_3.O(C_2H_5)_2 \rightleftharpoons 3RBF_2 + 2(C_2H_5)_2O$$

No attempt will be made here to describe the many applications of the boron halides and their derivatives in organic chemistry, where they can act as catalysts for esterifications, Friedel–Crafts acylations and alkylations, certain organic polymerizations and other types of reaction.

11.6 Carbon fluorides[10]

The carbon fluorides are unique among the fluorides of the non-metallic elements, both because they are so numerous and because of their unusual properties. Virtually all of the hydrocarbons, and also many of their derivatives, have fluorocarbon analogues. When all of the hydrogen in a hydrocarbon or one of its derivatives is replaced by fluorine, the products are referred to as perfluoro compounds, but when substitution is only partial the usual rules of organic nomenclature are applied.

Fluorocarbons exhibit high thermal and chemical stability. This is to be expected as the C—F bond is strong (*ca.* 480 kJ, or 116 kcal), compared with C—H (416 kJ, or 99 kcal), the weakest link being the C—C bond (347 kJ, or 83 kcal). The overall heat evolved in the reaction

$$>C{-}H + F_2 \rightarrow >C{-}F + HF$$

is about 480 kJ (115 kcal), compared with 80 kJ (20 kcal) for the corresponding chlorination reaction. Release of this amount of energy normally results in an uncontrolled exothermic reaction if a hydrocarbon is exposed to fluorine, and there is partial or complete breakdown of the carbon skeleton of the molecule with production of CF_4 and other simple fluorocarbons. It follows that, if it is desired to convert a hydrocarbon to a fluorocarbon by reaction with fluorine, some of the energy must be removed, or an alternative, less exothermic, reaction path must be used. This may be done in any of four ways:

(a) Direct fluorination of the hydrocarbon in an inert solvent, the evaporation of which will absorb heat. The fluorine may also be diluted with nitrogen, though in either case fluorination tends to be incomplete and the method is now seldom used.

(b) Fluorination of the hydrocarbon vapour by an F_2–N_2 mixture in a reactor packed with gold- or silver-plated copper turnings heated to 100–200°. The latter serve to remove heat, but it is probable that a surface layer of fluoride (AuF_3 or AgF_2) catalyses the conversion, as in method (c). This method will give moderate to high yields of many of the simpler fluorocarbons from the corresponding hydrocarbons or their derivatives, e.g.,

$$C_nH_{2n+2} \rightarrow C_nF_{2n+2}; \quad (C_2H_5)_2O \rightarrow (C_2F_5)_2O$$

and

$$(CH_3)_3N \rightarrow (CF_3)_3N + (CF_3)_2NF + CF_3NF_2$$

Yields become much lower when an attempt is made to fluorinate organic compounds with more complex structures, and in every case all unsaturation is eliminated (e.g. $C_6H_6 \rightarrow C_6F_{12}$).

(c) Fluorination with a metal fluoride in a higher valency state. The two compounds most commonly used are CoF_3 and AgF_2, which are prepared in the reactor by passing fluorine over the lower fluoride. The reactor is then purged with nitrogen, heated to 150–300° and the hydrocarbon vapour is passed. The essential reaction is

$$>C{-}H + 2CoF_3 \rightarrow >C{-}F + 2CoF_2 + HF + 210\ \text{kJ}\ (50\ \text{kcal})$$

(i.e. less heat is evolved than when fluorine is used). The CoF_3 may be regenerated when it has been completely reduced by further passage of fluorine, the conversion being operated as a batch process. The range of hydrocarbons that can be treated and the yields are much the same as in method (b).

(d) Electrochemical fluorination.[11] A solution of the organic compound in anhydrous hydrofluoric acid is electrolysed between nickel electrodes. Many organic compounds (e.g. acids, amines or ethers) give conducting solution (see Section 7.3), but if necessary a solid fluoride such as KF is added to increase the conductivity. With an applied voltage of 5–6 V, hydrogen is evolved at the cathode, but at the anode no fluorine appears. Instead, the solute is converted to its perfluoro analogue, possibly by a mechanism involving a higher nickel fluoride. This process is particularly valuable for the preparation of fully fluorinated organic acids, derivatives of which are widely used as surface active agents.

Fluorocarbons and their derivatives have remarkable physical properties, their melting and boiling points being fairly close to those of the hydrocarbon analogues, and much lower than might be expected from their molecular weights and by analogy with the corresponding chlorocarbons, where the latter are known. This may be illustrated by the boiling point values, given in Table 11.3 below.

Table 11.3 *Comparison of the boiling points of fluorocarbons and hydrocarbons*

Number of carbon atoms	1	2	3	4	8	16
n-Paraffin hydrocarbon b.p.	−161°	−89°	−44°	−0·5°	125°	287°
Fluorocarbon b.p.	−128°	−79°	−38°	−5°	103°	234°

Densities and viscosities are higher and surface tensions much lower than for the corresponding hydrocarbons. They are also insoluble in water, anhydrous hydrofluoric acid and alcohols, but dissolve in chlorocarbons.

Chemically the fluorocarbons are very inert. They resist oxidation in air and often have cracking temperatures above 400°. They are not hydrolysed by concentrated alkalis and are also unattacked by concentrated acids. Their chief applications are as coolants, sealing liquids, high-temperature lubricants and as reaction and dielectric media.

As soon as a functional group is introduced into a fluorocarbon the reactivity is enhanced, though it is different from that of the hydrocarbon analogue. Thus in perfluoroethers and perfluoroamines, which can be represented as $(R_F)_2O$ and $(R_F)_3N$, neither the oxygen nor the nitrogen atom has any appreciable donor function. The perfluoro acids, R_FCOOH, are very strong and, to take another example, CF_3I, the analogue of methyl iodide, behaves as a positive halogen compound, yielding CF_3H and HOI on alkaline hydrolysis. Many thousands of derivatives of a wide range of types are now known, including numerous organometallic compounds with fluoroalkyl or fluoroaryl groups (Chapter 21); otherwise discussion of their chemistry lies outside the scope of this book. Perfluoroaromatic compounds are usually prepared by fluorinating highly chlorinated aromatic compounds with a reagent such as KF in a polar solvent (e.g. $C_6Cl_6 \rightarrow C_6F_6$). Here too the chemical properties are much modified, though physical properties are similar.

Partial fluorination of an organic compound can be brought about by a variety of methods, including the addition of HF to double bonds and the replacement of chlorine by fluorine. The latter method is particularly valuable in the preparation of fluorochlorohydrocarbons, which are called Freons or Arctons. They were first made by Swarts by treating the chlorocarbon with SbF_3 and $SbCl_5$, the latter acting as a catalyst, though these reagents are now often used in association with HF. The Freons are stable, low-boiling liquids (e.g. CF_2Cl_2, b.p. −30°; $CFCl_3$, b.p. 25°; and CHF_2Cl, b.p. −41°), which are non-toxic, non-inflammable, non-corrosive and have thermodynamic properties which render them suitable for use in refrigeration. A further major use is as a propellent in sprays for insect control, etc.

Two important plastics are made on a large scale from chlorofluorohydrocarbons. The first, polytetrafluoroethylene, is produced by first preparing the monomer by the pyrolysis of CHF_2Cl in a silver or platinum tube at 650–800°.

$$CHCl_3 \xrightarrow[SbCl_5,\ HF]{SbF_3} CHF_2Cl \rightarrow C_2F_4 + 2HCl$$

The monomer (b.p. −76·6°) is polymerized in the presence of water and with a persulphate or organic peroxide as catalyst. The product is a white solid, which is stable to heat up to about 300°, is unaffected by most solvents and resists chemical attack by hot acids and alkalis. Its use in chemical plant (e.g. as a gasket material) is well established.

A second plastic, polymerized perfluorovinyl chloride $(CF_2{=}CFCl)_n$(Kel—F), is prepared from the Freon $CF_2ClCFCl_2$ by dechlorination:

$$CF_2ClCFCl_2 \xrightarrow{Zn/EtOH} CF_2{=}CFCl$$

The monomer (b.p. −28°) is again readily polymerized in the presence of a peroxide catalyst and the product, like polytetrafluoroethylene, is very inert. It has the advantage of being almost transparent and is more easily fabricated.

11.6.1. Carbon monofluoride.[12] When graphite is heated in fluorine to 500–700° it burns, the main product being CF_4. At 420–460°, however, the gas is taken into the graphite lattice and the product is a grey to white hydrophobic solid of the approximate composition $(CF)_n$, which, unlike graphite, is practically non-conducting. The temperature at which reaction occurs is lowered by the addition of HF to the fluorine, and also depends on the physical condition of the graphite. With coarse graphite and a partial HF pressure of 200 mm, fluorine is taken up at 250°. There is always considerable swelling, believed to be responsible for the rapid disintegration of graphite electrodes used in the electrolytic generation of fluorine. The difficulty is overcome by employing non-graphitic carbon for the electrodes and using a higher proportion of HF in the electrolyte, so as to lower the operating temperature of the cell.

Carbon monofluoride as prepared above is unreactive and is not attacked either by strong acids or by alkalis, nor does it react with hydrogen up to 400°. When heated to 400–500°, however, a rapid breakdown takes place with formation of carbon, CF_4 and other fluorocarbons.

The nature of carbon monofluoride was in doubt for some time. Ruff, who first prepared it, thought the structure to be ionic (i.e.

comparable with graphite salts (see Section 13.3), which, however, are rapidly attacked by water). The X-ray structure of the solid, in conjunction with other evidence, shows that this is not so. The carbon atoms are arranged in puckered layers (Fig. 11.7), with each carbon bonded to three other carbon atoms in the sheet and to one fluorine atom, which corresponds with sp^3 hybridization. The interplanar distance is 6·6 Å, compared with 3·35 Å in graphite itself, in which the sheets are planar. The C—C distance in the sheets

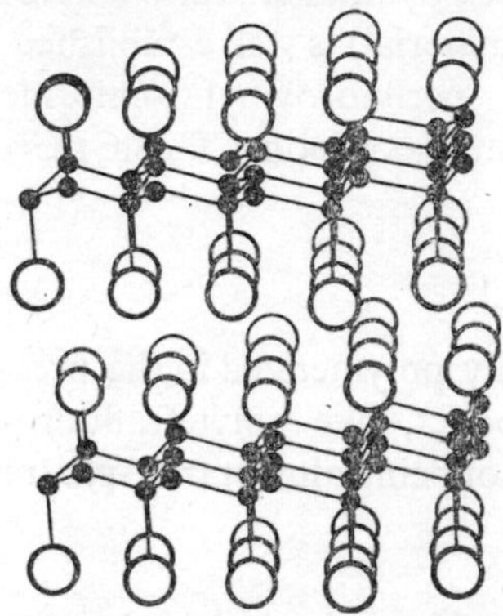

Figure 11.7 The structure of carbon monofluoride

(1·54 Å) equals that for single bonds, whereas in graphite it is 1·41 Å. The infrared spectrum of carbon monofluoride shows a strong broad band at 1215 cm^{-1}, which is due to a C—F stretching vibration and is observed in perfluorinated hydrocarbons in the range 1200–1320 cm^{-1}. In many samples of the monofluoride, fluorination is not complete, which leads to small changes in the lattice dimensions.

This evidence explains why carbon monofluoride differs from graphite in being a non-conductor. All of the four carbon valency electrons are used in σ-bonds. It also accounts for its comparatively high chemical inertness, since C—F bonds are present, and for its hydrophobic character. A second compound, tetracarbon monofluoride, can also be prepared by passing a fluorine–hydrogen fluoride mixture over powdered graphite at room temperature. The product is velvety-black in appearance and has a composition in the range C_4F–$C_{3\cdot6}F$, depending on the exact conditions of preparation. It is less stable to heat than the monofluoride, and vigorous decomposition sets in above about 100° with formation of carbon and fluorocarbons. Decomposition also occurs if an attempt is made to convert C_4F to CF by treatment with fluorine at 200–300°. In its resistance to attack by chemical reagents the compound is almost as unreactive as the monofluoride.

The structure of C_4F is like that of the monofluoride, but the layers are planar and not puckered and the distance between them is smaller (5·5 Å). Fluorine is again believed to be covalently bonded to carbon but, since this involves only every fourth carbon atom, it is not surprising to find that some electrical conductivity is retained—the actual value is roughly two powers of ten less than that of the original graphite. The uptake of chlorine and bromine by graphite up to a limiting composition C_8X is referred to in discussing intercalation compounds (Section 13.3). It differs from the uptake of fluorine in being largely reversible. The products are also reactive and there is no evidence for the formation of covalent bonds between carbon and these halogens.

11.7 Nitrogen halides[13, 14]

Nitrogen trichloride, the most familiar of the nitrogen halides, is formed as a highly explosive oil (b.p. 71°) when chlorine is passed into a solution of ammonium chloride in water. It is endothermic ($\Delta H_f^0 = 231$ kJ (55·4 kcal) mole^{-1}) and is readily hydrolysed by alkali to ammonia and hypochlorite, which is indicative of the bond polarity N^-—Cl^+. Both NBr_3 and NI_3 are known only as ammoniates; they are obtained readily from ammonia and the free halogen and are also highly explosive. The crystal structure of $NI_3.NH_3$ has been determined, and the lattice is found to contain zig-zag chains of irregular NI_4 tetrahedra linked by corner iodine atoms, with the ammonia molecules bonded to non-bridging iodine atoms. Both bromine and iodine display a similar acceptor function in other instances (e.g. in the adducts that they form with aliphatic and aromatic tertiary amines).

In contrast to the above, nitrogen trifluoride is a gas (b.p. −129°), which is slightly soluble in water and resistant to hydrolysis by acids and alkalis. It is also exothermic ($\Delta H_f^0 = -109$ kJ (−26·5 kcal) mole^{-1}). The reason for this difference in the thermodynamic stabilities of NF_3 and NCl_3 has been referred to earlier (Section 11.1). The trifluoride was first prepared by the electrolysis of molten $NH_4F.HF$, but is also formed, with other products, in the electrochemical fluorination of nitrogen-containing organic compounds such as pyridine (see Section 11.6), and in the controlled fluorination of ammonia. Yields of up to 60 per cent are obtained in the gas-phase reaction between fluorine and nitrous oxide:

$$N_2O + 2F_2 \xrightarrow{700^\circ} NF_3 + NOF$$

The nitrogen atom in NF_3 is devoid of donor properties and, since no *d* orbitals are available for back-bonding, it differs from PF_3 in being unable to act as a ligand to transition metals (see Section 20.6). The molecule, like NH_3, is pyramidal, but the dipole moment is only 0·23 D, compared with 1·46 D for NH_3. This is because the polarity due to the N—F bonds is opposed to that arising from the lone pair of electrons on nitrogen, whereas the two dipoles act in conjunction in the case of ammonia, as shown below:

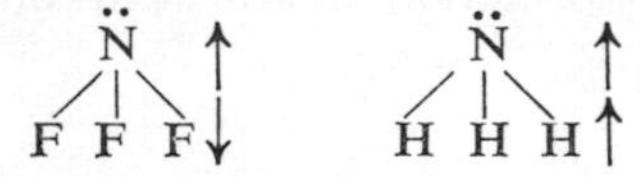

$$\underset{\mu = 0{\cdot}23\ \text{D}}{\text{F F F (N with lone pair} \uparrow\text{, bonds} \downarrow)} \qquad \underset{\mu = 1{\cdot}46\ \text{D}}{\text{H H H (N with lone pair} \uparrow\text{, bonds} \uparrow)}$$

The trifluoride is attacked by sodium only above the melting point of the latter. It is not, however, as inert as was originally thought. It does not react with hydrogen unless the mixture is sparked, when an explosion occurs; the observed stability in this case is kinetic rather than thermodynamic. It can also act as a fluorinating agent at elevated temperatures; for example,

$$3NF_3 + 2CrO_3 \xrightarrow{430^\circ} 2CrF_3 + 3NOF + \tfrac{3}{2}O_2$$

It also reacts with methyl magnesium bromide at −80° with quantitative formation of trimethylamine. Another striking reaction is that with fluorine in the presence of a fluoride ion acceptor, which yields salts of the fluoroammonium cation, NF_4^+:

$$NF_3 + F_2 + SbF_5 \xrightarrow[85\ \text{atm}]{200^\circ} NF_4SbF_6$$

Reaction with F_2 and AsF_5 occurs in a glow discharge at −80°, the product being NF_4AsF_6. Both salts are surprisingly stable, the hexafluoroantimonate decomposing only above 300°, when it yields NF_3, F_2 and SbF_5.

11.7.1. Tetrafluorohydrazine. Because of the strength of the N—F bond it is possible to prepare a wide variety of nitrogen–fluorine compounds which have no counterpart for the heavier halogens.

Tetrafluorohydrazine, N_2F_4 (b.p. −73°), was first made by Colburn and Kennedy in 1958 by passing NF_3 over copper heated to 350–400°. Defluorination of the trifluoride also occurs under similar conditions with other elements (e.g. C, Sb, Bi). Oxidation of difluoramine (NF_2H) in acid solution with $FeCl_3$ also gives the fluorinated hydrazine in high yield. The fluoramine can be made by first passing dilute fluorine through a cold aqueous solution of urea and then acidifying the solution with phosphoric acid, when hydrolysis of the *N,N*-difluoro urea takes place.

$$CO(NH_2)_2 + 2F_2 \xrightarrow{H_2O} NH_2CONF_2 + 2HF$$

$$NH_2CONF_2 + H_2O \xrightarrow{H^+} NF_2H + CO_2 + NH_4^+$$

Tetrafluorohydrazine dissociates reversibly to NF_2 radicals on heating, just as N_2O_4 dissociates to NO_2. The equilibrium may be studied by observing the changes in pressure of a sample of the gas heated at constant volume at temperatures up to 300°, where dissociation is complete. Dissociation is appreciable at room temperature and the chemistry of N_2F_4 is essentially that of the NF_2 radical. Alternative methods of studying the equilibrium are:

(a) Measurement of the strength of the electron spin resonance signal due to the NF_2 radical, as a function of temperature.

(b) Measurement of the intensity of the ultraviolet absorption band of NF_2, centred at about 2600 Å, as a function of temperature.

(c) By admitting samples of the gas from a vessel held at a series of temperatures into the mass spectrometer and observing the heights of peaks due to NF_2^+ and $N_2F_4^+$.

These methods give a value of 83·6 ± 4·2 kJ (20·0 ± 1·0 kcal) mole^{-1} for the N—N bond dissociation energy in N_2F_4, which, taken in conjunction with the heat of formation of N_2F_4 (8·4 ± 10·3 kJ (−2·0 ± 2·5 kcal) mole^{-1}), gives a value of 37·2 ± 10·3 kJ (8·9 ± 2·5 kcal) mole^{-1} for the heat of formation of NF_2. The difference in the heats of formation of NF_3 and NF_2 then gives the first N—F bond energy in NF_3 as 238 kJ (57 kcal) mole^{-1}. The average bond energy in NF_3 is, however, 276 kJ (66 kcal) mole^{-1}, so that the second and third N—F bonds must be stronger, with an average of 297 kJ (71 kcal) mole^{-1}. This difference in bond energies explains why it is possible to strip one fluorine from NF_3 under controlled conditions leaving NF_2, which dimerizes to N_2F_4 rather than breaking down further. In the case of NH_3 the energy needed to break the first N—H bond (434 kJ (104 kcal) mole^{-1}) is greater than the average for the next two (368 kJ (88 kcal) mole^{-1}), which probably explains why N_2H_4 cannot be prepared directly from NH_3.

Electron diffraction studies of the vapour of N_2F_4 at 25° show it to contain a mixture of *gauche* and *trans* conformers, while the NF_2 radical, which can be studied spectroscopically in a low-temperature nitrogen matrix, is like OF_2. The difluoramine radical behaves normally in combining with other free radicals. Thus with nitric oxide a deep purple coloured product $(NO)NF_2$ is formed which dissociates reversibly in the gas phase. Similar reactions occur with other free radicals produced either thermally or photochemically; for example,

$$(N_2F_4 \rightleftharpoons 2NF_2) + S_2F_{10} \xrightarrow{\Delta H} F_5SNF_2$$

$$+ FSO_2OOSO_2F \xrightarrow{\Delta H} FSO_2ONF_2$$

$$+ Cl_2 \xrightarrow{h\nu} ClNF_2$$

$$+ RC(O)\text{—}C(O)R \xrightarrow{h\nu} RCONF_2$$

$$+ C_2H_5I \xrightarrow{h\nu} C_2H_5NF_2$$

The radical also undergoes many hydrogen abstraction reactions of the normal type (e.g. $RSH + NF_2 \rightarrow HNF_2 + R_2S_2$; and $RC(O)H + NF_2 \rightarrow HNF_2 + RC(O)NF_2$). Addition to a variety of olefins and acetylenes also occurs readily; for example,

$$>C{=}C< + N_2F_4 \rightarrow >\underset{NF_2}{\underset{|}{C}}-\underset{NF_2}{\underset{|}{C}}<$$

Tetrafluorohydrazine will react with several Lewis acids which act as fluoride ion acceptors. Thus with SbF_5 the compound $N_2F_4.2SbF_5$ is obtained, and its ^{19}F n.m.r. spectrum in the NF region shows three different types of fluorine in equal abundance. This is in keeping with the formulation of the compound as a salt $(N_2F_3)^+(Sb_2F_{11})^-$, where there is restricted rotation about the N—N bond in

$$\left(\begin{matrix}F\\F\end{matrix}\!\!>N{=}N\diagup^{F}\right)^+$$

The characteristic spectrum of $(Sb_2F_{11})^-$ is also observed. The salt yields N_2F_4 when treated with KF.

11.7.2. Other nitrogen fluorides. The vapour-phase reaction between fluorine diluted with nitrogen and hydrazoic acid gives fluorine azide (FN_3), a gas (b.p. $-120°$) which, like the other halogen azides ClN_3, BrN_3 and IN_3, is explosive. The last three compounds are formed in the reaction between the free halogen and sodium or silver azide. Controlled vapour-phase decomposition of fluorine azide gives difluorodiazine (N_2F_2). The product is a mixture of the *cis* and *trans* isomers shown below (a and b); the third possible isomer (c) is unknown:

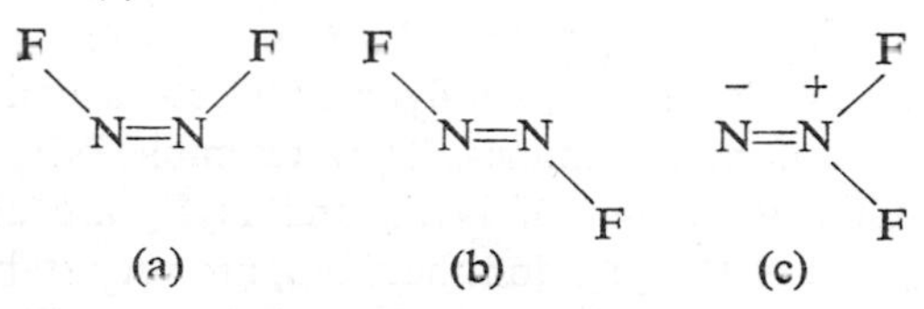

A mixture of the two isomers is also obtained by several other preparative methods. Some N_2F_2 is formed in the electrolysis of molten $NH_4F.HF$, where the main product is NF_3, but a more convenient route is the dehydrofluorination of difluoramine with KF, which occurs either in the dry state or in a concentrated aqueous solution at pH 8·6:

$$2KF + 2HNF_2 \rightarrow N_2F_2 + 2KHF_2$$

The isomers have been separated by fractional distillation and also by vapour-phase chromatography. The boiling and melting points are respectively: *cis*-N_2F_2, m.p. $-195°$, b.p. $-105·7°$; and *trans*-N_2F_2, m.p. $-172°$, b.p. $-111·4°$. The pure *trans* isomer goes over to an equilibrium mixture with approximately 90 per cent of the *cis* form when it is heated at 70–100°, the heat of isomerization deduced from the measured heats of formation of the *cis* and *trans* forms (68·5, 81·1 kJ (16·4, 19·4 kcal) $mole^{-1}$) being 12·5 kJ (3 kcal) $mole^{-1}$. There are small differences in the bond distances and angles for the isomers. The *cis* isomer is more active than the *trans* in its attack on glass and mercury. It also reacts with AsF_5 at room temperature to form the complex $N_2F_2.AsF_5$; no reaction occurs with the other isomer. Both, however, react with SbF_5 to give the same product, $N_2F_2.2SbF_5$. The arsenic compound has been shown to have the structure $N_2F^+AsF_6^-$; when treated with NaF in HF it gives pure *cis*-N_2F_2. Interestingly, the complex $N_2F_3^+Sb_2F_{11}^-$, referred to earlier in discussing N_2F_4, yields the *trans*-N_2F_2 in several reactions, including that with ferrocene in AsF_3 at $-8°$:

$$N_2F_3Sb_2F_{11} + 2(C_5H_5)_2Fe \rightarrow N_2F_2 + 2[(C_5H_5)_2Fe][SbF_6]$$

Like N_2F_4, difluorodiazine is a powerful fluorinating agent and can react explosively with organic compounds under some conditions. Difluoramine (NF_2H) and chlorodifluoramine ($ClNF_2$) are also highly unstable. The latter can be prepared very conveniently by the reaction

$$NHF_2 + Cl_2 + KF \rightarrow ClNF_2 + KF + HCl$$

The N—F bond is encountered in a range of organic compounds of various types, many of which are relatively very stable. This is especially true of fluoralkyl nitrogen fluorides (e.g. CF_3NF_2 and $(CF_3)_2NF$), which are formed in the catalytic or electrochemical fluorination of the corresponding amines. They resemble NF_3 in their inertness. The CF_3 analogues of N_2F_4 and N_2F_2 are also known (i.e. $N_2(CF_3)_4$ and $N_2(CF_3)_2$) and they, too, are very stable, $N_2(CF_3)_4$ showing no tendency to dissociate into $N(CF_3)_2$ radicals.

11.7.3. Nitrogen oxy-halides. The oxy-halides NOF, NOCl, NOBr, NO_2F and NO_2Cl are well known and will not be described here. The only recent addition to this list is trifluoramine oxide, ONF_3 (b.p. −85°), the analogue of OPF_3, which has been prepared by the reactions shown below:

$$2NF_3 + O_2 \xrightarrow[-190°]{\text{electrical discharge}} 2ONF_3$$

$$3NOF + 2IrF_6 \rightarrow ONF_3 + 2NOIrF_6$$

The oxygen and three fluorines have an almost tetrahedral arrangement round the central nitrogen. The compound can be heated to 300° in nickel or Monel vessels without decomposition; it resists hydrolysis by bases and is a strong oxidizing agent. Addition reactions to the double bond in some fluorinated olefins have been reported, the molecule behaving as if it cleaved to F and ONF_2. Stable solid 1 : 1 complexes are also formed with AsF_5 and SbF_5, which appear to be salts of the NF_2O^+ cation.

11.8 Halides of phosphorus, arsenic, antimony and bismuth[15–17]

There is a very extensive chemistry associated with substances in which a halogen is bonded to one of the heavier Group V elements, but this discussion will be restricted in the main to the binary halides and some of their simpler derivatives, the nature and reactions of which are different from those of the nitrogen halides.

The simple trichlorides can all be made by the direct reaction of a halogen with the element, provided the latter is in excess; otherwise, with excess of halogen, the pentahalide results if it exists. Other methods are, however, often more convenient; for example, arsenic trifluoride is usually prepared by the reaction shown below and is itself a good fluorinating agent:

$$As_4O_6 + 6CaF_2 + 6H_2SO_4 \rightarrow 4AsF_3 + 6H_2O + 6CaSO_4$$
$$PCl_3 + AsF_3 \rightarrow PF_3 + AsCl_3$$

There are also numerous mixed halides (e.g. PF_2Cl, PCl_2Br, etc.), some of which can be isolated, while others can be detected only by studies of the Raman, infrared and ^{31}P n.m.r. spectra of mixtures formed in redistribution reactions such as:

$$PCl_3 + PBr_3 \rightleftharpoons PCl_2Br + PClBr_2$$

Chlorofluorides are often made from chlorides by partial fluorination with reagents such as ZnF_2 or SbF_3.

All of the trihalides hydrolyse readily with water except PF_3, for which the reaction is slower. Apart from AsI_3, SbI_3 and BiI_3, which have layer lattices, and BiF_3, which is ionic, all have molecular lattices when in the solid state and pyramidal configurations as vapours.

Except for PF_3, all of the trihalides can act as halide ion acceptors, though the resulting halo anions usually have more complex structures than would be expected. Thus, for example, the salt $KSbF_4$ contains $[Sb_4F_{16}]^{4-}$ ions built up of SbF_5 units which share fluorines and in K_2SbF_5 the anion is octahedral with one stereochemical position occupied by an electron pair.

The trihalides give a very limited range of complexes with non-metallic acceptor atoms, but PF_3 in particular forms interesting adducts with some transition metals. Thus it is able to effect partial replacement of CO on $Ni(CO)_4$, and there is a group of derivatives in which PF_3 is the sole ligand (e.g. $Ni(PF_3)_4$, $Cr(PF_3)_6$), their stability being attributed to 'back-donation' involving interaction between occupied *d* orbitals of the transition metal and vacant *d* orbitals of phosphorus (see Section 20.6). Nitrogen has no *d* orbitals available for bonding and, accordingly, this type of compound is not formed by NF_3. Similar transition metal derivatives are formed by substituted phosphorus (III) halides such as

$$Mo[PF_2N(CH_3)_2]_3(CO)_3 \quad \text{and} \quad Ni(CH_3PF_2)_4$$

All of the diphosphorus tetrahalides have been prepared whereas only N_2F_4 is known. Unlike the latter, P_2F_4 is thermally stable and its dissociation to PF_2 radicals by heat has not been observed.

Phosphorus pentafluoride (b.p. $-84{\cdot}6°$) has a trigonal bipyramidal rather than a square pyramidal arrangement of fluorine atoms round phosphorus. This is consistent with its zero dipole moment. The ^{19}F n.m.r. spectrum shows the five fluorine atoms to be equivalent, which indicates rapid positional exchange among them. The structure of solid PF_5 is unknown. The pentachloride molecule also has the trigonal bipyramidal configuration in the vapour state and when melted under pressure, the melt being non-conducting. In the solid, however, there is an ionic lattice composed of tetrahedral $[PCl_4]^+$ and octahedral $[PCl_6]^-$. These ions are also present in the conducting solution in a polar solvent such as acetonitrile, though a carbon tetrachloride solution of the pentahalide does not conduct. The solid pentabromide contains $[PBr_4]^+$ and Br^- ions. It dissociates very readily in the vapour phase to PBr_3 and Br_2.

Arsenic pentafluoride is believed to have the same structure as its phosphorus analogue in the vapour phase; solid AsF_5 has not been investigated. The pentachloride is unknown but the $[AsCl_4]^+$ cation is found in association with large anions in several salts (e.g. $[AsCl_4]^+[AlCl_4]^-$ and $[AsCl_4]^+[SbF_6]^-$). Antimony pentafluoride, a viscous liquid, is polymeric. Its molecular weight is unknown, but association involves *cis*-fluorine bridge bonding with edge-to-edge sharing of fluorines between adjacent octahedra (Section 9.6). Antimony pentachloride has the trigonal bipyramidal configuration in the solid, liquid and vapour states, though it is believed to yield ions in acetonitrile. Solid bismuth pentafluoride, on the other hand, has a chain structure in which octahedral BiF_6 groups share opposite corners.

A number of mixed pentahalides can be made by adding a halogen to a trihalide. These are interesting because in some instances both ionic and molecular forms can be identified. A good example of such isomerism is provided by the gas PF_3Cl_2 (b.p. $7{\cdot}1°$), which is made by adding Cl_2 to PF_3 and is molecular. If, however, PCl_5 (i.e. $[PCl_4]^+[PCl_6]^-$) is treated with AsF_3 in $AsCl_3$ solution, the ionic isomer $[PCl_4]^+[PF_6]^-$ results. This is a solid which sublimes at 135° with some decomposition and partial conversion to the molecular form. Another example of the same sort is PCl_4F, the molecular form of which (m.p. $-59{\cdot}0°$) is transformed at room temperature to a white solid which is an electrolyte in acetonitrile. There are a few similar examples among the pentahalides of the remaining Group V elements. Thus, to quote a single instance, chlorination of AsF_3 gives AsF_3Cl_2, which is ionic in polar solvents but is partly converted to a molecular form if sublimed in vacuum.

For an account of the very extensive chemistry of the pentahalides of these elements, and particularly that of phosphorus (V) compounds with P—F bonds, the reader is referred to other sources.[17] It is, however, interesting to note the existence of the two hydrides PH_2F_3 (b.p. 3·8°) and PHF_4 (b.p. 39°), which are prepared by reaction of anhydrous HF with H_3PO_2 and H_3PO_3 respectively. Hexafluorophosphates are also of importance, the alkali metal salts being formed directly from the fluoride and PF_5. The anion is very stable in aqueous solution, but, like H_2SiF_6, the anhydrous acid cannot be prepared, though a crystalline hexahydrate is known. In mono- and difluorophosphoric acids (H_2PO_3F and HPO_2F_2) oxygen is replaced by fluorine in the PO_4^{3-} anion, but the chlorine analogues are unknown.

11.9 Halides of sulphur, selenium and tellurium[18–20]

It will be convenient to describe the halides and oxy-halides of these three elements on the basis of the three oxidation states that are encountered: +6, +4 and +2. Only the fluorides are known in the first of these, their boiling points or sublimation temperatures being shown below:

SF_6	−63·7° (subl.)	SeF_6	−47° (subl.)	TeF_6	−39° (subl.)
S_2F_{10}	29°		—	Te_2F_{10}	53°

Sulphur hexafluoride was first made in 1891 by Moissan by burning sulphur in fluorine, and this remains the standard preparative method. A little S_2F_{10} is formed in the reaction and can be removed when the gas is required for use as an electrical insulator, which is its main application, by pyrolysing at 400°. This converts S_2F_{10} to SF_4 and SF_6, and the former can then be scrubbed out with aqueous alkali, which does not hydrolyse SF_6.

Selenium and tellurium hexafluorides are made from the element and fluorine in the same way. Sulphur hexachloride is unknown and its preparation in the future seems unlikely as the S—Cl bond is weaker, and SCl_4 is only stable in the solid state at −80°, losing chlorine on melting.

The three hexafluorides have regular octahedral structures, with the following mean bond energies: S—F, 276 kJ (66 kcal); Se—F,

297 kJ (71 kcal) and Te—F, 339 kJ (81 kcal). Chemical reactivity increases in passing along the series. Sulphur hexafluoride resists hydrolysis by alkali, though this stability is due to kinetic rather than thermodynamic factors.

$$SF_6(g) + 3H_2O(g) \rightarrow SO_3(g) + 6HF(g): \Delta G^o = -200 \text{ kJ } (-48 \text{ kcal})$$

In contrast, TeF_6 is decomposed by water in a few hours. Although SF_6 shows a lack of reactivity with nucleophiles, it is attacked by some electrophiles; for example,

$$SF_6 + AlCl_3 \xrightarrow{225^\circ} \text{sulphur chlorides} + Cl_2 + AlF_3$$

$$SF_6 + SO_3 \rightarrow SO_2F_2$$

Alkali metals decompose it at approximately 200°.

Little is known about the chemistry of SeF_6, but TeF_6 behaves as a Lewis acid in forming Cs_2TeF_8 with CsF and adducts such as $(R_3N)_2TeF_6$ with trialkylamines. Perfluoroalkyl derivatives of SF_6 such as CF_3SF_5 and $(CF_3)_2SF_4$, both of which are formed with other products in the electrochemical fluorination of CS_2 or $(CH_3)_2S$, share its chemical inertness in large measure.

An important derivative of SF_6 is sulphur cholride pentafluoride, SF_5Cl (b.p. −21°), which can be prepared by either of the reactions shown below.

$$SF_4 + ClF \xrightarrow{380^\circ} SF_5Cl$$

$$SF_4 + CsF + Cl_2 \xrightarrow{110^\circ} SF_5Cl + CsCl$$

These can also be modified to give SF_5Br. Unlike SF_6 and S_2F_{10}, both SF_5Cl and SF_5Br are hydrolysed by alkali. They are useful in preparing a number of inorganic and organic pentafluorosulphur derivatives; for example,

$$SF_5Cl + H_2 \xrightarrow{h\nu} S_2F_{10}$$

$$SF_5Cl + O_2 \xrightarrow{h\nu} SF_5OOSF_5 + (SF_5)_2O$$

$$SF_5Cl + N_2F_4 \xrightarrow{h\nu} SF_5NF_2$$

In each case the primary photochemical reaction involves cleavage of the S—Cl bond. The last of the three products can also be prepared by the thermal reaction of N_2F_4 with elementary sulphur. At first sight the SF_5 radical might be expected to give a range of metallic and non-metallic derivatives comparable with those formed by CF_3 and other perfluoroalkyls. There is indeed some similarity, the compound $(SF_5)_2O$, for example, resembling $(CF_3)_2O$ in being thermally very stable, but SF_5 has marked oxidizing properties and converts elements such as phosphorus to their fluorides.

The main reaction used in preparing organic SF_5 derivatives from SF_5Cl is its addition to an unsaturated group, such as $>C{=}C<$ or $>C{=}O$. Addition occurs with olefins and fluoroolefins as well as with some acetylenes; for example,

$$SF_5Cl + C_2H_4 \rightarrow SF_5CH_2CH_2Cl + SF_5(CH_2CH_2)_nCl$$

Other examples of these addition reactions are given below; a considerable derivative chemistry has been built up in this way.

$$SF_5Cl + CH_2{=}C{=}O \rightarrow SF_5CH_2COCl\ (\rightarrow SF_5CH_2COOH)$$
$$SF_5Cl + ClCN \rightarrow SF_5N{=}CCl_2$$

The peroxide SF_5OOSF_5, formed photochemically from SF_5Cl and oxygen, cleaves readily at the O—O bond and can be used in syntheses. Reaction of the peroxide with N_2F_4, for example, yields SF_5ONF_2 and the product of its reaction with benzene is $C_6H_5OSF_5$.

Turning to sulphur (VI) oxy-halides, SO_2F_2 (b.p. −55°) and SO_2Cl_2 (b.p. 69°) are familiar compounds. The former can be prepared very conveniently from the chloride by reaction with sodium fluoride; SO_2Br_2 is unstable at room temperature. The chloro compound is extensively used in organic chemistry to introduce —Cl and —SO_2Cl groups into organic molecules. Towards reactive non-metals it behaves as an oxidative chlorinating agent.

Sulphur oxide tetrafluoride, SOF_4 (b.p. −49°), can be prepared by direct oxidation of SF_4 with NO_2 and O_2 at 200°, the NO_2 acting as an oxygen-carrier. It is readily hydrolysed by water to SO_2F_2, which is decomposed only under alkaline conditions:

$$SOF_4 + H_2O \rightarrow SO_2F_2 + 2HF$$

The structure has been shown by spectroscopic methods to consist of a trigonal bipyramid with two fluorine atoms and one oxygen in the equatorial plane.

Reaction of fluorine with sulphur trioxide at 180° in the presence of AgF_2 gives peroxydisulphuryl difluoride, FSO_2OOSO_2F (b.p.

57°), together with some FSO_2OF (see Section 11.10.3). The first of these compounds has also been made by electrolysis of a solution of an alkali fluorosulphonate in fluorosulphonic acid. It is remarkable because of the facile cleavage at the O—O bond to form OSO_2F radicals, which can be demonstrated by carrying the vapour mixed with nitrogen through a glass tube heated at one point to 120°. A brown colour appears in the heated zone, due to radical formation, and this disappears downstream as recombination occurs. Addition to various fluoroolefins has been observed, and SF_4 is also oxidized; for example,

$$FSO_2OOSO_2F + C_2F_4 \xrightarrow{20^\circ} FSO_2OC_2F_4OSO_2F$$

$$FSO_2OOSO_2F + SF_4 \longrightarrow FSO_2O(SF_4)OSO_2F$$

Interesting reactions also take place with various halides at, or a little above, room temperature, e.g.,

$$SnCl_4 \rightarrow SnCl(SO_3F)_3 \qquad KBr \rightarrow KBr(SO_3F)_4$$

$$CrO_2Cl_2 \rightarrow CrO_2(SO_3F)_2 \qquad KI \rightarrow KI(SO_3F)_4$$

It will be seen that the products from the alkali metal halides resemble polyhalides, with the highly electronegative radical behaving as a pseudo-halogen. This view is borne out by the existence of fluorosulphonates of chlorine, bromine and iodine which are analogous to the interhalogen compounds. Chlorine reacts with FSO_2OOSO_2F in a pressure vessel at 125° giving $Cl(OSO_2F)$, a liquid which reacts violently with water, liberating oxygen. Reaction with bromine and iodine occurs at room temperature and both the mono- and the tri-fluorosulphonates are formed. Excess of fluorine fluorosulphonate, which is described elsewhere (Section 11.10.3), also reacts with iodine to form $IF_3(OSO_2F)_2$:

$$I_2 + 6FOSO_2F \rightarrow 2IF_3(OSO_2F)_2 + S_2O_6F_2$$

A further point of resemblance to the interhalogens is the addition of $Br(OSO_2F)$ across the double bond of C_2F_4:

$$Br(OSO_2F) + C_2F_4 \xrightarrow{25^\circ} CF_2BrCF_2(OSO_2F)$$

11.9.1. Halides of the +4 state. The only stable binary sulphur compound of this type is SF_4 (b.p. −38°), which can be synthesized very conveniently in refluxing acetonitrile by the reaction shown below:

$$3SCl_2 + 4NaF \rightarrow SF_4 + S_2Cl_2 + 4NaCl$$

The structure of the tetrafluoride is based on a trigonal bipyramid, with two fluorine atoms and an electron pair occupying the equatorial sites. Unlike the hexafluoride, this compound is very readily hydrolysed. It reacts with a number of fluorides (e.g. SbF_5 and BF_3), which can function as fluoride ion acceptors, to form adducts which may be formulated as salts of the SF_3^+ cation. This view is supported by the fact that the infrared spectrum of the solid 1 : 1 adduct with BF_3 shows the presence of the BF_4^- ion. The tetrafluoride is also a valuable selective fluorinating agent for organic compounds, being able to replace the carbonyl oxygen by fluorine in a variety of compounds including aldehydes, ketones and carboxylic acids:[20a]

$$>C{=}O + SF_4 \rightarrow\ >CF_2 + SOF_2$$

Such reactions are generally carried out in an autoclave and require heat. Some simple examples are:

$$\begin{aligned} C_6H_5COOH + SF_4 &\rightarrow C_6H_5COF \rightarrow C_6H_5CF_3 \\ C_6H_5CHO &\rightarrow C_6H_5CHF_2 \\ CH{\equiv}CCOOH &\rightarrow CH{\equiv}CCF_3 \end{aligned}$$

The reagent has the advantage that it does not attack points of unsaturation in the molecule. Other reactions of the tetrafluoride include the synthesis of organoiminosulphur difluorides ($RN{=}SF_2$), by its reaction with organic and inorganic compounds having C—N multiple bonds; for example,

$$\left.\begin{aligned} RN{=}C{=}O + SF_4 &\rightarrow RN{=}SF_2 + COF_2 \\ RCN + SF_4 &\rightarrow RCF_2N{-}SF_2 \end{aligned}\right\} \text{(R = organic group)}$$

$$\begin{aligned} 2KSCN + 3SF_4 &\rightarrow 2CF_3N{=}SF_2 + 3S + 2KF \\ KOCN + 2SF_4 &\rightarrow CF_3N{=}SF_2 + SOF_2 + KF \end{aligned}$$

Various substituted derivatives of SF_4 are also known (e.g. CF_3SF_3 and $C_6H_5SF_3$) and have been studied in some detail.

In contrast to sulphur tetrafluoride, little is known about the selenium and tellurium analogues, the first of which boils at 106° while the second has a melting point of 130°. In keeping with the normal increase in basic character in passing to the heavier elements of the group, TeF_4 is less readily hydrolysed than the other tetrafluorides and will form a dihydrate. The structure of the anhydrous salt consists of a square pyramidal array of fluorines round each tellurium, with sharing of two fluorines to give infinite chains extending through the lattice. The other tetrahalides of selenium and tellurium ($SeCl_4$, $SeBr_4$, $TeCl_4$, $TeBr_4$ and TeI_4) and the complexes that they form will not be discussed here.

The chemistry of the lower halides of the three elements has a number of interesting points. It has long been known that a reactive gas is evolved if sulphur is heated to its melting point with silver (I) fluoride. This is now known to be a mixture of SF_4 with two isomeric forms of S_2F_2, thiothionyl fluoride, $S{=}SF_2$ (b.p. −10·6°), and disulphur difluoride, F—S—S—F (b.p. *ca.* −15°). The latter is less stable and changes to its isomer at about 30°. The two may be separated by trap-to-trap distillation.

Other routes to thiothionyl fluoride are also known; for example,

$$S_2Cl_2 + KSO_2F \rightarrow SSF_2 + SO_2ClF + KCl + SO_2$$
$$SCl_2 + KF \rightarrow SSF_2 + SF_4 + KCl$$

Both structures have been established by microwave spectroscopy. The first resembles that of OSF_2 and the second that of H_2O_2. Both compounds hydrolyse very readily, attack glass and also act as fluorinating agents, though their reactions have not been investigated in detail. Thiothionyl fluoride has been found to react with S_2Cl_2 to give F—S—S—F and Cl—S—S—F. The lower fluorides of selenium and tellurium appear not to have been prepared.

Sulphur monochloride (S_2Cl_2) and the dichloride (SCl_2) are familiar compounds; the latter loses chlorine very readily to form the monochloride. This behaviour is repeated with selenium, for, although both $SeCl_2$ and $SeBr_2$ can exist in the vapour state, they lose halogen very readily and form Se_2Cl_2 and Se_2Br_2. In contrast, $TeCl_2$ is much more stable. It is formed in the reaction of CCl_2F_2 with tellurium at 500° and is a black solid which melts at 175° without decomposition, and disproportionates with water to Te and TeO_2. The dibromide is very similar, the vapour consisting of molecules with the Br—Te—Br bond angle equal to 98°.

11.9.2. Nitrogen-sulphur-fluorine compounds.[21] Two compounds of this group, thiazyl fluoride ($N{\equiv}SF$) and thiazyl trifluoride ($N{\equiv}SF_3$) were referred to briefly in discussing polymers based on tetrasulphur tetranitride (Chapter 7). They are key substances in an important area of sulphur–fluorine chemistry, which has been extensively explored by Glemser and his co-workers in recent years.

Thiazyl fluoride (b.p. *ca.* 0°) may be made by several methods, one of the most convenient being that shown below:

$$Hg(N{=}SF_2)_2 \xrightarrow[\text{vacuum}]{100°} 2NSF + HgF_2$$

The mercurial is obtained by heating HgF_2 with $SF_2{=}NCOF$, which is a product of the reaction of SF_4 with $Si(NCO)_4$. Thiazyl fluoride polymerizes readily to $N_3S_3F_3$, but with this method of preparation it can be generated as required. It is a pungent gas which hydrolyses readily. The molecule has N—S and S—F bond distances of 1·446 and 1·646 Å with the N—S—F bond angle equal to 116° 52′. The chlorine analogue polymerizes too readily to be studied.

Numerous sulphur difluoride imides of the type $R{-}N{=}SF_2$ are known (e.g. $C_6H_5{-}N{=}SF_2$, $F_5S{-}N{=}SF_2$ and $F_3C{-}N{=}SF_2$), but the most interesting derivatives are the N-halo compounds $X{-}N{=}SF_2$ (X = Cl, Br, I), which are obtained by direct action of the halogen with $Hg(NSF_2)_2$. The N—X bond cleaves readily, so that additions occur across the double bonds of olefins.

Of the numerous related sulphur (VI) compounds we will select only thiazyl trifluoride ($N{\equiv}SF_3$) for discussion. It is a gas (b.p. −27·1°) which can be prepared by the reaction of AgF_2 with N_4S_4 in boiling CCl_4, and also by fluorinating NSF with AgF_2. The enthalpy of formation ΔH_f^0 is −340 kJ (−95 kcal) $mole^{-1}$, and the molecule shows some of the chemical stability of SF_6. The structure is approximately tetrahedral with a central sulphur atom and a very short N—S bond distance (1·416 Å), suggesting a bond order of just under three. The S—F bonds are of almost the same length as those in SF_6.

Thiazyl trifluoride is unattacked by sodium at 200° and hydrolysis by water is slow. Reaction with HF at 0° gives H_2NSF_5 and addition of ClF to the triple bond results in Cl_2NSF_5. A 1 : 1 adduct is formed with BF_3, but the infrared spectrum does not support its formulation as $NSF_2^+BF_4^-$. One of the fluorine atoms may be replaced by reaction with a dialkylamine:

$$N{\equiv}SF_3 + 2(C_2H_5)_2NH \xrightarrow[\text{ether}]{20°} N{\equiv}SF_2(N(C_2H_5)_2) + (C_2H_5)_2NH.HF$$

An unusual reaction with the lithium salt of hexamethyldisilazane has been reported which is shown below:

$$N{\equiv}SF_3 + LiN[Si(CH_3)_3]_2 \xrightarrow[-LiF]{0-20^\circ} (CH_3)_3SiN{=}SF_2{=}NSi(CH_3)_3$$

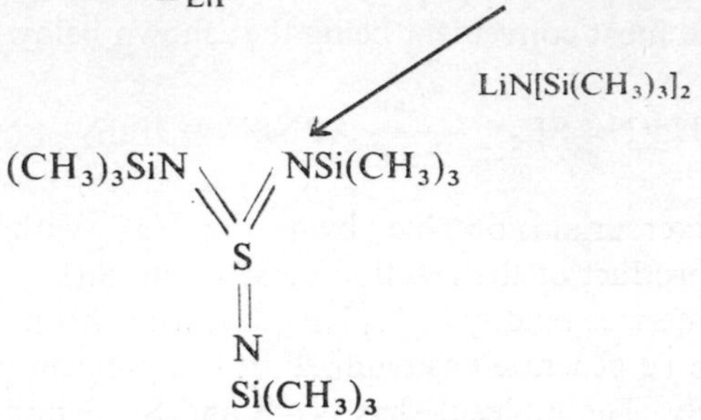

$$\downarrow LiN[Si(CH_3)_3]_2$$

$$[(CH_3)_3SiN{=}]_2S{=}NSi(CH_3)_3$$

The final product is the first monomeric compound so far reported in which sulphur has an oxidation number of +6 and a coordination number of 3 at room temperature. The reaction of NSF_3 with BCl_3, shown below, is also unusual, especially in the replacement of an S—F bond by S—Cl:

$$2NSF_3 + 3BCl_3 \xrightarrow{20^\circ} [N(SCl)_2]^+[BCl_4]^- + \tfrac{1}{2}N_2 + \tfrac{3}{2}Cl_2 + 2BF_3$$

11.10 Halogen oxides[22–24]

The general chemistry of the oxides of chlorine, bromine and iodine and the acids derived from them is fully discussed in many texts and will not therefore be described here. There are, however, a number of important recent advances in this field. This is especially true of fluorine-containing compounds, on which the main emphasis

Table 11.4 *Binary halogen–oxygen compounds*

F	*Cl*	*Br*	*I*
FOF	ClOCl	BrOBr	$(IO_2)_n$
FOOF	Cl_2O_3	BrO_2	I_4O_9
O_3F_2	ClO_2	Br_3O_8	I_2O_5
O_4F_5	Cl_2O_4	BrO_3*	I_2O_7*
O_5F_2	ClO_3, Cl_2O_6	Br_2O_7*	
O_6F_2	Cl_2O_7		

* Incompletely characterized.

is therefore placed. An overall picture of the binary halogen-oxygen compounds can be obtained from Table 11.4. We will deal fully with fluorine–oxygen compounds, but discuss only isolated points in the remainder of the field.

11.10.1. Oxygen fluorides. These compounds are called oxygen fluorides rather than fluorine oxides because fluorine is more electronegative than oxygen, whereas the reverse is true of the other halogens and oxygen. Oxygen difluoride, OF_2 (b.p. $-144{\cdot}8°$), differs from Cl_2O in being exothermic ($\Delta H_f^0 = -18$ kJ ($-4{\cdot}5$ kcal) mole^{-1}), the corresponding figure for Cl_2O being 94 kJ (22·4 kcal) mole^{-1}. It is not explosive, as is Cl_2O, though it behaves as a very strong oxidizing agent. It is produced by bubbling fluorine through a 2 per cent NaOH solution under controlled conditions, so as to minimize the second of the two reactions shown below:

$$2F_2 + 2NaOH \rightarrow OF_2 + 2NaF + H_2O$$
$$OF_2 + 2OH^- \rightarrow O_2 + 2F^- + H_2O$$

The gas is slightly soluble in water and Henry's law is obeyed, though slow hydrolysis occurs. This behaviour differs from that of Cl_2O and Br_2O, which yield HClO and HBrO respectively. There is no evidence for the formation of hypofluorous acid by reaction of OF_2 with water, though it has recently been prepared by another route described later.

Oxygen difluoride when pure is stable in glass vessels to at least 200°. When heated, it converts many metals to mixtures of oxide and fluoride. Phosphorus reacts at room temperature to give PF_5 and POF_3 and sulphur goes to a mixture of SO_2 and SF_4. There is an explosive reaction with H_2S at room temperature, but mixtures with hydrogen and hydrocarbons explode only on sparking. These examples show the gas to be considerably less reactive than fluorine.

The molecule has a symmetrical bent configuration (FOF angle = 103°; cf. ClOCl angle in Cl_2O = 111°). The single line in the ^{19}F n.m.r. spectrum shows the two fluorine atoms to be equivalent. When photolysed with ultraviolet light in an argon matrix at 4°K, a new infrared band appears which has been unambiguously identified as due to the OF radical. It disappears as the CsF window on which the matrix is deposited is warmed to 45°K because of recombination of OF and F to OF_2; the radical has thus only a transitory existence.

A much less stable oxide, O_2F_2 (m.p. $-145°$), is formed in good yield when a 1 : 1 F_2–O_2 mixture is passed through an electrical discharge at a pressure of about 12 mm in a vessel cooled by liquid air. The product condenses as an orange-yellow solid on the cold surface and decomposes rapidly to the elements at temperatures above about −100°. It is a powerful oxidizing agent, even at low temperatures. Thus at −150° SF_4 is converted to SF_6 and O_2, and N_2F_4 to NF_3 and O_2. Sulphur inflames at the temperature of liquid air and there are comparable reactions with other elements and compounds. Explosive reactions also occur with many organic compounds, but controlled addition to double bonds in perfluoro-olefins is possible; for example,

$$CF_3CF{=}CF_2 + O_2F_2 \xrightarrow[-180^\circ]{CF_3Cl} CF_3CF(OOF)CF_3 + CF_3CF_2CF_2(OOF)$$

Reaction of dioxygen difluoride with certain Lewis-acid fluorides leads to dioxygenyl salts (e.g. $BF_3 \rightarrow O_2^+BF_4^-$ (decomposes at 25°); and $SbF_5 \rightarrow O_2^+SbF_6^-$ (decomposes at 100°)). These are of the same type as O_2PtF_6, isolation of which led indirectly to the preparation of the first noble gas compound (see Section 13.1).

The structure of O_2F_2 resembles that of H_2O_2. The atoms are arranged in the order F—O—O—F, the O—F bonds lying in planes at an angle of 87° to one another. There is ready dissociation to fluorine and the OOF radical, and either in the liquid state or in solution in an inert solvent an electron spin resonance signal due to the OF_2 radical can be observed. It is more stable than the OF radical and is also formed on photolysis of a mixture of solid F_2 and O_2 in an argon matrix at 4°K.

The remaining oxygen fluorides (O_3F_2, O_4F_2, O_5F_2 and O_6F_2) are all produced by the method used for O_2F_2, but with different $O_2 : F_2$ ratios and progressively milder discharge conditions. All decompose very readily at low temperatures. Their structures have not been determined, the chemical reactions are largely unknown and characterization is by chemical analysis only. The compound O_3F_2 separates as a reddish solid (m.p. − 190°) when a 3 : 2 mixture of reactants is used with milder discharge conditions. The ^{17}O n.m.r. signal consists of a symmetrical doublet, which is difficult to reconcile with the obvious formulation F—O—O—O—F, ^{17}O n.m.r. signals from which should be a doublet with a 2 : 1 intensity ratio. It has been suggested that this substance is an O_2F_2–O_4F_2 mixture, which would give the observed ratio of oxygen to fluorine on decomposition. The compound O_4F_2 is formed similarly, using a 2:1 ratio of oxygen to fluorine. Its existence seems to be more probable, as it is the dimer of the O_2F radical, a strong e.s.r. signal due to which is observed for solutions of O_4F_2 in an inert solvent. Confirmation of the existence of O_5F_2 and O_6F_2, which are produced by the discharge method with 5:2 and 6:2 ratios of reactants, must await structural studies.

11.10.2. Hypofluorous acid. Early claims to have synthesized salts of fluorine oxy-acids by conventional methods are now thought to have been incorrect. The first indication of the existence of hypofluorous acid came in 1968 from experiments on the photolysis of mixtures of fluorine and water in a nitrogen matrix at 14–20°K, which gave a product with infrared bands attributable to HOF. Recently the compound has been prepared in milligram quantities by passing fluorine through a U-tube packed with wet Teflon (polytetrafluoroethylene) Raschig rings. The issuing gas is passed through U-tubes cooled to −50° and −80° to trap out water, and HOF is then condensed at −183°. It melts at *ca.* −117°, has a vapour pressure of 0·5 mm at −64°, and can be characterized by analysis and by the detection of the parent ion in the time-of-flight mass spectrometer. There is slow decomposition at room temperature to HF and O_2. Hydrolysis by water gives HF and H_2O_2; no salts have yet been prepared.

11.10.3. Covalent hypofluorites. The name hypofluorite is used here to denote compounds containing the OF group. The first such compound, O_2NOF, was made in 1934 by Cady, by the reaction of fluorine with 4N nitric acid, but the same product is obtained by fluorinating KNO_3:

$$F_2 + HNO_3 = HF + O_2NOF$$

It is an explosive gas (b.p. −45·9°), with a molecular structure very similar to that of nitric acid. The name fluorine nitrate is in common use, although nitroxyl hypofluorite or fluoroxide would be more descriptive. Explosive decomposition to NOF and O_2 occurs readily and unpredictably, but at low pressure there is a slow unimolecular decomposition at about 100° which gives NO_2F and O_2. Hydrolysis by water is slow, but with aqueous base the reaction is

$$2O_2NOF + 4OH^- \longrightarrow 2NO_3^- + 2F^- + 2H_2O + O_2$$

Few reactions have been studied.

A high yield of fluorine perchlorate, O_3ClOF (b.p. −15·9°), is obtained in the reaction when fluorine is passed into 70 per cent $HClO_4$ at 20°. This, too, may be made by direct fluorination of a metal perchlorate under controlled conditions. The essential difference from the acid is again that O—H has been replaced by O—F. The compound is explosive and has strong oxidizing properties, though these have been little studied. Fluorine fluorosulphonate, FSO_2OF (b.p. −31·3°), is also dangerously explosive, though rather more is known about it. Several synthetic methods are available. Sodium fluorosulphonate may be treated with fluorine at 200°, or SO_3 may be fluorinated at 200° in the presence of AgF_2. As would be expected, the compound has strong oxidizing properties. Because of the ready cleavage of the O—F bond, addition across the double bond of olefins and haloolefins occurs readily (e.g. $C_2F_4 \rightarrow C_2F_5OSO_2F$). Similar reactions are observed with fluorosulphonates of the other halogens (see Section 11.9). Sulphur tetrafluoride is oxidized to F_5SOSO_2F.

The compound F_5SOF (b.p. −35·1°), pentafluorosulphur hypofluorite or fluoroxypentafluorosulphur, was made originally by the reaction of F_2 and SOF_2 at 200° in the presence of AgF_2, but it may also be synthesized in higher yield from SOF_2 and F_2 in the presence of CsF; in the absence of the alkali metal salt this reaction gives SOF_4. It is more stable than the hypofluorites referred to above. There is quantitative addition across olefinic double bonds. Photolysis gives the peroxide F_5SOOSF_5 in low yield, and F_5SONF_2 is produced in either the thermal or the photochemical reaction with N_2F_4. The selenium analogue, F_5SeOF, which is also known, is rather more reactive, and it is interesting that fluorination of $KSeO_2F$ at −78° gives rise not only to F_5SeOF but also to $F_4Se(OF)_2$, which has a *trans* configuration of the O—F groups and is, incidentally, the only non-carbon-containing bis-fluoroxy compound known at present.

11.10.4. Carbon-containing hypofluorites. These fall into two groups, the perfluoroacyl hypofluorites of the type $R_FC(O)OF$ ($R_F = CF_3$, C_2F_5, C_3F_7), which are explosive, and fluoroalkyl hypofluorites and related compounds, which are more stable. Compounds of the first class are formed when the perfluoro acid, $R_FC(O)OH$, to which a little water has been added, is treated with fluorine:

$$R_FCOOH + F_2 \xrightarrow{H_2O} R_FCOOF + HF$$

Trifluoroacetyl hypofluorite (CF_3COOF), obtained from trifluoroacetic acid, decomposes at low pressure (30–80°) to CF_4 and OC_2 while the reaction with chlorine follows the equation,

$$CF_3COOF + Cl_2 \rightarrow CF_3Cl + CO_2 + ClF$$

Over thirty compounds of the second group are known. The simplest is trifluoromethyl hypofluorite, CF_3OF (b.p. −95°), which is best prepared by the reaction shown below:

$$COF_2 + F_2 \xrightarrow[20°]{CsF} CF_3OF$$

The salt $CsOCF_3$ is almost certainly an intermediate, and can be prepared in the crystalline form by passing COF_2 into a suspension of CsF in acetonitrile at 20°. The fluorides of potassium and rubidium react similarly, but not those of lithium and sodium. These salts decompose to M^IF and COF_2 at 80–100° and have X-ray powder diagrams which are very similar to those of the corresponding salts of the $[BF_4]^-$ ion, with which $[OCF_3]^-$ is isoelectronic. Spectroscopic evidence shows the structure of CF_3OF to be similar to that of CH_3OH (CF_3OH is unknown).

Trifluoromethyl hypofluorite decomposes to COF_2 and F_2 at 300–450°. The enthalpy of formation is −742 kJ (−177·3 kcal) and, unlike the perfluoroacyl hypofluorites, it is non-explosive, though a strong oxidizing agent. Reaction with mercury at room temperature gives HgF_2 and COF_2, and, in ultraviolet light the comparatively stable peroxide CF_3OOCF_3 is formed, fluorine from the primary photochemical decomposition ($CF_3OF \xrightarrow{h\nu} CF_3O\cdot + \cdot F$) reacting with the containing vessel. Hydrolysis by water occurs slowly. Sulphur dioxide at 250° gives CF_3OOSO_2F while, with SF_4, oxidation to CF_3OSF_5 is observed. Under some conditions explosions can occur with organic compounds, but it is possible to bring about additions to olefins, C_2H_4, for example, giving $CF_3OCH_2CH_2F$.

Examples of other compounds of this sort are shown below. We will confine the discussion to a few salient points.

CF_2ClOF, $CFCl_2OF$　　　　$FO(CF_2)_nOF$ ($n = 1, 2, 3, 4, 5$)
C_2F_5OF, C_3F_7OF
CCl_3CF_2OF, $O_2NCF_2CF_2OF$

The simple fluorochloro analogues of CF_3OF are formed by photolysis of the fluorochlorinated acetone with OF_2, the latter serving as a source of OF radicals. Thus the main product from

$CF_2ClC(O)CF_2Cl$ and OF_2 is CF_2ClOF. Carbon dioxide and fluorine react at room temperature in the presence of CsF to give $CF_2(OF)_2$ (b.p. −64°), which is similar to CF_3OF in its stability and reactions. A novel compound with both peroxide and trioxide linkages is formed when $CF_2(OF)_2$ is mixed with a 2 : 1 molar mixture of $CsOCF_3$ and CsF, and the system then warmed to −5°. The main product is CF_3OF, but smaller amounts of CF_3OOOCF_3 and $CF_3OOOCF_2OOCF_3$ are also obtained; the last of these compounds decomposes explosively at 40°.

Examples of the routes by which some of the multicarbon hypofluorites can be synthesized are shown below.

$$CF_3CF_2CH_2OH \xrightarrow{F_2} CF_3CF_2CF_2OF$$

$$CF_3C(O)CF_3.H_2O \xrightarrow{F_2} CF_3CF(OF)CF_3$$

$$(CF_3)_3COH \xrightarrow{F_2} (CF_3)_3COF$$

$$CF_3CF(NF_2)C(O)F \xrightarrow[-80°]{F_2,\ CsF} CF_3CF(NF_2)CF_2OF$$

$$F(O)COOC(O)F \xrightarrow[-95°]{F_2,\ KF} FOCF_2OOCF_2OF$$

The stability of these compounds varies considerably and a few are explosive, but in the main their reactions resemble those of CF_3OF.

11.10.5. Covalent hypochlorites. Alkyl hypochlorites have been known for over eighty years and may be produced by chlorination of a solution of sodium hydroxide in an alcohol. A few alkyl hypobromites have also been prepared, but all of these compounds are somewhat unstable. Only recently, however, has it been possible to prepare perfluoroalkyl hypochlorites by the general reaction

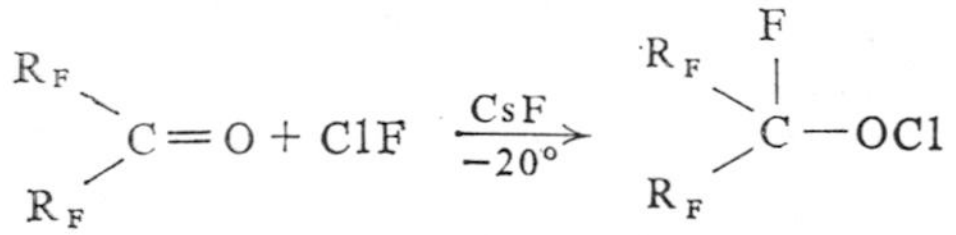

$$\begin{matrix} R_F \\ R_F \end{matrix}\!\!>C{=}O + ClF \xrightarrow[-20°]{CsF} \begin{matrix} R_F \\ R_F \end{matrix}\!\!>CF{-}OCl$$

R_F and R_F may both be atoms of fluorine, when the reaction parallels that between COF_2 and F_2, which gives CF_3OF. When R_F is CF_3 the product is $(CF_3)_2CFOCl$, and the same approach can be used to prepare difunctional derivatives; for example,

$$F(O)C(CF_2)_3C(O)F + 2ClF \xrightarrow[-20°]{CsF} ClO(CF_2)_5OCl$$

Chlorine monoxide, which is also a positive chlorine reagent, can be employed to convert COF_2 to CF_3OCl (and CsOCl) with CsF as catalyst. The salt $CsOCF_3$ (referred to in the last section) must be an intermediate in the last reaction since it is known to react with ClF to give CF_3OCl.

These new hypochlorites are more stable than their alkyl analogues: CF_3OCl decomposes slowly at 150°, whereas CH_3OCl does so at room temperature. They are strong oxidizing agents, with reactions similar to those of the hypofluorites. The trifluoromethyl compound is decomposed by ultraviolet light to CF_3OOCF_3, COF_2 and Cl_2, and with N_2F_4 it gives CF_3ONF_2.

Reaction of disodium perfluoropinacolinate with chlorine or bromine in CCl_3F at low temperature gives the hypochlorite or hypobromite, while, with ICl or IBr, a hypoiodite is formed:

$$[(CF_3)_2C(O)C(CF_3)_2O]Na_2 + 2X_2 \rightarrow (CF_3)_2C(OX)C(CF_3)_2OX + 2NaX \ (X = Cl, Br)$$

These compounds decompose at a little above room temperature.

An extension of this field has involved the syntheses of SF_5OCl by reaction of SOF_4 with ClF in the presence of CsF. The O—Cl bond is readily cleaved by ultraviolet irradiation and F_5SOOSF_5 and Cl_2 result. The same point of weakness in the molecule is apparent in the reaction with N_2F_4, from which F_5SONF_2 is formed.

11.10.6. Oxides of chlorine, bromine and iodine. The two endothermic and explosive oxides, Cl_2O and Br_2O, are obtained when the halogen vapour is passed over precipitated HgO, the iodine analogue being unknown. They behave chemically as anhydrides of the respective hypohalous acids. The Cl_2O molecule has a configuration like that

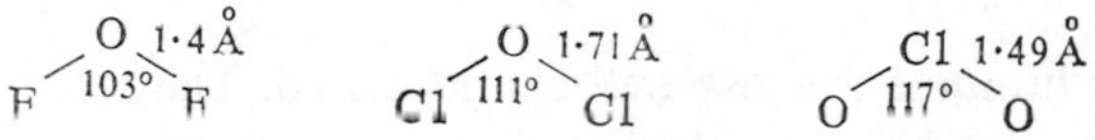

O 1·4 Å, F 103° F   O 1·71 Å, Cl 111° Cl   Cl 1·49 Å, O 117° O

of OF_2, though the bond angle is greater. Chlorine dioxide, the product of the reaction of sulphuric acid with a chlorate, or that of chlorine with silver nitrate, is likewise explosive.

$$2AgClO_3 + Cl_2 \xrightarrow{90^\circ} 2AgCl + 2ClO_2 + O_2$$

The configuration of the molecule is like that of Cl_2O, but the relative short O—Cl distance suggests some double bond character. The oxide is paramagnetic because of its odd electron, but has no detectable tendency to dimerize. With alkali it forms chlorite and chlorate (i.e. behaves as a mixed anhydride), but it is moderately soluble in water without reaction if the solution is kept in the dark; only on exposure to light is there slow decomposition to HCl and $HClO_3$. Chlorine dioxide is oxidized by $S_2O_6F_2$ (i.e. $FO_2S—O—O—SO_2F$) to chloryl fluorosulphonate which yields solvated ClO_2^+ ions in fluorosulphonic acid.

A new oxide of chlorine with the empirical formula Cl_2O_4 has recently been shown to be produced in good yield in the reaction of caesium perchlorate with chlorine fluorosulphonate at −45°:

$$CsClO_4 + ClSO_3F \rightarrow CsSO_3F + ClOClO_3$$

Its formulation as chlorine perchlorate is supported by the reaction with HCl or AgCl in which chlorine is set free:

$$MCl + ClOClO_3 \rightarrow MClO_4 + Cl_2$$

The infrared spectrum also shows features attributable to such a molecule. The oxide is a white solid (m.p. −117° ± 2°; b.p. 44·5°) which decomposes at room temperature to Cl_2O_6, Cl_2 and O_2.

Dichlorine hexoxide (Cl_2O_6) is formed in the reaction of ozone with ClO_2 and undergoes decomposition at its melting point (3·5°) to ClO_2 and O_2. In CCl_4 solution it is in equilibrium with the paramagnetic monomer ClO_3, but in the condensed phase association is believed to be complete. Reaction with organic matter may be explosive; with water perchloric and chloric acids are formed.

Dehydration of $HClO_4$ with P_2O_5 at −10° leads to Cl_2O_7, the anhydride of the acid, which is regenerated on the addition of water. This too is an explosive compound. Raman and infrared spectral studies on solid Cl_2O_7, its CCl_4 solution and on the vapour show the molecule to be made up of two ClO_3 groups joined by an oxygen bridge.

The oxides of bromine are less well characterized. The monoxide Br_2O, which has already been mentioned, decomposes above about −50°. The dioxide can be obtained by passing ozone into a solution of bromine in $CFCl_3$ at a low temperature. It decomposes violently to Br_2 and O_2 on warming, but, under controlled conditions, can give Br_2O and an unstable solid oxide which may be BrO_3 or Br_2O_7. Monomeric BrO_2 would be an odd electron molecule, but, since it gives no e.s.r. signal, it must be polymeric—possibly $O{=}Br—O—BrO_2$.

Iodine pentoxide (I_2O_5), which is produced when iodic acid is dehydrated at 240°, is also probably polymeric, as is shown by the infrared and mass spectra. A second oxide, usually formulated as I_2O_4, is formed in the partial hydrolysis of $(IO)_2SO_4$. Formerly it was thought to be iodosyl iodate ($IO^+IO_3^-$), but recent studies of the infrared spectrum indicate a polymeric structure made up of —I—O—I— chains to which IO_3 groups are covalently bonded. Reaction of ozone with iodine gives a third oxide, I_4O_9, which has been formulated as $I(IO_3)_3$, but is probably also polymeric. Iodine heptoxide, I_2O_7, an orange solid, is formed on treating periodic acid with oleum. It has not been fully characterized, but is probably polymeric.

11.10.7. Halogen oxyfluorides. In view of the similar high electronegativity of oxygen and fluorine, it is not surprising that oxyfluorides of the heavier halogens can exist. Chloryl fluoride, ClO_2F (b.p. −6°), which is formally related to ClF_5, is most conveniently prepared from bromine trifluoride and potassium chlorate:

$$6KClO_3 + 10BrF_3 \rightarrow 6KBrF_4 + 2Br_2 + 3O_2 + 6ClO_2F$$

Thermally, it is moderately stable and it forms a series of chloronium salts with fluoride ion acceptors (e.g. $ClO_2^+BF_4^-$ and $ClO_2^+SbF_6^-$). It also gives complexes with F^- ion donors (e.g. $Cs^+ClO_2F_2^-$). The bromine analogue, BrO_2F, formed from BrO_2 and F_2, is much less stable.

Fluorine perchlorate, which is mentioned earlier, is a hypofluorite, ClO_3OF, but perchloryl fluoride, ClO_3F (b.p. −46·7°), is a true oxyfluoride of chlorine (VII), and is also of great potential importance. Several preparative methods are available (e.g. electrolysis of $NaClO_4$ in anhydrous HF, and the reaction of elementary fluorine with $KClO_4$), but the best route is probably by the reaction of $KClO_4$ with HSO_3F containing a catalyst such as BF_3 or SbF_5. These are fluoride ion acceptors and the reaction mechanism may involve $ClO_3^+BF_4^-$ (or SbF_6^-) as an intermediate.

The structure of ClO_3F is very similar to that of the ClO_4^- ion. The compound does not decompose when heated in quartz until about 450°. It has a small positive free energy of formation and is very resistant to hydrolysis, with little reaction with water to 300° and only slow attack by alkali. Chemically it behaves as a strong oxidizing agent, and most combustible substances form shock-sensitive explosive mixtures with it. Its use as a high-energy fuel has been considered.

There is a remarkable reaction between ClO_3F and either liquid or aqueous ammonia. In the first instance $NaNH_2$ acts as a catalyst and the reaction is

$$ClO_3F + 3NH_3 \rightarrow NH_4F + NH_4NHClO_3$$

From the product the potassium salt K_2NClO_3 is obtained on addition of KOH. It is the potassium salt of the acid amide of perchloric acid (H_2NClO_3) and, not surprisingly, is explosive.

Perchloryl fluoride also reacts with benzene and certain substituted aromatic compounds in the presence of Friedel–Crafts catalysts to give derivatives with ClO_3 as a substituent, which confers properties rather similar to the nitro group. Further substitution in the aromatic ring is then possible; for example,

$$C_6H_6 + ClO_3F \xrightarrow{AlCl_3} C_6H_5ClO_3 \rightarrow C_6H_4(ClO_3)NO_2 \rightarrow C_6H_4(ClO_3)NH_2$$

The perchloryl aromatics are all explosive. There are also a number of instances where ClO_3F shows a specific fluorinating action; for example,

$$2ClO_3F + H_2C(COOEt)_2 + 2NaOEt \rightarrow F_2C(COOEt)_2 + 2NaClO_3 + 2EtOH$$

The bromine analogue of perchloryl fluoride, BrO_3F, has recently been made by the reaction of $KBrO_4$ with SbF_5 in anhydrous HF. Perbromates were for a long time thought to be non-existent. The first indication that this was not so came from a study of the β-decay of ^{83}Se incorporated into a selenate;

$$^{83}SeO_4^{2-} \rightarrow {}^{83}BrO_4^- + \beta^-$$

Enriched ^{82}Se was irradiated with thermal neutrons, oxidized to selenate and, after decay of the short-lived ^{83}Se, the product was found to be coprecipitated with rubidium perchlorate.

An aqueous solution of $NaBrO_3$ was later found to be oxidized to perbromate by xenon difluoride, but the salt is more readily prepared in quantity by oxidizing an alkaline solution of sodium bromate with fluorine. After removal of fluoride and bromate from the resulting solution, free perbromic acid is obtained by use of an ion exchange column. It is a strong monobasic acid, which can be concentrated to 3M, beyond which decomposition sets in. The potassium salt is stable to 280°, where it loses oxygen and gives bromate.

Iodine gives a series of oxyfluorides (IOF_3, IO_2F, IOF_5, IO_2F_3 and IO_3F), but their chemistry is at present relatively unknown. Periodyl fluoride (IO_3F), it may be noted, is formed when fluorine is passed into a solution of periodic acid in anhydrous hydrogen fluoride.

References for Chapter 11

1 Sharpe, A. G., *Quart. Rev.*, 1957, **11**, 49.
2 Sharpe, A. G., *Adv. Fluorine Chem.*, 1960, **1**, 29.
3 *Halogen Chemistry*, ed. V. Gutmann, Vol. 1, p. 1 (Academic Press, 1967).
4 Stein, L., in *Ref. 3*, p. 133.
5 Sharpe, A. G., *Quart. Rev.*, 1950, **4**, 115.
6 Musgrave, W. K. R., *Adv. Fluorine Chem.*, 1960, **1**, 1.
7 Wibenga, E. H., Havinga, E. E., and Boswijk, K. H., *Adv. Inorg. Chem. Radiochem.*, 1961, **3**, 133.
8 Popov, A. I., in *Ref. 3*, p. 225.
9 Coyle, T. D., and Stone, F. G. A., *Prog. Boron Chem.*, 1964, **1**, 83.
10 Banks, R. E., *Fluorocarbons and their Derivatives*, 2nd edn. (Macdonald, 1970).
11 Burdon, J., and Tatlow, J. C., *Adv. Fluorine Chem.*, 1960, **1**, 129.
12 Rüdorff, W., *Adv. Inorg. Chem. Radiochem.*, 1959, **1**, 223.
13 Colburn, C. B., *Adv. Fluorine Chem.*, 1963, **3**, 92; *Endeavour*, 1965, **24**, 138; *Chem. Britain*, 1966, **2**, 336.
14 Ruff, J. K., *Chem. Rev.*, 1967, **67**, 665.
15 Payne, D. S., *Quart. Rev.*, 1961, **15**, 173.
16 Kolditz, L., *Adv. Inorg. Chem. Radiochem.*, 1965, **7**, 1.
17 Schmutzler, R., *Adv. Fluorine Chem.*, 1965, **5**, 31; in *Ref. 3*, p. 31.
18 Cady, G. H., *Adv. Inorg. Chem. Radiochem.*, 1960, **2**, 105.
19 George, J. W., *Prog. Inorg. Chem.*, 1960, **2**, 72.
20 Williamson, S. M., *Prog. Inorg. Chem.*, 1966, **7**, 39.
20a Frith, W. C., *Angew. Chem.*, Intern. ed., 1962, **1**, 467.
21 Glemser, O., and Mews, R., *Adv. Inorg. Chem. Radiochem.*, 1972, **14**, 333.
22 Streng, A. G., *Chem. Rev.*, 1963, **63**, 607.
23 Turner, J. J., *Endeavour*, 1968, **27**, 42.
24 Schmeisser, M., and Brandle, K., *Adv. Inorg. Chem. Radiochem.*, 1963, **5**, 41.

12 Peroxides and peroxy-acids

12.1 Introduction

All peroxy compounds contain an O—O bond and are thus, in a sense, related to hydrogen peroxide. Metallic peroxides are of two types: the ionic compounds formed by the most electropositive elements and containing the O_2^{2-} ion (or, in the case of superoxides, the O_2^- ion) and peroxy compounds of the transition elements, lanthanides and actinides. In the latter the peroxy group is found in a variety of cationic and anionic species, occasionally as a bridging group, and also as a ligand bonded to the metal atom in a neutral molecule.

In the peroxy-acids either one or both of the hydrogen atoms of hydrogen peroxide are replaced by an acidic group, such as SO_3H in the peroxysulphuric acids. In a further class of peroxy compound, there is a covalent bond between one or both of the oxygen atoms and neutral groups, as, for example, in F_3COOCF_3 or F_5SOOSF_5.

12.2 Hydrogen peroxide

Pure hydrogen peroxide, an oily liquid with a pale blue colour, freezes at −0·9°, boils at 152°, and has a dielectric constant at 0° of 84·5, compared with 88·0 for water at the same temperature. As in the case of water, there is hydrogen bonding in the liquid phase and also in the crystals. The latter are tetragonal and contain a three-dimensional network of hydrogen-bonded molecules. The entropy of vaporization of the pure liquid is 113 J (27 cal) $mole^{-1}$ deg^{-1}. It behaves as a weak acid, the dissociation equilibrium being

$$2H_2O_2 \rightleftharpoons H_3O_2^+ + HO_2^-$$

The ionic product at 20° is $1{\cdot}55 \times 10^{-12}$ (i.e. the hydrogen ion concentration is roughly twelve times that in water). The compound is thermodynamically unstable with respect to decomposition to H_2O and O_2 ($\Delta H = -98$ kJ ($-23{\cdot}5$ kcal) mole^{-1}), and explosive decomposition may occur at the normal boiling point, though distillation at a reduced pressure is possible if special precautions are taken to exclude traces of impurities such as Fe^{3+}, which catalyse the decomposition. Both the anhydrous material and its aqueous solutions, which are the usual forms in which hydrogen peroxide is handled, are normally stabilized by the addition of substances such as urea.

In the vapour phase the molecule has been found spectroscopically to have the structure shown in Fig. 12.1. The two O—H bonds lie in planes that are at an angle of 101°, with O—O = 1·49 Å (which is almost the same as in O_2^{2-}) and O—H = 1·01 Å. This structure is con-

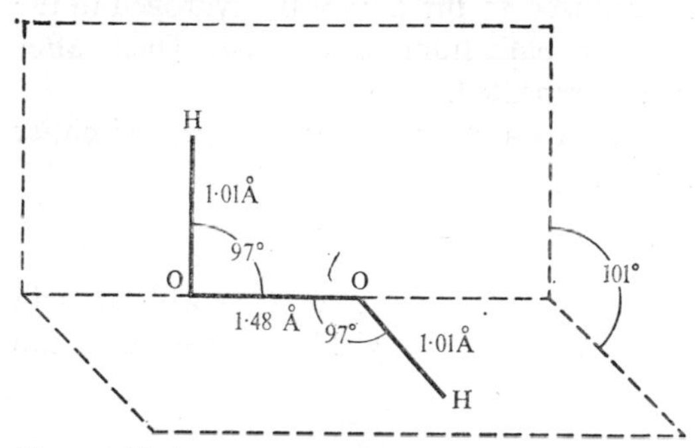

Figure 12.1

firmed by X-ray analysis of the crystalline adduct $CO(NH_2)_2.H_2O_2$, in which the urea molecule retains its normal configuration.

Hydrogen peroxide was formerly made on a technical scale by decomposing barium peroxide (BaO_2) with sulphuric acid. It is now prepared by electrolysis of ammonium sulphate solution in sulphuric acid with platinum electrodes and a high anode currect density. Hydrogen is evolved at the cathode and persulphate is formed at the anode:

$$2NH_4HSO_4 \rightarrow (NH_4)_2S_2O_8 + H_2$$

The persulphate solution is then hydrolysed by heating, to yield sulphate and hydrogen peroxide, which distils off at a reduced pressure and can be concentrated by low-pressure fractionation.

Increasing use is being made of a new process involving autoxidation, in which air is passed through a solution of 2-ethylanthraquinol in an organic solvent, oxidizing it to the quinone and hydrogen peroxide:

$$\text{(2-ethylanthraquinol: OH, OH, Et)} \xrightarrow{O_2} \text{(2-ethylanthraquinone: O, O, Et)} + H_2O_2$$

When the peroxide concentration has built up sufficiently, it is extracted with deionized water containing a stabilizer for the peroxide. After removal of small amounts of organic matter, the water solution is fractionally distilled to give high-strength hydrogen peroxide. The solvent phase containing the quinone is passed back to a hydrogenator, where it is reduced to the quinol by hydrogen in the presence of Raney nickel or palladium as catalyst. Then, after filtration, the air treatment is repeated.

Hydrogen peroxide behaves as a strong oxidizing agent in either acid or alkaline solution:

$$H_2O_2 + 2H^+ + 2e \rightleftharpoons 2H_2O \qquad E^0 = +1{\cdot}77\ V$$

Its reactions are familiar: those involving reaction with an acid halide such as chlorosulphonic acid to yield a peroxy-acid are referred to later.

12.3 Ionic peroxides[1]

Ionic peroxides are formed by the alkali and alkaline earth metals and are of three types: peroxides, superoxides and ozonides containing respectively O_2^{2-}, O_2^- and O_3^- ions. It is almost certain that peroxides of the type MO_2 formed by magnesium, zinc and cadmium are also ionic, though the classification of HgO_2 is less clear.

The two chief methods for preparing peroxides and superoxides are direct oxidation of the element with air or oxygen, and oxidation in liquid ammonia solution, the latter applying only to the alkali and alkaline earth elements, all of which are soluble. The alkali metal compounds will be described first, and here we find that lithium when heated in air or oxygen goes mainly to the normal monoxide, Li_2O, with probably a small amount of Li_2O_2. The peroxide is best made by gently heating $Li_2O_2.H_2O_2.2H_2O$, which is formed when H_2O_2 is added to LiOH solution. Sodium, when heated in air or oxygen, goes to the peroxide, Na_2O_2, the yellow colour of which is believed to be due to a trace of NaO_2; only at 400° and under an oxygen pressure of 150 atm is the superoxide, NaO_2, formed in good yield. Potassium, rubidium and caesium, on the other hand, give their superoxides by direct oxidation, and very careful control of the conditions is necessary if the reaction is to be arrested at the peroxide stage. These superoxides lose oxygen and go over to the peroxides when strongly heated. The hydroxides of rubidium and caesium also yield the superoxides when heated in oxygen.

There is some uncertainty as to the product formed when a solution of lithium in liquid ammonia is oxidized by oxygen at −30° to −80°, but it is probably mainly the peroxide. Sodium under these conditions gives the peroxide with some superoxide, but the three other alkali metals can be oxidized to either their peroxides or superoxides, with characteristic colour changes; for example,

$$\underset{\text{(blue)}}{\text{K in NH}_3} \rightarrow \underset{\text{(white)}}{K_2O_2} \rightarrow \underset{\text{(red)}}{K_2O_3} \rightarrow \underset{\text{(yellow)}}{KO_2}$$

The existence of K_2O_3, associated with the red colour, seems to be very doubtful. There is no evidence for the existence of the O_3^{2-} ion, but there may be mixed peroxide–superoxide phases. In these preparations side reactions such as the oxidation of ammonia to nitrite can also occur.

All of the peroxides are powerful oxidizing agents. For example, with moderate heating, organic compounds yield the alkali metal carbonate and metallic iron is converted to iron (VI). Sodium peroxide is used on a technical scale (e.g. in paper pulp bleaching). Hydrolysis by water gives H_2O_2, whereas hydrolysis of the superoxides gives both H_2O_2 and O_2:

$$M_2O_2 + 2H_2O = 2M^+ + 2OH^- + H_2O_2$$
$$2MO_2 + 2H_2O = 2M^+ + 2OH^- + H_2O_2 + O_2$$

The melting points of the peroxides are high (e.g. 675° for Na_2O_2), as also are the heats of formation, which are in the range 400–650 kJ (100–150 kcal) mole^{-1}. The superoxides contain an unpaired electron and are paramagnetic. Their melting points are rather lower (e.g. 380° for KO_2), as also are the heats of formation (e.g. 285 kJ (68 kcal) mole^{-1} for KO_2). Interatomic distances in the O_2^- and O_2^{2-} ions are 1·28 and 1·50 Å respectively. Sodium superoxide has a pyrites structure at low temperatures; both it and KO_2 adopt a cubic structure at higher temperatures, where the O_2^- rotates.

There is a greater variation in the structures of the alkali metal peroxides, arising from differences in the coordination of the cations, but the structures of K_2O_2, Rb_2O_2 and Cs_2O_2 are similar to the antifluorite structure, with cubic and tetrahedral coordination for the anions and cations respectively.

Ozonides of all of the alkali metals except lithium have been described, but only the potassium compound is well characterized. These compounds are prepared by treating the solid hydroxides with ozone:

$$3KOH + 2O_3 \rightarrow 2KO_3 + KOH.H_2O + \tfrac{1}{2}O_2$$

The pure compound can be extracted with liquid ammonia and forms orange-red needles. Potassium ozonide decomposes when heated to 60° to give KO_2 and O_2; there is a slow decomposition at room temperature. Reaction with water is vigorous and yields the hydroxide and oxygen. Solutions of the potassium salt in liquid ammonia are conducting, as would be expected from the ionic formulation $K^+O_3^-$. The paramagnetism of the salt corresponds with one unpaired electron per unit formula. Structural studies on potassium ozonide are not conclusive. There are similarities to the X-ray powder pattern of potassium azide, which imply that the O_3^- ion is linear. More recent work, however, has shown that the structure is probably monoclinic and isomorphous with KNO_2. This would mean that the ion, like NO_2^-, is bent, which is more probable.

Calcium peroxide octahydrate is precipitated when hydrogen peroxide is added to an alkaline solution of a calcium salt; this gives CaO_2 on careful dehydration. A small conversion of CaO to CaO_2 by heating in oxygen under pressure has been claimed. Strontium peroxide, on the other hand, can be made in an almost pure state from the monoxide and oxygen at 350–400° at 200–250 atm, and BaO_2 is obtained by heating BaO in air or oxygen at 1 atm and temperatures in the range 500–600°; at higher temperatures dissociation to the monoxide becomes appreciable. All three of these compounds crystallize with the calcium carbide structure. Their relative stabilities depend on the difference in lattice energy between MO_2 and MO, which is least for barium. Superoxides of the alkaline earth metals have not been made in a pure state, but small conversions have been reported when the peroxides are treated with 30 per cent hydrogen peroxide solution.

Magnesium peroxide can be prepared by adding hydrogen peroxide and alkali to a solution of a magnesium salt, but the products always contain water. This method likewise yields impure peroxides of zinc and cadmium, which always contain water and may contain hydroxide. Zinc peroxide containing about 80 per cent of ZnO_2 is also formed in the reaction of ethereal hydrogen peroxide with diethyl zinc:

$$2Zn(C_2H_5)_2 + 3H_2O_2 = 4C_2H_6 + 2ZnO_2 + H_2O + \tfrac{1}{2}O_2$$

Mercury peroxide (HgO_2), which exists in two crystalline forms, is produced by the action of concentrated hydrogen peroxide solution on mercury (II) oxide. The ionic radii of magnesium, zinc and cadmium are too small in relation to that of O_2^{2-} for the calcium carbide structure to be adopted, and these three compounds all have the pyrites structure. Only for one form of HgO_2 has the structure been determined. The symmetry is orthorhombic, with the mercury atoms and peroxide groups forming infinite zig-zag chains extending along the *c*-axis. This structure is probably best thought of as involving covalent bonding between the metal and oxygen, and it is likely that in ZnO_2 and CdO_2 the bonding is also largely covalent.

12.3.1. Peroxide hydrates and related compounds. A number of peroxide hydrates are known, for some of which structures have been determined. The most common preparative method is by precipitation from an alkaline solution containing the metal ion and hydrogen peroxide (e.g. by the addition of alcohol). Typical of such compounds are $Na_2O_2.8H_2O$ and $MO_2.8H_2O$ (M = Ca, Sr, Ba). In the alkaline earth octahydrates the lattice is found to be made up of metal ions, with peroxide ions and water molecules linked together into a network by hydrogen bonds. A considerable number of addition compounds between peroxides and hydrogen peroxide have also been made, and there is a substantial amount of structural evidence relating to them. While this will not be considered in detail here, it is of interest to note that hydrogen bonding again plays an important role. The compounds with one or two H_2O_2 molecules per peroxide ion have been most fully investigated and it is found that there are strong hydrogen bonds between H_2O_2 and the peroxide ions, so that a network is again built up which extends through the crystal.

12.4 Peroxy compounds of the transition elements, lanthanides and actinides[2]

Peroxy compounds of a high proportion of these elements are now known, though the literature of the subject is confused and there is a paucity of clear structural evidence. The peroxy group may be present in a cationic or anionic species and, in a few cases, in a neutral molecule. No attempt will be made to describe all of the compounds that have been reported. Instead, the main types will be illustrated by examples for which, as far as possible, some structural evidence is available.

12.4.1. Titanium, zirconium and hafnium. There are substantial resemblances between the peroxy compounds of these three elements. When a titanium (IV) solution is treated with H_2O_2 under approximately neutral conditions, addition of alcohol leads to precipitation of a yellow solid, $TiO_3.2H_2O$. This is found to have an oxidizing capacity, as measured by the liberation of iodine from an iodide, corresponding with the presence in the molecule of one peroxy group. Furthermore, in the initial solution prior to precipitation, it passes through both anionic and cationic exchange columns and is therefore uncharged. Both zirconium and hafnium form compounds of this type under comparable conditions, the most probable formulation being $M(OH)_3OOH$ ($M = Ti, Zr, Hf$), where the hydrogen atom of the OOH group would be expected to be very weakly acidic.

Addition of alkali and H_2O_2 to the above compound yields a peroxytitanate of the type $M^I_4Ti(O_2)_4$. The same compound is formed when a titanium (IV) solution is treated with H_2O_2 and alkali, and hydrates of the type $M^I_4Ti(O_2)_4.6H_2O$ may be isolated. The titanium salt decomposes readily at room temperature, with loss of oxygen, and also hydrolyses in solution to $Ti(OH)_3OOH$. Since analysis shows the presence of four peroxy groups per molecule, formulation of the anion as $\left[\mathrm{Ti}\left(\langle{}^{\mathrm{O}}_{\mathrm{O}}\right)_4\right]^{4-}$ is indicated, with each peroxy ligand contributing a charge of -2 to the complex. Salts of the tetraperoxyzirconate ion, $[Zr(O_2)_4]^{4-}$, are also known, but $Hf(OH)_3OOH$ will not dissolve in alkaline hydrogen peroxide. There are a few other anionic species which, though poorly characterized, appear to be related to the above. Among these are $[Ti(O_2)F_4]^{2-}$ and $[Ti(O_2)(C_2O_4)_2]^{2-}$. All such compounds are unstable and lose oxygen readily.

Addition of H_2O_2 to an acidified titanium (IV) solution gives an orange-red colour which is used as a sensitive analytical test both for titanium and for H_2O_2. The nature of the compound responsible for this colour has not been established, but it appears to contain one peroxy group per metal atom. The most probable structure is one that is similar to that of a titanyl derivative, with O_2 replacing O. Thus, in sulphuric acid solution, the reaction with H_2O_2 would be:

$$TiO(SO_4) + H_2O_2 \rightleftharpoons Ti(O_2)SO_4 + H_2O$$

The solid compound $Ti(O_2)SO_4.3H_2O$ has been isolated, and there are similar compounds containing the anions of other acids. The nature of the reaction of zirconyl and hafnyl salts with hydrogen peroxide in acid solutions is also very uncertain, and there is again a complete absence of structural evidence.

12.4.2. Vanadium, niobium and tantalum. Vanadium differs from titanium in forming polyacids, the degree of condensation of the anions of which is a function of pH. This is a complicating factor in considering its peroxy derivatives, but it is possible to recognize several relatively simple anionic types, some of which occur with niobium and tantalum.

Well-defined tetraperoxy compounds containing the $[V(O_2)_4]^{3-}$ anion are obtained from V_2O_5, an alkali hydroxide and H_2O_2 at below 0°. The solutions are blue and the salts (e.g. $K_3[V(O_2)_4]$) are precipitated by adding ethanol. The potassium salt is a vanadium (V) compound, and, since it is isomorphous with $K_3[Cr(O_2)_4]$, the structure of which is known from X-ray work (see Fig. 12.3), it presumably also has the same arrangement of ligands.

Three other simple types of peroxyvanadate ion occur in solution: $[HV(O)(O_2)_3]^{2-}$, $[HV(O)_2(O_2)_2]^{2-}$ and $[V(O)(O_2)]^{+}$. The first hydrolyses according to the equation below and forms the tetraperoxy ion with base and excess of hydrogen peroxide (hydrogen is present as the OH group):

$$[HV(O)(O_2)_3]^{2-}+H_2O \rightleftharpoons [HV(O)_2(O_2)_2]^{2-}+H_2O_2$$
$$[HV(O)(O_2)_3]^{2-}+H_2O_2+OH^- \rightleftharpoons [V(O_2)_4]^{3-}+2H_2O$$

Salts of the first two of these three ions have not been isolated, and the evidence for their existence in solution comes largely from spectrophotometry and cryoscopy. The third ion, $[VO(O_2)]^{+}$, which is formed on acidifying a metavanadate solution in the presence of H_2O_2, is stable in the presence of acid, the solution being red in colour. Several peroxyvanadates have been described in which there is more than one metal atom in the anion, but these will not be considered here.

For niobium and tantalum, alkali metal salts containing the $[M(O_2)_4]^{3-}$ have been made by adding ethanol to a solution containing niobium or tantalum pentoxide, hydrogen peroxide and an alkali metal hydroxide; they are reasonably stable and dissolve in water without decomposition. Acid converts them into salts containing the $[M(O)_2(O_2)]^-$ ion:

$$[M(O_2)_4]^{3-} + 2H_2O + 2H^+ \rightarrow [M(O)_2(O_2)]^- + 3H_2O_2$$

Other more complex species undoubtedly exist.

12.4.3. Chromium. When hydrogen peroxide is added to an acidified chromate solution, a deep blue colour is produced which is believed to be due to the compound $Cr(O)(O_2)_2$:

$$HCrO_4^- + 2H_2O_2 + H^+ \rightarrow Cr(O)(O_2)_2 + 3H_2O$$

The colour fades fairly rapidly because of decomposition, but the species responsible can be extracted into ether, in which the blue solution is much more stable. This is due to formation of an etherate, $R_2O.Cr(O)(O_2)_2$. The dimethylether adduct has been isolated, but explodes very readily.

Addition of pyridine and other nitrogen bases to the ether solution results in precipitation of crystalline violet solids in which the nitrogen base has displaced ether as the ligand (e.g. $C_5H_5N.Cr(O)(O_2)_2$). The structure of the pyridine adduct has been determined (Fig. 12.2). The chromium atom is at the centre of a pentagonal pyramid with an oxygen atom at the apex and two peroxy groups and the nitrogen atom of the ligand in the base. The peroxy groups are broadside-on to the metal atom with an O—O distance of 1·41 Å.

The compound is monomeric in benzene and the conductivity of a solution in dimethylformamide is very small, which shows it to be a neutral species. The magnetic properties are consistent with formulation as a chromium (VI) derivative. The presence of two peroxy groups per atom of chromium is confirmed by the reaction with neutral potassium permanganate, four equivalents of which are consumed per molecule of the complex. Compounds of this type have been made with a variety of bases; all should be handled with the greatest caution, as, under some conditions, they are explosive.

When an alkaline solution of a chromate is treated with H_2O_2,

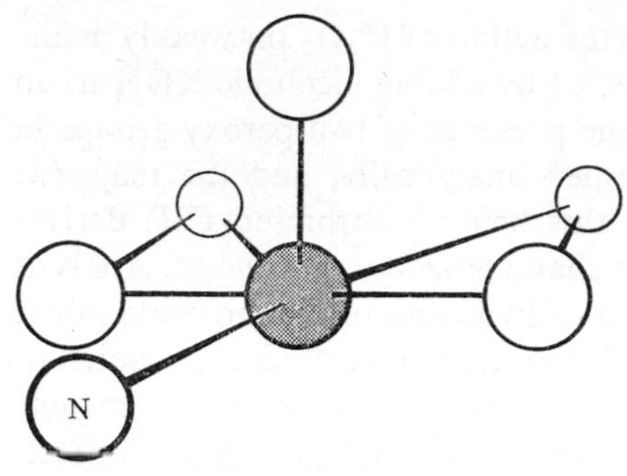

Figure 12.2

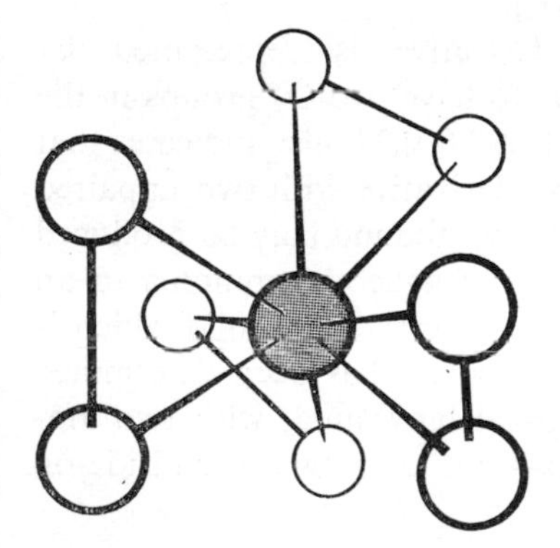

Figure 12.3

red-brown crystals of $M^I_3CrO_8$ deposit at 0°, in which there are four peroxy groups per metal atom. The paramagnetic susceptibility at 20° is 1·8 B.M., which is in keeping with the formulation $M^I_3[Cr(O_2)_4]$, with chromium in the +5 state. The salts are stable at room temperature and also for some days in aqueous solution; under neutral or alkaline conditions decomposition gives chromate and oxygen, but in acid solution the products are oxygen and chromium (III):

$$4[Cr(O_2)_4]^{3-} + 2H_2O = 4CrO_4{}^{2-} + 4OH^- + 7O_2$$
$$2[Cr(O_2)_4]^{3-} + 12H^+ = 2Cr^{3+} + 6H_2O + 5O_2$$

The structure of the ion of the potassium salt (Fig. 12.3), which is isomorphous with $K_3[V(O_2)_4]$, $K_3[Nb(O_2)_4]$ and $K_3[Ta(O_2)_4]$, shows four broadside-on peroxy groups arranged round the metal atom with an O—O distance of 1·47 Å.

A further series of compounds, the violet diperoxychromates, which are much less stable than the compounds described in the last paragraph, can be obtained by the action of H_2O_2 on weakly acidic alkali metal chromate solutions, or by adding alcoholic KOH to an ether solution of blue CrO_5. The presence of two peroxy groups in the molecule has been established analytically, and the magnetic properties are consistent with this being a chromium (VI) derivative. The instability is such that characterization is difficult. Analysis shows the presence of one atom of hydrogen in the molecule. Since the molecular conductivity of the ammonium salt in aqueous solution is about 100 ohm^{-1} cm^2, it is a uni-univalent electrolyte and the salts may be formulated as $M^I[Cr(O_2)_2(O)(OH)]$. This, however, is less well established than the formulae of the red peroxychromates of the type $M^I_3[Cr(O_2)_4]$.

The fourth type of peroxy derivative is represented by $Cr(O_2)_2.3NH_3$, which analysis shows to have two O_2 groups in the molecule. The paramagnetic moment of 2·8 B.M. also indicates that we are dealing with a chromium (IV) derivative with two unpaired electrons on the metal. This particular compound may be produced in several ways (e.g. by adding excess of aqueous ammonia to an ether solution of CrO_5). It is a red-brown crystalline solid which is only slightly soluble in water. The structure has been determined (Fig. 12.4) and is roughly a pentagonal bipyramid, with two ammonia molecules in the apical positions and two broadside-on

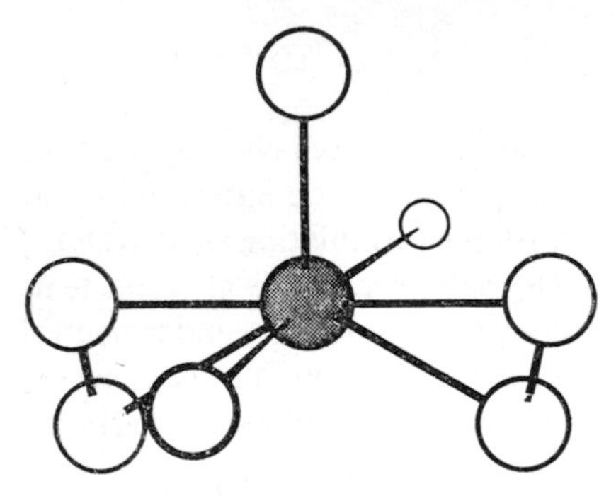

Figure 12.4

peroxy groups roughly in the equatorial plane. The O—O distance in the peroxy groups is 1·41 Å. The compound is stable at room temperature, but explodes when heated.

12.4.4. Molybdenum and tungsten. These two elements give three types of peroxy derivative, with O_2 : M ratios of 4 : 1, 2 : 1 and 1 : 1, and with the metal apparently always in the +6 state. There are also a number of less-clearly characterized peroxypolymolybdates and -tungstates.

Both molybdenum and tungsten give salts of the $[M(O_2)_4]^{2-}$ anion when excess of hydrogen peroxide is added to a neutral or slightly acid solution of a salt of the $[MO_4]^{2-}$ ion with a large cation. The molybdenum compounds are red and those of tungsten yellow. They are very unstable, exploding when heated or struck.

Acid solutions of alkali metal molybdates and tungstates, when treated with a high concentration of H_2O_2, give salts with anions of the type $[(O_2)_2M(O)OM(O)(O_2)_2]^{2-}$, where M = Mo or W. These are more stable than the tetraperoxy compounds, though still liable

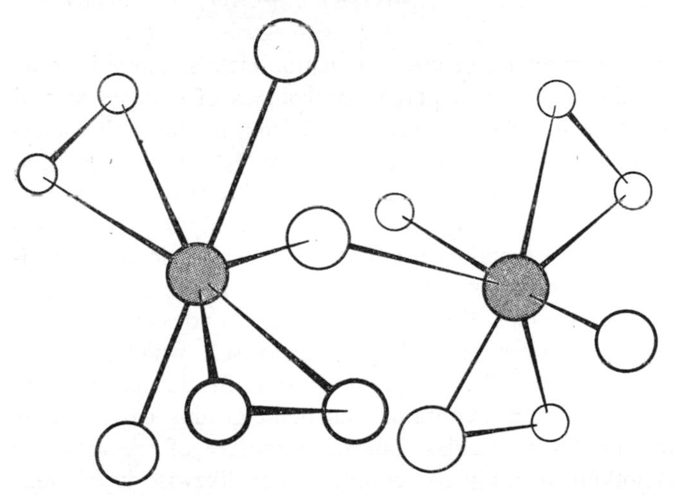

Figure 12.5

to explode when heated. The crystal structure of the tetrahydrate of the potassium salt has been determined (Fig. 12.5) and confirms the presence of an M—O—M bridge. The peroxy groups are broadside-on to the metal atom with the O—O distance equal to 1·50 Å.

The 1 : 1 peroxy species may be made by several methods, including treatment of the red 4 : 1 peroxy compounds with alkali, and decomposition of peroxymolybdates or -tungstates that are richer in peroxide. The acid $H_2[Mo(O)_3(O_2)]$ has also been prepared by dissolving molybdic acid in aqueous H_2O_2, evaporating excess water and drying at 110°. In peroxide-rich solutions it appears that there is a pH-dependent equilibrium between the 2 : 1 and the 4 : 1 ions:

$$[(O_2)_2M(O)OM(O)(O_2)_2]^{2-} + 2OH^- + 4H_2O_2 \rightleftharpoons 2[M(O_2)_4]^{2-} + 5H_2O$$

In peroxide-poor solutions at pH < 7 the 2 : 1 and 1 : 1 species are in equilibrium, while at high pH the 2 : 1 ion hydrolyses to the 4 : 1 species and molybdate or tungstate:

$$2[M(O)_3(O_2)]^{2-} + H_2O_2 \rightleftharpoons [(O_2)_2M(O)OM(O)(O_2)_2]^{2-} + H_2O$$

$$[(O_2)_2M(O)OM(O)(O_2)_2]^{2-} + 2OH^- + 4H_2O \rightleftharpoons [M(O_2)_4]^{2-} + MO_4^{2-} + 5H_2O$$

12.4.5. Peroxy compounds of other transition metals. There is some evidence for the existence of peroxy derivatives of manganese and rhenium, which will not be discussed. For iron, a poorly characterized solid produced by the action of 30 per cent H_2O_2 on a suspension of $Fe(OH)_2$ in alcohol, which has an approximately 1 : 1 ratio of peroxide to metal, is the only known derivative. Peroxy derivatives are, however, almost certainly involved in the catalytic decomposition of H_2O_2 by iron salts, and probably also in the operation of iron-containing enzymes such as peroxidase and catalase. The iron (II) oxygen-carriers in biological processes, which take up oxygen reversibly, are also of great interest in this connection. Haemoglobin itself, which contains four heme groups, can combine with up to four O_2 molecules, but the structure of the resulting complex is unknown. Only for cobalt, which likewise is of great biological importance, and iridium is there clear structural evidence of how oxygen is bonded, and the remainder of this section will be devoted to peroxy complexes of these two elements.

The best characterized peroxy derivatives of cobalt are cationic and binuclear, with a peroxy group bridging the two metal atoms.[3] Some were prepared by Werner, who established the stoichiometry and also, with great insight, proposed structures that have, in part, subsequently been confirmed by X-ray analysis. The cobalt compounds differ from the transition metal peroxides so far discussed in that a number may be synthesized with molecular oxygen, as well as by hydrogen peroxide.

The binuclear cobalt compounds are of two types, diamagnetic and paramagnetic, the paramagnetic compounds often being obtained by oxidation of the diamagnetic, in which two cobalt (III) atoms are present. Werner postulated that cobalt (III) and cobalt (IV) are present in the paramagnetic species, but the present view is that the cobalt atoms are equivalent and that it is the O_2^{2-} group that is oxidized, to O_2^-.

The preparation of these complexes is relatively simple. If, for example, air is passed through an ammoniacal solution of a cobalt (II) salt, the diamagnetic cation $[(H_3N)_5Co(O_2)Co(NH_3)_5]^{4+}$ first results, and this may then be further oxidized to the paramagnetic ion $[(H_3N)_5Co(O_2)Co(NH_3)_5]^{5+}$. The second oxidation step is brought about more effectively by ozone, hydrogen peroxide or persulphate. Similarly, a solution containing Co^{2+} and cyanide ions may be oxidized by air to the diamagnetic anion

$$[(NC)_5Co(O_2)Co(CN)_5]^{6-},$$

which hypobromite converts to the paramagnetic ion

$$[(NC)_5Co(O_2)Co(CN)_5]^{5-}.$$

We will not discuss preparative details here, but it may be noted that there are two other types of structure which are obtainable in much the same way:

(a) A dibridged structure, with one peroxy bridge, such as (a) below.
(b) A tribridged structure, with one peroxy bridge, such as (b) below. These are less common.

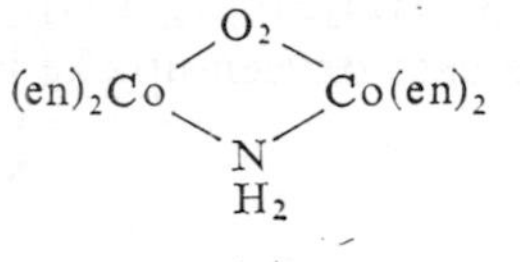

$$(en)_2Co\begin{smallmatrix} O_2 \\ \\ N H_2 \end{smallmatrix}Co(en)_2$$

(a)

$$(RH_2N)_3Co \begin{smallmatrix} O_2 \\ -OH- \\ NHR \end{smallmatrix} Co(NH_2R)_3$$

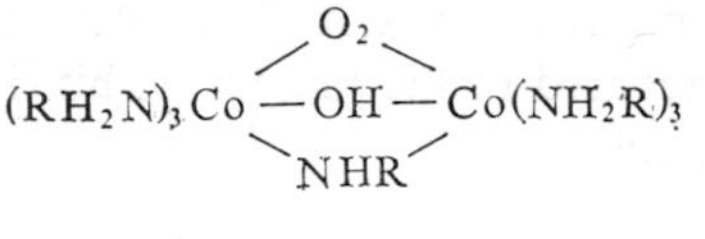

(b)

There is the usual approximately octahedral coordination in both these ions, and the main structural interest centres on the orientation of the peroxy group with respect to the metal atom and the location of the unpaired electron in the paramagnetic complexes.

The results of X-ray structural studies on two salts of the $[(H_3N)_5Co(O_2)Co(NH_3)_5]^{5+}$ cation were interpreted initially in terms of a symmetrical broadside-on orientation of the bridging group, as in Fig. 12.6. This would imply bonding involving interaction between the *d* orbitals of cobalt and the π electrons of the peroxy group, which is comparable with π bonding in platinum olefin complexes and, incidentally, very similar to the broadside-on orientation of peroxy groups in other transition metal peroxy compounds (see, for example, Figs. 12.2 and 12.3).

Later more precise studies on the nitrate and sulphate complexes show only one oxygen to be bonded to each metal atom (Fig. 12.7), the O—O axis being skewed to the Co—Co axis. The O—O distance is not much greater than that in O_2^-, the superoxide ion. The location of the odd electron in the paramagnetic species is less certain.

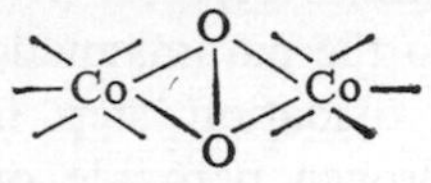

Figure 12.6

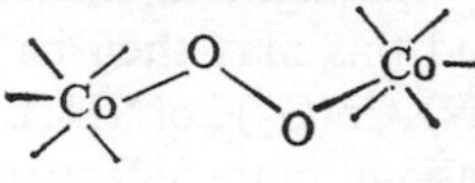

Figure 12.7

It is probably best to think of the +4 and +5 binuclear ions as derivatives of O_2^{2-} and O_2^- respectively, though the large ^{17}O hyperfine splitting in the +5 ion suggests that there is some delocalization of the unpaired electron on to the cobalt atoms. These peroxy salts all undergo slow decomposition at room temperature.

Iridium gives a well-characterized peroxy derivative, which is $(C_6H_5)_3PIr(O_2)(CO)Cl$, when $(C_6H_5)_3PIr(CO)Cl$ in benzene solution reacts reversibly with molecular oxygen. The oxygenated solid is stable at room temperature in an atmosphere of oxygen, but loses oxygen slowly in vacuum. Oxygen is also lost by the compound in solution if the pressure is reduced. It yields hydrogen peroxide when treated with aqueous acids and is monomeric in solution in an organic solvent. The analogous iodo compound, $(C_6H_5)_3PIr(O_2)(CO)I$, may be prepared in the same way, but in this case oxygen-uptake is not reversible.

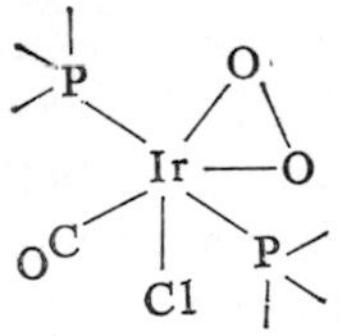

Figure 12.8

The structures of both the chloro- and the iodoperoxy complexes have been determined, and show an interesting difference in the O—O bond distance of the peroxy group, which is broadside-on in each case (Fig. 12.8). In the chloro compound it is 1·30 Å, which is close to the value for the O_2^- ion (1·28 Å), whereas in the iodo compound it is 1·51 Å. It has been suggested that the O_2 molecule donates a pair of electrons into a vacant *dsp*^{3} orbital of iridium in each case to form a π bond and that, in addition, there is back-donation from one of the filled iridium orbitals to oxygen. Back-bonding is influenced by the other ligands on the metal atom and is less with chlorine than with iodine: the former, accordingly, gives the weaker bond and oxygen-uptake is reversible.

12.5 Peroxy compounds of the main-group elements

In this section the main emphasis will be placed on peroxy-acids, which will be described on a group-by-group basis, but other types of covalently-bonded peroxide will also be referred to briefly in order to secure a more balanced presentation of the field.

12.5.1. Group III. The only elements in this group for which peroxy derivatives have been reported are boron and aluminium. No free perboric acid has been isolated, though the increase in acidity that occurs when hydrogen peroxide is added to a boric acid solution shows that a peroxy-acid may be present. In contrast, perborates are well known, those most fully studied being the alkali metal and ammonium salts. Sodium perborate ($NaBO_3.4H_2O$) is an important constituent of some commercial washing powders, where it functions as a mild bleaching agent. It may be prepared by the anodic oxidation of the borate ion in the electrolysis of a sodium borate–sodium carbonate solution. The role of carbonate is not understood, though little perborate is formed in its absence. The salt is also formed when boric acid is treated with sodium peroxide, or when hydrogen peroxide is added to a borate solution. The structure has been determined: the salt should be formulated as $Na_2[B_2(O_2)_2(OH)_4].6H_2O$, the anion containing two peroxy bridges as shown in Fig. 12.9.

This shows the only type of perborate that has been clearly identified. Other hydrates, and salts of this anion with other cations,

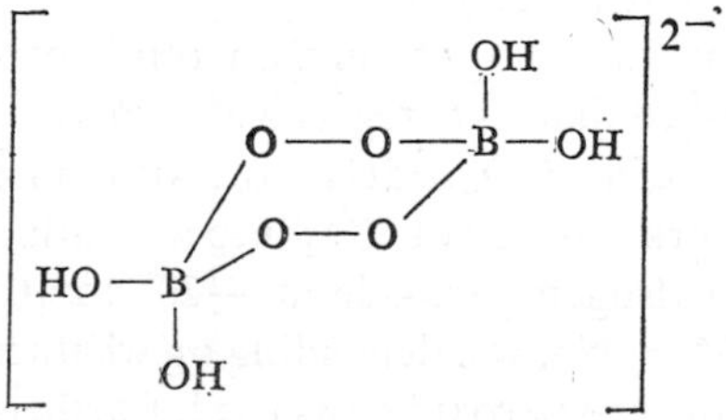

Figure 12.9

are known and there are also references in the literature to perborates with peroxide: boron ratios of 2 : 1 and 3 : 1, and to compounds with hydrogen peroxide of crystallization, but in no case is there any structural evidence.

A number of organoboron peroxides can be prepared either by nucleophilic substitution reactions or by autoxidation. Typical of the first method is the reaction of boron trichloride with alkyl hydroperoxides (see the next section):

$$BCl_3 + 3ROOH \longrightarrow B(OOR)_3 + 3HCl$$

The primary product of the autoxidation of a trialkylborane is a peroxyboronate; for example,

$$(C_2H_5)_3B \xrightarrow[30^\circ]{O_2} C_2H_5B(OOC_2H_5)_2$$

This may then rearrange to the alkoxy derivative. Aluminium forms no peroxy-acids or their derivatives, but nucleophilic substitution reactions with hydroperoxides have yielded organoaluminium peroxides, as in the case of boron.

12.5.2. Group IV. Most of the peroxy compounds of this group are derivatives of carbon. There are many types of organic peroxide, some of which are of major technical importance. These will be discussed briefly after first considering the peroxycarbonates.

Potassium peroxydicarbonate ($K_2C_2O_6$) is formed in the electrolysis of a concentrated solution of K_2CO_3 at -10°, with a high anode current density. It is described as being pale blue in colour, stable to heat up to 100–150°, and as liberating iodine immediately from a neutral potassium iodide solution in an amount equivalent to the whole of the active oxygen. There seems little doubt that the ion should be formulated as $[^-O(O)C—O—O—C(O)O^-]$, which corresponds with the formulation of peroxydisulphate. The structure has not, however, been confirmed by X-ray analysis.

There is considerable confusion in the literature on other peroxycarbonates. It has been established that CO_2 reacts with a mixture of Na_2O_2 and ice at below 0° to give $Na_2CO_4.1{\cdot}5H_2O$ or, with more CO_2, $Na_2C_2O_6$. In fairly recent work passage of CO_2 into a solution of NaOH or KOH in aqueous hydrogen peroxide at -20° to 10° has given M^IHCO_4 or $M^I_2C_2O_6$ ($M = Na, K$), depending on whether a 1 : 1 or 2 : 1 ratio of alkali to hydrogen peroxide was used. Furthermore, the peroxydicarbonate formed in this way is said to be identical with the electrolytic material. The peroxymonocarbonate is best formulated as shown in (a), where there is clearly the possibility

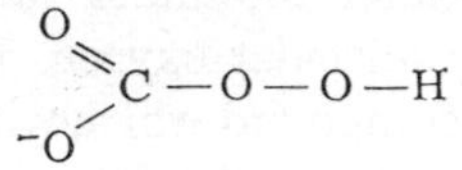

$$\underset{\text{(a)}}{\mathrm{O{=}C(O^-){-}O{-}O{-}H}} \qquad \underset{\text{(b)}}{\mathrm{^-O{-}C({=}O){-}O{-}C({=}O){-}O{-}O^-}}$$

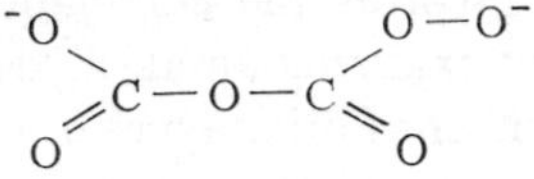

under suitable conditions of forming the salt $M^I_2CO_4$. At one time the peroxydicarbonate made from Na_2O_2 was thought to be an isomer of the electrolytic material and to have an ion with the structure shown in (b). This seems most unlikely in view of the ready conversion of M^IHCO_4 to $M^I_2C_2O_6$. The only reported difference between the electrolytic and the other material is the blue colour of the former. This may be due to an impurity, and it is of interest in this connection that $K_2C_2O_6$ prepared chemically gives a paramagnetic species when irradiated with ultraviolet light at −180°, presumably as a result of cleavage of the O—O bond.

Dialkyl peroxydicarbonates, which are esters of the hypothetical peroxydicarbonic acid, may be prepared from sodium peroxide and alkyl chlorocarbonates:

$$2ROC(O)Cl + Na_2O_2 \rightarrow ROC(O){-}O{-}O{-}C(O)OR + 2NaCl$$

Hydroperoxides may also be used in the syntheses of organic peroxycarbonates and carbonates; for example,

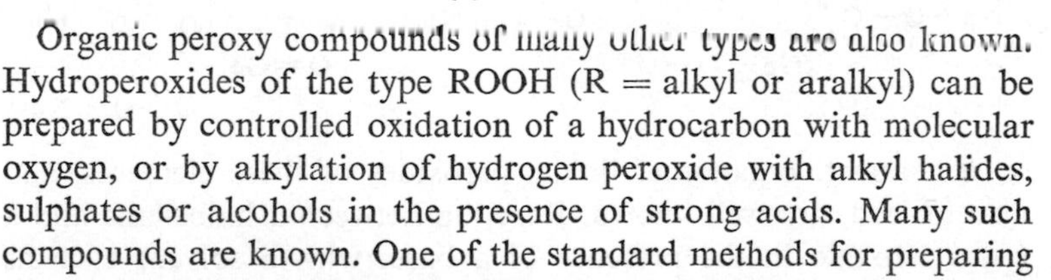

$$OCCl_2 \xrightarrow{(H_3C)_3C-O-O-H} ClC(O)OOC(CH_3)_3 \xrightarrow{(H_3C)_3C-O-O-H} OC[OOC(CH_3)_3]_2$$

$$ClC(O)OOC(CH_3)_3 \xrightarrow{RNH_2} OC(NHR)(OOC(CH_3)_3)$$

$$ClC(O)OOC(CH_3)_3 \xrightarrow{C_2H_5OH} OC(OC_2H_5)(OOC(CH_3)_3)$$

Organic peroxy compounds of many other types are also known. Hydroperoxides of the type ROOH (R = alkyl or aralkyl) can be prepared by controlled oxidation of a hydrocarbon with molecular oxygen, or by alkylation of hydrogen peroxide with alkyl halides, sulphates or alcohols in the presence of strong acids. Many such compounds are known. One of the standard methods for preparing symmetrical dialkyl peroxides is by reaction of dialkyl sulphate with hydrogen peroxide in the presence of an alkali; for example,

$$(C_2H_5)_2SO_4 + H_2O_2 \xrightarrow{NaOH} C_2H_5OOC_2H_5 + H_2SO_4$$

Organic peroxy-acids can be made from concentrated hydrogen peroxide and an organic acid in the presence of sulphuric acid, a process that involves the elimination of water or, under suitable conditions, by oxidizing an aldehyde with molecular oxygen. There are also a number of other types of peroxy compound which will not be described here. Compounds of this class are all relatively unstable and a number are dangerously explosive. The O—O bond is cleaved readily under controlled conditions by either heat or light to form free radicals, and this results in important applications of some of the organic peroxy compounds as catalysts in polymerization reactions.

12.5.3. Peroxy compounds of silicon, germanium, tin and lead. The hydroxides of these elements are weakly or very weakly acidic, and it is not surprising to find that their peroxy-acids are ill-defined. In the case of silicon, it would be expected that controlled reaction of a tetrahalide with anhydrous hydrogen peroxide would yield a peroxy derivative, which would be difficult to characterize. Such a reaction has not been reported, but in the case of germanium there is physicochemical evidence for the formation of two monoprotic acids when H_2O_2 is added to H_2GeO_3 in aqueous solution. A perstannic acid, formulated as $H_2Sn_2O_7.3H_2O$, is formed when a 30 per cent solution of H_2O_2 is added to freshly precipitated stannic acid, while sodium and potassium metastannates, similarly treated, give the salts $Na_2Sn_2O_7.3H_2O$ and $K_2Sn_2O_7.3H_2O$. Sodium and potassium metagermanates with 30 per cent hydrogen peroxide also give the sparingly soluble salts $Na_2Ge_2O_7.4H_2O$ and $K_2Ge_2O_7.4H_2O$, and a further compound Na_2GeO_5 may be isolated from the mother liquor after separating the sodium salt. The tentative structures assigned to these two compounds are shown below:

```
Na—O—O                  O—O—Na            O—O—Na
      \                /                 /
       Ge—O—Ge          O=Ge
      //       \\                        \
     O           O                        O—O—Na
```

In contrast, organometalloidal peroxides of these elements, and also of lead, are well known. Reaction between a halosilane and a hydroperoxide in the presence of a base, with a solvent such as petroleum ether, gives a range of well-defined products:

$$R_{4-n}SiX_n + nR'OOH \rightarrow R_{4-n}Si(OOR')_n + nHX$$

(R, R′ = alkyl, aralkyl, aryl; X = F, Cl, Br)

Symmetrical silyl organoperoxides are also formed in nucleophilic-substitution reactions between hydrogen peroxide and chlorosilanes in the presence of a base:

$$2R_3SiCl + H_2O_2 \xrightarrow{\text{pyridine}} R_3SiOOSiR_3 + 2HCl$$

Sodium peroxide can also be used, and polymeric species corresponding in type with the polysiloxanes, but with Si—O—O—Si groups in the structure, are known.

Very similar nucleophilic-substitution reactions take place with organogermanium and organotin halides, but these reactions will not be described here. Many of these compounds are sufficiently stable to be distilled under reduced pressure. Trialkyl and triaryl lead bromides will not react with hydroperoxides in the presence of a base, but a peroxy compound is formed in the reaction with the sodium salt of a hydroperoxide. The product decomposes very readily.

$$R_3PbBr + R'OONa \rightarrow R_3PbOOR' + NaBr$$

12.5.4. Group V. Nitrogen pentoxide reacts with anhydrous hydrogen peroxide at $-70°$, forming a mixture of pernitric acid (O_2NOOH) and nitric acid. The product, which has strong oxidizing properties, loses oxygen even at low temperatures, and neither the pure acid nor its salts have been isolated. Under favourable conditions, solutions in water or glacial acetic acid containing 70–80 per cent of the theoretical yield of acid are moderately stable, but weaker solutions hydrolyse readily.

Addition of HCl to a solution of sodium nitrite in aqueous H_2O_2 gives a brown colour, which disappears rapidly with evolution of oxygen. A more persistent yellow colour is obtained on adding sodium hydroxide to the brown solution. The brown colour is believed to be due to formation in solution of peroxynitrous acid (O=NOOH). The solution has strong oxidizing properties and there have been a number of studies of its reaction in the cold with various organic compounds. Aromatic compounds are, for example, converted to *o*-hydroxy and *m*-nitro derivatives. Such reactions, which can be explained in terms of an initial cleavage of the peroxy acid to OH and NO_2 radicals, provide indirect evidence for the formulation given above.

Two peroxyphosphoric acids are known either in the free state or as salts. The structures shown below in Figs. 12.10 and 12.11 are analogous to those of the two peroxysulphuric acids (*vide infra*), and are supported by observations on the Raman and infrared spectra of the solid salts or their solutions.

```
  OH         OH
  |          |
O=P—O—O—P=O
  |          |
  OH         OH
```

Figure 12.10

```
  OH
  |
O=P—O—O—H
  |
  OH
```

Figure 12.11

Solutions of permonophosphoric acid are obtained when phosphorus pentoxide is treated with aqueous H_2O_2. Potassium peroxydiphosphate is formed when a concentrated solution of potassium hydrogen phosphate is electrolysed at high anode current density in the presence of potassium fluoride. The salt $K_4P_2O_8$ which is isolated is readily converted to salts of other metals. In solution it is capable of oxidizing aniline to nitrosobenzene and nitrobenzene, while in strongly acid solution hydrolysis gives first H_3PO_4 and H_3PO_5 and then, from the latter, H_3PO_4 and H_2O_2. Peroxyarsenates have not been described, but peroxyantimonates result when an aqueous solution of $KSbO_3$ is treated with concentrated H_2O_2. The simplest of these salts is formulated as $KSbO_4.1{\cdot}7H_2O$, but it is non-crystalline and there is no evidence relating to its structure.

As for Groups III and IV, the organoperoxy derivatives of elements in this group are much more numerous. Chlorophosphates undergo the normal type of reaction with hydroperoxides in the presence of a base; for example,

$$(RO)_2P(O)Cl + R'OOH \rightarrow (RO)_2P(O)OOR' + HCl$$
$$(R = CH_3, C_2H_5; R' = C(CH_3)_3)$$

This is also one of several methods by which a variety of organoarsenic and organoantimony peroxides may be obtained. The compounds are of moderate though variable stability: a number may be stored at room temperature without appreciable decomposition.

12.5.5. Group VI. The two acids H_2SO_5 (permonosulphuric acid, or Caro's acid) and $H_2S_2O_8$ (perdisulphuric acid, or Marshall's acid) can be prepared in the pure crystalline state by gradual addition of the calculated amount of anhydrous H_2O_2 to $ClSO_2OH$:

$$H_2O_2 + ClSO_2OH \rightarrow H{-}O{-}O{-}SO_3H + HCl$$
$$H_2O_2 + 2ClSO_3H \rightarrow HO_3S{-}O{-}O{-}SO_3H + 2HCl$$

Their melting points are 65° and 45° respectively. Peroxydisulphates are made both in the laboratory and on a technical scale by electrolysing a well-cooled concentrated solution of potassium or ammonium sulphate in dilute sulphuric acid at a high anodic current density.

Acidified peroxydisulphate solutions hydrolyse rapidly, first to Caro's acid and then to hydrogen peroxide:

$$H_2S_2O_8 + H_2O \rightarrow H_2SO_4 + H_2SO_5$$
$$H_2SO_5 + H_2O \rightarrow H_2SO_4 + H_2O_2$$

One of the consequences of this hydrolytic reaction is that salts of Caro's acid cannot be isolated in the pure state and are normally contaminated by sulphate. Both acids, as well as their salts, are very strong oxidizing agents:

$$S_2O_8^{2-} + 2e \rightleftharpoons 2SO_4^{2-} \qquad E^0 = 2{\cdot}01\ V$$

Many metals dissolve in peroxydisulphate solutions without gas evolution and with formation of sulphates, while elements such as arsenic are converted to their oxy-acids. The configuration of the peroxydisulphate ion has been determined by X-ray analysis with several salts (Fig. 12.12).

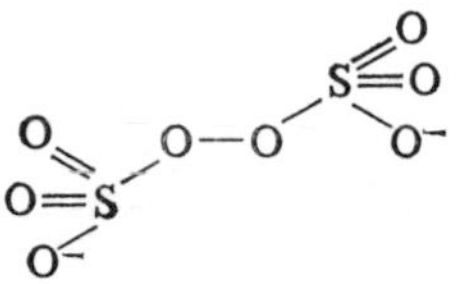

Figure 12.12

A quantitative yield of peroxymonoselenic acid (H_2SeO_5) is obtained in the reaction between chloroselenic acid ($ClSeO_2OH$) and anhydrous H_2O_2 at −10°. This decomposes, however, at >−10° and no peroxydiselenic acid is formed when a higher proportion of $ClSeO_2OH$ is used in the above reaction; nor, apparently, can the peroxydiselenates be prepared electrolytically. The reaction between anhydrous hydrogen peroxide and selenium dioxide at 0° gives the peroxyselenious acid (HOSe(O)OOH) which, though not isolated in the pure state, is stable at low temperatures: it is an isomer of selenic acid. Sulphur dioxide yields only sulphuric acid when treated with anhydrous H_2O_2, though SO_3 is converted to H_2SO_5. Selenyl chloride is only partly solvolysed by H_2O_2, the product ClSe(O)OOH being likewise unstable. Chlorotelluric acid is unknown, but treatment of potassium tellurate with 30 per cent H_2O_2

results in its quantitative conversion to the peroxytellurate, $K_2TeO_5.2H_2O$, from which the free acid is liberated on passage through an ion-exchange column. Attempts to prepare peroxy-ditellurates electrolytically have been unsuccessful.

A number of organosulphur peroxides have been reported. Benzene sulphonyl chloride, for example, reacts normally with sodium peroxide:

$$2C_6H_5SO_2Cl + Na_2O_2 \rightarrow C_6H_5S(O)_2OOS(O)_2C_6H_5 + 2NaCl$$

The peroxide $[CH_3S(O)_2O]_2$ has also been prepared electrolytically from methane sulphonic acid.

References for Chapter 12

1 Vannerberg, N. G., *Prog. Inorg. Chem.*, 1962, **4**, 125.
2 Connor, J. A., and Ebsworth, E. A. V., *Adv. Inorg. Chem. Radiochem.*, 1964, **6**, 279.
3 Sykes, A. G., and Weil, J. A., *Prog. Inorg. Chem.*, 1969, **13**, 1.

Miscellaneous topics in the chemistry of non-metals

13

13.1 Compounds of the noble gases[1–4]

Following the discovery of the noble gases in 1894–5, Ramsay and his co-workers made many unsuccessful attempts to induce them to form compounds. Subsequently others also failed, and it gradually came to be accepted that these elements were indeed chemically inert. This was seen to be in keeping with the presence in all of them of closed shells of electrons. Some of the earlier attempts to synthesize compounds did, however, come near to success. Thus, in 1933, Yost and Kaye[5] submitted a mixture of xenon and fluorine to an electrical discharge and, in the light of present knowledge, their failure to isolate a xenon fluoride must have been due only to experimental difficulties. It was also in 1933 that Pauling[6] made the remarkable prediction, based largely on considerations of ionic radii, that KrF_6, XeF_6 and xenic acid (H_4XeO_6) and its salts should be capable of existence. This, too, has been borne out in part by subsequent developments. The hydrates and other clathrates of the noble gases were discovered well before the isolation of the first of the true compounds. Their formation, however, involves quite different principles, and they will be described separately at the end of this section.

In 1962 Bartlett,[7] in the course of an investigation of the reaction of fluorine with platinum and its salts in glass or silica apparatus, observed the formation of a red solid, which proved to be the oxygenyl salt ($O_2^+PtF_6^-$), the oxygen having come from the container. It was subsequently obtained from molecular oxygen and PtF_6 at room temperature and Bartlett, realizing that this reaction involved the loss of an electron by O_2 and that the first ionization potential of O_2 (12·2 eV) was close to that of xenon (12·13 eV), investigated the reaction of the latter with PtF_6. There was again a rapid reaction at room temperature and the yellow product was formulated as $Xe^+PtF_6^-$.

Immediately following the announcement of this discovery it was confirmed at the Argonne National Laboratory, where xenon tetrafluoride was also prepared from the elements. Hoppe[8] and his co-workers in Germany also published their syntheses of the difluoride from xenon and fluorine in the electrical discharge. We will, however, abandon the historical approach in describing the many new discoveries that were crowded into the next few years in favour of a more systematic treatment.

13.1.1. Binary xenon fluorides. The three fluorides known with certainty to exist are XeF_2, XeF_4 and XeF_6. All can be made by heating suitable proportions of the two elements together in a nickel reactor, usually at 200–400° and at pressures up to 50 atm, excess of xenon favouring production of the difluoride and excess of fluorine that of the hexafluoride. A mixture always results because of the temperature- and pressure-dependent equilibrium that exists between the three compounds and their components. As a result only XeF_2 and XeF_6 can be isolated in a reasonably pure state, the tetrafluoride being always somewhat contaminated with the other two fluorides. The hexafluoride, conveniently made with an $Xe : F_2$ ratio of 1 : 20 at 300° at 50 atm, can be purified by allowing it to react with NaF, with which it forms the complex Na_2XeF_8. The other two compounds, which do not react, can be pumped away and the hexafluoride regenerated by heating the complex at 120°. The three fluorides can also be made in other ways. Irradiation of a xenon–fluorine mixture with ultraviolet light, for example, gives largely the difluoride, and other strong fluorinating agents such as O_2F_2 and CF_3OF can also replace fluorine in the reaction with xenon, though they are not as convenient.

The di-, tetra- and hexafluorides are all transparent crystalline solids. Some of their physical properties are tabulated below and it will be seen that all are volatile in vacuum.

The molecular structures of XeF_2 and XeF_4 are respectively linear and square-planar in both the vapour and the solid states. The hexafluoride probably has a distorted octahedral molecule in the vapour state, though there are uncertainties as to the structure of the solid, which exists in several crystalline forms. Bonding in these compounds is considered later.

All of the fluorides are stable at room temperature in the absence of moisture, though interconversion occurs on heating and in the presence of xenon and fluorine. All are fluorinating agents, the

Table 13.1

	XeF_2	XeF_4	XeF_6
M.p.	~140°	114°	49·5°
V.p. (mm) 25°	4·55	2·55	30
$-\Delta H_f^0$ (gas) (kJ mole^{-1})	108·5	216·0	294·5
(kcal mole^{-1})	25·9	51·5	70·4
Average Xe—F bond energy			
(kJ)	130·0	129·5	124·5
(kcal)	31·0	30·9	29·7

reactivity increasing from the di- to the hexafluoride. Thus hydrogen reacts with XeF_2 at 400°, with XeF_4 at 70–120° and with XeF_6 at room temperature. The difluoride reacts with certain olefins, converting C_2H_4, for example, to CH_2FCH_2F and CHF_2CH_3, but does not do so with perfluoropropene, which is converted by XeF_4 to perfluoropropane. The hexafluoride with perfluoropropene gives mainly CF_4 and C_2F_6. The trend in reactivity is also illustrated by the inertness of XeF_2 to NO_2, which both XeF_4 and XeF_6 convert to NO_2F. In all of these fluorinating reactions the fluoride is reduced to xenon. Differences in the reactions with water or alkali, and in the formation of complexes, are referred to below.

A 1 : 1 addition compound of XeF_2 and XeF_4 crystallizes when the two vapours are mixed, but an X-ray examination of the solid shows the two molecules to retain their identity in the solid, with no indication from interatomic distances of strong bonding between them. Formation of the octafluoride XeF_8 from the elements at 620° at 200 atm has been reported, but this observation has yet to be confirmed.

13.1.2. Xenon oxides, oxyfluorides and oxy-acids. Xenon difluoride is moderately soluble in water and initially is largely unchanged, since it can be extracted with carbon tetrachloride. The solution has strong oxidizing properties, liberating chlorine from hydrogen chloride and converting silver (I) to silver (II). There is, however, a slow hydrolysis to xenon, O_2 and HF, which occurs very rapidly in the presence of alkali. Reaction of the tetrafluoride with water is much more vigorous and is represented by the equation shown below, though $XeOF_2$ may be an intermediate:

$$3XeF_4 + 6H_2O \rightarrow XeO_3 + 2Xe + 1\tfrac{1}{2}O_2 + 12HF$$

Reaction of the hexafluoride with water is violent, the final product being XeO_3.

$$XeF_6 + 3H_2O \rightarrow XeO_3 + 6HF$$

Carefully controlled hydrolysis will, however, give $XeOF_4$, a colourless liquid which is stable at room temperature. A second, less stable intermediate, XeO_2F_2, has not been isolated as a product of the hydrolysis, though it can be made in good yield from XeO_3 and $XeOF_4$.

Xenon trioxide can be isolated by evaporating the solution resulting from the hydrolysis of either XeF_4 or XeF_6. It is a white solid which is dangerously explosive, the heat of formation ΔH_f^0 being 402 kJ (96 kcal) mole^{-1}. The structure has been determined by X-ray analysis (Fig. 13.1), and is similar to that of the isoelectronic IO_3^- ion, with the noble gas atom at the apex of a triangular pyramid and with the Xe—O distance equal to 1·76 Å and the O—Xe—O bond angle equal to 103°. Xenon oxide tetrafluoride

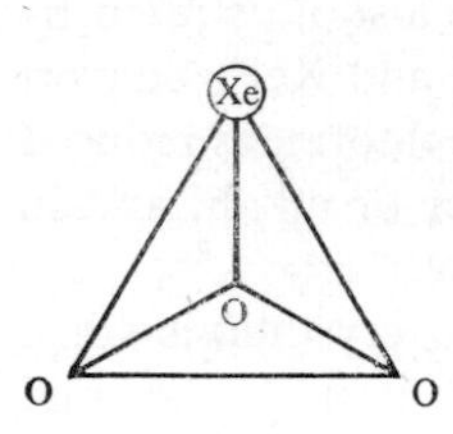

Figure 13.1

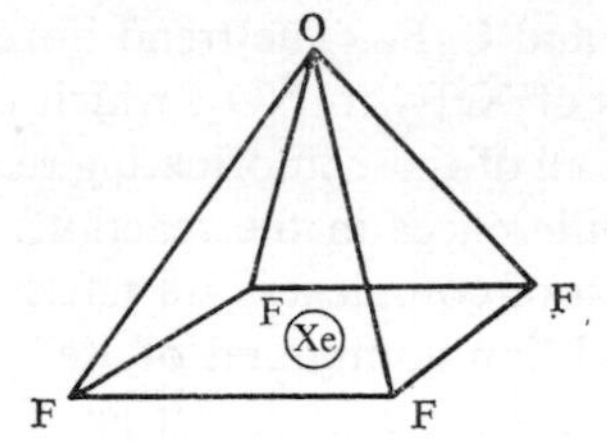

Figure 13.2

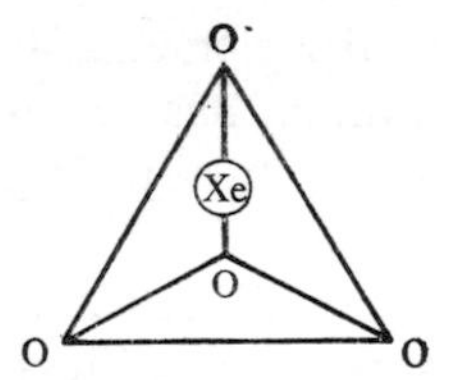

Figure 13.3

has been assigned a square pyramidal structure (Fig. 13.2) from observations of its Raman and infrared spectra, which are like that of IF_5. The structure of XeO_2F_2 is unknown. A second highly unstable and explosive xenon oxide, XeO_4, which can be made by treating barium perxenate (Ba_2XeO_6) with concentrated sulphuric acid at $-5°$, is fairly volatile, with a vapour pressure of 25 mm at 0°. The infrared spectrum shows that the molecule is probably tetrahedral (Fig. 13.3), as are the IO_4^- ion, with which the xenon compound is isoelectronic, and OsO_4.

Xenon trioxide is freely soluble in water and is the anhydride of xenic acid, which, however, is very weak. The aqueous solution, which has strong oxidizing properties, is almost non-conducting, and the Raman spectrum of a 2M solution of the oxide in water shows molecular XeO_3 to be present as a major component. On the other hand, exchange occurs with ^{18}O-enriched water, so that some interaction must be assumed. Monoalkali xenates of potassium, rubidium and caesium (M^IHXeO_4) can also be made by mixing solutions of the trioxide and alkali metal hydroxide. They are less stable than perxenates and decompose slowly at room temperature.

If potassium fluoride is added to an aqueous solution of XeO_3 containing a little HF the salt $KXeO_3F$ is left on evaporating the solution. The anion has been shown to be polymeric with XeO_3 units held together by Xe—F—Xe bridges. Caesium chloro- and bromoxenates are prepared similarly, using CsCl or CsBr. The fluoro compounds, when pure, are stable up to 200°, which is in marked contrast to other substances with Xe—O bonds; $CsXeO_3Cl$ decomposes at 150° but $CsXeO_3Br$ is very unstable. Sodium perxenate is best prepared by passing ozonized oxygen through a solution in 1M NaOH of XeO_3 prepared from XeF_4 or XeF_6, when hydrated Na_4XeO_6 separates, since it is sparingly soluble. Various other salts of the alkali and alkaline earth elements, and a few other elements, are also known; structural studies on several have confirmed the presence of the octahedral XeO_6^{4-} anion. Free perxenic acid is unknown and perxenates decompose in acid solution to xenon (VI) derivatives ($HXeO_4^-$ and XeO_3):

$$H_3XeO_6^- \rightarrow HXeO_4^- + \tfrac{1}{2}O_2 + H_2O$$

As already mentioned, the anhydride XeO_4 is formed on treating Ba_2XeO_6 with strong sulphuric acid. The perxenates are strong oxidizing agents, and are thermally stable to temperatures of 200–300°.

13.1.3. Xenon complexes. Xenon difluoride will react with fluoride ion acceptors to form complexes of three types: $2XeF_2.MF_5$ (M = As, Ru, Os, Ir, Pt), $XeF_2.MF_5$ (M = As, Ru, Os, Ir, Pt, Ta, Nb) and $XeF_2.2MF_5$ (M = Sb, Ru, Ir, Pt, Ta, Nb). Some are formed directly from the difluoride and the pentafluoride of the element in question, but in a number of cases BrF_3 is used as a solvent. In the 2 : 1 series the arsenic compound is found by single-crystal X-ray analysis to be $[Xe_2F_3]^+[AsF_6]^-$, and Raman and infrared spectra show the rest of the group to be similarly constituted. There is spectroscopic evidence that members of the second series should be formulated as $[XeF]^+[MF_6]^-$ while, in the third series, the antimony compound has been shown by X-ray analysis to be $[XeF]^+[Sb_2F_{11}]^-$, with a fluorine-bridged anion. Spectroscopic evidence again shows the other adducts of this type to be similarly constituted. The platinum metal compounds in each group are isomorphous.

It will be recalled that the first noble gas compound to be described was formulated as $XePtF_6$ by Bartlett. Reaction of xenon with PtF_6 and related hexafluorides gives products which vary in composition in the range $Xe(MF_6)_{1-2}$, and it is uncertain how these should be formulated and even if all are single substances. The difluoride forms a 1 : 1 molecular adduct with IF_5 in which X-ray work shows the separate molecules to exist (cf. $XeF_2.XeF_4$, Section 13.1.1).

Xenon tetrafluoride is reported to form an adduct with SbF_5, which is a good fluoride ion acceptor, but its nature is not fully established. It does not accept fluoride ions, as does XeF_6, and, as mentioned earlier, this difference is useful in separating XeF_6 from XeF_4. The hexafluoride forms numerous adducts with fluoride ion acceptors (e.g. $XeF_6.AsF_5$; $XeF_6.BF_3$; $XeF_6.PtF_5$), and the structures of the arsenic and platinum derivatives are known, both containing the XeF_5^+ ion, which is also present in the cubic form of solid XeF_6 ($XeF_5^+F^-$). There are also other adducts which have different stoichiometries. The alkali metal fluorides act as fluoride ion donors in forming highly reactive complexes of two types ($M^IF.XeF_6$ and $2M^IF.XeF_6$) with the hexafluoride, while NOF gives $2NOF.XeF_6$. There can be little doubt that these contain XeF_7^- or XeF_8^- ions, though their structures have not been reported. A few complexes of $XeOF_4$ with fluoride ion acceptors are also known (e.g. $XeOF_4.2SbF_5$), which are probably $XeOF_3^+$ salts, and, in addition, there are some complexes with fluoride ion donors (e.g. $CsF.XeOF_4$; $3KF.XeOF_4$), in which $XeOF_5^-$ or more complex fluorine-bridged ions are likely to be formed.

13.1.4. Other xenon compounds. Xenon dichloride has been obtained by passing a xenon–chlorine mixture containing a very large excess of xenon through a microwave discharge and condensing the product on a caesium iodide window at 20°K, so as to study the infrared spectrum. It appears to be unstable at room temperature and has not been prepared in macroscopic quantities. Both the di- and the tetrachloride, as well as $XeBr_2$, have been detected by Mössbauer spectroscopy as products of the β-decay of ^{129}I in $^{129}ICl_2^-$, $^{129}ICl_4^-$ and $^{129}IBr_2^-$.

Monosubstituted derivatives of XeF_2 are formed when xenon difluoride is treated with HSO_3F, $HClO_4$, $HOTeF_5$ or $HOC(O)CF_3$, the driving force for the reaction being

$$XeF_2 + HL \rightarrow FXeL + HF$$

provided by the highly exothermic formation of HF. All of these ligands, it will be noted, are highly electronegative. By using two molecular proportions of the anhydrous acid it is also possible to make the disubstituted derivatives, $Xe(OSO_2F)_2$, $Xe(ClO_4)_2$, $Xe(OTeF_5)_2$ and $Xe(OCOCF_3)_2$. The monosubstituted products are unstable and those of the second type even more so, though all have been analysed.

13.1.5. Krypton compounds. Krypton difluoride was first detected by means of the matrix isolation technique. A mixture of krypton, argon and fluorine held at 20°K was photolysed with ultraviolet light, and new bands in the infrared spectrum of the sample held at the low temperature were identified as due to KrF_2. Subsequently the compound, a white crystalline solid, was prepared in larger amounts by circulating a $Kr–F_2$ mixture through a low-temperature electrical discharge at −180°. It is highly reactive, and liberates oxygen from water:

$$KrF_2 + H_2O \rightarrow Kr + \tfrac{1}{2}O_2 + 2HF$$

Decomposition to krypton and fluorine occurs at room temperature, though an adduct with SbF_5 (which is probably $KrF^+ (Sb_2F_{11})^-$) is rather more stable. The value of ΔH_f^0 for the difluoride is −60·3 ± 3 kJ (−14·4 ± 0·8 kcal) mole^{-1}, compared with −108·5 kJ (−25·9 kcal) mole^{-1} for XeF_2, the average Kr—F bond energy being 49·0 kJ (11·7 kcal) compared with 130·0 kJ (31·0 kcal) for XeF_2. No other krypton compounds have been prepared.

13.1.6. Unstable species.[1] Up to this point noble gas chemistry has been discussed mainly in terms of compounds that are sufficiently stable to be isolated and characterized, but it is important to remember the existence of transient species, many of which can be detected only by spectroscopic techniques. Thus, for example, diatomic noble gas molecular ions such as Ar_2^+ and $HeAr^+$ have been observed mass spectrometrically, and there is similar evidence for species such as ArH^+ and KrH^+. The mass spectrometer also shows the transitory existence of positive ions formed with alkali metal atoms (e.g. $HeLi^+$, NeK^+, ArK^+, XeK^+). If a mixture of a noble gas with gases such as N_2, CO or CH_4 is introduced into the ionization chamber, ions such as XeN_2^+, $KrCO^+$ and $XeCH_3^+$ are observed. Without exception, however, these ions have only a very short lifetime and they cannot, therefore, be regarded as noble gas compounds in the normal sense.

13.1.7. Radon chemistry. The most abundant isotope of radon is ^{222}Rn, an α-emitter of half-life 3·82 days, which is produced in the α-decay of the naturally occurring radium isotope ^{226}Ra. It comes into radioactive equilibrium with its immediate short-lived decay products, shown below, in a few hours.

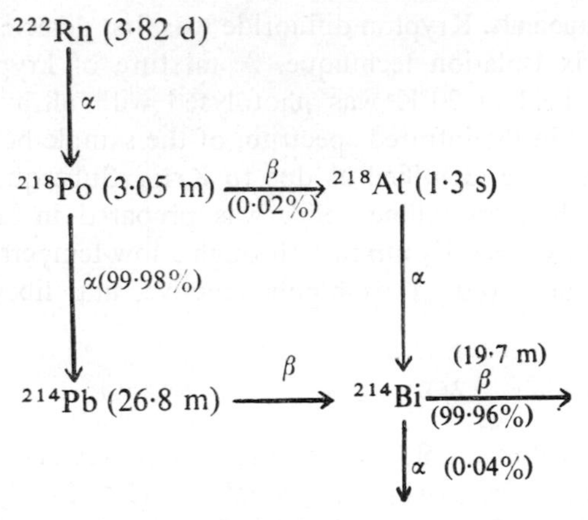

Both ^{214}Pb and ^{214}Bi emit γ-rays as well as β-rays, and the former are important in the tracer chemistry of radon since, unlike α- and β-rays, they are readily counted through the walls of metal or glass containing vessels. Radon is usually collected from a solution of a radium salt by pumping, freed from hydrogen and oxygen formed by radiolysis of water, and then sealed as gas in small containers. The boiling point is $-15°$, so that it can readily be transferred in the normal way in a vacuum line. One Curie of radon gas occupies a volume of 0·66 mm³, weighs 6·5 micrograms and produces $3{\cdot}7 \times 10^{10}$ disintegrations per second. This intense activity necessitates the use of very small quantities in chemical investigations, both because of the radiation hazard involved, and because of the decomposition of reagents and products under radioactive bombardment.

Ramsay, Rutherford, Soddy and others made a number of experiments in which they attempted to produce radon compounds, but all were negative and it became generally accepted that the element shared the chemical inertness of the other noble gases. The trend in stability of krypton and xenon compounds, however, leads to the expectation that compounds should be formed by radon. This was first shown to be so in 1962 when 5·1 microcuries of ^{222}Rn were sealed in a metal vessel with fluorine and heated to 400°. The radon was 'fixed' in an involatile form, presumed to be a fluoride. This experiment was repeated later with several Curies of radon, and the involatile product was shown to be decomposed by water (as is XeF_2) with liberation of the noble gas, and also to be reduced by hydrogen at 500°. Since the first ionization potential of radon (10·75 eV) is lower than that of Xe (12·13 eV), fluoride formation fits into the general pattern.

It might be expected that radon would also form compounds with oxygen, chlorine and perhaps other elements, and that these would be more stable than the xenon analogues, where known. These, however, have not yet been found. Radon reacts at room temperature with ClF, ClF_3, ClF_5, BrF_3, BrF_5 or IF_7 and the product remains in solution. Electromigration studies in either BrF_3 or HF—BrF_3 solution show the radon to migrate to the cathode, and to remain in the cathode solution, probably because it is reoxidized by the halogen fluoride as the ion is discharged at the electrode. These experiments prove that, although radon fluoride is unlikely ever to be analysed, it is a salt, most probably the difluoride. It is possible that the solutions of radon fluoride in halogen fluorides will find applications in the future in radiotherapy where, at present, radon itself in sealed containers is employed. A further potential application is in removing radon from the atmosphere of underground radium mines, where it is at present a health hazard, by using a circulating system to scrub the air with a halogen fluoride.

13.1.8. Bonding in noble gas compounds.[1] A full theoretical discussion of bonding in the noble gas compounds will be found elsewhere, and we will restrict ourselves here to a statement of the nature of the special problems presented and the general lines along which a solution has been sought. The first ionization potential of xenon is 12·13 eV, but for the gases above it the value becomes progressively greater (Kr, 13·99; Ar, 15·76; Ne, 21·56; He, 24·58 eV), so that formation of a unipositive ion becomes less probable. Since all of these gases have the electronic configuration ns^2 or ns^2np^6, formation of a covalent bond will require promotion of one or more electrons to a higher orbital. The energy required to do this is already high in the case of xenon, and even higher for the gases above it. These two factors both indicate a decreasing chance of compound formation in going from radon to helium.

Experimentally it is found that, with a few exceptions, only fluorine or oxygen can be bonded to xenon. Whether the resulting compound is exothermic or endothermic depends on whether the average energy of the bonds formed is greater or less than the energy needed to form an atom of the halogen or of oxygen. Largely because of the low bond dissociation energy of fluorine, it turns out that the xenon fluorides should be somewhat exothermic, and the chlorides and oxides endothermic, as they are. That such thermodynamically unstable compounds can be isolated must therefore be due to kinetic factors. Groups such as SO_3F are also highly electronegative, which is why they too can be bonded to xenon. Extrapolation of bond energy values for halides and oxides across the main groups of the periodic table also gives bond energy values for the noble gases which lead us to expect compound formation only by the heaviest members with fluorine and oxygen.

Bonding in the noble gas compounds has so far been considered only in relation to the fluorides, and special emphasis has been placed on whether a particular theoretical treatment can predict the geometry of the fluoride molecules correctly. One approach, the valence shell electron pair repulsion theory, considers the molecules as having localized electron pair bonds, the orientation of which is controlled by the interactions of the remaining electron pairs in the outer shell. Applying this to XeF_2, where there are ten electrons involved (eight from the xenon atom and two from the two fluorine atoms), it would be expected that the mutual repulsion of the three unshared pairs would force them into the equatorial plane of a trigonal bipyramid (Fig. 13.4), leaving a linear arrangement of the F—Xe—F bonds; this is in fact observed.

In the tetrafluoride, with twelve electrons, two pairs are not used in bonding and repulsive forces should lead to these occupying the *trans* positions in an octahedral arrangement (Fig. 13.5); this is the observed configuration. As has been pointed out, however, these configurations imply sp^3d^2 hybridization for XeF_4 and sp^3d or spd^3

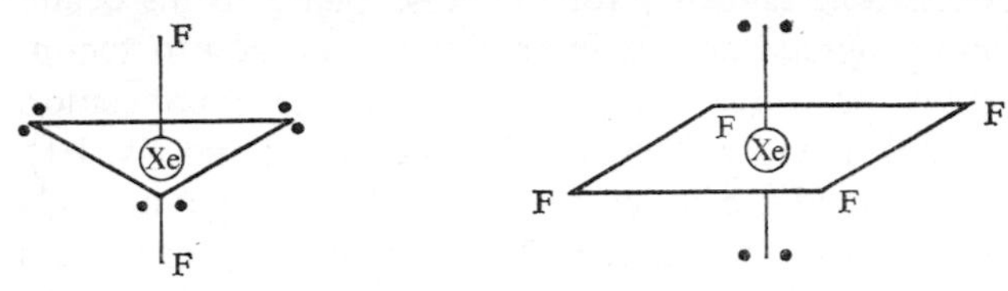

Figure 13.4 *Figure 13.5*

hybridization for XeF_2, and in each case the energy required to promote electrons to *d* orbitals is excessive, e.g. $5s^25p^6 \rightarrow 5s^25p^55d$ is approximately 10 eV.

In XeF_6 it is difficult to see how fourteen electrons can be reconciled with octahedral symmetry, and it is in this connexion that it is important to know the detailed structure of the hexafluoride. Any form of hybridization would, however, again raise the problem of the high energy needed to promote electrons.

In the main alternative theory, which is based on the use of molecular orbitals, only *p* orbitals are involved. For XeF_2, for example, a linear, three-centre, four-electron bond is postulated to cover the F—Xe—F structure. In XeF_4 and XeF_6 there are respectively two and three such three-centre bonds, and the molecules are expected to have square-planar and octahedral configurations, the Xe—F bonds having some ionic character. This approach to the problem removes the difficulty of *d* orbitals and, in the detailed treatment, also disposes satisfactorily of the electrons not involved in bonding.

13.1.9. Clathrates of the noble gases.[1, 9, 10] Clathrate compounds were so named in 1947 by Powell. They are solids in which small atoms or molecules (e.g. the noble gases, SO_2, H_2S, CH_4, HCl and CH_3OH) are trapped in relatively large cavities of the honeycomb structure formed by the crystallization of certain substances, including quinol and water. We will limit the discussion given here to clathrate compounds formed by the noble gases, noting that they illustrate all of the principles involved.

So-called hydrates of argon, krypton and xenon were first prepared in 1897–8 shortly after the elements themselves had been discovered. They are formed when crystallization of water is induced by cooling in the presence of the gas under pressure. The idealized composition is (N.G.).$6H_2O$, which corresponds with the filling of all available cavities in the host lattice, but the gas content is normally lower. There is a considerable tendency for the solid phase to dissociate and this becomes progressively less in passing from argon to xenon. Thus dissociation pressures for materials of the composition (N.G.)$5{\cdot}75H_2O$ at 0° are: Ar, 95·5 atm; Kr, 14·5 atm; and Xe, 1·15 atm. Heats of formation have been found for the last two compounds and are 58·2 and 69·9 kJ (13·9 and 16·7 kcal) mole^{-1} from the gas and ice.

Water forms clathrates not only with the noble gases but with a wide range of other molecules. Chlorine hydrate, $Cl_2(\sim 6H_2O)$, is a compound of this type, but other molecules (e.g. SO_2, N_2O, CH_3Cl and C_2H_2) behave similarly. One of the distinctive features common to all these hydrates is that water does not crystallize in the normal way. X-ray studies show that two structures can occur, each with cavities of two different sizes, and with diameters in the range 5·0–6·7 Å. Since the diameters of the atoms are: Ar, 3·4; Kr, 3·6; and Xe, 4·0 Å, it is understandable that these gases can be accommodated in the host lattice.

The quinol clathrates are very similar to the water clathrates both in their mode of formation and in their structures. A solution of quinol in water or benzene held under a noble gas pressure of 20–60 atm deposits on slow cooling a solid phase with the idealized composition $[C_6H_4(OH)_2]_3$(N.G.), in which the host lattice is formed by quinol molecules held together by hydrogen bonds. The structure adopted is different from that in the absence of the noble gas, and has cavities of the appropriate size. The compounds are always non-stoichiometric (i.e. all the cavities are not occupied). There is a considerable pressure of noble gas in equilibrium with the solid phase—3·4 atm at 2·5° in the case of the argon compound and less for the heavier gases. Clathrates of quinol are also formed from the molten material but so far there are no reports of their existence for helium and neon. Some very limited uses have been found for krypton clathrates prepared with the radioactive isotope ^{85}Kr.

13.2 Pseudohalogens

The term pseudohalogen was first applied by Birckenbach and Kellerman to certain univalent electronegative inorganic radicals which show a resemblance to the halogens in their chemical properties. The principal radicals coming within this category are cyanide, cyanate, thiocyanate, azidothiocarbonate, selenocyanate, tellurocyanate and azide. Several of these have been isolated in their dimeric forms, and all are known in the form of numerous derivatives.

The chemistry of these pseudohalogens is not, for the most part, of recent date. Cyanogen, for example, was first prepared in 1815 by Gay Lussac by the action of heat on mercuric or silver cyanide, and throughout the nineteenth century there was a sustained interest in this and related substances. The general resemblance to the halogens is evident from the combination of the pseudohalogens to form monobasic acids (e.g. HCN and HN_3), the formation of compounds similar in type to those formed by the halogens (e.g. $CO(N_3)_2$), and the formation of compounds with the halogens and among themselves (e.g. CNCl and CNSCN). It is now generally realized that there are a number of other electronegative radicals which satisfy some, if not all, of these criteria for classification as pseudohalogens. This is the case, for example, with the radicals NF_2, OSO_2F and SCF_3, and also with perfluoroalkyl or perfluoroaryl radicals, up to a point. We will, however, deal in this section only with the classical pseudohalogens, since many of the other species are described elsewhere in this book.

13.2.1. Cyanogen. This is the most readily accessible of the pseudohalogens, its general reactions being so well known as to need no detailed mention here. It is an inflammable gas (b.p. −25°) which is made on the laboratory scale by heating $Hg(CN)_2$, or by the reaction of the Cu^{2+} ion with cyanide in aqueous solution ($2Cu^{2+} + 4CN^- \rightarrow 2CuCN + C_2N_2$). It is also formed in the gas-phase oxidation of HCN (e.g. by NO_2). Polymerization of cyanogen to so-called paracyanogen occurs in the pure gas above about 600°, but is promoted by impurities and may take place at much lower temperatures. Basic hydrolysis gives cyanide and cyanate and resembles that of chlorine to chloride and hypochlorite, though secondary reactions lead to a range of other products (oxamide, ammonium oxalate, ammonium formate and urea). A number of the reactions of cyanogen involve addition across one or both of the multiple bonds in its molecule [N≡C—C≡N]. Thus H_2S gives NC—C(S)—NH_2 and H_2N—C(S)—C(S)NH_2, while aqueous hydroxylamine gives H_2N—C(=NOH)C(=NOH)—NH_2.

Hydrogen cyanide (b.p. 25·6°) is an associated liquid with a high dielectric constant which, in the anhydrous state, behaves as a protonic solvent system (see Chapter 7). In aqueous solution it is a very weak acid ($K_a = 2{\cdot}1 \times 10^{-9}$) and is therefore set free when soluble cyanides are treated with mineral acids, this being the normal laboratory preparative method. It is, however, an important industrial chemical and is manufactured by the catalytic oxidation of a mixture of ammonia and methane:

$$2CH_4 + 3O_2 + 2NH_3 \xrightarrow[>800^\circ]{} 2HCN + 6H_2O$$

The pure material polymerizes readily and is normally stored and transported with added stabilizers.

Cyanides of non-metals as well as of metals are well known and show a general resemblance to the corresponding halides: thus $Si(CN)_4$ and $B(CN)_3$ are readily hydrolysed. The cyanides of silver (I), lead (II), mercury (I) and copper (I) are also insoluble, as are the chlorides. The cyanides of sodium, potassium and rubidium crystallize at room temperature with the sodium chloride structure, while CsCN and TlCN have the caesium chloride structure. In all of these, the CN^- ion behaves as if it is freely rotating, with an effective radius of 1·92 Å, but there is a low-temperature modification of lower crystal symmetry in which the axis of CN^- has a fixed orientation, this behaviour being similar to that of the $C_2{}^{2-}$ ion in CaC_2, which also has a high- and low-temperature form. Lithium cyanide has a single orthorhombic form in which there is considerable polarization arising from the smaller size of the cation.

13.2.2. Oxycyanogen. Though the cyanates are very familiar, there has been no report so far of the isolation of the free pseudohalogen $(OCN)_2$. Electrolysis of a solution of potassium cyanate in methanol gives an anode solution which will liberate iodine from potassium iodide, and also dissolve copper, zinc or iron without gas evolution, and which therefore presumably contains $(OCN)_2$. The anhydrous acid HNCO (b.p. 23·5°) may be made by the thermal depolymerization of cyanuric acid, to which it reverts spontaneously; it gives a strongly acidic solution in water. There are also two series of esters corresponding with the two possible forms of the acid, HOCN and HNCO, though the free acid is in the iso form, HNCO, with a linear anion.

13.2.3. Thiocyanogen. This substance was first prepared in a pure state by the reaction of iodine or, better, bromine with an ethereal suspension of silver or lead thiocyanate. Evaporation of the solvent after filtering the silver halide gives the free pseudohalogen (m.p. 2–3°), which polymerizes irreversibly at room temperature to a brick-red amorphous solid. It has also been prepared by electrolysing an alcoholic solution of ammonium or potassium thiocyanate. Water decomposes it to a mixture of thiocyanic and hydrocyanic acids.

The free acid, which can be prepared by heating the potassium salt with $KHSO_4$, is in the iso form, HNCS, in the vapour state, and microwave spectroscopy shows the NCS group to be linear. There are, however, two well-defined series of esters. There are donor functions associated with both nitrogen and sulphur, and it is found that thiocyanate can accordingly act as a bridging group in certain complexes; for example,

```
Pr3P        S—C—N        Cl
    \      /      \      /
      Pt            Pt
    /      \      /      \
Cl          N—C—S        PPr3
```

In solid AgSCN there are also non-linear chains of silver atoms linked by SCN groups. Among the chemical reactions which suggest a similarity to the halogens are additions to olefins, ethylene, for example, giving $C_2H_4(SCN)_2$.

13.2.4. Selenocyanogen and tellurocyanogen. Selenocyanogen has been prepared by electrolysing a solution of potassium selenocyanate in methanol, or by thermal decomposition of lead tetraselenocyanate. A purer product is obtained by treating excess of an ether suspension of the silver salt with iodine:

$$2AgSeCN + I_2 \rightarrow 2AgI + (SeCN)_2$$

It is a yellow crystalline powder that is stable when dry and is soluble in C_6H_6, $CHCl_3$ or CCl_4. Hydrolysis occurs readily and is represented by the equation

$$2(SeCN)_2 + 3H_2O \rightarrow H_2SeO_3 + 3HSeCN + HCN$$

Tellurocyanogen has not yet been prepared in the pure state, though there are indications that it is produced in the electrolysis of a solution of potassium tellurocyanate in methanol.

13.2.5. Azidocarbondisulphide. $(SCSN_3)_2$ is an explosive white crystalline solid. The potassium salt, $KSCSN_3$, is formed in the reaction of a solution of potassium azide and carbon disulphide at 40°, the free pseudohalogen being then isolated by adding an oxidizing agent such as $FeCl_3$ or H_2O_2. It decomposes spontaneously at room temperature, probably according to the equation

$$(SCSN_3)_2 \rightarrow 2N_2 + 2S + (SCN)_2$$

With dilute aqueous alkali at −10° there is a reaction which is analogous to that between chlorine and cold alkali; for example,

$$(SCSN_3)_2 + 2KOH \rightarrow KSCSN_3 + KOSCSN_3 + H_2O$$

There is also evidence that $KOSCSN_3$ tends to decompose to form a new compound which is analogous to a chlorate:

$$3KOSCSN_3 \rightarrow 2KSCSN_3 + KO_3SCSN_3$$

13.2.6. Azides. The azide radical is unknown in the free state or as a dimer, but forms a number of derivatives which resemble those of the halogens. Sodium azide is prepared by reaction of N_2O with $NaNH_2$ at 190°; water produced gives some secondary decomposition as shown:

$$NaNH_2 + N_2O \rightarrow NaN_3 + H_2O$$
$$NaNH_2 + H_2O \rightarrow NaOH + NH_3$$

The free acid is obtained by distilling the sodium salt with 1 : 1 sulphuric acid and dehydrating the distillate with $CaCl_2$. It is endothermic ($\Delta H_f^0 = +294$ kJ (+70·3 kcal) mole^{-1}) and dangerously explosive. The boiling point of the anhydrous material is 37° and in water its acid strength is somewhat greater than that of acetic acid. The metallic azides have solubilities in water that parallel those of the halides: silver (I), lead (II), mercury (I), copper (I) and thallium (I) azides are insoluble. Azides of a number of heavy metals will explode on shock, and lead azide is used as a detonator filling. The alkali metal salts, on the other hand, are relatively stable. The azide radical can enter into the structure of basic salts (e.g. $Cu(OH)N_3$) or of complexes. In the latter it may occur as the azide of a complex ion (e.g. $[Cu(NH_3)_4](N_3)_2$), or as a group within the complex itself (e.g. $[Cu(NH_3)_2(N_3)_2]$, $K_2[Cu(N_3)_4]$).

In ionic crystals the ion is linear and symmetrical, with an N—N distance of 1·15 ± 0·02 Å, the ammonium salt being isomorphous with $NH_4[F—H—F]$. In covalent azides and in hydrazoic acid itself, however, the N_3 group is unsymmetrical. The molecular dimensions of CH_3N_3 and of HN_3 as determined by electron diffraction are shown below.

H_3C—(1·47 Å)—N—(1·24 Å)—N—(1·01 Å)—N (angle 120°)

N—(1·14 Å)—N—(1·25 Å)—N—(1·01 Å)—H (angle 116°)

13.2.7. Polyhalogenoids and related compounds. Various compounds are known in which pseudohalogens and halogens combine together. Examples are the halogen azides XN_3 (X = F, Cl, Br, I) and the cyanogen halides CNX (X = F, Cl, Br, I), which give cyclic trimers readily. What may be termed mixed pseudohalogens are also known (e.g. CNN_3, CNSCN and $ClSCSN_3$). The analogy between the halogens and pseudohalogens is also strengthened by the existence of polyhalide-like anions (e.g. $NH_4(SCN)_3$) and also of mixed polyhalides such as $CsI(CN)_2$.

13.3 Graphitic compounds[11–13]

Graphite has a layer-lattice type of structure consisting of sheets of carbon atoms linked in hexagonal array as in condensed aromatic skeletons, such as pyrene, so that each sheet constitutes a giant aromatic molecule (Fig. 13.6). Within the layer each carbon atom is bonded to three others, the C—C distance being 1·41 Å, which is

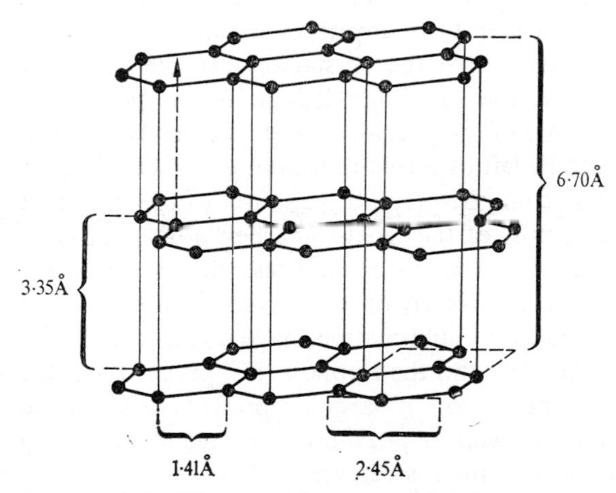

Figure 13.6 The graphite crystal lattice

slightly greater than in aromatic ring systems; the bond order is 1·33. The distance between the superposed carbon sheets is 3·35 Å. Within the sheets the carbon atoms form an sp^2 trigonal set of bonds. The remaining $2p_z$ orbitals, which are occupied by one electron per carbon atom, combine to give highly delocalized π-type aromatic orbitals running throughout each sheet. This extreme delocalization of the π-electrons is responsible for the highly anisotropic diamagnetism and for the electrical conductivity of graphite. The long distance between the sheets indicates that there is negligible orbital overlap and that interactions between the graphite layers are little more than van der Waals forces.

The structure thus accounts for the useful properties of graphite, its quasi-metallic electrical conductivity and its ready cleavage in a direction parallel to the layer planes. It is this second consequence of the graphite structure that is responsible for the useful lubricating properties of graphitic materials. The peculiarities of the structure are also reflected in the chemical properties, and lead to the formation of compounds in which atoms or molecules are introduced between the sheets. These are the so-called *intercalation* or *lamellar compounds*, which are discussed in this section.

All forms of carbon, apart from diamond, have been found to be graphitic in character. Variations in properties, and the apparently amorphous character of charcoals, arise from the varying size of unit crystallites, and the differing extents to which they are ordered within the particle. In the finest lamp black, or in norite, the crystallites may be only of the order of 40–50 Å across and 10 Å thick (i.e. each containing only two or three layers of carbon sheets a few hundred rings in extent). When the crystallites become extremely small, as in these ‘amorphous’ carbons, it is no longer possible to ignore atoms on the edges of the graphitic sheets. Each has only two neighbours and would be left as a free radical in an absolutely pure carbon. In fact, arising from the preparative source of the carbon or from exposure to air, most of the reactive carbon atoms are combined with hydrogen, oxygen or nitrogen. These chemically bound ‘impurities’, which are incompletely expelled even when the carbon is raised to its temperature of sublimation, have a considerable influence on both the chemical and the technological properties of the material. The so-called activation of carbon represents an empirical control of their nature, as well as providing a means of modifying the specific surface area and microstructure.

Graphite itself exists in two modifications, a hexagonal and a rhombohedral form. In the first, alternate layers are so displaced relative to each other that every second layer plane is superposable, as shown in Fig. 13.6. In the second modification every third layer is superposable, so that the types of stacking can be represented as (ABAB.....) and (ABCABC.....). The rhombohedral modification is often found in natural graphite. There is, however, no indication that the two forms differ in chemical reactivity or in their ability to form intercalation compounds.

The chemistry of graphitic compounds, the essential feature of which is that the carbon sheets shall remain intact, may be considered under four headings:

(a) Processes involving gross oxidation, in which new covalencies are formed, the sp^2 hybridization of carbon atoms within the sheets giving place to sp^3 hybridization. Concomitantly, the carbon sheet loses its aromatic character, becoming puckered rather than planar, as it is in graphite, and the electrical conductivity either disappears or is greatly reduced. Only two graphitic compounds of this sort are known. The first, carbon monofluoride, is described elsewhere (Section 11.6.1). Fluorine atoms, covalently bonded to carbon, are situated between the sheets. In the second, graphitic oxide, which is discussed below, oxygen atoms or oxygen-containing groups are σ bonded to carbon and again lie between the sheets.

(b) Alkali metal atoms may penetrate between the graphite sheets, to which they transfer electrons.

(c) The graphite sheets may act as electron donors, transferring electrons to certain acidic groups (e.g. HSO_4^- or ClO_4^-) which lie between the sheets. These are referred to as graphite salts.

(d) Molecules of metallic halides, etc., and related substances may be intercalated between the sheets, with only partial charge transfer.

13.3.1. Graphitic oxide. This is prepared by oxidizing graphite with potassium chlorate and a mixture of concentrated sulphuric and nitric acids. The colour changes gradually from black to brown or yellow and the product is usually freed from tenaciously held acid by washing with water. It retains the outward crystalline form of the original graphite but is found to have undergone considerable swelling. If all of the valencies of carbon were satisfied by bonding to oxygen the final C : O ratio for the material would be 2, but, in practice, well-oxidized samples give a value of 2·7 to 2·8 and the characteristic properties are in evidence when this figure is as high as 4. Furthermore, graphitic oxide also contains hydrogen in the form of OH and other groups in addition to water, so that its chemical nature is by no means simple.

Graphitic oxide is thermally unstable and deflagrates at about 200°, with formation of CO, CO_2, H_2O and carbon in the form of soot. X-ray analysis does not give complete structural information, but the observed reflexions enable the interplanar distance to be determined, and this is found to be about 6·0 Å in well-dried samples but to increase to at least 11 Å when the material is exposed to water. It is not possible to ascertain by X-ray analysis whether the layers are planar or not, but, if tetrahedral bonding of carbon is assumed, puckering of the rings must occur. As in the case of carbon monofluoride, the electrical conductivity of graphite is almost completely lost on its conversion to graphitic oxide: the electrical resistance of a pelleted sample of the dry material is greater than that of unreacted graphite by a factor of about 10^5–10^7, which implies the loss of aromatic character in the sheets.

The manner in which hydrogen and oxygen are combined with carbon is not yet fully established. Bands in the infrared spectrum which can be assigned to carbonyl, hydroxyl and perhaps also epoxy groups are observed, but it is not known if the carbonyl frequencies are associated with carboxyl groups or not. Chemical evidence favours the presence of OH and COOH groups. The former can be methylated with diazomethane, though the process is incomplete, and carboxyl groups can be esterified with methyl alcohol and hydrochloric acid. Acidic hydrogen can also be determined in an ion-exchange reaction with sodium ethylate. These roughly quantitative assays, however, leave a considerable part of the oxygen unaccounted for and the presence of ether linkages has been suggested, these being thought of as involving oxygen bonded to two carbon atoms which occupy *meta* positions in the same ring. That structural information is so imprecise appears to be an inherent difficulty with this particular material.

A number of other interesting observations on the behaviour of graphitic oxide have been made. The reversible uptake of water has been referred to already, but in alkaline media swelling proceeds even further, and the graphite lattice is broken up irreversibly to give a viscous suspension which is colloidal in nature. Acetone, dioxane, ether and various alcohols cause reversible swelling comparable with that observed with water, but benzene and other hydrocarbons produce either little or no swelling. There have also been a number of attempts to prepare organic derivatives of graphitic oxide, but they have led to no clear-cut results.

13.3.2. Alkali metal compounds. Graphite is wetted and immediately penetrated by liquid potassium, rubidium or caesium. At the same time, it swells and appears to disintegrate. When the excess of metal is evaporated off, a pyrophoric coppery-red mass remains which has the composition C_8M (M = K, Rb, Cs). If this material is heated further it is changed to a blue-black substance which approximates in composition to $C_{24}K$, and finally graphite is regenerated. The alkali metal can be washed out with mercury, showing that it is loosely held. Uptake of metal also occurs when graphite is exposed to the vapour. The heat of formation of the C_8M phase can be determined directly by introducing graphite into an excess of metal, and the following relatively low values are found: C_8K, 32·5 kJ (7·8 kcal); C_8Rb, 48·5 kJ (11·6 kcal); C_8Cs, 84·5 kJ (20·2 kcal) mole^{-1}. All of the products react violently with water.

Study of the isobaric breakdown curve of C_8K, coupled with X-ray examination, shows that five distinct phases can exist in which potassium is intercalated between the graphite sheets. In C_8K there are potassium atoms between every pair of sheets, while in $C_{24}K$ they are between every other pair. There are then three further phases, $C_{36}K$, $C_{48}K$ and $C_{60}K$, with three, four and five sheets between the metal layers, as shown below.

	Sequence	*c-axis*	*Stage*
C_8K	CKCKCK....	5·40 Å	1
$C_{24}K$	CCKCCKCCK....	5·40 + 3·35 = 8·75 Å	2
$C_{36}K$	CCCKCCCKCCCK....	5·40 + 2 × 3·35 = 12·10 Å	3
$C_{48}K$	CCCCKCCCCK....	5·40 + 3 × 3·35 = 15·45 Å	4
$C_{60}K$	CCCCCKCCCCCK....	5·40 + 4 × 3·35 = 18·80 Å	5

Introduction of potassium increases the normal interplanar distance of 3·35 to 5·40 Å, the corresponding figures for rubidium and caesium being 5·61 and 5·95 Å respectively. These formulae represent idealized compositions, and each stage has a homogeneity range within which the sequence of carbon and metal layers is retained, It is, however, possible for two stages to coexist.

X-ray examination shows that the graphite sheets are unaltered dimensionally by reaction with potassium. It also shows the metal atoms to be arranged in a definite way between the sheets, rather than randomly. Thus, in the first stage, they form a triangular net (Fig. 13.7) with each over the centre of a hexagon, so that it has twelve carbon atoms equidistant from it. This distance is 3·07 Å for C_8K and 3·24 Å for C_8Rb. This implies that the carbon planes are in the sequence AAA....., with the hexagons situated one above the other.

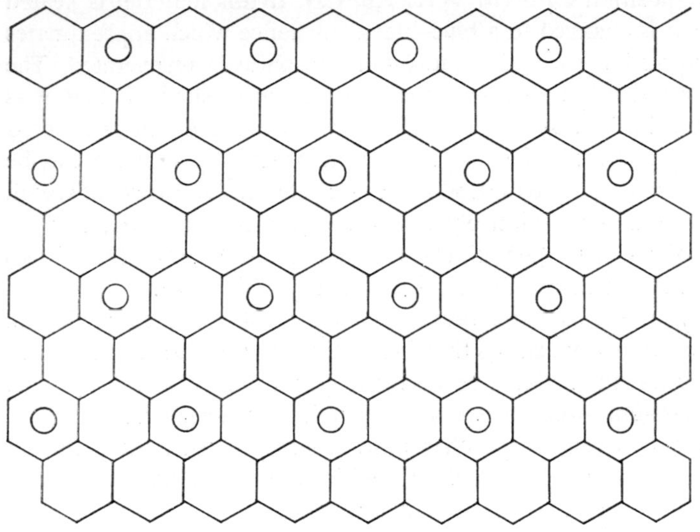

Figure 13.7 The triangular packing of the alkali atoms in C_8Me

The metal atoms are believed to occupy definite positions in the higher stages also, though their arrangement is not the same as in the first stage.

There are important differences between the physical properties of these lamellar compounds and those of graphite, which give a clue as to the nature of the bonding. The relatively large diamagnetic susceptibility of graphite, associated with the π electrons, is destroyed and replaced by a temperature-independent paramagnetism comparable with that shown by true metals. The electronic conduction is also much enhanced. These observations are consistent with the idea that the alkali metals, which have low ionization potentials, give up electrons to the conduction band of graphite, thus increasing the number of charge-carriers.

When graphite is heated with sodium at 400° the product has a deep violet colour and gives an X-ray diagram different from that of graphite. Very little metal is taken up, however, and compositions in the range $C_{60}Na$–$C_{68}Na$ have been reported. In the case of lithium, prolonged reaction at 500° leads only to the carbide Li_2C_2, and there is no evidence of lamellar compound formation. In contrast to the above, solutions of all of the alkali and alkaline earth metals in liquid ammonia react readily with graphite, and both metal and ammonia are intercalated. With an excess of metal, the graphite compound is deep blue in colour and very sensitive to air and moisture, though no longer inflammable. Compositions approximating to $C_{12}M(NH_3)_2$ are found (M = Li, Na, K, Rb, Cs, Ca, Sr, Ba). This corresponds with the first-stage potassium graphite compound C_8K. The distance between the carbon planes is about 6·5 Å in each case, compared with 5·4 Å for C_8K (i.e. does not appear to depend on which metal is present). Good evidence exists for a second-stage product $C_{26-29}M(NH_3)_{1\cdot7-4\cdot1}$ (M = Li, Na, K, Ca, Sr, Ba), and methylamine has also been shown to be intercalated with some of the alkali metals. Furthermore, tetramethyl- and tetraethylammonium or -phosphonium ions can replace the alkali metals. These products are formed when a solution of a salt of one of these ions in liquid ammonia is electrolysed with a graphite anode. The ion is probably ammoniated in the lattice, and the same is true of the product obtained when a solution of sodium amide in liquid ammonia is electrolysed with a graphite anode. The latter swells and disintegrates in each case.

The bonding in the ammoniates and amine derivatives appears to be of the same type as in the simple alkali metal–graphite compounds, since they have almost the same temperature-independent paramagnetism. The reason why lithium, sodium and the alkaline earth metals only intercalate in the presence of ammonia or an amine is not clear, but it must be related to the fact that all these metals already exist in liquid ammonia as solvated cations associated with solvated electrons. The metals cannot be recovered from the intercalated ammoniates, since decomposition occurs on raising the temperature, and hydrogen and the metal amide are produced.

13.3.3. Graphite salts. Whereas in the compounds just considered the graphite sheets acquire an essentially anionic character, charge-transfer from the partially filled graphite π orbitals to electron-accepting groups yields lamellar compounds in which the carbon skeleton is cationic. This is the case in the salts formed by graphite with certain oxy-acids. The first such salt to be prepared was graphite bisulphate, which results when graphite is suspended in sulphuric acid in the presence of small amounts of oxidizing agents such as HNO_3, CrO_3 or $(NH_4)_2S_2O_8$. The sample swells and at the same time becomes blue or purple in colour. It is difficult to free it from excess sulphuric acid as, on washing with water, it reverts to ordinary graphite; however, adhering acid can be removed by washing with phosphoric acid, leaving a material which contains about 0·32 gram of sulphuric acid per gram of graphite.

The properties of this substance can be explained in terms of the formation of graphite bisulphate, which is essentially salt-like in nature. The oxidizing agent takes electrons from the carbon layers, leaving them positively charged and able to bond anions, which become intercalated. From the consumption of oxidizing agent in its formation, the material appears to contain about one HSO_4^- ion per twenty-four atoms, whereas the total amount of sulphuric acid that can be washed out with water represents a ratio of one SO_4^{2-} ion to eight carbon atoms. The material is accordingly represented as $[C_{24}]^+HSO_4^-.2H_2SO_4$, implying that each ion carries two molecules of acid into the lattice. Treatment of graphite with other strong acids (e.g. $HClO_4$, CF_3COOH, H_2SeO_4 and HNO_3) in the presence of an oxidizing agent also leads to intercalation, both of the anion and of molecules of the acid, and the products again approximate in composition to $[C_{24}]^+X^-.2HX$. Weaker acids (e.g. CH_3COOH), on the other hand, do not give graphite salts. These compounds can also be produced electrolytically at a graphite anode. Thus the compound $[C_{24}]^+HF_2^-.4HF$ is formed by the anodic oxidation of graphite in the presence of anhydrous hydrogen fluoride, and a similar effect is observed with trichloracetic and sulphuric acids. The anode swells under these conditions and ultimately disintegrates.

The acid content of graphite salts can be decreased by cathodic reduction, by chemical reducing agents or, very strikingly, by treatment with a suspension of graphite, which is thereby converted to a graphite salt. X-ray examination of the products shows that there are successive stages in this reduction process, which are illustrated in Fig. 13.8 for the case of graphite perchlorate.

In the initial material where carbon atoms and anions are present in a ratio of 24 : 1 there is an ABAB..... sequence with a distance of 7·95 Å between the carbon planes. This varies by less than 0·3 Å for different anions. The five stages shown have been observed

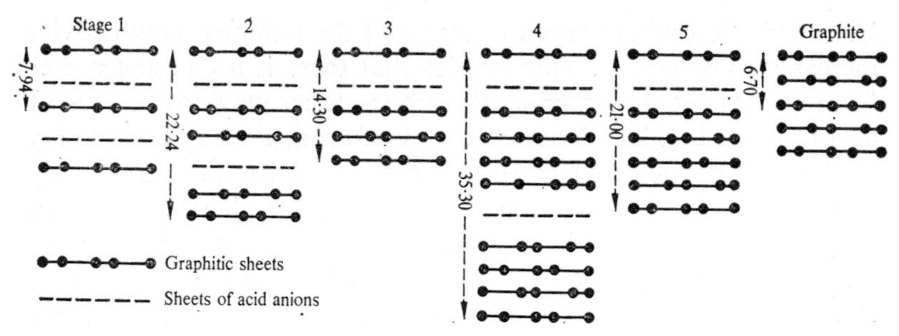

Figure 13.8 Successive stages of reduction of graphite perchlorate: reduction of graphite bisulphate proceeds in similar stages

though, as in the case of the alkali metal–graphite compounds, the representation is idealized.

An interesting dependence of the nature of the product on the concentration has been observed in the case of sulphuric acid which, when pure, gives a structure of the ABAB..... type (first stage). With 83 per cent acid and excess of oxidizing agent only the second stage is formed. At concentrations below 50 per cent no uptake of bisulphate can be detected by X-ray analysis.

The electrical conductivity of graphite is found to increase as a result of the uptake of bisulphate ions during electrolytic oxidation in sulphuric acid. This is associated with the loss of electrons by the graphite, leading to the production of 'holes' in its almost fully occupied band, with a consequential increase in the number of charge-carriers.

13.3.4. Graphite-halogen compounds. As well as acting anionically towards the alkali metals, and cationically towards acids, graphite combines in a highly selective manner with certain molecules. Thus it swells in contact with liquid bromine at ordinary temperatures, taking up approximately one Br_2 molecule per sixteen carbon atoms. Bromine is also taken up from the gas phase or from a solution in an organic solvent. In each case bromine can be almost completely removed from the product by reducing the external pressure, only 3–10 per cent being retained in what is referred to as a 'residue compound' (*vide infra*). X-ray measurements are best explained by assuming that this is a second-stage product with the AABAAB..... sequence, the interplanar distance being increased by the presence of bromine from 3·4 to 7·05 Å.

The anomalous diamagnetism of graphite is almost completely

lost at high bromine concentrations and the electrical conductivity increases considerably. This implies that there is a charge-transfer from the graphite to bromine, though the latter may not be present as a bromide ion. The inherent ability of aromatic π systems to act in this way is evidenced by the appreciable electronic conductivity found for the crystalline complex of bromine with the condensed polycyclic hydrocarbon perylene. The electrical conductivity of graphite also increases by a factor of ten when it is held in liquid chlorine for several days at −33°. It appears that a similar product is formed, though it decomposes above 0°. There is, however, no evidence that iodine can be taken into the graphite lattice, but iodine monochloride behaves in much the same way as bromine, forming a second-stage compound.

13.3.5. Lamellar compounds of metal halides. Graphite forms intercalation compounds with a remarkably wide range of metallic halides. This behaviour was first observed with iron (III) chloride, which is absorbed by graphite above 180°; reaction is accompanied by swelling of single crystals, and an expansion of the *c*-axis spacing to 9·4 Å, which would provide room for the incorporation of $FeCl_3$ molecules. The intercalated salt is neither hydrolysed nor hydrated by treatment with aqueous solutions, and is resistant towards reduction. Above about 300°, iron (III) chloride is lost by vaporization, and at about 400° graphite (or more probably a residue compound retaining a small concentration of iron (III) chloride) is reformed. The heat of sublimation of iron (III) chloride from the lamellar compound is 255 kJ (61 kcal) $mole^{-1}$, as compared with a value of 130 kJ (31 kcal) $mole^{-1}$ for the free anhydrous salt.

A range of compositions has been reported, but the intercalation isotherm of the chloride on graphite at 350° shows two compositions, $C_{12}FeCl_3$ and C_7FeCl_3, each of which is independent of pressure over a certain range. Single crystal X-ray analysis of a sample with the composition $C_{6\cdot7}FeCl_3$ shows that single incomplete layers of $FeCl_3$, which itself has a layer lattice, lie between successive parallel planes of carbon atoms. The material also contains free graphite, the proportion of which becomes greater when the $FeCl_3$ content is lower. For compositions richer in carbon than $C_{20}FeCl_3$ there is a more or less randomly varying number of carbon planes interposed between successive incomplete $FeCl_3$ sheets. On balance, the evidence is against there being a well-defined succession of phases of the type encountered in the alkali metal compounds and graphite salts, though the position is still far from clear.[14]

The X-ray investigation shows little correlation between atomic positions in the two kinds of layer, which suggests that there is neither covalent bonding nor coordination of the graphite π electrons to the iron atoms. It seems most probable that a charge transfer complex is formed, with the Fe^{3+} ions acting as electron acceptors. Small to moderate increases in conductivity relative to graphite have been reported, but the evidence is somewhat conflicting. The Mössbauer spectrum of a sample with the composition C_9FeCl_3 does, however, clarify the position. There is a single intense line centred at 0·80 mm sec^{-1}, compared with 0·70 for anhydrous $FeCl_3$. This can be explained on the assumption that the iron 3*d* orbitals receive charge from the graphite π-bond system, thus shielding the nucleus from the effect of *s* electrons. The strong peak is typical of high-spin iron (III).

Some thirty other metallic halides show similar intercalation in graphite under comparable conditions (Table 13.2). Others, including both non-metallic and metallic halides closely related to those

Table 13.2

$CuCl_2$	$SbCl_5$	$RuCl_3$
$CuBr_2$	$TaCl_5$	$RhCl_3$
$AuCl_3$	$CrCl_3$	$PtCl_4$
BCl_3	CrO_2Cl_2	YCl_3
$AlCl_3$	$MoCl_5$	$SmCl_3$
$GaCl_3$	WCl_6	$GdCl_3$
$InCl_3$	UCl_4	$YbCl_3$
$TlCl_3$	UO_2Cl_2	$DyCl_3$
$ZrCl_4$	$ReCl_4$	$EuCl_3$
$HfCl_4$	$CoCl_2$	

listed, appear to be inert. It is likely that bonding in the lamellar compound is very similar in a number of instances to that in the case of $FeCl_3$, but as yet there is insufficient detailed evidence to justify a generalization, and other factors may well be involved. Thus aluminium chloride is intercalated only in the presence of chlorine or oxygen (which, under the conditions of reaction, liberates some chlorine through the formation of AlOCl). The amount of chlorine entering the reaction is certainly less than would correspond to the formation of a graphite salt $C_n^+AlCl_4^-$. There is approximately one chlorine atom per three molecules of $AlCl_3$, and it has been suggested that the formation of ionic or charge-transfer compounds is favoured if discrete ions of the same sign are separated as much as possible by highly polarizable molecules. Other suitable electron acceptors can, in fact, bring about intercalation of aluminium chloride. Thus compounds of graphite with ($Br_2 + AlCl_3$) or ($FeCl_3 + AlCl_3$) can be prepared with a ratio of $AlCl_3$: electron acceptor of close to 3 : 1. All of the halide compounds resemble the graphite–iron (III) compound in appearance and as regards expansion of the original graphite in the direction of the *c*-axis. In most cases, too, it is possible to recover the halide from the lattice.

13.3.6. Analogues of the graphite-metal halide compounds. It is inherently probable that other polarizable molecules or ions could enter the graphite lattice in the same way as metal halides. This has been shown to be so in the case of a number of metal sulphides (e.g. Sb_2S_5, CuS, FeS_2, Cr_2S_3 and WS_2), which react when heated with graphite at 450–600°. It is significant that the presence of free sulphur promotes the reaction and that the metal–sulphur ratio in the products does not, in general, correspond to the compositions of the free metallic sulphides. Excess sulphur may act as the electron acceptor, forming sulphide ions, with the metal sulphide providing the polarizable spacing molecules. The analytical data are consistent with the general formula $C_n^+(MS_x)_{2-3}S^-$, where $n = 100$–400 for the CuS, Cr_2S_3 and WS_2 compounds. Only in the compounds of FeS_2 and Sb_2S_5 is there no excess sulphur, and in these the sulphide itself could presumably act as the electron acceptor.

Oxides of some of the transition metals in their highest oxidation states can also enter into lamellar compounds. Thus graphite is not oxidized by chromium (VI) oxide at 200°, but takes up the oxide to a composition $C_{6\cdot7}CrO_3$. Molybdenum trioxide is taken up similarly, though less extensively, and can be recovered by sublimation. As in the case of the compounds with some of the halides, more experimental evidence is needed before any reliable picture can be obtained of the bonding relationships in compounds such as these.

13.3.7. Boron nitride. Graphite occupies a special position as a host lattice for the formation of lamellar compounds because of its electronic structure, but the possibility exists that other compounds with layer lattices will behave similarly, though perhaps for different reasons. Boron nitride has a crystal structure which is very similar to that of graphite, with boron and nitrogen atoms forming plane hexagonal nets which lie parallel to one another in the lattice at a distance of 3·33 Å. The essential difference from graphite is that it does not conduct electricity, and has no delocalized π electron system in the sheets.

In spite of this difference, a number of metal halides (e.g. $FeCl_3$ and $AlCl_3$) were found to be taken up at 250–400°, with expansion of the lattice in the *c* direction, as observed by X-ray diffraction. The increase in weight was 10·6 per cent for $FeCl_3$ and 13·0 per cent for $AlCl_3$, and intercalation was attributed to bonding involving donation of electrons from nitrogen to the metal atom. Some other halides were intercalated in smaller amounts. Alternatively, in some cases (e.g. CuCl or $SbCl_3$, both of which are intercalated to a small extent) boron atoms in the sheets may act as electron acceptors. While these observations are not in doubt, it seems that the samples of boron nitride used were not pure. They were black, whereas boron nitride is white, and, when pure, does not intercalate either $FeCl_3$ or $AlCl_3$.

13.3.8. Residue compounds. Removal of the intercalated substance from a number of the lamellar compounds described above leaves a small amount of material which is very tenaciously held in a so-called residue compound. Graphite bisulphate and bromine residue compounds have been studied in some detail and several interesting points established. There is a marked structural difference from the original lamellar compound, with only a very small increase in the *c*-axis spacing of the graphite. This suggests that the residual ions or molecules are mainly associated with holes or imperfections in the graphite lattice and are not intercalated between the carbon planes. The conductivity of the graphite is also not restored fully to its normal value, and the chemical reactivity is somewhat greater than that of graphite. Furthermore, the residue compounds may be reconverted to the same lamellar compounds from which they were produced. An explanation of these observations is best based on the well-known fact that all lattices contain some imperfections: in the case of graphite these appear to have a special affinity for intercalated ions, which they retain or with which they react in an unknown manner.

13.4 Synthetic reactions of high-temperature species[15]

Many short-lived species produced in thermal and photochemical reactions or in the electrical discharge are now known to the chemist. Their identification depends largely on spectroscopic techniques, but only under special experimental conditions is it possible to use them in synthetic reactions. In the 1920's Harteck studied some of the reactions of oxygen and hydrogen atoms produced in a gas stream at low pressure, and short-lived gaseous free radicals produced thermally or photochemically were used by Paneth and others in a few syntheses. The idea of producing gaseous radicals or other species in high-temperature reactions under conditions such that they could be collected on a very cold surface, and subsequently allowed to undergo synthetic reactions, was used in some early investigations, but there has recently been a renewed interest in this approach, largely as the result of work published in 1963 by Skell and Weston on the reactions of condensed carbon vapour with a variety of organic compounds.

13.4.1. Reactions of carbon. When a carbon arc is operated in high vacuum it is possible to detect C_1, C_2, C_3, C_4 and higher polymers in the gas phase by mass spectrometry, these species being present in either the ground state or excited states, with C_1 predominating. If the arc is mounted in a vessel, the lower part of which is cooled by liquid nitrogen then, provided a high vacuum is maintained throughout, a second species (e.g. an olefin) can be introduced through a side tube and cocondensed with minimal decomposition due to the arc. Since evaporation of carbon occurs at a rate of up to 150 millimoles hr^{-1} under the conditions used, it is possible to identify products of its reaction with the added substance at low temperature, or on subsequent warming, by normal techniques: many reactions, indeed, can be carried through on a preparative scale.

A difficulty arises in this work because the condensed carbon contains several reactive species and it is not easy, at first sight, to see which is involved in the formation of a particular product. Skell and subsequent workers have overcome this, in part, in two ways. Firstly, sufficient time may be allowed for excited states to decay before adding the second reactant. This has allowed reactions of C_1 in the excited ^1D and ^1S states to be differentiated from those associated with the ground state (^3P). The half-life of the ^1S state on the cold surface has been estimated to be about 2 sec. Alternatively, using cocondensation, it may be apparent from the nature of the products that a species such as C_3 has been involved.

Much of Skell's work is of interest primarily to organic chemists and will not be considered in detail here. The following are, however, examples of simple reactions that have been found to occur.

$$C_1 + 2CH_3OH \xrightarrow{\text{cocondense}} (CH_3O)_2CH_2$$

$$C_1 + CF_3CF{=}CF_2 \longrightarrow CF_3CF{=}C{=}CF_2$$

$$C_3 + 2ROH \longrightarrow (RO)_2CHC{\equiv}CH$$

The reaction of carbon with a few inorganic substances has also been studied by cocondensation, without attempting to differentiate between products due to the excited and ground states. This may be illustrated by some reactions observed with non-metallic halides; for example,

$$C_1 + BCl_3 \rightarrow Cl_2C(BCl_2)_2 + ClC(BCl_2)_3$$

$$C_1 + GeCl_4 \rightarrow Cl_3C(GeCl_3) + Cl_2C(GeCl_3)_2$$

In the reaction with $SiCl_4$ the main product is $ClC{\equiv}CSiCl_3$, where apparently insertion of C_2 into the Si—Cl bond occurs; the reaction with B_2Cl_4 also gives $C(BCl_2)_4$ and other products derived from C_2.

13.4.2. Reactions with other atomic species. Clearly, in principle, many other atomic species can be studied in the same way as carbon. Under cocondensation conditions, the alkali metal atoms react in much the same way as at higher temperatures, but magnesium atoms when condensed with alkyl halides give alkyl magnesium halides. Boron evaporated at high temperature gives a highly reactive condensate which, if cocondensed with some non-metallic halides, is able to abstract chlorine on subsequent warming to room temperature (e.g. $B + PCl_3 \rightarrow BCl_3 + P_2Cl_4$ + polymer). Unlike carbon, silicon vaporizes mainly as Si, and is inserted into the Si—H bond of trimethylsilane at $-196°$:

$$Si + (CH_3)_3SiH \rightarrow [(CH_3)_3SiSiH] \xrightarrow{(CH_3)_3SiH} (CH_3)_3Si(SiH_2)Si(CH_3)_3$$

There are indications that germanium is also very reactive under these conditions, while vaporized tin will form alkyl tin halides at low temperatures with some alkyl halides, a reaction which requires much more drastic conditions when the massive metal is employed.

Experiments with the vapours of a number of transition metals have yielded some of the most interesting results in this field. Those of the second and third rows require excessively high temperatures for vaporization—often above 2500°—but Cr, Mn, Fe, Co, Ni, Pd, Cu, Ag and Au can all be evaporated in vacuum at useful rates (10–100 millimoles hr^{-1}) when heated to 1200–1600° in a crucible made of alumina and heated electrically with molybdenum wire. Mass spectrometry shows that the vapour consists largely of atoms, which can be condensed on a cold surface exactly as for carbon atoms, and allowed to react with a range of cocondensed materials. There is now abundant evidence that the reactivity of the metal atoms under these conditions is far greater than that of the massive material. This high reactivity can be exemplified by the reactions shown below with inorganic and organic ligands.

$$Cr + 6PF_3 \rightarrow Cr(PF_3)_6 \ (65\%)$$
$$Ni + 4PF_3 \rightarrow Ni(PF_3)_4 \ (100\%)$$
$$Cr + 2C_5H_6 \rightarrow Cr(C_5H_5)_2 \ (50\%) + H_2$$
$$Ni + PF_3 + PH_3 \rightarrow Ni(PF_3)_2(PH_3)_2 + Ni(PF_3)_3PH_3 + Ni(PF_3)_4$$

Several entirely new compounds have been synthesized in this way and there is clearly great scope for extending the observations, especially as the method lends itself to the preparation of useful quantities of material. Other reactions have involved dehalogenation by the reactive metal, cocondensation of copper vapour with BCl_3 at −196°, for example, giving up to 70 per cent yields of B_2Cl_4. This, in fact, is a better method for preparing the diboron halide than that involving passage of the trichloride vapour through a mercury discharge.

13.4.3. Reactions with molecular species. The cocondensation technique is not restricted to the study of the reactions of atoms; indeed much of the impetus for the present interest in this subject came from its early application to silicon difluoride, which is formed when the vapour of the tetrafluoride is streamed at a pressure of 0·1 mm over lump silicon heated to 1100°. The type of apparatus used is shown in Fig. 13.9.

Under the conditions used, the gas streaming from the heated zone contains roughly 50 per cent of SiF_2 admixed with unreacted SiF_4. This condenses unchanged on the cooled surface. The difluoride is also reasonably stable and unreactive in the gas phase, mass spectrometry showing a half-life of the order of 150 sec at 0·1 mm compared with a few milliseconds for $SiCl_2$ (produced

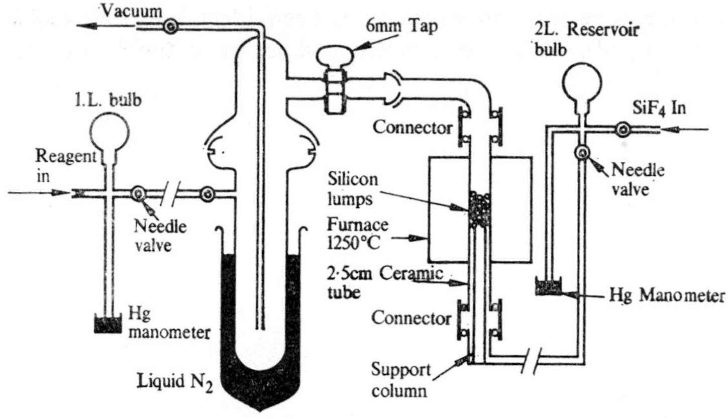

Figure 13.9 Apparatus for the preparation and reaction of SiF_2

similarly), ~1 sec for CF_2 and a few hundredths of a second for CH_2. Silicon difluoride deposits from the gas phase on to the walls of the containing material and also polymerizes in the condensed state to a plastic material, the Raman spectrum of which suggests a structure similar to that of polytetrafluoroethylene (Teflon). The two materials are, however, quite different chemically. Teflon is very inert and pyrolysis occurs only at *ca.* 600°, giving mainly the monomer, C_2F_4, whereas the SiF_2 polymer inflames in air, is readily hydrolysed to a mixture of silanes, and decomposes to a complex mixture of fluorosilanes with up to at least sixteen silicon atoms in the molecule when heated in vacuum at 200°.

Many of the low-temperature cocondensation reactions of SiF_2 give products with Si—Si bonds, which suggest that there is an initial polymerization on condensation, perhaps to reactive biradicals, but the exact reaction mechanism in most of the low-temperature reactions of this and other species is still largely unknown. The following are some of the reactions that have been observed.

With ethylene at −196° two volatile cyclic products (Figs 13.10 and 13.11) are formed in low yield.

H_2C——SiF_2
|　　　|
H_2C——SiF_2

Figure 13.10

　　CH_2
CH_2　SiF_2
|　　　|
CH_2　SiF_2
　　CH_2

Figure 13.11

With acetylene three products have been identified (Figs 13.12, 13.13 and 13.14), while with benzene a series of compounds of the

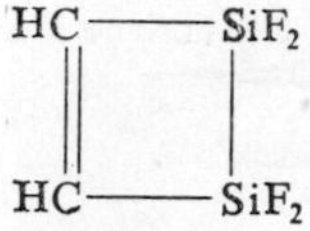

HC—SiF_2
HC—SiF_2

Figure 13.12

$H_2C{=}CHSiF_2SiF_2CH{=}CH_2$

Figure 13.13

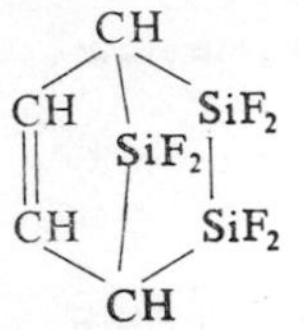

CH
CH SiF_2
SiF_2
CH SiF_2
CH

Figure 13.14

type $C_6H_6(SiF_2)_n$, $n = 2–6$, is obtained. These may all be identified by mass spectrometry and the most abundant, $C_6H_6Si_3F_6$, gives 1,4-cyclohexadiene on hydrolysis, so it is probably to be formulated as in Fig. 13.15. Cocondensation with hexafluorobenzene, on the

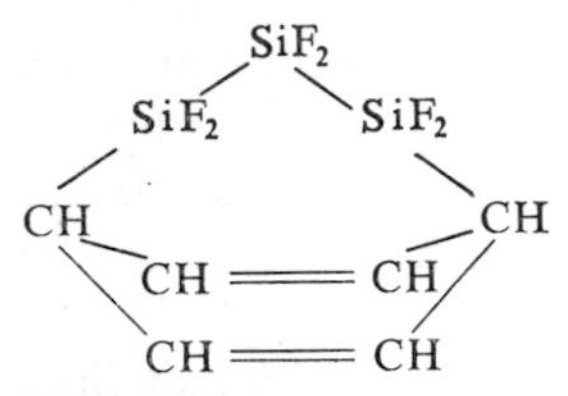

Figure 13.15

other hand, gives a product which, on warming to room temperature, decomposes to $C_6F_5(SiF_3)$ and the three isomers of $C_6F_4(SiF_3)_2$.

Several reactions of SiF_2 with inorganic molecules have been examined. The low-temperature reaction with BF_3 gives the mixed boron–silicon fluorides, $SiF_3SiF_2BF_2$ and $SiF_3(SiF_2)_2BF_2$, the ^{19}F n.m.r. spectra of which are distinctive, with small amounts of higher homologues in the series $SiF_3(SiF_2)_nBF_2$. The absence of SiF_3BF_2 is significant, indicating that the simplest reactive species may be the biradical $(SiF_2)_2$. When SiF_2 and BF_3 are incorporated into a krypton matrix at 20°K, infrared bands due to $SiF_3SiF_2BF_2$ appear on warming the matrix to 35°K, which is the temperature at which bands attributed to Si_2F_4 first appear in the infrared spectrum of matrix-isolated SiF_2.

Various other low-temperature reactions of SiF_2 have been investigated, among the more interesting being that with water, which gives $HSiF_2OSiF_2H$, and that with monogermane, from which $HSiF_2GeH_3$ and $HSiF_2SiF_2GeH_3$ are the main products. Some work has also been done with $SiCl_2$, formed when $SiCl_4$ is passed over silicon at low pressure at 1300°. It is much shorter lived: among its purely inorganic reactions under cocondensation conditions are that with BCl_3, giving $SiCl_3BCl_2$ in low yield (which contrasts with the reaction of SiF_2 and BF_3, mentioned above), and that with PCl_3, where the product is $SiCl_3PCl_2$. Exploratory work on other sub-halides is mentioned by Timms,[15] but only that on boron mono-fluoride will be described here.

Boron monofluoride can be obtained in high yield when the trifluoride is passed at low pressure over boron heated to 2000° in an apparatus similar to that used in making SiF_2 (Fig. 13.8):

$$B + BF_3 \longrightarrow BF\ (+\ BF_3)$$

When the mixture of BF and BF_3 is condensed at −196° and the product is then allowed to warm, a complex reaction occurs leading on the one hand to B_2F_4 and a new boron fluoride B_3F_5. The latter, which is obtained in better yield by cocondensing BF with B_2F_4, disproportionates above −50°:

$$4B_3F_5 \longrightarrow 2B_2F_4 + B_8F_{12}$$

The compound B_8F_{12}, the structure of which is unknown, decomposes above −10°, but reacts with various Lewis bases forming stable crystalline compounds of the type $(BF_2)_3BX$ ($X = CO$, PF_3, PCl_3, PH_3, $(CH_3)_2S$), the PF_3 compound having been shown to have a tetrahedral arrangement of PF_3 and three BF_2 groups round a central boron atom. In addition to the products referred to above, condensed BF gives a mixture of higher boron fluorides with up to at least fourteen boron atoms in the molecule containing, on the basis of spectroscopic evidence, $B—BF_2$ groups rather than BF.

The reaction of BF with several organic compounds has been investigated. On cocondensation with 2-butyne, for example, the 1,4-diboracyclohexadiene derivative results, while with 1-butene addition across the double bond gives $H_2C(BF_2)C(BF_2)HCH_2CH_3$. The production of BCl from BCl_3 and boron occurs only at temperatures in excess of 2000°, and experimental difficulties have so far prevented comparable studies with this species from being carried out.

Species of several other types have been obtained in high-temperature reactions. Thus gaseous silicon monoxide is evolved when a mixture of SiO_2 and silicon is heated in vacuum at 1300°, and is found to react with many unsaturated organic compounds on co-condensation at −196°, but the products, which are solids, have not been characterized. Germanium and tin monoxides may be prepared similarly, and gaseous boron monoxide (B_2O_2) is formed by heating B_2O_3 strongly with boron. When H_2S is passed over boron at 1300°, HBS is known to be produced in the gas phase and to be relatively long-lived, while SF_6 reacts with boron at 1300° forming a gaseous product $(FBS)_2$. This, when condensed at −196° and then allowed to warm to room temperature, gives a cyclic trimer $(FBS)_3$. Doubtless these and many other products produced by the high-temperature method will in time be examined in greater detail.

References for Chapter 13

1 Holloway, J. H., *Noble Gas Chemistry* (Methuen, 1968).
2 Malm, J. G., and Selig, H., *Chem. Rev.*, 1965, **65,** 199.
3 Claassen, H. H., *The Noble Gases* (D. C. Heath and Co., 1966).
4 Hyman, H. H. (ed.), *Noble Gas Compounds* (University of Chicago Press, 1963).
5 Yost, D. M., and Kaye, A. L., *J. Amer. Chem. Soc.*, 1933, **55,** 3890.
6 Pauling, L., *J. Amer. Chem. Soc.*, 1933, **55,** 1895.
7 Bartlett, N., *Proc. Chem. Soc.*, 1962, 218.
8 Hoppe, R., Dahne, W., Mattauch, H., and Rodder, K. M., *Angew. Chem.*, 1962, **74,** 903 [*Intern. Ed. English*, 1962, **1,** 599].
9 Cotton, F. A., and Wilkinson, G., *Advanced Inorganic Chemistry*, 2nd edn., p. 223 (Interscience, 1966).
10 Jeffrey, G. A., and McMullan, R. K., *Prog. Inorg. Chem.*, 1967, **8,** 43.
11 Rüdorff, W., *Adv. Inorg. Chem. Radiochem.*, 1959, **1,** 223.
12 Hennig, G. R., *Prog. Inorg. Chem.*, 1959, **1,** 125.
13 Croft, R. C., *Quart. Rev.*, 1960, **14,** 1.
14 See, for example, Bach, B., and Ubbelohde, A. R., *Proc. R. Soc. Lond.*, **A,** 1971, **325,** 437; *J. Chem. Soc.*, 1971, 3669.
15 Timms, P. L., *Adv. Inorg. Chem. Radiochem.*, 1972, **14,** 121.

Complexes of transition metals I: structure

14

14.1 Introduction

In this and the following chapters we shall be concerned with some aspects of the chemistry of the transition elements in which the general principles outlined in Chapters 3, 5 and 6 are extended to account for variations in properties along the series scandium–zinc, yttrium–cadmium and lanthanum–mercury. Even at the present time, there is not general agreement as to which elements should be classified as members of transition series. Some authors regard as transition elements only those which, in their ground states as neutral atoms, have partly filled *d* or *f* shells; others also include elements which have partly filled *d* or *f* shells in any common oxidation state. Neither of these definitions fits well into the natural classification of elements in terms of the permitted combinations of quantum numbers (2, 6, 10 and 14), and we have therefore preferred to consider the three series mentioned above as the transition elements and the lanthanides and actinides (the series of fourteen elements beginning with cerium and thorium) as inner transition elements.

All transition elements have the properties of hardness, high melting point and electrical conductivity that are characteristic of metals. All except the first members of each series and zinc are now known to exhibit more than one formal oxidation state. Nearly all form some coloured compounds (i.e. some compounds which absorb in the visible region of the spectrum); the intensity of such absorption is relatively low, and the frequency of the absorption maximum follows a regular sequence, that for some common ligands being

$$I^- < Br^- < Cl^- < F^- < H_2O < NH_3 < CN^-$$

(the so-called spectrochemical series—see Chapter 17) for complexes of any given oxidation state of any given metal. Nearly all transition elements also form some compounds that are paramagnetic (i.e. are attracted by a magnetic field); it is often found that the cyano complex is diamagnetic (i.e. repelled by a magnetic field) when halo complexes are paramagnetic, but never vice versa. For a given oxidation state, there is a slow decrease in ionic radius along a series, and analogous compounds of members of that series are usually isomorphous. There are, however, two major exceptions to this generalization: it does not hold for octahedral complexes in which the electronic configuration of the metal is d^4 (if the paramagnetic susceptibility of the compound indicates all these electrons are unpaired) or d^9; and it does not hold nearly so well for the less common coordination numbers such as 5, 7 and 8, for which the distribution of structures is more nearly a random one. Finally, there are regular variations in hydration energies of ions, lattice energies of compounds, and overall formation constants of complexes along the first transition series (the only one for which ample data are available), and marked kinetic inertness is associated with the electronic configurations d^3 and (provided the species concerned is diamagnetic) d^6.

These, and the variations in coordination number and geometry that we describe later in this chapter, are some of the experimental facts which any theory of bonding in transition metal complexes must try to account for. It will be observed that they are very diverse in nature, and in order to present a reasonably coherent account we shall therefore follow a description of structures and isomerism in transition metal complexes by an introduction to the main theories of bonding which have been applied in this field. We shall then consider magnetic, spectroscopic and thermochemical properties, and the kinetics and mechanism of reactions, concluding with accounts of selected groups of compounds classified by ligands and of the chemistry of the inner transition elements.

It will be noticed that in transition metal chemistry we tend to call any species other than the metal a ligand; some justifications for this will be given later (see Chapter 15). Meanwhile, however, it may be noted that the environment of the transition metal ion is often the same in what are conventionally called simple salts and complexes (e.g. in FeF_3 and $K_3[FeF_6]$). In this chapter we shall usually choose as illustrations compounds for which it would seem reasonable to write coordinate bonds from ligand to metal, but it has to be remembered that the geometry of a compound is not often by itself a guide to the nature of the bonding, and in any case our present purpose is to assemble the experimental material before going into its theoretical interpretation.

14.2 Werner's coordination theory

The basic features of the structures of coordination compounds, as it is still convenient to call them, were elucidated by Alfred Werner over seventy years ago. Werner, in discussing the nature of compounds such as $CoCl_3,6NH_3$ (as it was then written), postulated that neutral molecules or oppositely charged ions are grouped or coordinated round a central atom in what he termed the first sphere of attraction, the number of such ligands being the coordination number of the central atom. The latter had two sorts of valency. One, the primary valency, was identified with the oxidation state of the central atom; in $CoCl_3,6NH_3$, for example, this would be $+3$, the positive charge being balanced by that on the three chloride ions held in the outer coordination sphere. The ammonia molecules were envisaged as being held by secondary valencies round the central atom, having an octahedral distribution in space; the compound was thus formulated as $[Co(NH_3)_6]Cl_3$ (hexamminecobalt (III) chloride). It was also postulated that the number of coordinated ligands might vary as the central atom was changed, and that such ligands would not show the same reactivity as the same species in the free state.

The chief methods available to Werner for the investigation of complexes were the study of chemical reactivity, electrical conductivity and isomerism. The following examples, taken from his work, illustrate their application. By the action of oxidizing agents on solutions of cobalt (II) chloride in the presence of ammonia, three compounds are formed according to the conditions employed. One of these is the yellow $[Co(NH_3)_6]Cl_3$ already mentioned; all of the chloride in this compound is precipitated by cold aqueous silver nitrate. Another is the pink compound $CoCl_3.5NH_3.H_2O$; all the chloride in this is precipitated by cold silver nitrate, and a salt $Co(NO_3)_3.5NH_3.H_2O$ can be isolated from the reaction. This suggests the pink compound is $[Co(NH_3)_5(H_2O)]Cl_3$ (aquopentammine-cobalt (III) chloride). The third product, purple $CoCl_3.5NH_3$, yields only two moles of silver chloride on treatment with cold aqueous silver nitrate, and is thus $[Co(NH_3)_5Cl]Cl_2$ (chloropentammine-cobalt (III) chloride). In the series of complexes obtainable in the system K^+–Pt^{II}–Cl^-–NH_3, the values for the molar conductivities in aqueous solution of $K_2[PtCl_4]$, $K[PtCl_3(NH_3)]$, $[PtCl_2(NH_3)_2]$, $[Pt(NH_3)_3Cl]Cl$ and $[Pt(NH_3)_4]Cl_2$ (268, 107, 1, 116 and 261 ohm^{-1} cm^2 respectively) help in formulating the complexes, the contribution to the molar conductivity of any ion $I^{x\pm}$ being found to be about

60 x ohm^{-1} cm^2. The octahedral geometry of the cobalt (III) complexes and the planar geometry of the platinum (II) complexes was established with a high degree of probability by the studies of geometrical and optical isomerism, mentioned in Section 14.4.

Today there is an overwhelming amount of X-ray diffraction and other data in support of Werner's formulations; we shall now summarize this, and then end the chapter by a short account of all types of isomerism in metal complexes with special reference to recent work.

14.3 Coordination numbers and geometries in transition metal complexes[1–4]

14.3.1. Coordination number and structure. It has become clear during the last thirty years that the idea that a metal in a given oxidation state has a fixed coordination number and geometry is no longer tenable. Most transition metals are now known to exhibit more than one coordination number in some of their oxidation states, and even for a given ligand and a given metal ion the same coordination number may occur in different geometries.

Another fact that has come to light during recent years is that the environment of a metal atom is sometimes decidedly irregular. Thus iron (II) fluoride would generally be described as having the rutile structure; but of the six fluoride ions round the cation, two are at 1·99 Å and four at 2·12 Å. In vanadium pentoxide, the vanadium atom has oxygen atoms at distances of 1·54, 1·77, 1·88 (two), 2·02 and 2·81 Å. We could describe the environment of the metal atom in this compound as a very distorted octahedron, a distorted trigonal bipyramid with one non-bonded atom in an adjacent layer completing a very distorted octahedron, or an irregular tetrahedron with two more distant oxygen atoms, and so on. In these circumstances 'number of nearest neighbours' is not a suitable definition of coordination number, and Downs[5] has suggested its replacement by 'the number of vertices of the coordination polyhedron surrounding a particular atom, this polyhedron to encompass all other atoms or groups whose centres are nearer that atom than any other having an equivalent environment'. This is unambiguous, but it may lead to overemphasis of the importance of distant neighbours, since the extent of interaction between atoms decreases sharply with increase in interatomic distance. In this and the following chapters, therefore, we shall use coordination number to mean the number of neighbours at distances not greater than about 0·5 Å more than the distance of nearest neighbours; thus we shall describe the coordination of copper (II) in its halides as a distorted octahedron rather than as a square, though in such cases we shall usually cite interatomic distances so that the reader has a clear idea of how large the distortion is.

We shall now survey coordination numbers from two to twelve. No attempt will be made to be comprehensive; for further examples reference may be made to a number of sources.[1,4,6] Whilst it is clear that among the factors determining the coordination number and geometry of a metal atom or ion are

(a) size and steric effects,
(b) the electronic configuration, oxidation state and energetically significant orbitals on the metal,
(c) electrostatic and ligand field stabilization, and
(d) the nature of the ligand,

we are not yet at a stage at which the structures of transition metal compounds can often be predicted. It does, indeed, seem likely that the differences in energy between various possible structures are very small, and that in some instances the geometry of an ion or molecule is affected by a small change in environment. It has even been found recently that for the molecule $[Ni(PBzPh_2)_2Br_2]$ (Bz = benzyl, Ph = phenyl) and the ion $[Ni(CN)_5]^{3-}$ two different structures are found in the same crystal. Although, therefore, we shall discuss some aspects of the variations in coordination number and geometry in later chapters, we shall often be unable to do more than to draw attention to structures whose interpretation poses unsolved problems in theoretical chemistry.

14.3.2. Coordination number two. This coordination number is uncommon in transition metal chemistry and the few examples of it known at present are all complexes of d^{10} systems (e.g. $[CuCl_2]^-$, $[Ag(NH_3)_2]^+$ and $Hg(CN)_2$). In each of these species the configuration round the transition metal atom is linear, just as it is in two-coordination compounds of d^0 systems such as gaseous $BeCl_2$.

14.3.3. Coordination number three. This is also uncommon among transition metal compounds, and until recently the only known examples were those of d^{10} systems (e.g. $[Zn(OH)_3]^-$, which is planar, and $[HgI_3]^-$, which is very nearly so). Recently, though, it has been found that the compound $Fe[N(SiMe_3)_2]_3$ contains planar FeN_3 and

$FeNSi_2$ units, and the corresponding chromium compound is isomorphous with it. A pyramidal structure with a bond angle of 108° or less and a T-shaped structure like that of chlorine trifluoride have not yet been found among transition metal complexes.

14.3.4. Coordination number four. After six, this is the commonest coordination number. The configuration is usually a symmetrical or nearly symmetrical tetrahedron, but planar complexes are also found for metal ions having the electronic configurations d^7, d^8 and d^9; the unsymmetrical SF_4 structure has not been found among transition metal compounds.

Tetrahedral complexes have not yet been reported for ions of electronic configuration d^3 or d^4, but it is doubtful whether any special significance is to be attached to this statement. They are also relatively uncommon for metals of the second and third transition series; compounds of these metals often have six-coordinated structures where the corresponding compounds of metals of the first transition series exhibit tetrahedral coordination round the metal. Representative species having tetrahedrally coordinated metal atoms are as follows: $TiCl_4$, VO_4^{3-}, CrO_4^{2-}, MoO_4^{2-}, WO_4^{2-}, MnO_4^-, TcO_4^-, ReO_4^- and OsO_4 for d^0; VCl_4, MnO_4^{2-} and RuO_4^- for d^1; FeO_4^{2-} and RuO_4^{2-} for d^2; $[FeCl_4]^-$ for d^5; $[FeCl_4]^{2-}$ for d^6; $[CoCl_4]^{2-}$ for d^7; $[NiCl_4]^{2-}$ for d^8; $[CuCl_4]^{2-}$ (a somewhat flattened tetrahedron) for d^9; and $Ni(CO)_4$, $Pt(PF_3)_4$, $[Cu(CN)_4]^{3-}$, $[ZnCl_4]^{2-}$ and $[HgI_4]^{2-}$ for d^{10}.

Two complexes of Co^{II} (d^7) have been shown to contain planar

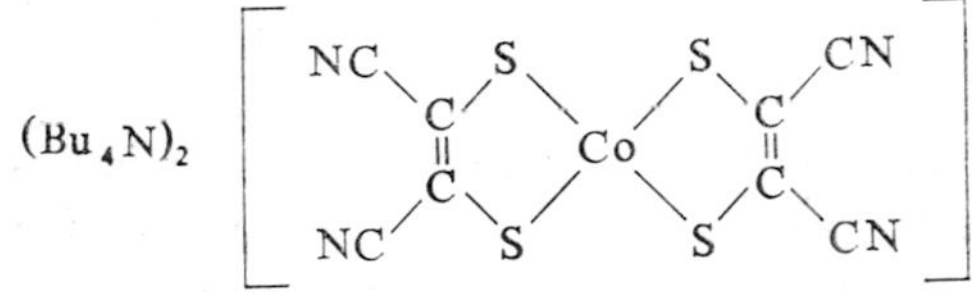

Figure 14.1

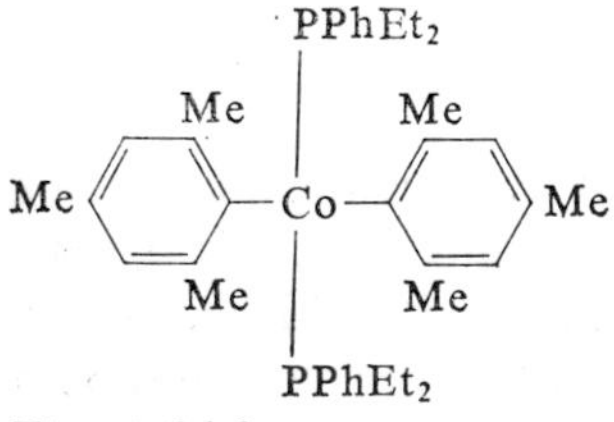

Figure 14.2

four-coordinated metal atoms: these are the maleonitriledithiolate complex (Fig. 14.1) and *trans*-dimesitylbis-diethylphenylphosphine cobalt (II) (Fig. 14.2). It seems possible that in these cases special electronic factors favour the planar structure. For the d^8 electronic configuration the planar structure is not uncommon in the case of Ni^{II} (e.g. in $[Ni(CN)_4]^{2-}$ and nickel bis-dimethylglyoximate) and is usual for derivatives of Pd^{II} and Pt^{II} (e.g. $[PdCl_4]^{2-}$ and $[PtCl_4]^{2-}$). It occurs also for Ag^{III} and Au^{III} in the ions $[AgF_4]^-$ and $[AuF_4]^-$. Most Cu^{II} (d^9) complexes have a distorted octahedral configuration round the metal atom, with four nearest neighbours in a plane and two more distant ones in directions perpendicular to it; however, electron diffraction measurements made on copper (II) nitrate vapour are compatible with a planar molecule and two bidentate nitrate groups, but a tetrahedral structure is not excluded. It might be expected that four-coordinated Ag^{II} complexes would have a planar configuration, but no detailed X-ray work on such compounds has yet been reported.

14.3.5. Coordination number five.[6] The two basic structures for five-coordination are the trigonal bipyramid and the square pyramid. These are easily interconvertible when ring systems are not involved by the bending process illustrated in Fig. 14.3, and when the ideal structures are somewhat distorted by the presence of bulky ligands the distinction between them is blurred. The ion $[Ni(CN)_5]^{3-}$ even occurs in both forms in crystalline $[Cr(en)_3][Ni(CN)_5],1{\cdot}5H_2O$: in the trigonal bipyramidal form the C(apical)—Ni—C(apical) angle is slightly distorted to 173°, Ni—C(apical) is 1·84 Å and Ni—C(equatorial) 1·91 or 1·99 Å; in the square-pyramidal form Ni—C(apical) is 2·17 Å and Ni—C(equatorial) 1·86 Å.[7]

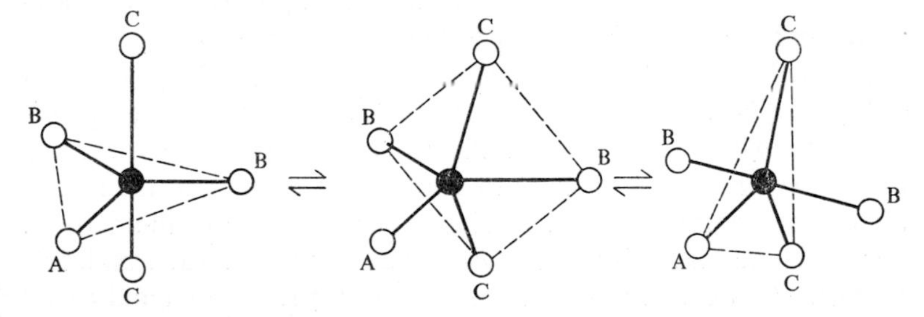

Figure 14.3 Interconversion of the trigonal bipyramid and the square pyramid and mechanism of exchange of apical and equatorial positions in the former

Among other species having the trigonal bipyramidal configuration are $NbCl_5$ (d^0); $MoCl_5$ (d^1); *N*-methylsalicylaldimine complexes of Mn^{II} (d^5), Co^{II} (d^7) and Zn^{II} (d^{10}) (Fig. 14.4); $Fe(CO)_5$, $[Co(CNCH_3)_5]^+$ and $[Pt(SnCl_3)_5]^{3-}$ (d^8); and $[CuCl_5]^{3-}$ (d^9). The first two of these are monomeric only in the vapour phase, and from

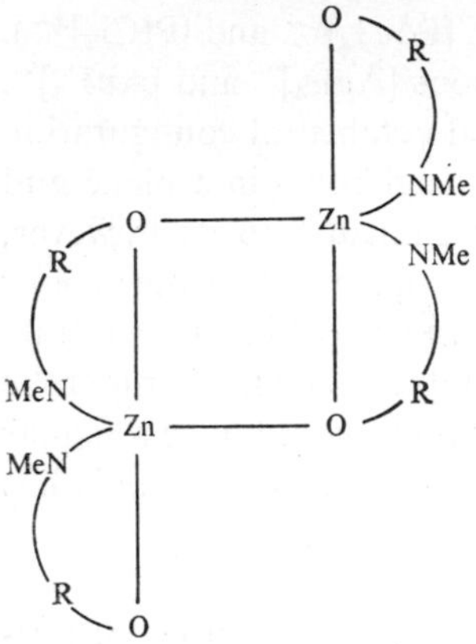

Figure 14.4

the early electron diffraction work on which the structures are based it is not known whether the apical and equatorial metal–chlorine distances are equal. The *N*-methylsalicylaldimine complexes (Fig. 14.4, R = *o*- —C_6H_4—CH=) are dimeric; the Zn—O—Zn bridges are unsymmetrical, but the other equatorial and apical bonds are not significantly different in length. There is, however, a significant difference between the apical and equatorial bond lengths in $Fe(CO)_5$ (1·806 and 1·833 Å respectively).

Species other than $[Ni(CN)_5]^{3-}$ which have approximately square-pyramidal structures include: $[VO(acac)_2]$ (acac = acetylacetonate anion) (d^1); $[(Ph_3P)_3RuCl_2]$ (d^6); $[Ni(triarsine)Br_2]$ (triarsine = $(CH_3)_2AsCH_2CH_2As(CH_3)CH_2CH_2As(CH_3)_2$) ($d^8$); copper (II) dimethylglyoximate (d^9), in which the fifth (apical) position is occupied by an oxygen atom belonging to another molecule; and the monohydrate of the zinc derivative of salicylaldehyde-ethylenediimine (d^{10}). In the first and last of these the oxygen atom and the water molecule respectively occupy the apical position; in the ruthenium and nickel complexes this is occupied by a halogen atom. In view of the complexities of these examples, it is difficult to say anything conclusive about bond lengths in most of them, but it is worth noting that the reported Ni—Br distances in the triarsine complex are quite different, being 2·37 and 2·69 Å for the basal and apical bonds respectively.

14.3.6. Coordination number six. As recently as 1965 it was believed that, with the exception of MoS_2 and WS_2 (which have layer lattices of sulphide ions with the metal ions surrounded by anions at the corners of a trigonal prism), all six-coordination compounds have the octahedral structure, though sometimes the octahedron is distorted tetragonally by electronic effects. Since it is not certain that the structure reported for MoS_2 is correct (when made in the laboratory, it has the $CdCl_2$ layer structure) we shall not discuss this compound further; the trigonal biprismatic structure has now been identified with certainty in Re^{VI} (d^1) tris-(*cis*-1,2-diphenylethene-1,2-dithiolate) (Fig. 14.5). In this compound the metal is surrounded by six equidistant sulphur atoms, the sides of the prism being nearly perfect

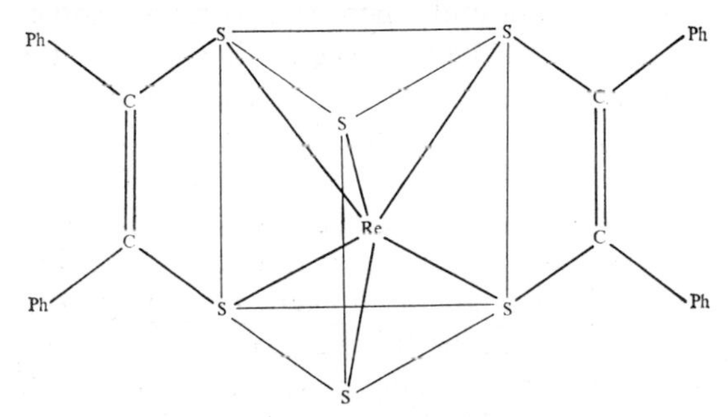

One C_2Ph_2 is not shown.

Figure 14.5

squares. It is interesting to note that the phenyl rings are twisted out of the planes of the chelate rings and appear not to be conjugated with them. There must thus be very considerable stability associated with trigonal biprismatic coordination in this instance. A further example of this structure has recently been described: in the complex cation $[Co(Co(OCH_2CH_2NH_2)_3)_2]^{2+}$, the oxygen atoms of the octahedral cobalt (III) ethanolamine complex are distorted round the middle cobalt (II) atom (d^7) in trigonal prismatic coordination. Preliminary reports of trigonal biprismatic coordination of a number of other metals, among them chromium (VI), molybdenum (VI), tungsten (VI) and zinc (II), have also appeared, and it seems certain that in the very near future our knowledge of the occurrence of this structure will be extended considerably.

The regular or very nearly regular octahedron is found for all electronic configurations from d^0 to d^{10} inclusive, typical examples being $[TiF_6]^{2-}$ (d^0), $[Ti(H_2O)_6]^{3+}$ (d^1), $[V(H_2O)_6]^{3+}$ (d^2), $[Cr(H_2O)_6]^{3+}$ (d^3), $[Mn(H_2O)_6]^{3+}$ (d^4), $[Fe(H_2O)_6]^{3+}$ (d^5), $[Fe(H_2O)_6]^{2+}$ (d^6), $[Co(H_2O)_6]^{2+}$ (d^7), $[Ni(H_2O)_6]^{2+}$ (d^8), $[Cu(NO_2)_6]^{4-}$ (d^9) and $[Zn(H_2O)_6]^{2+}$ (d^{10}). The only two of these that call for special comment are those of d^4 and d^9 systems. Most of the d^4 systems examined show substantial tetragonal distortion of the octahedron, and the same is true of very nearly all d^9 systems; the interpretation of these statements will be discussed in the next chapter. For $[Mn(H_2O)_6]^{3+}$ and $[Cu(NO_2)_6]^{4-}$, however, the symmetry of the structure appears to be established beyond doubt.

14.3.7. Coordination number seven.[4] For coordination numbers above six the differences between idealized geometries become even smaller than for five-coordination, and where structures are somewhat distorted from basic types by the presence of ring systems it is sometimes almost meaningless to describe them in terms of preconceived geometries. Moreover, for high coordination numbers the metal atoms involved are usually (though not invariably) those of the second and third transition series, and determination of the exact positions of ligand atoms of much lower atomic number is then impossible unless X-ray diffraction studies are supplemented by neutron diffraction work; up to the present time this has rarely been done.

Seven-coordinated structures are usually described in terms of three types: the pentagonal bipyramid, the capped trigonal prism (or one-face centred trigonal prism) and the capped octahedron (or one-face centred octahedron). The third of these was first reported many years ago, and was believed to be present in the ions $[ZrF_7]^{3-}$ and the isostructural $[NbOF_6]^{3-}$; more recently, however, it has been found that in the case of $[ZrF_7]^{3-}$ the report was incorrect, and since it now seems that there is no well-authenticated case of the capped octahedral structure, we shall confine attention to the first two.

The pentagonal bipyramidal structure (which is that of iodine heptafluoride) occurs in $[ZrF_7]^{3-}$ (d^0) and the bridged oxalate complex of titanium (III), μ-oxalatobis (oxalato) hexaquodititanium (III) (d^1) (Figs 14.6 and 7 respectively). In the latter, each titanium atom is coordinated by four oxygen atoms from oxalate and three from water; the water molecules occupy one equatorial and two apical positions round each titanium. It occurs also in the $[Fe(EDTA)(H_2O)]^-$ ion (d^5) (in which EDTA represents the ethylenediamine tetraacetate

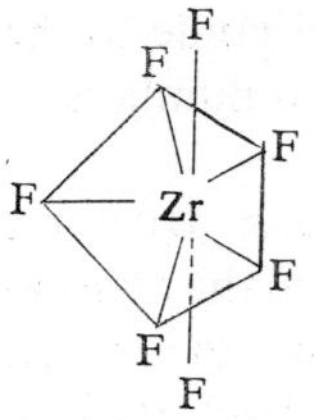

Figure 14.6 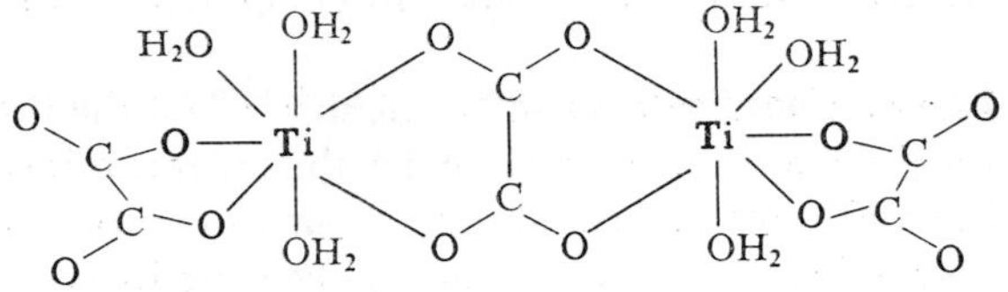 *Figure 14.7*

anion), with the water molecule occupying an equatorial position. If, however, the anion is replaced by that of diaminocyclohexane-tetraacetic acid, the structure changes to a capped trigonal prism, presumably owing to the steric effect of the cyclohexane ring. The balance between the two structures must be a fine one, for $[Mn(EDTA)(H_2O)]^{2-}$ (also a d^5 system) is also a capped trigonal prism.

The simplest examples of the capped trigonal prismatic structure are $[NbF_7]^{3-}$ and $[TaF_7]^{3-}$ (Fig. 14.8). In the former, the positions of the fluorine atoms have been confirmed by neutron diffraction; all lie within the range 1·91–1·96 Å of the metal atom. (The closest

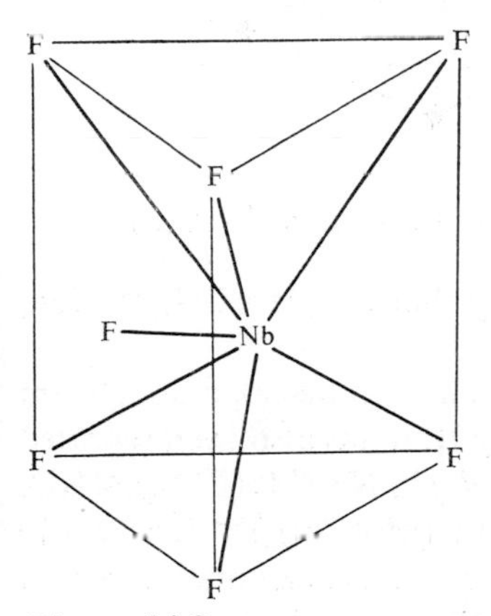

Figure 14.8

approaches of fluorine atoms, however, vary much more widely, extending from 2·36 to 2·91 Å.) Since niobium (V) and tantalum (V) are both d^0 like zirconium (IV), this difference of structure (like that of the manganese (II) and iron (III)–EDTA complexes) shows that for high coordination numbers the isoelectronic principle (that molecules or ions having the same number of electrons have the same structure) is no longer valid. There is, in fact, no reliable general guide to the structures of species of these less common coordination numbers.

14.3.8. Coordination number eight.[4, 8, 9] Of the three basic structures for eight-coordination (the cube, the square antiprism and the dodecahedron) only two commonly occur in complexes. The cube is well known in ionic salts, presumably because its symmetry lends itself to packing in infinite lattices, but individual molecules and ions usually adopt other structures in which, incidentally, repulsions between pairs of bonded electrons should be less. Recently, however, it has been found that the anion in $Na_3[PaF_8]$ is an almost perfect cube, and the corresponding compounds of uranium (V) and plutonium (V) are isostructural. The square antiprism and the dodecahedron are shown in Fig. 14.9(a) and (b) respectively; as with the structures for seven-coordination, there appears to be little to choose between them

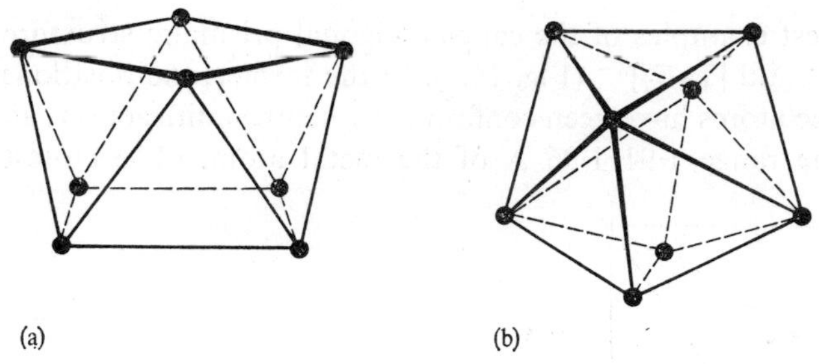

Figure 14.9 The square antiprism (a) and dodecahedron (b)

on energetic grounds, and it seems likely that the $[Mo(CN)_8]^{4-}$ ion, for example, can have either structure.

The square antiprism occurs in several d^0 complexes, among them $[TaF_8]^{3-}$, $[Zr(acac)_4]$ and $[Y(acac)_3.2H_2O]$; it has also been reported for $[ReF_8]^{2-}$ (d^1) and for $[Mo(CN)_8]^{3-}$ (d^1) in $Na_3[Mo(CN)_8].4H_2O$. Evidence concerning the structure of the $[Mo(CN)_8]^{4-}$ ion (d^2) in solution is conflicting; the Raman spectrum has been interpreted on the basis of both the antiprismatic and the dodecahedral structures, though the e.s.r. spectrum appears to support the latter; it has, however, recently been reported that the anion is a square antiprism in $H_4[Mo(CN)_8].6H_2O$ and in the analogous tungsten compound.

So far the square antiprism has not been found among the structures of compounds of the metals of the first transition series. The dodecahedron, however, occurs in a group of nitrates in which the ligand is bidentate: $Ti(NO_3)_4$ (d^0), $[Mn(NO_3)_4]^{2-}$ (d^5), $[Co(NO_3)_4]^{2-}$

(d^7) and $[Zn(NO_3)_4]^{2-}$ (d^{10}). It occurs also in $[Ti(diarsine)_2Cl_4]$ (diarsine=*o*-phenylenebis-dimethylarsine), $[Zr(C_2O_4)_4]^{4-}$, $[CrO_8]^{3-}$ (in which four peroxide ions act as bidentate ligands), $[Mo(CN)_8]^{3-}$ in the tetrabutylammonium salt and $[Mo(CN)_8]^{4-}$ in $K_4[Mo(CN)_8]$. $2H_2O$.

14.3.9. Coordination number nine.[4] The only structure found in complexes not containing polydentate chelating groups (which may largely determine the configuration) is the tricapped trigonal prism shown in Fig. 14.10. This occurs in $[Nd(H_2O)_9]^{3+}$ (in which the Nd—O distances are all nearly equal), in several other hydrated rare earth cations (also, incidentally, in $SrCl_2.6H_2O$, in which each strontium ion has three water molecules bonded to it alone and shares six others with adjacent metal ions in a cationic chain) and in $[ReH_9]^{2-}$ (d^0), which has been studied by neutron diffraction as well as by X-ray analysis (see Section 8.10.4).

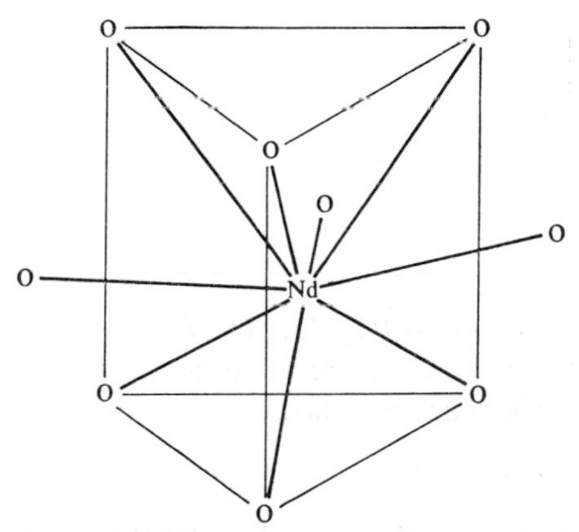

Figure 14.10

14.3.10. Higher coordination numbers.[4, 10–12] The only higher coordination numbers yet found in complexes are ten, eleven and twelve, and up to the present time coordination numbers above nine in molecules or ions have been identified only for the lanthanides and actinides. Examples of ten-coordination occur in $K_4Th(C_2O_4)_4.4H_2O$ (in which the arrangement of the ligands round the thorium is that of a bicapped square antiprism) and $La(NO_3)_2.2dipy$, in which the ten-coordination is provided by three bidentate nitrate ions and the two bidentate dipyridyl molecules; in this instance the structure is described as a bicapped dodecahedron.

Not surprisingly, the eleven-coordination of the thorium atom in $Th(NO_3)_4.5H_2O$ (eight-coordination by the bidentate nitrate ions plus three-coordination by some of the water molecules) is not describable in simple geometrical terms, though if the nitrate interactions are considered as single entities the geometry approximates to that of the capped trigonal prism ($[NbF_7]^{2-}$) structure. For twelve-coordination, the icosahedral structure shown in Fig. 14.11 has been established for cerium (III) and (IV) in the ions $[Ce(NO_3)_6]^{3-}$ and $[Ce(NO_2)_6]^{4-}$, and the same arrangement is found in $[Th(NO_3)_6]^{2-}$. Further examples will no doubt be discovered as the structural chemistry of the inner transition metals develops.

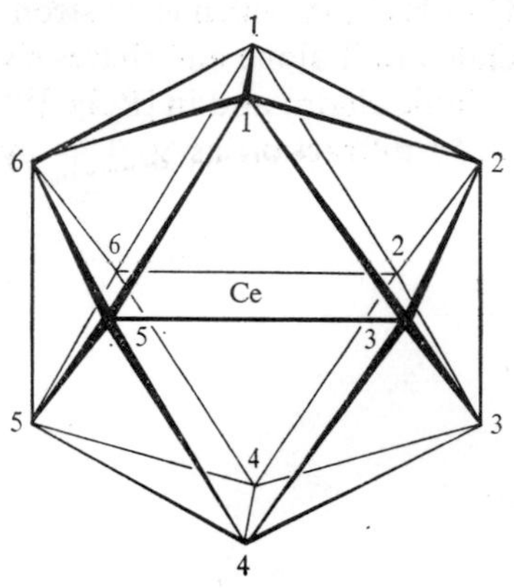

Figure 14.11 Six pairs of oxygen atoms (O_1, O_1; O_2, O_2, etc.) belong to six bidentate nitrate ions, the nitrogen and non-coordinated oxygen atoms of which are not shown

14.4 Isomerism in complexes[1, 2, 13, 14]

Examples of all of the types of isomerism mentioned below were discovered, mostly by Werner, during the last century, but we have chosen examples distinguished by relative simplicity or intensity of study. We have, too, used modern nomenclature throughout and have reproduced modern physicochemical work along with information obtained during the classical period of stereochemistry.

14.4.1. Geometrical isomerism. Geometrical isomers have the same structural framework, but differ in the spatial arrangement of the various substituent groups, the arrangement in a particular isomer being called its configuration. Configurational isomers can be isolated under ordinary conditions and are thus to be distinguished from different conformations of a molecule, which are in mobile equilibrium with one another. Whether a species exhibits evidence of configurational isomerism or merely of different conformations depends upon the activation energy for the interconversion process: the *cis* and *trans* isomers of 2-butene, for example, persist unchanged for long periods at room temperature, rotation about the double bond requiring a large amount of energy, whilst the staggered and eclipsed forms of ethane are very rapidly interconverted by rotation about the C—C bond. Clearly, however, there is an area in which such a distinction cannot be made rigidly, and at higher temperatures interconversion of geometrical isomers becomes much easier.

A regular tetrahedral molecule *Ma_2b_2 or Mabcd can exist in only one form, but a planar molecule Ma_2b_2 can exist in *cis* (a) and *trans* (b) forms, as shown below,

a b / M / b b and a b / M / b a

(a) (b)

A planar molecule Mabcd can exist in three forms, (c), (d) and (e), as shown below,

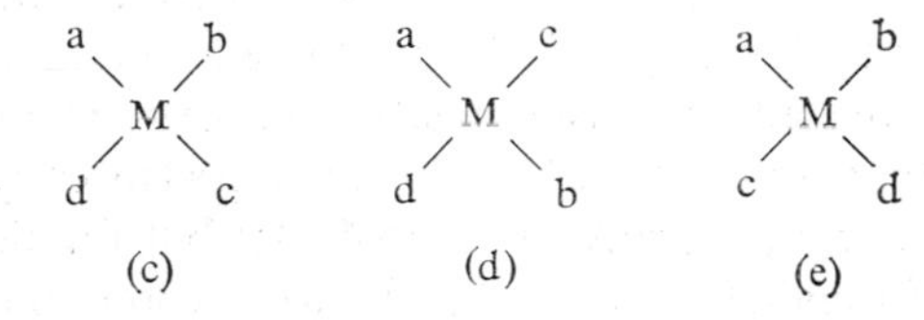

(c) (d) (e)

That there are two forms of many platinum complexes of the type $Pt^{II}a_2b_2$ (e.g. $[Pt(NH_3)_2Cl_2]$) and three isomers of the compound $[Pt(NH_3)(NH_2OH)(NO_2)(C_5H_5N)]NO_2$ eliminates the possibility of the groups round the platinum atom having a regular tetrahedral distribution; it does not *prove* the platinum atom and the four atoms bonded to it are coplanar, since, for example, a square pyramid, with the platinum atom at the apex, would lead to the same numbers of isomers. Furthermore, failure to isolate geometrical isomers may be just a question of technique, the more so since no general reliable methods are available.

As an example for further discussion we shall take the compound $[Pt(NH_3)_2Cl_2]$. The α-form of this substance is obtained by the action of ammonia on potassium tetrachloroplatinate (II) ($K_2[PtCl_4]$) as a

* Or ion (charges are omitted for clarity in this section).

yellow insoluble substance; ammonia converts it into $[Pt(NH_3)_4]Cl_2$, which on treatment with hydrochloric acid furnishes even less soluble $\beta[Pt(NH_3)_2Cl_2]$. Chemical identification of the isomers may be obtained by treating them with silver nitrate followed by oxalic acid, when the α compound gives a non-electrolyte and the β compound a dibasic acid. On the assumption that no change of structure occurs during these reactions and that oxalate can span only *cis* positions, the reactions are as shown in Fig. 14.12.

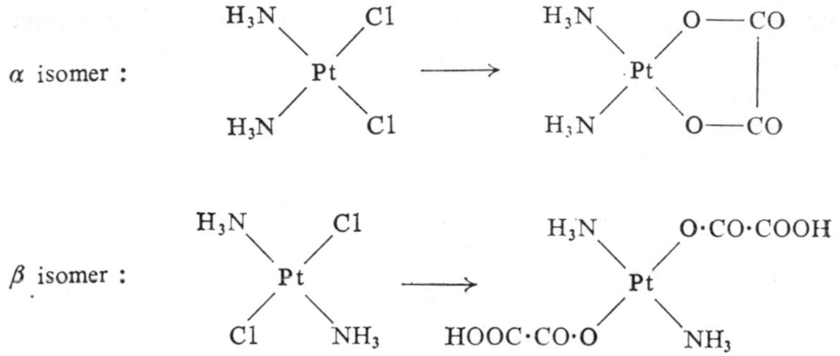

α isomer: H_3N, H_3N, Cl, Cl around Pt (cis) $\longrightarrow$ H_3N, H_3N around Pt with O—CO / O—CO chelate ring

β isomer: H_3N, Cl, Cl, NH_3 around Pt (trans) $\longrightarrow$ H_3N, O·CO·COOH, HOOC·CO·O, NH_3 around Pt

Figure 14.12

X-ray evidence confirms that these structures are correct. More rapid (if less convincing) physical methods for assignment of *cis* and *trans* configurations are provided by measurements of dipole moments (in solution in a non-polar organic solvent) and infrared spectroscopy. In the example just mentioned, no solvent is available; but for the analogous complexes of formula $[Pt(PEt_3)_2Cl_2]$ the *trans* compound is readily identified by its having a negligible dipole moment. Neither the Pt—Cl nor the Pt—NH_3 stretching frequency should appear in the infrared spectrum of *trans*-$[Pt(NH_3)_2Cl_2]$, which should thus have a simpler spectrum than that of the *cis* compound; as we have pointed out already, however, the vibrational spectra of solids are subject to several complicating factors, and measurements should be made in solution whenever possible (see Section 2.6).

Similar methods and considerations apply to the elucidation of the structures of geometrical isomers in which the configuration round the metal atom is octahedral. It may be noted that in the case of the compound $NH_4[Co(NH_3)_2(NO_2)_4]$, which is converted by oxalate into $NH_4[Co(NH_3)_2(NO_2)_2(C_2O_4)]$, there is strong evidence to suggest that a change of configuration takes place during the substitution of the nitrite ions by oxalate, and as a consequence chemical methods of assigning configurations are now to some extent discredited.

Among other elements to exhibit *cis–trans* isomerism is nickel in its complexes with unsymmetrical glyoximes, typical structures of which are shown in Fig. 14.13 ($R = CH_3$; $R' = CH_2C_6H_5$). Nickel

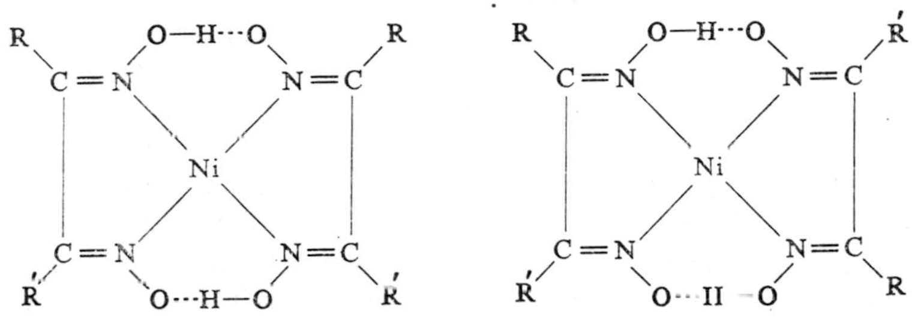

Figure 14.13

also exhibits another type of geometrical isomerism in certain complexes of formula $[NiL_2X_2]$, where L is a tertiary phosphine and X is chloride or bromide.

$[Ni(PRR'R'')_2X_2]$ has the *trans*-planar configuration if $R = R' = R'' = C_2H_5$ and $X = Br$, but is tetrahedral if $R = R' = R'' = C_6H_5$ and $X = Cl$; if $R = C_2H_5$, $R' = R'' = C_6H_5$ and $X = Br$, two isomers can be isolated and are interconvertible; if $R = CH_2C_6H_5$, $R' = R'' = C_6H_5$ and $X = Br$, molecules having the *trans*-planar and tetrahedral configurations actually occur in the same crystal, there being two of the tetrahedral configuration to every one of the planar configuration. Bond lengths are slightly less in the planar molecule.[15] In the case of nickel (II), the different configurations have different magnetic properties, planar complexes being diamagnetic and tetrahedral (and octahedral) complexes paramagnetic. We shall discuss this point further in Chapters 15 and 16.

14.4.2. Optical isomerism. Optical isomers differ only in the direction (right or dextro, and left or laevo) in which they rotate the plane of polarization of plane-polarized light. Such isomers (often called enantiomers) are related to one another as object and mirror images. For a finite molecule the formal criterion for enantiomorphism is the absence of an axis of rotatory inversion (possession of an *n*-fold axis of rotatory inversion means that the molecule is brought into coincidence with itself by rotation through $360°/n$ and inversion through its centre). For practical purposes this criterion normally corresponds to the absence of a centre or a plane of symmetry. Enantiomers are usually separated by forming a compound with one enantiomer (often a naturally occurring one) of a second optically active compound and utilizing differences in physical properties between the products. If, for example, a mixture of the dextro and laevo forms of an optically active base is treated with laevo-tartaric acid, the salts, which may be represented as (*d*-B)(*l*-A) and (*l*-B)(*l*-A), now differ slightly in all physical properties and can usually be separated by fractional crystallization. This process is called resolution; its reversal, the conversion of one optically active form into an inactive mixture of dextro and laevo forms, is termed racemization. Activation energies for racemization reactions vary widely and determine whether resolution is experimentally possible at a particular temperature. Thus failure to resolve a compound is not a proof that it has a centre or a plane of symmetry; a successful attempt at resolution does, however, establish beyond doubt the absence of these elements of symmetry. As we have mentioned already, rotation about a single bond, having a very low activation energy, takes place very readily at ordinary temperatures, and compounds which can acquire a centre or a plane of symmetry by such rotation are never resolvable.

Much elegant experimental work was done in the period before spectroscopic and diffraction methods came into use to elucidate molecular stereochemistry by resolution of suitably chosen compounds. A planar molecule Mabcd would not be resolvable, whereas a tetrahedral one would. Thus the resolutions of the compounds of Fig. 14.14 establish the tetrahedral configurations round phosphorus

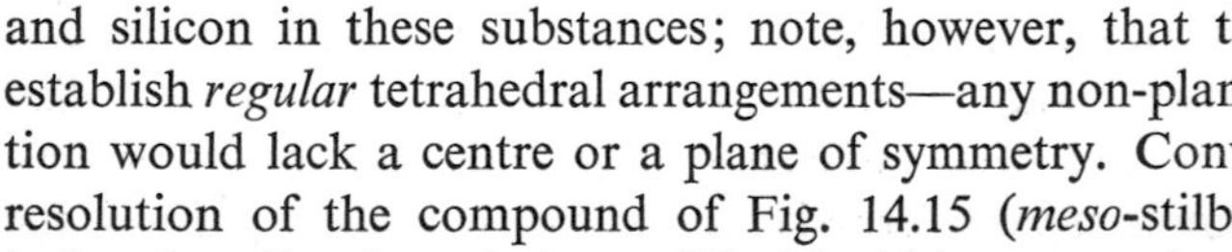

$$(CH_3)(C_2H_5)(C_6H_5)P \rightarrow O \quad \text{and} \quad C_6H_5CH_2{-}Si(C_2H_5)(C_3H_7){-}CH_2{\cdot}C_6H_4{\cdot}SO_3H$$

Figure 14.14

and silicon in these substances; note, however, that they do not establish *regular* tetrahedral arrangements—any non-planar distribution would lack a centre or a plane of symmetry. Conversely, the resolution of the compound of Fig. 14.15 (*meso*-stilbenediamino isobutylenediamino platinum (II) chloride) proves that the planes of the two rings are not perpendicular (since the ion would then have a plane of symmetry), but is not a conclusive demonstration of a planar configuration. Among cobalt (III) compounds, neither the *cis*

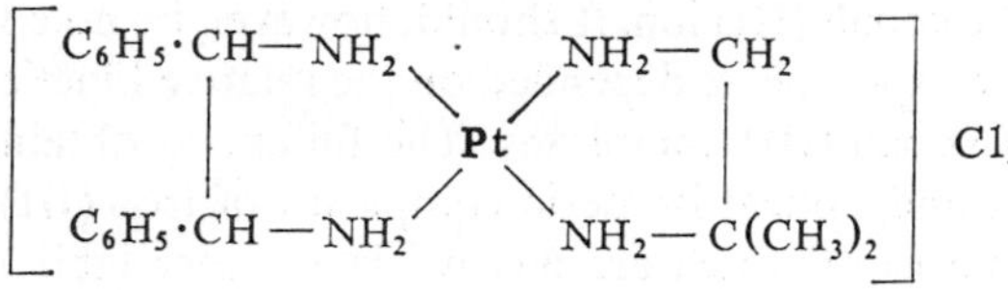

$$\left[\begin{array}{ccc} C_6H_5\cdot CH-NH_2 & & NH_2-CH_2 \\ | & Pt & | \\ C_6H_5\cdot CH-NH_2 & & NH_2-C(CH_3)_2 \end{array}\right]Cl_2$$

Figure 14.15

(Fig. 14.16(a)) nor the *trans* (Fig. 14.16(b)) form of the cation $[Co(NH_3)_4Cl_2]^+$ (present as a salt) should be resolvable. If, however, the four molecules of ammonia are replaced by two molecules of ethylenediamine (en), the rigidity conferred by the ring system should

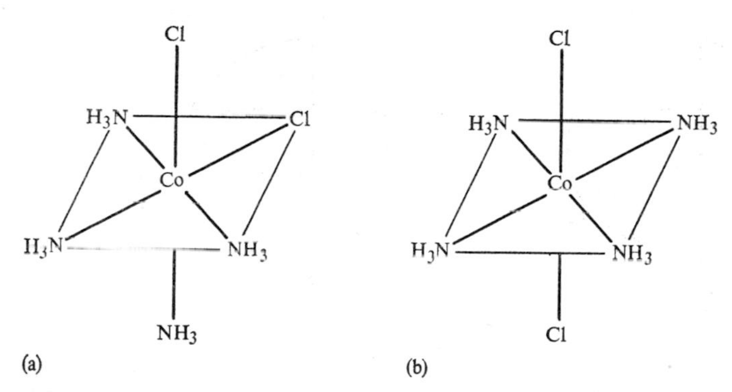

Figure 14.16

make the salt of the *cis* cation (Fig. 14.17(a)) resolvable while that of the *trans* cation (Fig. 14.17(b)) is not.

Confirmation of these predictions by Werner provided the strongest evidence short of full structure determination of the octahedral

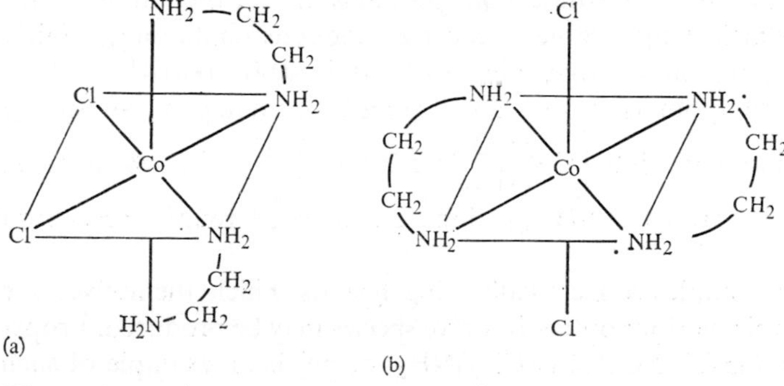

Figure 14.17

configuration round the cobalt (III) ion. It should, however, be noted that the success of these experiments depended on the relative kinetic inertness of nearly all cobalt (III) complexes. The failure to obtain analogous geometrical and optical isomeric complexes of iron (III) and aluminium (complexes of which are nearly always very labile) must not be taken to indicate that six-coordinated iron (III) and aluminium do not have an octahedral distribution of valencies; there is, indeed, convincing evidence from X-ray diffraction that they do.

Examination of the structure of the $[Co(en)_3]^{3+}$ ion shows that, although no geometrical isomerism is possible, the ion can exist in two forms (Fig. 14.18(a) and (b)), which are mirror images, and salts

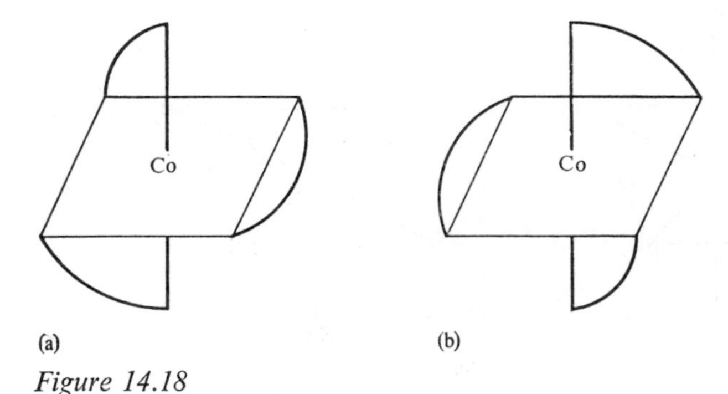

Figure 14.18

of the ion have been resolved into optically active forms. In general, X-ray diffraction (which confirms the structure of the ion) cannot distinguish between mirror images, but if the wavelength of the X-rays is so chosen as to produce an electronic transition of the cobalt atom an additional effect (anomalous X-ray dispersion) which permits the determination of absolute configurations arises; the study of this effect (which is applicable to most compounds containing a fairly heavy atom) shows that Fig. 14.18(a) is *D*-$[Co(en)_3]^{3+}$. Many similar compounds have been resolved, including a salt of the purely inorganic ion $\left[Co\left(\begin{matrix}HO\\HO\end{matrix}\!\!>Co(NH_3)_4\right)_3\right]^{6+}$, in which three bidentate $[(HO)_2Co(NH_3)_4]^+$ ions are chelated found the central atom.

When complexes are made using ligands which themselves are asymmetric, various optically active species may be produced. Propylenediamine ($H_2NCH_2CH(CH_3)NH_2$, or pn) is an example of such a ligand. If we denote the configuration of the whole ion by *D* or *L*, and that of the ligand molecules by *d* or *l*, the eight possible combinations are:

D-$[Co(l\text{-pn})_3]^{3+}$	*L*-$[Co(l\text{-pn})_3]^{3+}$
D-$[Co(l\text{-pn})_2(d\text{-pn})]^{3+}$	*L*-$[Co(l\text{-pn})_2(d\text{-pn})]^{3+}$
D-$[Co(l\text{-pn})(d\text{-pn})_2]^{3+}$	*L*-$[Co(l\text{-pn})(d\text{-pn})_2]^{3+}$
D-$[Co(d\text{-pn})_3]^{3+}$	*L*-$[Co(d\text{-pn})_3]^{3+}$

Such ions, like the salts of the different forms of an optically active base with *l*-tartaric acid, have slightly different thermodynamic properties, and some are formed preferentially in reactions which could produce random mixtures. For example when $[Co(l\text{-pn})_2Cl_2]^+$ reacts with *d*-pn, a mixture of *D*-$[Co(d\text{-pn})_3]^{3+}$ and *L*-$[Co(l\text{-pn})_3]^{3+}$, rather than mixed complexes, is obtained. The stereoselectivity found here and in related equilibria can be interpreted in terms of the packing of the chelate rings (which are, of course, puckered) round the metal atom, different arrangements leading to different amounts of repulsion between hydrogen atoms of different ligand molecules. The further discussion of stereoselectivity, however, lies beyond the scope of this book and a review article[16] should be consulted.

Two further related aspects of optical activity should be mentioned before we leave this subject—*optical rotary dispersion* and *circular dichroism*.[1, 2, 17] The effect of a given concentration of a particular optically active species on the plane of polarization of light is not constant, but depends upon the wavelength of the light, the direction as well as the extent of rotation being affected. This variation in sign and magnitude of rotation with wavelength, which is illustrated in Fig. 14.19, is known as optical rotatory dispersion. If the rotation rises to a maximum towards short wavelengths before changing sign, the compound is said to show a positive Cotton effect; the opposite behaviour constitutes a negative Cotton effect.

In order to discuss circular dichroism we must first consider further the interaction of polarized light and a molecule of an optically active substance. A beam of plane polarized light may be thought of as two coterminous beams of left- and right-handed circularly polarized light of equal amplitudes and in phase. Optical activity is then due to the different refractive indices of the medium for left- and right-handed circularly polarized light, n_l and n_r respectively. If the rotation per unit path length of the plane polarized light is α and its wavelength is λ,

$$\alpha = (n_l - n_r)\pi/\lambda$$

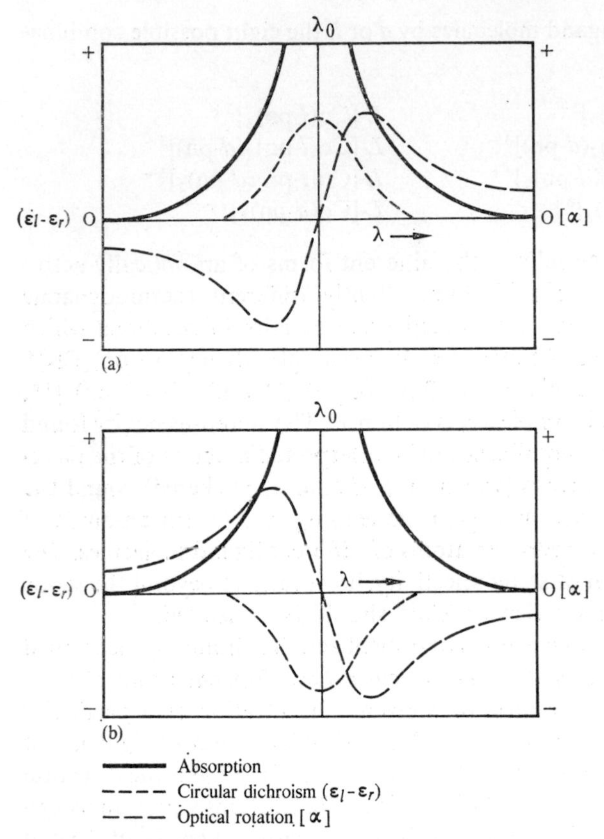

Figure 14.19 The Cotton effect, circular dichroism $\varepsilon_l - \varepsilon_r$, and optical rotatory dispersion [α] for an asymmetric compound having an absorption band centred at λ_0. (a) a positive Cotton effect (b) a negative Cotton effect

If $n_l > n_r$, the left-handed circularly polarized component is relatively delayed during passage through the optically active medium, and the plane of the emergent linearly polarized beam is rotated to the right. Since the refraction and absorption of light are interconnected, the condition $n_l > n_r$ in the longer-wavelength region in which an optically active medium showing a positive Cotton effect is transparent implies $\varepsilon_l > \varepsilon_r$ holds in the shorter-wavelength region where light is absorbed, ε_l and ε_r being the extinction coefficients of the medium for left- and right-handed circularly polarized light. A positive elliptical polarization or, as it is commonly called, circular dichroism $\varepsilon_l - \varepsilon_r$

then results; conversely, if $\varepsilon_r > \varepsilon_l$, a negative circular dichroism results. A positive Cotton effect thus denotes a positive circular dichroism as well as a particular type of variation of α with λ. Finally, it must be mentioned that the wavelength at which α is zero and the circular dichroism is a maximum or minimum coincides with that of a maximum in the electronic absorption spectrum of the optically active species. The relationship between the electronic absorption, the optical rotation α and the circular dichroism $\varepsilon_l - \varepsilon_r$ is shown in Fig. 14.19.

For closely related molecules or ions the optical rotatory dispersion and circular dichroism effects are similar, and can be used to correlate the configurations of related optically active molecules and hence to follow the steric course of certain reactions. Thus comparisons with *D*-$[Co(en)_3]^{3+}$ have been used to assign absolute configurations to some other cobalt (III)–ammine complexes. The relationship of the Cotton effects to electronic absorption is, however, a very difficult subject, and we shall not pursue it here beyond remarking that data on optical rotatory dispersion and circular dichroism are sometimes useful in the interpretation of electronic spectra.

14.4.3. Ionization, hydrate and coordination isomerism. These three names, all of which originated with Werner, refer to cases of isomerism in which there is ligand-exchange between a complex ion and the remainder of the compound.

Ionization isomerism involves interchange of a ligand anion with an associated anion outside the complex. An example is provided by the compounds $[Co(NH_3)_5Cl]SO_4$ and $[Co(NH_3)_5(SO_4)]Cl$, obtained by the reactions

$$[Co(NH_3)_5(H_2O)]Cl_3 \xrightarrow{\text{heat}} [Co(NH_3)_5Cl]Cl_2 \xrightarrow{Ag_2SO_4} [Co(NH_3)_5Cl]SO_4$$

and

$$[Co(NH_3)_5Cl]Cl_2 \xrightarrow{\text{conc. } H_2SO_4} [Co(NH_3)_5(SO_4)]HSO_4 \xrightarrow{BaCl_2} [Co(NH_3)_5(SO_4)]Cl$$

These substances give the usual reactions of ionic sulphate and chloride respectively. A further example is provided by the compounds $[Pt(NH_3)_3Cl]I$ and $[Pt(NH_3)_3I]Cl$.

In *hydrate isomerism* water replaces the ligand anion of ionization isomerism. The best-known example is that of the hydrated chromium (III) chlorides. When hydrogen chloride is passed into a freshly prepared solution of chromium (III) alum, violet $[Cr(H_2O)_6]Cl_3$ is obtained; addition of ether saturated with hydrogen chloride to the filtrate after removal of this compound gives light green $[Cr(H_2O)_5Cl]Cl_2.H_2O$; and crystallization of a hot solution obtained by reducing chromium (VI) oxide with concentrated hydrochloric acid gives $[Cr(H_2O)_4Cl_2]Cl.2H_2O$. These isomers in freshly prepared aqueous solution react with silver ion to precipitate all, two-thirds and one-third of their chloride respectively.

Coordination isomerism is an extreme case of ionization isomerism which arises when both anion and cation are complexes (e.g. $[Co(NH_3)_6][Cr(CN)_6]$ and $[Cr(NH_3)_6][Co(CN)_6]$, both made by precipitation from solutions containing the appropriate ions). Within one complex containing two metal atoms in different environments coordination position isomerism may occur; for example, in

$$\left[(NH_3)_4Co\begin{matrix} H \\ O \\ \diagup\;\diagdown \\ \\ \diagdown\;\diagup \\ O \\ H \end{matrix}Co(NH_3)_2Cl_2\right]Cl_2$$

and

$$\left[Cl(NH_3)_3Co\begin{matrix} H \\ O \\ \diagup\;\diagdown \\ \\ \diagdown\;\diagup \\ O \\ H \end{matrix}Co(NH_3)_3Cl\right]Cl_2$$

14.4.4. Linkage isomerism.[18] Ligands with two or more possible sites of attachment to a metal are known as ambident ligands; in principle they include NO_2^-, SCN^-, CN^-, $S_2O_3^{2-}$, CO, $CO(NH_2)_2$, $CS(NH_2)_2$ and $(CH_3)_2SO$, but up to the present only the first four of these have been shown to form linkage isomers (i.e. isomers differing only in the method of attachment of one or more ligands). Isomeric nitrito and nitro compounds, containing ONO and NO_2 radicals respectively, have been known for many years; their preparation is summarized by the scheme shown on the next page.

Thiocyanate forms both M–SCN (thiocyanato) and M–NCS (isothiocyanato) complexes, generally with class 'b' and class 'a' metals respectively (see Section 6.4.3). Zinc, for example, bonds to nitrogen and mercury to sulphur. In a few cases both thiocyanato and isothiocyanato complexes are known for the same metal (e.g. [dipy

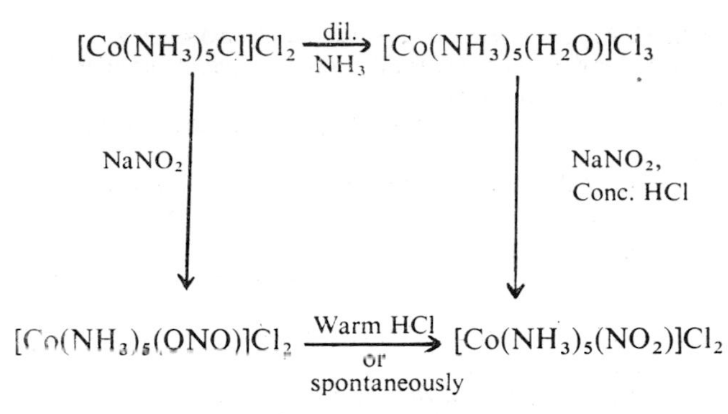

$Pd(SCN)_2]$ and $[dipyPd(NCS)_2]$, $[Mn(CO)_5(SCN)]$ and $[Mn(CO)_5(NCS)]$).

Linkage isomers are readily identified by means of infrared spectroscopy. The symmetrical NO_2 stretching frequency, for example, is quite different in nitro and nitrito groups bonded to a metal atom, being at 1310 and 1065 cm^{-1} respectively in the pentammine cobalt (III) complexes. Similarly, the C—S stretching frequency is much higher in M—N=C=S (*ca.* 820 cm^{-1}) than in M—S—C≡N (*ca.* 700 cm^{-1}).

14.4.5. Polymerization isomerism. This is not strictly isomerism; its inclusion under this heading arose originally because of the impossibility of determining molecular weights of insoluble non-volatile compounds. There are three compounds of empirical formula $Pt(NH_3)_2Cl_2$, *cis*- and *trans*-$[Pt(NH_3)_2Cl_2]$, and an insoluble green compound $[Pt(NH_3)_4][PtCl_4]$ made by mixing solutions of $K_2[PtCl_4]$ and $[Pt(NH_3)_4]Cl_2$, which are yellow-brown and colourless respectively. In the green compound, commonly known as Magnus's Green Salt, parallel square anions and cations are stacked in chains with Pt—Pt bonding. Another example of polymerization isomerism is provided by $[Co(NH_3)_6][Co(NO_2)_6]$ and the non-electrolyte $[Co(NH_3)_3(NO_2)_3]$ which is formed, together with other products, by oxidation of an ammoniacal solution of a cobalt (II) salt containing nitrite with air or hydrogen peroxide in the presence of charcoal as catalyst (a technique widely used in the synthesis of cobalt (III) complexes, though its mechanism is not known).

References for Chapter 14

1 Cotton, F. A., and Wilkinson, G., *Advanced Inorganic Chemistry*, 2nd edn. (Interscience, 1966).

2 Kettle, S. F. A., *Coordination Compounds* (Nelson, 1969).
3 Cartmell, E., and Fowles, G. W. A., *Valency and Molecular Structure*, 3rd edn. (Butterworths, 1966).
4 Muetterties, E. L., and Wright, C. M., *Quart. Rev. Chem., Soc.*, 1967, **21,** 109.
5 Downs, A. J., in *New Pathways in Inorganic Chemistry*, eds. E. A. V. Ebsworth, A. G. Maddock and A. G. Sharpe (Cambridge University Press, 1968).
6 Muetterties, E. L., and Schunn, R. A., *Quart. Rev. Chem. Soc.*, 1966, **20,** 245.
7 Raymond, K. N., Corfield, P. W. R., and Ibers, J. A., *Inorg. Chem.*, 1968, **7,** 1362.
8 Parish, R. V., *Coord. Chem. Rev.*, 1966, **1,** 439.
9 Lippard, S. J., *Prog. Inorg. Chem.*, 1967, **8,** 109.
10 Akhtar, M. N., and Smith, A. J., *Chem. Comm.*, 1969, 705.
11 Al-Karaghouli, A. R., and Wood, J. S., *J. Amer. Chem. Soc.*, 1968, **90,** 6548.
12 Beineke, T. A., and Delgaudio, J., *Inorg. Chem.*, 1968, **7,** 718.
13 Wilkins, R. G., and Williams, M. J. G., in *Modern Coordination Chemistry*, eds. J. Lewis and R. G. Wilkins (Interscience, 1960).
14 Basolo, F., and Pearson, R. G., *Mechanisms of Inorganic Reactions*, 2nd edn. (Wiley, 1967).
15 Kilbourn, B. T., and Powell, H. M., *J. Chem. Soc.*, **A,** 1970, 1688.
16 Dunlop, J. H., and Gillard, R. D., *Adv. Inorg. Chem. Radiochem.*, 1966, **9,** 185.
17 Gillard, R. D., *Prog. Inorg. Chem.*, 1966, **7,** 215.
18 Burmeister, J. L., *Coord. Chem. Rev.*, 1968, **3,** 225.

15 Complexes of transition metals II: bonding

15.1 Introduction

In Section 14.1 we mentioned the principal aspects of transition metal chemistry to which a theory of bonding should be applicable: structures, electronic spectra, magnetic properties, distortions from symmetrical structures, thermodynamic properties and kinetics. We now give an account of the most important theories of bonding in transition metal compounds in relation to these subjects. In doing so, we shall in this chapter follow the historical order of the development of theories of bonding, but subsequently we shall consider selected properties in more detail without regard to historical factors. In general, we shall limit the theoretical treatment to the minimum needed to deal with the experimental material; for fuller and more rigorous accounts of theory several excellent texts[1–6] are available.

The first generalization about the electronic configurations of transition metal complexes was Sidgwick's *effective atomic number rule*, according to which the central atom or ion in the complex accepts pairs of electrons from the ligands until its total number of electrons is that of the next noble gas. The neutral atom of iron, for example, has twenty-six electrons, so Fe^{2+} has twenty-four; acceptance of six pairs of electrons from six CN^- ions, as in $[Fe(CN)_6]^{4-}$, then brings the total number of electrons to thirty-six, the same as that of krypton; the simplest carbonyl of iron is $Fe(CO)_5$, a derivative of iron (0), and so the metal again attains a noble gas configuration. Although the effective atomic number rule is still an invaluable guide to the formulae of carbonyls and organometallic compounds, there are so many exceptions to it (e.g. $[Fe(CN)_6]^{3-}$ and $[FeCl_4]^-$, and the hexaaquo complexes of all but d^6 ions) that it is no longer useful in transition metal chemistry in general.

For two decades from 1930, Pauling's *valence bond theory*[7] was the standard medium for the discussion of transition metal complexes;

this concentrated on structure and magnetic properties, but had little to offer in connexion with electronic spectra, thermodynamics or kinetics. Furthermore, with the realization that most structures can be accounted for on the basis of repulsions between pairs of electrons just as well as on the basis of hybridized orbitals (see Section 5.2), the importance of describing structures (as distinct from predicting them—a much less common operation) began to diminish. Nevertheless, so much of the terminology and so many of the ideas of valence bond theory are still retained in some form that an acquaintance with it remains essential.

Next came *crystal field theory*, an electrostatic approach first developed many years earlier, but largely neglected by chemists, who were at that time generally more involved with the covalent bond. Crystal field theory is concerned with the effect of the external electric field due to the ligands on the relative energy levels of the *d* orbitals of the central atom or ion; in a regular octahedral field, for example, the orbitals split into a group of three of lower energy and a group of two of higher energy. The nature of the splitting determines the distribution of electrons among orbitals and hence leads to the interpretation of magnetic and spectroscopic properties, and is often very useful in the discussion of distorted structures, thermodynamic properties and kinetics. Because of its essential simplicity, crystal field theory has had an enormous influence in inorganic chemistry during the last twenty years, and for complexes for which the electrostatic model is most nearly valid (e.g. complex halides) it is still the treatment most widely used.

Just as molecular orbital theory has displaced electrostatic and valence bond theories in dealing with non-ionic compounds of typical elements, however, *ligand field theory* has now become the standard approach to the study of transition metal complexes in general. In this theory, the orbitals of the ligands are combined with those of the metal in the manner of molecular orbital theory; many of the conclusions and predictions of crystal field theory are not thereby affected substantially, but the arbitrary concentration on the metal orbitals is avoided and it is possible to deal more satisfactorily with complexes containing ligands such as CN^-, CO and organic groups, in many of which there is strong evidence for π bonding between metal and ligand. Because of its wider applicability, ligand field theory is now considered to be in principle the most generally useful and most nearly valid theory of transition metal complexes, and it is becoming common practice to use this title to include the

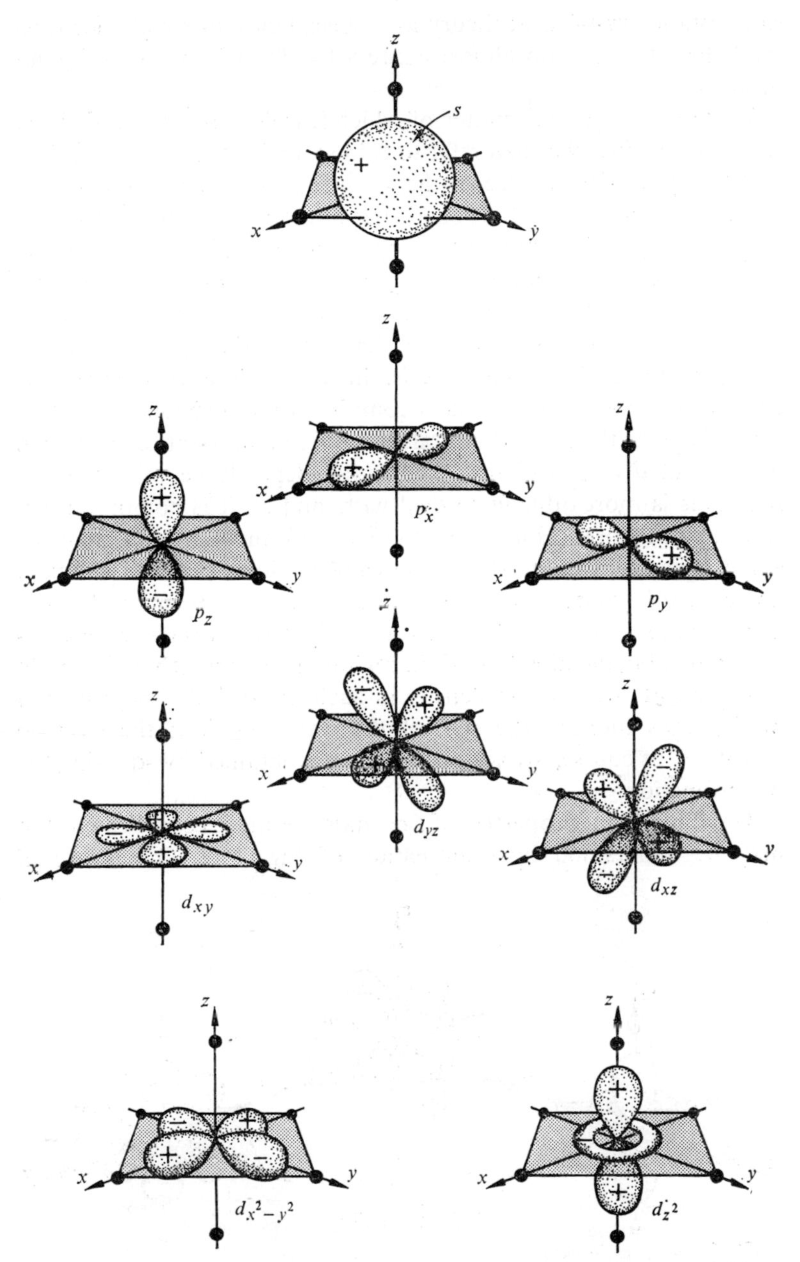

Figure 15.1 Directional properties of *s*, *p* and *d* orbitals

electrostatic crystal field theory as a special case in which electrons are believed to remain almost entirely localized in metal or ligand orbitals.

There is one general matter of which it is desirable to remind the reader here, since we shall refer to it in several different places. This is the angular dependence of the wave-functions for different electrons. The *s* orbital has full spherical symmetry; the three degenerate p_x, p_y and p_z orbitals are mutually perpendicular. In the case of the *d* orbitals it is not possible to choose five functions which are independent and have the same shape, and it is customary to introduce the d_{z^2} orbital having a different shape from the d_{xy}, d_{xz}, d_{yz} and $d_{x^2-y^2}$ orbitals. (This difference is, in fact, only apparent: the d_{z^2} orbital can be written as a linear combination of orbitals having the same shape as the other *d* orbitals but not independent of them (e.g. $d_{z^2-x^2}$ and $d_{z^2-y^2}$ orbitals). The radial dependence of the wave-functions is more difficult to deal with, and so it is common practice in non-theoretical accounts to reproduce the wave-functions for a hydrogen atom and to state that their directional properties remain unchanged in more complex systems. Thus the conventional boundary contours shown in Fig. 15.1 are only a very rough representation indicating the direction of the region in space which will contain nearly all of the electron density in such an orbital. The plus and minus signs indicate the sign of the wave-function; the electron density is, of course, always positive and is obtained by squaring the wave-function.

The directional properties of *f* orbitals are more complicated, but may be understood by examination of Fig. 15.2. One *f* orbital

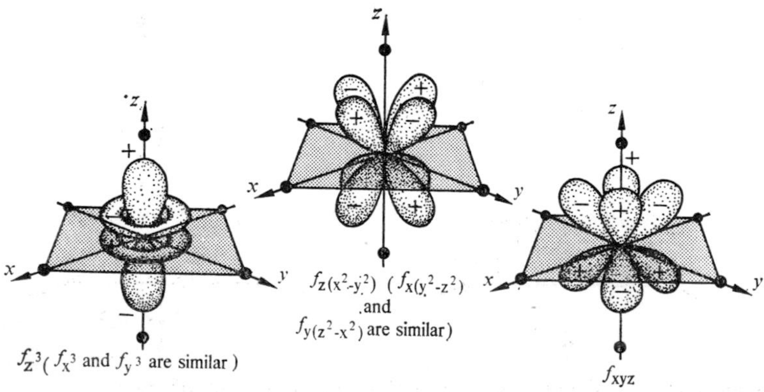

Figure 15.2 Directional properties of *f* orbitals

(denoted f_{xyz}) has eight lobes pointing between the x-, y- and z-axes; three more ($f_{z(x^2-y^2)}$, $f_{x(y^2-z^2)}$ and $f_{y(z^2-x^2)}$) have eight lobes directed along two of the axes; and the remaining three (f_{x^3}, f_{y^3} and f_{z^3}) are somewhat analogous to the d_{z^2} orbital (though the wave-function changes sign at the nucleus) and are conventionally represented as directed along one axis with two more regions of electron density (corresponding to two different signs for the wave-function) above and below the plane of the other two axes.

15.2 Valence bond theory

In valence bond theory applied to transition metal complexes we examine how the d orbitals can be hybridized with other orbitals of the same atom to give new orbitals directed according to the symmetry of the complex. These orbitals are then considered to overlap ligand orbitals and each of the resulting orbitals is occupied by a pair of electrons donated by a ligand.

Let us consider the case of an octahedral complex as an example. Hybridization of the s, p_x, p_y, p_z, $d_{x^2-y^2}$ and d_{z^2} orbitals (preferably the d orbitals of principal quantum number one less than that of the s and p orbitals) leads to the formation of six hybrid d^2sp^3 bonds directed towards the six ligands, and overlap of these with six ligand orbitals leads to formation of six σ bonds. The d_{xy}, d_{yz} and d_{xz} atomic orbitals, being directed between the axes on which the ligands are placed, play no part in σ bonding, but they may overlap p or d orbitals on the ligands to form π bonds. If the ligand has vacant orbitals and there are electrons in the d_{xy}, d_{yz} and d_{xz} orbitals, such π bonding will help to remove negative charge from the central metal atom and is often referred to as 'back-donation'; its importance was first recognized by Pauling in connexion with his *electroneutrality principle*, according to which charge tends to get distributed within a molecule or ion so that no atom has a resultant charge greater than about $\pm e$.

The valence bond diagram for an octahedral complex of Cr^{III} (d^3) can then be represented as:

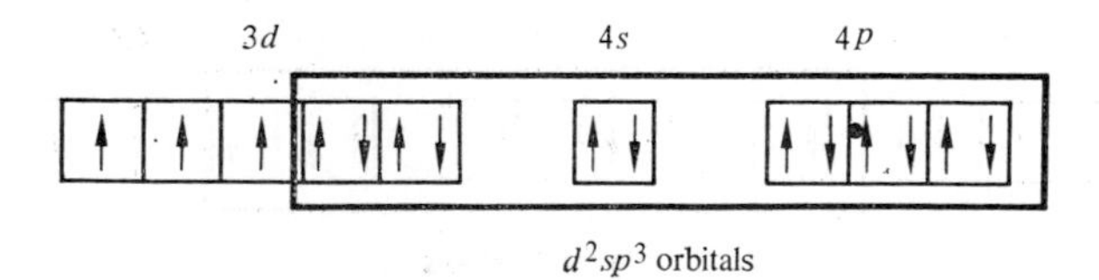

Hund's rule applies to the non-bonding *d* electrons, and the complex has a paramagnetic susceptibility corresponding to the presence of three unpaired electrons. If, however, we consider an octahedral complex of Cr^{II} or Mn^{III} (d^4), two possibilities are found to exist: either the complex may have four unpaired electrons (as shown by magnetic susceptibility) or it may have only two. Octahedral complexes of Mn^{II} or Fe^{III} (d^5) similarly fall into two groups having five and one unpaired electrons respectively, whilst those of Fe^{II} and Co^{III} (d^6) have four or zero unpaired electrons. Such types of complex are best known by the self-explanatory terms high-spin (or, somewhat ambiguously and less satisfactorily, spin-free) and low-spin (or spin-paired) respectively. In the early days of valence bond theory, however, high-spin complexes, in which the magnetic moment of the free ion remained unchanged, were called *ionic complexes*, whereas those in which the moment was modified were called *covalent complexes*. Thus in $[FeF_6]^{3-}$, a high-spin complex, the electronic configuration of the Fe^{3+} ion would be represented as the same as it is in the gas-phase ion by means of the diagram:

3*d* 4*s* 4*p*

In $[Fe(CN)_6]^{3-}$, a low-spin complex, it would be represented as:

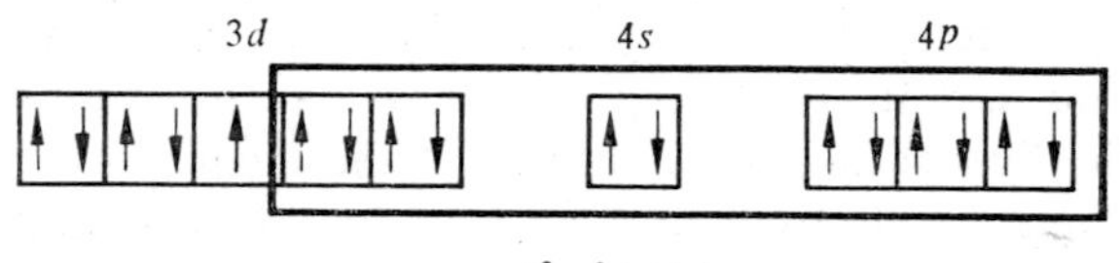

d^2sp^3 orbitals

Later, the view that all complexes contained covalent bonds came to prevail, and it was then believed that high-spin complexes utilized *d* orbitals of the same quantum number as the *s* and *p* orbitals for the formation of hybrid bonds; high-spin complexes were then called outer-orbital complexes (low-spin complexes being inner-orbital complexes), and the diagram for the electronic configuration of Fe^{3+} in $[FeF_6]^{3-}$ became:

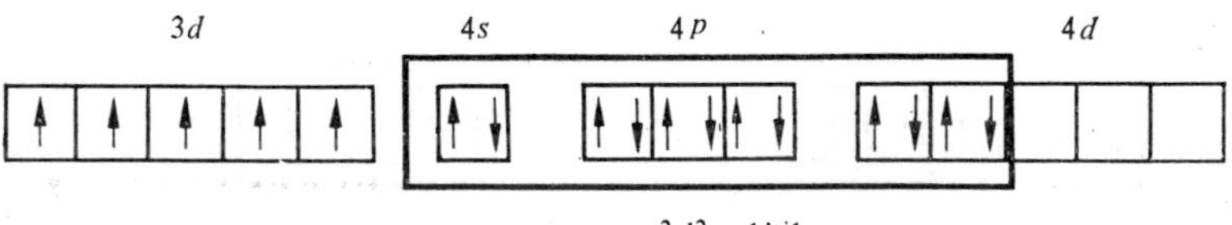

sp^3d^2 orbitals

(Only a combination of the $d_{x^2-y^2}$ and d_{z^2} orbitals with one s and three p orbitals will give an octahedral arrangement.) The distinction between inner- and outer-orbital complexes is, however, solely one of magnetic properties, and does not imply anything about bond strength (though it has often been taken to do so).

The combinations of orbitals required for hybridization to give a particular geometry have been worked out by a number of authors, and some of the results are collected in Table 15.1. Although it is often possible to find examples of a particular geometrical configuration, it would certainly not have been possible to predict the occurrence of all coordination numbers and configurations described in the last chapter. Nevertheless, it is interesting to note, for example, that cubic eight-coordination requires the use of f orbitals and that this configuration has only recently been found in a complex of protactinium. Nor are other theories of valency much more successful in predicting the structures of compounds of high coordination number; and since, as we showed in the last chapter, there is apparently often little difference in energy between different structures, prediction in this field is, and is likely to remain, a very severe test of any theory.

The formation of complexes in which the coordination number of

Table 15.1 *Sigma-bonding orbitals for some geometrical configurations*

Coordination number	*Geometry*	*Sigma-bonding orbitals*
2	Linear	s, p_z or d_{z^2}, p_z
3	Equilateral triangle	s, p_x, p_y or d_{z^2}, p_x, p_y
4	Tetrahedral	s, p_x, p_y, p_z or $s, d_{xy}, d_{xz}, d_{yz}$
	Square	$d_{x^2-y^2}, s, p_x, p_y$ or $d_{x^2-y^2}$ d_{z^2}, p_x, p_y
5	Trigonal bipyramid	$s, p_x, p_y, p_z, d_{z^2}$
6	Octahedron	$s, p_x, p_y, p_z, d_{x^2-y^2}, d_{z^2}$
8	Square antiprism	$d_{xy}, d_{xz}, d_{yz}, d_{x^2-y^2}, d_{z^2}, p_x, p_y, p_z$
	Dodecahedron	$d_{x^2-y^2}, d_{z^2}, d_{xz}, d_{yz}, s, p_x, p_y, p_z$
	Cube	$s, p_x, p_y, p_z, d_{xy}, d_{xz}, d_{xy}, f_{xyz}$

the transition metal is other than six may be illustrated by a few examples. The square planar configuration of many Ni^{II}, Pd^{II} and Pt^{II} (d^8) complexes, all of which are diamagnetic (i.e. have no unpaired electrons), may be represented in the typical case of $[Ni(CN)_4]^{2-}$ as the hybridization of the $3d_{x^2-y^2}$, $4s$, $4p_x$ and $4p_y$ orbitals to give dsp^2 hybrid orbitals, calculated and found to result in the planar configuration:

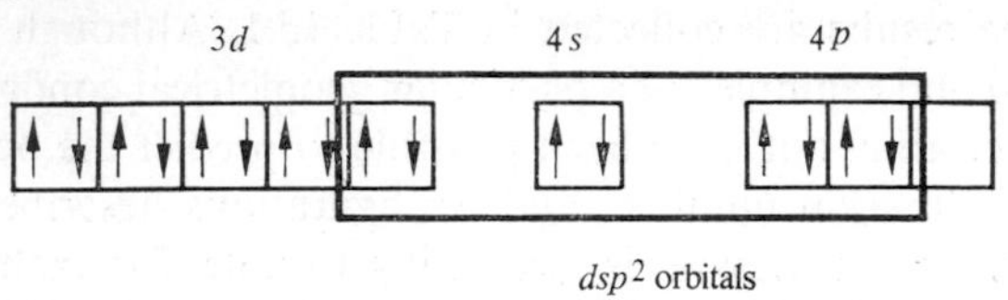

dsp^2 orbitals

Similarly, a tetrahedral complex of Ni^{II} (d^8) would involve occupation of the $4s$ and $4p$ orbitals by electrons from the ligands; the eight electrons of the metal ion could then occupy all the $3d$ orbitals, giving two unpaired electrons:

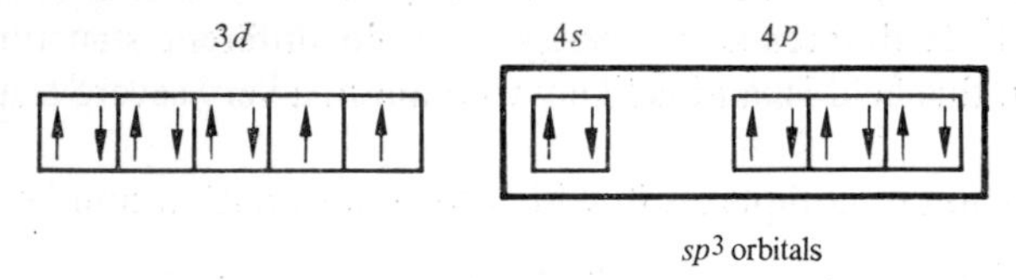

sp^3 orbitals

This correlation of stereochemistry and magnetic properties in the case of nickel (II) complexes was undoubtedly one of the successes of valence bond theory. There is, however, no correlation implied here between either of these properties and the electronic spectra of the complexes, nor can we see why some ligands give rise to high-spin and some to low-spin complexes. To understand these matters we must turn to later theories.

15.3 Crystal field theory

In crystal field theory, a complex is regarded as composed of a metal ion surrounded by anions or negative ends of polar ligands, and the bonding is assumed to be purely electrostatic. We shall confine attention to octahedral, tetrahedral, tetragonal and square complexes; the treatment of more complex systems is difficult, and few useful results have yet been obtained.

15.3.1. Octahedral complexes. Suppose six ligands are brought towards the positively charged central metal ion along the x-, y- and

z-axes. There will be a net electrostatic attraction, but owing to repulsion between the ligand electrons and electrons in the d orbitals of the metal there will be a general increase in the energy of these orbitals. However, not all d electrons will be affected equally; those in the $d_{x^2-y^2}$ and d_{z^2} orbitals (which are directed along the x-, y- and z-axes) will be repelled more than those in the d_{xy}, d_{yz} and d_{xz} orbitals, which are directed between these axes. The former group of orbitals are, for reasons which will be indicated later, designated e_g orbitals, the latter group t_{2g} orbitals (the respective names d_γ and d_ε are also used sometimes).

It is obvious that the three t_{2g} orbitals are degenerate. This is, however, also true of the two e_g orbitals: the d_{z^2} orbital can be envisaged as a linear combination of $d_{z^2-x^2}$ and $d_{z^2-y^2}$ orbitals which, by symmetry, must behave as the $d_{x^2-y^2}$ orbital does. The relative amounts by which the energies of the t_{2g} and e_g orbitals are affected can be seen by considering the following argument. Suppose a d^{10} cation is placed at the centre of a hollow sphere of radius equal to the metal–ligand distance, with a total charge equal to that of the six ligands spread uniformly over its surface. In this symmetrical environment the five d orbitals are still degenerate. Then imagine the total charge on the sphere being collected into six point-charges lying on the surface of the sphere but also at the apexes of the octahedron associated with the x-, y- and z-axes. Such redistribution of charge cannot affect the total energy of the system when all the d orbitals are filled symmetrically, and therefore the combined energy of the e_g orbitals must be raised by the same amount as that by which the combined energy of the t_{2g} orbitals is lowered. The difference in energy between the t_{2g} and e_g orbitals is known as Δ_o (the subscript denotes the octahedral environment) or sometimes (for historical reasons) as $10Dq$. Thus the three t_{2g} orbitals are each $0{\cdot}4\ \Delta_o$ or $4Dq$ lower, and the two e_g orbitals each $0{\cdot}6\ \Delta_o$ or $6Dq$ higher, than the centre of gravity of the set of levels. This pattern of splitting is general for any splitting of energy levels so long as the forces are purely electrostatic and the set of levels being split is well removed from other sets of energy levels with which they might interact.

The relationship between the energy levels of d orbitals in the free metal ion, in a spherically symmetrical field and in an octahedral crystal field, is summarized in Fig. 15.3. The difference between the energy levels in the free ion and in a spherically symmetrical field depends upon the nature of the metal ion, the nature of the ligand and the internuclear distance. It is important to understand that although the energies of the *d* orbitals are raised when the ion is put into any negative field, the total energy of the system is lowered by the much larger effect of the electrostatic interaction of the negative charges or dipoles with the positive charge on the cation. The magnitude of Δ_o is usually quite small in comparison with the total energy of interaction, but it is on the splitting of the *d* orbitals that crystal field theory concentrates attention; this point is sometimes forgotten when small variations in Δ_o are being discussed for systems for which the total interaction energies may be quite different.

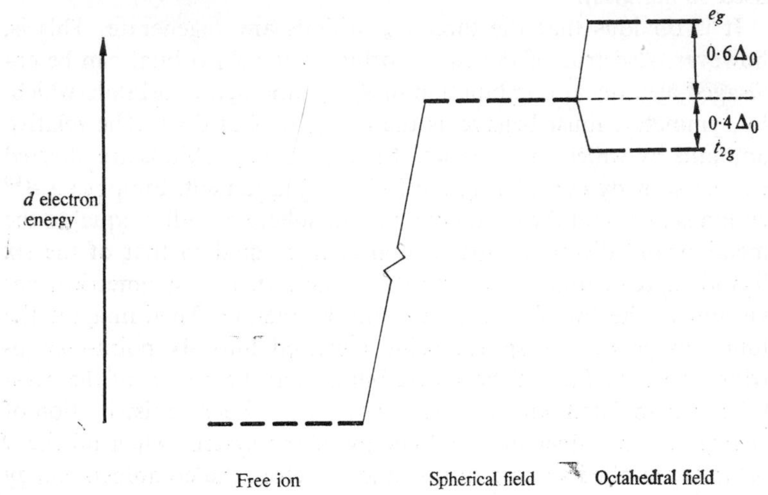

Figure 15.3 The effects on *d* orbitals of a spherical and of an octahedral crystal field

The great significance of Δ_o is that it can often be determined experimentally from the absorption spectrum of the complex. In simple cases the frequency at the maximum of the lowest frequency band in the visible or ultraviolet absorption spectrum corresponds to promotion of an electron from a t_{2g} orbital to an e_g orbital. For many transition metal complexes this band in the visible region is the origin of the colour of the complex; the $[Ti(H_2O)_6]^{3+}$ ion, for example, has a broad band with a maximum at 20,400 cm^{-1} (4900 Å), corresponding to an increase in energy of 244 kJ (58 kcal) $mole^{-1}$ for this promotion. Such a transition would be spectroscopically forbidden in the free Ti^{3+} ion (the selection rule for electronic spectra is $\Delta l = \pm 1$, where l is the second quantum number), but occurs in the complex because the vibration of the ligand molecules about the central ion destroys the perfect octahedral symmetry and broadens what would otherwise be a line into a band; the intensity of the absorption, however, remains low (i.e. the colour is pale). The quantity 0·4 Δ_o (8160 cm^{-1}, 98 kJ, or 23 kcal) is called the crystal field stabilization energy (CFSE) or, more generally in order to accommodate the same quantity for non-electrostatic complexes treated by ligand field theory, the ligand field stabilization energy (LFSE). For ions containing several electrons, the interpretation of the spectra is often more complicated, for reasons which we shall discuss later, and comparison with values of Δ_o estimated theoretically is often helpful in assigning bands. Nevertheless, a large number of values of Δ_o for different ions and different ligands are now known with a fair degree of certainty.

There is a striking general pattern in the relative values of Δ_o for different ligands. Almost irrespective of the nature of the metal ion, Δ_o increases along the series:

$I^- < Br^- < Cl^- < F^- < OH^- < C_2O_4^{2-} < H_2O <$ pyridine $<$ $NH_3 <$ ethylenediamine $<$ dipyridyl $<$ *o*-phenanthroline $< CN^-$

This is the so-called spectrochemical series. It should be noted that since it is an experimental series it incorporates all effects of the ligands in splitting the *d* orbitals (π bonding, for example), and that its validity in no way depends upon the validity or otherwise of crystal field theory. For a given ligand and a given oxidation state of the central metal ion, it is found that Δ_o varies somewhat for metals in the same transition series (e.g. over the range 8000 cm^{-1} to 14,000 cm^{-1} for $[M(H_2O)_6]^{2+}$ in the first series). For a given ligand, it is higher for the higher oxidation state of the same metal (e.g. 9400 cm^{-1} for $[Fe(H_2O)_6]^{2+}$ and 13,700 cm^{-1} for $[Fe(H_2O)_6]^{3+}$). Finally, Δ_o is about 30–50 per cent greater for members of the second transition series than for corresponding members of the first transition series, and greater again by the same margin on going from the second to the third transition series (e.g. values for $[Co(NH_3)_6]^{3+}$, $[Rh(NH_3)_6]^{3+}$ and $[Ir(NH_3)_6]^{3+}$ are 23,000, 34,000 and 41,000 cm^{-1} respectively).

Let us now consider the interpretation of the magnetic properties of octahedral complexes on the basis of crystal field theory. For the electronic configurations d^1, d^2 and d^3 the t_{2g} orbitals are half-filled progressively and the electrons remain unpaired; for d^4, however, two possibilities exist. If Δ_o is small, the fourth electron goes into an e_g orbital and all electrons are unpaired; if, however, Δ_o is large, it may cause less absorption of energy to pair the fourth electron with one of the t_{2g} electrons than to put it in an e_g orbital. Similarly for d^5, the two possible electronic configurations are $(t_{2g})^3(e_g)^2$, with five unpaired electrons, and $(t_{2g})^5$, with only one unpaired electron. These are the high-spin and low-spin configurations mentioned earlier. Which configuration is found depends upon the total octahedral stabilization energy and the energy of repulsion between the paired electrons. Two kinds of repulsion are actually involved, the electrostatic repulsion between like charges and a quantum mechanical exchange interaction which favours parallel spins for electrons in the same set of *d* orbitals. For the present, however, we shall not separate these effects and shall consider only the total mean pairing energy per electron, *P*, which can be obtained approximately from spectroscopic data. The mean pairing energies per electron vary irregularly from metal to metal (as do the Δ_o values) for a given ligand.

Fig. 15.4 shows the occupation of the 3*d* orbitals of Fe^{3+} in a high-spin (weak field) and in a low-spin (strong field) complex (the numerical values for the magnetic moments are discussed in the following chapter). In Fig. 15.5 are shown the high-spin and low-spin electron configurations for d^4, d^5, d^6 and d^7 ions in octahedral fields; it will be remembered that the distinction does not arise for d^1, d^2 and d^3 ions in octahedral fields, and the same is true for d^8, d^9 and d^{10}, in

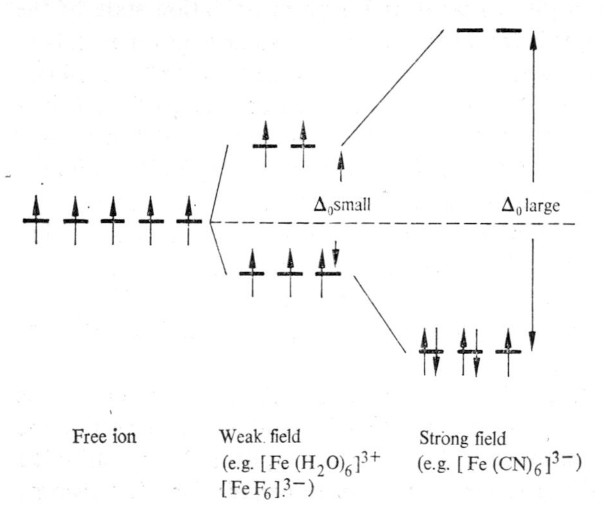

Figure 15.4 Occupation of the 3*d* orbitals in weak and strong octahedral fields, illustrated for Fe^{3+} ($3d^5$) complexes

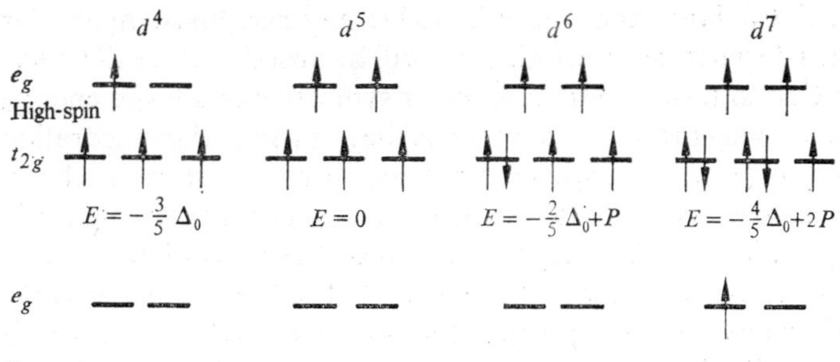

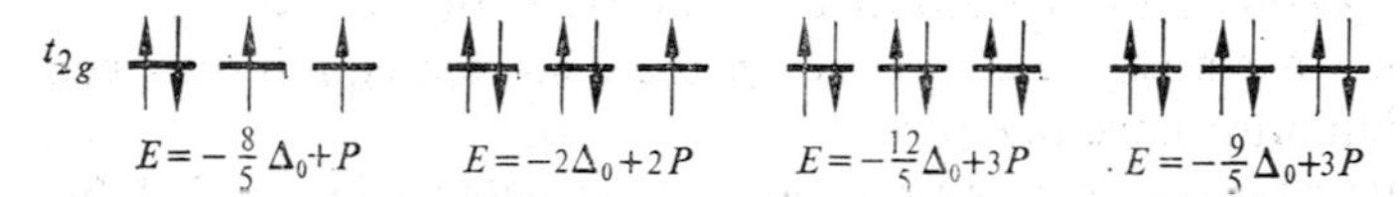

Figure 15.5 Configurations, crystal fields stabilization energies, and pairing energies of d^4, d^5, d^6 and d^7 ions in high-spin and low-spin ground states in octahedral fields

which the distribution of electrons is $(t_{2g})^6(e_g)^2$, $(t_{2g})^6(e_g)^3$ and $(t_{2g})^6(e_g)^4$ respectively. Also shown in Fig. 15.5 are the energies of the high-spin and low-spin configurations of d^4, d^5, d^6 and d^7 ions relative to the energy of the ion in a spherical field, expressed in terms of Δ_o and P. In each case it can be seen that the high-spin and low-spin configurations have equal energy if Δ_o is equal to P; or, to put the matter differently, whether the complex is of the high-spin or of the low-spin type depends on whether Δ_o is less or greater than P.

15.3.2. Distorted octahedral structures: the Jahn-Teller effect. When the e_g orbitals are occupied to different extents (as in high-spin d^4 and d^9 systems), it would be expected that the cation would interact to a different extent with ligands along one axis and those along the other two axes. For example, for the high-spin d^4 case, if there is an electron in the d_{z^2} orbital and the $d_{x^2-y^2}$ orbital is empty, cation–anion interaction along the z-axis should be less than along the x- and y-axes, leading to a larger interionic distance along the z-axis. The octahedron is then said to have undergone tetragonal distortion. A similar effect is to be expected where occupation of the t_{2g} orbitals is unsymmetrical, but it should be much smaller in this case, since these orbitals are directed between ligands.

These expectations are generally, though not invariably, fulfilled, as is shown later. They are expressed more rigorously in a theorem enunciated by Jahn and Teller in 1937: any non-linear molecular system in a degenerate electronic state will be unstable and will undergo distortion to form a system of lower symmetry and lower energy, thereby removing the degeneracy. Thus the high-spin d^4 configuration is doubly degenerate, since in the absence of a distortion the e_g electron can occupy either the d_{z^2} or the $d_{x^2-y^2}$ orbital; the high-spin d^6 configuration is triply degenerate, since the t_{2g} orbital which is doubly occupied can be d_{xy}, d_{xz} or d_{yz}. It should be noted that the Jahn–Teller theorem only predicts the occurrence of a distortion; it does not predict its nature or its magnitude. There is, however, one general guide to the nature of distortions: if the undistorted configuration has a centre of symmetry, the distorted equilibrium configuration must have one too.

Let us now consider some octahedral structures in which Jahn–Teller effects occur. In CrF_2, each Cr^{2+} has four F^- at 2·00 Å and two at 2·43 Å; similarly, in CuF_2 each Cu^{2+} has four F^- at 1·93 Å and two at 2·27 Å. In these cases it is inferred that the electron density is higher in the d_{z^2} orbital. The reverse situation is found in K_2CuF_4, in which the Cu^{2+} has two F^- at 1·95 Å and four at 2·08 Å, indicating that in this case the electronic configuration of the metal ion is $(t_{2g})^6(d_{x^2-y^2})^2(d_{z^2})^1$. In FeF_2 (in which the cation has the high-spin d^6 configuration) the Fe^{2+} ion has two F^- at 1·99 Å and four at 2·12 Å; but in MnF_2 and NiF_2 (in which the configurations of the cations are high-spin d^5 and d^8 respectively) the environment of the cation is, within very narrow limits, regular octahedral.

The Jahn–Teller effect is also observed in excited states of some complexes which in their ground states are regular, or very nearly regular, octahedra, such as $[Ti(H_2O)_6]^{3+}$, $[Fe(H_2O)_6]^{2+}$ and $[CoF_6]^{3-}$; the second and third of these ions are high-spin complexes, so that in each case a promotion of an electron to the e_g level results in unsymmetrical occupation of the e_g orbitals, and in each case the promotion gives rise to two bands (or one band with a shoulder) in the electronic spectrum.

Clear though these examples are, there are several puzzling observations that we can mention only briefly here. In MnF_3, for example, there are not two but three different Mn–F distances (1·79, 1·91 and 2·09 Å); and in $[Mn(H_2O)_6]^{3+}$ and $[Cu(NO_2)_6]^{4-}$ the octahedra are symmetrical. Attribution of this symmetry to a dynamic Jahn–Teller effect is a possible explanation, but if it is true the ions should become unsymmetrical at low temperatures, and this has been found not to be so, at least down to −195°C, for $[Mn(H_2O)_6]^{3+}$.

On balance, however, we see that the simple crystal field theory makes a notable contribution to the interrelation of stereochemistry, magnetic properties and electronic spectra in the case of octahedral complexes; we shall consider the last two topics further in the two following chapters, and further applications of crystal field theory to octahedral complexes in Chapters 18 and 19.

15.3.3. Tetrahedral complexes. The splitting of the *d* orbitals depends markedly upon the symmetry of the ligand field round the central ion. For a tetrahedral field the *d* orbitals are again split into two groups of different energies, but in this case the d_{xy}, d_{xz} and d_{yz} orbitals have the higher energy since their directions lie closer to the ligands than those of the d_{z^2} and $d_{x^2-y^2}$ orbitals. For the same ion and ligand at the same distance as in the octahedral case, it can be shown that Δ_t, the difference between the two sets of orbitals in a tetrahedral field, is given by

$$\Delta_t = \tfrac{4}{9}\,\Delta_o$$

(This relationship would not be expected to be quantitatively very useful in practice, since a change of coordination number from six to four would nearly always be accompanied by a change in interatomic distance.) The effects on *d* orbitals of a spherically symmetrical and a tetrahedral crystal field are then as represented in Fig. 15.6.

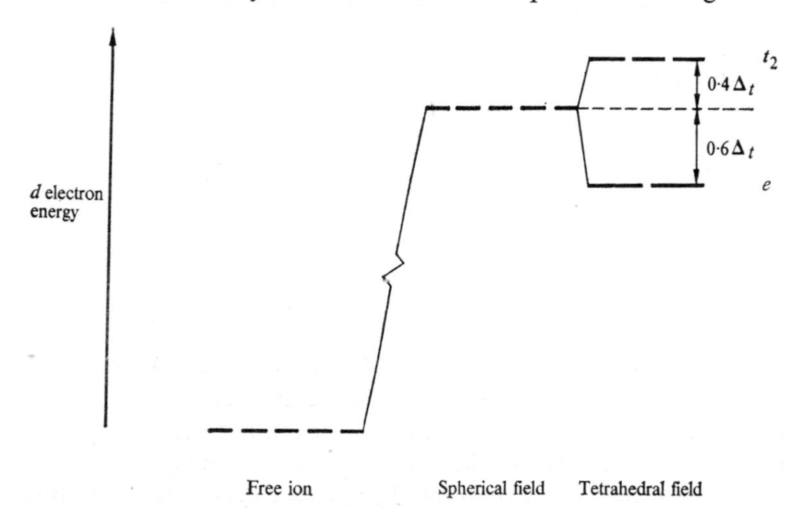

Figure 15.6 The effects on *d* orbitals of a spherical and of a tetrahedral field

The spectrochemical series follows the same order as for octahedral fields, but low-spin tetrahedral complexes have not yet been obtained; apparently the crystal field stabilization energies are never sufficient to bring about spin-pairing. Nor have substantial distortions yet been found in tetrahedral complexes, but this may be because relatively few have been examined.

Crystal field stabilization energies for high-spin tetrahedral and octahedral configurations of the same species at the same distance are shown in Table 15.2. Two points of interest about the values for tetrahedral fields may be noted. First, the difference between the stabilizations is least for d^0, d^5 and d^{10} (for which it is zero in all cases), and next smallest for d^1, d^2, d^6 and d^7; most of the examples of tetrahedral coordination given in Section 14.3.4 belong to one of these systems. Secondly, since Δ_t is much lower than Δ_o, complexes of the same ion with ligands occupying nearby positions in the spectrochemical series have their absorption maxima for *d–d* spectra at lower frequencies and higher wavelengths for tetrahedral complexes than for octahedral complexes. Thus high-spin octahedral cobalt (II) complexes such as $[Co(H_2O)_6]^{2+}$ absorb in the blue region of the spectrum and are pink, whilst tetrahedral cobalt (II) complexes such as $[CoCl_4]^{2-}$ absorb in the red and are blue in colour.

Table 15.2 *Crystal field stabilization energies for high-spin tetrahedral and octahedral configurations (other things being equal)*

Number of d electrons	*CFSE*	
	Tetrahedral	*Octahedral*
0, 5, 10	0	0
1, 6	$0{\cdot}6\ \Delta_t = 0{\cdot}27\ \Delta_o$	$0{\cdot}4\ \Delta_o$
2, 7	$1{\cdot}2\ \Delta_t = 0{\cdot}53\ \Delta_o$	$0{\cdot}8\ \Delta_o$
3, 8	$0{\cdot}8\ \Delta_t = 0{\cdot}35\ \Delta_o$	$1{\cdot}2\ \Delta_o$
4, 9	$0{\cdot}4\ \Delta_t = 0{\cdot}18\ \Delta_o$	$0{\cdot}6\ \Delta_o$

15.3.4. Tetragonal and square complexes. We have already mentioned the tetragonal distortion of an octahedral complex arising from the Jahn–Teller effect. Such a distortion has the effect of destroying the degeneracy of the e_g orbitals; if, for example, the metal ion has four near and two distant ligands, the d_{z^2} orbital is of lower energy than the $d_{x^2-y^2}$ orbital. If the difference in energy between them is large enough, a d^8 complex (e.g. of Ni^{II}) will be diamagnetic with the electronic configuration $(d_{xz})^2(d_{yz})^2(d_{xy})^2(d_{z^2})^2$. In the limit of tetragonal distortion, two of the ligands are removed to infinity and a square-planar complex results. The relative changes in energy when an octahedral complex is successively distorted are shown in Fig. 15.7, from which we can see that planar complexes of electronic configurations d^7, d^8 and d^9 (the only ones at present known) should have 1, 0 and 1 unpaired electrons respectively. Fig. 15.7, however, does not really tell us much about the likelihood of forming planar complexes. As we have pointed out before, the crystal field stabilization energy is only a small part of the total interaction energy in the complex. It may also be doubted whether a truly electrostatic system would result in planar coordination, and it is indeed noteworthy that ligands which often give planar complexes are also good π-bonding ligands. We shall therefore not discuss planar coordination further in terms of crystal field theory; fuller treatments are given in more specialized works.[3, 4, 8]

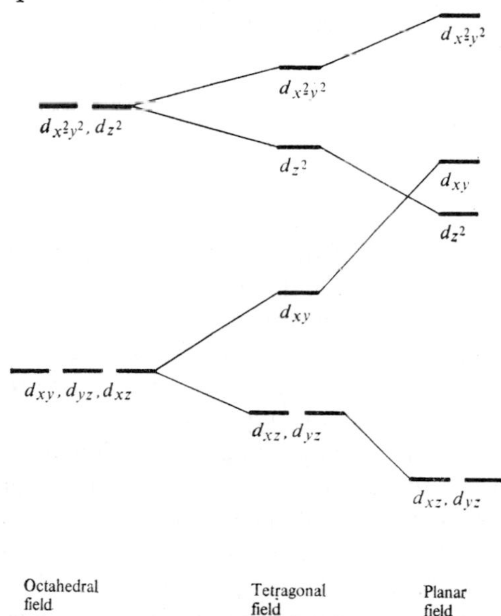

Figure 15.7 Crystal field splittings of the d orbitals in octahedral, tetragonal and planar complexes

15.4 Ligand field theory

The existence of stable complexes formed from uncharged metal atoms and neutral ligands (e.g. the metal carbonyls) is of itself a proof that any general treatment of complex formation must make provision for covalent bonding between metal and ligand. Even for complexes which can be treated satisfactorily by crystal field theory, there is often evidence of electron delocalization between metal and ligand (e.g. in the hyperfine structure of the e.s.r. spectrum of the $[IrCl_6]^{2-}$ ion, which shows that the single unpaired electron is only about 70 per cent an iridium (IV) 4*d* electron, and in the ^{19}F n.m.r. spectra of transition metal fluorides, which show that some of the electron spin density is transferred from metal to ligand). Ligand field theory, which is a combination of crystal field theory and molecular orbital theory, is therefore in principle the most satisfactory of the theories of bonding discussed in this chapter. Unfortunately, it is also the most elaborate, and it is less able than crystal field theory to provide a simple working model in terms of which quantitative predictions may be made by the non-theoretician. Many of the simple calculations of crystal field theory may be incorporated into ligand field theory, of course, but it is not necessarily advantageous to do this. In this section, therefore, we shall concentrate on presenting the ligand field theory treatment of octahedral complexes so as to bring out its particular merits; in subsequent chapters we shall use whichever theory is more convenient for the discussion of the experimental material under consideration.

Let us consider first an octahedral complex with no π bonding. Of the nine orbitals in the valence shell of the metal atom or ion, six (the *s*, *p* and $d_{x^2-y^2}$ and d_{z^2} orbitals) are suitable for σ bonding, and the other three (d_{xy}, d_{xz} and d_{yz}), which are directed between the ligands, are not. The six individual σ orbitals of the ligands are combined to form six orbitals of the correct symmetries for each one to combine with one of the metal orbitals: an *s*-type orbital to combine with the *s* orbital of the metal, three *p*-type orbitals to combine with the *p* orbitals of the metal and two *d*-type orbitals to combine with the $d_{x^2-y^2}$ and d_{z^2} orbitals of the metal. The molecular orbitals formed are shown schematically in Fig. 15.8, from which it is seen that the t_{2g} orbitals remain unaffected. The symbols shown against the molecular orbitals are derived from group theory; their meanings are as follows. The letter *A* denotes an energy level which is singly degenerate, corresponding to a single orbital having the full symmetry of the system; *E* denotes an energy level which is doubly degenerate and corresponds to a pair of orbitals differing only in directional properties; and *T* denotes a triply degenerate energy level corresponding to a set of three orbitals differing only in directional properties. The subscript numeral, if there is one, denotes the symmetry of the set of orbitals; subscript one means the wave-functions do not change sign on rotation about the cartesian axes and subscript two that they do not change sign on rotation about axes diagonal to the cartesian axes. The subscripts g and u indicate whether the orbitals are centrosymmetric (g from the German *gerade* = even) or antisymmetric (u from *ungerade* = uneven). A set of three p orbitals corresponds to a triply degenerate energy level T_{1u}, a set of two $d_{x^2-y^2}$ and d_{z^2} orbitals to a doubly degenerate energy level E_g, and a set of three d_{xy}, d_{xy} and d_{yz} orbitals to a triply degenerate energy

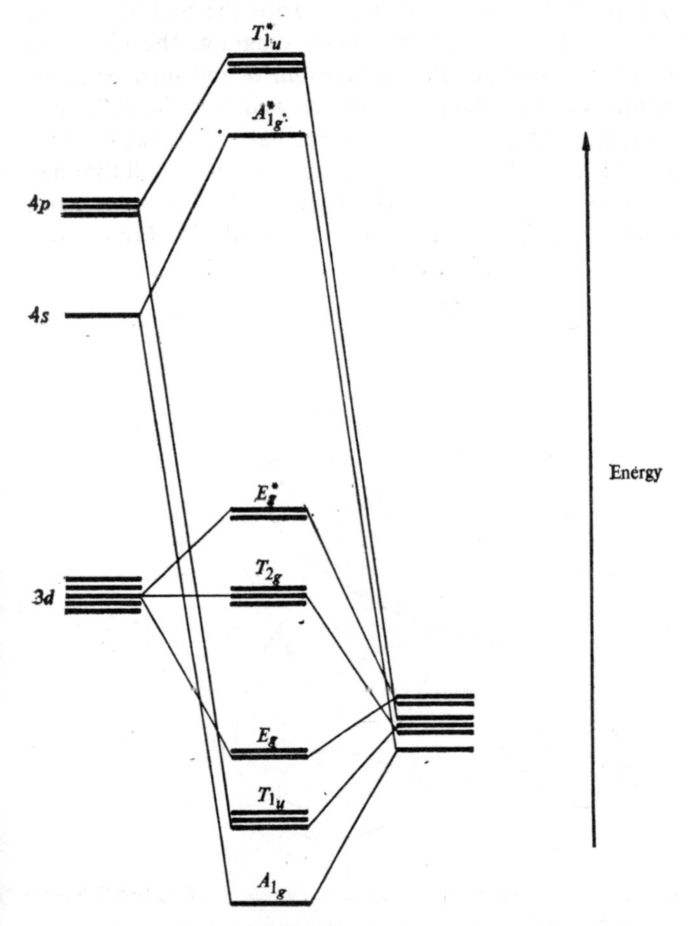

Figure 15.8 Molecular orbital energy level scheme for an octahedral complex with no π bonding

level T_{2g}. (The same symbols with small letters in place of capitals are used, as we have already seen, for individual orbitals.) Asterisks denote antibonding orbitals.

In a diagram such as Fig. 15.8, a molecular orbital near in energy to only one of the atomic orbitals used in its formation has very much the character of that orbital. Thus in the situation represented in this diagram, the six bonding orbitals have more the character of ligand orbitals than metal orbitals. Conversely, in Fig. 15.8, the antibonding orbitals are more nearly metal orbitals, and any electrons in them more nearly metal electrons. The T_{2g} orbitals are purely metal orbitals and play no part in the bonding. Thus the middle of Fig. 15.8 is the octahedral crystal field theory diagram, the electrons originally in the *d* orbitals of the metal atom or ion now being in T_{2g} non-bonding or E_g* antibonding orbitals, and Δ_o is the difference between the T_{2g} and E_g* energy levels. There has, of course, been no change in Δ_o, which is anyway an experimental quantity; all that has changed is the interpretation of it. Although attempts are now being made to calculate the relative energy levels for octahedral (and other) complexes, it is not yet possible to do so accurately.

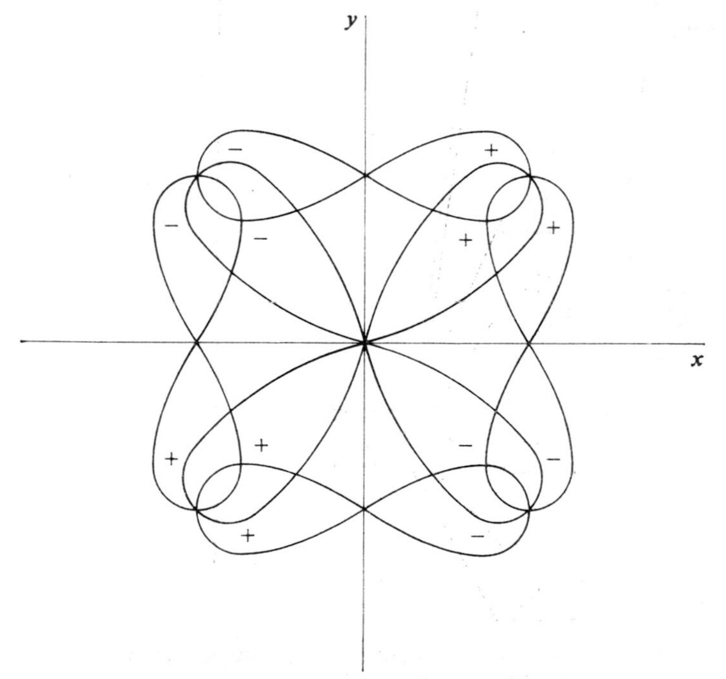

Figure 15.9 Interaction of a d_{xy} orbital with ligand $p\pi$ orbitals

If the ligands have π orbitals of the correct symmetry, filled or unfilled, it is necessary to consider their interaction with the T_{2g} orbitals of the metal. The ligand π orbitals may be p_π orbitals (as in a halide ion), d_π orbitals (as in phosphines) or molecular orbitals (as in CO, CN^- or ethylene). The interaction of the d_{xy} orbital with a set of p_π orbitals is represented in Fig. 15.9. Any π bonding modifies the energy level diagram shown in Fig. 15.8 somewhat. Unfortunately, most ligands can conceivably act in different ways depending on, for example, whether filled p orbitals or empty d orbitals are involved in π bond formation, and we are then driven to infer from observed values of Δ_o in what capacity the ligand is acting. Thus in the halide complexes Δ_o is small (and smallest for iodide), from which it appears that the filled p orbitals are used for π bonding. In this case the ligand orbitals are lower in energy than the metal t_{2g} orbitals, and the central part of Fig. 15.8 is modified to Fig. 15.10(a); although iodide exerts the weakest electrostatic field, its taking part in the strongest π bonding could outweigh this factor. Where the ligands lead to high values of Δ_o, as in the case of phosphines, arsines and sulphides, it is inferred that their empty d orbitals are used for π bonding, leading to the situation shown in Fig. 15.10(b). For cyanide and carbon monoxide, the very high values of Δ_o are attributed to the use of the empty ligand antibonding π orbitals, and these ligands are often described as moderate donors and very good π acceptors.

It is clear that if other factors are equal, π bonding in which the ligands use empty orbitals should be a maximum when the metal has the largest possible number of t_{2g} electrons; for ions capable of existing in both high- and low-spin configurations, the latter would be better for π bonding in these particular circumstances. Thus we can say that with phosphorus and sulphur ligands low-spin Co^{2+}, Co^{3+}, Fe^{2+}, Ni^{2+} and Cu^{2+} should be strong π bond formers. The occurrence of these ions as sulphide minerals has been held to support this suggestion.[8] There is, however, no method of assessing quantitatively the extent of these π interactions, since all experimental data relate to the total interaction of metal and ligand, and although we shall often use the concept of π bonding in transition metal chemistry in a qualitative sense we are not yet in a position to do more.

It is, however, instructive to reconsider the effective atomic number rule again in the light of ligand field theory. The rule holds well for carbonyl, cyanide and unsaturated hydrocarbon complexes. We have argued already that carbonyls, in particular, must be treated by

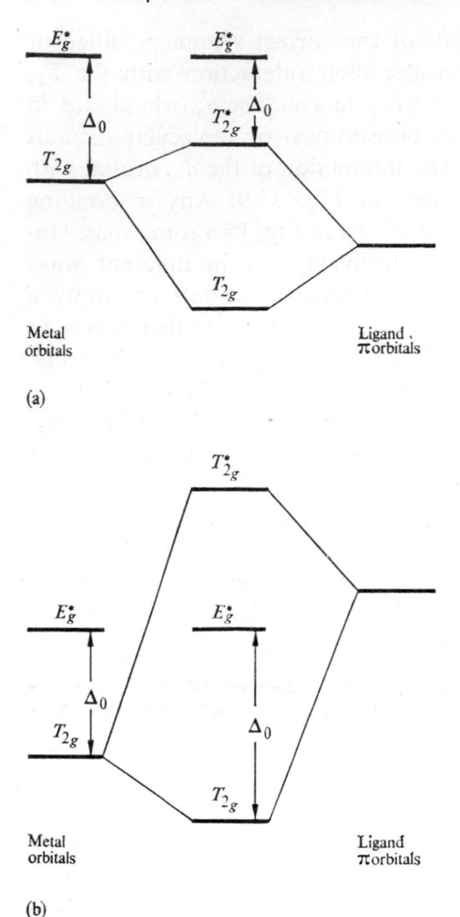

Figure 15.10 The effects of π bonding on Δ_o: (a) ligand π orbitals of lower energy than metal T_{2g} orbitals (b) ligand π orbitals of higher energy than metal T_{2g} orbitals

ligand field theory, and we have attributed the high values of Δ_o in octahedral carbonyls to π bonding involving the use of ligand antibonding orbitals. A very high Δ_o means that an electron in an E_g* orbital will be particularly unstable (i.e. the E_g* will be strongly antibonding). Thus the formulae of the carbonyls appear to be determined by two tendencies, that of the metal to have the maximum number of t_{2g} electrons (so as to give the maximum amount of π bonding) and that of the E_g* orbitals to be empty (to avoid antibonding interactions). For non-π bonding ligands the presence or absence of electrons in the non-bonding t_{2g} orbitals is immaterial, and the effective atomic number rule is not valid.[9]

Molecular orbital energy level schemes have also been given for tetrahedral and other geometries, but we cannot discuss them here. We should, however, draw attention to the fact that the tetrahedron has no centre of symmetry, and in consequence the classification of molecular orbitals as g or u is inappropriate in this system (see further Section 17.3).

References for Chapter 15

1 Orgel, L. E., *An Introduction to Transition Metal Chemistry: Ligand Field Theory*, 2nd edn. (Methuen, 1966).
2 Jones, M. M., *Elementary Coordination Chemistry* (Prentice-Hall, 1964).
3 Cotton, F. A., and Wilkinson, G., *Advanced Inorganic Chemistry*, 2nd edn. (Interscience, 1966).
4 Ballhausen, C. J., *Introduction to Ligand Field Theory* (McGraw-Hill, 1962).
5 Day, M. C., and Selbin, J., *Theoretical Inorganic Chemistry*, 2nd edn. (Reinhold, 1969).
6 Kettle, S. F. A., *Coordination Compounds* (Nelson, 1969).
7 Pauling, L., *The Nature of the Chemical Bond*, 3rd edn. (Cornell University Press, 1960).
8 Burns, R. G., *Mineralogical Applications of Crystal Field Theory* (Cambridge University Press, 1970).
9 Mitchell, P. R., and Parish, R. V., *J. Chem. Education*, 1969, **46**, 811.

Complexes of transition metals III: magnetic properties

16

16.1 Magnetic susceptibility[1–3]

The use of magnetic susceptibility measurements to decide between different electronic configurations was mentioned in the last chapter; in this section we go into the background of such measurements and discuss magnetic properties in more detail. Magnetic properties and electronic spectra are closely connected, and there is some overlap between this chapter and the next. We should draw attention to the fact that in the next four paragraphs the cgs, rather than the SI, system is used.

When a substance is placed in a magnetic field of strength H the magnetic induction or density of lines of force B within the substance is given by

$$B = H + 4\pi I$$

where I is the intensity of magnetization or magnetic moment per unit volume. Alternatively, we can divide by H and write

$$P = 1 + 4\pi\kappa$$

where P is the permeability and κ the susceptibility per unit volume. Susceptibility is, in practice, usually expressed per unit mass, and given the symbol χ; and the molar susceptibility is then χ_m, the product of the gram susceptibility and the molar weight.

If P is less than unity (i.e. I, κ and χ are negative), the substance causes a reduction in the intensity of the magnetic field and, in an inhomogeneous field, moves to the region of lowest field strength; it is then said to be diamagnetic. Values of χ for diamagnetic substances are very small (about -1×10^{-6}) and are usually independent of field strength and temperature. All substances have diamagnetic susceptibility (which arises from the induction of a small magnetic moment opposing the effect of the field when any substance is placed in a magnetic field), and it is necessary to attempt to correct for this in determining paramagnetic susceptibilities, which result from the presence of unpaired electrons. Otherwise, however, numerical values of diamagnetic susceptibilities are of little interest in inorganic chemistry, and we shall not discuss them further.

If P is greater than unity the substance is said to be paramagnetic; it then causes an increase in the intensity of the field, and in an inhomogeneous field it moves to the region of highest field strength. Paramagnetic susceptibilities are usually much larger than diamagnetic susceptibilities (χ is about 1 to 100 $\times 10^{-6}$); they are independent of field strength, but, because the ordering of the magnetic moments that give rise to them is resisted by thermal motion, they depend inversely upon temperature. Ferromagnetic and antiferromagnetic substances are special classes of paramagnetic materials. For the former, P is much greater than unity and χ may be as high as 10^4; they are both field strength- and temperature-dependent, and show the properties of remanence and hysteresis. For antiferromagnetic substances, χ is about the same or somewhat less than for most paramagnetic materials; it is temperature-dependent and sometimes also field strength-dependent. Ferro- and antiferromagnetism are cooperative phenomena (i.e. they arise from the interaction of individual paramagnetic species with one another) and are therefore more likely to occur when there is a high concentration of paramagnetic species. A distinction is therefore made between magnetically concentrated and magnetically dilute materials. The former are exemplified by metals, alloys and compounds like oxides and fluorides in which the metal cations are combined with small anions; typical magnetically dilute materials are aqueous solutions of paramagnetic substances and salts in which the paramagnetic ions are shielded from one another by large numbers of ligands.

Curie's law,

$$\chi_M = C/T$$

where C is the Curie constant and T the absolute temperature, holds for many paramagnetic substances; for others, the Curie–Weiss law,

$$\chi_M = C/(T + \theta)$$

holds. (This second law is sometimes written with the denominator $T - \theta$, thereby changing the sign of θ.) In this, the constant θ is known as the Curie temperature; it is readily obtained by plotting

$1/\chi_M$ against T. It is customary to express the properties of a paramagnetic substance in terms of μ, its 'effective magnetic moment', in Bohr magnetons. The value of the Bohr magneton β is given by

$$\beta = \frac{eh}{4\pi mc}$$

where e is the electronic charge, h Planck's constant, c the velocity of light and m the mass of the electron. A statistical mechanical treatment of the relationship between the molar susceptibility and temperature lends to the expression

$$\chi_M = \frac{N\beta^2\mu^2}{3kT}$$

in which N is Avogadro's number, k the Boltzmann constant and the other symbols have the same meaning as before. Thus

$$\mu = 2{\cdot}84\sqrt{\chi_M T}$$

This expression shows μ as independent of temperature; this is true, of course, only if the Curie law holds. Nevertheless, owing to uncertainties in the interpretation of θ, it is common practice to work out μ on the basis of this equation (and hence of the Curie law), and to state the temperature of the measurements; herein lies the reason for the use of the description of μ as the 'effective magnetic moment'.

The fact that a substance is diamagnetic indicates the absence of unpaired electrons, and in the cases of the compounds of empirical formula $GaCl_2$ and H_2PO_3 (now known to be $Ga^+[GaCl_4]^-$ and $H_4P_2O_6$) their diamagnetism provided a useful clue to their structures; in general, however, it is with paramagnetism, ferromagnetism and antiferromagnetism that chemists are most concerned.

16.2 Paramagnetism[1-5]

The origin of paramagnetism is the spin and orbital angular momentum possessed by extranuclear electrons. Although the picture of the electron as a particle travelling round the nucleus is no longer accepted as correct, the usefulness of this model in a qualitative introduction to atomic spectroscopy or magnetism remains, and we shall employ it here. Quantum mechanics leads to the conclusion that the magnetic moment associated with the spin of a single electron, μ_s, is given in Bohr magnetons (B.M.) by the equation

$$\mu_s = g\sqrt{s(s+1)}$$

where s is the absolute value of the spin quantum number and g is a constant called the gyromagnetic ratio (the ratio of magnetic moment to angular momentum when both are expressed in their respective quantum units) or the Landé splitting factor (it was first derived in atomic spectroscopy by Landé). For a free electron $g = 2{\cdot}0023$ (the departure from 2·00 arises from a small relativistic effect which does not concern us here), and the 'spin-only' moment of a single electron ($s = \frac{1}{2}$) is thus

$$\mu_s = 2{\cdot}00\sqrt{\tfrac{1}{2}.\tfrac{3}{2}} = 1{\cdot}73 \text{ B.M.}$$

The magnetic moment associated with the orbital angular momentum of a single electron, on the other hand, is given by

$$\mu_l = \sqrt{l(l+1)}$$

where l is the subsidiary quantum number.

It will be recalled that the interpretation of atomic spectra requires that four quantum numbers be available for the description of any electron in any atom. These are: a principal quantum number n, which may have values 1, 2, 3, 4,. . .; a subsidiary or orbital quantum number l, which may be 0, 1, 2, 3,. . . $(n - 1)$, electrons having $l = 0$, 1, 2 and 3 being denoted s, p, d and f electrons respectively; a third or magnetic quantum number m of value $-l$, $-(l - 1)$, . . ., 0, . . ., $(l - 1)$, l; and a spin quantum number s of value $+\frac{1}{2}$ or $-\frac{1}{2}$. According to the Pauli principle, no two electrons in the same atom may have the same values for all four quantum numbers. For an atom containing several electrons, the total orbital angular momentum (in units of $h/2\pi$) is given by L, this being the vector sum of the individual subsidiary quantum numbers l_1, l_2, l_3, etc. The total spin quantum S is similarly the resultant of the individual spin quantum numbers; in this case it is $n/2$, where n is the number of unpaired electrons. For an individual electron, the resultant of the orbital and spin contributions to the angular momentum is obtained by adding l and s to give a so-called inner quantum number j, which can thus have the values $l + \frac{1}{2}$ and $l - \frac{1}{2}$. (It is the numerical value of j that determines the total angular momentum; for $l = 0$ there is only one energy level.) L and S may be combined to give a total inner quantum number J with possible values $L + S$, $L + S - 1$, . . ., $L - S$ (i.e. there are $2S + 1$ possible combinations of L and S if $S < L$ or $2L + 1$ if $L < S$). These combinations have slightly different energies, so for each value of L there is always a multiplet of slightly different energies corresponding to the different values of J.

Whether in fact L and S combine in this way, in what is called *LS* or Russell–Saunders coupling, depends on the relative magnitudes of the various types of interaction possible. For two electrons, these are between

(a) the two spins (s_1s_2),
(b) the two orbital momenta (l_1l_2),
(c) the spin of one electron and the orbital momentum of the same electron (s_1l_1), termed spin-orbit coupling and
(d) the spin of one electron and the orbital momentum of the other (s_1l_2).

Normally the last of these is negligible, and the two extreme situations are

(i) $s_1s_2 > l_1l_2 > s_1l_1$
(ii) $s_1l_1 > s_1s_2, l_1l_2$

Of these case (i) is *LS* coupling; it is found to apply to the lighter elements up to the end of the first transition series, and to the lower energy levels of the heavier elements. Case (ii) is the so-called *jj* coupling, in which j values for individual electrons, instead of L and S values for all the electrons, combine to give J; it seldom occurs in the pure form, but it holds in part for the heavier elements, particularly in excited states. In what follows we are referring to systems for which *LS* coupling is found to prevail.

The components of the multiplet mentioned above (or 'states' or 'terms') are conventionally represented by S, P, D, F, G, $H, \ldots$ according to whether L is 0, 1, 2, 3, 4, 5, ... The value of S is shown by a superscript, the 'multiplicity' of the term, which is $2S + 1$, and the value of J is added as a subscript. It was found by Hund from analysis of atomic spectroscopic data that, for the lowest energy term,

(a) S has the highest value allowed by the Pauli principle,
(b) L then has the highest value possible and
(c) J then has the value $L - S$ if the electron shell is less than half full and $L + S$ if it is more than half full.

If it is half full and S has the maximum value, L must be zero and $J = S$. The energy difference between adjacent states of J values J' and $(J' + 1)$ is $(J' + 1)\lambda$, where λ is the spin-orbit coupling constant. For the d^2 configuration, for example, spin-spin interaction separates the energy levels into a set having the spins of the same sign ($S = \frac{1}{2} + \frac{1}{2} = 1$, so the multiplicity is three—these are triplets) and a set having the spins of opposite sign ($S = \frac{1}{2} - \frac{1}{2} = 0$, so the multiplicity is one—these are singlets); the former are lower in energy. Orbital coupling splits these levels further; for d electrons $l = 2$ and if $S = 1$ the maximum value of L (which may be obtained by algebraic summation of values of m, since this corresponds to vectorial addition of l) is then $2 + 1 = 3$. For $L = 3$, $S = 1$ the term symbol is ^3F. This is split by spin-orbit coupling into states $^3F_2$, $^3F_3$ and $^3F_4$, of which the first is the lowest; the differences between successive pairs are 3λ and 4λ respectively.

In a magnetic field spectroscopic states are split again to give $(2J + 1)$ levels, each separated by $g\beta H$, where g is the Landé splitting factor for the system, β the Bohr magneton and H the field strength.

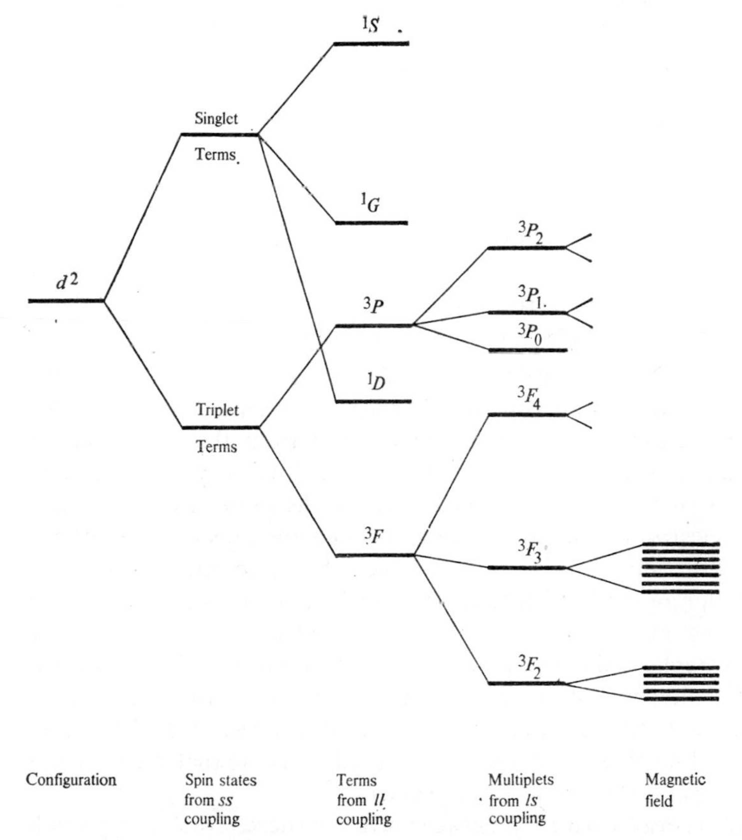

Figure 16.1 Splitting of the states of a gaseous d^2 ion (not to scale)

For a field of 10,000 gauss, $g\beta H$ is about 1 cm^{-1}; since this corresponds to only 0·012 kJ (0·003 kcal) $mole^{-1}$, magnetic fine structure, like spectroscopic fine structure, is concerned with extremely small amounts of energy compared with energy changes in chemical processes. The overall splitting pattern is shown in Fig. 16.1.

The extent of occupation of energy levels of different J values depends on their separation and on the value of the thermal energy available, kT; at 300°K, kT is approximately 200 cm^{-1}, or 2·6 kJ (0·6 kcal) $mole^{-1}$.

If the separation of these energy levels is large compared to kT, theory shows that

$$\mu = g\sqrt{J(J+1)}$$

where

$$g = 1 + \frac{S(S+1) - L(L+1) + J(J+1)}{2J(J+1)}$$

The magnetic moments of the lanthanide tripositive ions, for which λ is about 10^3 cm^{-1}, show almost ideal behaviour in this respect if their room temperature magnetic moments are compared with calculated values, as may be seen from Table 16.1. For Ce^{3+}, for example, with one unpaired electron, the ground state is $^2F_{\frac{5}{2}}$, g is $\frac{6}{7}$, $S = \frac{1}{2}$, $L = 3$, $J = \frac{5}{2}$ and μ (calc.) for the free ion is 2·54 B.M., whilst the values found in solid compounds range between 2·3 and 2·5 B.M. This good agreement is believed to arise from the very effective screening of the unpaired $4f$ electrons from environmental effects by the overlying s and p electrons (this is also the cause of the sharpness of the electronic spectroscopic bands in compounds of the lanthanides). For Sm^{3+} and Eu^{3+}, for which the agreement is poor, the separation of levels of different J is smaller, and energy levels above the lowest are populated to some extent. (The situation is actually more complicated than that we have described here; it is now known that the moments of the lanthanide ions are temperature-dependent, and at low temperatures the agreement between calculated and measured moments is not nearly so good, showing that the presence of the ligands does have some effect.)

If the energy separation between levels of successive J is very small relative to kT (as it is for some, though not all, ions of metals of the first transition series), coupling of L and S is negligible and they inter-

Table 16.1 *Magnetic moments of lanthanide tripositive ions*

Ion	*Ground state*	*g*	μ *Calc.*	*Exptl.**
La^{3+}	$^1S_0$	—	0	0
Ce^{3+}	$^2F_{\frac{5}{2}}$	$\frac{6}{7}$	2·54	2·3–2·5
Pr^{3+}	$^3H_4$	$\frac{4}{5}$	3·58	3·4–3·6
Nd^{3+}	$^4I_{\frac{9}{2}}$	$\frac{8}{11}$	3·62	3·5–3·6
Pm^{3+}	$^5I_4$	$\frac{3}{5}$	2·68	—
Sm^{3+}	$^6H_{\frac{5}{2}}$	$\frac{2}{7}$	0·84	1·5–1·6
Eu^{3+}	$^7F_0$	—	0	3·4–3·6
Gd^{3+}	$^8S_{\frac{7}{2}}$	2	7·94	7·8–8·0
Tb^{3+}	$^7F_6$	$\frac{3}{2}$	9·72	9·4–9·6
Dy^{3+}	$^6H_{\frac{15}{2}}$	$\frac{4}{3}$	10·63	10·4–10·5
Ho^{3+}	$^5I_8$	$\frac{5}{4}$	10·60	10·3–10·5
Er^{3+}	$^4I_{\frac{15}{2}}$	$\frac{6}{5}$	9·57	9·4–9·6
Tm^{3+}	$^3H_6$	$\frac{7}{6}$	7·63	7·1–7·4
Yb^{3+}	$^2S_{\frac{7}{2}}$	$\frac{8}{7}$	4·50	4·4–4·9
Lu^{3+}	$^1F_0$	—	0	0

* At ordinary temperatures.

act independently with the external field, producing orbital and spin moments,

$$\mu_L = \sqrt{L(L+1)}$$

and

$$\mu_S = g\sqrt{S(S+1)} = 2\sqrt{S(S+1)}$$

Then

$$\mu = \sqrt{4S(S+1) + L(L+1)}$$

This is the equation for transition metal ions in general. When the ground states of transition metal ions or lanthanide ions are *S* states (for which $L = 0$), there can be no orbital contribution and

$$\mu = \sqrt{4S(S+1)}$$

or, if we write *n* for the number of unpaired electrons,

$$\mu = \sqrt{n(n+2)}$$

This relationship holds as expected for compounds of manganese (II)

Table 16.2 *Magnetic moments of some first transition series ions in high-spin configurations at ordinary temperatures*

Ions	*Ground term*	μ			
		Calculated from			*Exptl.*
		$g\sqrt{J(J+1)}$	$\sqrt{4S(S+1)+L(L+1)}$	$\sqrt{4S(S+1)}$	
Ti^{3+}	$^2D_{\frac{3}{2}}$	1·55	3·01	1·73	1·7–1·8
V^{3+}	$^3F_2$	1·63	4·49	2·83	2·8–3·1
V^{2+}, Cr^{3+}	$^4F_{\frac{3}{2}}$	0·70	5·21	3·87	3·7–3·9
Cr^{2+}, Mn^{3+}	$^5D_0$	0	5·50	4·90	4·8–4·9
Mn^{2+}, Fe^{3+}	$^6S_{\frac{5}{2}}$	5·92	5·92	5·92	5·7–6·0
Fe^{2+}, Co^{3+}	$^5D_4$	6·71	5·50	4·90	5·0–5·6
Co^{2+}	$^4F_{\frac{9}{2}}$	6·63	5·21	3·87	4·3–5·2
Ni^{2+}	$^3F_4$	5·59	4·49	2·83	2·9–3·9
Cu^{2+}	$^2D_{\frac{5}{2}}$	3·55	3·01	1·73	1·9–2·1

and iron (III) in their high-spin configurations; however, as may be seen from Table 16.2, it also holds much better for several other species than the relationship which takes into full account the contribution from the orbital angular momentum. (Moments calculated from the formula $\mu = g\sqrt{J(J+1)}$, it may be noted, are, except for the d^5 configuration, in even worse agreement with the values for solid compounds.) This is interpreted qualitatively as showing that the ligand field is greater than the spin-orbit interaction for transition metal ions in general, and that, in addition, it is often sufficient to 'quench' the orbital contribution partly or completely by restricting the orbital motion of the electrons. We shall, however, return to this point shortly.

Spin coupling takes place when many ions of metals of the first transition series having the configurations d^4, d^5, d^6 and d^7 are placed in very strong ligand fields (notably those of cyanide, dipyridyl, and phosphorus- and sulphur-containing ligands); this is the formation of low-spin complexes (or strong field, inner orbital, or covalent complexes in other terminologies) discussed in the last chapter. If we consider the application of a field of increasing strength to two states of different energies whose relative values are reversed on going from a weak field to a strong one, we can envisage a certain region of field strength for which the separation is of the order of kT; a chemical equilibrium mixture of the two states will then be observed. This is quite common for nickel (II) complexes, not only in solution but also in the solid state (see Section 14.4.1).

The 'quenching' of the orbital contribution to the magnetic moment can be understood on the basis of crystal field theory. For an electron to have orbital angular momentum about an axis the orbital which it occupies must be transformable into an entirely equivalent and degenerate orbital by rotation about the axis in question. In a free atom or ion the d_{xy} and $d_{x^2-y^2}$ orbitals are related by rotation through 45° and the d_{xz} and d_{yz} orbitals by rotation through 90°, about the z-axis. An electron in any of these orbitals therefore has orbital angular momentum. A d_{z^2} electron, on the other hand, has no angular momentum about the z-axis.

In an octahedral or tetrahedral complex the d_{xz} and d_{yz} orbitals remain rotationally and energetically equivalent, but the d_{xy} and $d_{x^2-y^2}$ orbitals do not, being now separated by Δ. A d_{z^2} electron has no orbital angular momentum, and the contribution of a d_{xy} or $d_{x^2-y^2}$ electron is now completely quenched by the ligand field. In an octahedral field, the total orbital angular momentum becomes zero if the ion has the ground state configuration $t_{2g}{}^3$ or $t_{2g}{}^6$, and among high-spin complexes only the configurations d^1, d^2, d^6 and d^7 ($t_{2g}{}^1$, $t_{2g}{}^2$, $t_{2g}{}^4e_g{}^2$ and $t_{2g}{}^5e_g{}^2$ respectively) have any orbital angular momentum. Consideration of tetrahedral species in the same manner shows that in this case only the (high-spin) configurations d^3, d^4, d^8 and d^9 ($e^2t_2{}^1$, $e^2t_2{}^2$, $e^4t_2{}^4$ and $e^4t_2{}^5$ respectively) should have orbital moments. For d^7, it is seen that the orbital moment should be quenched in a high-spin tetrahedral complex but not in a high-spin octahedral complex. It is found that the pink octahedral complexes of cobalt (II) such as $[Co(H_2O)_6]^{2+}$ do in fact have magnetic moments considerably higher than those of the blue tetrahedral complexes such as $[CoCl_4]^{2-}$ (about 5·0 and 4·4 B.M. respectively). Since the value for a tetrahedral complex is above the spin-only value, however, another factor must also be involved.

This is spin-orbit coupling, which opposes the quenching of orbital angular momentum by coupling it with spin angular momentum, which is not directly influenced by the ligand field. This mixing in of higher energy states with different magnetic properties from the ground state has the effect (in ways described in more specialized works) of modifying μ_s, the value calculated on the spin-only formula, according to the equation

$$\mu = \mu_s\left(1 - \alpha\frac{\lambda}{\Delta}\right)$$

where α is a constant which depends on the spectroscopic ground state of the ion in the ligand field, Δ is the separation between the ground energy level and the level being mixed in, and λ is the spin-orbit coupling constant which (as mentioned earlier) is obtained from the differences in energy between states of different J values. For high-spin complexes λ is related to the spin-orbit coupling constant per electron ζ by the equation

$$\zeta = \pm 2S\lambda$$

where S is the value for the free ion. For low-spin complexes S is the value in the complex. For comparisons between different ions, values of ζ rather than λ should be used so as to take into account different numbers of electrons; ζ increases steadily along the series Ti^{2+} to Ni^{2+} and Ti^{3+} to Ni^{3+}, values for the tripositive ions being on average about 15 per cent greater than for the dipositive ions; ζ is always positive, but λ takes the positive sign for a shell less than half full and the negative sign for a shell more than half full. There is no spin-orbit coupling for the high-spin d^5 configuration; the ^6S state which arises from this configuration of the gaseous ion is unaffected by a ligand field, so there is no higher level with which interaction could take place. A few values for ζ and λ for ions of the first transition series are given in Table 16.3. For $[Ni(H_2O)_6]^{2+}$, Δ_o obtained spectroscopically is 8500 cm^{-1} and α is 4. Substitution of $\lambda = -325$ cm^{-1} leads to $\mu = 3{\cdot}3$ B.M.; the experimental value is 3·2 B.M. For first transition series ions the effect of spin-orbit coupling is usually one of about $\pm(0{\cdot}2 - 0{\cdot}4)$ B.M. in μ.

Table 16.3 *Some values for ζ and λ*

Ion	Ti^{2+}	V^{2+}	Cr^{2+}	Mn^{2+}	Fe^{2+}	Co^{2+}	Ni^{2+}
ζ (cm^{-1})	121	167	230	347	410	533	649
λ (cm^{-1})	60	56	57	0	−102	−177	−325

For ions of metals of the second and third transition series, ζ is much greater than for those of the first transition series elements (e.g. the approximate values for Ru^{4+} and Os^{4+} are 1600 and 6400 cm^{-1}). Furthermore, most compounds of the heavier transition elements are of the low-spin type; not only are the $4d$ and $5d$ orbitals spatially larger than the $3d$ orbitals, so that double occupation of them produces less repulsion, but values for Δ for analogous compounds are usually larger by 30–50 per cent for ions of the second transition series than for analogous ions of the first series, and there is a similar increase between the second and third series. Thus there are very few high-spin complexes of the heavier transition elements, and when complexes of these elements do have unpaired electrons, their magnetic moments at ordinary temperatures (when $\lambda \gg kT$) are well below the spin-only values. Furthermore, the moments are temperature-dependent (i.e. the Curie law is not obeyed). This is true also for compounds of ions with very low spin-orbit coupling constants, but in these cases ($\lambda < kT$) the effect is apparent only at very low temperatures. The variation of μ with kT/λ is complex and depends upon both the electronic configuration and the stereochemistry, and for a discussion of the so-called Kotani diagrams, which represent the variations, more specialized works must be consulted. We must, however, emphasize the need for magnetic moments of compounds of the later transition series to be measured over a temperature range if they are to be interpreted to serve any useful purpose; it will be apparent from the foregoing discussion that magnetochemistry is a much more difficult subject than it appeared to be when we were considering the classification of complexes in the last chapter.

We should mention here the temperature-independent paramagnetism shown by many systems (such as CrO_4^{2-}, MnO_4^- and many low-spin cobalt (III) complexes) which do not contain unpaired electrons in the ground state. This arises from spin-orbit coupling of the ground state to a paramagnetic excited state, well removed from the ground state, under the influence of a magnetic field; the degree of mixing in of the higher state depends linearly upon the field strength but this means, of course, that the susceptibility is independent of the field strength. The effect is independent of temperature because the thermal population of the excited state is very nearly zero and does not change appreciably with change of temperature.

So far we have adopted a crystal field approach to paramagnetism. On ligand field theory, what we have assumed to be metal orbitals are molecular orbitals derived from metal and ligand orbitals, and there is no reason to assume that the ligand contribution to the moment is equal to that part of the metal contribution which it has replaced. It is customary, therefore, in applying ligand field theory to paramagnetic complexes, to apply an empirical correlation to λ for the free ion, and this consists of multiplying it by an orbital contribution reduction factor k, which is generally less than unity and often about 0·7.

16.3 Ferromagnetism and antiferromagnetism[2, 4]

In the compounds discussed so far we have assumed that paramagnetic species do not interact with one another. This is not always the case; 'exchange' or 'cooperative' effects may arise when the paramagnetic species are very close together (as in metals) or when the intervening atoms can transmit magnetic interactions (as in many oxides, fluorides and chlorides). In ferromagnetic materials such as metallic iron the spins on each of the atoms couple together to form a resultant unit cell moment. If we imagine that the unit cell moment is perpendicular to one of the six faces of the cubic unit cell, there are domains within which the unit cells are aligned, but in the whole crystal there is a random distribution over the six orientations. The effect of an applied magnetic field is then to align the domains in its direction. Even a moderately strong field may suffice for this purpose; at higher field strengths the susceptibility can increase no further, and the crystal has reached magnetic saturation. Hysteresis arises because thermal energy is not always able to randomize the alignment of domains when the field is removed; at the lowest temperature at which it is able to do so (the Curie point) ferromagnetism is replaced by ordinary paramagnetism. In Fe_3O_4, an inverse spinel (Section 3.4), half the Fe^{3+} ions occupy tetrahedral holes and half octahedral holes; their spins cancel (the interaction between spins in different sites is usually antiferromagnetic), but the alignment of the spins of the Fe^{2+} ions leads to ferromagnetism. The bulk magnetic susceptibility of Fe_3O_4 corresponds to a magnetic moment indicating one high-spin Fe^{2+} per formula weight. Compounds which are ferromagnetic owing to incomplete cancelling of coupled spins are sometimes described as ferrimagnetic.

In antiferromagnetic materials there is coupling between paramagnetic species, usually through ligand atoms, with the magnetic moments so aligned as to produce no resultant magnetic moment. In this case thermal motion up to a certain temperature (the Néel temperature) results in an increased susceptibility; above this temperature, when normal paramagnetism replaces antiferromagnetism, the susceptibility decreases with further rise in temperature. The χ/T patterns for a simple paramagnetic, a ferromagnetic and an antiferromagnetic material are shown in Fig. 16.2. Among typical antiferromagnetic compounds are MnO, $KNiF_3$ and K_2OsCl_6. For the last of these μ is 1·4 B.M. per osmium atom at the ordinary temperature, but this value increases to 1·9 B.M. in dilute solid solution in

K_2PtCl_6 (which is diamagnetic) owing to the separation of the paramagnetic species from one another.

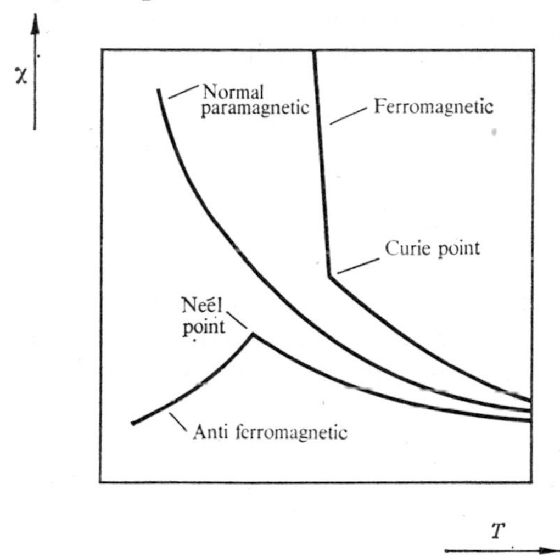

Figure 16.2 χ/T variation for different materials

16.4 Electron spin resonance spectroscopy (electron paramagnetic resonance spectroscopy)[5–8]

A magnetic field splits spectroscopic states into levels separated by $g\beta H$. In the simplest case, that of a molecule or an ion containing an unpaired electron in an S state ($L = 0$), the field makes the energies of the electron for its two alignments corresponding to the spin angular momentum quantum numbers $+\frac{1}{2}$ and $-\frac{1}{2}$ differ by

$$E = g\beta H$$

where g is the Landé splitting factor, β the Bohr magneton and H the field strength. At thermal equilibrium the lower state is slightly more populated than the upper one. If radiation of frequency ν such that

$$h\nu = g\beta H$$

is applied to the system, there is a small absorption of energy; by substitution in this expression it may be shown that for a magnetic field strength of 3000 gauss, such as is commonly employed, ν is about 9000 megacycles sec^{-1} or the wavelength 3 cm (i.e. in the microwave region). In many respects electron spin resonance is formally similar to nuclear magnetic resonance, but since the magnetic moment of an electron is much greater than that of a nucleus the energy change is much larger; as in nuclear magnetic resonance spectroscopy (Section 2.8) it is usual to keep ν constant and to vary H, but in electron spin resonance it is usual to plot the first derivative (i.e. the slope) of the absorption curve against the field strength so that shoulders on the absorption curve appear as peaks.

For a free electron g has the value 2·0023. In many free radicals the value of g, obtained from ν, β and H, is close to that for the free electron, but for metal ions g values are often appreciably different. The value of g depends in general upon the orientation of the species containing the unpaired electron with respect to the magnetic field. Because of free molecular motion, g measured in solution or in the gas phase is averaged over all orientations. In a crystal, if the paramagnetic species is located on a site of perfectly cubic symmetry, the g value is independent of the crystal orientation and is said to be isotropic. For any site of lower symmetry, however, g depends upon the orientation; in any particular direction it can be expressed in terms of three components, g_x, g_y and g_z in mutually perpendicular directions. The z direction is defined as being coincident with the highest-fold rotation axis of symmetry (which can be determined by X-ray methods) and g_z is termed $g_{\parallel}$, the g value when the z-axis is parallel to the external magnetic field. When, as is often the case, the paramagnetic species is located on a site of tetragonal symmetry, g_x and g_y are equal and are termed $g_{\perp}$. Small distortions from cubic symmetry which would escape detection by X-ray diffraction can often be detected by electron spin resonance spectroscopy.

When we discussed spin-orbit coupling in Section 16.2, we considered the change in μ_s, the 'spin-only' magnetic moment. We could equally well have considered μ_s to be a constant and discussed the change in g from the value of 2·0023 for a free electron. In this case we should have written

$$g = g_s\left(1 \pm \alpha \frac{\lambda}{\Delta}\right)$$

where the other symbols have the same meaning as before. Values for chromium (III) complexes, for example, are all low: for Cr_2O_3 in solid solution in Al_2O_3, $g = 1{\cdot}988$; for $[Cr(H_2O)_6]^{3+}$ in aluminium alum, $g = 1{\cdot}976$; and for $[Cr(CN)_6]^{3-}$, $g = 1{\cdot}993$. Here, spin-orbit coupling mixes an excited state with the ground state; the amount of mixing is seen to be inversely proportional to Δ (i.e. the difference between the states). Since g and Δ are both measurable, λ can be determined and compared with that for the free ion; the disagreement is evidence for covalent bonding in complexes. For copper (II) complexes, in which the geometry is usually that of a distorted octahedron owing to the Jahn–Teller effect, $g_{\parallel}$ and $g_{\perp}$ differ by 0·1–0·3; detailed interpretation of g values can then lead to information on the distribution of the unpaired electron, but this subject lies beyond the scope of this book and specialized works must be consulted for further information.

Another aspect of electron spin resonance spectroscopy of interest in inorganic chemistry is the hyperfine splitting that occurs when the paramagnetic species is near a nucleus with a spin I, when interaction between the nuclear spin and the electron spin causes the absorption to be split into $2I + 1$ components. The classic case of hyperfine splitting is that of the hexachloriridate (IV) anion, $[IrCl_6]^{2-}$, studied in solid solution in a single crystal of $Na_2[PtCl_6].6H_2O$. The quartet hyperfine structure due to the iridium nucleus of $I = \frac{3}{2}$ is split further by the chlorine nuclei ^{35}Cl and ^{37}Cl, each also of $I = \frac{3}{2}$, to give a complex pattern; this establishes electron-transfer between the iridium and chlorine atoms, and from detailed examination of the coupling it is concluded that the unpaired electron spends only about 70 per cent of its time on the iridium nucleus. Similar delocalization of unpaired electrons has been established for bis-(salicylaldiminato) copper (II) and several other compounds.

References for Chapter 16

1 Lewis, J., *Science Prog.*, 1962, **50,** 419.
2 Earnshaw, A., *Introduction to Magnetochemistry* (Academic Press, 1968).
3 Kettle, S. F. A., *Coordination Compounds* (Nelson, 1969).
4 Figgis, B. N., and Lewis, J., *Prog. Inorg. Chem.*, 1964, **6,** 37.
5 Phillips, C. S. G., and Williams, R. J. P., *Inorganic Chemistry* (Oxford University Press, 1966).
6 Carrington, A., *Chem. Britain*, 1968, **4,** 301.
7 Drago, R. S., *Physical Methods in Inorganic Chemistry* (Reinhold, 1965).
8 Goodman, B. A., and Raynor, J. B., *Adv. Inorg. Chem. Radiochem.*, 1970, **13,** 135.

17

Complexes of transition metals IV: electronic spectra

17.1 Introduction

The connection between the visible or ultraviolet spectra of transition metal complexes and the splitting of the *d* orbitals into two groups in an octahedral field has been indicated in Chapter 15. We now turn to this subject in more detail, first discussing the spectra in general and then considering those for a few relatively simple systems in more detail. Before doing so, however, we should draw attention to the fact that not all bands in the visible and ultraviolet spectra of transition metal complexes are *d–d* (or ligand field) spectra. Some arise from electron transfer (which may be in either direction) between metal ion and ligand; such *charge-transfer spectra* are of high intensity and usually occur at somewhat higher frequencies than *d–d* transitions. We shall say more about them in Section 17.4. Moreover, transitions may occur which involve electrons being excited from one ligand orbital to another; these are usually observed in the ultraviolet and are often almost the same in complexes as in the free ligand. It should also be mentioned that some oxy-anions have weak bands in the same region as *d–d* transitions; they include NO_2^- and NO_3^-. The anions Cl^-, ClO_4^- and SO_4^{2-} do not absorb in this region and are therefore preferable as 'counter-ions' when the spectra of complex cations are being studied.

There are two properties of an absorption band in the visible or ultraviolet to which we shall refer: its *'frequency' of maximum intensity* ν_{max}, usually expressed in cm^{-1}, and its *molar extinction coefficient* ε_{max}, defined by the equation

$$\log_{10}(I_0/I) = cl\varepsilon$$

where I_0/I is the ratio of the intensity of the incident radiation I_0 to that of the emergent radiation I, and c and l are the molar concentration and the path length in centimetres respectively. Reference is sometimes also made to the *half band width* $\Delta\nu_{\frac{1}{2}}$; this is the width of the absorption band at half height (i.e. where $\varepsilon = \varepsilon_{max}/2$).

Absorption spectra rarely approach the ideal situation of a number of symmetrical and separate bands, and in order to separate overlapping bands it is usual to treat the bands as simple Gaussian error curves.

17.2 Selection rules

There are two selection rules for electronic transitions in complexes; these are similar to the selection rules in atomic spectroscopy. They are:

(a) Transitions between states of different multiplicity, *S*, are forbidden. Usually this means that the number of unpaired electrons must not be changed, but this is not quite the same thing. For example, the transition $s^2 \rightarrow s^1p^1$ is spin-allowed so long as the spins of the two electrons in the s^1p^1 state are $+\frac{1}{2}$ and $-\frac{1}{2}$ (i.e. the state is a singlet); transition to the triplet state, in which both spins have the same sign, is forbidden.

(b) Transitions within a given set of *p* or *d* orbitals (i.e. transitions involving only redistribution of electrons in the given sub-shell) are forbidden if the molecule or ion has a centre of symmetry. A more formal statement of this rule, first put forward by Laporte, is that in a molecule which has a centre of symmetry, transitions between two *g* states or between two *u* states (see Section 15.4 for the significance of *g* and *u*) are forbidden.

As we have pointed out in connection with molecular spectra (Chapter 2), forbidden transitions often occur, but the probability of their doing so is much less than that of allowed transitions, and the intensities of the spectral lines (in the case of atoms) or bands (in the case of compounds) associated with them are relatively low. Transitions breaking rule (a), breaking rule (b) and breaking both rules are described as spin (or multiplicity)-forbidden, Laporte-forbidden, and spin (or multiplicity)- and Laporte-forbidden respectively. Since the regular octahedron has a centre of symmetry, *d–d* transitions in regular octahedral complexes are Laporte-forbidden. Many octahedral complexes would therefore be colourless if it were not for *vibronic coupling* of normal vibrations of the octahedron (some of which slightly distort its symmetry) with electronic absorption. Since there are many vibrational transitions (which have relatively low energies) that can be coupled with any electronic transition, the lines of atomic spectra are broadened into bands in the case of complexes. As it is, *d–d* absorption is weak. When, as in high-spin d^5 complexes, such as $[Mn(H_2O)_6]^{2+}$, any *d–d* transition must necessarily involve a change in multiplicity as well as a breach of the Laporte rule, the intensity is particularly low. Thus the colour of a dilute solution of a manganese (II) salt can scarcely be seen. Conversely, substitution in the octahedron, by destroying the centre of symmetry, results in increased intensity of absorption. For tetrahedral coordination, the Laporte restriction does not apply, since the tetrahedron has no centre of symmetry. Thus tetrahedral complexes are more intensely coloured than octahedral complexes of the same ion (compare, for example, $[CoCl_4]^{2-}$ and $[Co(H_2O)_6]^{2+}$). Multiplicity-forbidden bands can often be recognized by their sharpness. Multiplicity-allowed $t_{2g} \rightarrow e_g$ transitions lead to the production of an excited state in which the e_g electron lengthens the equilibrium metal–ligand distance, but since electronic transitions are much faster than atomic vibrations (the Franck–Condon principle), the excited state is not one of equilibrium interatomic distance but one which is vibrationally excited. In solution, for example, the complex then interacts to different extents with different solvent molecules, and a broad band results. On the other hand, a $t_{2g} \rightarrow t_{2g}$ transition does not result in a substantial change in the metal–ligand equilibrium distance, and the product of the transition is a low vibrational energy level of an excited state whose potential energy curve is very like that of the ground state; consequently the band is very narrow.

For the ions of transition metals of the second and third series it was seen that spin-orbit coupling is important in the interpretation of magnetic properties. This is certainly also true of their electronic spectra, which contain bands of intensities considerably greater than those of corresponding ions of the first transition series metals; but spectra of later transition series ions are not well understood at the present time, and in this account we shall confine attention to elements of the first transition series.

17.3 Spectra and energy level diagrams[1–10]

The identification of the 20,400 cm^{-1} (4900 Å) band of the d^1 $[Ti(H_2O)_6]^{3+}$ ion with the energy Δ_o required to promote an electron from a t_{2g} orbital to an e_g orbital was mentioned in Section 15.3 (owing to a Jahn–Teller effect in the excited state, a shoulder is present on the band, but we can ignore this in the following discussion). The spectroscopic term for the ground state of the gaseous Ti^{3+} ion is ^2D, and the corresponding descriptions of the $(t_{2g})^1$ and $(e_g)^1$ configurations are T_{2g} and E_g (the nomenclature was discussed in Section 15.4). As we showed earlier, the T_{2g} state lies 0·4 Δ_o below, and the E_g state 0·6 Δ_o above, the values of the energies of the five d orbitals in a spherical field. As Δ_o increases, the energy for the transition (which spectroscopists usually write as $E_g \leftarrow T_{2g}$ in absorption and $E_g \rightarrow T_{2g}$ in emission) increases. We can represent this by the simple diagram shown in Fig. 17.1.

For a d^9 ion in an octahedral field (a rare occurrence owing to the Jahn–Teller effect), the value of Δ_o corresponds to the promotion of

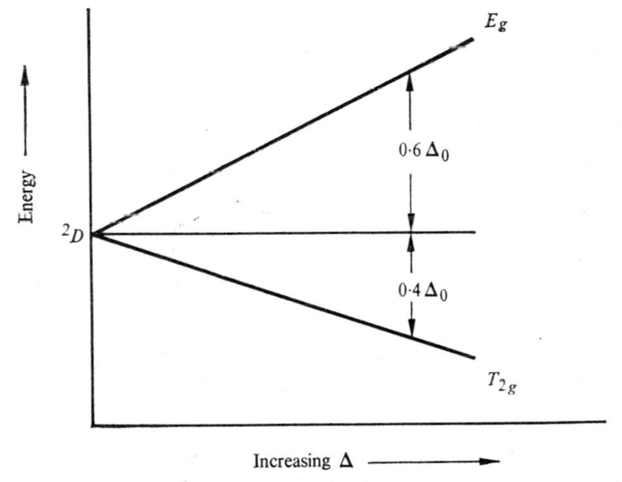

Figure 17.1 Energy level diagram for a d^1 ion

an electron from a t_{2g} orbital to an e_g orbital as before, but in this case the state of lower energy is E_g and that of higher energy is T_{2g}. This is so because the ground state $(t_{2g})^6(e_g)^3$ is doubly degenerate (it can be $(t_{2g})^6(d_{x^2-y^2})^2(d_{z^2})^1$ or $(t_{2g})^6(d_{x^2-y^2})^1(d_{z^2})^2$) and the excited state $(t_{2g})^5(e_g)^4$ is triply degenerate (it can be $(d_{xy})^2(d_{yz})^2(d_{xz})^1(e_g)^4$ or $(d_{xy})^2(d_{yz})^1(d_{xz})^2(e_g)^4$ or $(d_{xy})^1(d_{yz})^2(d_{xz})^2(e_g)^4$). Thus the diagram for a d^9 ion is inverted relative to that for a d^1 ion. Alternatively, the situation may be described by saying that for the d^9 configuration there is a hole in an e_g orbital in the ground state and a hole in a t_{2g} orbital in the excited state. The diagram is also inverted by a change of configuration from octahedral to tetrahedral (though at a given metal–ligand distance Δ_t is only $\frac{4}{9}$ Δ_o). All of this information is summarized in the *Orgel diagram* shown for the Ti^{3+} ion in Fig. 17.2. (For a d^1 system all terms necessarily have the same multiplicity and this is therefore not shown.)

By the argument already given, the same form of diagram would hold for a d^9 ion, though in this case the octahedral case would be on the left-hand side and the tetrahedral case on the right-hand

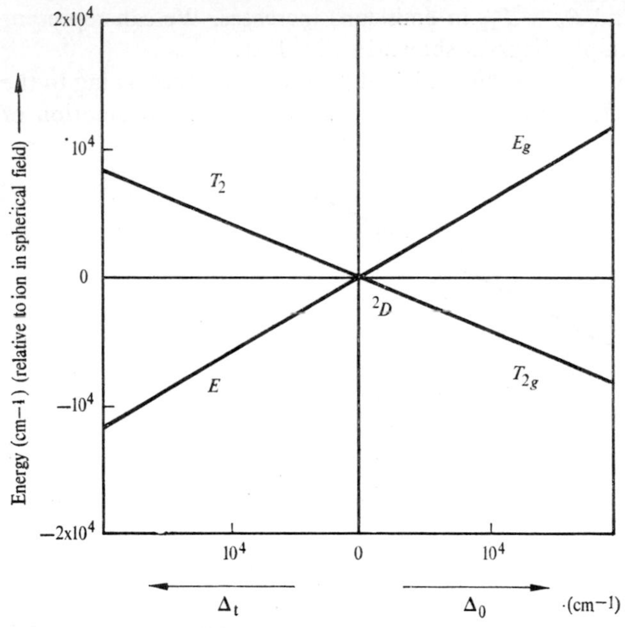

Figure 17.2 Orgel diagram for octahedral and tetrahedral complexes of a $3d^1$ ion (e.g. Ti^{3+})

side. Furthermore, a high-spin d^6 ion in an octahedral or a tetrahedral field is analogous to a d^1 ion in the same environment, and the same is true of a high-spin d^4 ion and a d^9 ion. The case of the high-spin d^4 ion is complicated by the Jahn–Teller effect, but for the high-spin d^6 ion in an octahedral complex the only allowed *d–d* transition is from $(t_{2g})^4(e_g)^2$ to $(t_{2g})^3(e_g)^3$, i.e. $^5E_g \leftarrow {}^5T_{2g}$; as in the case of the d^1 ion $[Ti(H_2O)_6]^{3+}$, the spectrum of $[Fe(H_2O)_6]^{2+}$ (high-spin d^6) consists of a single band (somewhat distorted by a Jahn–Teller effect in the excited state). For such systems, Δ is given by the frequency of the absorption maximum. We can thus give a single generalized Orgel diagram for d^1, d^4 (high-spin), d^6 (high-spin) and

d^9 ions for octahedral and tetrahedral fields; this is shown in Fig. 17.3.

For configurations other than d^1, d^9, and high-spin d^4 and d^6, the interpretation of electronic spectra is much less straightforward, and it is necessary to consider not only the splitting of the d orbitals in the ligand field, but also the interelectronic repulsions among the d electrons themselves. The need to do this is clearly shown by the fact that for these configurations there are more bands in the spectra than can be accounted for by the splitting of the d orbitals. For most

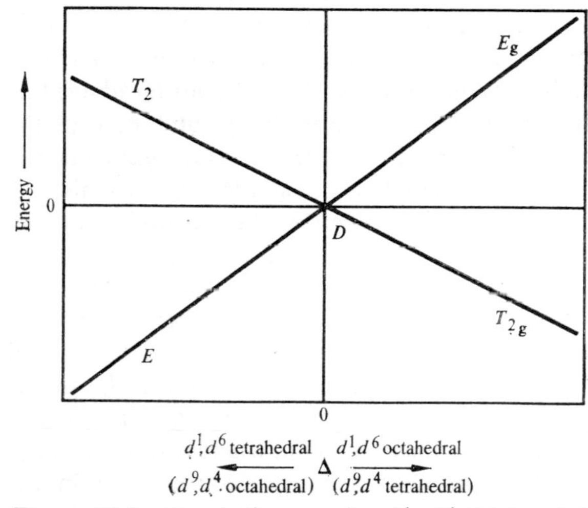

Figure 17.3 Orgel diagram for d^1, d^4 (high-spin), d^6 (high-spin) and d^9 ions in an octahedral or a tetrahedral field. (For ions in parentheses the g subscripts should be transferred to the left-hand side)

complexes, Δ is comparable with the differences between the energy levels of the free ion (obtained from atomic spectroscopy) from which the complexes have been derived. The strength of the ligand field is then described as intermediate, whereas if Δ is considerably smaller than the differences between the energy levels of the free ion the field is said to be weak. Conversely, a strong field is one for which the effect on the metal d electrons is much greater than the interelectronic repulsions. In the *weak field* approach to the electronic spectra of complexes we therefore consider first the free ion energy levels and then how these are modified somewhat by the effect of the ligand field; in the *strong field* approach we begin by describing the energy levels in terms of the ways in which the t_{2g} and e_g orbitals are occupied, and then make allowance for the effect of the electronic repulsions. The approach to intermediate field complexes (i.e. most real ones) may then be made by either method, the calculated energy level diagrams which serve as guides in the interpretation of the spectra being known as Orgel (weak field) and Tanabe and Sugano (strong field) diagrams respectively. To go into these methods in detail would be beyond the scope of this book, but we can give some idea of modern work by considering further the spectra of d^2 and high-spin d^8 complexes in octahedral fields.

Consider a d^2 gaseous ion with its electrons unpaired (i.e. in a triplet state). The ground state according to Hund's rules is ^3F ($l = 2$ and $l = 1$ for the individual electrons, so $L = 3$). Not far above this in energy is another triplet state ^3P. For the ground state of the octahedral d^2 complex there are three degenerate arrangements, the two electrons being placed in any two of the three t_{2g} orbitals; the symbol $^3T_{1g}(F)$ may be used to describe this state, denoting a triply degenerate *gerade* state of multiplicity three derived from the ^3F state. (There is also an excited state in which the two electrons are paired in one t_{2g} orbital, but transitions to this state are multiplicity-forbidden). The next state in terms of increasing energy is the excited state $(t_{2g})^1(e_g)^1$. If an electron is promoted from the d_{xz} or d_{yz} orbitals, leaving the remaining electron in the d_{xy} orbital, the promoted electron will go into the d_{z^2} orbital, where it will encounter less interelectronic repulsion from the d_{xy} electron than in the $d_{x^2-y^2}$ orbital, which is nearer to the d_{xy} orbital. Similarly, an electron promoted from the d_{xy} orbital will go to the $d_{x^2-y^2}$ orbital; but promotion of an electron from the d_{xy} orbital can leave the remaining electron in either the d_{xz} or the d_{yz} orbitals. There are thus three degenerate arrangements for $(t_{2g})^1(e_g)^1$

$$[(d_{xy})^1(d_{z^2})^1, (d_{xz})^1(d_{x^2-y^2})^1 \text{ and } (d_{yz})^1(d_{x^2-y^2})^1]$$

and this set gives the state $^3T_{2g}(F)$. The arrangements

$$(d_{xy})^1(d_{x^2-y^2})^1, (d_{xz})^1(d_{z^2})^1 \text{ and } (d_{yz})^1(d_{z^2})^1$$

are higher in energy; they are also degenerate and produce the state $^3T_{1g}(P)$. Other possible arrangements corresponding to $(t_{2g})^1(e_g)^1$ would involve reversing one of the electron spins; this would give a singlet state and transition to it would be multiplicity-forbidden. Lastly, a two-electron transition to produce the excited state $(d_{z^2})^1(d_{x^2-y^2})^1$ gives rise to a singly degenerate state $^3A_{2g}(F)$. The three bands in the spectra of octahedral vanadium (III) complexes are then assigned to the transitions

$$^3T_{2g}(F) \leftarrow {}^3T_{1g}(F),\ {}^3T_{1g}(P) \leftarrow {}^3T_{1g}(F) \text{ and } {}^3A_{2g}(F) \leftarrow {}^3T_{1g}(F)$$

in increasing frequency. (These are not all observed for the $[V(H_2O)_6]^{3+}$ ion, the one of highest frequency being overlapped by a charge-transfer band, but they are all observed for V^{3+} ions incorporated into an Al_2O_3 lattice.) For octahedral $Ni^{2+}(d^8)$ complexes a similar discussion can be given, though in this case the order of the states derived from the ^3F configuration is reversed and the transitions become, in order of increasing frequency,

$$^3T_{2g} \leftarrow {}^3A_{2g},\ {}^3T_{1g}(F) \leftarrow {}^3A_{2g} \text{ and } {}^3T_{1g}(P) \leftarrow {}^3A_{2g}$$

The spectra of $[Ni(H_2O)_6]^{2+}$ and $[Ni(en)_3]^{2+}$ are reproduced in Fig. 17.4; it will be seen that the middle peak in the spectrum of the

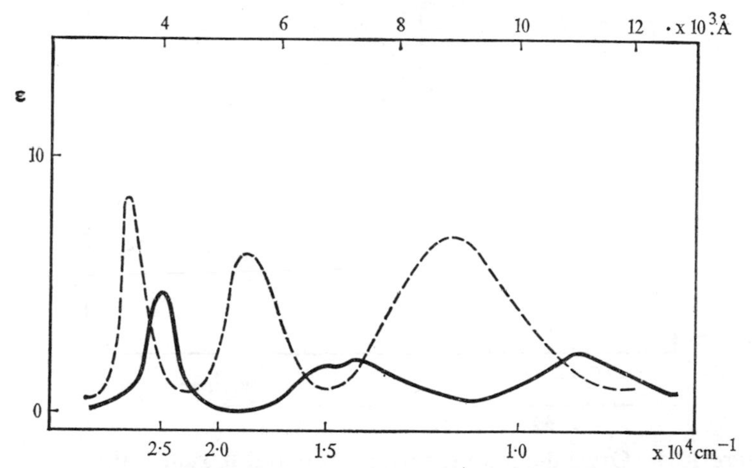

Figure 17.4 Absorption spectra of $[Ni(H_2O)_6]^{2+}$ (———) and $[Ni(en)_3]^{2+}$ (- - - - -)

aquo ion is actually a doublet; it is believed this arises from spin-orbit coupling between the $^3T_{1g}(F)$ and a nearby ^1E_g level; in the stronger field of the ethylenediamine complex these levels are further apart and no mixing occurs.

This is the weak field approach, and the discussion given above may be followed on the Orgel diagram for V^{3+} in Fig. 17.5. If we had used the strong field approach, we should have begun by considering the energy levels of the configurations $(t_{2g})^2$, $(t_{2g})^1(e_g)^1$

and $(e_g)^2$ when we allow for some interaction between individual electrons. The correlation diagram for a d^2 ion in weak, intermediate and strong octahedral fields is shown in Fig. 17.6, in which, because of the relative energy levels of different states, many states are shown to which transitions are forbidden. Using the same selection rules as in the weak field approach, we see that the same assignment of bands can be made.

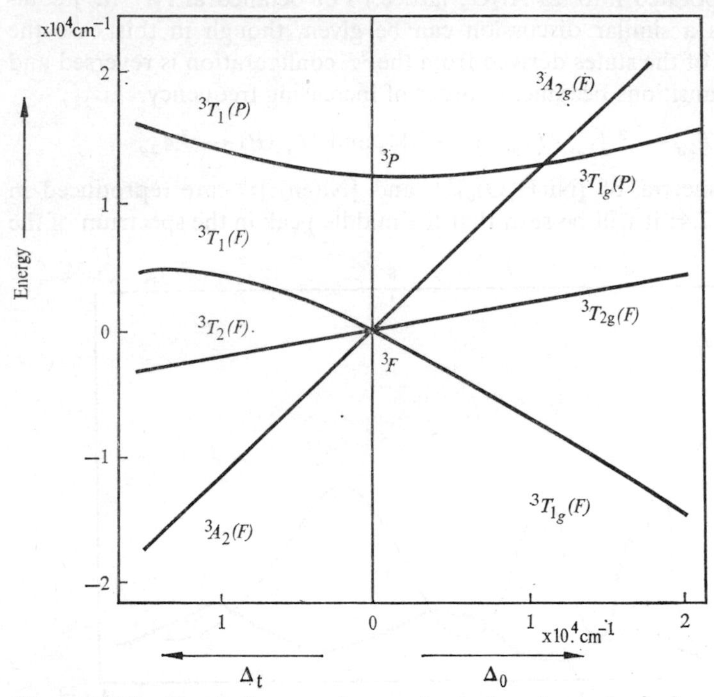

Figure 17.5 Orgel diagram for octahedral and tetrahedral complexes of $V^{3+}(d^2)$

Just as the spectra of d^1 and d^6 and of d^4 and d^9 ions are closely related, so are those of d^2 and d^7 and of d^3 and d^8 ions, and a second generalized Orgel diagram (for maximum multiplicity terms only) is given in Fig. 17.7. Thus the spectrum of $[Cr(H_2O)_6]^{3+}$, present in violet solutions of chromium (III) salts (green solutions contain complexes), contains three bands corresponding to the transitions

$$^4T_{2g} \leftarrow {}^4A_{2g},\ {}^4T_{1g}(F) \leftarrow {}^4A_{2g},\ \text{and}\ {}^4T_{1g}(P) \leftarrow {}^4A_{2g}$$

As we have already mentioned, a high-spin d^5 ion should have no *d–d* transitions, and all such bands in the spectra of Mn^{2+} and Fe^{3+} complexes are of very low intensity (the extinction coefficients are about one-hundredth of those for most complexes of ions of the first transition series). There are numerous closely spaced energy

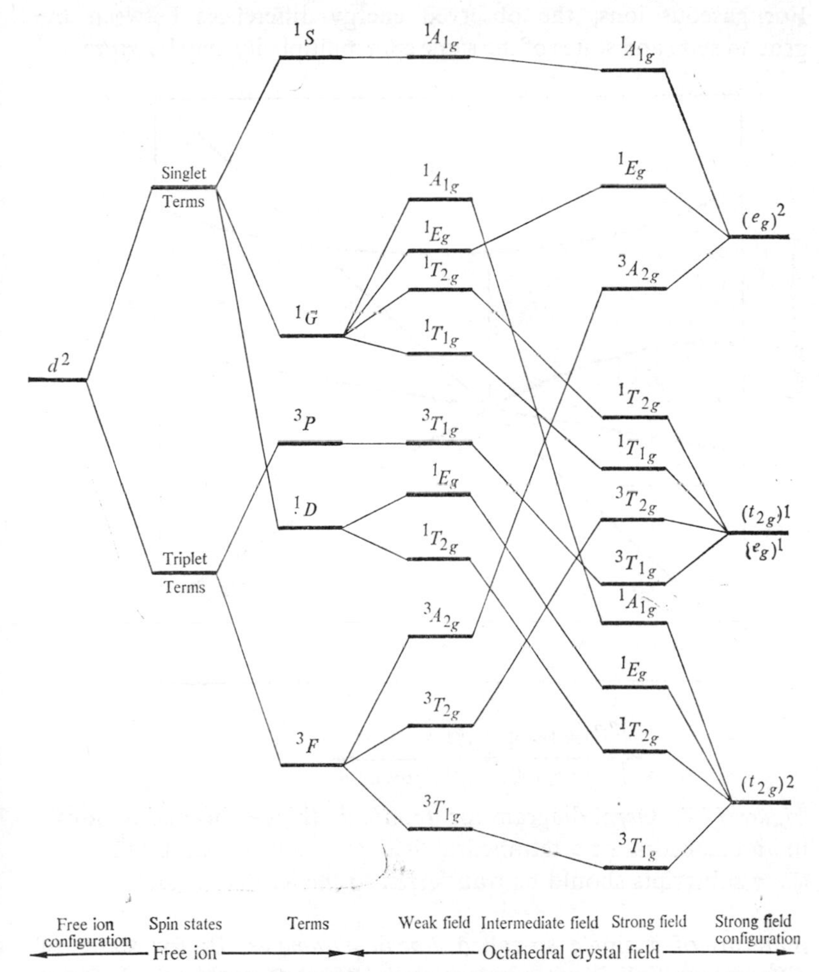

Figure 17.6 Correlation diagram for a d^2 ion

levels derived from the excited states ^4G, ^4P, ^4D and ^4F of the free ion, and a complicated spectrum results. Most of the bands are fairly broad, but those which result from the transfer of an electron from one t_{2g} orbital to another are, for reasons given earlier, very sharp.

The qualitative assignment of spectral bands is an impressive achievement, but of course the ultimate aim is to predict them, if not from first principles, at least from other spectroscopic data. This necessitates a knowledge of the interelectronic repulsion parameters. For gaseous ions, the observed energy differences between the ground state and states of the same spin multiplicity may be expressed

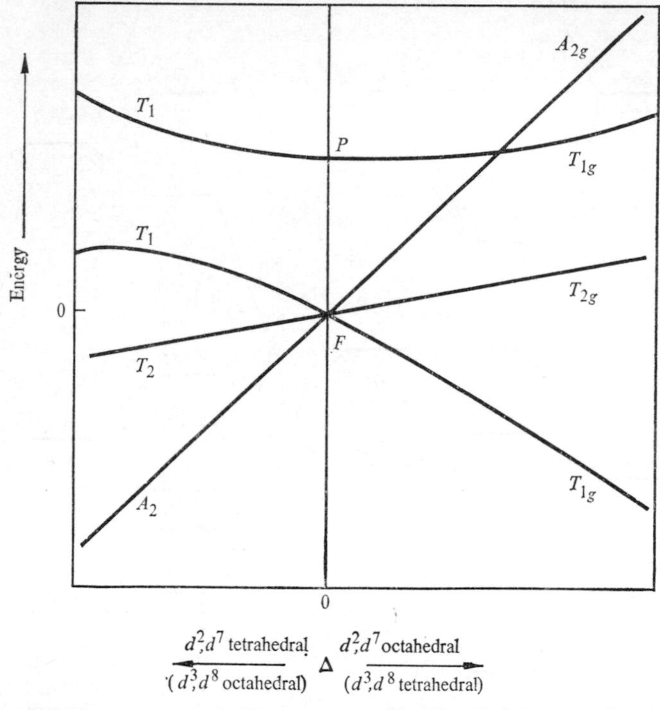

Figure 17.7 Orgel diagram for d^2, d^3, d^7 (high-spin) and d^8 ions in an octahedral or a tetrahedral field. (For ions in parentheses the *g* subscripts should be transferred to the left-hand side)

in terms of a single so-called *Racah parameter*, *B*; for states of different multiplicities two parameters, *B* and *C*, are required, and it is found that *C* is usually about four times as large as *B*. In the calculations of frequencies of transitions in terms of Δ, *B* and *C* the values of *B* and *C* for the free ion are used in the first place; these, however, are found to lead to results which are only roughly correct, and the process proceeds by adjustment of the *B* and *C* parameters. We cannot go into these calculations here, but one point of great general interest emerges: the values of the parameters are always lower in a complex than in the free ion. If B is the value in the free ion and B' that in a complex, for example,

$$B'/B = \beta \sim 0{\cdot}6\text{–}0{\cdot}9$$

the value of β depending upon both the ion and the ligand. For a series of complexes of the same metal ion with different ligands, there is a general order of values of β which is

$$I^- > Br^- \geqslant CN^- \geqslant Cl^- > OH^- > NH_3 > H_2O > F^-$$

This series has been interpreted as indicating the tendency of the electrons of metal ions to move to the ligands or for the mean distance of the d electrons from the nucleus to increase. This swelling of the d electron cloud, as it may be envisaged, has given rise to the name *nephelauxetic* (from the Greek for 'cloud-expanding') series for the progression above. It should be noted that it is quite different from the spectrochemical series, and the two assess different properties of the ligand field. A similar series may be derived for cations; in this case it is

$$Pt^{4+} > Pd^{4+} > Rh^{3+} > Co^{3+} > V^{3+} > Cr^{3+} > Fe^{3+} > Fe^{2+} > Co^{2+} > Ni^{2+} > Mn^{2+}$$

As for anions, this represents fairly well intuitive ideas about the likelihood of covalent bonding. The interpretation of electronic spectra may thus be said to provide further evidence for electron delocalization in transition metal complexes.

We have mentioned earlier that for d^1, d^4 (high-spin), d^6 (high-spin) and d^9 octahedral and tetrahedral complexes Δ_o and Δ_t are given directly by the one d–d band in the spectrum (Jahn–Teller splitting being neglected). For other ions, Δ can be obtained only if account is taken of interelectronic repulsions and, as we have seen, the repulsion parameters for free ions are changed somewhat on complex formation. The evaluation of Δ for complexes of d^2, d^3, d^5, d^7 and d^8 ions (and for low-spin d^4 and d^6 ions) is therefore dependent upon the quantitative interpretation of the spectra in terms of Δ and the Racah parameters, and the values obtained are subject to some uncertainty. However, values of Δ_o for the hexaquo and some other complexes of the common oxidation states of metals of the first transition series are now known with a reasonable degree of reliability, and we shall make considerable use of them in discussing the energetics of complex formation by these elements. Experimental data for tetrahedral complexes are scanty, but for $[CoCl_4]^{2-}$ and $[NiCl_4]^{2-}$ at least Δ_t (experimental) is fairly close to the value of $\frac{4}{9}\Delta_o$ calculated on a purely electrostatic model; for oxy-anions, however, π bonding and spin-orbit coupling make interpretation of spectra very difficult, and the theoretical description of the colour of permanganate, for example, remains an unsolved problem. Concerning spectra of complexes of planar and more complex geometries, not enough is known at the present time to justify a discussion in this book.

17.4 Charge-transfer spectra[1, 4, 9, 11]

Most of the very intense bands found in the spectra of transition metal complexes arise from transfer of an electron from a ligand orbital to a metal orbital or vice versa. This process does not infringe the usual selection rules, and such charge-transfer bands have extinction coefficients in the range 500–2000 compared with below about 100 for *d–d* transitions. Charge-transfer bands usually occur in the near ultraviolet; often there is overlap between the end of a charge-transfer band and a *d–d* absorption, and when this happens it is impossible to obtain the full *d–d* spectrum of the complex.

The high extinction coefficient is usually a reliable indication of a charge-transfer band, but evidence is also available from other sources. In the series of complexes of formula $[Co(NH_3)_5X]^{2+}$, where X is a halogen, strong absorption in the ultraviolet takes place at progressively higher frequencies for X = I, Br, Cl, F; for X = F the spectrum is, in fact, nearly the same as for X = NH_3. This suggests that the shift to higher frequencies is associated with increasing difficulty of transferring an electron from the ligand to the metal (i.e. of reduction of the cobalt (III) by the ligand). The same process is believed to occur in the highly coloured compounds formed by interaction of iron (III) and thiocyanate or phenols. If the frequency of the transition is low enough, reduction of the metal occurs. Thus solutions of copper (II) in the presence of high concentrations of fluoride remain blue; in the presence of high concentrations of chloride or bromide dark solutions are produced; and iodide effects reduction to copper (I) iodide.

A majority of charge-transfer complexes of transition metals involve ligand to metal electron transfer, as is to be expected from the availability of non-bonding or antibonding orbitals of the metal; the reverse process does, however, take place. Metal to ligand charge-transfer occurs in the red complexes of iron (II) with dipyridyl or *o*-phenanthroline; it is obviously favoured if the metal is in a low oxidation state and the ligand has a low-lying antibonding orbital.

This simple view of charge-transfer reactions involves looking at a complex from the electrostatic viewpoint. Considered in terms of ligand field theory, the electron transfer is between two molecular orbitals, one of which has more ligand character than metal character, and the other more metal character than ligand character. Theoretical treatments of charge-transfer spectra make use of this model, but discussion of the spectra in these terms is beyond the scope of this book.

Before leaving this subject we should, however, refer briefly to complexes containing a transition metal in two different oxidation states. Perhaps the best known of such *mixed valence compounds* is Prussian blue, potassium iron (III) hexacyanoferrate (II), $KFe^{III}[Fe^{II}(CN)_6]$. In this compound, high-spin iron (III) is octahedrally coordinated by the nitrogen atoms of six cyanide ions, and low-spin iron (II) by six carbon atoms. In the excited state the former is reduced to high-spin iron (II) and the latter oxidized to low-spin iron (III). Cyanide plays an important part in this transition, its π-bonding system greatly facilitating the transfer of charge. Several other mixed valence compounds of iron are known. Whether they give charge-transfer spectra depends upon the separation of the two types of iron: iron (II) iron (III) phosphates do show charge-transfer bands, but complex halides of the type $[Fe^{II}(NCCH_3)_6][Fe^{III}Br_4]_2$, in which there is no easy way of relaying charge, do not.

Many other metals form intensely coloured mixed valence compounds, typical examples being $[Pt^{II}(NH_3)_2Br_2][Pt^{IV}(NH_3)_2Br_4]$ and $Cs_2[Au^{I}Cl_2][Au^{III}Cl_4]$ and, among derivatives of typical elements, $(NH_4)_4[Sb^{III}Br_6][Sb^{V}Br_6]$; the absence of platinum (III), gold (II), and antimony (IV) in these compounds is shown by their diamagnetism.

References for Chapter 17

1 Orgel, L. E., *Ligand Field Theory*, 2nd edn. (Methuen, 1966).
2 Day, M. C., and Selbin, J. H., *Theoretical Inorganic Chemistry*, 2nd edn. (Reinhold, 1969).
3 Cotton, F. A., and Wilkinson, G., *Advanced Inorganic Chemistry*, 2nd edn. (Interscience, 1966).

4 Phillips, C. S. G., and Williams, R. J. P., *Inorganic Chemistry* (Oxford University Press, 1966).
5 Drago, R. S., *Physical Methods in Inorganic Chemistry* (Reinhold, 1965).
6 Kettle, S. F. A., *Coordination Compounds* (Nelson, 1969).
7 Dunn, T. M., in *Modern Coordination Chemistry*, eds. J. Lewis and R. G. Wilkins (Interscience, 1960).
8 Sutton, D., *Electronic Spectra of Transition Metal Complexes* (McGraw-Hill, 1968).
9 Lever, A. B. P., *Inorganic Electronic Spectroscopy* (Elsevier, 1968).
10 Carlin, R. L., *J. Chem. Education*, 1963, **40**, 135.
11 Robin, M. B., and Day, P., *Adv. Inorg. Chem. Radiochem.*, 1967, **10**, 247.

18 Complexes of transition metals V: thermochemistry and the stabilization of oxidation states

18.1 Introduction

In the early chapters of this book we discussed in a general way the energetics of formation of ionic salts (Chapter 3), molecules (Chapter 5) and complexes in aqueous media (Chapter 6). We now discuss these subjects again with special reference to transition metal chemistry, in which the crystal field stabilization energy term imposes an interesting and general variation upon what would otherwise be a regular progression along each series. In this chapter, however, we shall first consider salts and complexes together, for reasons mentioned in Chapter 16, and defer consideration of molecules to a later stage; there are very few uncharged species containing transition metals for which thermochemical data are available, and in most of these the metal may be regarded as having either a d^0 or a noble gas electronic configuration, so that there is no crystal field stabilization energy. The relative stability of different oxidation states is in principle a thermochemical matter, and we shall treat this subject quantitatively in a few simple cases; often, however, the data necessary for a precise analysis are unavailable. Where this is so, physical measurements of a non-thermochemical nature (e.g. vibrational spectra) can sometimes be used to develop a qualitative understanding of the role of certain ligands in making possible the isolation of compounds of transition metals in their less common formal oxidation states. Such an understanding, though hardly a substitute for a rigid thermochemical discussion, is nevertheless capable of providing useful guidance in preparative chemistry, and for this reason we shall consider both quantitative and qualitative aspects of the stabilization of oxidation states here. As in the last two chapters, we shall pay particular attention to octahedral high-spin complexes of ions of elements of the first transition series, since these have been the most thoroughly investigated and their behaviour is well understood.

18.2 Crystal field stabilization energies: high-spin octahedral systems[1–6]

Let us consider as a first example the high-spin dipositive ions of calcium and members of the first transition series in an octahedral environment. For a splitting of the d orbitals by an amount Δ_o the crystal field stabilization energy is, as we saw in Section 15.3, zero for d^0, d^5 and d^{10} configurations, 0·4 Δ_o for d^1 and d^6, 0·8 Δ_o for d^2 and d^7, 1·2 Δ_o for d^3 and d^8, and 0·6 Δ_o for d^4 and d^9. Thus from d^0 to d^{10} successive crystal field stabilization energies are 0, 0·4, 0·8, 1·2, 0·6, 0, 0·4, 0·8, 1·2, 0·6 and 0 Δ_o (see Fig. 18.1). So long as we are

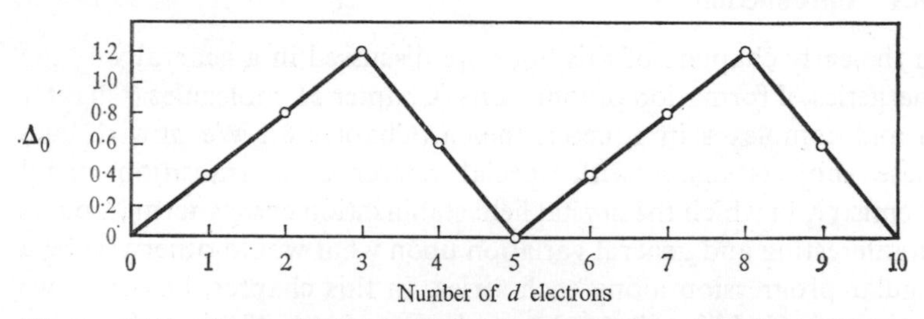

Figure 18.1 Crystal field stabilization energies of high-spin d^n ions in octahedral fields

considering only the extra stability which results from the splitting of the d orbitals in high-spin complexes, we shall neglect the pairing energy term for the d^6–d^{10} configurations, because even in the gaseous ions pairing in these systems is inevitable, and the term 'high-spin' denotes that the number of paired electrons in the actual compound being considered is the same as in the gas-phase ion. Strictly speaking, this neglect of the pairing energy term is not quite justified, for P is found to be somewhat smaller in complexes than in free ions, but the effect of this is not great in the systems we are considering now. We shall also neglect the consequences of Jahn–Teller effects in the d^4 and d^9 systems; these confer slight extra stability on such systems.

As we saw in Chapter 3, the lattice energy of, say, a difluoride can be obtained either from X-ray and compressibility data, or from the Born cycle. In the cases of CaF_2, MnF_2 and ZnF_2 (d^0, d^5 and d^{10} systems), the values obtained by the two methods agree closely, supporting their formulation as ionic salts. (For reasons discussed in Chapter 3, the fact that CaF_2 has the fluorite structure whilst the other two compounds have the rutile structure is, from the lattice energy viewpoint, unimportant.) It may reasonably be expected that for ions other than d^0, d^5 and d^{10} the lattice energies of difluorides would lie on the line joining the values for CaF_2, MnF_2 and ZnF_2 if it were not for the crystal field effect, and hence the difference between the observed (Born cycle) and intrapolated lattice energies for CrF_2, FeF_2, CoF_2, NiF_2 and CuF_2 (ScF_2, TiF_2 and VF_2 are not known) is a thermochemical measure of the crystal field stabilization energy. Similar treatments may be given for the other dihalides; and the lattice energies of all the fluorides and chlorides are shown in Fig. 18.2. The discussion may also be extended to oxides (where these are stoichiometric) and to some other salts, similar double-humped curves being obtained in every case. (The limited data available for compounds of tripositive d^0 to d^{10} ions show the same kind of variation.)

Since, as we saw earlier in this section, the variation of crystal field stabilization energy with the number of d electrons is also a

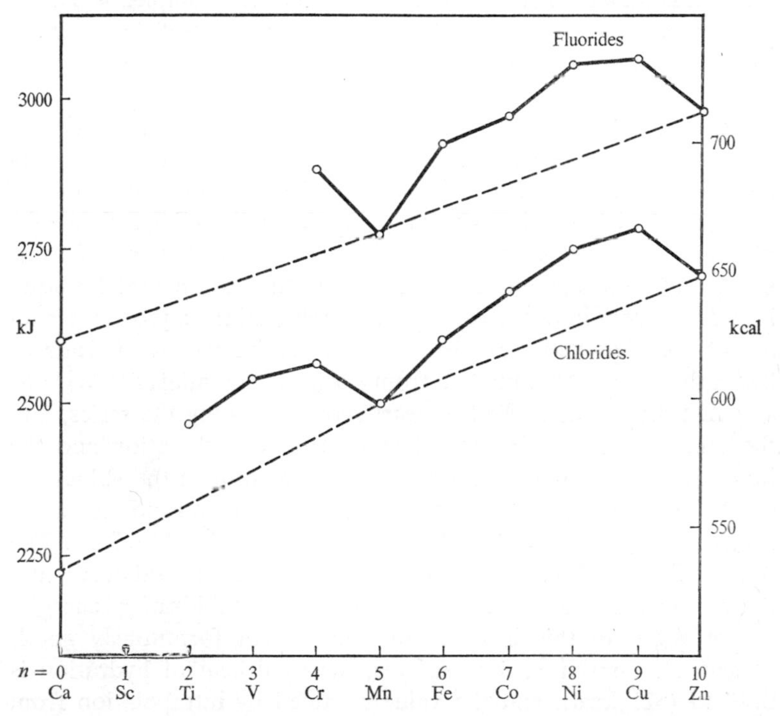

Figure 18.2 Lattice energies of difluorides and dichlorides of d^n ions

double-humped curve, the general similarity of the two diagrams is very strong evidence for the reality of the crystal field stabilization energy effect. If, however, we compare the stabilization energies measured by the difference between Born cycle and intrapolated lattice energies with those evaluated from spectroscopic values of Δ_o (obtained as described in the last chapter), we find that there is only rough agreement between the two sets of data. For the nickel halides, for instance, the thermochemical stabilization energies are 20–40 per cent greater than the spectroscopic ones. This is partly because the Born cycle lattice energies are not very accurate, involving as they do uncertainties in the other terms in the cycle (particularly in heats of formation of fluorides), but partly also for a more fundamental reason. We mentioned in Chapter 17 that when an electron in an octahedral complex is promoted from a t_{2g} to an e_g orbital the excited state of the complex which results does not have the equilibrium metal–ligand distance, since promotion of the electron is much faster than the distortion of the complex when the electron enters the e_g orbital: Δ_o (spectroscopic) therefore *ought not* to be the same as Δ_o (thermochemical), which refers to systems in equilibrium.

As a second example of the application of the concept of crystal field stabilization energy in transition metal chemistry we may consider the heats of hydration of the dipositive ions of metals of the first transition series. Crystallographic studies of hydrated salts of these ions, and comparison of the electronic spectra of these solids and their aqueous solutions, show that octahedral hexaquo ions are present in both; for dipositive ions, all the aquo complexes are of the high-spin type. In this case relative values of heats of hydration are readily obtained from electrode potentials for M^{2+}/M systems; the entropy effect is nearly constant along the series, and the other terms involved (the latent heats of sublimation and the ionization potentials) are all known. Again we take the values for Ca^{2+}, Mn^{2+} and Zn^{2+} as reference points, and intrapolate values for other dipositive ions in the absence of crystal field effects. The observed values (Fig. 18.3) again fall on a double-humped curve, and thermochemical and spectroscopic values of stabilization energies are in good (in this instance to some extent fortuitously good) agreement. For Ni^{2+}, for example, the total heat of hydration is 2180 kJ (521 kcal), and the value obtained by intrapolation from those for Mn^{2+} and Zn^{2+} (1950 and 2130 kJ, or 466 and 510 kcal, respectively) is 2060 kJ (492 kcal), whence the thermochemical

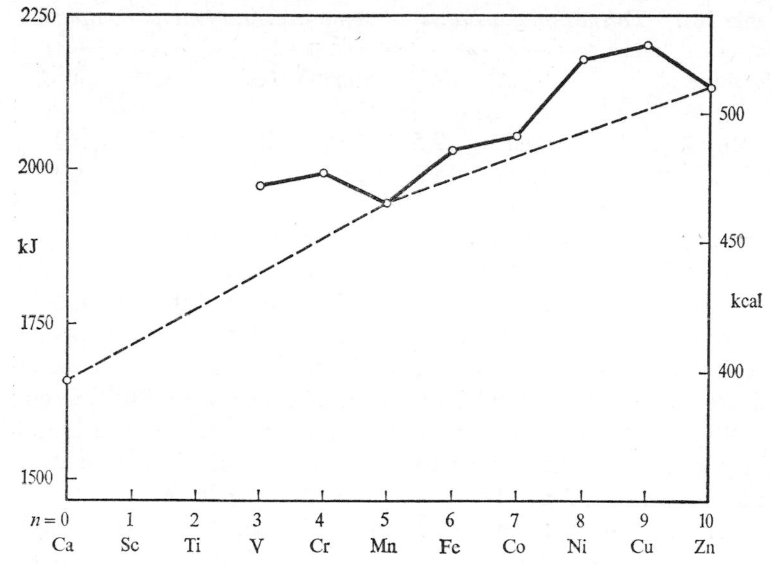

Figure 18.3 Heats of hydration of dipositive d^n ions (absolute values)

crystal field stabilization energy is 120 kJ (29 kcal). For $[Ni(H_2O)_6]^{2+}$, Δ_o is 8500 cm^{-1} and so the spectroscopic crystal field stabilization energy, 1·2 Δ_o, is 125 kJ (30 kcal).

Complex formation in aqueous solution involves displacement of water molecules by ligands, and the equilibrium constant for this process is related to thermodynamic quantities by the equations

$$\Delta G^o = \Delta H^o - T\Delta S^o = RT \ln K = 2{\cdot}303\ RT \log_{10} K$$

For a given ligand and a given cationic charge among metals of the first transition series, ΔS^o is nearly constant and hence the variation in log K is parallel to that in ΔH^o. For dipositive ions, data are very scanty in the first half of the scandium–zinc series, but plots of log K against the number of d electrons from manganese to zinc generally follow the pattern of the heats of hydration so long as only high-spin complexes are formed. Thus for ethylenediamine and ethylenediaminetetraacetate ion as ligands, the values for the tris- and mono-complexes respectively are shown in Table 18.1.

If Δ_o were constant for a given ligand and did not depend (as it does to an appreciable extent) upon the nature of the cation, there should be a general relationship of crystal field stabilization

Table 18.1 *Formation constants for some transition metal complexes*

Complex	Mn^{2+}	Fe^{2+}	Co^{2+}	Ni^{2+}	Cu^{2+}	Zn^{2+}
$M(en)_3^{2+}$ (ln *K*)	5·7	9·5	13·8	18·6	18·7	12·1
$M(EDTA)^{2-}$ (ln *K*)	13·8	14·3	16·3	18·6	18·7	16·1

energies for high-spin complexes of the ions Mn^{2+}–Zn^{2+} (viz. $Mn^{2+} < Fe^{2+} < Ni^{2+} > Cu^{2+} > Zn^{2+}$). If we then regard the total interaction energy as given by the imposition of the crystal field stabilization energy on a steadily increasing electrostatic energy of interaction from Mn^{2+} to Zn^{2+}, we should get a series of curves of similar shape for the total interaction of these ions with different ligands, and the difference between the curves for a ligand L and water would then be a measure of the energy of displacement of water by the ligand L. The values of log *K* in Table 18.1 are in fact proportional to such an energy of displacement, and the rough similarity in the pattern of variation of log *K* and of the heat of hydration with the number of *d* electrons is striking testimony to the similarities of the interactions of the ions with water, ethylenediamine and the ethylenediaminetetraacetate ion.

In practice Cu^{2+} usually forms rather more stable complexes than Ni^{2+}; this is generally attributed to the strong Jahn–Teller effect for Cu^{2+} complexes, as a consequence of which comparison with other ions in the series Mn^{2+}–Zn^{2+} is not strictly valid. The modified sequence $Mn^{2+} < Fe^{2+} < Ni^{2+} < Cu^{2+} > Zn^{2+}$ is part of a larger one first formulated by Irving and Williams, and empirically correlated by them with the sum of the first two ionization potentials of the metals; we cannot, however, discuss this interesting relationship further here.

In concluding this section we should again draw attention to the smallness of the crystal field stabilization energy term relative to the total metal ion–ligand interaction. With the maximum value of $1{\cdot}2\ \Delta_o$ for a high-spin octahedral complex, the crystal field stabilization energy contribution to the total lattice energy of NiF_2 or the total heat of hydration of Ni^{2+} is only about 5 per cent. The predictive uses of calculations of such stabilization energies are therefore greatest when like is being compared with like, and we are considering only small changes such as the replacement of one ligand by another or the interchange of cations between different environments. We shall now go on to discuss the latter subject.

18.3 Crystal field stabilization energies: tetrahedral systems and site preferences [7, 8]

As mentioned in Chapter 15, all tetrahedral complexes of first transition series elements are of the high-spin type. Because the splitting pattern for octahedral complexes is reversed in tetrahedral complexes, relative values of crystal field stabilization energies are different for these species, the values for systems containing different numbers of d electrons being: d^0, d^5 and d^{10}, 0; d^1 and d^6, 0·6 Δ_t; d^2 and d^7, 1·2 Δ_t; d^3 and d^8, 0·8 Δ_t; d^4 and d^9, 0·4 Δ_t. Since, so long as other factors are equal, $\Delta_t = \frac{4}{9}\Delta_o$, it follows that crystal field stabilization energies are always less in tetrahedral than in high-spin octahedral complexes containing the same number of d electrons; when other factors are equal this consideration should thus cause ions having configurations other than d^0, d^5 and d^{10} to prefer six-coordination. There is, in fact, little to be said about crystal field stabilization energies in tetrahedral complexes, since quantitative data are not available for the lattice energies of the compounds, or the formation constants or heats of formation of these complexes. Furthermore, many tetrahedral complexes certainly involve π bonding and hence would not be accurately described in terms of the simple electrostatic model we have used so far.

If we express all crystal field stabilization energies in terms of Δ_o, the values for ions in octahedral and tetrahedral complexes in which other factors are equal are given in Table 18.2. It would appear from this table that it is least likely for d^3 and d^8 ions to be found in tetrahedral environments. Up to the present time no

Table 18.2 *Crystal field stabilization energies for high-spin complexes (in terms of Δ_o)*

Configuration	d^0	d^1	d^2	d^3	d^4	d^5	d^6	d^7	d^8	d^9	d^{10}
Octahedral	0	0·4	0·8	1·2	0·6	0	0·4	0·8	1·2	0·6	0
Tetrahedral	0	0·27	0·53	0·35	0·18	0	0·27	0·53	0·35	0·18	0

tetrahedral d^3 complex appears to have been prepared, and in general d^8 ions certainly prefer octahedral (or planar) coordination; nickel salts, for example, form hexammine rather than tetrammine complexes. But although they are less common than octahedral complexes, tetrahedral complexes of nickel (II) are well established, the simplest of them being the blue $[NiCl_4]^{2-}$ anion. In crystalline nickel chloride, on the other hand, the metal ion is octahedrally coordinated in the cadmium chloride type of lattice. We must emphasize again the smallness of the crystal field stabilization energy relative to the total interaction energy. For a ligand at the weak field end of the spectrochemical series (such as Cl^-) Δ_o for the nickel (II) complex is about 7000 cm^{-1}, and hence 0·85 Δ_o, the extra stabilization energy in an octahedral field, is approximately 70 kJ (17 kcal), only a few per cent of the total Ni^{2+}–Cl^- interaction. Thus lattice or solvation energy factors can easily overcome the crystal field stabilization energy factor in this instance, and $[NiCl_4]^{2-}$ may be isolated as a tetramethylammonium salt or characterized in solution in nitromethane.

Where consideration of the relative values of crystal field stabilization energies in octahedral and tetrahedral fields has been particularly useful is in the interpretation of the distribution of normal and inverse spinel structures (Sections 3.4 and 4.4). In the normal structure $A^{II}[B_2^{III}]O_4$ the tetrahedral sites are occupied by the dipositive ions and the octahedral sites (denoted by the square brackets) by the tripositive ions; in the inverted structure the distribution is represented by the formula $B^{III}[A^{II}B^{III}]O_4$. In spinel itself A = Mg and B = Al. If at least one of the cations is that of a transition metal the inverted structure is often, though by no means always, found: thus $Zn^{II}Fe_2^{III}O_4$, $Fe^{II}Cr_2^{III}O_4$ and $Mn^{II}Mn_2^{III}O_4$ are normal, whilst $Ni^{II}Al_2^{III}O_4$, $Co^{II}Fe_2^{III}O_4$ and $Fe^{II}Fe_2^{III}O_4$ are inverted. For each of these compounds consideration of the crystal field stabilization energy factor is instructive, though on occasion we must be prepared to modify the data in Table 18.2 to take into account the fact that if other things are equal Δ_o for a tripositive ion is substantially greater than Δ_o for a dipositive ion of the same metal. Thus, among the compounds having the normal structure, in the first neither Zn^{2+} nor Fe^{3+} has any stabilization energy; in the second Cr^{3+} has a much greater additional stabilization energy in an octahedral site than Fe^{2+}; and in the third only the Mn^{3+} ion has any stabilization energy, this being (as usual) greater in an octahedral than in a tetrahedral site. Among the compounds having the inverted structure, in the first only Ni^{2+} has any stabilization energy (and it has a strong preference for octahedral sites); in the second only Co^{2+} and in the third only Fe^{2+} have any stabilization energy, and both have slight preferences for octahedral sites.

We can, however, cite two examples to show that here, as elsewhere, other factors are not necessarily equal, and predictions based solely on crystal field stabilization energy considerations sometimes prove incorrect. For $Fe^{II}Al_2{}^{III}O_4$ only Fe^{2+} has any stabilization energy, and hence this compound should have the inverted spinel structure; it is in fact a normal spinel. For $Ni^{II}Mn_2{}^{III}O_4$ both ions have stabilization energies, and the greater site preference factor for the d^8 ion is offset by a greater Δ_o for the triply charged d^4 ion. If we assume (in the absence of data for oxides) that the values of Δ_o for Ni^{2+} and Mn^{3+} in the oxide lattice are the same as in the $[Ni(H_2O)_6]^{2+}$ and $[Mn(H_2O)_6]^{3+}$ ions (8400 and 21,000 cm^{-1} respectively) we can show that the preference for octahedral sites is slightly greater for Mn^{3+} than for Ni^{2+}. The compound would thus be expected to be a normal spinel, whereas it is inverted. This may be the consequence of a Jahn–Teller effect for Mn^{3+} (the unit cell is not quite cubic) or it may arise from a small difference between the Madelung constants for the normal and inverted structures; whatever the reason, it is clear that when only small site preference energies are involved, crystal field theory is not a sufficient basis for the prediction of structures.

18.4 Crystal field stabilization energies: low-spin octahedral systems

In Table 18.3 are given the high-spin and low-spin octahedral crystal field stabilization energies for the electronic configurations (d^4–d^7 inclusive) for which the distinction is meaningful. Also shown

Table 18.3 *Crystal field stabilization energies for high-spin and low-spin d^4–d^7 ions in octahedral fields*

		d^4	d^5	d^6	d^7
Crystal field stabilization energy	High-spin	$0{\cdot}6\ \Delta_o$	0	$0{\cdot}4\ \Delta_o$	$0{\cdot}8\ \Delta_o$
	Low-spin	$1{\cdot}6\ \Delta_o$	$2\ \Delta_o$	$2{\cdot}4\ \Delta_o$	$1{\cdot}8\ \Delta_o$
Pairing energy	High-spin	0	0	P	$2P$
	Low-spin	P	$2P$	$3P$	$3P$

are the pairing energies (which represent destabilization) in terms of the average pairing energy per electron, P. As we noted in Chapter 15, whether a high-spin complex is formed depends on whether P is greater than Δ_o. Although values of P (or of the Racah parameters, which are another way of expressing the interelectronic repulsions) are obtainable for gaseous ions from atomic spectra, it is found that such values are somewhat too high for the ions in complexes, and in principle therefore P and Δ_o should both be obtained by analysis of the electronic spectrum of the appropriate low-spin complex. Since P, as well as Δ_o, depends on the nature of both metal ion and ligand, a full treatment of low-spin complexes would involve the interpretation of a very large number of spectra, some of them of considerable complexity. The limited data at present available suggest that for a given ion, however, P varies with variation in the nature of the ligand less than Δ_o does, so it is possible to make some qualitative comments about low-spin complexes in general.

The pairing energies in gaseous Fe^{3+} and Co^{3+}, for example, are 30,000 and 21,000 cm^{-1} respectively, and Δ_o values for the $[Fe(H_2O)_6]^{3+}$ and $[Co(H_2O)_6]^{3+}$ ions are 13,700 and 18,200 cm^{-1} respectively. The difference between the two quantities represents what may be thought of as the resistance to spin-pairing; clearly it is very much greater for iron (III) than for cobalt (III), and this may be correlated with the observation that whereas common complexes of iron (III) (except for the cyanide) are high-spin in character, those of cobalt (III) (except for the fluoride) are low-spin in character. The point made earlier that free ion pairing energies are higher than those in complexes is, incidentally, illustrated by the low-spin character of $[Co(H_2O)_6]^{3+}$, despite the apparent increase in energy on spin-pairing.

When spin-pairing has occurred, there is a large increase in the crystal field stabilization energy for d^5 and d^6 systems. This is often correlated with the high formation constants of low-spin iron (III), iron (II) and cobalt (III) complexes, but it is only one factor (and probably a relatively small one) in a complicated situation. Low-spin complexes of these ions should have shorter metal–ligand distances than their hypothetical high-spin counterparts because of the removal of electrons from e_g to t_{2g} orbitals; on an electrostatic model this means a higher electrostatic interaction energy, and on a covalent model it means a stronger σ bond. Furthermore, the electrons now in t_{2g} orbitals become available, in ligand field theory, for π bonding. We have no means of reliably sorting out the effects of high Δ_o, π bonding and ordinary electrostatic or covalent σ bonding, and can only reiterate the caution given previously against attaching too much significance to the crystal field stabilization energy term in isolation.

As we have seen in considering magnetic properties and electronic spectra, very much less is known about complexes of the metals of the second and third transition series than about those of members of the first series, and we are not yet in a position to discuss the thermochemistry of most of these elements in any detail.

18.5 Molecular compounds of transition metals

We shall confine attention in this section to some of the few molecular halides of transition metals for which thermochemical data are available, these being the only compounds for which reliable bond energy data are available. Heats of formation of many other compounds are known, of course, but when two or more types of bond are present it is impossible to obtain values for their energies from a single observation; information about bond lengths and force constants may then be used in a qualitative discussion but numerical values cannot be assigned.

The formation of any molecular halide MX_n from solid M and gaseous X_2 may be represented by the following series of changes:

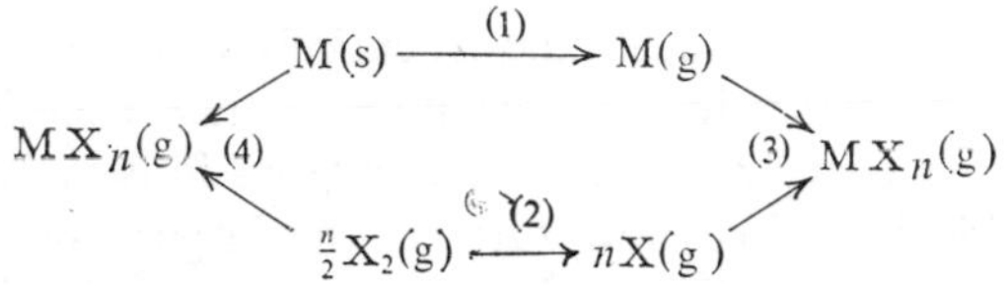

$$\begin{array}{ccccc} & M(s) & \xrightarrow{(1)} & M(g) & \\ \swarrow & & & & \searrow \\ MX_n(g)\ (4) & & & & (3)\ MX_n(g) \\ \nwarrow & & & & \nearrow \\ & \frac{n}{2}X_2(g) & \xrightarrow{(2)} & nX(g) & \end{array}$$

In this (1) is the latent heat of sublimation (or, better expressed, the heat of atomization) of M, (2) is the appropriate number times the dissociation energy of the halogen, (3) is n times the M—X bond energy in MX_n and (4) is the standard heat of formation of gaseous MX_n so long as X is fluorine or chlorine and M is a solid at 25°C. Obviously the cycle may be modified slightly if X is bromine or iodine or M is not a solid. Since values of (1) and (2) are known, measurement of the heat of formation of a gaseous halide enables the bond energy to be evaluated.

Among a periodic group of typical elements it is almost invariably found that bond energies decrease as the atomic number of M increases. The values for the M—Cl bond in Group IV tetrachlorides, for example, are C—Cl, 327 kJ (78 kcal); Si—Cl, 391 kJ (93 kcal); Ge—Cl, 342 kJ (82 kcal); Sn—Cl, 320 kJ (76 kcal); and Pb—Cl,

244 kJ (58 kcal). Among similar groups of transition metal halides, however, the bond energy increases with increase in atomic number of the metal. Thus in the series of tetrachlorides of titanium, zirconium, and hafnium the bond energies are Ti—Cl, 430 kJ (103 kcal); Zr—Cl, 488 kJ (117 kcal); and Hf—Cl, 519 kJ (124 kcal). No data are available for a similar series of three fluorides, but the same sequence of bond energies is shown in the pairs VF_5 and NbF_5, and MoF_6 and WF_6, the values being V—F, 468 kJ (112 kcal)*; Nb—F, 582 kJ (139 kcal)*; Mo—F, 448 kJ (107 kcal); and W—F, 502 kJ (120 kcal). Force constants for stretching vibrations of analogous halides also increase on going from compounds of metals of the second transition series to those of metals of the third series. Stronger π bonding by the heavier elements, which is suggested by other data, appears to provide a satisfactory explanation; but it should also be noted that because of the lanthanide contraction compounds of niobium and tantalum, and of molybdenum and tungsten, are almost isodimensional; this is in contrast to analogous compounds of, say, tin and lead, between which there is an appreciable increase in bond length.

18.6 Oxidation states of transition metals

We have so far in this chapter dealt almost entirely with one oxidation state at a time, discussing, for example, lattice energies of difluorides, hydration energies of dipositive gaseous ions and bond energies in tetrachlorides. We now turn to the larger and more difficult problems of why nearly all transition metals exhibit a range of oxidation states, how the relative stabilities of these states vary from one metal to another and how, for a given metal, the relative stabilities of different oxidation states can be modified by a change of environment (i.e. by choice of ligand).

The general question of why transition metals usually show variability of oxidation state is made more difficult by the fact that they form three types of compound, for each of which a different thermochemical cycle is appropriate. Most compounds of metals of the first transition series are those of low positive oxidation states and may be considered to a first approximation in terms of an electrostatic model; for these the Born cycle, or a modification of it, provides a good basis for discussion. For metals of the second and

* These are mean bond energies, equatorial and apical bonds in a trigonal bipyramid not being quite equivalent.

third transition series, on the other hand, high oxidation states are more stable; many compounds of such oxidation states are molecular in character, and a cycle of the type given in the last section for the formation of a covalent halide of formula MX_n is then appropriate. Compounds in which metals of the second and third series exhibit low oxidation states often contain clusters of metal atoms (no doubt partly, at least, because of the high heats of atomization of these elements), and a third type of cycle is then necessary. We shall confine attention here to compounds containing a single transition metal atom. As examples of the analytical treatment of change of oxidation state in terms of thermochemical parameters we shall consider four conversions; MX_n into MX_{n+1} (both ionic); MX_n into MX_{n+1} (both molecular); MX_n (ionic) into MX_n (molecular); and MX_n (ionic) into MX_{n+1} (molecular). Only the first will be considered in detail.

(a) $MX_n \longrightarrow MX_{n+1}$ (both ionic)

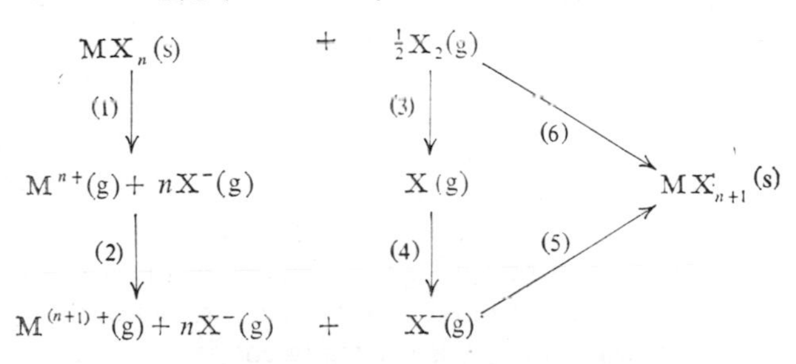

For this change the appropriate cycle is as follows. Stage (1) involves absorption of the lattice energy of MX_n, stage (2) the $(n + 1)$th ionization potential of M, stages (3) and (4) the dissociation energy and electron affinity of X_2 and X respectively, and stage (5) the liberation of the lattice energy of MX_{n+1}. For X = F the overall effect of stages (3) and (4) is liberation of 260 kJ (62 kcal), and for X = Cl the corresponding quantity is 235 kJ (56 kcal). If we ignore the small contribution of $T\Delta S^{\circ}$, we see, therefore, that whether stage (6) is possible depends upon whether the sum of the increase in lattice energy on going from MX_n to MX_{n+1} plus approximately 250 kJ (60 kcal) for X = F or Cl is greater than the $(n + 1)$th ionization potential of M. If the energy change for stage (6) is nearly zero MX_n and MX_{n+1} will be about equally stable, though the position of the equilibrium can, of course, be affected by continuous removal of X_2.

The lattice energy of MX_{n+1} will be greater than that of MX_n for three reasons: the increase in the product of the charges on the ions, the increase in Madelung constant (see Section 3.7) and the decrease in interionic distance, the radius of $M^{(n+1)+}$ being slightly less than that of M^{n+}. It is instructive to consider the dependence of the increase in lattice energy on the nature of X. Let us consider only the coulombic attraction and assume that the Madelung constants of the compounds of formula MX_n and MX_{n+1} (A and A' respectively) are independent of the structures for a given formula type. Then

$$U(MX_{n+1}) - U(MX_n) \propto \frac{A'(n+1)}{r(M^{(n+1)+}) + r(X^-)} - \frac{An}{r(M^{n+}) + r(X^-)}$$

The increase in lattice energy will clearly be greatest when $r(X^-)$ is smallest (i.e. X = F). Thus several metal fluorides can be converted by fluorine into higher (essentially ionic) fluorides although chlorine has no action on the corresponding chlorides; familiar examples are MnF_2, CoF_2 and AgF. The greater liberation of energy for X = F in the process $\frac{1}{2}X_2(g) \rightarrow X^-(g)$ plays some part in this, but the decisive influence is that of the lattice energy, which results from the size of the fluoride ion. Conversely, for a given $r(X^-)$, the increase in lattice energy is seen to be greatest when $r(M^{(n+1)+})$ and $r(M^{n+})$ are smallest. Since ionic radii decrease along a transition series, the increase in the coulombic lattice energy factor should most stabilize MX_{n+1} with respect of $MX_n + \frac{1}{2}X_2$ for later members of the series. This conclusion is not altered if we take into account repulsion energy, for the value of the Born exponent is constant along the series.

We have seen that for the dihalides the crystal field stabilization energy is only 5 per cent or less of the total lattice energy, but in the problem we are now discussing we should consider the difference in crystal field stabilization energy between MX_n and MX_{n+1}. Let us consider as an example the case of $n = 2$. Obviously M^{3+} has one less d electron than M^{2+}. If the cations are in high-spin configurations and are octahedrally coordinated in both MX_2 and MX_3 (as they are if M is a member of the first transition series), and if Δ_o' for MX_3 is taken as $1{\cdot}5\ \Delta_o$, Δ_o being the value for MX_2, the increase in crystal field stabilization energy is obtained as shown in Table 18.4.

We see that for several electronic configurations of M^{2+}, the change in crystal field stabilization energy will be zero or very small. For d^4, d^5, d^9 and d^{10} M^{2+} ions, it will be appreciable if Δ_o is about

Table 18.4 *Crystal field stabilization energies in di- and trihalides*

Number of d electrons in M^{2+}	*Crystal field stabilization energy*		
	MX_2	MX_3	$MX_3 - MX_2$
1,6	$0{\cdot}4\ \Delta_o$	0	$-0{\cdot}4\ \Delta_o$
2,7	$0{\cdot}8\ \Delta_o$	$0{\cdot}4\ \Delta_o' = 0{\cdot}6\ \Delta_o$	$-0{\cdot}2\ \Delta_o$
3,8	$1{\cdot}2\ \Delta_o$	$0{\cdot}8\ \Delta_o' = 1{\cdot}2\ \Delta_o$	0
4,9	$0{\cdot}6\ \Delta_o$	$1{\cdot}2\ \Delta_o' = 1{\cdot}8\ \Delta_o$	$1{\cdot}2\ \Delta_o$
5,10	0	$0{\cdot}6\ \Delta_o' = 0{\cdot}9\ \Delta_o$	$0{\cdot}9\ \Delta_o$

7000 cm^{-1} (84 kJ, or 20 kcal), a typical value for a halide complex of a dipositive ion of the first transition series, and it could suffice to make possible the existence of a trihalide which might otherwise be unobtainable. A detailed examination[9] of Born cycle and estimated lattice energy data, however, shows that the increase in crystal field stabilization energy exercises only a minor effect on the increase in the difference of the lattice energies of MX_2 and MX_3 as cationic radii decrease along the transition series. The conclusion that, so far as lattice energy considerations are concerned, MX_{n+1} should become more stable with respect to $MX_n + \frac{1}{2}X_2$ as we go from scandium to zinc, remains unchanged.

Since such an order is completely at variance with the experimental facts (no element after iron forms a trichloride), the variation in the third ionization potential must be an overriding factor; that this is so is shown by the data assembled in Table 18.5, in which are given the difference in lattice energies of MCl_2 and MCl_3 (estimated where necessary for the latter), the third ionization potentials, and ΔH^0 values for the reaction

$$MCl_2(s) + \tfrac{1}{2}Cl_2 = MCl_3(s)$$

For these reactions $T\Delta S^0$ arises largely from the loss of the entropy of gaseous chlorine and is nearly constant, so that the variation in ΔG^0 is almost identical with that in ΔH^0. We see that now the effect of the ionization potential has been taken into account, thermochemical calculations give results in complete agreement with experiment, even though the layer structures adopted by the dichlorides and trichlorides imply that the electrostatic model is not completely valid. There is, however, more point in such calculations than the demonstration that the existence of already well-known

Table 18.5 $MCl_2(s) + \frac{1}{2}Cl_2 = MCl_3(s)$ (*values in kJ, in parentheses in kcal*)

M		Sc	Ti	V	Cr	Mn
$U(MCl_3) - U(MCl_2)$	ΔH^0	−2498 (−597)	−2632 (−629)	−2737 (−654)	−2916 (−697)	−3004 (−718)
$\frac{1}{2}Cl_2 + e \rightarrow Cl^-$	ΔH^0	−234 (−56)	−234 (−56)	−234 (−56)	−234 (−56)	−234 (−56)
$M^{2+} \rightarrow M^{3+} + e$	ΔH^0	+2393 (+572)	+2657 (+635)	+2833 (+677)	+2990 (+715)	+3260 (+779)
Overall	ΔH^0	−339 (−81)	−209 (−50)	−138 (−33)	−160 (−38)	+22 (+5)
	$T\Delta S^0$	−25 (−6)	−20 (−5)	−25 (−6)	−30 (−7)	−30 (−7)
	ΔG^0	−314 (−75)	−189 (−45)	−113 (−27)	−130 (−31)	+52 (+12)
Est. uncertainty in	ΔG^0	±30 (±7)	±10 (±2)	±20 (±5)	±20 (±5)	±30 (±7)

M		Fe	Co	Ni	Cu	Zn
$U(MCl_3) - U(MCl_2)$	ΔH^0	−2787 (−666)	−2878 (−688)	−2888 (−690)	−2965 (−709)	−3100 (−741)
$\frac{1}{2}Cl_2 + e \rightarrow Cl^-$	ΔH^0	−234 (−56)	−234 (−56)	−234 (−56)	−234 (−56)	−234 (−56)
$M^{2+} \rightarrow M^{3+} + e$	ΔH^0	+2962 (+708)	+3243 (+775)	+3402 (+813)	+3556 (+850)	+3837 (+917)
Overall	ΔH^0	−59 (−14)	+131 (+31)	+280 (+67)	+357 (+85)	+503 (+120)
	$T\Delta S^0$	−30 (−7)	−25 (−6)	−20 (−5)	−30 (−7)	−25 (−6)
	ΔG^0	−29 (−7)	+156 (+37)	+300 (+72)	+387 (+92)	+528 (+126)
Est. uncertainty in	ΔG^0	±10 (±2)	±60 (±15)	±60 (±15)	±60 (±15)	±60 (±15)

compounds can be accounted for; a slightly positive ΔG^0 indicates that the higher chloride might be obtainable under specific conditions or in the form of a complex. It is interesting to note that complex chlorides of manganese (III) and (IV) have been known for some time, and that convincing evidence for the existence of a complex chloride ion of cobalt (III), $[CoCl_6]^{3-}$, has recently been reported.

We have said nothing here about the *reasons* for the variation in the third ionization potential from scandium to zinc, since ionization potentials are primary thermochemical data; we shall, however, comment briefly on this variation in Section 18.8.

(b) $MX_n \rightarrow MX_{n+1}$ (both molecular)

In this case, if we consider gaseous molecules, the appropriate cycle is

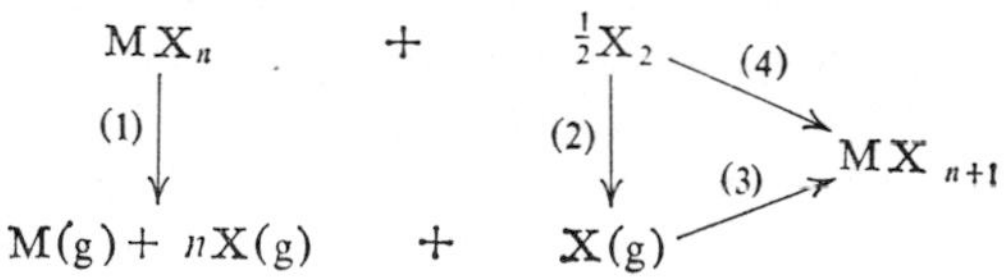

$$\begin{array}{ccccc} MX_n & + & \frac{1}{2}X_2 & \xrightarrow{(4)} & MX_{n+1} \\ \downarrow(1) & & \downarrow(2) & \nearrow(3) & \\ M(g) + nX(g) & + & X(g) & & \end{array}$$

This is much less easy to deal with than an electrostatics problem, for bond energies relate, by definition, to formation of bonds from ground state atoms, and clearly the ground state of M is rather remote from its two different valence states in MX_n and MX_{n+1}. Among the typical elements it is found that mean bond energies decrease with increase in oxidation state (e.g. from PF_3 to PF_5), and this may be considered to be the resultant of two factors, the promotion energy absorbed in the production of a higher valence state, and different degrees of overlap of orbitals in the two different valence states; unfortunately, neither of these factors is susceptible to direct experimental measurement. However, if we make the reasonable assumption that the bond energy in MX_{n+1} will be less than that in MX_n, we see that the question of whether the total energy of $(n + 1)$ weaker bonds (stage 3) is greater than the sum of that of n stronger bonds (stage 1) plus half the dissociation energy of X_2 (stage 2) is a nice one. In these circumstances $T\Delta S^0$ may play a decisive part, especially if one of the halides is in fact a solid, or if T is very high, in determining ΔG^0 for stage (4).

(c) MX_n (ionic) $\rightarrow MX_n$ (molecular)

This case is most simply considered as the conversion of an ionic solid to a molecular vapour at the same temperature by means of the cycle

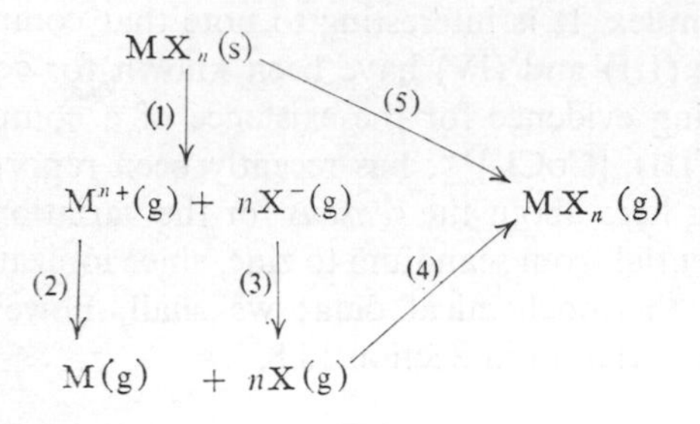

We see that the energy change of stage (5) is determined by the relative magnitudes of [the lattice energy of MX_n (stage 1) plus n times the electron affinity of X (stage 3)] and [the sum of the first n ionization potentials of M (stage 2) plus n times the bond energy in MX_n (stage 4)]. We shall not discuss this problem further, but it will be seen that the interplay of the factors that determine whether MX_n is ionic or covalent is a subtle matter, and it is not surprising that the distinction between the two extreme bond types is not maintained in practice as a sharp one.

(d) MX_n (ionic) $\longrightarrow$ MX_{n+1} (molecular)

This is the combination of conversions (c) and (b), since we may envisage the change as taking place in two stages:

$$MX_n \text{ (ionic)} \longrightarrow MX_n \text{ (molecular)}$$
$$MX_n \text{ (molecular)} + \tfrac{1}{2}X_2 \longrightarrow MX_{n+1} \text{ (molecular)}$$

This represents a very common case in practice (the conversion of a salt-like lower halide to a volatile higher halide), and it is noteworthy that the physicochemical treatment of it presents so difficult a problem.

18.7 Oxidation states of transition metals in aqueous media[4, 5]

The factors which determine the magnitudes of standard electrode potentials and the influence of pH, precipitation and complex formation on the relative stabilities of oxidation states were discussed at length in Chapter 6, and we have already discussed the hydration energies of M^{2+} ions earlier in this chapter. Because of the relative ease with which water is reduced or oxidized, the range of oxidation states attainable in aqueous media is less than in the solid state: thus scandium (II) and titanium (II) would reduce water, and nickel (III), copper (III) and zinc (III) would oxidize it. E^0 for the system M^{2+}/M is determined by the algebraic sum of the hydration energy of M^{2+}, the first two ionization potentials of M, and the latent heat of sublimation of M, E^0 for the system

$$H^+(aq) + e \rightleftharpoons \tfrac{1}{2}H_2$$

being taken as zero. The sum of the first two ionization potentials (values for which are given in Section 18.8) increases successively from scandium (1866 kJ, or 446 kcal) to zinc (2640 kJ, or 630 kcal) except that the value for copper is higher than that for zinc. Hydration energies, as we have seen, increase fairly regularly, most of the departure from regularity being accounted for by the crystal field stabilization energy term. Latent heats of sublimation, however, vary erratically between 281 kJ (67 kcal) for manganese and 515 kJ (123 kcal) for vanadium, with the exception of that for zinc which has the much lower value of 125 kJ (30 kcal). The observed variation in E^0 for M^{2+}/M for those elements for which measurements can be made is: V, −1·18; Cr, −0·9; Mn, −1·18; Fe, −0·44; Co, −0·28; Ni, −0·25; Cu, +0·34; and Zn, −0·76 V. In so far as this sequence can be said to be reflected in any single one of the variables we have mentioned, it is probably in ionization potential, with the proviso that the very low latent heat of sublimation of zinc places this element in a special position. But it must be made clear that with so many variables no simple pattern of behaviour is to be expected.

If we consider the variation in the M^{3+}/M^{2+} standard potential, the problem is simplified by the fact that the latent heat of sublimation of M is no longer relevant, and we are concerned only with the third ionization potential of M and the hydration energies of M^{3+} and M^{2+}. The measured values for this potential are: V, −0·26; Cr, −0·41; Mn, +1·51; Fe, +0·77; and Co, +1·97 V. These values, and the more negative but unknown ones for scandium and titanium and more positive but unknown ones for nickel, copper and zinc, show a variation which is generally very similar to that in the third ionization potentials given in Table 18.5. This suggests that though there is a steady increase in the difference between the heats of hydration of M^{3+} and M^{2+} (as there was in the difference between the lattice energies of MCl_3 and MCl_2), this factor is outweighed by the ionization potential term. The most marked exception to this is chromium: although the third ionization potential of chromium is 159 kJ (38 kcal) greater than that of vanadium, chromium (II) is a slightly more powerful reducing agent than vanadium (II). Now V^{2+} and Cr^{2+} are d^3 and d^4 ions respectively, and if we turn back to Table 18.4 we see that for octahedrally coordinated ions the gain in crystal field stabilization energy on going from an M^{2+} complex to an M^{3+} complex is given approximately by zero for a d^3 ion and 1·2 Δ_o (Δ_o being the value for the M^{2+} ion) for a d^4 ion. For $[Cr(H_2O)_6]^{2+}$, Δ_o is 14,100 cm^{-1} (169 kJ, or 40 kcal). Thus, if for the present we disregard the difference between spectroscopic and thermochemical values, the heat of hydration of Cr^{3+} should be greater than that of Cr^{2+} by about 200 kJ (50 kcal) more than the heat of V^{3+} is greater than that of V^{2+} because of crystal field stabilization energy considerations.

Another value which calls for comment is that of the Co^{3+}/Co^{2+} potential, which is much nearer to that of Fe^{3+}/Fe^{2+} than the respective third ionization potentials of these elements would suggest. This can be correlated with the fact that, unlike the halides we discussed in the last section and unlike the aquo complexes of the other tripositive ions mentioned in this, $[Co(H_2O)_6]^{3+}$ is a low-spin complex and thus has considerable extra stabilization energy; for a low-spin d^6 ion the crystal field stabilization energy is $2{\cdot}4\ \Delta_o - 3P$. It has been calculated that if $[Co(H_2O)_6]^{3+}$ were a high-spin ion E^0 for Co^{3+}/Co^{2+} would be about $+2{\cdot}8$ V (i.e. the same as that of the $\frac{1}{2}F_2/F^-$ system).

It was shown in Chapter 6 that for the general system

$$ML_n{}^{3+} + e \rightleftharpoons ML_n{}^{2+}$$

the formation of a more stable complex by M^{3+} (the common state of affairs) lowers E^0 for the $ML_n{}^{3+}/ML_n{}^{2+}$ system relative to that for $M(H_2O)_n{}^{3+}/M(H_2O)_n{}^{2+}$; the formation of a more stable complex by M^{2+} has the opposite effect. For the Fe^{3+}/Fe^{2+} system, complexing by cyanide, oxalate or EDTA lowers E^0, and complexing by 2,2′-dipyridyl and *o*-phenanthroline raises it (see Table 18.6). In each case the iron (III) and iron (II) complexes are both high-spin (H_2O and EDTA as ligands) or both low-spin (CN^-, *o*-phenanthroline and 2,2′-dipyridyl as ligands). Although it may seem obvious that a negatively charged ligand should complex the more highly charged cation more strongly, it must be remembered that there is really little difference between the metal—OH_2 bond being destroyed and the metal—anion bond being created; furthermore, the entropy effect can seldom be neglected in aqueous reactions involving ions. Stabilization of the lower oxidation state, as by *o*-phenanthroline and 2,2′-dipyridyl, must involve non-electrostatic interactions, and it is generally agreed that such ligands stabilize the lower oxidation state by virtue of their ability to absorb some or all of the extra negative charge in their relatively low energy antibonding orbitals; the low-spin d^6 configuration for Fe^{2+} provides the maximum number of t_{2g} electrons to take part in such π bond formation. But since we cannot assess π bonding quantitatively we shall defer further consideration of this matter until Section 18.9.

Table 18.6 *Redox potentials for some Fe^{3+}/Fe^{2+} couples*

Ligand	E^0 (V)
o-phenanthroline	+1·10
2,2′-dipyridyl	+0·96
H_2O (at $[H^+] = 1$)	+0·77
CN^-	+0·36
$EDTA^{4-}$	−0·10

We should, however, draw attention here to the outstanding example in the first transition series of the stabilization of an oxidation state by complexing. Comparison of the standard potentials for the $[Co(H_2O)_6]^{3+}/[Co(H_2O)_6]^{2+}$ and $[Co(NH_3)_6]^{3+}/[Co(NH_3)_6]^{2+}$ systems, +1·97 and +0·10 V respectively, shows that the overall formation constant of $[Co(NH_3)_6]^{3+}$ must be greater than that of $[Co(NH_3)_6]^{2+}$ by a factor of about 10^{32}. In commenting upon this difference, which is very large by the standards of formation constants of complexes in aqueous solution, it is not enough to say that $[Co(NH_3)_6]^{3+}$ is a low-spin complex whilst $[Co(NH_3)_6]^{2+}$ is a high-spin one; the same is true of the analogous aquo complexes. The formation constant of $[Co(NH_3)_6]^{3+}$ is really determined by the difference between the total interaction energy of gaseous Co^{3+} with water and its interaction energy with six molecules of ammonia plus the hydration energy of the resulting cation; the formation constant of $[Co(NH_3)_6]^{2+}$ is determined by an analogous difference, and E^0 for the $[Co(NH_3)_6]^{3+}/[Co(NH_3)_6]^{2+}$ system by the net difference between these two differences.

We cannot calculate this net difference accurately, but it is interesting to enquire to what extent differences in crystal field stabilization energies contribute to it. Even this cannot be done quantitatively in the absence of a knowledge (which we do not yet have) of the pairing energies for Co^{3+} and Co^{2+} in all the ions involved, but there is something very simple about the electronic spectra of the ions we are discussing that sheds light on the problem. The spectra of $[Co(H_2O)_6]^{2+}$ and $[Co(NH_3)_6]^{2+}$ show that Δ_o is nearly the same in these two complexes, and since the crystal field stabilization energy for a high-spin d^7 octahedral complex is $0{\cdot}8\ \Delta_o - 2P$, the difference in crystal field stabilization energies between the two ions is very small except in the unlikely event of the pairing energies in them being substantially different. For $[Co(H_2O)_6]^{3+}$, on the other hand, Δ_o is 4200 cm^{-1} less than in $[Co(NH_3)_6]^{3+}$, and since both are low-spin d^6 octahedral complexes for which the crystal field stabilization energy is $2{\cdot}4\ \Delta_o - 3P$, the difference in spectroscopic crystal field stabilization energies (taking P as constant) is as much as 10,080 cm^{-1}, equivalent to a difference in E^0 of about 1·2 V. Thus much of the difference in E^0 can be accounted for by the difference in crystal field stabilization energies, which in turn arises from the difference in Δ_o for the cobalt (III) complexes and their low-spin character (also a consequence of values for Δ_o). Although it would be an oversimplification, we might say that a large part of the modification of $E^0(Co^{3+}/Co^{2+})$ by complexing with ammonia can be interpreted on the basis of the observations that whereas $[Co(H_2O)_6]^{2+}$ and $[Co(NH_3)_6]^{2+}$ are both pink, $[Co(H_2O)_6]^{3+}$ is blue but $[Co(NH_3)_6]^{3+}$ is orange.

18.8 Ionization potentials of transition elements[5]

It is clear from the discussions in Sections 18.6 and 18.7 that ionization potentials play a very important part in transition metal chemistry, and we end the quantitative treatment of oxidation states of transition metals by considering these quantities further. Values for the first three potentials of elements of the first transition series are assembled in Table 18.7. Since we have been most concerned with third ionization potentials, we shall discuss these first.

For a transition metal ion which has the configuration d^n beyond the argon configuration, the total energy may be considered in two parts, the electrostatic attraction of the positively charged argon core for the d electrons, and the interactions among the d electrons themselves. The second part may in turn be divided into a repulsion energy between electrons (approximately proportional to the

Table 18.7 *Ionization potentials of metals of the first transition series (values in kJ, in parentheses in kcal)*

Ionization potential	Sc	Ti	V	Cr	Mn	Fe	Co	Ni	Cu	Zn
1	631	656	650	653	717	762	758	736	745	906
	(151)	(157)	(155)	(156)	(171)	(181)	(181)	(176)	(178)	(217)
2	1235	1309	1414	1592	1509	1561	1644	1752	1958	1734
	(295)	(313)	(338)	(380)	(361)	(373)	(393)	(419)	(468)	(414)
3	2393	2657	2833	2990	3260	2962	3243	3402	3556	3837
	(572)	(635)	(677)	(715)	(779)	(708)	(775)	(813)	(850)	(917)

number of pairs of electrons) and the quantum mechanical exchange energy (which stabilizes the ion and is approximately proportional to the number of pairs of parallel spins). If the proportionality constants for the coulombic repulsion and exchange energies are J and K, and the energy of attraction of each d electron to the argon core is $-U$, the total energy $E(d^n)$ of the d shell is given approximately by

$$E(d^n) = -nU + J\,{}^nC_2 - Km$$

where ${}^nC_2$ is the number of pairs of electrons in the d shell and m the number of pairs of parallel spins. The ionization potential I is then given by

$$\begin{aligned} I = E(d^{n-1}) - E(d^n) &= U + J[{}^{n-1}C_2 - {}^nC_2] + K\delta m \\ &= U - (n-1)J + K\delta m \end{aligned}$$

where δm is the decrease in the number of pairs of parallel spins. The overall increase of any ionization potential along the series shows that the rise in U with increase in n outweighs that in $(n-1)J$, and if we assume a smooth variation in each of these quantities with n, $U - (n-1)J$ should increase smoothly with n. The variation in $K\delta m$ is then superimposed on this increase. The number of pairs of parallel spins m and δm on ionization are shown for different d^n configurations in Table 18.8.

From the table we see that the loss of exchange energy rises to a maximum at d^5, drops to zero at d^6 and rises to another maximum at d^{10}. There should thus be an abrupt break in the plot of ionization potential against the number of d electrons between d^5 and d^6. For the

Table 18.8 *Numbers of pairs of parallel spins m for different d^n configurations and changes in m on ionization*

Configuration	*m*	*δm*
d^0	0	—
d^1	0	0
d^2	1	1
d^3	3	2
d^4	6	3
d^5	10	4
d^6	10	0
d^7	11	1
d^8	13	2
d^9	16	3
d^{10}	20	4

third ionization potential we are concerned with the removal of an electron from an M^{2+} ion, and for all transition elements the *s* electrons have been lost in the ground state of such an ion. There is indeed a marked break in ionization potential between the values for Mn^{2+} and Fe^{2+}. (There are similar breaks between Tc^{2+} and Ru^{2+} and between Re^{2+} and Os^{2+} in the later series.) A similar discontinuity is found in second ionization potentials between the same configurations (i.e. between the ionization potentials of Cr^+ and Mn^+, Mo^+ and Tc^+, and W^+ and Re^+). For the first ionization potentials, however, there is no break between vanadium and chromium, niobium and molybdenum, or tantalum and tungsten. The first ionization potential, though, always involves the loss of an *s* electron, so the principles discussed here do not apply; but they do apply to the second ionization potential if the ground state of the M^+ ion is d^n rather than $d^{n-1}s^1$. This is usually the case (e.g. for Fe^+ the ground state is d^7 rather than d^6s^1), and even when it is not, the difference between the configurations is very small and therefore unimportant.

For the first transition series the discontinuity in the fourth ionization potential comes between iron and cobalt, and in the fifth ionization potential between cobalt and nickel, as expected on the basis of the variation of δm with n. This simple treatment therefore provides a useful generalization in the study of oxidation states.

Once again, however, we have shown that we can predict the pattern of an irregularity superimposed on a major effect which we cannot treat quantitatively. Thus although we appear to understand the effect of a change in n and hence in m on the exchange energy factor in the magnitude of ionization potentials, we cannot yet calculate the much larger contributions from the other factors. This situation is like those we have encountered in discussing crystal field stabilization energy contributions to lattice and hydration energies. The thought that we can discuss 5 per cent of a total interaction energy quantitatively but cannot make more than a rough estimate of the remaining 95 per cent is a sobering one. Fortunately, much of chemistry is concerned with the small term that can be treated quantitatively on the basis of a theory of valency. Thus while there is little prospect of calculating E^0 for an M^{3+}/M^{2+} system accurately from ionization potentials and hydration energies evaluated theoretically, the study by means of crystal field (or, better, ligand field) theory of variations in E^0 produced by change of ligand is potentially a rewarding subject.

18.9 Qualitative survey of the stabilization of oxidation states[4, 10, 11]

The range of oxidation states discussed so far has been restricted by the wish to give a treatment which, in principle at least, is of a quantitative nature. A survey of formal oxidation states actually obtainable shows that most transition metals, other than the first and last members of each series, exhibit three or more. Some of the less familiar ones are illustrated for each triad of transition metals in Table 18.9. This list is by no means exhaustive, and it should be made clear that the description 'less common' refers to the oxidation state of an individual element (+IV, for example, is a common oxidation state for rhodium and iridium, but uncommon for cobalt; there is a general tendency to form stable higher oxidation states with increase in atomic number in any triad).

We mentioned in Chapter 15 that there is evidence for some electron delocalization, even in a simple fluoride like MnF_2, and that the hyperfine structure of the e.s.r. spectrum of the ion $[IrCl_6]^{2-}$ shows that the unpaired electron spends only some 70 per cent of the time on the iridium atom. For some of the ligands that feature prominently in Table 18.9 it is very doubtful to what extent they should be taken as really having the formal charge conventionally

Table 18.9 *Less common formal oxidation states of transition metals (see the text for comments on the validity, or otherwise, of oxidation states assigned here)*

Metals	*Examples*
Sc, Y, La	O in $[Sc(dipy)_3]$ and $[Y(dipy)_3]$; +II in LaI_2
Ti, Zr, Hf	O in $[Ti(dipy)_3]$; +II in $TiCl_2$ and $ZrCl_2$
V, Nb, Ta	−I in $[V(CO)_6]^-$; O in $V(CO)_6$; +I in $[\pi\text{-}C_5H_5Ta(CO)_4]$; +II, III in Ta_6I_{14}
Cr, Mo, W	−II in $[Cr(CO)_5]^{2-}$; −I in $[Cr_2(CO)_{10}]^{2-}$; O in $Cr(CO)_6$; +I in $[(C_6H_6)_2Mo]^+$; +II in Mo_6Cl_{12}; +IV in $[CrF_6]^{2-}$; +V in $[CrOF_4]^-$
Mn, Tc, Re	−III in $Mn(NO)_3CO$; −I in $[Re(CO)_5]^-$; O in $Mn_2(CO)_{10}$; +I in $[Mn(CN)_6]^{5-}$; +II in $[Re(diarsine)_2Cl_2]$; +V in $[MnO_4]^{3-}$
Fe, Ru, Os	−II in $[Fe(CO)_4]^{2-}$; O in $Fe(CO)_5$ and $Fe(PF_3)_5$; +I in $[Fe(H_2O)_5NO]^+$; +II in $[Os(CN)_6]^{4-}$; +IV in $[Fe(diarsine)_2Cl_2]^{2+}$; +VI in $[FeO_4]^{2-}$; +VII in $[RuO_4]^-$; +VIII in OsO_4
Co, Rh, Ir	−I in $[Co(CO)_4]^-$; O in $Co_2(CO)_8$; +I in $[Co(CN)_3CO]^{2-}$; +II in $[Rh(dipy)_2Cl]^+$; +IV in $[CoF_6]^{2-}$; +V in $[IrF_6]^-$; +VI in RhF_6 and IrF_6
Ni, Pd, Pt	−I in $[Ni_2(CO)_6]^{2-}$; O in $Ni(CO)_4$, $[Ni(CN)_4]^{4-}$ and $Pd(PF_3)_4$; +I in $[Ni_2(CN)_6]^{4-}$; +III in $[(Et_3P)_2NiBr_3]$; +IV in $[NiF_6]^{2-}$; +V in PtF_5; +VI in PtF_6
Cu, Ag, Au	+III in $[CuF_6]^{3-}$ and $[AgF_4]^-$; +V in $[AuF_6]^-$
Zn, Cd, Hg	+I in $[Cd_2]^{2+}$ and $[Hg_2]^{2+}$ salts

assigned to them (e.g. zero for CO, PF_3, dipyridyl or diarsine (*o*-phenylenebis-dimethylarsine), +I for NO^+ and −I for CN^-). Furthermore, some apparently simple halides in Table 18.9 have properties or structures which indicate that the assignment of oxidation states is not a straightforward matter. Thus LaI_2 is a metallic conductor, and should be written $La^{3+}(I^-)_2(e^-)$; Mo_6Cl_{12} contains a metal cluster cation of formula $[Mo_6Cl_8]^{4+}$ and four Cl^- ions; and Ta_6I_{14} is similarly $[Ta_6I_{12}]^{2+}(I^-)_2$.

It might be thought that the simplest method of finding out

whether, for example, the cyanide ion in Ni^0, Ni^I, and Ni^{II} complex cyanides is the same as in an alkali metal cyanide would be to measure accurately the C—N bond length in appropriate compounds. This has not been done, but even if it had been, the interpretation of the results might well be complicated by the different geometries of the ions and the consequent bonding of the carbon to differently hybridized nickel atoms in different oxidation states: $[Ni(CN)_4]^{4-}$ is tetrahedral, $[Ni_2(CN)_6]^{4-}$ is a metal–metal bonded ion and $[Ni(CN)_4]^{2-}$ is planar. The interpretation of the variation in infrared C≡N stretching frequency is similarly open to some question. This objection cannot, however, be raised to the comparison of the infrared C≡O stretching frequency in the isoelectronic and isostructural species $[V(CO)_6]^-$, $Cr(CO)_6$ and $[Mn(CO)_6]^+$; the values of 1860, 2000 and 2090 cm^{-1} respectively indicate, provided we make the reasonable assumption of similar potential energy curves, that the C≡O bond strength increases (i.e. there is less electron transfer from the metal to the ligand antibonding orbital) as the oxidation state of the metal increases. Conversely, we infer that in the species containing a metal in the lowest oxidation state, more of the electron density is back-donated to the ligand. Thus the range of actual charges on the metal atom in the series $[V(CO)_6]^-$, $Cr(CO)_6$ and $[Mn(CO)_6]^+$ is probably considerably less than the range of formal oxidation states of the metals in these species. A similar conclusion may be reached for the ions $[Fe(CN)_6]^{4-}$ and $[Fe(CN)_6]^{3-}$, for which the infrared C≡N stretching frequencies are 2044 and 2125 cm^{-1}; in general, however, the variation of the C≡N stretching frequency in complex cyanides is much less than that of the C≡O frequency in carbonyls, suggesting that the cyanide ion (as might be expected from its negative charge) is a poorer acceptor of electrons than carbon monoxide.

The overall position thus appears to be that oxide and fluoride (whether in salts, because of lattice energy effects, or in molecules, because of strong covalent bond formation) are particularly suitable for the production of high oxidation states, whilst ligands like CO, dipyridyl and other unsaturated organic systems with low energy π antibonding orbitals are particularly suitable for the production of low oxidation states. Much depends, of course, upon the choice of suitable experimental conditions. For the conversion of elements into compounds of high oxidation states, fluorine at high temperatures, ozone, persulphate in aqueous media, and electrolytic oxidation at low temperatures and high current densities are the most general methods. For the preparation of compounds of metals in low oxidation states, by far the most widely used method is reduction with alkali metals in liquid ammonia (see Chapter 7). Many examples of the uses of these reagents, and more detailed information about many of the species listed in Table 18.9, are given in other chapters of this book.

References for Chapter 18

1 Orgel, L. E., *An Introduction to Transition Metal Chemistry: Ligand Field Theory*, 2nd edn. (Methuen, 1966).
2 George, P., and McClure, D. S., *Prog. Inorg. Chem.*, 1959, **1**, 381.
3 Basolo, F., and Pearson, R. G., *Mechanisms of Inorganic Reactions*, 2nd edn. (Wiley, 1967).
4 Phillips, C. S. G., and Williams, R. J. P., *Inorganic Chemistry* (Oxford University Press, 1966).
5 Johnson, D. A., *Some Thermodynamic Aspects of Inorganic Chemistry* (Cambridge University Press, 1968).
6 Day, M. C., and Selbin, J., *Theoretical Inorganic Chemistry*, 2nd edn. (Reinhold, 1969).
7 Dunitz, J. D., and Orgel, L. E., *Adv. Inorg. Chem. Radiochem.*, 1960, **2**, 1.
8 Greenwood, N. N., *Ionic Crystals, Lattice Defects, and Nonstoichiometry* (Butterworths, 1968).
9 Nelson, P. G., and Sharpe, A. G., *J. Chem. Soc.*, **A**, 1966, 501.
10 Nyholm, R. S., and Tobe, M. L., *Adv. Inorg. Chem. Radiochem.*, 1963, **5**, 1.
11 Cotton, F. A., and Wilkinson, G., *Advanced Inorganic Chemistry*, 2nd edn. (Interscience, 1966).

Complexes of transition metals VI: kinetics and mechanisms of reactions

19

19.1 Introduction[1–5]

For the purposes of this chapter, nearly all reactions of transition metal complexes may be subdivided into substitution processes (in which there is a change in the coordination of the metal atom without a permanent change in its oxidation state) and redox reactions (in which the oxidation state of the metal changes, with or without a change in its coordination). There are also, however, a few examples of intramolecular isomerizations which may be held to constitute a special group; these are discussed in Section 19.2. Substitution reactions of metal carbonyls are mentioned in Chapter 20.

Whether a particular reaction involves the transition metal atom directly is not always immediately apparent, and tracer methods may be needed to settle this question. Thus in the reaction

$$[(NH_3)_5Co(CO_3)]^+ + 2H_3O^+ \rightarrow [(NH_3)_5Co(H_2O)]^{3+} + 2H_2O + CO_2$$

the use of ^{18}O-labelled water as solvent shows that the oxygen in the final aquo complex is derived entirely from carbonate, from which it is clear that the reaction really involves proton attack on the oxygen atom of the Co–O–C system followed by breakage of the O—C bond:

$$[(NH_3)_5CoOCO_2]^+ + H^+ \rightarrow [(NH_3)_5Co(OH)CO_2]^{2+} \rightarrow$$

$$[(NH_3)_5CoOH]^{2+} + CO_2 \xrightarrow{H^+} [(NH_3)_5Co(OH_2)]^{3+} + CO_2$$

In the exchange of iron between $[Fe(CN)_6]^{4-}$ and $[Fe(CN)_6]^{3-}$, the kinetic inertness of both ions (as measured by the slowness of their exchange of labelled cyanide) shows that the process is not really one of metal transfer but one of electron transfer between the two complex ions.

We shall assume here an elementary knowledge of kinetics and of current views on the nature of substitution at a saturated carbon atom. The latter subject had been well understood for about twenty years before substitution in metal complexes became a widely investigated field of study, and inevitably organic chemical thinking has exerted a great influence on the interpretation of the kinetics of inorganic reactions. It is, therefore, useful to discuss some difficulties in the elucidation of the mechanisms of reactions (particularly of substitution) of complexes of metals in general, and transition metals in particular, before going on to specific cases.

The distinction between labile and kinetically inert complexes was drawn in Section 1.4.4, in which we mentioned the rapidity of ligand exchange with $[Fe(H_2O)_6]^{3+}$ and $[Ni(CN)_4]^{2-}$ and its slowness with $[Cr(H_2O)_6]^{3+}$ and $[Fe(CN)_6]^{4-}$, and also stated that there was no general connection between thermodynamic stability with respect to dissociation and kinetic inertness. For the first transition series the two cations whose complexes are generally kinetically inert are Cr^{3+} and Co^{3+} (the high-spin complex $[CoF_6]^{3-}$ is an exception in this case). Complexes of other ions (including the hydrated ions) usually react so rapidly (at temperatures for which measurements can be made on aqueous solutions) that no distinction can be made by conventional methods of following reactions between those which require, say, 10^{-2} sec and those which require only 10^{-10} sec for near-completion.

The scope of chemical kinetics has, however, been much increased in recent years by developments in the study of fast reactions. For reactions in solution much use has been made of relaxation methods, particularly by Eigen and his collaborators. The principle underlying such methods is as follows. The position of equilibrium in a system depends not only upon concentrations of reactants and products, but also on variables such as temperature, pressure and electric field intensity. If one of these variables is suddenly changed, the position of equilibrium changes, and the approach to the new equilibrium can be followed using a cathode-ray oscillograph to record spectroscopic or conductance changes. Alternatively, the variable can be altered periodically (e.g. by the application of an alternating high density electrical field or ultrasonic waves), the reaction then lags behind the change in the physical quantity, and a displacement of equilibrium results; the change of position can be related to rate constants.

The first-order rate constants for replacement of a water molecule from a hydrated cation by a ligand are shown in Fig. 19.1. Within a group of typical elements, the rate constant increases with increasing size of cation (i.e. with decreasing hydration energy). This suggests that breaking of the metal ion–water bond is particularly important in the rate-determining step. Comparison of the data for Li^{+}, Mg^{2+} and Ga^{3+} (which have nearly the same crystal radii) shows that increasing formal charge retards substitution, a conclusion which also supports the bond-breaking as the rate-determining step suggestion. So does the relative independence of the rates on the nature of the substituting ligand. For dipositive cations of the

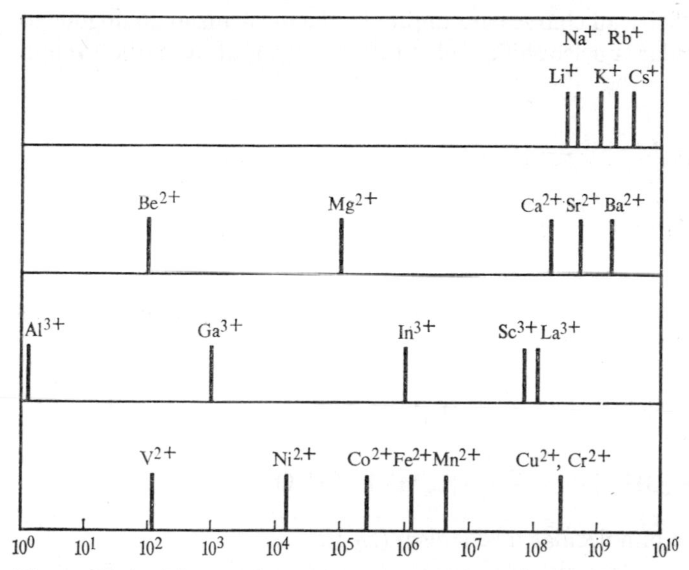

Figure 19.1 First-order rate constants (sec^{-1}) for substitution of water of the inner coordination sphere of metal ions

first transition series, however, there is no correlation with ionic radius, and electronic factors must be involved. We return to this point later.

Relatively few measurements have been made at low temperatures, mainly because few complexes are soluble in non-aqueous solvents of low freezing point. Where they are soluble, formation of ion-pairs and triple ions is common and adds to the difficulties of interpreting results; even in aqueous media, ion-pair formation sometimes plays a decisive part in a reaction. Further, because liquid water is a good donor of concentration 55M, direct measurement of its role in a process is impossible. Such a role may be kinetic (e.g. in intermediate replacement of a ligand by water, followed by substitution of the water molecule) or thermodynamic (e.g. by helping to produce a reactive intermediate when, by virtue of its basic properties, it removes a proton from a ligand). We may illustrate the uncertainties introduced into the interpretation of kinetic data by these considerations with a few simple examples.

For a ligand replacement reaction of the general type

$$[L_nMX] + Y = [L_nMY] + X$$

(for simplicity all charges are omitted), the mechanism analogous to unimolecular nucleophilic substitution (S_N1) at a carbon atom would be

$$[L_nMX] \xrightarrow[k_1]{\text{slow}} [L_nM] + X$$

$$[L_nM] + Y \xrightarrow{\text{fast}} [L_nMY]$$

It would, however, be impossible without further evidence to distinguish this mechanism from the following one, in which k_1 is reduced by a factor of 55:

$$[L_nMX] + H_2O \xrightarrow[k_1]{\text{slow}} [L_nM(OH_2)] + X$$

$$[L_nM(OH_2)] + Y \xrightarrow{\text{fast}} [L_nMY] + H_2O$$

For the bimolecular mechanism (S_N2),

$$[L_nMX] + Y \xrightarrow[k_1]{\text{slow}} [L_nMXY]$$

$$[L_nMXY] \xrightarrow{\text{fast}} [L_nMY] + X$$

the kinetic data would be equally compatible with ion-pair formation (if both reactants are ions) followed by a unimolecular reaction of the ion-pair:

$$[L_nMX] + Y \underset{k_{-1}}{\overset{k_1}{\rightleftharpoons}} [L_nMX]Y$$

$$[L_nMX]Y \xrightarrow[k_2]{\text{slow}} [L_nMY] + X$$

This leads to

$$\frac{d}{dt}[L_nMY] = \frac{k_1k_2[L_nMX][Y]}{k_{-1}+k_2} = k[L_nMX][Y]$$

where

$$k = \frac{k_1k_2}{k_{-1}+k_2}$$

Detailed investigation of such a reaction can lead to a value for k_1/k_{-1}, the equilibrium constant for ion-pair formation; this can be compared with the value obtained by static methods. Several reactions studied in this way in non-aqueous media have shown good agreement between values obtained by the two different methods, and it is now widely believed that no genuine bimolecular substitution at an octahedrally coordinated metal atom has yet been identified.

The part played by proton abstraction from a complex may be shown for the reaction

$$[Co(NH_3)_5Cl]^{2+} + OH^- = [Co(NH_3)_5OH]^{2+} + Cl^-$$

an apparently S_N2 process since the rate law is

$$-\frac{d}{dt}[Co(NH_3)_5Cl] = k[Co(NH_3)_5Cl][OH]$$

The hydroxyl ion is by far the most reactive reagent for attack on most cobalt (III) complexes in aqueous solution: $[Co(NH_3)_6]^{3+}$, for example, though thermodynamically very unstable with respect to decomposition by acid, is kinetically stable to hot sulphuric acid, but it is rapidly decomposed by cold aqueous base. In the example given above, it is believed that rapid proton abstraction is the first step in the mechanism:

$$[Co(NH_3)_5Cl]^{2+} + OH^- \underset{k_{-1}}{\overset{k_1}{\rightleftharpoons}} [Co(NH_3)_4(NH_2)Cl]^+ + H_2O$$

$$[Co(NH_3)_4(NH_2)Cl]^+ \underset{k_2}{\overset{\text{slow}}{\longrightarrow}} [Co(NH_3)_4(NH_2)]^{2+} + Cl^-$$

$$[Co(NH_3)_4(NH_2)]^{2+} + H_2O \xrightarrow{\text{fast}} [Co(NH_3)_5OH]^{2+}$$

This leads to the rate law

$$\frac{d}{dt}[Co(NH_3)_5OH] = \frac{k_1k_2[Co(NH_3)_5Cl][OH]}{k_{-1}[H_2O]^2 + k_2[H_2O]}$$
$$= k[Co(NH_3)_5Cl][OH]$$

where

$$k = \frac{k_1k_2}{k_{-1}[H_2O]^2 + k_2[H_2O]}$$

This is an example of the S_N1CB (S_N1 conjugate base) mechanism. Strong circumstantial evidence, from a different source, of its essential correctness is provided by the fact that if ammonia in the example given is replaced by pyridine or cyanide, neither of which contains a N—H bond, the rate of the reaction is no longer pH-dependent.

In substitution at a carbon atom, there is a clear-cut correlation between the kinetics and the stereochemical course of the reaction. The octahedron, however, presents several stereochemical possibilities (involving both geometrical and optical isomerism) depending on how the reagent attacks and on the nature of the transition state. No simple generalization is then possible, and each case has to be discussed on its own merits.

Finally, we should mention that, as in nearly all other branches of kinetics, rate laws often show that more than one reaction is taking place simultaneously. In the chromium(III)–thiocyanate reaction, for example, the rate expression is given by

$$(k_1 + k_2[H^+]^{-1} + k_3[H^+]^{-2})[Cr(H_2O)_6{}^{3+}][NCS^-]$$

This is interpreted as indicating that three important paths, involving $[Cr(H_2O)_6]^{3+}$, $[Cr(H_2O)_5(OH)]^{2+}$ and $[Cr(H_2O)_4(OH)_2]^+$ are involved in the reaction.

As in the measurement of formation constants of complexes (Section 6.4), it is customary to avoid the consequences of change in ionic strength during a reaction by working in media of high ionic strength. This point must be borne in mind when quantitative comparisons are made of rate data, not all of which may relate to the same experimental conditions.

19.2 Substitution reactions of octahedral complexes[1–3]

19.2.1. Replacement of coordinated water. The general reaction of displacement of water by an anion is often known as *ananation*. We have discussed the use of fast reaction techniques to investigate rates of ananation of a variety of cations in the last section, where we also discussed the ambiguities which can arise in the interpretation of kinetic data. To simplify the problem let us consider replacement of water in a species containing five non-labile ligands such as $[Co(NH_3)_5H_2O]^{3+}$, and let us reverse the experimental procedure and attempt to infer kinetic behaviour from a postulated mechanism. This is

$$[L_5M(H_2O)] \underset{k_{-1}}{\overset{k_1}{\rightleftharpoons}} [L_5M] + H_2O$$

$$[L_5M] + Y \xrightarrow{k_2} [L_5MY]$$

Since Y competes with solvent water for the active intermediate $[L_5M]$, the rate of formation of $[L_5MY]$ can be dependent on the concentration of Y. On the other hand, there should be some high concentration of Y at which the rate of replacement of water no longer depends on the concentration of Y. The rate of formation of $[L_5MY]$ at this concentration should be equal to the rate of formation of $[L_5M]$ and also equal to the rate of exchange of water between $[L_5M(H_2O)]$ and the solvent. Thus the rate of formation of $[L_5M]$ is given by

$$\frac{d}{dt}[L_5M] = k_1[L_5M(H_2O)] - k_{-1}[L_5M][H_2O] - k_2[L_5M][Y]$$

According to the steady-state approximation, the concentration of the very reactive $[L_5M]$ remains small and constant during the reaction $\left(\text{i.e. } \frac{d}{dt}[L_5M] = 0 \text{ at the steady state}\right)$. Thus,

$$[L_5M] = \frac{k_1[L_5M(H_2O)]}{k_{-1}[H_2O] + k_2[Y]}$$

and

$$\frac{d}{dt}[L_5MY] = \frac{k_1k_2[L_5M(H_2O)][Y]}{k_{-1}[H_2O] + k_2[Y]}$$

If $k_{-1}[H_2O] \gg k_2[Y]$,

$$\frac{d}{dt}[L_5MY] = \frac{k_1k_2}{k_{-1}}[L_5M(H_2O)][Y]$$

and a second-order reaction will be observed. On the other hand, if $k_2[Y] \gg k_{-1}[H_2O]$

$$\frac{d}{dt}[L_5MY] = k_1[L_5M(H_2O)]$$

giving first-order kinetics with the overall first-order constant equal to that for the dissociation of the aquo complex; furthermore, if Y represents a group of reagents, the same limiting rate should apply to all. If $k_2[Y] \sim k_{-1}[H_2O]$, the kinetic behaviour may be complicated, though for very low and very high concentrations of Y the kinetics will approach those for second-order and first-order processes respectively. However, at the high concentrations of Y necessary to achieve first-order kinetics, ion-pairing becomes important and the second-order ion-pairing mechanism outlined in the last section takes over.

Ion-pairing can, however, be avoided by studying substitution of an anion into an anion. For entry of a wide range of anions into the complex $[Co(CN)_5(H_2O)]^{2-}$ a dissociative mechanism has been established. Under these conditions

$$\frac{d}{dt}[Co(CN)_5Y] = k[Co(CN)_5(H_2O)]$$

where

$$k = \frac{k_1k_2[Y]}{k_{-1}[H_2O] + k_2[Y]}$$

or

$$\frac{1}{k} = \frac{k_{-1}[H_2O]}{k_1k_2[Y]} + \frac{1}{k_1}$$

Thus values of k_1 and k_{-1}/k_2 can be obtained from a graph of $1/k$ against $1/[Y]$ for various concentrations of Y. For $[Co(CN)_5(H_2O)]$, k_1 was found to be approximately constant at constant ionic strength, and was very nearly the same as the rate constant for the exchange of coordinated water with $H_2{}^{18}O$. The values of k_2/k_{-1} permit a comparison of the reactivity of all the ligands examined (including water) towards the intermediate $[Co(CN)_5]^{2-}$, and the following sequence has been established:

$$OH^- > N_3^- > SCN^- > I^- > NH_3 > Br^- > S_2O_3^{2-}$$

An analogous mechanism has been established for the reactions

$$[Co(NH_3)_4(SO_3)X] + Y = [Co(NH_3)_4(SO_3)Y] + X$$

where $X = SCN^-$, $Y = NH_3$; $X = NO_2^-$, $Y = NH_3$; $X = OH^-$, $Y = NH_3$; $X = NH_3$, $Y = OH^-$, CN^-, NO_2^- or SCN^-; and $X = OH^-$, $Y = CN^-$. Within this group, different rate laws are obtained for different reactions, in accordance with the treatment given for substitution into $[L_5M(H_2O)]$; for details, however, the original paper[6] must be consulted. Until this work was reported, it was customary to interpret the apparently special position of the $[Co(CN)_5(H_2O)]^{3-}$ ion in terms of Co—CN π bonding stabilization of the five-coordinated intermediates; clearly, however, such an explanation cannot apply with equal force to $[Co(NH_3)_4(SO_3)]^-$, and it seems more likely that the dissociative mechanism is of general occurrence but that in many instances, for the reason given already, faster reaction pathways are available.

19.2.2. Hydrolysis of cobalt (III) ammine complexes. The replacement of a ligand X in $[Co(NH_3)_5X]$ by water or hydroxide is commonly termed hydrolysis of the complex, and it is generally described by the rate law

$$-\frac{d}{dt}[Co(NH_3)_5X] = k_A[Co(NH_3)_5X] + k_B[Co(NH_3)_5X][OH^-]$$

In this expression k_B is usually about 10^4–10^8 k_A, so that the first term is more important in acidic, and the second term in alkaline, solution; k_A and k_B are generally called the rate constants for acid and base hydrolysis respectively.

The rate law for acid hydrolysis at low pH thus becomes

$$-\frac{d}{dt}[Co(NH_3)_5X] = k_A[Co(NH_3)_5X]$$

(If X is the anion of a weak acid, a term $k_{H^+}[Co(NH_3)_5X][H^+]$ is added.) As we have shown previously, such a rate law is compatible with either a slow dissociation of the complex into $[Co(NH_3)_5]^{3+}$ and X or replacement of X by H_2O as the rate-determining step. In order to try to decide between these alternatives, the rates of hydrolysis of a series of complexes of formula $[Co(AA)_2Cl_2]^+$, where AA is a substituted ethylenediamine, were examined. For replacement of a single chloride ion at pH 1 the order found for values of k_A was

$$\begin{matrix} CH_2NH_2 \\ | \\ CH_2NH_2 \end{matrix} < \begin{matrix} CH_3CHNH_2 \\ | \\ CH_2NH_2 \end{matrix} < \begin{matrix} CH_3CHNH_2 \\ | \\ CH_3CHNH_2 \end{matrix} \ll \begin{matrix} (CH_3)_2CNH_2 \\ | \\ (CH_3)_2CNH_2 \end{matrix}$$

Such an acceleration of substitution by bulky ligands suggests that the dissociative mechanism is operative; although introduction of methyl groups must have some inductive effect, the variation in base strengths among the diamines is very much less than the variation in rate constants for the hydrolysis of their cobalt (III) complexes, and it seems reasonable to attribute the kinetic effect mainly to steric factors. Now since steric factors favour S_N1 reactions, this is evidence for the dissociative mechanism. Further evidence for this mechanism is provided by:

(a) a general inverse correlation between the rate of replacement of X in $[Co(NH_3)_5X]$ and the formation constant of the $[Co(NH_3)_5X]$ complex from $[Co(NH_3)_5(H_2O)]^{3+}$ and X, and

(b) the decrease in the rate of the exchange reaction $[Co(NH_3)_5(H_2O)]^{3+} + H_2{}^{18}O = [Co(NH_3)_5(H_2{}^{18}O)]^{3+} + H_2O$ at high pressures.

For a rate-determining step involving an increase in the number of particles,

$$[Co(NH_3)_5(H_2O)]^{3+} \rightarrow [Co(NH_3)_5]^{3+} + H_2O$$

increase in pressure should retard the reaction, whereas for a rate-determining S_N2 step such as the formation of a seven-coordinated intermediate an acceleration should be found. The increase in rate is, however, less than calculated for a pure S_N1 reaction.

It is now believed that there is, in fact, a continuous range of mechanisms between the limiting S_N1 and S_N2 cases, and current practice is moving towards a less rigid classification of mechanisms. If the intermediate formed in the rate-determining step has a lower coordination number than the reacting complex, the mechanism is described as dissociative (*D*); conversely, if the intermediate has a higher coordination number, the mechanism is associative (*A*). If the reaction is one involving no change in coordination number, and ligand exchange occurs with no kinetically detectable intermediate, the mechanism is described as interchange (*I*). For a series of *I* reactions

$$[L_5MX] + Y \rightarrow [L_5MY] + X$$

if the rate is more sensitive to variation in the nature of Y than of X, it is said to be association-controlled and the mechanism is I_a; conversely, if the rate depends more on the nature of X, the mechanism is I_d.

Let us now turn to the kinetics of base hydrolysis. At pH values above about 5, the rate law for the reaction

$$[Co(NH_3)_5X]^{3+} + OH^- = [Co(NH_3)_5OH]^{2+} + X$$

is given by

$$-\frac{d}{dt}[Co(NH_3)_5X] = k_B[Co(NH_3)_5X][OH^-]$$

which is compatible with either the S_N2 or the S_N1 CB mechanism. We have mentioned earlier that if ammonia is replaced as ligand by pyridine, rapid base hydrolysis does not occur, and that this observation is indirect support for the S_N1 CB mechanism. The rate law requires that the establishment of the pre-equilibrium

$$[Co(NH_3)_5X] \rightleftharpoons [Co(NH_3)_4(NH_2)X] + H^+$$

should be very rapid compared with the rate of the overall process and, further, that the concentration of the intermediate amido complex should be very low (otherwise a limiting rate would be reached at too low a concentration of hydroxyl ion). Both of these requirements appear to be met. Exchange studies show that the interchange of hydrogen and deuterium between $[Co(NH_3)_5Cl]^{2+}$ and alkaline D_2O is much faster than chloride liberation, and cobalt ammine complexes are such weak acids that the concentration of the conjugate base is low at all hydroxyl ion concentrations. Finally, the conjugate base mechanism is supported by a study of the $^{18}O : {}^{16}O$ ratios in water, the hydroxyl ion and the complex $[Co(NH_3)_5OH]^{2+}$ produced from $[Co(NH_3)_5X]^{2+}$ where X is Cl^-, Br^- or NO_3^-. The equilibrium constant for the reaction

$$H_2{}^{16}O + {}^{18}OH^- \rightleftharpoons H_2{}^{18}O + {}^{16}OH^-$$

is 1·040, so that water and the hydroxyl ion, being labelled to different extents, can be distinguished. In the S_N1CB mechanism the $[Co(NH_3)_4(NH_2)]^{2+}$ ion reacts with water to give $[Co(NH_3)_5(OH)]^{2+}$, whilst in the S_N2 mechanism substitution of X by OH^- proceeds directly; the $^{18}O : {}^{16}O$ ratio should therefore be the same as in water and in the hydroxyl ion respectively for these mechanisms, and it turns out to be the same as in water.

19.2.3. Racemization of octahedral complexes. We shall not discuss the subject of change or retention of stereochemical configuration on substitution in this book, but one aspect of stereochemical change calls for comment here. This is the racemization of optically active octahedral complexes, which may occur by an intermolecular or an intramolecular mechanism. The rate of loss of optical activity of a methanolic solution of D-*cis*-$[Co(en)_2Cl_2]^+Cl^-$ at 35° is equal to the rate of exchange of coordinated chloride and labelled chloride ion in the solution, and the product is a mixture of 70 per cent *trans*-$[Co(en)_2Cl_2]^+Cl^-$ and 30 per cent of the racemic *cis* isomer. These observations suggest that the complex dissociates to a symmetrical five-coordinated intermediate and that recombination with chloride ion at different sites takes place at slightly different rates. (In aqueous solution, on the other hand, replacement of one chlorine atom by water takes place essentially with retention of configuration, showing that a similar intermediate is not formed under these conditions.) For $[Ni(phen)_3]^{2+}$ and $[Ni(dipy)_3]^{2+}$ the rate of racemization has been found to be the same as the rate of exchange with ^{14}C-labelled ligand, showing that either the bis complex is symmetrical or that it racemizes more rapidly than it reacts with the ligand.

For $[Cr(C_2O_4)_3]^{3-}$, $[Co(C_2O_4)_3]^{3-}$, and (rather surprisingly in view of the behaviour of the nickel complexes) $[Fe(phen)_3]^{2+}$ and $[Fe(dipy)_3]^{2+}$, the rates of racemization are greater than those of ligand exchange. Two possible mechanisms for racemization without exchange (i.e. intramolecular racemization) are:

(a) the breaking of one bond of one chelating group, formation of a symmetrical five-coordinated intermediate and re-formation of the broken bond; and
(b) a twisting process (e.g. one in which a trigonal prism is formed and reconverted into an octahedron).

For racemization in solution there is so far convincing evidence only for the bond-breaking mechanism. Thus $[Cr(C_2O_4)_3]^{3-}$ undergoes acid-catalysed exchange of *all* its oxygen atoms with $H_2{}^{18}O$ much faster than it exchanges oxalate with $^{14}C_2O_4{}^{2-}$ (which requires both bonds to one ligand to be broken), and at almost the same rate as it undergoes racemization. The compound cobalt(III) trisacetylacetylacetonate has also been shown to racemize by a bond-breaking mechanism in an ingenious experiment based on the use of n.m.r. spectroscopy to distinguish CD_3 and CH_3 groups. Isomerization of the type

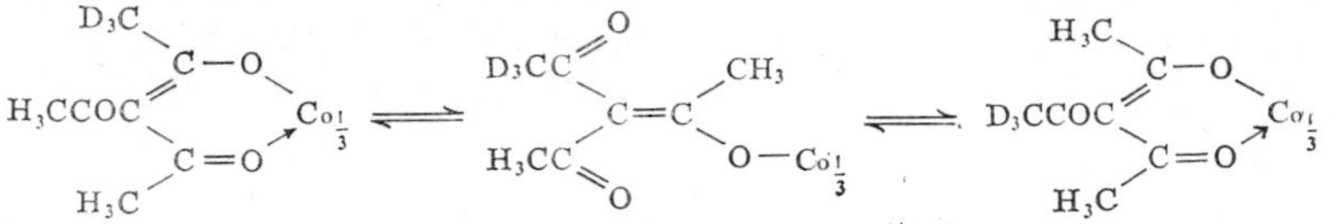

$$\left[\begin{matrix}D_3C \\ \; C{-}O \\ H_3CCOC \quad\quad Co \\ \; C{=}O \\ H_3C\end{matrix}\right]_{\frac{1}{3}} \rightleftharpoons \left[\begin{matrix}O \\ D_3CC \quad CH_3 \\ C{=}C \\ H_3CC \quad O{-}Co \\ O\end{matrix}\right]_{\frac{1}{3}} \rightleftharpoons \left[\begin{matrix}H_3C \\ \; C{-}O \\ D_3CCOC \quad\quad Co \\ \; C{=}O \\ H_3C\end{matrix}\right]_{\frac{1}{3}}$$

cannot take place by a twisting process and must involve a ring opening and closing mechanism; isomerization and racemization are found to occur at the same rate.

Racemization of tris-oxalato complexes also takes place in the solid state and for $K_3[Co(C_2O_4)_3].3H_2O$ the rate has been shown to increase at high pressures, indicating a decrease in volume in the transition state. Whether this is caused by the formation of a trigonal prismatic structure or by transference of a water molecule into the complex ion is not known, however.

19.2.4. Correlation of rates of substitution of octahedral complexes with electronic configuration. It will be clear from the foregoing discussion that we have as yet only a very incomplete knowledge of the transition states in substitution reactions. Nevertheless, we can draw some general conclusions about reactivity and electronic configuration. Cobalt (III) complexes have featured by far the most frequently in our accounts of both isomerism and kinetics of substitution; other octahedral complexes mentioned have been those of chromium (III), low-spin iron (II), and nickel (II). The rates of complex formation from aquo ions of some dipositive metals of the first transition series were shown by fast reaction techniques to increase in the order

$$V^{2+} < Ni^{2+} < Co^{2+} < Fe^{2+} < Mn^{2+} < Cr^{2+} < Cu^{2+}$$

(all of these are high-spin species). We may now ask whether there is any explanation of these facts in terms of theories of valency.

Consideration of the available data in terms of valence bond theory shows that labile complexes are generally either of the outer orbital type (Section 15.2) or, if of the inner orbital type, are derivatives of d^1 or d^2 cations. There is, however, no way of predicting relative labilities in terms of valence bond theory. Crystal field theory, on the other hand, is capable of being used as a basis for calculating changes of crystal field stabilization energy when the transition state is formed if we assume the geometry of the transition state and the constancy of all other variable factors. These are very large assumptions, but bearing in mind the success of crystal field theory in treating variations in thermochemical properties, it is interesting to see the results of calculations carried out on this basis by Basolo and Pearson. They evaluated the crystal field stabilization energies for high-spin and low-spin d^n complexes having the five-coordinated square pyramidal and the seven-coordinated pentagonal bipyramidal structures, with the same spin-multiplicity and metal–ligand distances as the reacting octahedral complexes. Their results are given in Table 19.1, in which a negative value denotes a loss of

Table 19.1 *Changes in crystal field stabilization energy on converting an octahedral complex into a square pyramid (CN5) or bipyramid (CN7) (in units of* Δ_o*)*

Number of d	*High-spin complexes*		*Low-spin complexes*	
electrons	CN5	CN7	CN5	CN7
0	0	0	0	0
1	+0·06	+0·13	+0·06	+0·13
2	+0·11	+0·26	+0·11	+0·26
3	−0·20	−0·43	−0·20	−0·43
4	+0·31	−0·11	−0·14	−0·30
5	0	0	−0·09	−0·17
6	+0·06	+0·13	−0·40	−0·85
7	+0·11	+0·26	+0·11	−0·53
8	−0·20	−0·43	−0·20	−0·43
9	+0·31	−0·11	+0·31	−0·11
10	0	0	0	0

crystal field stabilization energy. It is seen that the configuration to lose most stabilization energy on either model for the transition state is low-spin d^6; other configurations affected substantially, whatever the substitution mechanism assumed, are d^3 and d^8. The qualitative agreement with our list of the least labile species (complexes of low-spin Co^{III} and Fe^{II} (d^6), Cr^{III} (d^3), Ni^{II} (d^8) and V^{II} (d^3)) is very striking; quantitatively, however, agreement with observation is less satisfactory, complexes of nickel (II), for example, being generally far more labile than those of chromium (III), for which the changes in crystal field stabilization energies are the same. Nevertheless, the correlation of relative rates for complexes differing only in the number of *d* electrons they contain with electronic configuration is a very considerable achievement. The particular merits of calculations of this kind are their relative simplicity, the fact that they bear directly on the fundamental quantities of chemical kinetics (the stabilities of transition states) and the comparative ease with which predictions can be tested by experiment. More will undoubtedly be done along these lines in the future.

19.3 Substitution reactions of planar complexes[1, 3]

Most work on planar complexes has been carried out using those of platinum (II); the limited data for complexes of palladium (II) and nickel (II) indicate a general similarity to complexes of platinum (II), but reaction rate constants are much larger. It would seem reasonable that for a planar complex there should be little steric opposition to the bonding of an entering ligand before the substituted ligand has left (i.e. to an S_N2 mechanism), and this proves to be so. For example, the rate constants for the replacement of one chlorine atom by water in each of the species $[PtCl_4]^{2-}$, $[Pt(NH_3)Cl_3]^-$, $Pt(NH_3)_2Cl_2$ and $[Pt(NH_3)_3Cl]^+$ are almost identical; for an S_N1 type of reaction the rate constants would be expected to vary substantially with the variation in charge on the complex. For reactions of the general type

$$[L_nPtCl_{4-n}] + Y = [L_nPtCl_{3-n}Y] + Cl^-$$

the usual form of the rate law is

$$-\frac{d}{dt}[L_nPtCl_{4-n}] = k'[L_nPtCl_{4-n}] + k''[L_nPtCl_{4-n}][Y]$$

The second term in this expression corresponds to the S_N2 mechanism we have mentioned; the first represents a two-stage reaction in which one Cl^- is first replaced by H_2O, probably also by a rate-determining S_N2 step, and the H_2O is then more rapidly replaced by the ligand Y. This term, it may be noted, appears only when the solvent is, like water, itself a good ligand.

The particular feature of substitutions at a platinum (II) atom that excites general interest is the so-called *trans* effect. When the reaction

$$[LPtX_3] + Y \rightarrow [LPtX_2Y] + X$$

takes place, it is found that the proportions of *cis* and *trans* isomers in the product depend markedly upon the nature of L. The effect of a coordinated group on the rate of substitution of ligands opposite to it in a complex is called its *trans* effect; among common ligands the order of *trans* effects is

$$H_2O < OH^- < NH_3 < Cl^- < Br^- < I^- \sim NO_2^- \\ \sim PR_3 \sim H \ll CO \sim C_2H_4 \sim CN^-$$

Thus when $[PtCl_4]^{2-}$ is treated with ammonia, the product is $[PtCl_3(NH_3)]^-$; the further action of ammonia yields *cis*-$[Pt(NH_3)_2Cl_2]$, an ammonia molecule *trans* to a chlorine being more active than one *trans* to ammonia. Conversely, successive substitution of chloride into the $[Pt(NH_3)_4]^{2+}$ ion yields *trans*-$[Pt(NH_3)_2Cl_2]$.

It is not yet clear how the *trans* effect operates, though for the ligands with the greatest effect it seems certain that π bonding between platinum and carbon monoxide, ethylene or cyanide ion plays an important part. Since the substitution is an S_N2 process, the most reasonable model for the transition state is probably a trigonal bipyramid with the *trans* ligand in the same plane as the entering and leaving groups.

As we have defined it, the *trans* effect is concerned only with kinetics. Reference is sometimes made to a static *trans* effect which affects physical properties such as bond lengths, vibrational frequencies and chemical shifts in n.m.r. spectroscopy. It is perhaps not surprising that there is no close correlation between the kinetic and the static *trans* effects, and we shall not discuss the latter further here; for a fuller account a detailed review[1] is available. Nor shall we discuss the evidence for *trans* effects in complexes of other elements, including octahedral complexes of cobalt; for this subject also a review article[7] may be consulted.

19.4 Oxidation–reduction reactions[1–3, 8–10]

19.4.1. Outer sphere reactions. The simplest examples of oxidation–reduction reactions are those between species which differ only in charge, such as $[Fe(CN)_6]^{4-}$ and $[Fe(CN)_6]^{3-}$ or $[IrCl_6]^{3-}$ and $[IrCl_6]^{2-}$. Such reactions are commonly investigated by tracer methods, using flow methods and rapid quenching if necessary; for very fast reactions, the broadening of e.s.r. or n.m.r. spectra can sometimes be utilized. In one novel method, the rate of loss of optical activity on mixing solutions of a *D*-complex of one oxidation state and the *L*-complex of another oxidation state (both complexes being kinetically inert) gives the rate of electron transfer by the reaction

$$D\text{-}[Os(dipy)_3]^{2+} + L\text{-}[Os(dipy)_3]^{3+} \rightleftharpoons L\text{-}[Os(dipy)_3]^{2+} + D\text{-}[Os(dipy)_3]^{3+}$$

Where both reactants are non-labile, e.g. in the case of $[Fe(CN)_6]^{4-}$ and $[Fe(CN)_6]^{3-}$, a close approach of the metal atoms is impossible, and the electron transfer must take place by a *tunnelling* or *outer sphere* mechanism. Although for an isotopic change the equilibrium constant is nearly unity and ΔG^0 is nearly zero, activation energy is required to overcome the electrostatic repulsion between ions of like charge, to distort the coordination of both species and to modify the solvent structure around both species.

The second-order rate constants for a number of exchange reactions are given in Table 19.2. All of these are believed to occur by

Table 19.2 *Rates of electron exchange reactions with outer sphere mechanisms in water at 25°*

Reactants	*Rate constant* (M^{-1} sec^{-1})
$[Fe(phen)_3]^{2+}$, $[Fe(phen)_3]^{3+}$	10^5
$[Os(dipy)_3]^{2+}$, $[Os(dipy)_3]^{3+}$	10^4
$[IrCl_6]^{3-}$, $[IrCl_6]^{2-}$	10^3
$[Fe(CN)_6]^{4-}$, $[Fe(CN)_6]^{3-}$	740 (0°)
$[Co(phen)_3]^{2+}$, $[Co(phen)_3]^{3+}$	1
$[Co(NH_3)_6]^{2+}$, $[Co(NH_3)_6]^{3+}$	$<10^{-9}$

the outer sphere mechanism, though this is not certain in the cases of the two reactions involving cobalt complexes, since the cobalt (II) species undergo ligand exchange. These results may be supplemented by those for some reactions between non-labile complexes to form non-labile complexes; for example,

$$[Os(dipy)_3]^{2+} + [Mo(CN)_8]^{3-} = [Os(dipy)_3]^{3+} + [Mo(CN)_8]^{4-}$$

for which k is 10^9 M^{-1} sec^{-1}, near the limit of diffusion control. The very wide variation in these rate constants may be understood by considering the electronic configurations of the transition metal ions in the reacting species. In $[Fe(CN)_6]^{4-}$ and $[Fe(CN)_6]^{3-}$, for example, the Fe—C bond distances are nearly the same, since the iron atoms differ only in that the iron (II) ion has one more t_{2g} electron. On crystal field theory this is in a non-bonding orbital; on ligand field theory it is at least partly dispersed over the ligands on account of π bonding. When electron transfer between the ions takes place, the products are at first in excited states, since according to the Franck–Condon principle electron transfer is much faster than atomic movement. If, however, the geometries of the reacting complexes are only slightly different from what they would be in the transition state, the activation energy for the electron exchange is low and the reaction is rapid. For $[Co(NH_3)_6]^{2+}$ and $[Co(NH_3)_6]^{3+}$, on the other hand, the Co—N bond lengths (2·11 and 1·96 Å) are appreciably different and, furthermore, the cobalt (II) complex is a high-spin one and the cobalt (III) complex a low-spin one, the electronic configurations of the metal ions being $(t_{2g})^5(e_g)^2$ and $(t_{2g})^6$ respectively. After electron transfer these presumably become $(t_{2g})^5(e_g)^1$ and $(t_{2g})^6(e_g)^1$, neither of which is the ground state of the ion. In this case, therefore, the exchange reaction has a high activation energy.

The $[Fe(CN)_6]^{4-}$–$[Fe(CN)_6]^{3-}$ exchange reaction is catalysed by alkali metal ions, the effect being greatest for caesium and smallest for lithium. The very large cation Ph_4As^+ has little effect, however. These results suggest that a partly desolvated cation accelerates exchange by helping to overcome electrostatic repulsion by formation of a transition state such as

$$[Fe(CN)_6]^{4-} \;.....\; M^+ \;.....\; [Fe(CN)_6]^{3-}$$

A small cation holds its hydration sheath too strongly, whilst a very large one does not bring the anions into close proximity. It is interesting to note that the MnO_4^{2-}–MnO_4^- exchange reaction is also subject to alkali metal ion catalysis, the order of effectiveness being the same as for the $[Fe(CN)_6]^{4-}$–$[Fe(CN)_6]^{3-}$ reaction.

Outer-sphere reactions between complexes of different metals (e.g. the $[Os(dipy)_3]^{2+}$–$[Mo(CN)_8]^{3-}$ reaction mentioned earlier) are usually faster than outer sphere exchange reactions between different oxidation states of the same element. For such reactions the decrease in energy when excited states of products are converted into ground states can appear as the free energy of the reaction (ΔG^0 must be negative, or the reaction would not take place). This is tantamount to saying that for such reactions the structure of the transition state is more like that of the reactants; hence the activation energy is lowered and the rate is increased. It has been shown theoretically that there should be a relationship between the rates of such reactions and their standard free energies, and there is considerable evidence in support of this theory; its discussion, however, lies outside the scope of this book.

19.4.2. Inner sphere reactions. Many oxidation–reduction reactions have been shown to occur by a *ligand-bridging* or *inner sphere* mechanism in which substitution of the coordination shell of one of the metal ions occurs. The classic example of such a reaction is that between $[Co(NH_3)_5Cl]^{2+}$ and $[Cr(H_2O)_6]^{2+}$ in acidic solution, first investigated by Taube. The net reaction is found to be

$$[Co(NH_3)_5Cl]^{2+} + [Cr(H_2O)_6]^{2+} + 5H_3O^+ = [Co(H_2O)_6]^{2+} + [Cr(H_2O)_5Cl]^{2+} + 5NH_4^+$$

Now since $[Co(NH_3)_5Cl]^{2+}$ and $[Cr(H_2O)_5Cl]^{2+}$, typical complexes of low-spin cobalt (III) and of chromium (III), are non-labile (unlike the two aquo ions in the equation), the only reasonable explanation for the nature of the products is that a bridged intermediate is formed and breaks down to give the chloro complex of chromium (III):

$$[(NH_3)_5CoCl]^{2+} + [Cr(H_2O)_6]^{2+} \rightarrow [(NH_3)_5Co^{III}Cl\text{–}Cr^{II}(H_2O)_5]^{4+} + H_2O$$

$$\downarrow \; H^+ \mid H_2O$$

$$5NH_4^+ + Co^{2+}(aq) + [ClCr(H_2O)_5]^{2+}$$

If the reaction is carried out in the presence of $^{36}Cl^-$ in the solution, none of this isotope appears in the chromium (III) complex; this fact provides further support for the bridging mechanism. Because the change in oxidation states of the metal ions is accompanied by transfer of a chlorine atom, the process is often referred to as an atom transfer reaction.

Similar reactions occur when the chloride in the above example is replaced by other halide ions, sulphate, phosphate, acetate, succinate, oxalate and maleate. Among halide ions, the effectiveness for bridging purposes (as measured by relative reaction rates) is $F^- < Cl^- < Br^- < I^-$, in accordance with the expected order of ability to transmit an electron and undergo covalent bond-breakage. For the organic ions mentioned, oxalate and maleate (which contain conjugated systems) are considerably more effective than acetate and succinate.

Chromium compounds have played a large part in the establishment of the outer sphere mechanism because of the lability of complexes of the lower oxidation state and the kinetic inertness of complexes of the higher oxidation state. Lability of the lower oxidation state complex is not, however, essential for the demonstration of an outer sphere mechanism; a lower coordination number in the lower oxidation state may serve instead. Thus when the non-labile cobalt (II) complex, $[Co(CN)_5]^{3-}$ (obtained by addition of a cobalt (II) salt to excess of cyanide in the absence of air), reacts with $[Fe(CN)_6]^{3-}$, a binuclear complex, $[(NC)_5Co(NC)Fe(CN)_5]^{6-}$, is produced and may be isolated as the barium salt; hydrolysis of the ion, which is very slow, gives $[Co(CN)_5(H_2O)]^{2-}$ and $[Fe(CN)_6]^{4-}$, showing that the bridging mechanism for electron transfer need not necessarily result in atom transfer. This is found to be so for the $Cr^{2+}(aq)$–$[IrCl_6]^{2-}$ reaction also; when solutions of the reactants are mixed, a dark colour, believed to be due to $[Cl_5Ir^{IV}Cl–Cr^{II}(H_2O)_5]$, is produced, but then the colour becomes lighter and it is found that the main products are $[IrCl_6]^{3-}$ and $Cr^{3+}(aq)$.

The same reductant may appear to react by both inner and outer sphere mechanisms. We have already mentioned one inner sphere reaction of $[Co(CN)_5]^{3-}$; other examples are its reactions with cobalt (III) complexes of formula $[Co(NH_3)_5X]^{2+}$, where $X = Cl^-$, N_3^-, SCN^- or OH^-. For each of these the stoichiometry of the reaction is

$$[Co(NH_3)_5X]^{2+} + [Co(CN)_5]^{3-} = [Co(CN)_5X]^{3-} + 5NH_3 + Co^{2+}$$

and the rate in the presence of excess of cyanide is proportional only to $[Co(NH_5)X][Co(CN)_5]$. These are typical inner sphere reactions. If, however, $X = NH_3$, the stoichiometry of the reaction changes and $[Co(CN)_6]^{3-}$ is formed according to the equation

$$[Co(NH_3)_6]^{3+} + [Co(CN)_5]^{3-} + CN^- = [Co(CN)_6]^{3-} + Co^{2+} + 6NH_3$$

and the rate is proportional to $[Co(NH_3)_6][Co(CN)_5][CN]$. It is inferred that $[Co(CN)_5]^{3-}$ in the presence of cyanide is in equilibrium with a low concentration of $[Co(CN)_6]^{4-}$ and that this reacts with $[Co(NH_3)_6]^{3+}$ by the outer sphere mechanism. Since $[Co(CN)_5]^{3-}$ is rapidly formed from Co^{2+}(aq) or $[Co(NH_3)_6]^{2+}$ and cyanide, it may be seen that Co^{2+}(aq) will catalyse the reaction

$$[Co(NH_3)_6]^{3+} + 6CN^- = [Co(CN)_6]^{3-} + 6NH_3$$

This reaction is extremely slow in the absence of a catalyst, but the combination of rapid substitution into the labile cobalt (II) species followed by a rapid outer sphere redox reaction produces a fast overall reaction. Similar catalysis of the $[Co(NH_3)_5X]^{2+}$–CN^- reactions, this time involving rapid substitution and inner sphere redox reactions, also takes place. A closely related observation is the catalysis of the racemization of $[Co(en)_3]^{3+}$ by added $[Co(en)_3]^{2+}$.

The catalysis of reactions of chromium (III) complexes by Cr^{2+}(aq) undoubtedly depends upon similar series of reactions. It has long been known that the dissolution of anhydrous $CrCl_3$ by water is catalysed by a trace of chromium (II) chloride solution, and the chromium (III)-containing product in solution is $[Cr(H_2O)_5Cl]^{2+}$; clearly a chloride bridged intermediate is involved.

There are many other reactions which are believed to proceed by the inner sphere mechanism. Where all the species involved are too labile for tracer methods to be applicable or for the mechanism to be inferred from the nature of the products, dependence of the rate of reaction upon the concentration of an ion present in the solution can provide useful information. Caution must be exercised in reaching conclusions from such evidence, since, as we have seen, ions can catalyse outer sphere reactions such as the $[Fe(CN)_6]^{4-}$–$[Fe(CN)_6]^{3-}$ exchange as well as take part in inner sphere reactions. There is some evidence to show that change of anion makes much more difference to the rates of inner sphere reactions (e.g. of $[Cr(NH_3)_5X]^{2+}$–Cr^{2+}(aq), where X = F, Cl, Br or I) than to the rates of outer sphere reactions (e.g. of the $[Co(en)_3]^{2+}$–$[Co(en)_3]^{3+}$ exchange catalysed by F^-, Cl^-, Br^- or I^-). It has therefore been inferred that the F^-, Cl^- or Br^- catalysed exchanges between Fe^{2+} and Fe^{3+}, which proceed at about the same rate (approximately ten times that of the uncatalysed reaction described in Section 1.4), are all catalysed outer sphere reactions. This conclusion, however, has been challenged in the case of the chloride ion catalysed reaction, for which it is maintained that detailed interpretation of the kinetics shows that the principal reaction involves atom transfer between $FeCl^{2+}$(aq) and Fe^{2+}(aq). Despite extensive study over a period of several years, the nature of the apparently simple Fe^{2+}–Fe^{3+} exchange reaction is still uncertain.

References for Chapter 19

1 Basolo, F., and Pearson, R. G., *Mechanisms of Inorganic Reactions*, 2nd edn. (Wiley, 1966).
2 Sykes, A. G., *Kinetics of Inorganic Reactions* (Pergamon, 1966).
3 Benson, D., *Mechanisms of Inorganic Reactions in Solution* (McGraw-Hill, 1968).
4 Kettle, S. F. A., *Coordination Compounds* (Nelson, 1969).
5 McAuley, A., and Hill, J., *Quart. Rev. Chem. Soc.*, 1969, **23,** 18.
6 Halpern, J., Palmer, R. A., and Blakley, L. M., *J. Amer. Chem. Soc.*, 1966, **88,** 2877.
7 Pratt, J. M., and Thorp, R. G., *Adv. Inorg. Chem. Radiochem.*, 1969, **12,** 375.
8 Taube, H., *Adv. Inorg. Chem. Radiochem.*, 1959, **1,** 1.
9 Halpern, J., *Quart. Rev. Chem. Soc.*, 1961, **15,** 207.
10 Sykes, A. G., *Adv. Inorg. Chem. Radiochem.*, 1967, **10,** 153.

20

Transition metal carbonyls and related compounds

20.1 Introduction

Most transition metals form complexes with a wide variety of uncharged molecules such as carbon monoxide, substituted phosphines, and unsaturated or aromatic hydrocarbons. In many of these, the metal is in zero or another low oxidation state and, as we have already explained, π bonding between the metal atom and the ligands is believed to play an important part in stabilizing these complexes. The formation of metal–carbon σ and π bonds in a metal carbonyl is shown in Fig. 20.1. In this chapter we shall consider the metal carbonyls, anions derived from them, some of their substitution products, and complexes formed by a few other ligands such as NO^+, N_2 and CN^- (all of which are isoelectronic with CO), and PF_3 and PMe_3. The closely related subject of organometallic

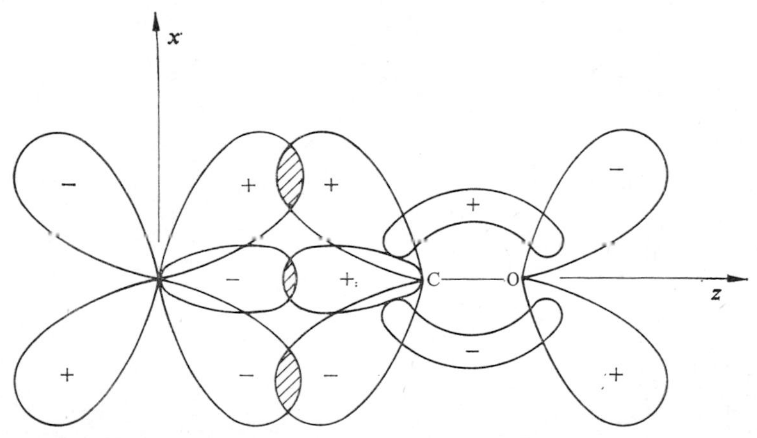

Figure 20.1 The formation of σ and π bonds between a transition metal and CO. Only one t_{2g} metal orbital, and one π and one π^* ligand orbital, are shown

Table 20.1 *The binary carbonyls*

Electrons needed to attain noble gas configuration	13	12	11	10	9	8
First transition series	$V(CO)_6$ blue solid	$Cr(CO)_6$ white solid (sublimes)	$Mn_2(CO)_{10}$ yellow solid (m.p. 154°)	$Fe(CO)_5$ yellow liquid (b.p. 103°) $Fe_2(CO)_9$ orange solid (sublimes) $Fe_3(CO)_{12}$ black solid (sublimes)	$Co_2(CO)_8$ orange solid (m.p. 51°) $Co_4(CO)_{12}$ black solid $Co_6(CO)_{16}$ black solid	$Ni(CO)_4$ colourless liquid (b.p. 43°)
Second transition series		$Mo(CO)_6$ white solid (sublimes)	$Tc_2(CO)_{10}$ white solid	$Ru(CO)_5$ colourless liquid (m.p. −22°) $Ru_2(CO)_9$* $Ru_3(CO)_{12}$ orange solid (m.p. 154°)	$Rh_2(CO)_8$* $Rh_4(CO)_{12}$ orange solid $Rh_6(CO)_{16}$ black solid	
Third transition series		$W(CO)_6$ white solid (sublimes)	$Re_2(CO)_{10}$ white solid (m.p. 177°)	$Os(CO)_5$ colourless liquid (m.p. −15°) $Os_2(CO)_9$ orange solid $Os_3(CO)_{12}$ yellow solid (m.p. 224°)	$Ir_2(CO)_8$† yellow solid $Ir_4(CO)_{12}$ yellow solid $Ir_6(CO)_{16}$ red solid	

* Not isolated. † Identity not confirmed.

compounds of transition metals is discussed in the following chapter.

The metal carbonyls (together with organometallic compounds of transition metals) are remarkable for the way in which, almost without exception, they conform to the effective atomic number rule (see Chapters 15 and 18); $V(CO)_6$ is the only important exception among simple binary carbonyls. Thus a glance at the formulae of the simplest carbonyls listed in Table 20.1 shows that, except for $V(CO)_6$, mononuclear carbonyls contain a number of carbon monoxide molecules such that if each donates two electrons to the metal atom, the latter acquires a total number of electrons equal to that of an atom of the next noble gas. When the atomic number of the transition metal is odd, the simplest carbonyl is always (except for that of vanadium) binuclear, and its diamagnetism suggests that a metal–metal bond is present. Thus the atomic number of iron is 26, and this plus five pairs of electrons gives the iron atom in $Fe(CO)_5$ an effective atomic number of 36. Cobalt, the next element, acquires eight electrons from carbon monoxide molecules and then bonds to another cobalt atom to give the same effective atomic number, the simplest carbonyl being $Co_2(CO)_8$. Useful though the effective atomic number rule (or, as it is sometimes called, the eighteen-electron rule, which refers to the nd, $(n + 1)s$ and $(n + 1)p$ orbitals) is in predicting formulae, however, it does not predict structures. Thus $Co_2(CO)_8$ could conform to the rule if it had any of the

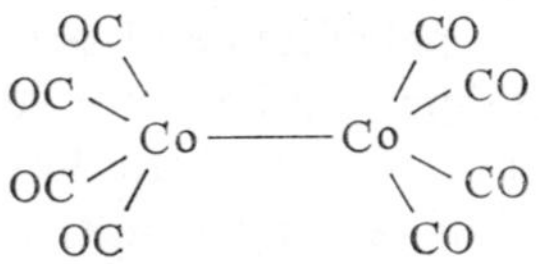

Figure 20.2

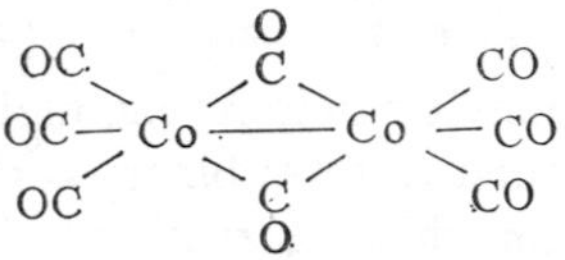

Figure 20.3

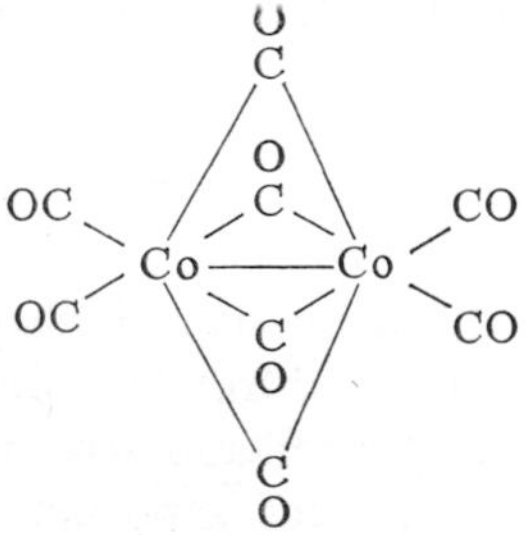

Figure 20.4

structures shown in Figs 20.2, 20.3 and 20.4, and a bridging carbonyl group were considered to contribute a single electron to the valence shell of each of two metal atoms. Figure 20.4 looks unlikely on steric grounds, and large numbers of bridging carbonyls between two atoms are in practice not found, but $Co_2(CO)_8$ may actually have either of the structures of Figs 20.2 or 20.3, and there is very little difference in energy between them.

Complexes containing NO as a ligand nearly always appear to contain it as NO^+, and in such complexes, and in those of PF_3 and analogous compounds of Group V elements, the effective atomic number rule is obeyed with the same consistency as by the carbonyls. Cyanide as a ligand, however, is a much less effective π-acceptor because it already has a negative charge, and the rule does not provide a satisfactory basis for the rationalization or prediction of the formulae of cyanide complexes; as we pointed out in Chapter 18, however, the valence shells of transition metals in stable cyanide complexes seldom if ever contain *more* than a noble gas number of electrons.

Table 20.1 is not a comprehensive collection of the formulae of all reported carbonyls; new polynuclear species are being discovered every year. Nor does the table include heteronuclear carbonyls such as $(OC)_5MnCo(CO)_4$ and $FeRu_2(CO)_{12}$, some of which are mentioned later, or unstable intermediates like $Ni(CO)_3$, which is formed during substitution reactions of $Ni(CO)_4$ and may be isolated by photolysis of the latter compound in an inert matrix at 15°K. A few metals form carbonyl halides (e.g. copper, palladium and gold) or carbonylate anions (e.g. niobium and tantalum) but have not yet been found to form binary carbonyls; it is, however, doubtful how significant this observation is in the absence of recent attempts to identify the missing carbonyls at low temperatures using modern techniques. It is noteworthy that although neither $Pd(CO)_4$ nor $Pt(CO)_4$ has yet been prepared, $Pd(PF_3)_4$ and $Pt(PF_3)_4$, analogous to $Ni(PF_3)_4$, are well established.

20.2 Preparation of metal carbonyls[1, 2]

The following are the chief methods used in the synthesis of binary carbonyls. Of these compounds, $Fe(CO)_5$, $Ni(CO)_4$, $Mo(CO)_6$ and $W(CO)_6$ are now available commercially. The formation and decomposition of the nickel compound is, of course, of great importance in the extraction of the metal.

20.2.1. Direct synthesis. Carbon monoxide will react with several metals, especially if they are finely divided and have a clean, oxide-free surface. Nickel reacts at, or a little above, room temperature and at atmospheric pressure of carbon monoxide; $Fe(CO)_5$ and $Co_2(CO)_8$ can be made conveniently at 150–250° at 200 atm, and high pressure reactions have also given carbonyls of molybdenum, tungsten, rhodium and ruthenium, though other methods are better.

20.2.2. Preparation from a metal compound, carbon monoxide and a reducing agent (reductive carbonylation). These reactions involve the interaction of a metal compound and carbon monoxide in the presence of a reducing agent, usually at elevated temperatures and pressures; sometimes the reducing agent is excess of carbon monoxide. Among other reducing agents that have been used are hydrogen, sodium, magnesium, aluminium, zinc, copper, aluminium alkyls, lithium aluminium hydride and sodium benzophenone ketyl. Some typical examples are as follows.

Carbon monoxide as reductant and carbonylating agent may be used for the preparation of $Os(CO)_5$ and $Re_2(CO)_{10}$ from OsO_4 and Re_2O_7 respectively; for example,

$$OsO_4 + CO \xrightarrow{250^\circ \text{ at } 350 \text{ atm}} Os(CO)_5\ (+\ Os_3(CO)_{12}) + CO_2$$

Copper, silver and zinc are often used as halogen or sulphur acceptors in preparing carbonyls from metal halides (usually iodides) or sulphides; for example,

$$RuI_3 + CO + Ag \xrightarrow{175^\circ \text{ at } 250 \text{ atm}} Ru(CO)_5 + AgI$$

$$CoS + CO + Cu \xrightarrow{200^\circ \text{ at } 200 \text{ atm}} Co_2(CO)_8 + Cu_2S$$

These reactions proceed via the formation and dehalogenation of a carbonyl halide; sometimes the halide is so stable as to resist decomposition, and it then appears as the product of the reaction (e.g. $Re(CO)_5Cl$ from K_2ReCl_6, CO and Cu at 230° at 50 atm). Similar reactions using other reducing agents include

$$CrCl_3 + CO + LiAlH_4 \xrightarrow[\text{ether}]{115^\circ \text{ at } 70 \text{ atm}} Cr(CO)_6 + LiCl + AlCl_3$$

$$Cr(acac)_3 + CO + Mg \xrightarrow[\text{pyridine}]{160^\circ \text{ at } 300 \text{ atm}} Cr(CO)_6$$

$$CoCO_3 + CO + H_2 \xrightarrow[\text{pet. ether}]{150^\circ \text{ at } 300 \text{ atm}} Co_2(CO)_8$$

In the last reaction the hydride $HCo(CO)_4$, which is thermally unstable, is an intermediate.

The best method for the preparation of vanadium carbonyl is reduction with sodium in 'diglyme' (diethyleneglycol dimethyl ether) followed by acidification:

$$VCl_3 + Na + CO \xrightarrow[\text{diglyme}]{100° \text{ at } 150 \text{ atm}} [V(CO)_6]^- \xrightarrow{H^+} V(CO)_6$$

For manganese carbonyl, sodium benzophenone ketyl may be employed with advantage:

$$Na + Ph_2CO \rightarrow Ph_2\dot{C}ONa$$

$$MnCl_2 + 2Ph_2\dot{C}ONa \rightarrow (Ph_2\dot{C}O)_2Mn + 2NaCl$$

$$2(Ph_2\dot{C}O)_2Mn + 10CO \xrightarrow[\text{tetrahydrofuran}]{200° \text{ at } 200 \text{ atm}} Mn_2(CO)_{10} + 4Ph_2CO$$

Finally, we may mention the use of iron pentacarbonyl; for example,

$$WCl_6 + 3Fe(CO)_5 \xrightarrow{100°} W(CO)_6 + 3FeCl_2 + 9CO$$

20.2.3. From mononuclear or binuclear carbonyls. The classic examples are the production of $Fe_2(CO)_9$ by irradiation of $Fe(CO)_5$ in acetic acid solution with ultraviolet light at or below room temperature, and of $Fe_3(CO)_{12}$ by heating $Fe_2(CO)_9$ at 50°. The binuclear carbonyl decomposes according to the equation

$$Fe_2(CO)_9 \rightarrow Fe(CO)_5 + Fe(CO)_4$$

and trimerization of $Fe(CO)_4$ gives $Fe_3(CO)_{12}$ in low yield. A better preparation involves oxidation of $Na[HFe(CO)_4]$ in methanol with manganese dioxide; this gives the anion $[HFe_3(CO)_{11}]^-$, which when treated with acid yields $Fe_3(CO)_{12}$, H_2 and iron (III) salts.

Polynuclear carbonyls of other metals are generally made by similar methods; e.g. $Ru_3(CO)_{12}$ and $Os_3(CO)_{12}$ by spontaneous decomposition of the monocarbonyls (they may thus be prepared by slight modification of methods used for the monocarbonyls), $Co_4(CO)_{12}$ by heating $Co_2(CO)_8$ at 70°. The action of potassium in tetrahydrofuran solution on $Co_4(CO)_{12}$ gives the salt $K_4[Co_6(CO)_{14}]$, which is oxidized by iron (III) chloride to $Co_6(CO)_{16}$. Reductive carbonylation of $RhCl_3$ gives $Rh_4(CO)_{12}$ and $Rh_6(CO)_{16}$ without appreciable yield of $Rh_2(CO)_8$. The last compound has been shown to be formed from $Rh_4(CO)_{12}$ and carbon monoxide under high pressures at low temperatures, a species with an infrared absorption very like that of the bridged form of $Co_2(CO)_8$ being present in the mixture of products; but it has never been obtained in the pure state. The ruthenium carbonyl $Ru_2(CO)_9$ has been characterized in a similar way.

20.2.4. From complex cyanides. Although of no preparative importance, the formation of $Ni(CO)_4$ by a disproportionation of a nickel (I) complex is of considerable scientific interest. Reduction of a solution of the well-known complex $K_2[Ni(CN)_4]$ with potassium amalgam gives a red solution containing the $[Ni_2(CN)_6]^{4-}$ ion; this reacts with carbon monoxide to give the ion $[Ni_2(CN)_6(CO)_2]^{4-}$, which in water disproportionates roughly according to the equation

$$2[Ni_2(CN)_6(CO)_2]^{4-} + 2H^+ \rightarrow 3[Ni(CN)_4]^{2-} + Ni(CO)_4 + H_2$$

Cobalt carbonyl may be made by the action of carbon monoxide at atmospheric pressure on an alkaline solution of potassium pentacyanocobaltate (II); $K[Co(CO)_4]$ is formed, and on acidification this yields the hydride $HCo(CO)_4$, which decomposes to $Co_2(CO)_8$ and hydrogen.

Partial substitution of cyanide in a number of other complexes by carbon monoxide has been reported, but other metal carbonyls have not yet been made by this method.

20.2.5. Heteronuclear carbonyls. It is convenient at this point also to indicate methods for the synthesis of heteronuclear carbonyls; the following will serve as examples.

(a) $Na[Mn(CO)_5] + Re(CO)_5Cl \rightarrow (OC)_5MnRe(CO)_5 + NaCl$
$Na[Co(CO)_4] + Mn(CO)_5Br \rightarrow (OC)_5MnCo(CO)_4 + NaBr$

Such reactions are usually carried out in tetrahydrofuran, in which the sodium salts are soluble.

(b) $[Re(CO)_6]^+[Co(CO)_4]^- \xrightarrow{60^\circ} (OC)_5ReCo(CO)_4 + CO$

(c) $Mn_2(CO)_{10} + Re_2(CO)_{10} \xrightarrow[\text{hexane}]{\nu h} 2(OC)_5ReMn(CO)_5$

$$3Fe(CO)_5 + Ru_3(CO)_{12} \xrightarrow{110^\circ} FeRu_2(CO)_{12} + Fe_2Ru(CO)_{12} + 3CO$$

20.3 Structures of metal carbonyls[3, 4]

The structures of the mononuclear carbonyls $V(CO)_6$, $Cr(CO)_6$, $Fe(CO)_5$ and $Ni(CO)_4$ have been established by X-ray diffraction, infrared spectroscopy and (except for the vanadium compound) electron diffraction; the last method has also been used for $Mo(CO)_6$ and $W(CO)_6$. All of the hexacarbonyls have regular octahedral structures, and nickel carbonyl has a regular tetrahedral structure. In the trigonal bipyramidal molecule of iron pentacarbonyl, however, the most recent electron diffraction study suggests that the Fe—C apical bond is slightly shorter than the equatorial one, the lengths being 1·806 and 1·833 Å respectively; it is interesting to note that in trigonal bipyramidal molecules of typical elements it is usually the equatorial bond that is shorter, but the significance of this observation is not clear.

X-ray diffraction studies have shown that in the solid state $Mn_2(CO)_{10}$, $Fe_2(CO)_9$ and $Co_2(CO)_8$ have the structures shown in Figs 20.5, 20.6 and 20.7 respectively; $Tc_2(CO)_{10}$ and $Re_2(CO)_{10}$ are isomorphous with $Mn_2(CO)_{10}$. The structure of $Mn_2(CO)_{10}$ may be described as two staggered octahedra sharing one corner; in each of the octahedra the four equatorial Mn—CO bonds are bent inwards

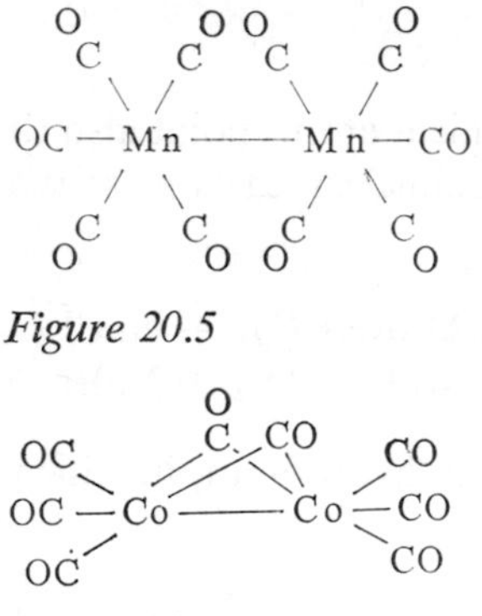

Figure 20.5

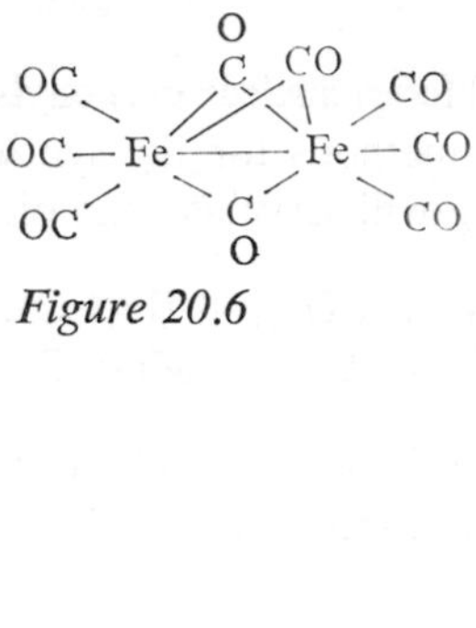

Figure 20.6

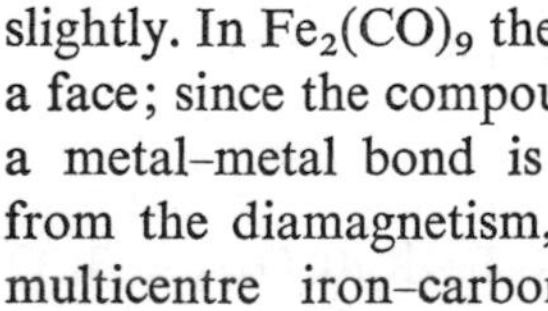

Figure 20.7

slightly. In $Fe_2(CO)_9$ the structure consists of two octahedra sharing a face; since the compound is diamagnetic it is usually inferred that a metal–metal bond is present. This does not necessarily follow from the diamagnetism, since coupling of unpaired spins through multicentre iron–carbonyl–iron bonding could occur: but the shortness of the Fe—Fe distance (2·46 Å, almost the same as that in metallic iron) probably indicates a metal–metal bond; Mn—Mn in $Mn_2(CO)_{10}$, in which there must be a metal–metal bond, is as great as 2·92 Å. A puzzling property of $Fe_2(CO)_9$, in view of its molecular structure, is its insolubility in all organic solvents. Dicobalt octacarbonyl has approximately the structure of $Fe_2(CO)_9$ with one of the bridging carbonyls removed; again the diamagnetism of the compound and the Co—Co distance (2·52 Å) are generally taken to indicate a metal–metal bond.

Metal–carbon distances in $Fe_2(CO)_9$ and $Co_2(CO)_8$ fall into two groups, metal-bridging carbonyl distances being about 0·1 Å longer than metal terminal carbonyl distances. Such a difference is compatible with the concept of two-electron donation by terminal carbonyls, and one-electron donation (to each of two metal atoms) by bridging carbonyls, though the possible existence of π bonds of different strengths makes quantitative interpretation impossible. That the extent of π bonding to terminal and bridging carbonyls is different is clearly shown by the carbonyl stretching frequencies. Carbon monoxide itself has a stretching frequency of 2143 cm^{-1}; neutral metal carbonyls known to have no bridging carbonyl groups have stretching frequencies in the range 2125–2000 cm^{-1}; and $Fe_2(CO)_9$ and $Co_2(CO)_8$, in addition to showing bands in this region, also show carbonyl absorption at 1830 and 1860 cm^{-1} respectively. In general, carbonyl absorption in the 1900–1800 cm^{-1} region is indicative of the presence of bridging carbonyl groups in uncharged species, though the presence of other groups may result in the lowering of the stretching frequencies of terminal carbonyl groups into this region (in carbonylate anions such as $[Co(CO)_4]^-$ and $[Fe(CO)_4]^{2-}$ very low carbonyl stretching frequencies of 1883 and 1788 cm^{-1} respectively result from the strong metal–carbon π bonding which stabilizes the low oxidation state of the metal—see Chapter 18). In a few neutral species believed to contain carbonyl groups bonded to three metal atoms, stretching frequencies of 1800 cm^{-1} or less are found. Absorption in the terminal or the bridging region often consists of several closely spaced bands, even when measurements are made in solution; we cannot, however, discuss the details of the spectra here, and so we shall generally refer to terminal carbonyl or bridging carbonyl absorption without specifying the number or origin of the bands present.

The infrared absorption spectrum of $Co_2(CO)_8$ in saturated hydrocarbon solvents is dependent on temperature, and both bridged and non-bridged structures are present—at room temperature in nearly equal proportions; ΔH for the conversion of the bridged into the non-bridged structure is +5·5 kJ (1·3 kcal), but this is accompanied by an increase in entropy, so that there is very little difference in free energy between the two forms.

Of the three trinuclear carbonyls at present known, crystalline $Os_3(CO)_{12}$ and $Ru_3(CO)_{12}$ have the structure shown in Fig. 20.8, whilst $Fe_3(CO)_{12}$ has the structure shown in Fig. 20.9; the infrared spectrum of the last compound in solution shows a complex dependence on temperature, and it appears that under these conditions the structure of Fig. 20.9 is in equilibrium with another form of the compound.

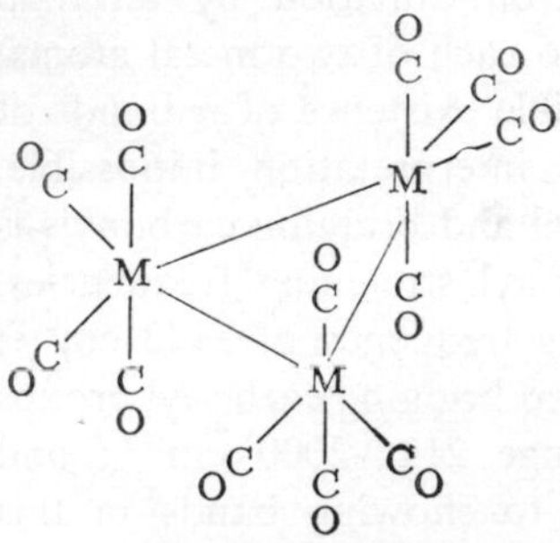

Figure 20.8

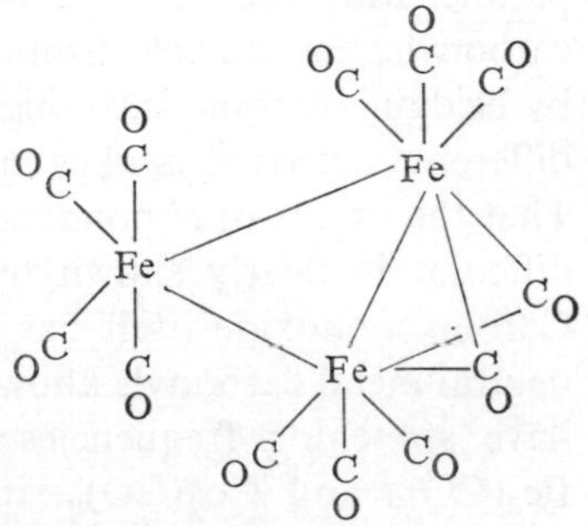

Figure 20.9

A similar change of structure occurs among the tetracarbonyls of members of the cobalt group: $Ir_4(CO)_{12}$ has the non-bridged structure shown in Fig. 20.10, whilst $Co_4(CO)_{12}$ and $Rh_4(CO)_{12}$ have the structure shown in Fig. 20.11. The only hexametalcarbonyl whose structure has yet been determined is $Rh_6(CO)_{16}$, which con-

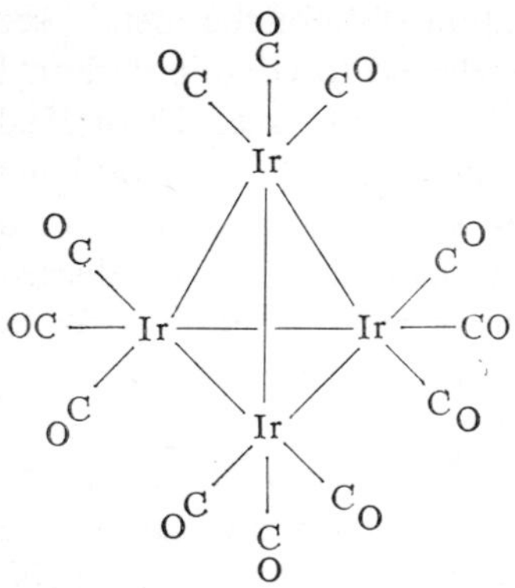

Figure 20.10

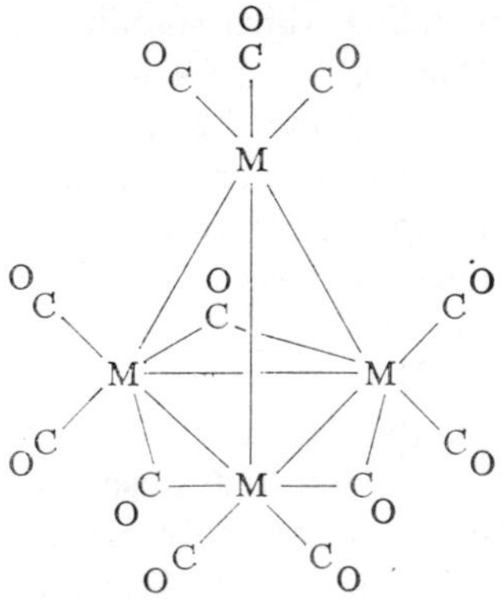

Figure 20.11

tains a Rh_6 octahedron, two terminal carbonyl groups on each metal atom and four carbonyl groups each bridging three metal atoms (Fig. 20.12). Thus the environment of each metal atom is the same, the configuration round it being tetragonal antiprismatic. The Rh—Rh distance is 2·78 Å between adjacent atoms, compared with 2·70–2·80 (weighted average 2·73) Å in the tetracarbonyl. For the tetracarbonyl, if we assume that each metal atom receives six electrons from carbonyl groups (either three pairs or two pairs and two singles) and three electrons from the three metal atoms to which it is bonded, the total number of electrons associated with each rhodium atom is $45 + 6 + 3 = 54$ (i.e. the same as for xenon). For $Rh_6(CO)_{16}$, however, similar assumptions (with a carbonyl group

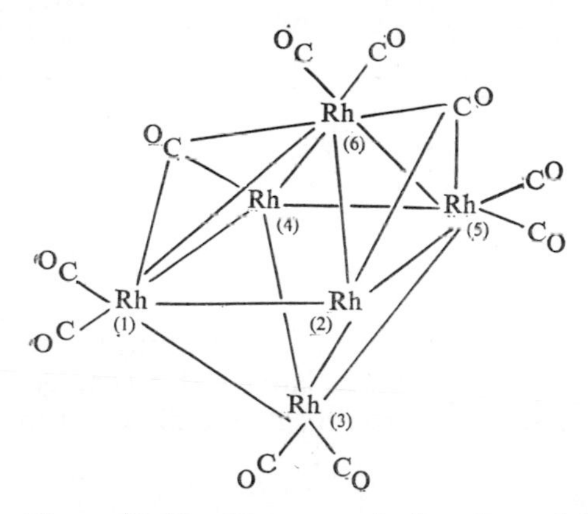

Figure 20.12 (Four terminal and two bridging CO groups omitted)

bridging three metal atoms being taken as contributing two-thirds of an electron to each) lead to the conclusion that the compound does not obey the effective atomic number rule, which would require the species to be $[Rh_6(CO)_{16}]^{?+}$. Probably multicentre bonding is involved in the Rh_6 octahedron. The development of laser Raman spectroscopy makes it possible to study metal–metal bond vibrations in polynuclear carbonyls, and some measurements (which show these vibrations to be in the range 250–100 cm^{-1}) have already been reported; unfortunately, however, reliable assignments have not yet been made in most cases and discussion of force constants is therefore not yet rewarding at the level of this book.

The structures of a few heteronuclear carbonyls are also known. X-ray data show that $Mn_2Fe(CO)_{14}$ has the linear Mn—Fe—Mn grouping and the structure in Fig. 20.13. Infrared spectroscopy suggests that $Fe_2Ru(CO)_{12}$ has the structure of Fig. 20.9, whilst $FeRu_2(CO)_{12}$ has the non-bridged structure of Fig. 20.8.

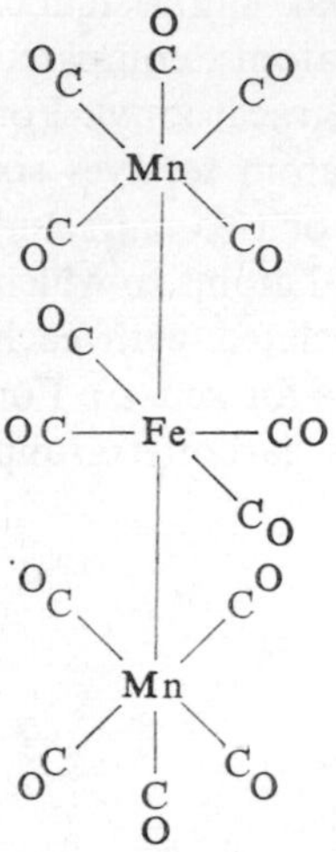

Figure 20.13

The structures of several carbonylate anions and carbonyls containing inorganic substituents are mentioned later in this chapter; those of carbonyls containing organic substituents are discussed in the following chapter.

20.4 Physical and chemical properties of the carbonyls[1]

The metal carbonyls are all crystalline solids with the exceptions of $Fe(CO)_5$, $Ru(CO)_5$, $Os(CO)_5$ and $Ni(CO)_4$, which are liquids. They melt or decompose at low temperatures, and are either insoluble in water (e.g. $Ni(CO)_4$) or decomposed by it (e.g. $Fe(CO)_5$). With the exception of $Fe_2(CO)_9$, all are more or less soluble in organic solvents. As we have mentioned already, thermal or photochemical decomposition of the simpler carbonyls under mild conditions often yields polynuclear carbonyls, but the ultimate action of heat is formation of metal and carbon monoxide. The carbonyls vary greatly in their resistance to atmospheric oxidation: $Co_2(CO)_8$ decomposes in air at ordinary temperatures, forming a carbonate; $Fe(CO)_5$ and $Ni(CO)_4$ are very easily oxidized, their vapours forming explosive mixtures with air; and $Cr(CO)_6$, $Mo(CO)_6$ and $W(CO)_6$

are quite stable in air. This stability is, however, kinetic rather than thermodynamic in character; the bond energies in $Cr(CO)_6$, $Fe(CO)_5$ and $Ni(CO)_4$ (obtained from the heats of combustion, the bond energies in the carbonyl groups being assumed to be the same in all three compounds) are 113, 117 and 146 kJ (27, 28 and 35 kcal) respectively.

The exchange of carbon monoxide with some metal carbonyls has been studied using ^{14}CO; $Ni(CO)_4$ and $Co_2(CO)_8$ exchange rapidly, and $Fe(CO)_5$, $Cr(CO)_6$ and $Mn_2(CO)_{10}$ very slowly. The rate of exchange of the nickel compound is independent of the concentration of ^{14}CO and depends on the first power of that of $Ni(CO)_4$; the mechanism is thus

$$Ni(CO)_4 \rightarrow Ni(CO)_3 + CO \qquad \text{(slow)}$$
$$Ni(CO)_3 + {}^{14}CO \rightarrow Ni(CO)_3({}^{14}CO) \qquad \text{(fast)}$$

Another ligand such as triphenylphosphine may replace ^{14}CO without change in the kinetics of the reaction. The intermediate $Ni(CO)_3$ has been isolated, as we mentioned in Section 20.1, by the photolysis of $Ni(CO)_4$ in an inert matrix at low temperatures. A point of interest about the exchange of CO in $Co_2(CO)_8$ is that all eight molecules are equivalent in reactivity in solution; this is understandable in terms of the equilibrium between the ordinary (bridged) structure and the non-bridged structure under these conditions.

One of the earliest known reactions of iron pentacarbonyl was its conversion into the $[Fe(CO)_4]^{2-}$ anion by the action of aqueous sodium hydroxide. All neutral metal carbonyls can be converted into anions, a molecule of carbon monoxide being eliminated; it can be considered as being replaced by two electrons (in a doubly charged anion such as $[Fe(CO)_4]^{2-}$), one electron and a hydrogen atom (as in $[HFe(CO)_4]^-$) or two hydrogen atoms (in the uncharged carbonyl hydride, $H_2Fe(CO)_4$). There are now, indeed, more carbonylate anions than neutral carbonyls known. Protonation by strong acids will convert a few carbonyls into cations (e.g. $Fe(CO)_5$ into $[HFe(CO)_5]^+$) and other cations can be obtained in different ways. The action of chlorine, bromine or iodine on some carbonyls gives carbonyl halides; for example,

$$2Mo(CO)_6 + 2Cl_2 \rightarrow [Mo(CO)_4Cl_2]_2 + 4CO$$
$$Mn_2(CO)_{10} + Br_2 \rightarrow 2BrMn(CO)_5$$
$$Fe(CO)_5 + I_2 \rightarrow I_2Fe(CO)_4 + CO$$

The recent preparation of a carbonyl fluoride by the reaction

$$4Ru(CO)_5 + 4F_2 \rightarrow [Ru(CO)_3F_2]_4 + 8CO$$

(a similar reaction occurs between $Mo(CO)_6$ and fluorine) is of particular interest, since it has been widely believed that carbonyl fluorides do not exist. Most other reactions of metal carbonyls involve replacement of carbon monoxide by Lewis bases (e.g. tertiary phosphines, PF_3, pyridine, NO, *o*-phenanthroline and ethylenediamine) or organic molecules or radicals; the latter will be discussed in the following chapter. Meanwhile, we shall complete the description of the inorganic chemistry of the carbonyls by discussing in turn anionic and cationic carbonyl complexes, Lewis base complexes, carbonyl hydrides, carbonyl halides and complexes with non-transition elements as ligands. For more detail than we can include here the recent reviews by Abel and Stone,[1, 3] on which much of the material in Sections 20.5–9 is based, are particularly recommended.

20.5 Anionic and cationic carbonyl complexes[1]

There are three important methods for the conversion of metal carbonyls into anions, but none of them is of universal applicability.

(a) The action of hydroxide ion or nitrogenous bases (ammonia, amines, pyridine); for example,

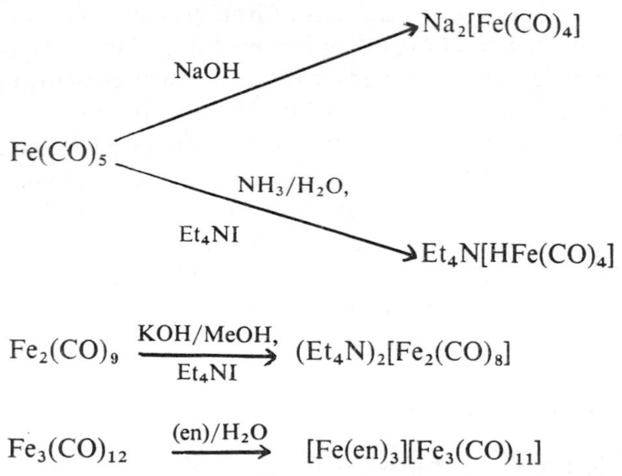

These reactions appear to involve nucleophilic attack by hydroxyl ion on the carbon atom of a carbonyl group, followed by electron

transfer to the metal atom and formation of carbon dioxide, which is converted into carbonate:

$$(OC)_4Fe(CO) \xrightarrow{OH^-} \left[(OC)_4Fe\text{—}C \begin{matrix} \diagup\!\!\diagup O \\ \diagdown OH \end{matrix} \right]^- \longrightarrow [Fe(CO)_4]^{2-} + H^+ + CO_2$$

They are, however, more complicated than this simple scheme suggests: in the presence of aqueous triethylamine, for example, $Fe(CO)_5$ yields $Et_3NH[HFe_3(CO)_{11}]$, and $Fe_3(CO)_{12}$ with pyridine in ultraviolet light forms $[Fe(py)_6][Fe_4(CO)_{13}]$. The anion in the last-mentioned compound is derived from a carbonyl not yet isolated, $Fe_4(CO)_{14}$.

The carbonyl $Co_2(CO)_8$ is converted by pyridine and many other bases into $[Co(py)_6][Co(CO)_4]_2$ and analogous compounds. In this case the initial reaction is probably the disproportionation

$$Co_2(CO)_8 + py \rightarrow [Co^{I}(CO)_4py]^+[Co^{-I}(CO)_4]^-$$

This is followed by another electron transfer reaction and loss of carbon monoxide, the whole process being represented by the equation

$$3Co_2(CO)_8 + 12py \rightarrow 2[Co(py)_6]^{2+} + 4[Co(CO)_4]^- + 8CO$$

(b) Reduction of carbonyls with alkali metals, alkali metal amalgams or borohydrides; for example,

$$Co_2(CO)_8 \xrightarrow[THF]{Na/Hg} Na[Co(CO)_4]$$

$$Mn_2(CO)_{10} \xrightarrow[THF]{Na/Hg} Na[Mn(CO)_5]$$

$$Re_2(CO)_{10} \xrightarrow[THF]{NaBH_4} Na_2[Re_4(CO)_{16}]$$

$$Cr(CO)_6 \xrightarrow{Na/NH_3} Na_2[Cr(CO)_5]$$

$$Cr(CO)_6 \xrightarrow[THF]{NaBH_4} Na[HCr_2(CO)_{10}]$$

$$Cr(CO)_6 \xrightarrow[NH_3]{NaBH_4} Na_2[Cr_2(CO)_{10}]$$

Molybdenum and tungsten carbonyls react in the same way as the chromium compound.

(c) Many substituted carbonyl anions can be obtained by displacement of carbon monoxide from a metal carbonyl; for example,

$$Me_4NI + Mo(CO)_6 \rightarrow Me_4N[Mo(CO)_5I] + CO$$
$$Mo(CO)_6 + NaB_3H_8 \rightarrow Na[Mo(CO)_5(B_3H_8)] + CO$$
$$Fe(CO)_5 + [Mn(CO)_5]^- \xrightarrow{h\nu} [FeMn(CO)_9]^- + CO$$
$$2Fe(CO)_5 + [Mn(CO)_5]^- \rightarrow [Fe_2Mn(CO)_{12}]^- + 3CO$$

We should mention here that carbonyl anions of niobium and tantalum are known, though no neutral carbonyls have yet been obtained. Sodium salts, in which the cation is complexed by diglyme, are obtained by the reaction

$$MCl_5 + Na + CO \xrightarrow[\text{diglyme}]{100° \text{ at } 200 \text{ atm}} [Na(diglyme)_3]^+[M(CO)_6]^-$$

Cationic carbonyl complexes may be obtained in the form of complex halides by the action of carbon monoxide and a Lewis acid such as $AlCl_3$ or BF_3; for example,

$$Mn(CO)_5Cl + CO + AlCl_3 \rightarrow [Mn(CO)_6]^+[AlCl_4]^-$$

Although few unsubstituted carbonyls react in this way a large number of organically substituted carbonyls do so.

Cationic complexes are also obtained by protonation of carbonyls in strong acids, and n.m.r. spectroscopy shows that the proton becomes bonded to the metal atom; for example,

$$Fe(CO)_5 \xrightarrow{HCl/BCl_3} [FeH(CO)_5]^+[BCl_4]^-$$

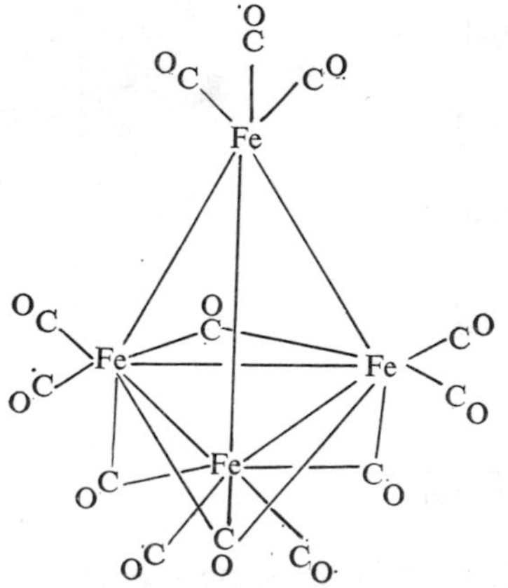

Figure 20.14

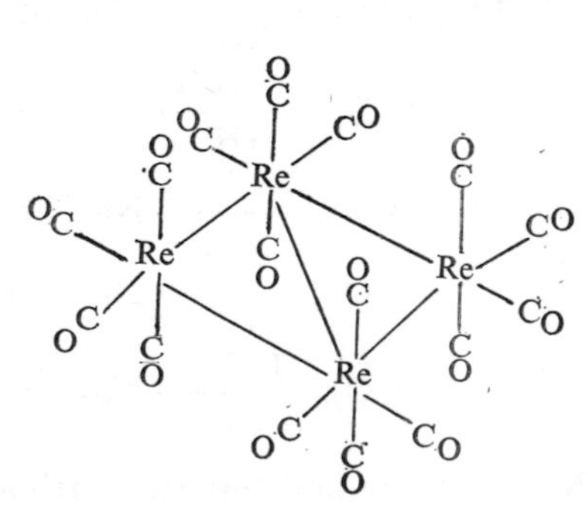

Figure 20.15

The structures of carbonyl anions and cations, in so far as they are known, are the same as those of neutral carbonyls with which they are isoelectronic: for example $[V(CO)_6]^-$ and $[Mn(CO)_6]^+$ are octahedral like $Cr(CO)_6$; $[Fe(CO)_4]^{2-}$ and $[Co(CO)_4]^-$ are tetrahedral like $Ni(CO)_4$; and $[MnFe_2(CO)_{12}]^-$ has the same structure as $Fe_3(CO)_{12}$. (These conclusions are based mainly on infrared and Raman spectroscopy.) Structures of a few anions which have no isoelectronic neutral carbonyls have been determined by X-ray diffraction. In $[Fe_4(CO)_{13}]^{2-}$ (Fig. 20.14), terminal, bridging and triply bridging carbonyl groups are all present. In $[Re_4(CO)_{16}]^{2-}$ (Fig. 20.15), only terminal carbonyl groups are present. Anions containing hydrogen are mentioned in Section 20.7.

20.6 Lewis base derivatives of carbonyls[1]

These compounds are nearly always made by direct replacement of carbon monoxide. Most substituents are compounds with π-acceptor as well as σ-donor properties (ammonia and ethylenediamine are exceptions), but apart from PF_3 and NO they are poorer π-acceptors than CO itself. This is inferred from the decrease in $\nu(CO)$ on substitution (e.g. in the series $Ni(CO)_4$, $Me_3PNi(CO)_3$ and $(Me_3P)_2Ni(CO)_2$, the highest carbonyl stretching frequency is 2128, 2063 and 1994 cm^{-1} respectively). By this criterion (which is necessarily imperfect, since the stretching frequency depends also on other factors) the sequence of π bonding ability is approximately

$$NO \sim CO \sim PF_3 > PCl_3 > P(OR)_3 > PPh_3 > py > R_3N$$

Nitric oxide acts as a three-electron ligand and coordinates as NO^+, which is isoelectric with CO; complexes containing this ligand are discussed in Section 20.10. The other species mentioned so far are either two-electron donors or, if they contain two donor atoms, four-electron donors which displace two molecules of carbon monoxide; in these cases substitution results in no major structural change.

Large numbers of triphenylphosphine and similar substituted metal carbonyls are known; $Ni(CO)_2(PPh_3)_2$ is of great importance as a catalyst for the polymerization of olefins and acetylenes (e.g. butadiene to cyclooctadiene and acetylene to benzene and styrene). Analogous compounds can be obtained by the action of triphenylphosphine on iron pentacarbonyl. Dicobalt octacarbonyl gives two products with a $Ph_3P : Co$ ratio of 1 : 1; one is $[Co_2(CO)_6(PPh_3)_2]$, the other the salt $[Co(CO)_3(PPh_3)_2][Co(CO)_4]$ in which the cation has the expected trigonal bipyramidal structure. It is noteworthy that a platinum complex, $Pt(CO)_2(PPh_3)_2$, can be obtained by the action of carbon monoxide on $Pt(PPh_3)_4$.

Since the first discovery that PF_3 (which until then was not considered a donor molecule) would replace CO in nickel carbonyl, a large number of its complexes have been prepared by treating carbonyls with phosphorus trifluoride under pressure and either heating the mixture or exposing it to ultraviolet radiation. This method yields a mixture of substituted products; fully substituted PF_3 complexes are obtained by the action of phosphorus trifluoride on finely divided metals or metal vapours at high pressures, sometimes in the presence of iodine as a catalyst, or by heating a metal halide in phosphorus trifluoride in the presence of copper or zinc. Typical preparations are:

$$Ni + 4PF_3 \xrightarrow[\text{trace } I_2]{400\text{ atm}, 150°} Ni(PF_3)_4$$

$$PtCl_2 + 4PF_3 + Cu \xrightarrow{100\text{ atm}, 100°} Pt(PF_3)_4 + CuCl_2$$

$$2Co + H_2 + 8PF_3 \xrightarrow{200\text{ atm}, 200°} 2HCo(PF_3)_4$$

Among PF_3 complexes now known are $Cr(PF_3)_6$, $Fe(PF_3)_5$, $Ni(PF_3)_4$, $Pd(PF_3)_4$ and $Pt(PF_3)_4$; cobalt forms $[Co_2(PF_2)_2(PF_3)_6]$, PF_2 groups replacing bridging carbonyls. Compounds and ions analogous to other carbonyl derivatives have also been obtained (e.g. $[HCo(CO)(PF_3)_3]$, $[Co(PF_3)_4]^-$, $[H_2Fe(PF_3)_4]$ and $Fe(PF_3)_4I_2$). The nickel compound $Ni(PF_3)_4$, which may be described as a typical complex, is a colourless liquid boiling at 71° (nickel carbonyl boils at 43°). The PF_3 in this compound is less reactive than in the free state, and the complex can be steam-distilled without decomposition. This appears to be a kinetic effect, since the P—F bond length is very similar to that in the free ligand. Analogous complexes are formed by PCl_3 and $P(CF_3)_3$.

A number of complexes in which phosphine itself replaces carbon monoxide have been reported recently. Thus $Cr(CO)_6$ reacts with phosphine to yield successively $Cr(CO)_5(PH_3)$, *cis*-$Cr(CO)_4(PH_3)_2$, *cis*-$Cr(CO)_3(PH_3)_3$ and *cis*-$Cr(CO)_2(PH_3)_4$. Other compounds prepared include $Fe(CO)_4(PH_3)$, $Ru_3(CO)_8(PH_3)_4$ and $Ni(CO)_3(PH_3)$; complexes are also formed by trimethylphosphine. The compound $Ni(PH_3)_4$ has recently been made by the action of phosphine on bis(1,5 cyclooctadiene) nickel at $-40°$ and characterized by its vibrational spectrum; it decomposes at $-30°$.

Ammonia and amines, as we have already seen, often cause metal carbonyls to disproportionate and form salts containing carbonylate anions, but a few simple substitution products can be obtained (e.g. $Cr(CO)_5(NH_3)$ and $Cr(CO)_3(NH_3)_3$). When, however, the nitrogen is part of an aromatic system capable of accepting some back-donation from the metal, the tendency to bring about disproportionation is reduced. Thus although ethylenediamine reacts with nickel carbonyl to give the tris-ethylenediamine nickel (II) salt of a carbonylate anion, probably $[Ni_4(CO)_9]^{2-}$, *o*-phenanthroline gives $Ni(CO)_2$(phen).

20.7 Carbonyl hydrides[1, 5]

We have already mentioned that some metal carbonyls dissolve in alkali (with the loss of a molecule of carbon monoxide in the case of mononuclear carbonyls) and form carbonylate anions. Iron pentacarbonyl, for example, gives a solution which is colourless in the absence of air and which on treatment with dilute mineral acid evolves iron carbonyl hydride ($H_2Fe(CO)_4$) as an evil-smelling gas which can be condensed to a white solid melting at $-70°$. It decomposes rapidly above $-10°$, giving hydrogen and products of decomposition of the $Fe(CO)_4$ radical. A number of other carbonyl hydrides, among them $HMn(CO)_5$, $HRe(CO)_5$, $H_2Ru(CO)_4$, $H_2Os(CO)_4$ and $HCo(CO)_4$, can be obtained by acidification of solutions containing carbonylate anions obtained as described in Section 20.5. Other methods of preparation include:

$$Fe(CO_4)I_2 \xrightarrow[\text{THF}]{NaBH_4} H_2Fe(CO)_4$$

$$Mn_2(CO)_{10} + H_2 \xrightarrow[200\ \text{atm}]{200°} 2HMn(CO)_5$$

$$Co + 4CO + \tfrac{1}{2}H_2 \xrightarrow[50\ \text{atm}]{150°} HCo(CO)_4$$

The most stable of these compounds with respect to thermal decomposition are $HMn(CO)_5$ and $H_2Os(CO)_4$, both colourless liquids which are stable up to about 80° and 100° respectively. Most reactions of $HMn(CO)_5$ involve replacement of the hydrogen atom (e.g. it reacts with Me_2Cd and with GeH_4 to give $Cd[Mn(CO)_5]_2$ and $H_2Ge[Mn(CO)_5]_2$ respectively), but Ph_3P displaces a molecule of CO to form $HMn(CO)_4(PPh_3)$. Substitution of triphenylphosphine in other carbonyl hydrides enhances the stability: $HCo(CO)_3(PPh_3)$ and $HV(CO)_5(PPh_3)$, for example, are well-defined species, whereas $HCo(CO)_4$ decomposes above −20° and $HV(CO)_6$ has not yet been prepared.

The acidities of mononuclear carbonyl hydrides vary very widely: $HCo(CO)_4$, though only slightly soluble in water, is as strong as a mineral acid in methanol; for $H_2Fe(CO_4)_2$ pK_1 and pK_2 are 4·5 and 14 respectively; and for $HMn(CO)_5$ pK is 7. It seems clear that the hydrogen atoms in all these compounds are bonded to the metal, though only for $HMn(CO)_5$ is the structure known in detail: the molecule is octahedral, with the Mn—H distance equal to 1·60 Å. Two physical properties may be used to show that metal–hydrogen bonds are present in the other compounds; the metal–hydrogen stretching vibration, which occurs at 2200–1700 cm^{-1}, is lowered by the theoretical factor of $\sqrt{2}$ on deuteration, and transition metal hydrides have characteristic high-field ^1H n.m.r. spectra in the range τ 12–45 (17·5 for $HMn(CO)_5$ and 20 for $HCo(CO)_4$). The structure of the substituted carbonyl hydride $(Ph_3P)_3Rh(CO)H$ has also been determined by X-ray crystallography; this molecule is a trigonal bipyramid with hydrogen and carbon monoxide occupying the apical positions.

Many di- or polynuclear carbonyl hydrides have been reported, but their formulae are not always related to those of known carbonylate anions, and many of the structures assigned to them rest only on spectroscopic evidence. We may, however, mention $HRe_2Mn(CO)_{14}$ (Fig. 20.16) and the anion $[HCr_2(CO)_{10}]^-$ (Fig. 20.17). In each case

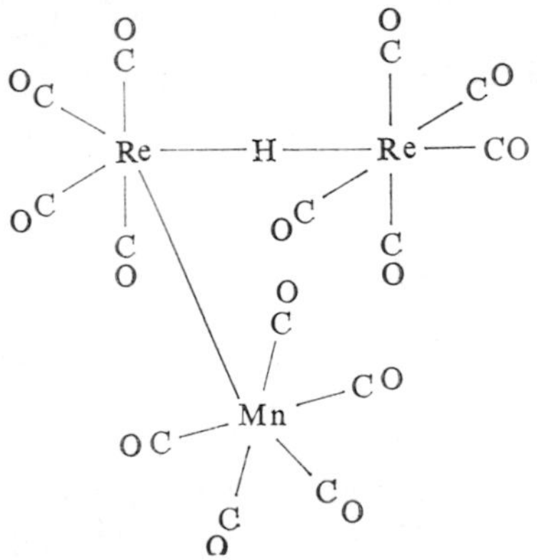

Figure 20.16

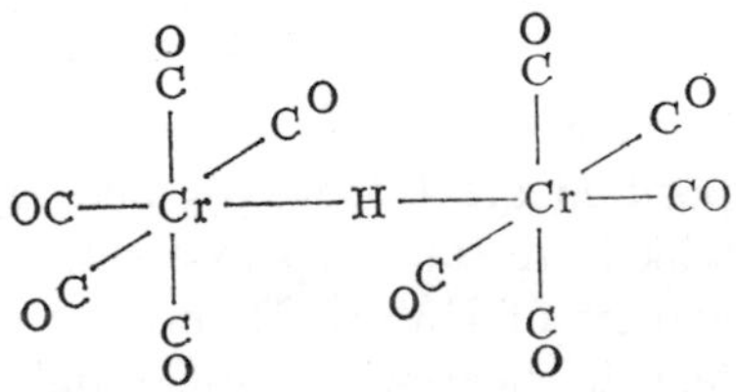

Figure 20.17

spectroscopic and diffraction data are consistent with a linear symmetrical hydrogen bridge and eclipsed carbonyl groups on the metal atoms bonded to hydrogen. The structure of $[HFe_3(CO)_{11}]^-$ is closely related to that of $Fe_3(CO)_{12}$ (Fig. 20.9), and can be regarded as the latter with one bridging carbonyl replaced by hydrogen. In $[HFe_2(CO)_8]^-$ a hydride ion similarly replaces a bridging carbonyl group in the $Fe_2(CO)_9$ (Fig. 20.6) structure. Among several carbonyl hydrides obtained by dissolution of metal carbonyls in alkali to give unidentified coloured products, and subsequent acidification, are $H_3Mn_3(CO)_{12}$ (from $Mn_2(CO)_{10}$), and $H_4Os_4(CO)_{12}$ and $H_2Os_4(CO)_{13}$ (from $Os_3(CO)_{12}$). Evidently considerable molecular rearrangements can take place in what might seem to be a simple process.

In addition to the triphenylphosphine-substituted carbonyl hydrides already mentioned, many others are known. Indeed, the combination of hydride, carbonyl and triphenylphosphine as ligands for the platinum metals is so favourable that several compounds of this type can be synthesized easily, a typical example being

$$Os(CO)_5 \xrightarrow[Ph_3P]{130°} (Ph_3P)_2Os(CO)_3 \xrightarrow[130° \text{ at } 120 \text{ atm}]{H_2/THF} H_2Os(CO)_2(PPh_3)_2$$

Addition of hydrogen to a triphenylphosphine carbonyl halide will be mentioned in the following section.

20.8 Carbonyl halides[1, 6]

Like other derivatives of metal carbonyls, carbonyl halides can exist as neutral molecules (e.g. $[Mo(CO)_4Cl_2]_2$), as anions (e.g. $[Mo(CO)_5I]^-$) and as cations (e.g. $[Fe(CO)_5Br]^+$). These three species are obtained in the following typical reactions:

$$Mo(CO)_6 \xrightarrow{Cl_2} [Mo(CO)_4Cl_2]_2$$

$$Mo(CO)_6 \xrightarrow[\text{diglyme}]{Et_4NI} Et_4N[Mo(CO)_5I]$$

$$Fe(CO)_5 \xrightarrow[\text{anhydrous HCl, then } BCl_3]{Br_2 \text{ in}} [Fe(CO)_5Br][BCl_4]$$

We shall restrict attention here to the neutral carbonyl halides.

The interaction of a metal carbonyl and a halogen is a general method for the preparation of carbonyl halides, though only two fluorides ($Ru_4(CO)_{12}F_8$ and a polymer of $Mo(CO)_2F_4$) have so far been obtained. Typical examples are:

$$Fe(CO)_5 \xrightarrow{I_2} Fe(CO)_4I_2$$

$$Mn_2(CO)_{10} \xrightarrow[40^\circ]{Br_2} Mn(CO)_5Br \underset{CO}{\overset{\text{pet. ether, } 120^\circ}{\rightleftharpoons}} [Mn(CO)_4Br]_2$$

$$Ru(CO)_5 \xrightarrow{F_2} [Ru(CO)_3F_2]_4$$

More frequently, however, carbonyl halides are prepared by the interaction of a metal halide and carbon monoxide; for example,

$$RuCl_3 \xrightarrow[65^\circ \text{ at } 10 \text{ atm}]{CO/CH_3OH} [Ru(CO)_3Cl_2]_2$$

$$RhCl_3.3H_2O \xrightarrow[100^\circ \text{ at } 1 \text{ atm}]{CO} [Rh(CO)_2Cl]_2$$

$$AuCl_3 \xrightarrow[C_2Cl_4 \text{ solution at } 120^\circ]{CO} Au(CO)Cl$$

It will be noted (see Table 20.1) that gold and platinum do not form binary carbonyls. Substitution of triphenylphosphine for some of the carbonyl groups greatly enhances the stability of carbonyl halides, just as it does that of carbonyl hydrides; thus although $Co(CO)_4I$ is unstable, $(Ph_3P)Co(CO)_3I$ can be made by the remarkable reaction

$$[(Ph_3P)Co(CO)_3]^- \xrightarrow{2CF_3I} (Ph_3P)Co(CO)_3I + I^- + C_2F_6$$

Triphenylphosphine carbonyl halides of rhodium and iridium may be prepared by interaction of the metal halide (or a complex halide) and triphenylphosphine in a variety of organic solvents, the solvent serving as the source of the carbonyl group:

$$(NH_4)_2IrCl_6 \xrightarrow[\text{reflux in ethylene glycol}]{Ph_3P} (Ph_3P)_2Ir(CO)Cl$$

$$IrCl_3.3H_2O \xrightarrow[\text{reflux in dimethylformamide}]{Ph_3P} (Ph_3P)_2Ir(CO)Cl$$

The product of these reactions ('Vaska's compound') is a highly reactive complex, the chemistry of which is discussed later.

The carbonyl halides are white, yellow or orange solids which are soluble in most organic solvents and decomposed by water. They readily undergo substitution of carbon monoxide by Lewis bases such as pyridine, diarsine and triphenylphosphine; for example,

$$Fe(CO)_4I_2 + 2Ph_3P \rightarrow Fe(CO)_2(PPh_3)_2I_2 + 2CO$$
$$[W(CO)_4Cl_2]_2 + 4Ph_3P \rightarrow 2W(CO)_3(PPh_3)_2Cl_2 + 4CO$$

Carbonyl halides often dimerize, always (so far as is known at present) by means of halogen bridges. If we assume a bridging halogen contributes one electron to the bond with one metal atom and forms a two-electron coordinate link to the other, the metal atoms in these compounds normally have a noble gas electronic configuration. Among structures which have been established by X-ray crystallography are those of $[Mn(CO)_4Br]_2$ (Fig. 20.18), $[Ru(CO)_3Br_2]_2$ (Fig. 20.19) and $[Ru(CO)_3F_2]_4$ (Fig. 20.20).

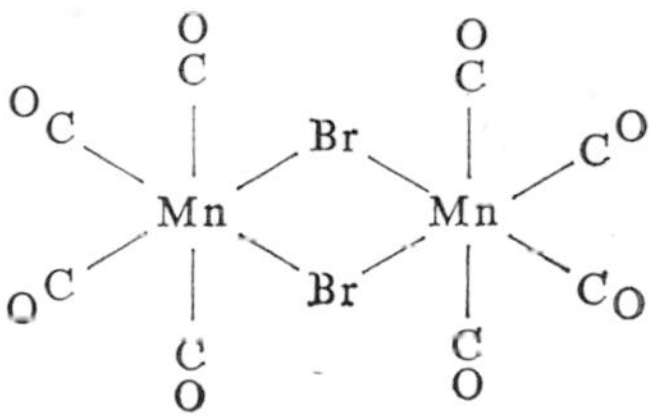

Figure 20.18

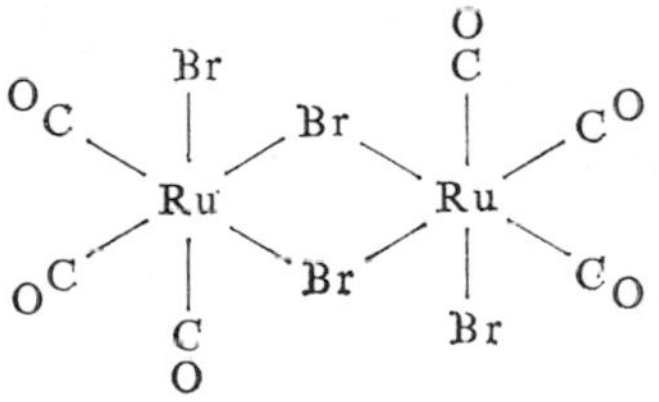

Figure 20.19

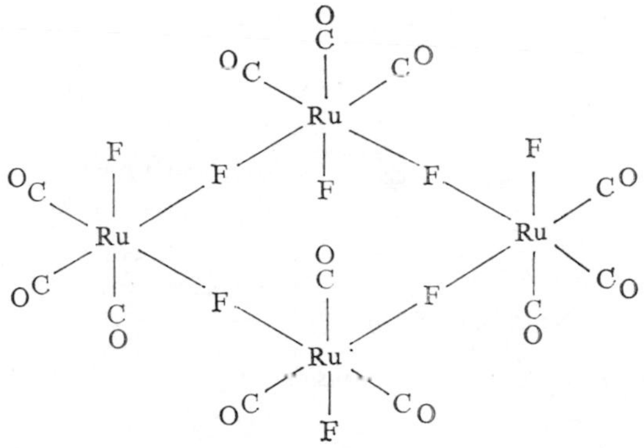

Figure 20.20

The halide $Ru(CO)_4I_2$ is octahedral, with the iodine atoms occupying *cis* positions, and so is $Fe(CO)_4I_2$ when freshly prepared; irradiation of the latter, however, even in the infrared spectrometer, produces the *trans* isomer.

'Vaska's compound' (*trans*-$(Ph_3P)_2Ir(CO)Cl$) and the analogous rhodium complex, unlike most carbonyl halides, contain metal atoms with valence shells of sixteen electrons, and this probably accounts in large measure for their activity as hydrogenation catalysts and for the great range of reactions which they undergo. The iridium compound is remarkable for its reversible uptake of H_2, O_2 and SO_2 to give crystalline 1 : 1 adducts which can be decomposed by lowering the pressure; for example,

$$(Ph_3P)_2Ir(CO)Cl \underset{\text{vacuum}}{\overset{C_6H_6\text{ solution}, O_2}{\rightleftharpoons}} O_2Ir(PPh_3)_2(CO)Cl$$

In the oxygen adduct, oxygen atoms occupy *cis* octahedral positions; the O—O distance of 1·30 Å suggests that oxygen is present as O_2^- rather than O_2^{2-}. Some of the many other reactions of Vaska's compound are shown in Fig. 20.21. We have described the reactions

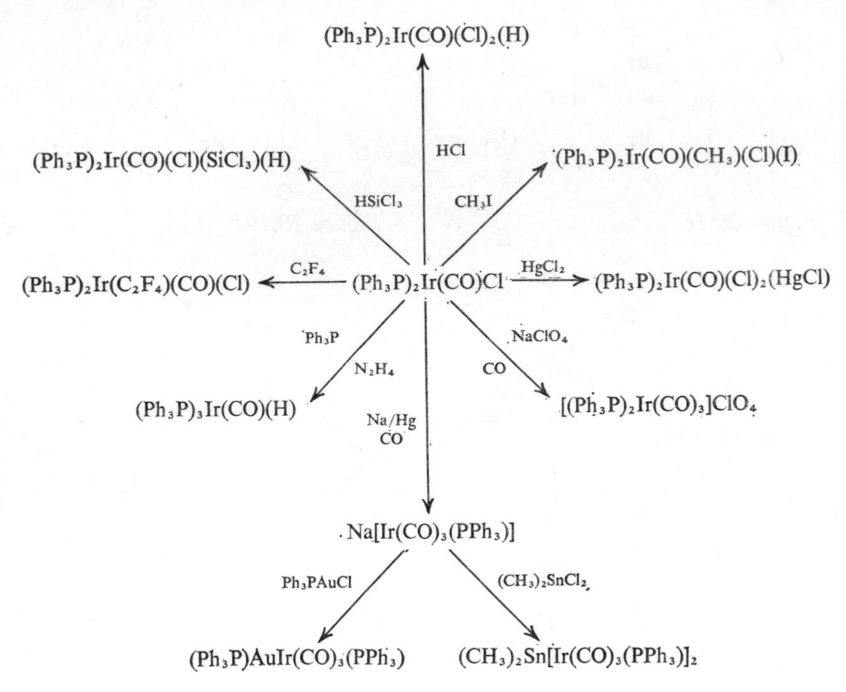

Figure 20.21

of Vaska's compound in some detail here because it is a carbonyl halide; but it should be added that many triphenylphosphine complexes containing rhodium and iridium show similar reactivity and catalytic activity.

20.9 Miscellaneous derivatives of metal carbonyls

Under this heading we shall briefly describe three types of carbonyl derivative not referred to previously: carbonyl carbide and carbonyl sulphur compounds, carbonyl complexes in which Group IV elements are ligands, and carbonyl complexes in which the metal carbonyl residue is bonded to an element of the copper or the zinc group.

20.9.1. Carbide and sulphur complexes.[1-3] The first of these to be discovered was $CFe_5(CO)_{15}$, which is obtained as a minor product in the reactions of $Fe_3(CO)_{12}$ and acetylenes. An X-ray crystallographic examination of this substance shows it to have five $Fe(CO)_3$ groups at the corners of a square pyramid, the carbon atom being just below the centre of the base and at approximately equal distances from each of the iron atoms. When $Ru_3(CO)_{12}$ is heated in n-octane, $CRu_6(CO)_{17}$ is obtained. This consists of an octahedron of ruthenium atoms with the carbon atom in the centre; four of the metal atoms are bonded to three terminal carbonyl groups each, the other two each having two terminal carbonyl groups and sharing one bridging carbonyl group (Fig. 20.22).

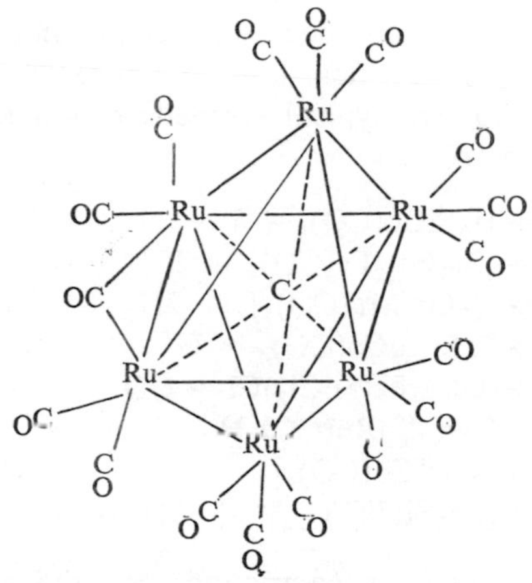

Figure 20.22

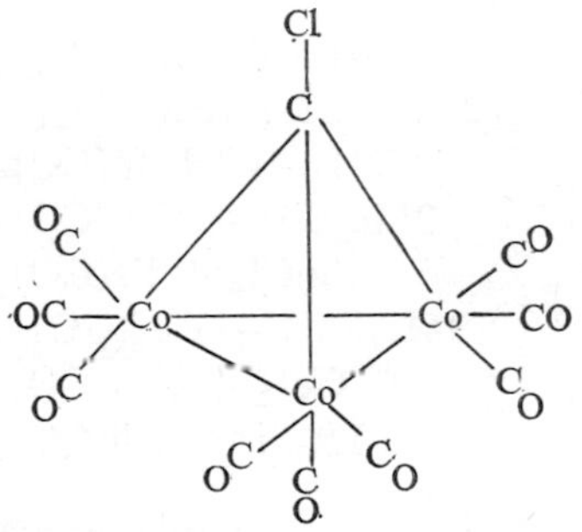

Figure 20.23

Interaction of dicobalt octacarbonyl and carbon tetrachloride in ethanol yields the complex $ClCCo_3(CO)_9$, the structure of which is shown in Fig. 20.23. When the analogous bromine compound is heated in boiling toluene it loses bromine to form $(OC)_9Co_3C.CCo_3(CO)_9$

and $(OC)_9Co_3C(CO)CCo_3(CO)_9$. The sulphur compound $SCo_3(CO)_9$, obtained by the action of ethylmercaptan on dicobalt octacarbonyl at 0°, has the sulphur atom bridging three cobalt atoms in place of the CCl group in Fig. 20.23, and the isostructural heteronuclear complex $SFeCo_2(CO)_9$ may also be obtained.

Many other carbide and sulphur complexes related to these substances have been characterized, very often by X-ray diffraction after unsuccessful attempts at chemical analysis. Although $ClCCo_3(CO)_9$ and its derivatives are adequately represented by conventional electronic formulae, the interpretation of the structures of the more complex species, like that of the complex halide ions containing metal clusters (e.g. $[Nb_6Cl_{12}]^{2+}$ and $[Mo_6Cl_8]^{2+}$), poses formidable problems in valence theory.

20.9.2. Complexes containing Group IV elements as ligands.[1, 7] Silyl-substituted metal carbonyls are considerably more stable than the corresponding alkyl compounds; $H_3SiCo(CO)_4$, for example, decomposes only above 80°, whereas $H_3CCo(CO)_4$ decomposes below room temperature. It has been suggested that this is at least partly due to π bonding between the Group IV element and the metal atom, the π bonding character of the silyl group in other compounds (e.g. $(H_3Si)_3N$) being well established. Many compounds in which silicon, germanium, tin or, less often, lead is bonded to a metal carbonyl residue are now known; typical synthetic methods are indicated by the following examples:

$$SiH_3I + [Co(CO)_4]^- \rightarrow H_3SiCo(CO)_4 + I^-$$
$$2Et_3PbCl + [Fe(CO)_4]^{2-} \rightarrow (Et_3Pb)_2Fe(CO)_4 + 2Cl^-$$
$$GeH_4 + 2HMn(CO)_5 \rightarrow H_2Ge[Mn(CO)_5]_2 + 2H_2$$
$$2Me_3SiH + Co_2(CO)_8 \rightarrow 2Me_3SiCo(CO)_4 + H_2$$
$$SnCl_4 + Fe(CO)_5 \rightarrow (OC)_4Fe(SnCl_3)Cl + CO$$
$$3Me_3SnH + Ru_3(CO)_{12} \rightarrow 3Me_3SnRu(CO)_4H$$
$$GeI_2 + Co_2(CO)_8 \rightarrow I_2Ge[Co(CO)_4]_2$$
$$Ph_3Si^- + Ni(CO)_4 \rightarrow [Ph_3SiNi(CO)_3]^- + CO$$

Polynuclear complexes may result from compounds containing one or two bonds between Group IV elements and transition metals undergoing further reactions or thermal decomposition; for example,

$$Fe(CO)_5 \xrightarrow{GeI_4} Fe(CO)_4(GeI_3)I \xrightarrow[\text{heat}]{Fe(CO)_5} [(OC)_4FeGeI_2]_2$$

$$2Me_3SiRu(CO)_4H \rightarrow [Me_3SiRu(CO)_4]_2 + H_2$$

These products appear from chemical evidence to contain a four-membered ring with alternating germanium and iron atoms and a Si–Ru–Ru–Si chain respectively. Iodine and sodium amalgam cleave the Ru—Ru bond according to the equations

$$[Me_3SiRu(CO)_4]_2 + I_2 \rightarrow 2Me_3SiRu(CO)_4I$$
$$[Me_3SiRu(CO)_4]_2 + 2Na \rightarrow 2Na[Me_3SiRu(CO)_4]$$

The anion produced in the second reaction will attack many halides to produce new metal–metal bonds, typical examples being

$$(Ph_3P)AuCl + [Me_3SiRu(CO)_4]^- \rightarrow (Me_3Si)(Ph_3PAu)Ru(CO)_4 + Cl^-$$

$$Mn(CO)_5I + [Me_3SiRu(CO)_4]^- \rightarrow (Me_3Si)[(OC)_5Mn]Ru(CO)_4 + I^-$$

Structures of several of these Group IV element–metal carbonyl complexes have been determined. In general they are what would be expected from the methods of synthesis, and in the absence of data for single covalent radii of many of the elements involved it is difficult to comment usefully on bond lengths. There are, however, some compounds with remarkable structures, of which we will mention two. In $Sn[Fe(CO)_4]_4$ (Fig. 20.24), which is obtained as a by-product in the reaction between iron pentacarbonyl and tri-n-butyltin chloride, the closest distance of approach of two iron atoms is 2·87 Å, much more than in $Fe_2(CO)_9$. The diamagnetism of the compound is therefore attributed to pairing of unpaired spins in three-centre orbitals involving a vacant *d* orbital on the tin atom. The compound $Fe_2(CO)_6(GeMe_2)_3$, prepared by heating the carbonyl $Fe_3(CO)_{12}$ with dimethylgermane at 65°, has the $Fe_2(CO)_9$ structure

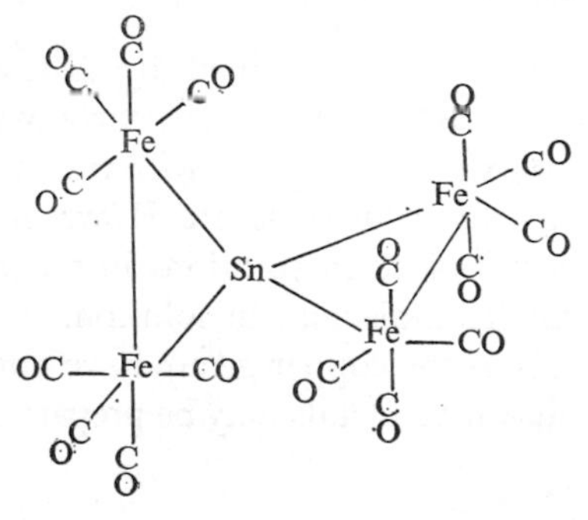

Figure 20.24

with Me_2Ge groups replacing the three bridging carbonyls. Again the Fe—Fe distance (2·74 Å) is appreciably greater than that in the carbonyl, and the manner in which the electron spins on the iron atoms are paired is uncertain. It may be noted that other compounds containing a Group IV element bonded to a transition metal are formed not only from organo-substituted metal carbonyls but also from trialkylphosphine complexes and halides; for example,

$$2Ph_3GeLi + (Et_3P)_2PtCl_2 \rightarrow (Et_3P)_2Pt(GePh_3)_2 + 2LiCl$$
$$PtCl_2 + 2SnCl_3^- \rightarrow [PtCl_2(SnCl_3)_2]^{2-}$$

Compounds in which Group III elements (notably boron, gallium and indium) act as ligands in substituted carbonyl complexes are also known.

20.9.3. Complexes in which the metal carbonyl residue is bonded to a metal of the copper or the zinc group.[1] Mercury (II) derivatives of metal carbonyls were prepared for the first time many years ago: interaction of iron carbonyl hydride and mercuric chloride in acetone yields successively $(ClHg)_2Fe(CO)_4$ and $HgFe(CO_4)_4$, whilst mercury combines directly with dicobalt octacarbonyl under pressure at 180° to yield $Hg[Co(CO)_4]_2$. The same compound is obtained by heating mercury with cobalt powder or cobalt bromide under a high pressure of carbon monoxide, and zinc and cadmium compounds may be prepared by analogous methods. Derivatives of other carbonyls may be obtained by the reactions

$$[Na(diglyme)_3][Ta(CO)_6] \xrightarrow{EtHgCl} EtHgTa(CO)_6$$
$$Zn + Mn_2(CO)_{10} \xrightarrow[diglyme]{120^\circ} Zn[Mn(CO)_5]_2$$

The structure of $Hg[Co(CO)_4]_2$ has been determined by X-ray crystallography; the Co–Hg–Co system is linear and the carbonyl groups perpendicular to the CoHgCo axis are staggered with respect to one another. The zinc compound has the same structure. However, whereas the latter substance is largely ionized in solutions in polar solvents, the mercury compound remains molecular in solution.

Derivatives of simple ions of metals of the copper group have not yet been made, but derivatives of unipositive metals may be prepared by reactions such as

$$(Ph_3P)AuCl + [Mn(CO)_5]^- \rightarrow (Ph_3P)AuMn(CO)_5 + Cl^-$$

Whether the presence of π-acceptor groups like Ph_3P is essential for the formation of stable compounds containing copper, silver or gold bonded to a metal carbonyl residue is not yet certain.

20.10 Nitrosyl complexes of transition metals[1, 3, 8]

Nitric oxide is a paramagnetic molecule with an electron in an antibonding orbital. This electron is relatively easily lost with formation of the NO^+ ion and an increase in the N—O stretching frequency from 1878 cm^{-1} in NO to 2200–2400 cm^{-1} in nitrosonium salts. Most complexes of nitric oxide and transition metals are best considered to be those of the NO^+ ion, three electrons being transferred to the metal atom; M—N back π-bonding then takes place in exactly the same way as for carbon monoxide. Because of its positive charge, however, coordinated NO is a better π-acceptor than coordinated CO, and the N—O stretching frequency in complexes of NO^+ is some 300–500 cm^{-1} lower than that in salts such as $NO^+BF_4^-$.

Mononuclear nitrosyls and nitrosyl carbonyls are restricted to the following compounds: $Co(NO)(CO)_3$, $Fe(NO)_2(CO)_2$, $Mn(NO)_3CO$ and $Co(NO)_3$ (isoelectronic with $Ni(CO)_4$); $Mn(NO)(CO)_4$ (isoelectronic with $Fe(CO)_5$); and $V(NO)(CO)_5$ (isoelectronic with $Cr(CO)_6$). In addition a binuclear species $Mn_2(NO)_2(CO)_7$ (isoelectronic with $Fe_2(CO)_9$) and a number of nitrosyl complexes containing organic groups or triphenylphosphine as substituents have been prepared. Nitric oxide displaces carbon monoxide from $V(CO)_6$, $(Ph_3P)_2Mn_2(CO)_8$, $Fe_2(CO)_9$ and $Co_2(CO)_8$ to give $V(NO)(CO)_5$, $Mn(NO)(CO)_4$, $Fe(NO)_2(CO)_2$ and $Co(NO)(CO)_3$ respectively; the further action of nitric oxide on the manganese and cobalt compounds yields $Mn(NO)_3(CO)$ and $Co(NO)_3$. All of these substances are solids of low melting point or liquids which are thermally rather unstable and are decomposed by air and by water. In the reaction of $Fe(NO)_2(CO)_2$ with alkali in methanol, $[Fe(NO)(CO)_3]^-$ is formed, but under comparable conditions $Co(NO)(CO)_3$ gives $[Co(CO)_4]^-$, $Co(OH)_2$ and other cobalt-free products.

The limited evidence available is consistent with tetrahedral structures for $Fe(NO)_2(CO)_2$ and $Co(NO)(CO)_3$ and a trigonal bipyramidal structure (with NO in the equatorial plane) for $Mn(NO)(CO)_4$; $(Ph_3P)_2Mn(NO)(CO)_2$ also has a trigonal bipyramidal structure, the two triphenylphosphine molecules occupying the apical positions. Since $Co(NO)_3$ shows two N—O stretching frequencies in the infrared, it must be pyramidal rather than planar, but the detailed structure is not known.

Volatile diamagnetic nitrosyl halides of formula $Fe(NO)_3X$ are formed by the action of nitric oxide on iron carbonyl halides in the presence of finely divided iron as a halogen-acceptor. These readily lose NO to give $[Fe(NO)_2X]_2$, in which the halogen atoms act as bridges. Analogous compounds of cobalt and nickel may be formed by reactions similar to those involved in the high pressure synthesis of carbonyls; for example,

$$CoX_2 + Co + 4NO \rightarrow 2Co(NO)_2X$$
$$4NiI_2 + 2Zn + 8NO \rightarrow 2[Ni(NO)I]_4 + 2ZnI_2$$

The ease of formation of these compounds increases in the sequences Ni < Co < Fe and X = Cl < Br < I. Nitrosyl chloride and nickel carbonyl in liquid hydrogen chloride, on the other hand, give $Ni(NO)_2Cl_2$, which is probably monomeric and tetrahedral. Nitrosyl halides are also formed by some metals which, so far as is known, do not form nitrosyls or nitrosyl carbonyls. Thus molybdenum and tungsten (but not chromium) carbonyls react with nitrosyl chloride:

$$M(CO)_6 + 2NOCl \xrightarrow[CH_2Cl_2]{20^\circ} M(NO)_2Cl_2 + 6CO$$

Palladium (II) chloride in methanolic solution yields $Pd(NO)_2Cl_2$, and nitrosyl halide molecules or anions are formed also by several other transition metals.

Two other NO^+ derivatives of iron may be mentioned briefly here. The species formed in the brown-ring test for nitrate is $[Fe(H_2O)_5NO]^{2+}$. The equilibrium

$$[Fe(H_2O)_6]^{2+} + NO \rightleftharpoons [Fe(H_2O)_5NO]^{2+} + H_2O$$

is reversible, and the brown complex may be destroyed by blowing nitrogen through the solution to remove nitric oxide. In this species the N—O stretching frequency is 1745 cm^{-1}, and the magnetic moment is 3·9 B.M., corresponding to the presence of three unpaired electrons; formally, therefore, the ion is a high-spin d^7 complex of Fe^I and NO^+, but the N—O stretching frequency indicates very strong π bonding and the intense brown colour strongly suggests Fe^I–NO^+ charge transfer. Sodium nitroprusside (nitrosopentacyanoferrate (II)) is prepared by the action of nitric acid or sodium nitrite on the hexacyanoferrate (II). In the former process the overall reaction is

$$[Fe(CN)_6]^{4-} + 4H^+ + NO_3^- \rightarrow [Fe(CN)_5NO]^{2-} + CO_2 + NH_4^+$$

In the latter process, two successive equilibria are involved:

$$[Fe(CN)_6]^{4-} + NO_2^- \rightleftharpoons [Fe(CN)_5NO_2]^{4-} + CN^-$$
$$[Fe(CN)_5NO_2]^{4-} + H_2O \rightleftharpoons [Fe(CN)_5NO]^{2-} + 2OH^-$$

These are driven to completion by adding barium chloride to the reaction mixture and blowing a current of carbon dioxide through the hot solution to remove the hydrogen cyanide liberated by the reaction

$$2[Fe(CN)_6]^{4-} + 2NO_2^- + 3Ba^{2+} + 3CO_2 + H_2O \rightarrow 2[Fe(CN)_5NO]^{2-} + 2HCN + 3BaCO_3$$

The formulation of the complex anion as a NO^+ derivative of iron (II) is supported by its diamagnetism, a N—O stretching frequency of 1939 cm^{-1} and a N—O distance of 1·13 Å. The purple colour obtained from nitrosopentacyanoferrate (II) and sulphide is due to the ion $[Fe(CN)_5(NOS)]^{4-}$, analogous to $[Fe(CN)_5NO_2]^{4-}$.

Complexes containing the NO^- ion strictly lie outside the scope of this chapter, but we should point out that they are known. For example there are two isomeric series of salts of the formula $[Co(NH_3)_5NO]X_2$, coloured black and red respectively, which are obtained by the action of nitric oxide on ammoniacal solutions of cobalt (II) salts. The black series contains the monomeric cation $[Co(NH_3)_5NO]^{2+}$, in which a very low N—O stretching frequency of 1170 cm^{-1} and a long N—O bond (variously reported as 1·26 or 1·41 Å) suggests the presence of NO^-. The red series are derivatives of hyponitrite, the structure of the dimeric cation being

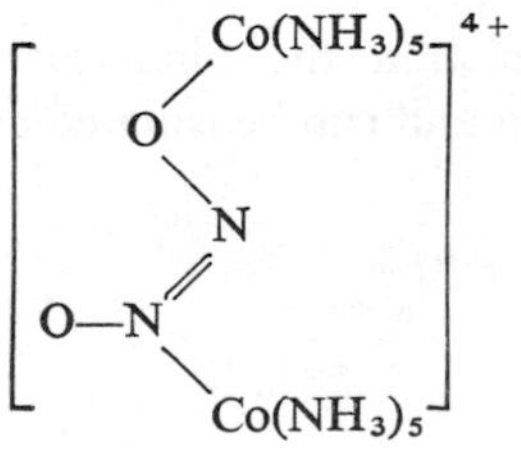

```
         Co(NH3)5 ⁴⁺
        /
      O
        \
          N
         //
  O—N
        \
         Co(NH3)5
```

in which hyponitrite acts as an oxygen-donor at one end of the ion and as a nitrogen-donor at the other. The *cis* configuration of the hyponitrite anion presumably arises from steric factors; the *trans* configuration would lead to interactions between the pentammine-cobalt (III) groups.

20.11 Complexes of molecular nitrogen[9]

These substances (often called nitrogenyl or dinitrogen complexes, to distinguish them from those containing the nitride ion) have been known for only a few years. In 1965, Allen and Senoff obtained salts containing the $[Ru(NH_3)_5N_2]^{2+}$ cation by the action of hydrazine hydrate on various compounds of tri- and tetrapositive ruthenium, amongst them ruthenium trichloride and ammonium hexachloro-ruthenate (IV). Many other complexes containing one or two (but not, so far, more) molecules of coordinated nitrogen have now been prepared, and it is clear that N_2 acts as a σ-donor and π-acceptor in the same way as the isoelectronic molecule CO, though the complexes formed are much less stable than carbonyls. Much of the interest in this field centres on the possibility of developing new methods for nitrogen fixation; up to the present time, however, no method has been found for the reduction of nitrogen in the complexes described here (though this has been achieved by systems involving an organo-titanium complex under powerfully reducing conditions[10]).

Most, though not all, nitrogenyl complexes have triphenyl-phosphine and halide or hydride as other ligands in the complex. The following examples illustrate methods for their preparation.

(a) The action of nitrogen on a metal complex; for example,

$$CoCl_2 + Ph_3P \xrightarrow[EtOH]{NaBH_4} (Ph_3P)_3CoH_3 \xrightarrow{N_2} (Ph_3P)_3CoH(N_2)$$

$$[Ru(NH_3)_5H_2O]^{2+} \xrightarrow{N_2} [Ru(NH_3)_5(N_2)]^{2+}$$

It is noteworthy that in the latter reaction the binuclear ion $[(NH_3)_5RuN_2Ru(NH_3)_5]^{4+}$ is also formed and can be isolated as its fluoroborate.

(b) Reactions of coordinated azide; for example,

$$[Ru(NH_3)_5Cl]^{2+} + N_3^- \xrightarrow[NH_3]{MeSO_3H} [Ru(NH_3)_5N_2]^{2+}$$

$$(Ph_3P)_2Ir(CO)Cl + RCON_3$$

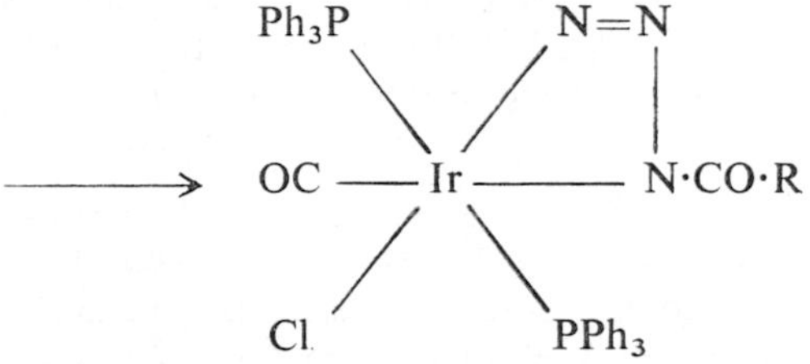

Ph_3P N=N

$\longrightarrow$ OC — Ir — N·CO·R

Cl PPh_3

$$\xrightarrow[EtOH]{CHCl_3} (Ph_3P)_2IrCl(N_2)$$

$$[RuCl(NO)(diarsine)_2]Cl_2 \xrightarrow{N_2H_4} [RuCl(N_3)(diarsine)_2]$$

$$\xrightarrow{NO^+PF_6^-} [RuCl(N_2)(diarsine)_2][PF_6]$$

(c) Reactions of coordinated ammonia; for example,

$$[Os(NH_3)_5(N_2)]^{2+} + HNO_2 \rightarrow cis\text{-}[Os(NH_3)_4(N_2)_2]^{2+} + 2H_2O$$

The most stable nitrogenyl complexes are those of the heavier members of the iron and cobalt groups, some of which are unaffected by dry air and can be heated to 100–200° before thermal decomposition takes place. Most, however, are rapidly oxidized by air and decompose when heated gently. Coordinated nitrogen is easily displaced by other ligands; in the case of the compound $(Ph_3P)_3CoH(N_2)$, an orange solid, for example, displacement by hydrogen, ethylene or ammonia is reversible, whilst that by carbon monoxide, a stronger ligand, is irreversible. Some reactions of $(Ph_3P)_2IrCl(N_2)$, a yellow solid, are shown below:

$(Ph_3P)_2IrCl(N_2)$

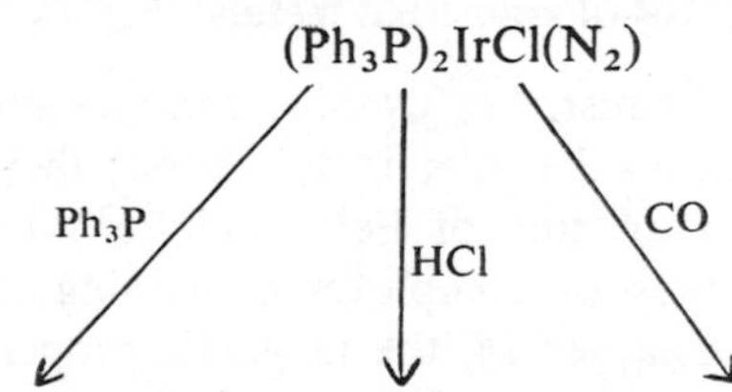

$(Ph_3P)_3IrCl + N_2$ $(Ph_3P)_2IrHCl_2 + N_2$ $(Ph_3P)_2Ir(CO)Cl + N_2$

Nitrogenyl complexes show an asymmetric infrared N≡N

stretching frequency in the range 2230–1920 cm^{-1} (the Raman stretching frequency in N_2 is 2331 cm^{-1}). The decrease in $\nu(N \equiv N)$ on coordination is rather greater than that for CO in compounds of analogous types. This suggests that N_2 is a better π-acceptor than CO, but owing to its higher ionization potential it would be expected to be a poorer σ-donor; as we have stated already, in its overall effect N_2 is a poorer ligand than CO. Structures of three nitrogenyl complexes have been determined by X-ray crystallography; the ion $[Ru(NH_3)_5(N_2)]^{2+}$ is octahedral, the molecule $(Ph_3P)_3CoH(N_2)$ is trigonal bipyramidal, with the three triphenylphosphine groups occupying the equatorial positions. In both cases the $M—N \equiv N$ system is linear, and the $N \equiv N$ distance is 1·10 Å, the same as in molecular nitrogen. A similar value (1·12 Å) is found in the $[(H_3N)_5RuN_2Ru(NH_3)_5]^{4+}$ ion, which is linear; as would be expected from this structure, the $N \equiv N$ stretching frequency is infrared-inactive, but it occurs at 2100 cm^{-1} in the Raman spectrum. This method of bridging may be contrasted with that found in bridging carbonyls, in which carbon donates to two atoms. With this exception, however, the structural chemistry of the coordinated N_2 group is very like that of the coordinated isoelectronic CO group.

Values of ΔH^0 for the reactions in aqueous solution

$$[Ru(NH_3)_5(H_2O)]^{2+} + N_2 \rightarrow [Ru(NH_3)_5(N_2)]^{2+} + H_2O$$

and

$$[Ru(NH_3)_5(N_2)]^{2+} + [Ru(NH_3)_5(H_2O)]^{2+} \rightarrow [(H_3N)_5RuN_2Ru(NH_3)_5]^{4+} + H_2O$$

have recently been determined as —42 kJ (—10 kcal) and —46 kJ (—11 kcal) respectively.[11] There are at present, unfortunately, no related thermochemical data with which to compare them.

20.12 Cyanide complexes of transition metals[12, 13]

Many aspects of the chemistry of cyanide complexes of transition metals have been mentioned earlier in this book; they include the thermodynamics of complexing of Fe^{2+} and Fe^{3+} by cyanide in Chapter 6, the structures of complexes containing molybdenum, tungsten or nickel in Chapter 14, the magnetic properties of complex cyanides and the position of the ion in the spectrochemical series in Chapter 15, the charge-transfer spectrum of Prussian blue in Chapter 17, π bonding by cyanide and its effect on the stabilization of oxidation states in Chapter 18, and some aspects of the kinetics and mechanism of the formation of cobalt (III) complexes in Chapter 19. The formal similarity between cyanide complexes and carbonyls is often very marked, perhaps the most striking difference being that when cyanide acts as a bridging ligand it nearly always does so by both carbon and nitrogen acting as donor atoms.

Formal similarities notwithstanding, however, it must be remembered that cyanide, unlike carbon monoxide and nitrogen, is an ion. Just as NO^+ acting as a ligand shows greater back-bonding ability than CO, so the negatively charged CN^- shows less, and the range of C≡N stretching frequencies observed in complex cyanides is quite small (2000–2170 cm^{-1}, compared with 2080 cm^{-1} in KCN). Furthermore, most of the chemistry of complex cyanides is that of aqueous solutions or ionic solids. Quantitative treatment of the former is made difficult by the kinetic inertness of many complex cyanide ions, those of d^3 and d^6 systems in particular; the factors underlying this kinetic inertness were discussed in Chapter 19.

Although, therefore, CN^- and CO can replace one another in a few systems (the best known examples being $[Mo(CO)_3(CN)_3]^{3-}$, $[W(CO)_3(CN)_3]^{3-}$, $[Fe(CO)(CN)_5]^{3-}$, $[Ni_2(CO)_2(CN)_6]^{4-}$ and $[Ni(CO)_2(CN)_2]^{2-}$), the similarities in the case of their complexes are largely confined to the structural side.

Mention may also be made here of complex acetylides.[14] The $(C{\equiv}CH)^-$ ion is isoelectronic with cyanide, but, since it is a much stronger base, its complexes must be made in non-aqueous media by the action, for example, of alkali metal acetylides on thiocyanates in liquid ammonia ($K_2[Ni(C{\equiv}CH)_4]$ and $K_4[Fe(C{\equiv}CH)_6]$ may be made in this way). These substances are violently explosive, but the corresponding derivatives of phenylacetylene are somewhat more stable. All are converted into complex cyanides by the action of aqueous potassium cyanide. In view of the fact that cobalt (II) forms only a pentacyano complex $[Co(CN)_5]^{3-}$ (which dimerizes in the solid state in alkali metal salts), it is interesting to note that acetylene and methylacetylene yield the salts $K_4[Co(C{\equiv}CH)_6]$ and $Na_4[Co(C{\equiv}CMe)_6]$ respectively. Mixed carbonyl acetylide complexes can also be made, for example, by the typical reaction

$$Mo(CO)_3(NH_3)_3 + 3KC{\equiv}CR \rightarrow K_3[Mo(CO)_3(C{\equiv}CR)_3] + 3NH_3$$

References for Chapter 20

1 Abel, E. W., and Stone, F. G. A., *Quart. Rev. Chem. Soc.*, 1970, **24,** 498.
2 Johnston, R. D., *Adv. Inorg. Chem. Radiochem.*, 1970, **13,** 471.
3 Abel, E. W., and Stone, F. G. A., *Quart. Rev. Chem. Soc.*, 1969, **23,** 325.
4 Churchill, M. R., and Mason, R., *Adv. Organometallic Chem.*, 1967, **5,** 93.
5 Green, M. L. H., and Jones, D. J., *Adv. Inorg. Chem. Radiochem.*, 1965, **7,** 115.
6 Calderazzo, F., in *Halogen Chemistry*, ed. V. Gutmann, Vol. 3 (Academic Press, 1967).
7 Young, J. F., *Adv. Inorg. Chem. Radiochem.*, 1968, **11,** 91.
8 Cotton, F. A., and Wilkinson, G., *Advanced Inorganic Chemistry*, 2nd edn. (Interscience, 1966).
9 Fergusson, J. E., and Love, J. L., *Rev. Pure and Appl. Chem.*, 1970, **20,** 33.
10 van Tamelen, E., *Accounts Chem. Research*, 1970, **3,** 361.
11 Armor, J., and Taube, H., *J. Amer. Chem. Soc.*, 1970, **92,** 6170.
12 Griffith, W. P., *Quart. Rev. Chem. Soc.*, 1962, **16,** 188.
13 Chadwick, B. M., and Sharpe, A. G., *Adv. Inorg. Chem. Radiochem.*, 1966, **8,** 83.
14 Nast, R., *Chemical Society Special Publication* No. 13 (1959).

Organometallic compounds of transition metals

21

21.1 Introduction[1–5]

No field of chemistry has expanded more rapidly in the past twenty years than that of the organometallic chemistry of the transition metals. Although the formation of potassium ethylenetrichloroplatinate (II), $K[PtCl_3(C_2H_4)]$, by the action of ethylene on a solution of potassium chloroplatinate (II) was noted by Zeise as long ago as 1830, and trialkylplatinum halides (e.g. $[(CH_3)_3PtI]_4$) were prepared by Pope and Peachey in the early 1900's, it is only comparatively recently that extensive research into organic complexes of the transition metals has been carried out. The chance discovery of ferrocene (see Section 21.7) in 1951 may be said to mark the beginning of a period of intense activity, originating partly in the scientific interest of compounds of transition metals with both saturated and unsaturated organic systems, and partly in the discovery that some of these compounds are of great utility as catalysts in organic chemistry.

The chemistry of organometallic derivatives of transition metals shows many similarities to that of the carbonyls and their derivatives discussed in the last chapter. However, whereas in the carbonyls the carbon atom uses its lone pair of electrons to provide the primary bond, in the case of complexes of transition metals with olefins and systems capable of providing more than two electrons the π electrons themselves are involved in the primary bond formation. In contrast to the carbonyls, in which the metal atom is in the nodal plane of the π electrons, bonding in the olefin complexes (which we may take as the simplest example) is perpendicular to the nodal plane. The primary bonding is reinforced, as in the carbonyls, by back-donation from a filled transition metal t_{2g} orbital to a π^* orbital of the ligand; this total interaction is shown in Fig. 21.1, which may be compared with Fig. 20.1 at the beginning of the preceding chapter. Both

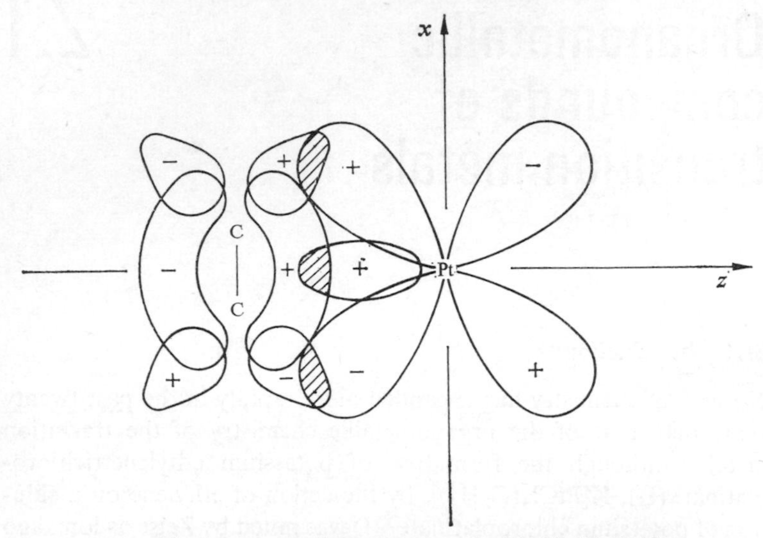

Figure 21.1 Platinum–ethylene bonding in $[PtCl_3(C_2H_4)]^-$, showing interaction of a $5d\,6s\,6p^2$ orbital with the π orbital of the ethylene and of the d_{xz} orbital with the π^* orbital of ethylene

removal of bonding electrons from ethylene (to consider the simplest system) and donation into its π^* orbital weaken the carbon–carbon bond, and it is therefore to be expected that the C═C stretching frequency should be less than it is in ethylene (1623 cm^{-1}); values in the range 1460–1560 cm^{-1} are usually found for complexed olefins (e.g. in $[PtCl_3(C_2H_4)]^-$ at 1511 cm^{-1}). The carbon–carbon bond length in this ion is 1·37 Å compared with 1·32 Å in ethylene. The bonding of other unsaturated hydrocarbons to transition metals can be discussed similarly on the basis of a simple molecular orbital theory, but for details more specialized works[1, 2] must be consulted.

In the following account we shall first consider σ-bonded derivatives of alkyl (including fluoroalkyl) and aryl groups. In a σ bond the metal and the alkyl or aryl group each provide one electron, and these derivatives are sometimes, therefore, classified as those of one-electron donors. In classifying further hydrocarbon complexes it is generally convenient to consider the number of electrons formally available (without reference to whether they are involved or not, which can only be inferred from the structure of the complex) from the neutral hydrocarbon or hydrocarbon radical. Thus we have, with increasing numbers of formally available electrons, the following: one-electron donors (alkyl and aryl groups); two-electron donors (monoolefins or, to give them their modern name in organic chemistry, monoalkenes); three-electron donors (enyl groups, e.g. π-allyl); four-electron donors (dienes); five-electron donors (dienyl groups, e.g. π-cyclopentadienyl); six-electron donors (trienes and arenes); seven-electron donors (trienyl groups, e.g. cycloheptatrienyl); and eight-electron donors (tetrenes).

As in the chemistry of metal carbonyls, the effective atomic number rule (or eighteen-electron rule) is a useful basis for the discussion of formulae, though there are now many exceptions to it, especially among complexes at the beginnings and ends of the transition series. In many cases its application in terms of numbers of formally available electrons is straightforward; in π-$C_5H_5Mo(CO)_2\pi$-C_3H_5, for example, the π-cyclopentadienyl and π-allyl groups contribute five and three electrons respectively, the carbonyl groups two each and the molybdenum atom six, making a total of eighteen. In biscyclooctatetrene iron, however, addition of available electrons would give the iron atom a total of $8 + (2 \times 8) = 24$, but an X-ray diffraction investigation of the structure indicates that the iron atom is near enough for bonding only to three double bonds in one ring and two in the other; the ligands are therefore acting as a triene and a diene respectively, and the eighteen-electron rule is then obeyed. As with binuclear carbonyls, metal–metal bonding is often invoked to account for the diamagnetism of complexes containing two metal atoms; in such cases counting electrons then results in the eighteen-electron rule being obeyed. In $[\pi$-$C_5H_5Fe(CO)_2]_2$, for example, there are two bridging and two terminal carbonyl groups; these and the π-cyclopentadienyl groups would bring the number of electrons in the $4d$, $5s$ and $5p$ orbitals of each iron atom to seventeen, whilst coupling of the unpaired spins by formation of an iron–iron bond would make each metal atom have eighteen electrons in these orbitals. In $[\pi$-$C_5H_5Mo(CO)_3]_2$, on the other hand, there are no bridging carbonyl groups, and the dimer must be held together by a metal–metal bond; again the eighteen-electron rule is obeyed. Most (though not all) exceptions to the rule occur in complexes of nickel, copper and zinc and their homologues (metals of the zinc group rarely, if ever, form π complexes). In general it is fair to say that although the eighteen-electron configuration is sometimes not attained, it is probably very seldom exceeded, and it often provides a useful signpost to formulae or structures.

21.2 Preparation of organometallic compounds of transition metals

In this section we shall describe briefly some of the most general methods for the synthesis of organometallic complexes of transition metals. In so doing, however, we do not wish to obscure the part played by chance discovery in the development of this branch of chemistry. Many substances referred to here or elsewhere were first obtained as unexpected products which were identified by a combination of physical methods or, occasionally, by X-ray diffraction alone. Many of the compounds discussed in this chapter have catalytic properties, and it is not uncommon for a ligand to undergo chemical change during complex formation.

21.2.1. From a metal or metal compound and the ligand. Few ligands (or hydrocarbons from which they are derived) react directly with transition metals to give organometallic derivatives, since these substances are seldom stable at the high temperatures required to bring about reactions of the metals. One exception, however, is bis-(π-cyclopentadienyl) iron, $(\pi\text{-}C_5H_5)_2Fe$ (*ferrocene*), which is formed in low yield by passing cyclopentadiene over reduced iron at 300°.

Displacement reactions (e.g. of solvent molecules, coordinated anions and, especially, of carbon monoxide) are very commonly used. When ethylene is passed through a solution of potassium tetrachloroplatinate (II) the compound $K[PtCl_3(C_2H_4)].H_2O$ (Zeise's salt) is precipitated. Less volatile olefins will displace ethylene from this complex. Aqueous rhodium trichloride and olefins react readily; for example,

$$C_2H_4 + RhCl_3(aq) \longrightarrow [(CH_2{=}CH_2)_2Rh(\mu\text{-}Cl)_2Rh(CH_2{=}CH_2)_2]$$

Many olefins form complex silver ions in aqueous media, but solid products cannot be isolated from these solutions. They are, however, readily obtained from certain silver salts, especially the fluoroborate $AgBF_4$, by direct combination; aromatic hydrocarbons react similarly.

The reaction with metal carbonyls is greatly assisted by the volatility of carbon monoxide, which is thereby continuously displaced from many carbonyl complexes at ordinary temperature or on gentle heating or ultraviolet irradiation. It is worth noting that thermal and photochemical reactions often proceed by different mechanisms (basically S_N2 and S_N1 respectively) and hence may lead to different products. As in the case of substitutions by inorganic ligands, it is usually very difficult to substitute all the carbonyl groups in a metal carbonyl, and hence organo-substituted carbonyls have been isolated in large numbers. Typical examples of reactions involving displacement of carbon monoxide are the following:

$$\text{1,3,5-}Me_3C_6H_3 + Mo(CO)_6 \xrightarrow{160^\circ} (Me_3C_6H_3)Mo(CO)_3$$

$$C_5H_6 + Fe(CO)_5 \xrightarrow{130^\circ} (C_5H_6)Fe(CO)_3 \longrightarrow \pi\text{-}C_5H_5Fe(CO)_2H \xrightarrow{\text{heat}} [\pi\text{-}C_5H_5Fe(CO)_2]_2$$

$$\text{Butadiene} + Fe(CO)_5 \xrightarrow[\text{n-hexane}]{40^\circ} (H_2C{=}CH{-}CH{=}CH_2)Fe(CO)_4$$

$$\text{Butadiene} + Fe(CO)_5 \xrightarrow{140^\circ,\ 20\,\text{atm}} (CH_2{=}CH{-}CH{=}CH_2)Fe(CO)_3$$

21.2.2. From a metal salt, the ligand and a reducing agent. This is an extension of the method used frequently in the preparation of metal carbonyls. Some typical examples are:

$$CrCl_3 \xrightarrow[AlCl_3,\ Al]{C_6H_6} [(\pi\text{-}C_6H_6)_2Cr]^+ [AlCl_4]^-$$

$$[(\pi\text{-}C_6H_6)_2Cr]^+ \xrightarrow{S_2O_4^{2-}} (\pi\text{-}C_6H_6)_2Cr$$

$$[(\pi\text{-}C_6H_6)_2Cr]^+ \xrightarrow{H_2O} (\pi\text{-}C_6H_6)_2Cr + Cr^{2+}$$

$$cis\text{-}(Ph_3P)_2PtCl_2 \xrightarrow[N_2H_4,\ H_2O;\ EtOH]{trans\text{-}PhCH{=}CHPh} (Ph_3P)_2Pt \leftarrow \begin{matrix} Ph \quad H \\ C \\ \| \\ C \\ H \quad Ph \end{matrix}$$

$$(py)_2MnCl_2 \xrightarrow[\text{in dimethylformamide}]{C_5H_6,\ Mg,\ CO,\ H_2} (\pi\text{-}C_5H_5)Mn(CO)_3$$

A modification of this method is illustrated by

$$FeCl_3 \xrightarrow[\text{ether}]{C_5H_5MgBr} (\pi\text{-}C_5H_5)_2Fe$$

21.2.3. From a halide and an organometallic derivative of a non-transition metal. Alkali metal derivatives of hydrocarbons and Grignard reagents are particularly useful for reactions of this kind; for example,

$$TiCl_4 \xrightarrow[-80°,\ \text{ether}]{MeLi} TiMe_4 \text{ (decomposes at } -78°)$$

$$(Ph_3P)_2NiCl_2 \xrightarrow{C_6F_5Li} (Ph_3P)_2Ni(C_6F_5)_2$$

$$PtCl_4 \xrightarrow[\text{benzene}]{MeMgI} [Me_3PtI]_4$$

$$NiCl_2 + 2CH_2{=}CHCH_2MgBr \xrightarrow[-10°]{\text{ether}} (\pi\text{-}C_3H_5)_2Ni$$

(product drawn as Ni bonded to two allyl groups, $HC\langle^{CH_2}_{CH_2}$ and $H_2C{\cdots}CH{\cdots}H_2C$)

$$CrCl_2 \xrightarrow[\text{THF}]{NaC_5H_5} (\pi\text{-}C_5H_5)_2Cr$$

21.2.4. From complex anions of transition metals and halides. This method is closely related to the previous one; typical examples are:

$$Na[Mn(CO)_5] + MeI \longrightarrow MeMn(CO)_5$$

$$[\pi\text{-}C_5H_5Fe(CO)_2]_2 \xrightarrow[THF]{Na/Hg} Na[\pi\text{-}C_5H_5Fe(CO)_2] \xrightarrow{C_6F_6} C_6F_5Fe(CO)_2\pi\text{-}C_5H_5$$

$$Na[Mn(CO)_5] \xrightarrow{CH_2{=}CHCH_2Cl} (CO)_5MnCH_2CH{=}CH_2$$

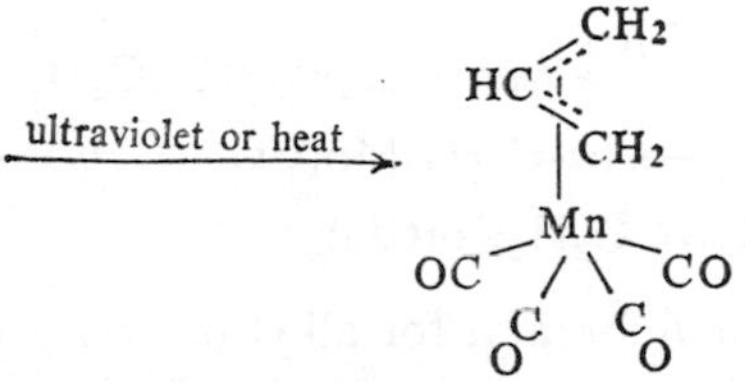

$$\xrightarrow{\text{ultraviolet or heat}} (\pi\text{-}C_3H_5)Mn(CO)_4$$

(structure: CH_2, HC, CH_2 π-allyl bonded to Mn bearing four CO groups)

Other syntheses are mentioned in the discussion of groups of complexes.

All organometallic compounds are thermodynamically unstable with respect to oxidation by air, and in many cases oxidation takes place at a significant rate at the ordinary temperature. Much preparative work therefore has to be carried out *in vacuo*, under nitrogen or under argon.

21.3 Derivatives of one-electron donors (σ-bonded compounds)[1, 2, 3]

Binary compounds of alkyls and transition metals (other than zinc, cadmium and mercury) are highly unstable; $TiMe_4$, for example, one of the very few such compounds yet characterized, decomposes above $-78°$. Alkyl titanium halides (which are important polymerization catalysts) and titanium aryls are slightly more stable with respect to thermal decomposition. They may be made by reactions such as:

$$TiCl_4 + Me_3Al \longrightarrow MeTiCl_3 + Me_2AlCl$$

(The alkyl aluminium halide is converted into the complex $Na[AlMe_2Cl_2]$ with sodium chloride, and the alkyl titanium compound can then be separated by distillation.)

$$TiCl_4 \xrightarrow[\text{ether, } -80°]{PhLi} Ph_4Ti \qquad WCl_6 \xrightarrow[\text{ether}]{Me\,Li} WMe_6$$

These substances are very readily oxidized by air, rapidly hydrolysed by water and decompose thermally above about −10°. Dimethylmanganese, obtained from manganese (II) iodide and methyl lithium, is similarly unstable and readily detonates; since it is insoluble in ether it is probably polymeric.

If, however, π bonding ligands such as carbon monoxide, π-cyclopentadienyl, cyanide and substituted phosphines are present in the molecule, stable alkyl and aryl derivatives of transition metals can easily be made by the general methods given in Sections 21.2.3 and 21.2.4, and also by insertion reactions such as:

$$trans\text{-}(Et_3P)_2PtClH \xrightarrow{C_2H_4} trans\text{-}(Et_3P)_2PtCl(C_2H_5)$$

$$(\pi\text{-}C_5H_5)Mo(CO)_3H + CH_2N_2 \rightarrow (\pi\text{-}C_5H_5)Mo(CO)_3(CH_3) + N_2$$

$$HCo(CO)_4 + C_2H_4 \rightleftharpoons C_2H_5Co(CO)_4$$

Substitution of fluoroalkyl or fluoroaryl for alkyl or aryl groups also increases the thermal stability of the compounds: $CF_3Co(CO)_4$, for example, can be distilled without decomposition at 91°, whereas $CH_3Co(CO)_4$ is stable only below −20°. Typical preparations of these complexes are illustrated by the reactions:

$$CF_3COCl + Na[Mn(CO)_5] \xrightarrow{THF} CF_3COMn(CO)_5 \xrightarrow[\text{ultraviolet}]{\text{heat or}} CF_3Mn(CO)_5 + CO$$

(Perfluoroalkyl halides do not give perfluoroalkyl derivatives on direct treatment with metal carbonylates.)

$$(CF_3)I + Fe(CO)_5 \xrightarrow{50°} (CF_3)Fe(CO)_4I + CO$$

$$(Et_3P)_2PtF_2 + 2C_2F_4 \rightarrow (Et_3P)_2Pt(C_2F_5)_2$$

$$2[Co(CN)_5]^{3-} + C_2F_4 \rightarrow [(NC)_5CoCF_2CF_2Co(CN)_5]^{6-}$$

$$Na[(\pi\text{-}C_5H_5)Fe(CO)_2] + C_6F_6 \rightarrow (\pi\text{-}C_5H_5)Fe(CO)_2(\sigma\text{-}C_6F_5) + NaF$$

Alkyl and perfluoroalkyl derivatives show considerable differences in chemical reactions as well as in thermal stability. For example, alkyl complexes often undergo insertion reactions, especially carbonylation:

$$MeMn(CO)_5 + CO \xrightarrow{\text{pressure}} MeCOMn(CO)_5$$

(Tracer studies using ^{14}CO show that the CO of the acetyl group formed is one of those originally present in $MeMn(CO)_5$.) Fluoroalkyl complexes do not usually give this reaction. Nor do cleavage reactions of the metal–carbon bond by hydrogen halides, for example,

$$cis\text{-}(Et_3P)_2PtMe_2 \xrightarrow{HCl} cis\text{-}(Et_3P)_2PtMeCl$$

take place with perfluoroalkyl derivatives.

The difficulty of discussing the enhanced stability of the perfluoroalkyl derivatives arises partly from uncertainty about the kind of stability (e.g. thermodynamic or kinetic, and with respect to what reactions) we are considering. Among qualitative factors which have been put forward in explanation of the experimental facts are:

(a) Withdrawal of electrons by the fluorine atoms makes the metal–carbon bond more polar, thus increasing the ionic contribution to the bond energy;
(b) The higher resulting positive charge on the metal atom contracts the *d* orbitals, leading to reduced non-bonding interactions and better overlap between metal bonding *d* orbitals and suitable ligand orbitals; and
(c) π bonding contributes to the metal–carbon bond in the perfluoroalkyl derivatives.

Some evidence in support of the last suggestion is provided by infrared spectroscopic and X-ray data. In $CF_3Mn(CO)_5$, for example, the C—F stretching frequency is about 100 cm^{-1} less than in CF_3X (where X = Cl, Br, I), and the Rh—C(C_2F_4) distance is 0·15 Å shorter than the Rh—C(C_2H_4) distance in Rh(acac)(C_2F_4)(C_2H_4). Further, in (π-C_5H_5)Rh(CO)(C_2F_5)I the Rh—C(C_2F_5) distance is only 0·11 Å larger than the Rh—C(CO) distance despite the different hybridizations of the carbon atoms attached to the metal, which would tend to make this difference larger.

A few points about the structures of individual alkyl complexes may be made here. Trimethylplatinum chloride is tetrameric, with the structure shown in Fig. 21.2. It was for some time believed that $(PtMe_4)_4$ had the same structure, but the compound investigated was actually $[Me_3Pt(OH)]_4$, and tetramethylplatinum has not yet been prepared. Another polymeric complex is $[(n\text{-}Pr)_2AuCN]_4$ (Fig. 21.3), which contains a twelve-membered planar ring. The cyclopentadienyl radical may function as a one-electron donor or a five-electron donor, sometimes in the same molecule. When (π-C_5H_5)Fe$(CO)_2$Cl reacts with NaC_5H_5, the compound having the structure shown in Fig. 21.4 is formed. The presence of a σ-cyclopentadienyl group is shown by the infrared spectrum of the compound and by X-ray diffraction.

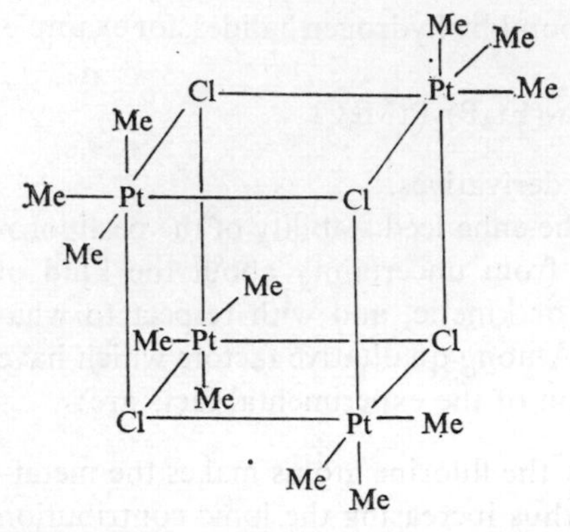

Figure 21.2

At ordinary temperatures, however, its proton magnetic resonance spectrum shows only a single resonance for the protons of the σ-bonded ring. At −80° this single resonance is split into three of relative intensities 1 : 2 : 2. It is deduced that the σ-bonded ring at

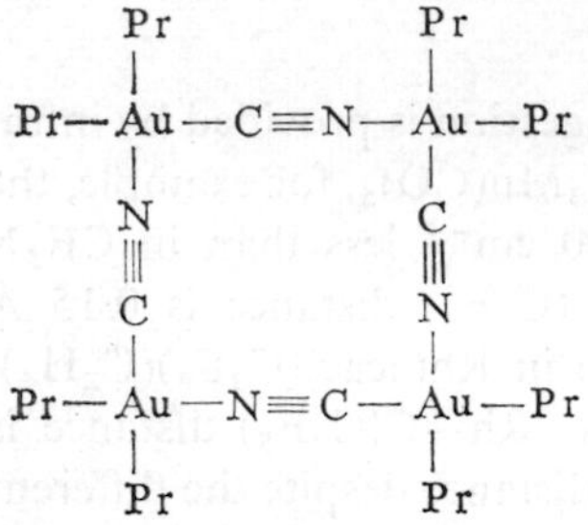

Figure 21.3

ordinary temperatures is rapidly changing the carbon atom bonded to the iron atom by the mechanism shown in Fig. 21.5, the activation energy for this process being about 13–26 kJ (3–6 kcal).

Finally, mention should be made of σ-derivatives of acetylenes

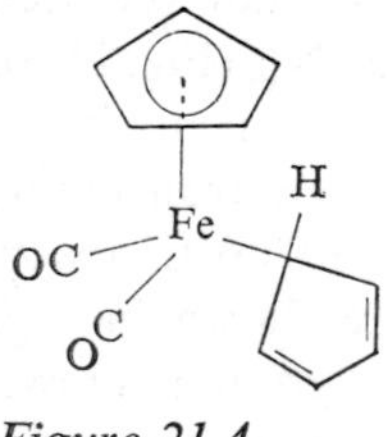

Figure 21.4

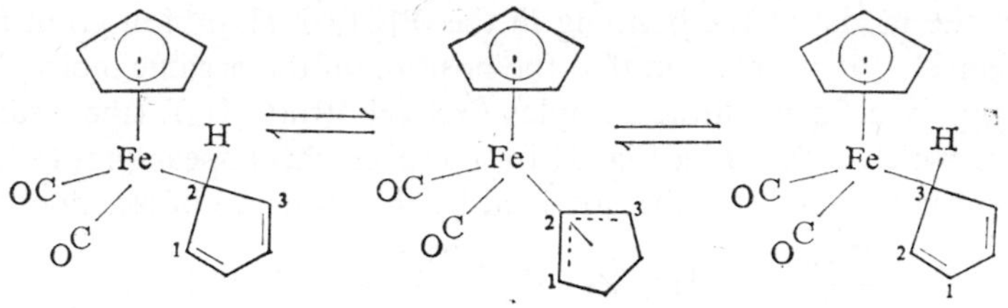

Figure 21.5

(alkynes).[1, 6] Acetylides of copper, silver and gold and a few other metals are obtained by interaction of salts of these metals and acetylenes or alkali metal acetylides (in the presence of reducing agents if the oxidation state of copper or gold is greater than one). Derivatives of acetylene itself are polymeric explosive solids; those of phenylacetylene are more stable; and those of acetylenes when tertiary phosphines or similar substances are also present as ligands are crystalline compounds which melt without decomposition. For the platinum complexes $(Et_3P)_2Pt(C{\equiv}CR)_2$ the C≡C stretching frequencies are a little lower (by 100–150 cm^{-1}) than those in the parent acetylenes; the C≡C distance in $(Et_3P)_2Ni(C{\equiv}CPh)_2$ of 1·18 Å is, however, very close to that in typical acetylenes. Copper, silver and gold complexes containing acetylide and tertiary phosphine are usually polymeric (e.g. $(PhC{\equiv}CCuPMe_3)_4$); X-ray diffraction studies indicate that in the polymeric units π bonding from the triple bond and metal–metal bonding, as well as σ bonding between the acetylide ion and the metal, are involved. For details of these complicated structures, however, a reference[1] should be consulted.

Acetylide ion in complexes such as $K_2[Ni(C{\equiv}CH)_4]$ is formally a two-electron donor isoelectronic with carbon monoxide and cyanide ion; compounds of this type were described briefly in Section 20.12.

21.4 Derivatives of potential two-electron donors (monoolefins)[1, 3, 7]

The preparation and structures of some complexes of monoolefins have already been described in Sections 21.1 and 21.2. Diolefins which are not conjugated act as chelating bifunctional two-electron donors and are dealt with in this section; conjugated diolefins are considered to be potential four-electron donors and are discussed in Section 21.6. Some unconjugated olefins isomerize to conjugated compounds on complex-formation, and other changes also occur (e.g. cyclohexene often yields π-cyclohexenyl complexes).

On the model of the bonding in the $[PtCl_3(C_2H_4)]^-$ ion given in Section 21.1, it might seem that the position of the organic molecule was rigidly defined. In the complex $(\pi\text{-}C_5H_5)Rh(C_2H_4)_2$ (the structure of which is shown in Fig. 21.6), however, this appears not to be so. This compound may be obtained by the sequence of reactions

$$RhCl_3 \xrightarrow[EtOH]{C_2H_4} [(C_2H_4)_2RhCl]_2 \xrightarrow{C_5H_5Na} (\pi\text{-}C_5H_5)Rh(C_2H_4)_2$$

The temperature-dependence of the proton magnetic resonance spectrum of this complex indicates that the ethylene molecules rotate

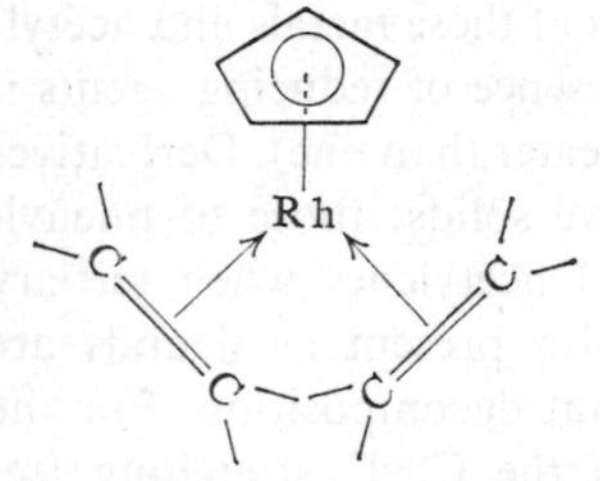

Figure 21.6

about the metal–ethylene σ bond axis with an energy barrier of about 25 kJ (6 kcal). The explanation of this observation is that ethylene in different positions of rotation can be bonded to the metal by d_{xz}, d_{yz} and d_{xy} orbitals during the operation of rotation and that the energy barrier is a measure only of the difference in the strength of the π bonding in different positions. The complex $(\pi\text{-}C_5H_5)Rh(C_2H_4)_2$ exchanges olefin with labelled ethylene only slowly, but $(acac)Rh(C_2H_4)_2$, which has a similar structure, exchanges rapidly. Ethylene in $[PtCl_3(C_2H_4)]^-$ and olefins in many other complexes also undergo rapid exchange reactions.

Tetrafluoroethylene forms complexes with several transition metals. For this ligand it would be expected that σ bonding (donor) properties should be much weaker and π bonding (acceptor) properties much stronger than in ethylene. Among typical preparations are those of $(C_2F_4)Fe(CO)_4$ from iron pentacarbonyl and tetrafluoroethylene, and of $(Ph_3P)_2Ni(C_2F_4)$ by displacement of ethylene from $(Ph_3P)_2Ni(C_2H_4)$. If in such species the π bond is regarded as mainly responsible for holding the ligand on to the metal, the best representation of the structure becomes:

F_2C–CF_2 bridged to $Fe(CO)_4$ or $(Ph_3P)_2Ni$ bridged to CF_2–CF_2

Support for such a formulation is provided in the case of the complex $(Ph_3P)_2Pt(CF_2{=}CFCF_3)$ or

$$(Ph_3P)_2Pt\begin{array}{l} \diagup CF_2 \\ \;\; | \\ \diagdown CF \\ \;\;\; | \\ \;\;\; CF_3 \end{array}$$

by the absence of any C=C stretching frequency in the infrared absorption spectrum, but a detailed X-ray diffraction study of one of these complexes has not yet been reported. In the analogous complex of tetracyanoethylene, however, the 'olefinic' carbon–carbon bond length is 1·51 Å, almost as long as for a typical single bond. In this case, therefore, the complex should really be formulated as a dialkyl derivative rather than an olefinic complex.

The mechanism of the palladium chloride catalysed oxidation of olefins to aldehydes or ketones is discussed in Section 21.9.

21.5 Derivatives of potential three-electron donors (π-allyl complexes)[1, 3]

The simplest three-electron donating ligand is the allyl group π-bonded to a transition metal by means of a conjugated system. This group may also, of course, function as a σ-donor in the same way as an alkyl group, and σ complexes are often formed as intermediates in the preparation of π-allyl complexes from metal halides and allyl Grignard reagents, or from anions and allyl halides. Representative reactions additional to the preparations of $(\pi\text{-}C_3H_5)_2Ni$ (from nickel chloride and allyl magnesium bromide) and $(\pi\text{-}C_3H_5)Mn(CO)_4$ (from sodium manganesepentacarbonylate and allyl chloride) mentioned in Section 21.2 are:

$$(\pi\text{-}C_5H_5)_2Ni + C_3H_5MgBr \rightarrow (\pi\text{-}C_5H_5)(\pi\text{-}C_3H_5)Ni$$

$$[Co(CN)_5]^{3-} \xrightarrow{C_3H_5MgBr} [(NC)_5CoCH_2CH{=}CH_2]^{3-} \underset{+CN^-}{\overset{-CN^-}{\rightleftharpoons}} [(NC)_4Co(\pi\text{-}C_3H_5)]^{2-}$$

Allylic (or enyl) complexes may also be obtained from olefins and metal halides or from allyl halides and metal carbonyls; for example,

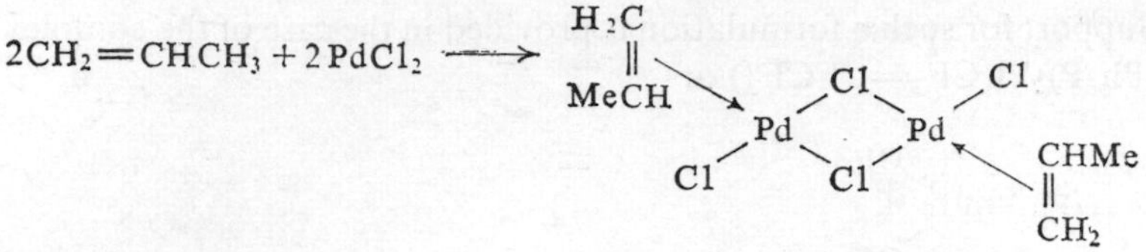

$$2CH_2{=}CHCH_3 + 2PdCl_2 \longrightarrow [(\text{MeCH}{=}\text{CH}_2)PdCl_2]_2$$

$$\xrightarrow{-2HCl} [\pi\text{-}C_3H_5PdCl]_2$$

$$CH_2{=}CHCH_2I + Fe(CO)_5 \longrightarrow (\pi\text{-}C_3H_5)Fe(CO)_3I$$

Palladium forms a number of π-allyl complexes stable in air.

Attempts to make perfluoroallyl complexes have not been successful; when alkali metalcarbonylates and perfluoroallyl chloride inter-

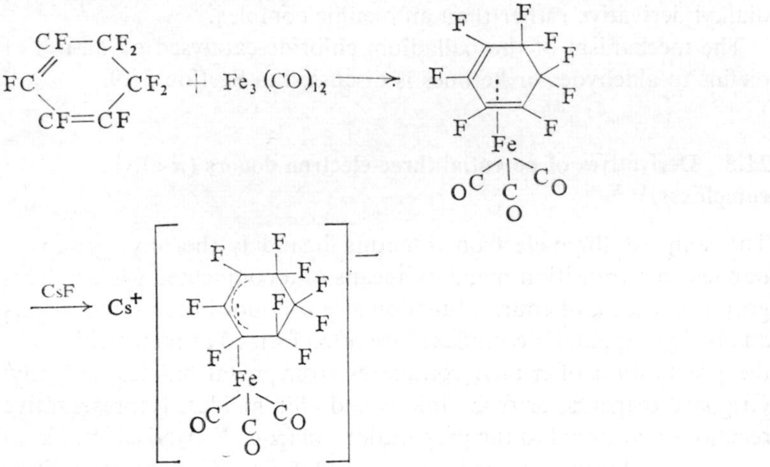

$$C_6F_8 + Fe_3(CO)_{12} \longrightarrow C_6F_8Fe(CO)_3 \xrightarrow{CsF} Cs^+[C_6F_9Fe(CO)_3]^-$$

Figure 21.7

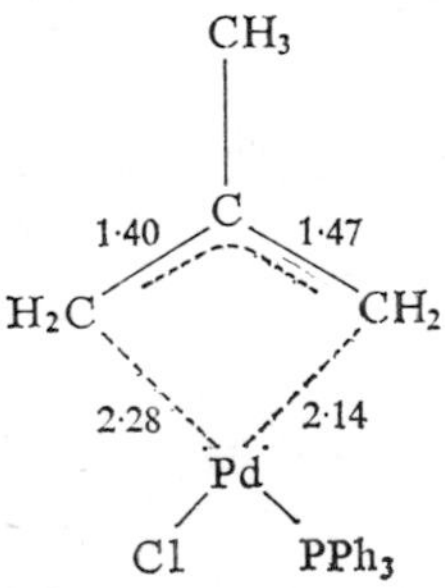

Figure 21.8

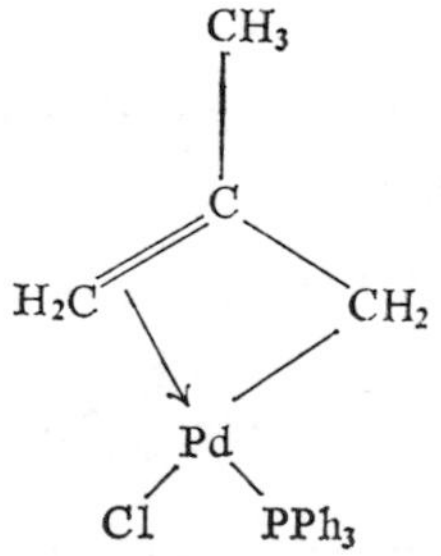

Figure 21.9

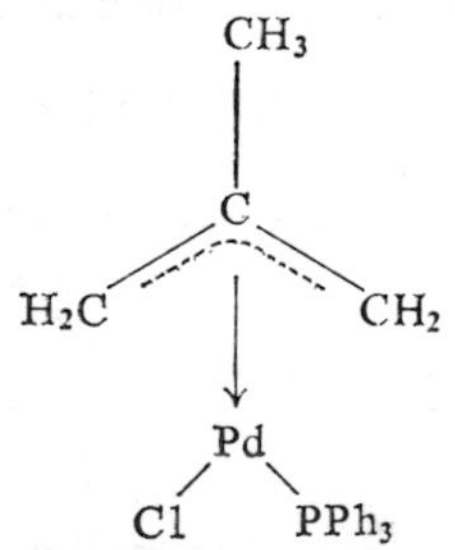

Figure 21.10

act, rearrangement takes place and σ-propenyl complexes are formed. Perfluoro-π-enyl complexes of more complicated systems are, however, known (for example, see Fig. 21.7).

The detailed structures of π-allyl complexes do not lend themselves to simple description. In $(\pi\text{-}C_3H_4CH_3)(Ph_3P)PdCl$ (Fig. 21.8), for example, obtained by the action of triphenylphosphine on $[(\pi\text{-}C_3H_4CH_3)PdCl]_2$, the allyl group appears from both proton magnetic resonance studies and X-ray diffraction to be bonded unsymmetrically to the metal atom, the observed C—C distances indicating a structure intermediate between those shown in Figs 21.9 and 21.10.

The proton magnetic resonance spectra of some π-allyl complexes exhibit temperature-dependence. Those of $(\pi\text{-}C_3H_5)_2M$ (M = Ni, Pd, Pt), for example, show three peaks of relative intensity 1 : 2 : 2 at low temperatures, corresponding to the protons numbered 1, 2 and 3 in Fig. 21.11. At high temperatures, however, there are

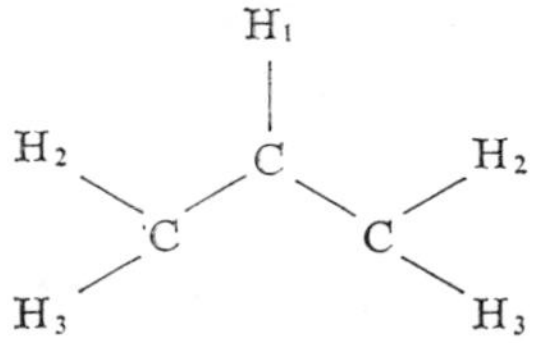

$$\begin{array}{c} H_1 \\ | \\ H_2\diagdown\quad C \quad\diagup H_2 \\ \quad C \diagup\quad\diagdown C \\ H_3\diagup\qquad\qquad\diagdown H_3 \end{array}$$

Figure 21.11

only two peaks, of relative intensity 1 : 4, showing that, within the relatively long time-scale of the proton magnetic resonance method, the *syn* and *anti* protons have become equivalent; this could occur by means of a series of $\pi \rightleftharpoons \sigma$ conversions or by rotation of the CH_2 groups about the C—C axis.

Complexes of a number of other π-enyl ligands are known, including two of the π-triphenylcyclopropenyl group; when nickel carbonyl reacts with triphenylcyclopropenyl bromide, for example, the following reaction occurs:

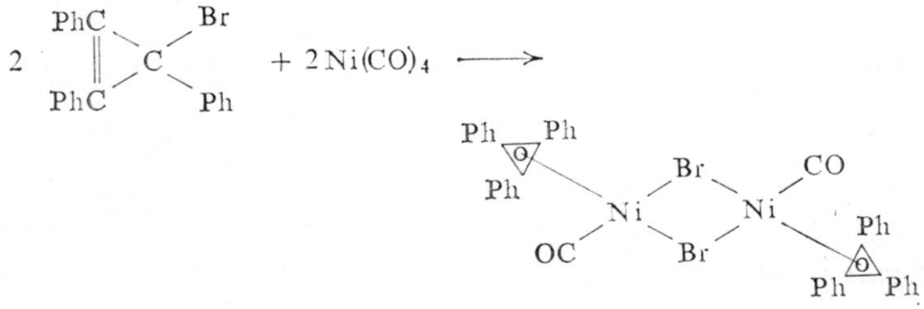

$$2\ \begin{array}{c} PhC \\ \| \\ PhC \end{array}\!\!\!>\!C\!<\!\!\begin{array}{c} Br \\ \\ Ph \end{array} + 2Ni(CO)_4 \longrightarrow [(\pi\text{-}Ph_3C_3)Ni(CO)Br]_2$$

(product: two Ni atoms bridged by two Br; each Ni bears one CO and one π-triphenylcyclopropenyl group, Ph at each ring carbon)

21.6 Derivatives of potential four-electron donors[1, 3]

Under this heading we shall consider briefly complexes of acetylenes, dienes and cyclic dienes. Since there are two π bonds in acetylene it should be feasible to complex a metal atom with each of them. The interaction of acetylenes and organometallic compounds frequently leads to the formation of complex compounds in which extensive changes have taken place (see, for example, Section 20.9), but a few relatively simple species containing an acetylene bonded to two metal atoms are known. Diphenylacetylene, for example, displaces the bridging carbonyls from $Co_2(CO)_8$ when these substances are heated together in benzene. As in the case of adducts of C_2F_4, however, the order of the carbon–carbon bond is changed considerably during the reaction, and its length in the product (1·46 Å) shows the structure of the product to be intermediate between those shown in Figs 21.12 and 21.13.

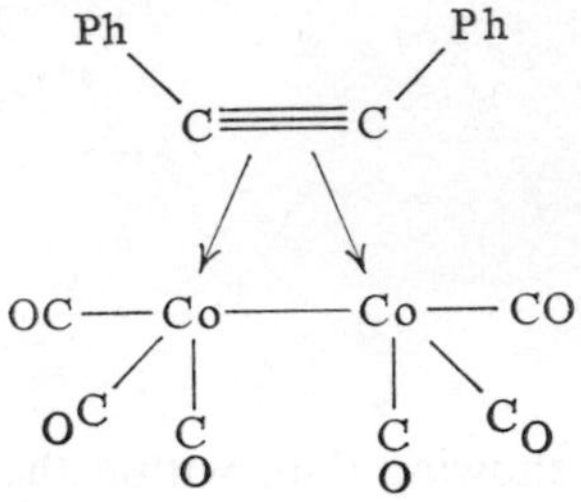

Figure 21.12

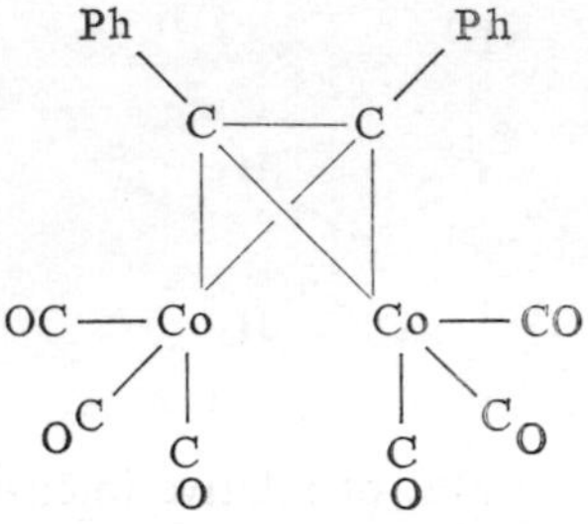

Figure 21.13

Butadiene iron tricarbonyl was first made in 1930, by the action of butadiene on iron pentacarbonyl under pressure. The product is a yellow-brown oil which is soluble in most organic compounds and is slowly oxidized by air. The complexed ligand is difficult to hydrogenate and does not give the Diels–Alder reaction. The structure of the complex is as shown in Fig. 21.14, with the ligand in the *cis* con-

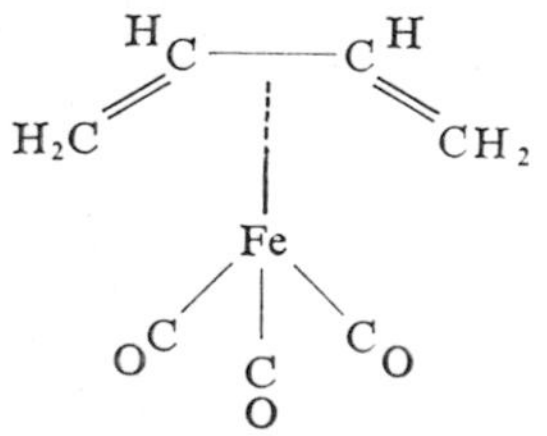

Figure 21.14

figuration, the plane of four carbon atoms being nearly parallel to that of the carbonyl groups. The metal atom is equidistant from each of the four carbon atoms in the diene and the carbon–carbon bond lengths are all 1·45 Å, suggesting a π complex rather than a complex containing two carbon–metal bonds. It is noteworthy that iron carbonyls isomerize non-conjugated dienes to form carbonyl complexes of the conjugated compounds.

In platinum complexes butadiene may function as a two-electron donor bonded to only one metal atom (as in $K[(C_4H_6)PtCl_3]$), or as a bridging ligand forming two-electron bonds to two metal atoms (as

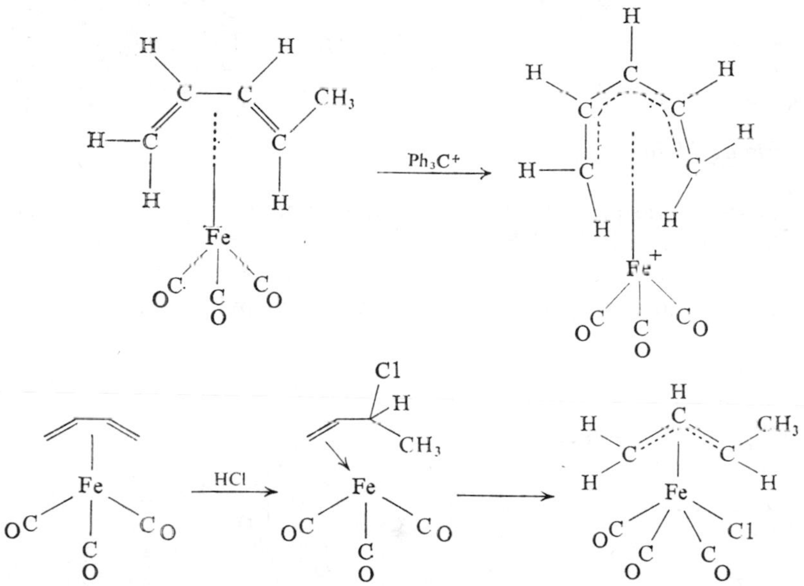

Figure 21.15

in $K_2[Cl_3PtC_4H_6PtCl_3]$); generally, however, butadiene and its substitution products are bonded to a single metal atom.

Hydride abstraction (e.g. by $Ph_3C^+BF_4^-$) or proton addition (e.g. by HCl) convert complexes of conjugated olefins into cations complexed by five-electron π-dienyl donors or π-enyl derivatives (complexes of three-electron donors) respectively, as shown in Fig. 21.15. Similar changes occur if olefin complexes are treated with the same reagents, the products then being π-allyl and alkyl complexes respectively.

Cyclobutadiene complexes are of particular interest in that their

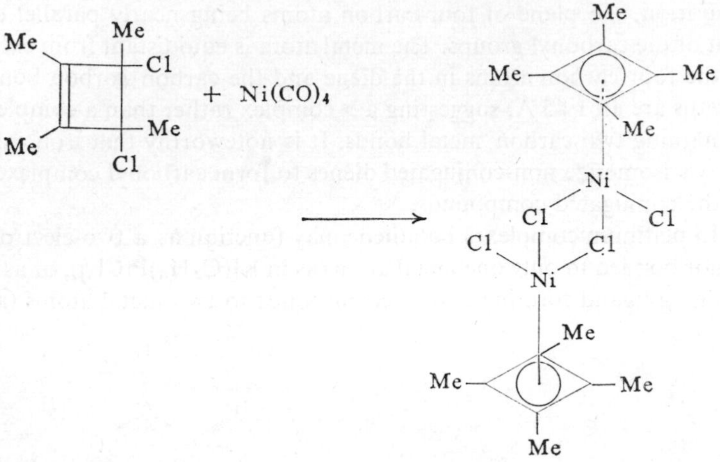

Figure 21.16

existence was predicted on theoretical grounds some years before the first one was isolated by the reaction shown in Fig. 21.16. Many other cyclobutadiene complexes are now known, among them the unsubstituted π-cyclobutadiene iron tricarbonyl prepared by the

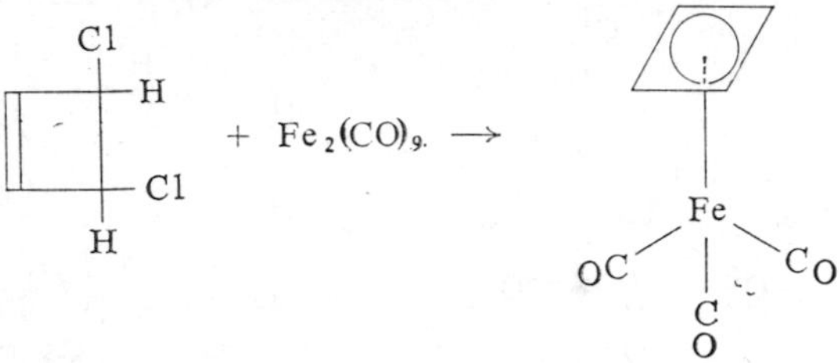

Figure 21.17

reaction shown in Fig. 21.17. The palladium complex of tetraphenylcyclobutadiene (which is analogous to the nickel complex mentioned above) is readily obtained from diphenylacetylene and palladium (II) chloride. It reacts with a wide range of metal carbonyls and π-cyclopentadienyl complexes, transferring one C_4Ph_4 group to another metal; for example,

$$[C_4Ph_4PdCl_2]_2 + (\pi\text{-}C_5H_5)_2Co \rightarrow (C_4Ph_4)Co(\pi\text{-}C_5H_5)$$

X-ray diffraction studies of complexes of methyl- and phenyl-substituted cyclobutadienes show that the C_4 ring is square-planar; the substituents are slightly displaced out of the plane of the C_4 ring on the side remote from the metal atom.

Cyclobutadiene iron tricarbonyl is a remarkably stable substance, and a number of electrophilic substitutions may be carried out on the ring; acetyl chloride and aluminium chloride give an acetyl derivative, and formaldehyde and hydrogen chloride introduce a chloromethyl group. When the complex is oxidized by cerium (IV) salts, the unstable free hydrocarbon is released and may be characterized by its reaction with substituted acetylenes to form 'Dewar isomers' of aromatic hydrocarbons (e.g. Fig. 21.18).

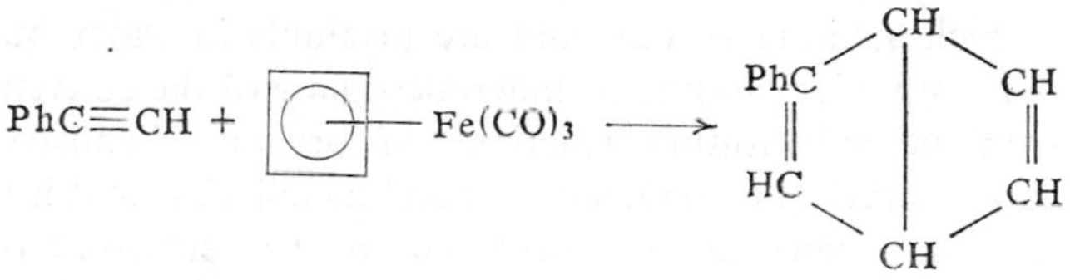

$$PhC{\equiv}CH + C_4H_4Fe(CO)_3 \longrightarrow$$

PhC, CH, CH, CH, HC, CH

Figure 21.18

Other potential four-electron donors include some substances in which one C=C bond is replaced by C=O (e.g. cinnamic aldehyde and acrolein). These substances have been shown to react with metal carbonyls to yield complexes, but the products have so far been little studied.

21.7 Derivatives of potential five-electron donors (including π-cyclopentadienyl complexes)[1, 3, 4, 8, 9]

Ligands formally contributing five electrons when bonded to a transition metal are called dienyl ligands; much the most important of them is the cyclopentadienyl group, C_5H_5 (to which we shall confine our attention), but derivatives of cyclohexa- and cycloheptadienyl groups, C_6H_7 and C_7H_8, and of the non-cyclic pentadienyl group, C_5H_5, are also known.

Cyclopentadiene itself has been known for over fifty years to form metallic derivatives with alkali metals; these, the cyclopentadienides, contain the $C_5H_5^-$ anion. Most of the compounds with which we are concerned in this section are essentially covalent derivatives in which the cyclopentadienyl group is π-bonded to a transition metal (the manganese derivative is a possible exception); but we shall also have occasion to mention some σ-cyclopentadienyl compounds in which one carbon atom of the ring is bonded to the metal by a two-electron bond. The π-cyclopentadienyl metal bond in many complexes has a high kinetic stability towards thermal decomposition and oxidation, and very large numbers of π-cyclopentadienyl derivatives are known. Alone among transition metal complexes of this kind, the parent bis-(π-cyclopentadienyl) metal complexes are known by trivial names analogous to that of ferrocene, $(\pi\text{-}C_5H_5)_2Fe$, the first of these compounds to be prepared.

The first two preparations of ferrocene were both accidental, the compound being obtained by the interaction of cyclopentadienyl magnesium iodide and iron (III) chloride, and by the action of cyclopentadiene on reduced iron at 300°. Ferrocene forms orange crystals (m.p. 174°) which sublime at 100° and are insoluble in water but soluble in many organic solvents. It undergoes many of the conventional electrophilic substitution reactions of organic chemistry, notably acylation, alkylation, halogenation and metallation; but it is more resistant than benzene to hydrogenation. The structure of ferrocene in the solid state is shown by X-ray diffraction to be a sandwich in which the two cyclopentadienyl groups are staggered with respect to one another; an electron diffraction study of the vapour shows that in this phase the rings are eclipsed. All C—C distances in the ring are $1{\cdot}40 \pm 0{\cdot}02$ Å, almost exactly the same as in benzene. The inter-ring distance is 3·32 Å, and the iron–carbon distance 2·04 Å. The compound is diamagnetic. The discussion of the bonding in ferrocene lies beyond the scope of this book, but it should be said that a molecular orbital treatment compatible with eighteen electrons in the nine available bonding orbitals accounts for the structure and physical properties such as ionization potential and magnetic susceptibility.

Ferrocene is thermally stable to almost 500° in the absence of oxygen, and it is unattacked by boiling sodium hydroxide solution or by hydrochloric acid. It is, however, easily oxidized (e.g. by silver ion or dilute nitric acid) to the blue ferricinium ion, $[(\pi\text{-}C_5H_5)_2Fe]^+$, several salts of which have been isolated. This oxidation is reversible, and E^0 for the system

$$[(\pi\text{-}C_5H_5)_2Fe]^+ + e \rightleftharpoons (\pi\text{-}C_5H_5)_2Fe$$

is about $+0{\cdot}5$ V.

All metals of the first transition series from titanium to nickel inclusive are now known to form neutral bis-(π-cyclopentadienyl) complexes, and those of vanadium, chromium, cobalt and nickel are isomorphous with ferrocene. Bis-(π-cyclopentadienyl) derivatives of zirconium, technetium, ruthenium, osmium, rhodium and iridium have also been prepared; those of ruthenium and osmium have been shown to have a structure similar to that of ferrocene, but with the rings in the eclipsed conformation.

The general method for the preparation of these compounds is by the action of freshly prepared sodium cyclopentadienide on a halide or other soluble salt in tetrahydrofuran, 1,2-dimethoxyethane or other ethers. Alternatively, the action of cyclopentadiene on a metal halide in a basic solvent (diethylamine or piperidine) may be used; for example,

$$NiCl_2 + 2C_5H_6 \xrightarrow{Et_2NH} (\pi\text{-}C_5H_5)_2Ni + 2Et_2NH_2Cl$$

The properties of manganocene are described later. The neutral complexes of all of the other first-row transition elements are more readily oxidized than ferrocene, the stability order being Ni $>$ Co $>$ V $\gg$ Cr $>$ Ti. If the rings in the two complexes are taken to be identical, the iron-ring and nickel-ring bond energies in ferrocene and nickelocene are 290 and 240 kJ (70 and 57 kcal) respectively. The ions $[(\pi\text{-}C_5H_5)_2Ni]^+$ and $[(\pi\text{-}C_5H_5)_2Co]^+$ are well established; the latter species is very resistant to reduction (it may be noted that the cobalt atom has the eighteen-electron configuration) and its derivatives are formed during the attempted preparation of cobaltocene unless reducing conditions are employed. It is also unaffected by hot concentrated sulphuric and nitric acids; presumably the positive charge hinders the usual electrophilic attack on the ring. Cobaltocene and nickelocene are both more prone to undergo reactions in which the aromatic character of one of the π-cyclopentadienyl rings is destroyed (e.g. Fig. 21.19).

Rhodocene, $(\pi\text{-}C_5H_5)_2Rh$, is even more easily oxidized than cobaltocene, and can be obtained only by the action of sodium vapour on compounds of the $[(\pi\text{-}C_5H_5)_2Rh]^+$ ion. It behaves as a radical and dimerizes rapidly at room temperature to give a product in which the atom attains the eighteen-electron configuration shown in Fig. 21.20. The iridium compound behaves similarly.

Manganocene has similar physical properties to ferrocene and it has the same structure (so, indeed, has magnesium cyclopentadienide) but it is rapidly hydrolysed by water and in solid solution in the magnesium compound it has a magnetic moment of 5·9 B.M., indicating the presence of five unpaired electrons. The same moment is found for the pink form of the compound, which is stable from 160°C

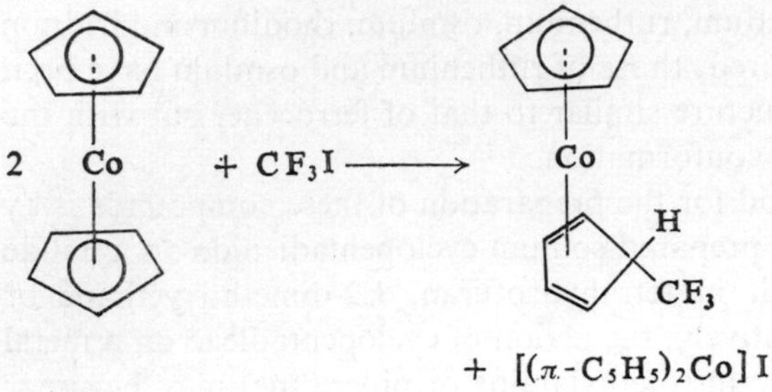

2 Co + CF_3I → Co (H, CF_3) + $[(\pi\text{-}C_5H_5)_2Co]I$

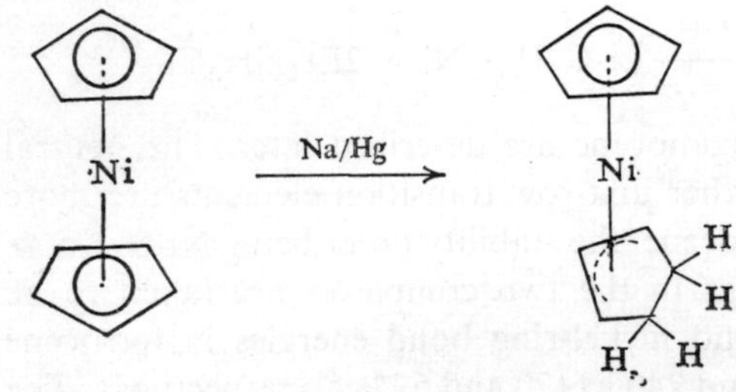

Ni $\xrightarrow{Na/Hg}$ Ni (H, H, H, H)

Figure 21.19

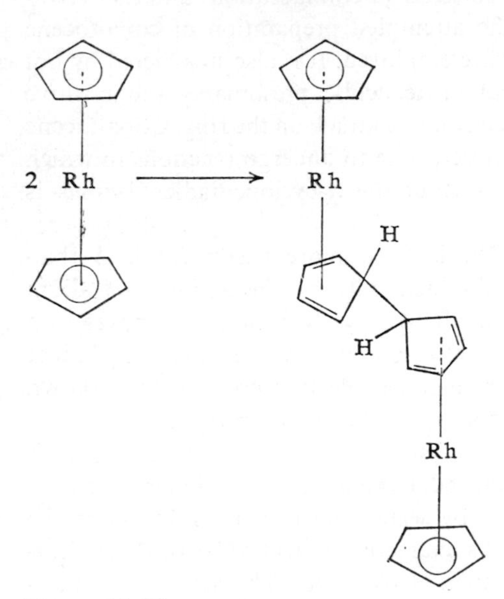

Figure 21.20

to the melting point of 173°C, but at lower temperatures the brown form is antiferromagnetic. The mass spectrum of the manganese and magnesium compounds are similar but different from those of other transition metal bis-(π-cyclopentadienyl) derivatives. On the basis of this evidence it is commonly accepted that manganocene is, like magnesium cyclopentadienide, an ionic solid; the structural similarity to ferrocene probably means no more than that the sandwich structure is the best way of packing the ions. Whilst these arguments are not compelling, they are at least strong evidence for some marked difference in bonding between manganocene and the formally analogous compounds of other metals of the first transition series.

A number of cyclopentadienyl derivatives of the lanthanides and actinides are known; they include $La(C_5H_5)_3$ and several other tris compounds of lanthanides, $Eu(C_5H_5)_2$, $Yb(C_5H_5)_2$, $Th(C_5H_5)_4$, $U(C_5H_5)_4$, $Np(C_5H_5)_4$, $Pu(C_5H_5)_3$ and $Am(C_5H_5)_3$. Their structures, however, are unknown, and we shall not discuss them further here.

Among complexes containing a π-cyclopentadienyl group and other ligands, the π-cyclopentadienyl carbonyl complexes are the most important. With very few exceptions, these compounds, unlike the π-cyclopentadienyl complexes themselves, obey the eighteen-electron rule, as may be seen from the formulae $(\pi\text{-}C_5H_5)V(CO)_4$, $[(\pi\text{-}C_5H_5)Cr(CO)_3]_2$, $(\pi\text{-}C_5H_5)Mn(CO)_3$, $[(\pi\text{-}C_5H_5)Fe(CO)_2]_2$, $(\pi\text{-}C_5H_5)Co(CO)_2$ and $[(\pi\text{-}C_5H_5)Ni(CO)]_2$; compounds of the same formula types are also formed by metals of the second and third transition series. A few cations and anions derived from binuclear species are also known (e.g. $[(\pi\text{-}C_5H_5)Cr(CO)_4]^+$ and $[(\pi\text{-}C_5H_5)Cr(CO)_3]^-$), together with a few more complex polynuclear species such as $[(\pi\text{-}C_5H_5)Fe(CO)]_4$, $[(\pi\text{-}C_5H_5)Rh(CO)]_3$ and $(\pi\text{-}C_5H_5)_3Ni_3(CO)_2$.

The π-cyclopentadienyl carbonyl complexes are usually made by the action of cyclopentadiene on metal carbonyls, or of carbon monoxide on π-cyclopentadienyl complexes. These reactions are not simple ligand replacements, since CO and $C_5H_5^-$ do not carry the same charge; in the reaction

$$2C_5H_6 + Co_2(CO)_8 \rightarrow 2(\pi\text{-}C_5H_5)Co(CO)_2 + 4CO + H_2$$

for example, the hydrogen produced from the cyclopentadiene reduces some of this substance. On the other hand, the method used for $[(\pi\text{-}C_5H_5)Ni(CO)]_2$ is a straightforward ligand exchange:

$$(\pi\text{-}C_5H_5)_2Ni + Ni(CO)_4 \rightarrow [(\pi\text{-}C_5H_5)Ni(CO)]_2 + 2CO$$

(Prolonged interaction in this case gives the trinuclear species mentioned earlier.) The complex π-methylcyclopentadienyl manganese tricarbonyl is made from the pyridine complex $(py)_2MnCl_2$, magnesium, methylcyclopentadiene and carbon monoxide; it is used, together with lead tetraalkyls, in antiknock mixtures. The methylcyclopentadienyl complex is a liquid and is therefore more suitable than the solid unsubstituted derivative.

The structures of π-cyclopentadienyl carbonyls pose the same problems as those of the metal carbonyls. X-ray diffraction shows that the dimeric iron and nickel compounds have two bridging carbonyl groups, their structures in the solid state being shown in Figs 21.21

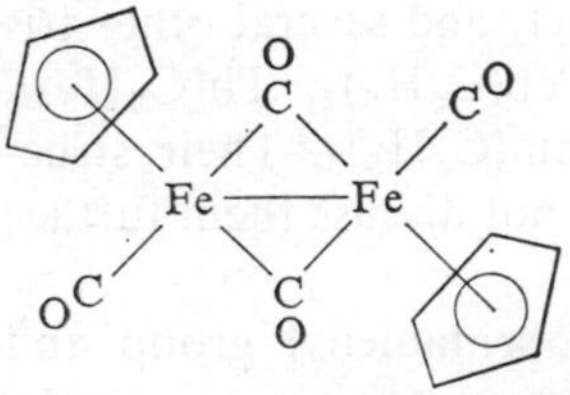

Figure 21.21

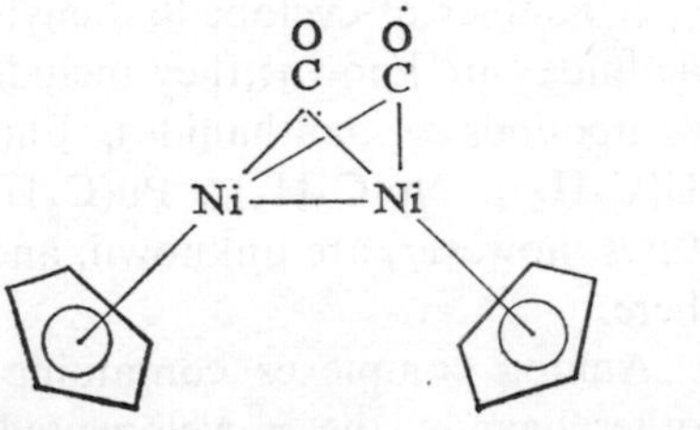

Figure 21.22

and 21.22; as in the metal carbonyls, the presence of a metal–metal bond is inferred from the diamagnetism and the interatomic distance, which in $[(\pi\text{-}C_5H_5)Fe(CO)_2]_2$ is almost the same as in $Fe_2(CO)_9$. The infrared spectrum of $[(\pi\text{-}C_5H_5)Cr(CO)_3]_2$, however, shows no bridging carbonyl stretching frequency, and the analogous molybdenum complex has been shown by X-ray analysis to be $(\pi\text{-}C_5H_5)(CO)_3Mo\text{–}Mo(CO)_3(\pi\text{-}C_5H_5)$. The osmium complex, $[(\pi\text{-}C_5H_5)Os(CO)_2]_2$, also shows no bridging carbonyl frequency in the infrared spectrum, whilst for the ruthenium complex the solid has a bridged structure. However, in cyclohexane at 30° bridged and unbridged forms are present in roughly equal proportions. At low temperatures in solution the *trans* form of $[\pi\text{-}C_5H_5Fe(CO)_2]_2$ is largely converted into the *cis* form; at room temperature and above the *cis* form is in equilibrium with the unbridged form, though the extent of the conversion is smaller than for the ruthenium complex under the same conditions. The increased preference for metal–metal bonding with increase in atomic number, which was found among the carbonyls, is therefore also found here; its most likely origin is the stronger metal–metal bonding in the heavier transition elements, as shown by their higher heats of atomization.

The structures of $[(\pi\text{-}C_5H_5)Rh(CO)]_3$ and $(\pi\text{-}C_5H_5)_3Ni_3(CO)_2$

(with all cyclopentadienyl rings omitted) are shown in Figs 21.23 and 21.24 respectively; the nickel compound involves carbonyl groups bridging three metal atoms.

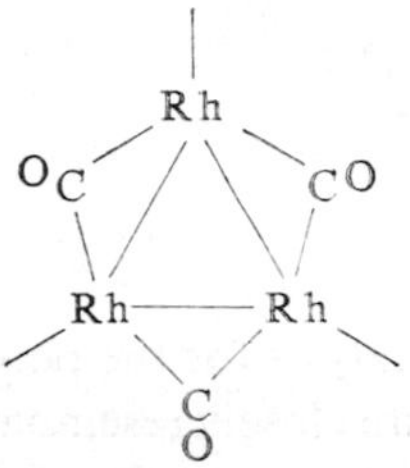

Figure 21.23

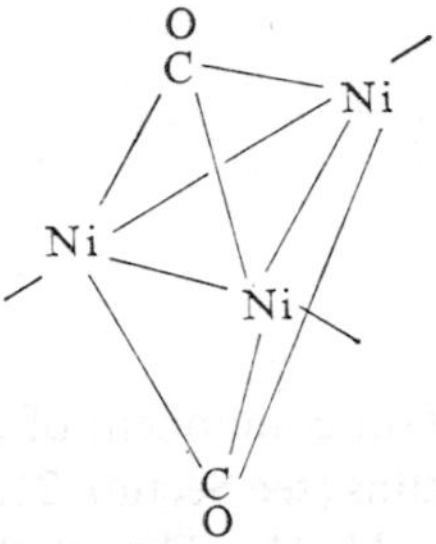

Figure 21.24

The (π-C_5H_5)$Mo(CO)_3$ and (π-C_5H_5)$Fe(CO)_2$ groupings are relatively inert, and a large number of compounds containing them have been prepared. A few representative reactions of the molybdenum complexes are shown in Fig. 21.25.

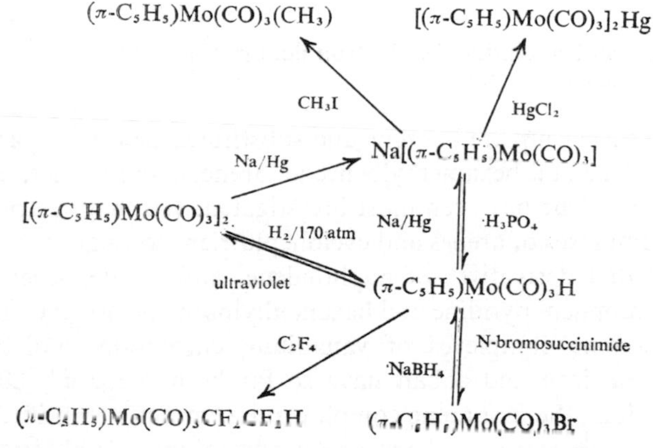

Figure 21.25

Other π-cyclopentadienyl metal compounds include hydrides, halides and nitrosyls. In (π-C_5H_5)$_2MoH_2$, obtained from molybdenum pentachloride, sodium cyclopentadienide and sodium borohydride in tetrahydrofuran, the two rings are inclined at an angle of 34°, the structure being shown in Fig. 21.26. The best-known halides are those of titanium, particularly (π-C_5H_5)$_2TiCl_2$, which is readily prepared from titanium tetrachloride and sodium cyclopentadienide,

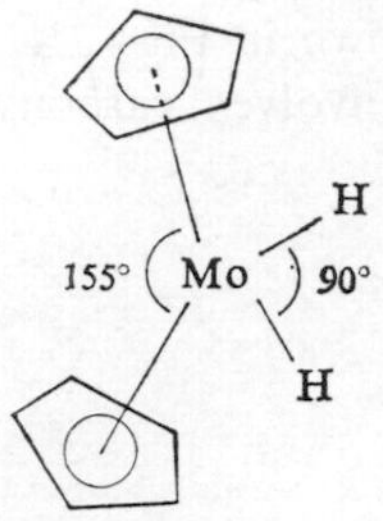

Figure 21.26

and is an important component of the Ziegler catalysts for the polymerization of olefins (see Section 21.9). Its structure closely resembles that of $(\pi\text{-}C_5H_5)_2MoH_2$. The most stable nitrosyl is that of nickel, $(\pi\text{-}C_5H_5)NiNO$, which is readily obtained by the action of nitric oxide on nickelocene or $[(\pi\text{-}C_5H_5)Ni(CO)]_2$.

Finally, we should remind the reader that there is a very large organic chemistry of ferrocene and some other π-cyclopentadienyl derivatives,[1, 10] but the discussion of substituted π-cyclopentadienyl derivatives lies outside the scope of this book.

21.8 Derivatives of potential six-electron donors (including benzene complexes)[1, 3, 4, 8, 9]

The ligands in this group are benzene and substituted benzenes, particularly mesitylene and hexamethylbenzene (arenes), and trienes, of which cycloheptatriene has been most investigated. We shall restrict attention to complexes of arenes and cycloheptatriene, though it may be mentioned that derivatives of naphthalene, anthracene, several other trienes, thiophen, pyridine and hexamethylborazine are known.

Neutral bis-arene complexes of vanadium, chromium, molybdenum, tungsten, iron and cobalt have so far been prepared. Dibenzene chromium, the first arene complex to be correctly identified, is best made by the action of benzene on chromium (III) chloride in the presence of aluminium and aluminium chloride, the resulting cation $[(\pi\text{-}C_6H_6)_2Cr]^+$ being reduced to the neutral complex by dithionite or allowed to disproportionate in the presence of aqueous alkali:

$$3CrCl_3 + 2Al + AlCl_3 + 6C_6H_6 \rightarrow 3[(\pi\text{-}C_6H_6)_2Cr][AlCl_4]$$

$$2[(\pi\text{-}C_6H_6)_2Cr]^+ + 4OH^- + S_2O_4{}^{2-} \rightarrow 2(\pi\text{-}C_6H_6)_2Cr + 2H_2O + 2SO_3{}^{2-}$$

or

$$2[(\pi\text{-}C_6H_6)_2Cr]^+ \xrightarrow{H_2O} (\pi\text{-}C_6H_6)_2Cr + 2C_6H_6 + Cr^{2+}$$

The vanadium, molybdenum and tungsten compounds are obtained by analogous methods. Alkyl substitution in the ring substantially increases stability, and for iron and cobalt only hexamethylbenzene derivatives have been prepared; conversion of the cations $[(\pi\text{-}C_6Me_6)_2Fe]^{2+}$ and $[(\pi\text{-}C_6Me_6)_2Co]^+$ to the neutral complexes requires powerful reducing agents such as sodium in liquid ammonia.

The neutral complexes are crystalline solids which sublime *in vacuo* at about 100°C and are mostly thermally stable up to 200–300°C. They are soluble in organic solvents but not in water. In both the vapour and the solid, dibenzene chromium is a sandwich molecule; all the C—C distances in the ring are 1·42 Å (slightly more than in benzene itself), and the Cr—C distances are 2·15 Å. The metal-ring bond energies in dibenzene chromium and dibenzene molybdenum are approximately 170 and 210 kJ (40 and 50 kcal) respectively. Most of these dibenzene complexes oxidize in air, the cobalt complex especially rapidly. Dibenzene chromium forms a number of 1 : 1 molecular complexes with acceptors such as tetracyanoethylene; such behaviour is in accord with its easy oxidation to salts of the $[(\pi\text{-}C_6H_6)_2Cr]^+$ ion. Salts of this ion are kinetically stable to further oxidation.

Arene metal carbonyls may be obtained by displacement of carbon monoxide from metal carbonyls; for example,

$$Cr(CO)_6 + C_6H_6 \rightarrow (\pi\text{-}C_6H_6)Cr(CO)_3 + 3CO$$

In contrast to dibenzene chromium, the product does not undergo oxidation in air. The electron-withdrawing effect of the $Cr(CO)_3$ group is shown by the low reactivity of the benzene in this compound towards electrophilic reagents; susceptibility to attack by nucleophilic reagents, on the other hand, is increased. Thus $(\pi\text{-}C_6H_5Cl)Cr(CO)_3$ is readily attacked by methoxide ion to yield $(\pi\text{-}C_6H_5OMe)Cr(CO)_3$.

The structure of benzene chromium tricarbonyl is shown in Fig. 21.27. The C—C distances in the ring are all 1·40 Å and the Cr—C (ring) distance is 2·25 Å, 0·1 Å more than in dibenzene chromium. Hexamethylbenzene chromium tricarbonyl has a similar structure;

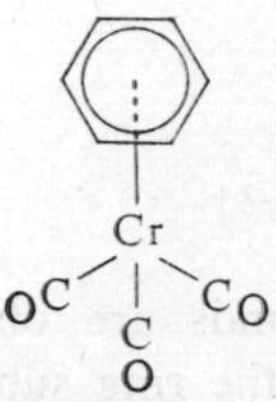

Figure 21.27

the methyl and ring carbons in the aromatic molecule are essentially coplanar.

It is interesting to note that aromatic complexes of chromium of the type described here were obtained some fifty years ago by Hein in an examination of the action of phenyl magnesium bromide on chromium (III) chloride, but were formulated as salts of poly (σ-bonded phenyl) chromium complexes. They are now recognized to be salts of π-benzene, π-biphenyl or bis-(π-biphenyl) chromium cations, probably formed via σ-bonded intermediates.

It may also be mentioned here that benzene forms 1 : 1 adducts with several silver (I) salts (e.g. $AgClO_4$, $AgBF_4$ and $AgAlCl_4$) and also forms a complex $C_6H_6CuAlCl_4$. In these compounds, however, the metal ion is near to only one bond in the ring, from which it is inferred that the aromatic hydrocarbon is here acting only as a two-electron donor.

Cycloheptatriene may act as a four-electron donor (in $C_7H_8Fe(CO)_3$) or as a six-electron donor. When it displaces benzene from (π-C_6H_6)$Mo(CO)_3$ it acts in the latter capacity; the structure of the resulting complex is shown in Fig. 21.28, the positions of the

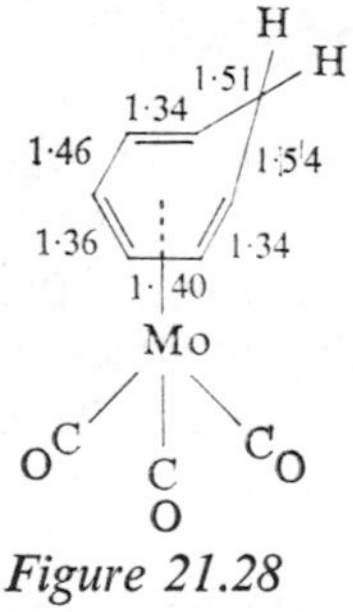

Figure 21.28

carbonyl groups when viewed from above the ring being staggered with respect to those of the double bonds. Within experimental error the distances between the metal atom and the six nearest ring carbon atoms are equal. The conversion of this cycloheptatriene complex into a cycloheptatrienyl (seven-electron donor) one is discussed in the next section.

21.9 Derivatives of potential seven-electron donors[1, 3]

The only actual seven-electron donor system to have been examined in any detail is the π-cycloheptatrienyl (C_7H_7) group. The neutral cycloheptatriene complex mentioned in the last section may be converted into a salt of the $[(\pi\text{-}C_7H_7)Mo(CO)_3]^+$ cation by the action of triphenylmethyl fluoroborate:

$$C_7H_8Mo(CO)_3 + Ph_3CBF_4 \rightarrow [(\pi\text{-}C_7H_7)Mo(CO)_3]^+BF_4^- + Ph_3CH$$

This change may be reversed by treatment with sodium borohydride. The π-benzene ligand may be displaced from $(\pi\text{-}C_5H_5)(\pi\text{-}C_6H_6)Cr$ (which is made from chromium (III) chloride by interaction with phenyl and cyclopentadienyl magnesium bromides) by the action of cycloheptatriene in the presence of aluminium chloride. In this case hydrogen is lost, and the resulting cationic complex can be reduced to $(\pi\text{-}C_5H_5)(\pi\text{-}C_7H_7)Cr$ by the same method as is used for the reduction of $[(\pi\text{-}C_6H_6)_2Cr]^+$ to $(\pi\text{-}C_6H_6)_2Cr$ (see Fig. 21.29). The

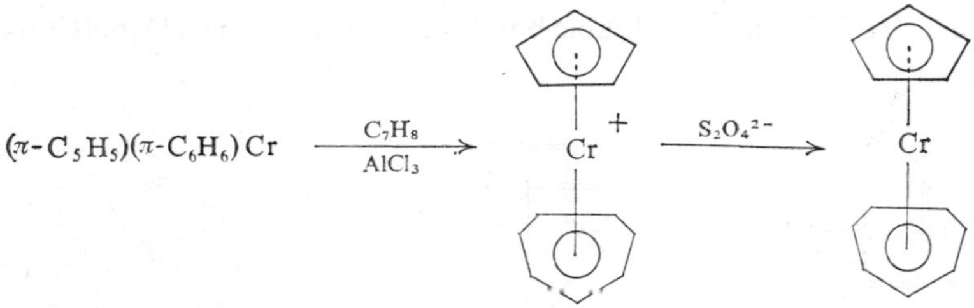

$$(\pi\text{-}C_5H_5)(\pi\text{-}C_6H_6)Cr \xrightarrow[AlCl_3]{C_7H_8} [(\pi\text{-}C_5H_5)(\pi\text{-}C_7H_7)Cr]^+ \xrightarrow{S_2O_4^{2-}} (\pi\text{-}C_5H_5)(\pi\text{-}C_7H_7)Cr$$

Figure 21.29

methylcycloheptatrienyl complex cation $[(\pi\text{-}C_5H_5)(\pi\text{-}C_7H_6CH_3)Cr]^+$ results from a ring expansion when $(\pi\text{-}C_5H_5)(\pi\text{-}C_6H_6)Cr$ reacts with acetyl chloride in the presence of aluminium chloride; it may be reduced to a neutral complex by dithionite.

No complex containing two π-cycloheptatrienyl groups bonded to the same metal atom is known at present, and it is noteworthy that all complexes containing one of these groups and a π-benzene or π-cyclopentadienyl group are derivatives of metals at the beginning of the transition series; corresponding complexes of later members of the series would, of course, have several electrons in antibonding orbitals. It should, however, be possible to obtain species such as $(\pi\text{-}C_3H_5)(\pi\text{-}C_7H_7)Fe$. It is interesting to note that when azulene (Fig. 21.30) forms complexes with metals it acts, so far as is known at present, as a 5 + 3 electron donor or a 5 + 5 electron donor.

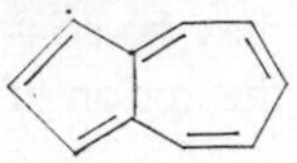

Figure 21.30

The only π-cycloheptatrienyl derivatives whose structures have yet been determined are $(\pi\text{-}C_5H_5)(\pi\text{-}C_7H_7)V$ and $(\pi\text{-}C_7H_7)V(CO)_3$, made from cycloheptatriene and $(\pi\text{-}C_5H_5)V(CO)_4$ and $V(CO)_6$ respectively. In each of these the seven-membered ring is planar and the carbon–carbon distances are equal within the experimental error. In general π-cycloheptatrienyl complexes closely resemble arene and π-cyclopentadienyl derivatives.

21.10 Derivatives of potential eight-electron donors[1]

Cyclooctatetrene (C_8H_8) acts as a ligand in several different ways. In C_8H_8CuCl only one double bond is coordinated; in $C_8H_8Fe(CO)_3$

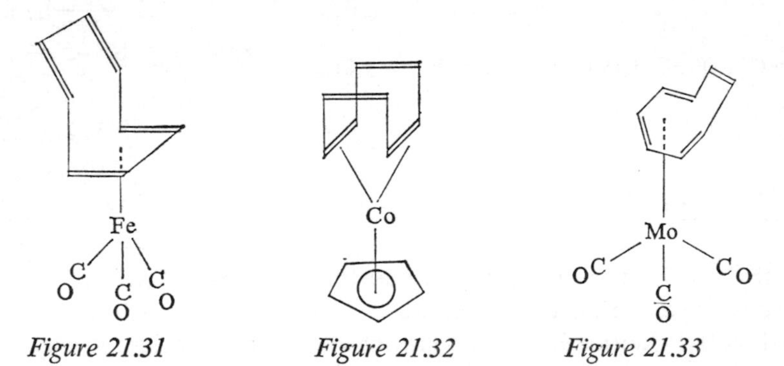

Figure 21.31 *Figure 21.32* 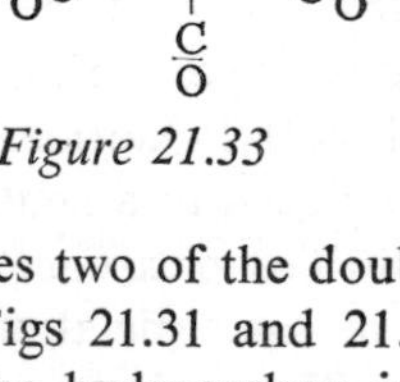 *Figure 21.33*

and $C_8H_8(\pi\text{-}C_5H_5)Co$ the coordination involves two of the double bonds and the structures are as shown in Figs 21.31 and 21.32 respectively; in $C_8H_8Mo(CO)_3$ (Fig. 21.33) the hydrocarbon is a six-electron ligand; and in $(C_8H_8)_2Fe$ one ligand acts as a six-electron, and one as a four-electron, donor. In $(C_8H_8)_2Ti$ (obtained from the alkoxide $Ti(OC_4H_9)_4$, $AlEt_3$ and cyclooctatetrene), however, one ring is symmetrically π-bonded; the other ring is in the boat form with only four carbon atoms involved in the π bonding. In this way the compound, like those mentioned earlier in this section, obeys the eighteen-electron rule.

Finally, we should mention that $U(C_8H_8)_2$, a derivative of $C_8H_8^{2-}$, formally a ten-electron donor, has recently been made by the interaction of $K_2C_8H_8$ and uranium tetrachloride; an X-ray diffraction study shows that the rings are planar and eclipsed, and that the C—C bond lengths are all equal. The mass spectrum, which shows high-intensity peaks corresponding to the ions $(C_8H_8)_2U^+$ and $(C_8H_8)U^+$, establishes that there is strong bonding between the metal and the rings, and it seems clear that in this instance the *f* orbitals of the uranium must be involved.[11]

21.11 Homogeneous catalysis by organometallic compounds of transition metals[1, 3, 12–14]

As stated in Section 21.1, some of the impetus of recent research in organometallic chemistry has come from the value of transition metal derivatives as catalysts, and we end this chapter with a few illustrations of homogeneous catalysis in which these substances are involved.

Since compounds containing organic ligands (or similar species such as triphenylphosphine) and hydrogen attached to the same metal play an important part in this subject, it is desirable first to remind the reader of the types of transition metal hydride derivatives now known.[15] The only complexes known at the present time in which hydrogen is the only ligand are $[TcH_9]^{2-}$ and $[ReH_9]^{2-}$, which are obtained by reduction of TcO_4^- and ReO_4^- with potassium in ethylenediamine; the tricapped trigonal prismatic structure of these ions is described in Chapter 8. Carbonyl hydrides were discussed in Section 20.7, a carbonyl halide hydride, $(Ph_3P)_2Ir(CO)(Cl)(H)_2$ (obtained by the action of hydrogen on 'Vaska's compound'), was mentioned in Section 20.8, and some cyclopentadienyl metal hydrides and carbonyl hydrides were described in Section 21.7. Other hydride species include those of tertiary phosphine metal complexes, derivatives of amine complexes and hydrido cyano complexes. A representative preparation of one member of each of these classes is illustrated below:

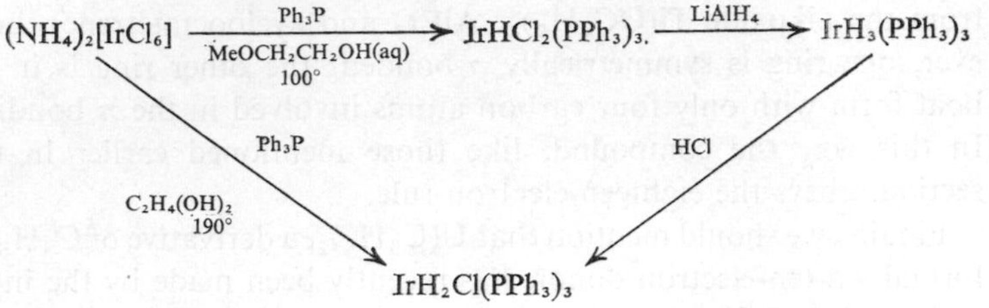

$$(NH_4)_2[IrCl_6] \xrightarrow[\substack{MeOCH_2CH_2OH(aq)\\100^\circ}]{Ph_3P} IrHCl_2(PPh_3)_3 \xrightarrow{LiAlH_4} IrH_3(PPh_3)_3$$

$$(NH_4)_2[IrCl_6] \xrightarrow[\substack{C_2H_4(OH)_2\\190^\circ}]{Ph_3P} IrH_2Cl(PPh_3)_3 \qquad IrH_3(PPh_3)_3 \xrightarrow{HCl} IrH_2Cl(PPh_3)_3$$

$$[Rh(en)_2Cl_2]^+ \xrightarrow[H_2O]{NaBH_4} [Rh(en)_2HCl]^+$$
(isolated as the BPh_4^- salt)

$$2[Co(CN)_5]^{3-} + H_2 \longrightarrow 2[Co(CN)_5H]^{3-}$$
(isolated as the Cs/Na salt)

Neutron diffraction, infrared spectroscopy (especially in conjunction with deuteration studies) and proton magnetic resonance spectroscopy are the structural techniques most used in the study of these hydride-containing species; all have been mentioned earlier. Sometimes, as in carbonylate anions containing hydrogen (Section 20.7), the hydrogen bridges two groups; other examples include borohydride and boron hydride complexes such as $(\pi\text{-}C_5H_5)_2TiH_2BH_2$ and $HMn_3(CO)_{10}(BH_3)_2$ and, probably, several species involved in catalytic reactions.

The two particular reactions of transition metal hydride complexes that are important in connection with catalysis are addition to olefins (often a reversible process) and addition to cyclic systems such as ethylene oxide and its derivatives, followed by a carbonyl insertion reaction:

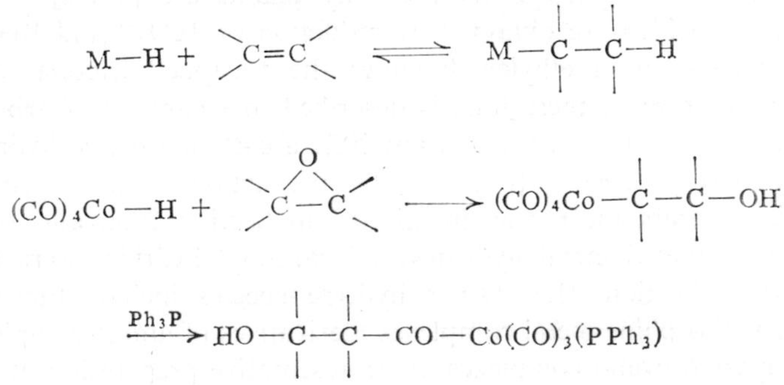

$$M{-}H + \rangle C{=}C\langle \rightleftharpoons M{-}\overset{|}{\underset{|}{C}}{-}\overset{|}{\underset{|}{C}}{-}H$$

$$(CO)_4Co{-}H + \rangle C\overset{O}{-}C\langle \longrightarrow (CO)_4Co{-}\overset{|}{\underset{|}{C}}{-}\overset{|}{\underset{|}{C}}{-}OH$$

$$\xrightarrow{Ph_3P} HO{-}\overset{|}{\underset{|}{C}}{-}\overset{|}{\underset{|}{C}}{-}CO{-}Co(CO)_3(PPh_3)$$

21.11.1. Isomerization reactions. There are many instances of the isomerization of olefins or olefinic compounds under the influence of transition metal compounds as catalysts. Some of these are pent-1-ene into *cis*- and *trans*-pent-2-ene ($RhCl_3$ in methanol or H_2PtCl_6 + $SnCl_2$), hex-1-ene into hex-2-enes and hex-3-enes ($Fe_3(CO)_{12}$), penta-1,4-diene into *trans*-penta-1,3-diene $Fe(CO)_3$ ($Fe(CO)_5$), allyl alcohol into propionaldehyde ($Fe(CO)_5$ or $HCo(CO)_4$), propylene oxide into acetone ($Co_2(CO)_8$ in methanol) and isopropylmagnesium bromide into the n-propyl compound ($TiCl_4$). In these processes it is believed that the double bond reacts with the transition metal complex, giving an intermediate which decomposes with regeneration of the double bond in a new position. Possible mechanisms for such changes are shown in Fig. 21.34 (a) and (b).

(a)

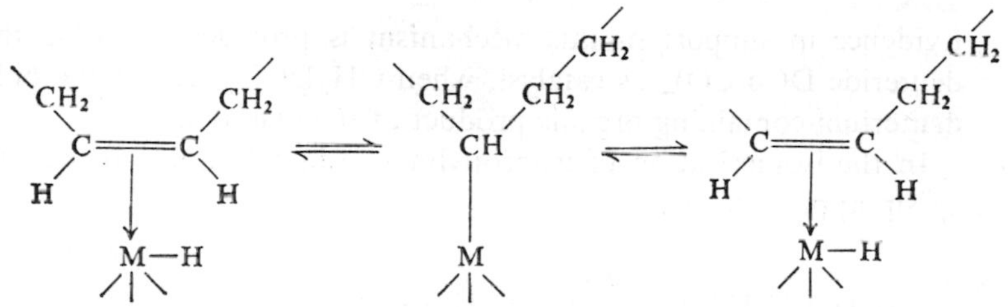

(b)

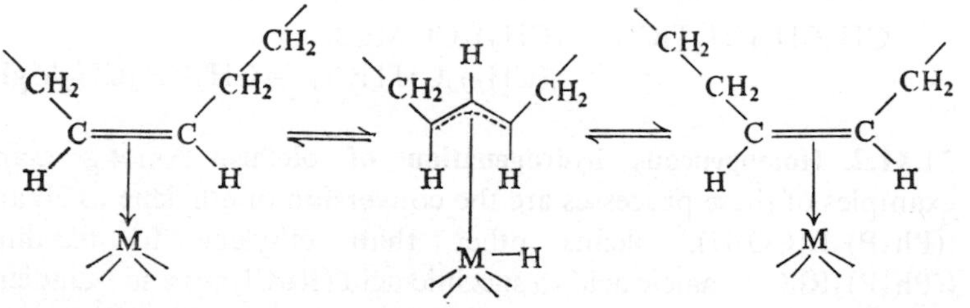

Figure 21.34

Isomerization normally leads to the isomer or mixture of isomers predicted by thermodynamic considerations. When a large amount of catalyst is present, however, it has to be remembered that complex formation between the various possible products and the catalyst has to be considered. Thus many catalysts will bring about the transformation of cycloocta-1,5-diene into the conjugated 1-3-diene, but the reverse change is effected by molybdenum carbonyl because only the 1,5-diene forms a $C_8H_{12}Mo(CO)_4$ complex.

For the isomerization of allyl alcohol to propionaldehyde by $HCo(CO)_4$ the following mechanism has been suggested:

$$CH_2{=}CHCH_2OH + HCo(CO)_4 \longrightarrow CH_2{=}CHCH_2OH + CO$$

$$\downarrow$$

$$HCo(CO)_3$$

$$\longrightarrow CH_3{-}\underset{\underset{\displaystyle Co(CO)_3}{|}}{\overset{\overset{\displaystyle H}{|}}{C}}{-}\underset{\underset{\displaystyle H}{|}}{\overset{\overset{\displaystyle H}{|}}{C}}{-}OH \longrightarrow CH_3CH{=}CHOH \longrightarrow CH_3CH_2CHO$$

$+ HCo(CO)_3$
(which reacts with more allyl alcohol)

Evidence in support of this mechanism is provided by using the deuteride $DCo(CO)_4$ as catalyst, when CH_2DCH_2CHO is the only deuterium-containing organic product of the reaction.

In the isomerization of isopropylmagnesium bromide the mechanism appears to be:

$$(CH_3)_2CHMgBr \xrightarrow{TiCl_4} (CH_3)_2CHTiCl_3 \rightarrow TiCl_3H + CH_3CH{=}CH_2$$

$$CH_3CH{=}CH_2 + TiCl_3H \rightleftharpoons CH_3CH_2CH_2TiCl_3$$

$$CH_3CH_2CH_2TiCl_3 + (CH_3)_2CHMgBr \rightleftharpoons (CH_3)_2CHTiCl_3 + CH_3CH_2CH_2MgBr$$

21.11.2. Homogeneous hydrogenation of olefins. Among many examples of these processes are the conversion of ethylene to ethane ($(Ph_3P)_3Ir(CO)H$), olefins other than ethylene to paraffins ($(Ph_3P)_3RhCl$), maleic acid to succinic acid ($RuCl_2aq$) and butadiene to *cis*- and *trans*-but-2-enes and but-1-ene ($[Co(CN)_5H]^{3-}$). The mode of action of $(Ph_3P)_3RhCl$ has been investigated in detail. In solution in benzene, which is a suitable solvent for the hydrogenation of olefins by this method, the complex loses a molecule of triphenylphosphine and readily and reversibly absorbs a molecule of hydrogen, forming the (solvated) complex $(Ph_3P)_2RhH_2Cl$, which rapidly reduces olefins, including ethylene. The solution of $(Ph_3P)_2RhCl$ will also absorb ethylene, yielding $(Ph_3P)_2Rh(C_2H_4)Cl$; this compound, however, does not react with hydrogen at an appreciable rate. Other olefins, by way of contrast, do not form complexes with

$(Ph_3P)_2RhCl$. Thus ethylene, by blocking the catalyst, prevents hydrogenation of itself or any other olefin.

The transition state in the interaction of the dihydride complex with ethylene appears to be as shown in Fig. 21.35. This breaks down

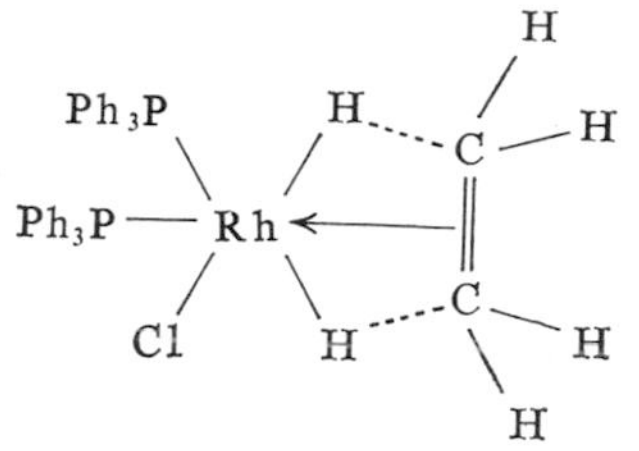

Figure 21.35

to $(Ph_3P)_2RhCl$, which can recombine with hydrogen, and ethane. If, for example, the catalyst is exposed to a mixture of hydrogen and deuterium and then ethylene, only C_2H_6 and $C_2H_4D_2$ (with no C_2H_5D) are formed. The delicate balance between composition and catalytic activity is shown by the fact that although the triphenylarsine and triphenylstibine analogues of the complex show similar reactions to the triphenylphosphine derivative, they are not nearly such good catalysts.

21.11.3. The dimerization, oligomerization and polymerization of olefins and acetylenes. This is a large subject and we can mention only a few of the processes which have been described: the dimerization of ethylene to butenes ($RhCl_3$ solution), the dimerization ($Ni(CO)_2[P(OPh)_3]_2$) or trimerization ($(\pi\text{-}C_3H_5)_2Ni$) of butadiene, the conversion of acetylene into benzene ($(Ph_3P)_2Ni(CO)_2$) or cyclooctatetrene ($Ni(CN)_2$), and the polymerization of olefins (by the Ziegler–Natta catalysts—see later).

The dimerization of ethylene has been studied in detail and appears to occur as follows:

$$C_2H_4 \xrightarrow[H_2O]{RhCl_3} CH_3CHO + [(C_2H_4)_2RhCl_2]^- \xrightarrow{+HCl}$$

$$[(CH_3CH_2)Rh(C_2H_4)Cl_3]^- \rightarrow [CH_3CH_2CH_2CH_2RhCl_3]^- \xrightarrow[-HCl]{C_2H_4}$$

$$[(C_2H_4)_2RhCl_2]^- + CH_3CH_2CH{=}CH_2$$

Whether butadiene dimerizes or trimerizes depends on the nickel catalyst employed: trimerization results only when all the groups attached to the catalyst are readily displaced. Thus butadiene in the presence of $Ni(CO)_2[P(OPh)_3]_2$ gives cycloocta-1,5-diene via an intermediate which may be represented as shown in Fig. 21.36.

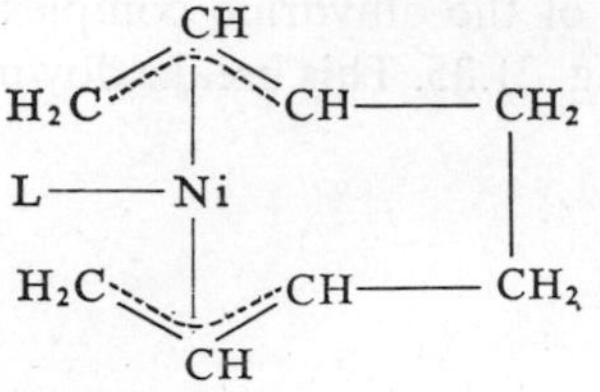

CH
H_2C CH—— CH_2
L—— Ni
H_2C CH—— CH_2
CH

Figure 21.36

Warming of this intermediate gives a L—Ni (cyclooctadiene) complex from which butadiene displaces cyclooctadiene; the process is therefore catalytic. If the original nickel complex contains only readily replaceable groups (e.g. (π-$C_3H_5)_2Ni$), treatment with butadiene gives a cyclododecatriene nickel complex with a formula as shown in Fig. 21.37. Butadiene displaces cyclododecatriene, and again a catalytic process results. Similar principles have been invoked to account for the behaviour of different catalysts in the polymerization of acetylene.

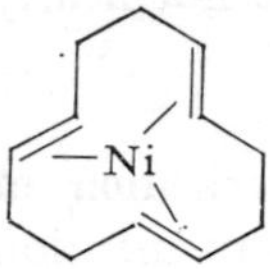

Ni

Figure 21.37

In the polymerization of ethylene and other simple olefins, aluminium alkyls are particularly effective catalysts. They may, however, be replaced by mixtures of dialkyl aluminium halides and transition metal compounds, especially those of titanium. Most such systems are heterogeneous catalysts, but there are a few homogeneous systems known (e.g. (π-$C_5H_5)_2TiCl_2Al(C_2H_5)_2$ in dichloromethane). This compound has the structure shown in Fig. 21.38. There is, however, as yet no generally accepted view on how the

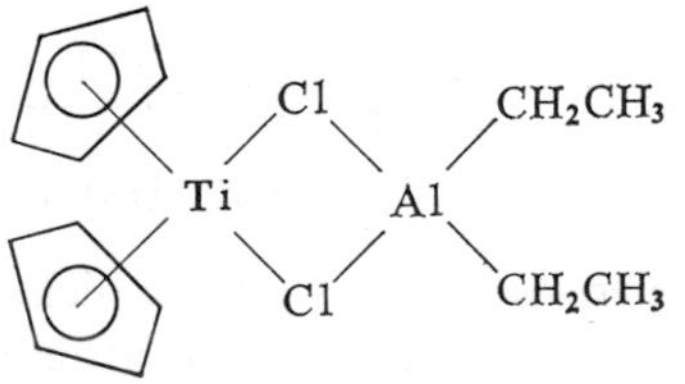

Cl CH_2CH_3
Ti Al
Cl CH_2CH_3

Figure 21.38

catalyst functions. Since neither of the related compounds, $(\pi\text{-}C_5H_5)_2TiCl_2Al(C_2H_5)Cl$ nor $(\pi\text{-}C_5H_5)_2TiCl_2AlCl_2$, is a catalyst, and choice of solvent is important, it seems that the requirements for catalytic activity in this process are very critical.

21.11.4. Oxidation of olefins. The aerial oxidation of olefins to aldehydes or ketones using palladium (II) as a catalyst is the Hoechst–Wacker process; it involves the formation and decomposition of the palladium analogue of the well-known $[PtCl_3(C_2H_4)]^-$ ion. We shall illustrate the process with reference to the conversion of ethylene into acetaldehyde. The palladium (O) formed on decomposition of the $[PdCl_3(C_2H_4)]^-$ ion is oxidized back to palladium (II) by copper (II) and, since the copper (I) produced can be reconverted into copper (II) by air, the whole reaction is catalytic with respect to palladium:

$$C_2H_4 + [PdCl_4]^{2-} + H_2O \rightarrow Pd^{\circ} + 2HCl + CH_3CHO + 2Cl^-$$
$$Pd^{\circ} + 2CuCl_2 + 4Cl^- \rightarrow [PdCl_4]^{2-} + 2[CuCl_2]^-$$
$$2[CuCl_2]^- + \tfrac{1}{2}O_2 + 2HCl \rightarrow 2CuCl_2 + H_2O + 2Cl^-$$

Overall

$$C_2H_4 + \tfrac{1}{2}O_2 \xrightarrow{PdCl_2/CuCl} CH_3CHO$$

The kinetics of this process have been studied in some detail. The rate law is

$$-\frac{d}{dt}[C_2H_4] = \frac{k[PdCl_4{}^{2-}][C_2H_4]}{[H^+][Cl^-]^2}$$

There is an isotope effect $k_H/k_D = 4{\cdot}0$ when the reaction is carried out in D_2O, but the effect on the rate of replacing C_2H_4 by C_2D_4 is very small. Only CH_3CHO is formed from C_2H_4 when the reaction is run in D_2O. These observations are interpreted in terms of the mechanism:[16]

$$C_2H_4 + [PdCl_4]^{2-} \rightleftharpoons [PdCl_3(C_2H_4)]^- + Cl^-$$
$$[PdCl_3(C_2H_4)]^- + H_2O \rightleftharpoons [Pd(C_2H_4)(OH_2)Cl_2] + Cl^-$$
$$[Pd(C_2H_4)(OH_2)Cl_2] \rightleftharpoons [Pd(C_2H_4)(OH)Cl_2]^- + H^+$$

$$(H_2C{=}CH_2)PdCl_2(OH) + H_2O \xrightarrow{slow} [HOCH_2CH_2{-}PdCl_2(OH_2)]^-$$

$$\left[\begin{array}{c} \text{Cl} \quad\quad\quad \text{H} \\ | \quad\quad\quad\quad | \\ \text{Cl—Pd—CH}_2\text{—C—O—H} \\ | \quad\quad\quad\quad | \\ \text{OH}_2 \quad\quad\quad \text{H} \end{array}\right]^- \xrightarrow{\text{fast}} 2Cl^- + Pd^0 + CH_3CHO + H_3O^+$$

The first three equilibria are labile, and the third is introduced because of the inverse dependence of the overall rate on the hydrogen ion concentration. The observed isotope effect on change of solvent is understandable from the fact that weak acids containing deuterium instead of hydrogen are generally weaker than their hydrogen analogues by a factor of about 4. The last step must be fast because of the near-identity in rates of oxidation of C_2H_4 and C_2D_4, and hence the addition of hydroxide ion to the coordinated olefin is the rate-determining step.

In acetic acid solution containing sodium acetate, vinyl acetate is formed, the first breakdown step now being:

$$\left[\begin{array}{c} \text{Cl} \\ | \\ \text{Cl—Pd—CH}_2\text{CH}_2\text{OAc} \\ | \\ \text{OAc} \\ \text{H} \end{array}\right]^- \longrightarrow 2Cl^- + HOAc + Pd^0 + H^+ + CH_2{=}CHOAc$$

21.11.5. Hydroformylation and related carboxylation reactions. The hydroformylation of olefins (the 'oxo' reaction) is their conversion into aldehydes by the action of hydrogen and carbon monoxide under pressure, usually in the presence of $Co_2(CO)_8$:

$$\gt C{=}C\lt + CO + H_2 \xrightarrow{Co_2(CO)_8} H{-}\overset{|}{\underset{|}{C}}{-}\overset{|}{\underset{|}{C}}{-}CHO$$

Temperatures used are 90–200°C and pressures are 100–400 atm. Under these conditions $Co_2(CO)_8$ is converted into $HCo(CO)_4$, and strong evidence for this as the important cobalt-containing species in the overall reaction is provided by its ability to bring about hydroformylation at ordinary temperatures and pressures. The rate of the overall reaction is first-order with respect to olefin concentration and approximately first-order with respect to the amount of cobalt present; it increases with increasing pressure of hydrogen and falls with increasing pressure of carbon monoxide.

The carbon monoxide insertion reaction

$$MeMn(CO)_5 + CO \longrightarrow MeCOMn(CO)_5$$

was mentioned in Section 21.3, when it was also stated that tracer studies show that the CO of the acetyl group was one of those originally present in methylmanganesepentacarbonyl. Since decarboxylation of acetylmanganese pentacarbonyl containing ^{14}CO in the acetyl group produces no ^{14}CO, reversal of the carboxylation reaction must also involve an intramolecular change, probably a migration of a methyl group from the carbonyl carbon atom to the metal in decarboxylation and the reverse process during the uptake of carbon monoxide. These reactions are closely related to hydroformylation, the mechanism of which can be envisaged as:[16]

$$HCo(CO)_4 \rightleftharpoons HCo(CO)_3 + CO$$
$$HCo(CO)_3 + C_2H_4 \rightleftharpoons HCo(CO)_3C_2H_4$$
$$HCo(CO)_3(C_2H_4) \rightarrow C_2H_5Co(CO)_3$$
$$C_2H_5Co(CO)_3 + CO \rightarrow C_2H_5Co(CO)_4$$
$$C_2H_5Co(CO)_4 \rightleftharpoons C_2H_5COCo(CO)_3$$

and either $C_2H_5COCo(CO)_3 + H_2 \rightarrow C_2H_5CHOCo(CO)_3H \rightarrow C_2H_5CHO + HCo(CO)_3$

or

$$C_2H_5COCo(CO)_3 + HCo(CO)_4 \rightarrow C_2H_5CHO + Co_2(CO)_7$$
$$Co_2(CO)_7 + H_2 \rightleftharpoons HCo(CO)_3 + HCo(CO)_4$$

This process is frequently followed by reduction of the aldehyde to an alcohol; some reduction of the olefin to a saturated hydrocarbon also usually takes place. For the reduction of propionaldehyde to n-propyl alcohol by hydrogen in the presence of $Co_2(CO)_8$, the rate law at 150° is

$$\text{Rate} = k[C_2H_5CHO][Co][H_2][CO]^{-2}$$

where [Co] represents cobalt in any form in the system. The inverse dependence upon $[CO]^2$ again suggests that $HCo(CO)_3$ rather than $HCo(CO)_4$ is the active entity, and further suggests that the cobalt atom must be only four-coordinated before reaction with hydrogen takes place. The mechanism is then

$$HCo(CO)_4 \rightleftharpoons HCo(CO)_3 + CO$$
$$HCo(CO)_3 + RCHO \rightleftharpoons RCH(OH)Co(CO)_3$$
$$RCH(OH)Co(CO)_3 + H_2 \rightarrow RCH_2OH + HCo(CO)_3$$

If the reaction

$$RCH(OH)Co(CO)_3 + CO \rightleftharpoons RCH(OH)Co(CO)_4$$

takes place and $RCH(OH)Co(CO)_4$ is unreactive towards H_2 or

$HCo(CO)_4$, this would account for the retarding effect of carbon monoxide being greater than would follow from the $HCo(CO)_4 \rightleftharpoons HCo(CO)_3 + CO$ equilibrium alone.

The reaction

$$CO + H_2O \xrightarrow[400^\circ]{Fe} CO_2 + H_2$$

is a well-known heterogeneous one. Nevertheless, it can be made to proceed by a homogeneous mechanism. At high pressures of carbon monoxide the ion $[HFe(CO)_4]^-$, produced by the reaction

$$Fe(CO)_5 + 3NaOH \longrightarrow NaHFe(CO)_4 + Na_2CO_3 + H_2O$$

can act as a catalyst. The reactive entity is the dimeric species $[HFe(CO)_4]_2{}^{2-}$, which can lose hydrogen according to the equation

$$[HFe(CO)_4]_2{}^{2-} \longrightarrow [Fe_2(CO)_8]^{2-} + H_2$$

The ion $[Fe_2(CO)_8]^{2-}$ then reacts with water and carbon monoxide:

$$[Fe_2(CO)_8]^{2-} + CO + H_2O \longrightarrow CO_2 + [HFe(CO)_4]_2{}^{2-}$$

This iron-containing system can also catalyse the hydroformylation reaction, though the usual product is an alcohol rather than an aldehyde. Nickel carbonyl will also catalyse some similar reactions such as the conversion of olefins to acids:

$$\gt C{=}C\lt + CO + H_2O \xrightarrow[250^\circ \text{ at } 200 \text{ atm}]{Ni(CO)_4} HC{-}C{-}COOH$$

Although the nickel carbonyl is converted into the metal as this reaction proceeds (indicating that it is the agent for the transfer of carbon monoxide), it may be regenerated by the action of acid under high pressures of carbon monoxide, and continuous operation is thus possible. Whilst it is likely that acyl nickel complexes are involved in the process, no compounds of this type have yet been isolated.

References for Chapter 21

1 Green, M. L. H., *Organometallic Compounds*, Vol. II (Methuen, 1968).
2 Coates, G. E., Green, M. L. H., Powell, P., and Wade, K., *Principles of Organometallic Chemistry* (Methuen, 1968).
3 Pauson, P. L., *Organometallic Chemistry* (Arnold, 1967).

4 King, R. B., *Transition Metal Organometallic Chemistry* (Academic Press, 1969).
5 Cotton, F. A., and Wilkinson, G., *Advanced Inorganic Chemistry*, 2nd edn. (Interscience, 1966).
6 Nast, R., *Chemical Society Special Publication* No. 13, p. 103 (1959).
7 Quinn, H. W., and Tsai, J. H., *Adv. Inorg. Chem. Radiochem.*, 1969, **12,** 217.
8 Fischer, E. O., and Fritz, H. P., *Adv. Inorg. Chem. Radiochem.*, 1959, **1,** 56.
9 Wilkinson, G., and Cotton, F. A., *Prog. Inorg. Chem.*, 1959, **1,** 1.
10 Rosenblum, M., *The Chemistry of the Iron Group Metallocenes* (Wiley, 1965).
11 Zalkin, A., and Raymond, K. N., *J. Amer. Chem. Soc.*, 1969, **91,** 5667.
12 Wender, I., and Pino, P. (eds.), *Organic Syntheses via Metal Carbonyls* (Interscience, 1968).
13 Bird, C. W., *Transition Metal Intermediates in Organic Synthesis* (Academic Press, 1966).
14 Candlin, J. P., Taylor, K. A., and Thompson, D. T., *Reactions of Transition Metal Complexes* (Elsevier, 1968).
15 Green, M. L. H., and Jones, D. J., *Adv. Inorg. Chem. Radiochem.*, 1965, **7,** 115.
16 Basolo, F., and Pearson, R. G., *Mechanisms of Inorganic Reactions*, 2nd edn. (Wiley, 1967).

22

Inner transition elements I: the lanthanides

22.1 Introduction[1-4]

Much of the revival of interest in the chemistry of the lanthanide or rare earth elements has arisen from the discovery of the actinides and the use of the earlier series as a model for the interpretation and prediction of the properties of members of the later one. There are, nevertheless, other features of general interest about this series of elements. Some of these concern the effects of a slow but steady decrease in atomic and ionic sizes with increase in atomic number (the so-called lanthanide contraction): the metals in their normal tripositive oxidation state resemble one another much more closely than members of any other series of elements, and the separation of some of the lanthanides was until recently a matter of great difficulty. Some elements of the series, however, exhibit oxidation states other than three, and the occurrence of these less common oxidation states can be partly correlated with electronic configurations, such correlation often being expressed in the statement that extra stability with respect to oxidation or reduction is associated with the configurations f^0, f^7 and f^{14}.

We have mentioned some aspects of the chemistry of the lanthanides in previous chapters: the structures of some compounds containing lanthanide ions of high coordination number in Chapter 14 and the magnetic properties of the tripositive ions in Chapter 16; in addition, the discussion of the equilibrium

$$MX_n(s) + \tfrac{1}{2}X_2 \rightleftharpoons MX_{n+1}(s)$$

in Chapter 18 is highly relevant to the question of unusual ionic oxidation states in the halides of these strongly electropositive elements. The metallic nature of LaI_2 and the interpretation of the role of ligands such as dipyridyl in stabilizing low oxidation states were also mentioned in Chapter 18. In this chapter we shall concentrate on the electronic configurations of the elements, the separation of their ions, their general chemical properties and their oxidation states. Strictly speaking, lanthanum is not a member of the series of fourteen inner transition elements, but just as we often found it desirable to mention calcium in discussing the metals of the first transition series, so we shall often consider lanthanum here along with the lanthanides proper. Yttrium, having an ionic radius of 0·88 Å (about equal to those of dysprosium and holmium), is often classed with the lanthanides, with which it occurs naturally; scandium forms a smaller ion, but like yttrium it has only one oxidation state, and it, too, should be mentioned here. The differences between scandium and the lanthanides are differences of degree; not surprisingly, scandium (ionic radius 0·68 Å) is more prone to complex formation, its hydroxide is a weaker base, its chloride more volatile and its nitrate more easily decomposed by heat, than the corresponding compounds of yttrium, lanthanum and the lanthanides. Although it forms compounds of formula ScO, $ScCl_{1\cdot5}$, $ScBr_{1\cdot5}$ and $ScI_{2\cdot2}$, none of these is a genuine derivative of an oxidation state lower than three; ScO and $ScI_{2\cdot2}$ are metallic conductors, and $ScCl_{1\cdot5}$ and $ScBr_{1\cdot5}$ appear to contain metal clusters. There are at present no features of special interest in the chemistry of scandium or (with the exception of a stable volatile complex $Cs^+[Y(CF_3COCHCOCF_3)_4]^-$) of yttrium, and in the rest of this chapter we shall restrict attention to lanthanum and the following fourteen elements.

22.2 Electronic configurations

The configurations of lanthanum and the following fourteen elements, determined by atomic spectroscopy, are given, together with the configurations of M^{2+}, M^{3+} and M^{4+} ions and the radii of the metals and M^{3+} ions, in Table 22.1. It should be said that since simple compounds of these elements often crystallize with complicated structures (e.g. in LaF_3 each cation has $5F^-$ at 2·36 Å and $6F^-$ at 2·70 Å, and in La_2O_3 each cation has $3O^{2-}$ at 2·38, $1O^{2-}$ at 2·45 and $3O^{2-}$ at 2·72 Å), the assignment of individual values is more than usually uncertain. Although, therefore, the relative magnitudes of the radii of the M^{3+} ions (which are based mainly on oxides) are certainly correct, absolute values (insofar as they can be said to exist) may be different from those given in Table 22.1.

The lanthanide contraction is attributed to the same factor as the

Table 22.1 *Electronic configurations and radii of lanthanum and the lanthanides*

Atomic number	Name	Symbol	Electronic configuration*				Radii	
			M	M^{2+}	M^{3+}	M^{4+}	M	M^{3+}
57	Lanthanum	La	$5d^1 6s^2$		[Xe]		1·87	1·061
58	Cerium	Ce	$4f^2 6s^2$		$4f^1$	[Xe]	1·83	1·034
59	Praseodymium	Pr	$4f^3 6s^2$		$4f^2$	$4f^1$	1·82	1·013
60	Neodymium	Nd	$4f^4 6s^2$	$4f^4$	$4f^3$	$4f^2$	1·81	0·995
61	Promethium	Pm	$4f^5 6s^2$		$4f^4$		—	0·979
62	Samarium	Sm	$4f^6 6s^2$	$4f^6$	$4f^5$		1·79	0·964
63	Europium	Eu	$4f^7 6s^2$	$4f^7$	$4f^6$		2·04	0·950
64	Gadolinium	Gd	$4f^7 5d^1 6s^2$		$4f^7$		1·80	0·938
65	Terbium	Tb	$4f^9 6s^2$		$4f^8$	$4f^7$	1·78	0·923
66	Dysprosium	Dy	$4f^{10} 6s^2$		$4f^9$	$4f^8$	1·77	0·908
67	Holmium	Ho	$4f^{11} 6s^2$		$4f^{10}$		1·76	0·894
68	Erbium	Er	$4f^{12} 6s^2$		$4f^{11}$		1·75	0·881
69	Thulium	Tm	$4f^{13} 6s^2$	$4f^{13}$	$4f^{12}$		1·74	0·869
70	Ytterbium	Yb	$4f^{14} 6s^2$	$4f^{14}$	$4f^{13}$		1·94	0·858
71	Lutecium	Lu	$4f^{14} 5d^1 6s^2$		$4f^{14}$		1·74	0·848

* Only electrons outside the [Xe] core are shown; data for M^{2+}, M^{3+} and M^{4+} relate to ions in solids.

decrease in size along a series of ordinary transition metal ions (i.e. the imperfect shielding of one electron by another in the same set of orbitals). As we go along the lanthanide series, the nuclear charge and number of $4f$ electrons both increase, but owing to the shapes of the f orbitals there is little screening from the nucleus of an electron in one f orbital by electrons in other f orbitals; hence each of these electrons experiences an increasing effective nuclear charge, and the ionic size decreases.

The importance of the lanthanide contraction is not confined to the chemistry of the inner transition elements. The interpolation of this series of metals between lanthanum and hafnium results in a near-identity of atomic and ionic radii of zirconium and hafnium, niobium and tantalum, molybdenum and tungsten, etc., as a consequence of which metals of the third transition series resemble the corresponding metals in the second series much more closely than the latter resemble the corresponding metals in the first series. There is, of course, a large difference in nuclear charge between, say, zirconium and hafnium, and the ionization potentials of these elements are substantially different; but properties which do not involve change of oxidation state (e.g. complex formation, solubilities of salts and structures of compounds) are very nearly the same for these two elements and for niobium and tantalum, etc.

The variation in atomic radii between lanthanum and lutecium is less regular than that in ionic radii. All the metals except europium crystallize in either the slightly distorted hexagonal close-packed or the cubic close-packed structure; europium has the body-centred cubic structure, so that instead of having twelve equidistant nearest neighbours it has eight (at 1·994 Å); the value quoted in Table 22.1 is an estimate for twelve-coordination. This minor structural change is, however, completely overshadowed by the fact that the atomic radii of europium and ytterbium are much larger than those of the other elements. These two metals are the two which form the most stable M^{2+} oxidation state, and it therefore seems that in the metals they contribute fewer electrons to the conduction band than the other lanthanide elements; this leads to weaker interatomic bonding, which finds expression not only in the larger atomic radii of these metals, but also in their lower heats of sublimation (180 and 150 kJ (43 and 36 kcal) for Eu and Yb respectively, compared with values of 250–460 kJ (60–110 kcal) for most other lanthanides). (It should, however, be said that samarium, which also has a low heat of sublimation (205 kJ, or 49 kcal) and a well-defined dipositive oxidation state, shows no anomaly in atomic radius.)

Promethium does not occur naturally in appreciable concentrations; information on this element is derived from studies of isotopes separated from fission products, the most stable of which is ^{147}Pm, which has a half-life of 2·6 years.

22.3 Separation of lanthanum and the lanthanides

The most important source of the lanthanide elements is monazite sand, predominantly phosphates of the lighter lanthanides but containing also thorium, silicon and the heavier lanthanides. Sulphuric acid takes the metals into solution as sulphates; after dilution, addition of a previously prepared mixture of lanthanide oxides raises the pH and results in precipitation of thorium and other tetrapositive metals (e.g. titanium and zirconium) that may be present, salts of these elements being more readily hydrolysed than those of tripositive lanthanides. Cerium may now be separated by precipitation with sodium hydroxide followed by drying in air at 100°; under these conditions cerium forms CeO_2, which is a weaker base than the trihydroxides and is not dissolved by dilute nitric acid. Alternatively, or subsequently, the general separation of the lanthanides from one another by the ion-exchange method may be employed.

A typical ion-exchange resin is a sulphonated polystyrene, which may be denoted HR, since it corresponds to an insoluble strong acid. When a solution containing the M^{3+} cations is passed down the column, the equilibrium

$$M^{3+}(aq) + 3HR(s) \rightleftharpoons MR_3(s) + 3H^+(aq)$$

is set up. The position of the equilibrium is found to lie slightly more to the right for the lighter lanthanides (i.e. those which have the larger unhydrated ions), but this effect is not large enough for a satisfactory separation. If, however, the column is eluted with a weakly acidic solution of a suitable complexing agent (e.g. citrate or EDTA) the partial separation brought about by preferential adsorption on the column is reinforced by the larger effect of preferential complex formation by the heavier (i.e. smaller) cations. Thus on treatment of the column with, for example, citrate at pH 5, the cations are displaced in reverse order of their atomic numbers.

(Further reference to this method is made in the following chapter in connexion with the separation of the actinides.)

In place of ion-exchange, separation by solvent extraction may be employed. Extraction by tri-*n*-butyl phosphate in kerosene from aqueous nitric acid solution occurs preferentially for heavier cations (i.e. those which form more stable complexes). Laborious fractionation methods such as fractional crystallization and fractional thermal decomposition of oxy-salts are now obsolete, but it is still instructive to note the principles on which such methods are based. Crystallization depends on a balance between the lattice energy of a solid and the sum of the solvation energies of the ions with which it is in equilibrium (see Section 6.2); thermal decomposition depends on the difference in lattice energy between the oxy-salt and the oxide, and this difference in turn depends on the size of the cation in a series of compounds of the same structure (see Chapters 11 and 12). The special merits of the ion-exchange and solvent-extraction methods are the relatively high separation factors (distribution coefficients between resin and citrate eluant, and between nitric acid and tributyl phosphate, differ by factors of about two between successive elements) and their suitability for continuous operation; they still depend on equilibria, the relative positions of which are functions of relative ionic sizes.

The separation of europium calls for special mention; because Eu^{2+} is much more stable with respect to oxidation than any other dipositive lanthanide ion in aqueous solution, it is easily separated by reduction of Eu^{3+} with zinc dust and coprecipitation with barium sulphate, from which it is separated by subsequent oxidation. The concentration of europium in monazite sand is, in fact, very low; the element occurs in higher concentrations in the dipositive state in some alkaline earth minerals (notably strontianite, $SrCO_3$), though owing to lack of demand it is not isolated from these sources.

22.4 Properties of lanthanum and the lanthanides: the tripositive oxidation state

All of the metals are obtained by reduction of halides or oxides. For the lighter lanthanides, reduction of the chlorides at 1000° or over with sodium or calcium is used; chlorides of the heavier lanthanides are too volatile for this method to be successful, and the fluorides are reduced with magnesium at high temperatures. It is, however, noteworthy that trihalides of samarium, europium and ytterbium are readily reduced only to the dihalides; these elements are obtained by reduction of their oxides with lanthanum.

The metals are silvery-white in appearance and very reactive, though the heavier ones, owing to the ready formation of an oxide coating, are kinetically less reactive than the lighter ones. For the reaction

$$M^{3+}(aq) + 3e \rightleftharpoons M$$

E^0 varies steadily from $-2{\cdot}52$ V for M = La to $-2{\cdot}25$ V for M = Lu; this is, of course, a very small variation and it shows that changes in the heats of sublimation are compensated by changes in ionization potentials and hydration energies. The metals combine with oxygen when heated, forming M_2O_3 except in the case of cerium, which forms MO_2. With fluorine, cerium, praseodymium and terbium give MF_4, and complex fluorides of the M^{IV} state may be prepared for some other elements (see the next section); other halogens give trihalides. With ammonia or nitrogen, nitrides MN are formed. The action of hydrogen at 300–400° gives hydrides MH_2 and MH_3 whose properties have been discussed in Chapter 8, and combination with carbon at high temperatures yields carbides MC_2. The C—C distance of 1·32 Å in LaC_2 (which has the CaC_2 structure), and the electrical conductivity of this compound, indicate that the carbide contains La^{3+}, a reduction product of the acetylide ion, and electrons in the conduction band (CaC_2, it should be noted, has a C—C distance of 1·20 in $C_2{}^{2-}$ and is an insulator). Europium and ytterbium dissolve in liquid ammonia to form M^{2+} and solvated electrons (which subsequently reduce the ammonia) in a manner similar to the alkali and alkaline earth metals.

The normal oxidation state of lanthanum and all the lanthanides is the tripositive one, and we shall describe compounds of the metals in this oxidation state first. Neither La^{3+} nor Lu^{3+} shows absorption bands in the ultraviolet or visible region, but all the other tripositive ions do so, an ion with n $4f$ electrons usually, though not invariably, having an absorption spectrum similar to that of the $(14\text{-}n)4f$ electron-containing species (e.g. Ce^{3+} and Yb^{3+} derivatives are colourless, Pr^{3+} and Tm^{3+} derivatives are green). The bands corresponding to these f—f transitions, being Laporte-forbidden if the species is centrosymmetric (see Chapter 17), are of low intensity. They are also very sharp (quite unlike those for d—d transitions) and very little affected by complexing agents; crystal field stabilization energy terms are only a small fraction of those for ions of the ordinary transition series. The apparently complex magnetic properties of the M^{3+} ions are satisfactorily accounted for on the basis of Russell–Saunders coupling, but since this subject was dealt with fully in Chapter 16 and is of specialized rather than general significance, we shall not go into it again here.

The trihydroxides of the elements are strong bases, $La(OH)_3$ being the strongest base of all hydroxides of tripositive ions; it absorbs atmospheric carbon dioxide. Solubility and basic strength decrease with increasing atomic number and decreasing ionic radius; heating $Yb(OH)_3$ and $Lu(OH)_3$ with concentrated NaOH solution gives $Na_3Yb(OH)_6$ and $Na_3Lu(OH)_6$, but the other hydroxides are not amphoteric. The trifluorides are sparingly soluble, even in dilute mineral acid; they do not form complexes in aqueous media. The chlorides, on the other hand, which are made by the action of carbonyl chloride, carbon tetrachloride or ammonium chloride on the oxides at moderate temperatures, are usually deliquescent. Among oxy-salts, nitrates and sulphates are soluble, and carbonates, phosphates and oxalates only sparingly soluble, in water. The insolubility of their oxalates in dilute mineral acid is a characteristic feature of the lanthanides. In hydrated salts the ions are usually associated with more than six molecules of water (e.g. $[Nd(H_2O)_9]^{3+}$), and hence, for example, the sulphates do not form alums. Large numbers of double salts are, however, formed, and some of these (such as those of formula $M_2(SO_4)_3.3Na_2SO_4.12H_2O$) have been used for the partial separation of the lanthanides, since the compounds of the heavier lanthanides are considerably more soluble than those of the lighter metals.

Although it is often stated that the tripositive lanthanide ions form few complexes, this statement is not really correct. Complex formation by these ions in aqueous solutions produces no striking change in properties, and many anions will not form complexes at all in aqueous media—but this often means only that they cannot compete successfully with water as a ligand. Complexes of the types $[M(en)_4]X_3$ and $Et_4N[M(S_2CNEt_2)_4]$, in which the metal ions are completely coordinated by nitrogen and sulphur respectively, can be obtained in non-aqueous media. It is, nevertheless, true to say that all the most stable complexes involve chelating ligands such as acetylacetonate, citrate and ethylenediaminetetraacetate.

As we have already indicated, the crystal structures of even binary compounds of the lanthanides tend to be complicated, and we shall not describe them in detail here. It is, however, interesting to note that compounds of ions of the heavier metals often adopt structures having lower coordination numbers than the corresponding compounds of the lighter metals: the chlorides $LaCl_3$–$GdCl_3$ inclusive, for instance, have the UCl_3 layer structure, in which the metal is nine-coordinated, whilst $DyCl_3$–$LuCl_3$ inclusive have the $AlCl_3$ layer structure with six-coordination for the metal. Similarly, the oxides M_2O_3 have seven-coordinated metal atoms for La_2O_3, and six-coordinated metal atoms for Eu_2O_3–Lu_2O_3 inclusive; Ce_2O_3–Sm_2O_3 inclusive are dimorphic. Although in LaF_3 the La^{3+} ion is best described as eleven-coordinated (5F at 2·36 Å forming a trigonal bipyramid round the cation plus 6F at 2·70 Å forming a trigonal prism around it), La^{3+} eight-coordinated by fluoride ion occurs in $KLaF_4$, which has a disordered fluorite structure. For further details of the structures of these and similar compounds the account by Wells[5] should be consulted.

22.5 The dipositive oxidation state

The majority of the lanthanides form at least one halide of formula MX_2, but if M is lanthanum, cerium, praseodymium or gadolinium the halide is a metallic conductor and hence should be formulated $M^{3+}(X^-)_2(e)$. Up to the present time only neodymium, samarium, europium, dysprosium, thulium and ytterbium have been found to form saline dihalides, and of these elements only samarium, europium and ytterbium have an extensive chemistry in this oxidation state. For reasons explained in Chapter 18, the position of the equilibrium

$$MX_n(s) \rightleftharpoons MX_{n-1}(s) + \tfrac{1}{2}X_2$$

is most likely to lie on the right-hand side when X = I, and the iodides are the most easily obtained compounds of these metals in the dipositive state. When M = Eu, thermal decomposition of the trichloride, -bromide or -iodide gives a dihalide, but for M = Sm or Yb only the triiodide decomposes, the trichloride and tribromide requiring reduction with hydrogen at a high temperature.

There is, however, another route to the dihalides by the reverse of the disproportionation reaction which is their normal mode of decomposition:

$$2MX_3(s) + M(s) \rightleftharpoons 3MX_2(s)$$

This is the method which has been used at about 500° for the preparation of $NdCl_2$ and NdI_2, $DyCl_2$, and TmI_2, the only compounds of neodymium (II), dysprosium (II) and thulium (II) yet known. For all three of these elements this oxidation state is very powerfully reducing and cannot exist in aqueous media. The structures of $NdCl_2$ and TmI_2 are known, the compounds being isomorphous with $PbCl_2$ and CdI_2 respectively.

Samarium, europium and ytterbium can be obtained in the dipositive state in aqueous solution by electrolytic reduction, reduction with an alkali metal amalgam or (in the case of europium only) metallic iron or zinc. The Eu^{3+}/Eu^{2+} half-cell has a standard potential of $-0{\cdot}43$ V, very close to that of the Cr^{3+}/Cr^{2+} system; semiquantitative estimates of E^0 for Sm^{3+}/Sm^{2+} and Yb^{3+}/Yb^{2+} give values of $-1{\cdot}5$ and $-1{\cdot}1$ V respectively. It follows from these values that solutions of europium (II) can be used for the preparation of many europium (II) compounds, but that solutions of samarium (II) and ytterbium (II) are so unstable with respect to decomposition of water that pure samarium (II) and ytterbium (II) compounds are unlikely to be obtained in aqueous media. Europium (II) has recently attracted some attention as a one-electron reducing agent.

In addition to dihalides, lower oxides of samarium, europium and ytterbium can be obtained, reduction of M_2O_3 by the metal itself or by lanthanum being the method employed. In their structural chemistry, Sm^{2+}, Eu^{2+} and Yb^{2+} (ionic radii 1·11, 1·10 and 0·93 Å respectively) closely resemble Ca^{2+}, Sr^{2+} and Ba^{2+}: the monoxides all have the NaCl structure; SmF_2 and EuF_2 have the CaF_2 structure; $SmBr_2$ and $EuBr_2$ are isomorphous with $SrBr_2$; and YbI_2 is isomorphous with CaI_2 (and hence with CdI_2).

It would be interesting to discuss the variation in stability of the dipositive oxidation state of the lanthanides from a thermochemical viewpoint, but unfortunately experimental values for the third ionization potentials of these elements are available for only a few of them; lattice energies for the trihalides (the complicated structures of which were mentioned earlier) are also inaccessible. Nevertheless, it seems likely from the very smooth variation in interatomic distances in, say, fluorides and oxides that there is a smooth variation in the difference between the lattice energies of MX_3 and MX_2, and that the third ionization potentials therefore play a decisive part in determining the stabilities of the trihalides with respect to decomposition into dihalides and halogens. Without going into detail here, it can be said that the likely variation in ionization potentials among the lanthanides can be discussed in the same way as for ordinary transition elements (see Chapter 18), i.e. in terms of the effects of increasing nuclear charge and loss of exchange energy on ionization. In this way it can be shown that the loss of exchange energy on removal of an *f* electron from the configuration f^n rises to a maximum at $n = 7$, drops to zero at $n = 8$ and then rises to another maximum. If we are considering the change $M^{2+} \rightarrow M^{3+}$, $n = 7$ corresponds to M = Eu. Thus there should be a sharp drop in the third ionization potential between europium and gadolinium, making $Gd^{3+} + e$ much more stable with respect to Gd^{2+} than $Eu^{3+} + e$ is with respect to Eu^{2+}. Whilst this argument is reasonably convincing so far as preparation of a lower halide by reduction or decomposition is concerned, it is not, of course, a satisfactory treatment of the preparation of a dihalide from a trihalide and a metal, which poses a much more complicated problem. To consider this matter in detail is beyond the scope of this book; but it should be said that, with the aid of some estimated thermochemical data, a discussion of the problem that correctly predicts which elements can form dihalides has recently been given.[6]

22.6 The tetrapositive oxidation state

Only cerium has a tetrapositive oxidation state capable of existence in the presence of water (and then only because the oxidation of water by cerium (IV), though thermodynamically possible, is very slow); the other elements to exhibit this oxidation state do so only in fluorides and (in some cases) oxides. It will therefore be convenient to deal with cerium first.

The only binary compounds of cerium (IV) yet known are CeO_2 (made by heating $Ce(OH)_3$ in air or by oxidation of a solution of a cerium (III) salt with hypochlorite) and CeF_4 (made from CeF_3 and F_2). Several solid complexes are known, however, among them $(NH_4)_2[Ce(NO_3)_6]$ and $(pyH)_2[CeCl_6]$. The solution chemistry of cerium (IV) is rather like that of zirconium (IV) insofar as hydrolytic cationic polymerization and formation of an acid-insoluble phosphate are concerned, but cerium (IV) is also a powerful oxidizing agent. In M perchloric acid, E^0 for the reaction

$$Ce^{4+} + e \rightleftharpoons Ce^{3+}$$

is +1·70 V. Values of E^0 in M HNO_3, H_2SO_4 or HCl are lower, being +1·61, 1·44 and 1·28 V respectively, from which it is deduced that solutions in at least three of the acids contain the cerium in the form of a complex ion. Cerium (IV) sulphate is a familiar reagent in titrimetric analysis, in which it is often used in place of permanganate or dichromate; as a one-electron oxidant it is less prone to undergo side-reactions than the more familiar reagents.

Praseodymium forms PrO_2 (made by heating Pr_2O_3 in oxygen at 500° and 100 atm pressure) and a number of fluoro compounds, including impure PrF_4, Na_2PrF_6 and K_2PrF_6, all made in the dry way. An oxide of approximate composition Pr_6O_{11} is obtained by heating the nitrate, carbonate or hydroxide of praseodymium (III) in air; several other phases intermediate between $PrO_{1.5}$ and $PrO_{2.0}$ also exist in different temperature ranges. Praseodymium (IV) compounds liberate oxygen on treatment with acids, and there are no oxy-salts of the metal in this oxidation state. Terbium (IV) is represented by TbO_2 and TbF_4 (and some ill-defined complexes of the latter) and there is also an intermediate oxide of approximate composition Tb_4O_7; again no oxy-salts are known. For neodymium and dysprosium, fluorination of mixtures of composition 3CsCl : MF_3 yields mixtures believed to contain Cs_3NdF_7 and Cs_3DyF_7 respectively; these are the only derivatives yet made of the metals in a higher oxidation state.

The oxides CeO_2, PrO_2 and TbO_2 all have the fluorite structure; CeF_4 and TbF_4 are isomorphous with UF_4; and $(NH_4)_2[Ce(NO_3)_6]$, as mentioned in Chapter 14, contains a twelve-coordinated metal atom, the six nitrate ions all acting as bidentate ligands.

The discussion in the preceding section concerning the variation in ionization potential with atomic number can be extended to cover the fourth ionization potential. The abrupt fall in the ionization potential for the configuration f^8 then means that terbium should have a much lower fourth ionization potential than gadolinium (i.e. be more easily oxidized). Thus we see the exchange energy factor provides a good explanation of the relative stabilities of europium (II) and terbium (IV). It does not, however, suffice to account for all the other 'anomalous' oxidation states, and until more thermodynamic data are available for all the lanthanides a full thermochemical treatment will be impossible. In the meantime, the qualitative idea of the stability of the half-filled shell remains a useful half-truth.

References for Chapter 22

1 Topp, N. E., *The Chemistry of the Rare-Earth Elements* (Elsevier, 1965).
2 Spedding, F. H., and Daane, A. M. (eds.), *The Rare Earths* (Wiley, 1961).
3 Cotton, F. A., and Wilkinson, G., *Advanced Inorganic Chemistry*, 2nd edn. (Interscience, 1966).
4 Phillips, C. S. G., and Williams, R. J. P., *Inorganic Chemistry* (Oxford University Press, 1966).
5 Wells, A. F., *Structural Inorganic Chemistry*, 3rd edn. (Oxford University Press, 1962).
6 Johnson, D. A., *J. Chem. Soc.*, **A**, 1969, 2578.

Inner transition elements II: the actinides

23

23.1 Introduction[1–3]

At the time of the first presentation of the Periodic Table in its long form the only known elements after actinium were thorium, protactinium and uranium, the chemistry of which resembles that of hafnium, tantalum and tungsten much more closely than that of cerium, praseodymium and neodymium. For many years, therefore, these elements were regarded as members of an ordinary transition series. With the isolation of neptunium and plutonium (which are quite unlike rhenium and osmium), the possibility of there being a second inner transition series began to be discussed, and after the isolation of americium, curium and later elements it became generally accepted that the very heavy elements are part of a series analogous to the lanthanides, the last member of which is element number 103, lawrencium. The recent preparation of element number 104, rutherfordium, and the demonstration that it is almost certainly the homologue of hafnium, now makes it desirable to consider elements numbers 90 to 103 (thorium to lawrencium) inclusive as members of this inner transition series even though the thorium atom in its ground state has no $5f$ electrons.

23.2 Occurrence and preparation of the actinides

The atomic numbers, names, probable electronic structures of the ground state gaseous atoms and ionic radii in the tripositive and tetrapositive states of actinium and the later elements are given in Table 23.1. Thorium and uranium occur naturally in relatively large amounts; actinium and protactinium are present in very small amounts as decay products; and neptunium and plutonium are also found in traces as a result of the slight natural conversion of ^{238}U under the influence of slow neutrons. Neptunium and plutonium are, however, almost entirely made by neutron irradiation of uranium in a pile, as described in Chapter 1.

The very high flux of neutrons in a pile leads to the production of small amounts of higher members of the series by a process of multiple neutron capture. Thus, for example, if a ^{239}Pu target is used the following series of reactions will occur:

$$^{239}Pu(n,\gamma)^{240}Pu(n,\gamma)^{241}Pu(n,\gamma)^{242}Pu(n,\gamma)^{243}Pu \xrightarrow[5\ hr]{\beta} {}^{243}Am$$

$$^{243}Am(n,\gamma)^{244}Am \xrightarrow[26\ min]{\beta} {}^{244}Cm$$

$$^{244}Cm(n,\gamma)^{245}Cm(n,\gamma) \ldots\ldots {}^{249}Cm \xrightarrow[65\ min]{\beta} {}^{249}Bk$$

The process of synthesis involves successive neutron captures until the accumulation of mass renders the nucleus unstable, when β-decay occurs. The berkelium isotope shown above will itself capture neutrons, and the isotopes ^{253}Cf and ^{254}Cf result in turn. As a consequence of multiple neutron capture minute amounts of the higher transuranic elements accumulate with plutonium in the uranium rods during their irradiation in the reactor. It is also interesting to note that einsteinium and fermium were first detected as by-products of a thermonuclear explosion, the result of multiple neutron capture in the very high concentration of neutrons produced.

The transuranic elements may also be synthesized with the aid of high-voltage machines such as the cyclotron; for example,

$$^{238}U(\alpha,n)^{241}Pu \xrightarrow[1\ yr]{\beta} {}^{241}Am \xrightarrow[500\ yr]{\alpha} {}^{237}Np$$

$$^{239}Pu(\alpha,n)^{242}Cm \xrightarrow[5\ mth]{\alpha} {}^{238}Pu$$

Numerous other reactions of this sort have been carried out. The following illustrate the production of some of the higher members of the series: $^{239}Pu(\alpha,2n)\ ^{241}Cm$; $^{243}Am(\alpha,n)\ ^{246}Bk$; $^{242}Cm(\alpha,2n)\ ^{244}Cf$; $^{249}Cf(d,3n)\ ^{248}Es$; $^{249}Cf(\alpha,2n)\ ^{251}Fm$; $^{253}Es(\alpha,n)\ ^{256}Mv$.

An important advance in the technique for producing the very heavy elements entails the use of heavy ion bombardments, especially with ions derived by stripping off all electrons from ^{11}B, ^{12}C,

Table 23.1 *Electronic configurations and radii of actinium and later elements*

Atomic number	*Name*	*Symbol*	*Probable electronic configuration**	*Radii†* M^{3+}	M^{4+}
89	Actinium	Ac	$6d^17s^2$	1·08	
90	Thorium	Th	$6d^27s^2$		0·99
91	Protactinium	Pa	$5f^26d^17s^2$		0·96
92	Uranium	U	$5f^36d^17s^2$	1·00	0·93
93	Neptunium	Np	$5f^57s^2$	0·99	0·92
94	Plutonium	Pu	$5f^67s^2$	0·97	0·90
95	Americium	Am	$5f^77s^2$	0·96	0·89
96	Curium	Cm	$5f^76d^17s^2$	0·95	
97	Berkelium	Bk	$5f^86d^17s^2$	0·93	
98	Californium	Cf	$5f^{10}7s^2$	0·92	
99	Einsteinium	Es	$5f^{11}7s^2$		
100	Fermium	Fm	$5f^{12}7s^2$		
101	Mendelevium	Mv	$5f^{13}7s^2$		
102	Nobelium	No	$5f^{14}7s^2$		
103	Lawrencium	Lw	$5f^{14}6d^17s^2$		
104	Rutherfordium‡	Rf	$5f^{14}6d^27s^2$		

* Only electrons outside the [Rn] core are shown.
† Data from trifluorides and dioxides.
‡ After early claims to have made this element the name kurchatovium was suggested.

^{14}N and ^{16}O.[4] The bombardment of a target by such ions after acceleration in a cyclotron or similar accelerator results in the gain in a single reaction of considerably more nuclear charge and mass than in α-particle bombardment; examples of its application are

$$^{238}U(^{16}O,4n)^{250}Fm \qquad ^{238}U(^{14}N,p + 4n)^{247}Cf$$
$$^{252}Cf(^{11}B,6n)^{257}Lw$$

There are several important factors which limit the range and amounts of the heaviest elements that can be produced by the methods described. If, for example, ^{252}Cf were required as a target it would take 5–10 years to produce about one milligram by multiple neutron capture from one kilogram of ^{239}Pu at a neutron flux of 3×10^{14} neutrons cm^{-2} sec^{-1}; and the yield of product from such a target would be smaller by many orders of magnitude. Fortunately, however, the decay properties and yield of the product can be predicted, and with the aid of ion-exchange resins the separation of a minute amount of product is comparatively easy. Many of the heavier isotopes are also short-lived in the sense of radioactive decay of the normal type (usually α-particle emission) or may undergo spontaneous fission; the longest-lived isotope of fermium, for example, is ^{254}Fm, with half-lives of 3·3 hours and 220 days for α-decay and fission respectively. Thus (to mention the most stable isotopes) ^{241}Am and ^{244}Cm are available in gram quantities, ^{249}Bk, ^{249}Cf and ^{251}Cf in milligrams and ^{253}Es in micrograms; for elements beyond einsteinium the highest amounts yet obtained are a few atoms, and these elements can be studied only in solution.

The present limit to the elements known and studied chemically is rutherfordium, three isotopes of which, ^{257}Rf, ^{259}Rf and ^{261}Rf, are obtained by bombarding ^{249}Cf with ^{12}C and ^{13}C ions and ^{248}Cm with ^{18}O ions respectively; all decay by α-emission, with half-lives of a few seconds. It seems likely that a few more elements may be prepared, but still shorter half-lives may make chemical characterization impossible.

23.3 General properties of the actinides

Although the general resemblance in electronic configuration between the actinides and the lanthanides is clear, there are several ways in which the later inner transition series differs from the earlier one. In the first place, it is only with the second element after actinium that the ground state of the neutral atoms contains *f* electrons; the point at which the binding energy of a 5*f* electron becomes lower than that of a 6*d* electron is thus different from the corresponding point of change for 4*f* and 5*d* electrons. At the beginning of the actinide series the energies of the 5*f*, 6*d*, 7*s* and 7*p* orbitals are comparable, and covalent bonding can involve any of them. That the 5*f* orbitals have greater spatial extension than the 4*f* orbitals is proved by the e.s.r. spectrum of UF_3 in a CaF_2 lattice, which shows fine structure attributable to the interaction of the electron spin of the U^{3+} ion with the fluorine nuclei; NdF_3, the corresponding compound in the lanthanide series, shows no such effect.

The range of oxidation states exhibited by the actinides (particularly the lighter ones) is very much greater than that found among the lanthanides, and high oxidation states involving covalent bonding to oxygen are well established for uranium, neptunium, plutonium and americium. In the second half of the series, the tripositive state becomes the dominant oxidation state, however, and the two metals before the last actinide (mendelevium and nobelium, the homologues of thulium and ytterbium) both have dipositive states; in the case of nobelium this is, indeed, the stable oxidation state, E^0 for the No^{III}/No^{II} system being approximately +1·4 V. Americium, however, the element corresponding in position to europium, is not known to exist as Am^{2+} in any stable compound. All oxidation states known at the present time are shown in Table 23.2; apparently low oxidation states of thorium are discussed in Section 23.5.

The electronic spectra of the actinide ions are sharp like those of

Table 23.2 *Oxidation states of actinium and later elements*

Ac	Th	Pa	U	Np	Pu	Am	Cm	Bk	Cf	Es	Fm	Md	No	Lw	Rf
												2	2		
3		3	3	3	3	3	3	3	3	3	3	3	3	3	
	4	4	4	4	4	4	4	4							4
		5	5	5	5	5									
			6	6	6	6									
				7	7										

the lanthanides, but their intensities are somewhat greater. Though the spectra are less influenced by ligand fields than those of ordinary transition metal ions, complexing does affect them slightly, so that there are small crystal field stabilization energy terms for actinide ions. The magnetic properties of the actinide ions are extremely difficult to interpret, and it seems that there is a partial breakdown of the Russell–Saunders coupling scheme. The general shape of the curve of molar magnetic susceptibility against the number of f electrons is, however, like the corresponding curve for the lanthanides, though the oxidation states involved are quite different; for the actinides the f^1 configuration, for example, is represented by neptunium (VI) or protactinium (IV) rather than thorium (III).

The crystal chemistry of the actinides, in the limited range of compounds that can be studied, shows a general resemblance to that of the lanthanides, a similar contraction in ionic size taking place along the series; as in the case of the lanthanides, the decrements in radius between successive elements become smaller towards the end of the series. Thus all dioxides have the fluorite structure, and only very small variations, or no variations, in structure type are found among the trifluorides, tetrafluorides, tetrachlorides or hexafluorides. The metals, themselves, however, show very complex behaviour; polymorphism is common (plutonium exists in six forms), and many of the structures are peculiar to one element. As for the lanthanides, for a detailed account of the structures of the elements and their compounds the description by Wells[5] should be consulted.

Because the distribution of different oxidation states between the members of the actinide series is so uneven and the techniques of investigation are so different for the earlier and later members, it is unsatisfactory to review their chemistry in terms of oxidation states in the way that served for the lanthanides. We shall, therefore, deal with elements separately or in small groups, and will include also brief accounts of the chemistry of actinium and rutherfordium as the elements before and after the actinides proper. It has to be said that whilst our knowledge of the physical chemistry of uranium, neptunium and plutonium is very extensive, quantitative information about the later actinides is necessarily extremely scanty, and no general discussion of the chemistry of this group of elements in terms of variations in independently determined physicochemical data can yet be given.

23.4 Actinium[1,6]

Actinium occurs in pitchblende (though in much smaller concentrations than radium); it is formed by α-decay of protactinium-231 and was first detected as a radioactive material which accompanied the lanthanide fraction in the chemical separation of the main elements in the mineral. It is now made in milligram amounts by neutron irradiation of ^{226}Ra, the naturally occurring radium isotope:

$$^{226}\mathrm{Ra}(n,\gamma)^{227}\mathrm{Ra} \xrightarrow{\beta} {}^{227}\mathrm{Ac}$$

This isotope decays by two paths shown on the next page.

The separation of actinium is complicated by the difficulty of detecting the element by its radiation. In the decay scheme shown

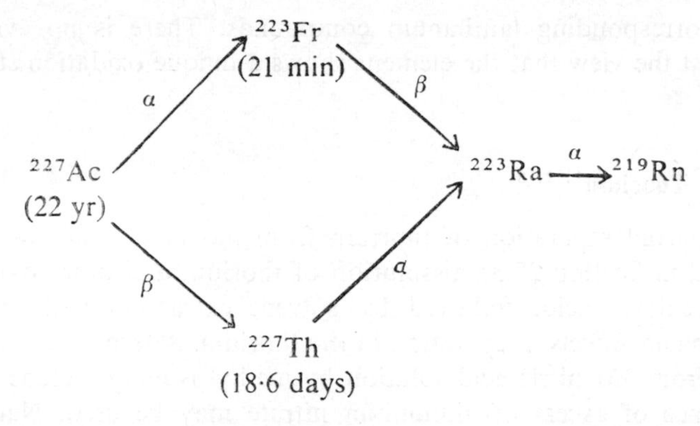

for ^{227}Ac, α-decay occurs to the extent of 1·2 per cent, and was, indeed, detected only after the discovery of francium. The β-particle emitted in the main mode of disintegration is of very low energy and cannot be counted by the usual equipment. The element is, therefore, usually detected by measuring the radioactivity of its daughter elements. This accentuates the need to have the original actinium preparation free from radioactive contamination. Separation of actinium from neutron-irradiated radium involves not only the separation from the target radium but also that from daughter elements of radium and actinium (Th, Po, Bi, Pb), long-lived isotopes of which accumulate during the irradiation. This is achieved by ion-exchange or by preferential solvent extraction into benzene with thenoyltrifluoroacetone (TTA) at pH 6, under which conditions most of the impurities remain in the aqueous layer. The actinium is then recovered in an aqueous phase by washing with 6M hydrochloric acid. Repetitions of this cycle give carrier-free actinium. Ten isotopes are known, the longest-lived of which, ^{227}Ac, of half-life 22 years, is used in all chemical work. The study of even small amounts of actinium is made difficult by the intense γ-radiation of its decay products.

Precipitation of AcF_3 from solutions and reduction with lithium at 1200° gives the silvery-white metal, which closely resembles lanthanum in both physical and chemical properties; it is, however, slightly more basic, as might be expected from its greater ionic radius. Thus it is more strongly absorbed on ion-exchange resins, is less easily extracted from nitric acid solution by tributyl phosphate, and forms oxy-halides AcOX (X = F or Cl) rather than oxides when the halides are heated in water vapour at 1000°. The oxide Ac_2O_3 is produced by igniting the insoluble oxalate in oxygen. The known structures of actinium compounds are the same as those of the corresponding lanthanum compounds. There is no evidence against the view that the element shows a unique oxidation state of three.

23.5 Thorium[1, 2, 6]

The partial separation of thorium from monazite sand was mentioned in Section 22.3; dissolution of thorium and other oxides in hydrochloric acid, followed by solvent extraction with tributyl phosphate, effects a separation of the thorium. Alternatively, extraction from 3M nitric acid solution by methyl isobutyl ketone in the presence of excess of aluminium nitrate may be used. Naturally occurring thorium consists mainly of ^{232}Th, of half-life 10^{10} years; the chemistry of the element is therefore studied almost entirely by conventional methods.

The metal is obtained by reduction of the tetrafluoride with calcium in the presence of zinc chloride as a flux, or by electrolysis of a fused mixture of ThF_4, KCl and NaCl. It may be obtained very pure by thermal decomposition of the vapour of the tetraiodide on a hot tungsten wire.

The reactivity of metallic thorium is comparable to that of the lanthanide metals, but the chemistry is almost entirely that of thorium (IV). E^0 for the Th^{IV}/Th system is $-1{\cdot}9$ V (compared with $-1{\cdot}7$ and $-1{\cdot}5$ V for Hf^{IV}/Hf and Zr^{IV}/Zr respectively), and there is no evidence for a lower oxidation state in aqueous media. A golden-yellow iodide, ThI_3, may be obtained by heating the tetraiodide with the metal, and a sulphide, ThS, is obtained by an analogous method; the conductivity of these substances, however, indicates that they are really derivatives of thorium (IV) with electrons in conduction bands.

Thorium dioxide is converted by carbon tetrachloride at 600° into the anhydrous tetrachloride, which is soluble in water in the presence of hydrochloric acid, and from solutions of which the hydroxide and hydrated fluoride may be precipitated. Ignition of the hydroxide yields ThO_2; the anhydrous fluoride is best made by the action of hydrogen fluoride on the oxide at 600°. The nitrate and sulphate are readily soluble, but the iodate, oxalate and phosphate are insoluble even in dilute mineral acid. A peroxy derivative $Th(O_2)_2SO_4.3H_2O$ of unknown structure is also insoluble in dilute acid. The remarkable borohydride $Th(BH_4)_4$ is a white solid (m.p. 204°) which is stable in dry air and can be sublimed *in vacuo* at 150°; it is obtained by the interaction of aluminium borohydride and thorium tetrafluoride:

$$4Al(BH_4)_3 + 3ThF_4 \rightarrow 3Th(BH_4)_4 + 4AlF_3$$

As might be expected from the high charge, the thorium (IV) ion is extensively hydrolysed in aqueous solution, the main products being $Th(OH)^{3+}$ and $Th(OH)_2{}^{2+}$ (and polymers thereof). A cation chain of composition $Th(OH)_2{}^{2+}$ is found in solid $Th(OH)_2CrO_4.H_2O$ and $Th(OH)_2SO_4$; in it there are two parallel rows of hydroxyl groups with thorium atoms on alternate sides, so that each thorium atom is in contact with four hydroxyl groups and each hydroxyl group with two thorium atoms, giving a structure which may be represented as follows:

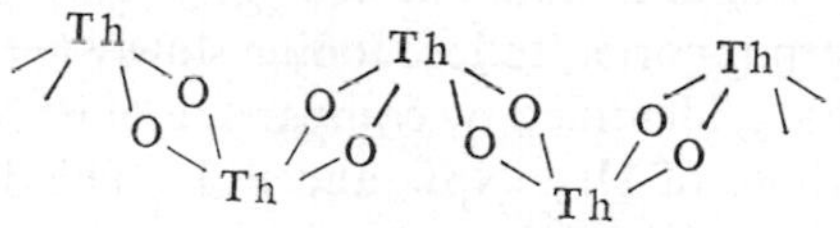

```
  Th             Th             Th
 / \ O        O / \ O        O / \
     O \   / O       O \   / O
        Th              Th
```

Complex formation with anions (e.g. F^-, Cl^-, $NO_3{}^-$, $C_2O_4{}^{2-}$) also takes place in aqueous media.

23.6 Protactinium[1, 6–8]

The naturally occurring isotope ^{231}Pa (half-life $3{\cdot}2 \times 10^4$ years) is present to the extent of about one part in 10^7 in pitchblende, being formed by α-decay of ^{235}U; it is a powerful α-emitter and for this reason (and also because of the extreme tendency of protactinium compounds to hydrolyse in solution, with consequent adsorption on precipitates or containing vessels) is very difficult to work with. The isotope ^{233}Pa, which is a β-emitter of half-life 27 days, is safer to handle and has been used for some work on this element; it is best obtained by a (n,γ) reaction on ^{232}Th in an atomic pile, followed by β-decay of the ^{233}Th which is first formed. Up to 100 g of ^{231}Pa has been separated from pitchblende; among many stages involved in this operation are extraction from HNO_3/HF solution by tributyl phosphate in kerosene, anion-exchange separation from hydrogen chloride solution and precipitation as a peroxide $Pa_2O_9.3H_2O$ from dilute sulphuric acid solution; the action of heat on this compound yields successively hydrated and anhydrous Pa_2O_5.

The action of hydrogen fluoride on Pa_2O_5 gives $PaF_5.2H_2O$, which when heated forms the oxyfluoride Pa_2OF_8; the anhydrous pentafluoride is obtained by fluorination of PaF_4 at 700°, the PaF_4

(a brown solid which can be handled in the atmosphere without hydrolysis or oxidation occurring) being made by passing hydrogen fluoride and hydrogen over the pentoxide at 500°. Lithium at 1300° reduces the tetrafluoride to metallic protactinium, which is reactive and, like uranium, forms a trihydride when heated in hydrogen. The yellow pentachloride, which sublimes *in vacuo* at 180° and is rapidly hydrolysed, is obtained by heating the pentoxide in thionyl chloride vapour; reduction with hydrogen at 800° gives the greenish-yellow tetrachloride. The black pentaiodide may be prepared by iodine exchange between protactinium pentachloride and silicon tetra-iodide; like the pentachloride, it is very unstable towards water but is thermally much more stable than the corresponding uranium compound. Reduction of PaI_5 with aluminium at 400° gives dark-green PaI_4, which has recently been reported to lose iodine slowly *in vacuo* at 300° with formation of PaI_3, identified by comparison of its X-ray powder photograph with those of UI_3, NpI_3 and PuI_3. This is the only preparation of a protactinium (III) compound yet described. Reduction of Pa_2O_5 in hydrogen at high temperatures eventually yields PaO_2, but a number of other oxide phases are formed during this reaction.

Most of the aqueous solution chemistry of protactinium is that of anionic complexes of protactinium (V), though solutions containing protactinium (IV) having absorption spectra very similar to those of cerium (III) ($4f^1$) can be obtained by reduction with zinc amalgam. Such solutions undergo rapid oxidation by air. Protactinium is best kept in aqueous solution as fluoride, oxalate, sulphate, citrate or tartrate complexes. Large numbers of fluoride complexes have been characterized; among them are Na_3PaF_8 (containing a cubic $[PaF_8]^{3-}$ ion), K_2PaF_7 (in which the anion is an infinite chain, each metal atom sharing two pairs of fluorines with other metal atoms) and $RbPaF_6$ (in which there is also an infinite chain anion with each metal sharing two pairs of fluorine atoms, the protactinium atom being eight-coordinated as compared with nine-coordinated in K_2PaF_7). It may be noted that whilst $KPaF_6$, KUF_6, $RbUF_6$ and $CsPaF_6$ are isostructural with $RbPaF_6$, $CsUF_6$ contains a discrete $[UF_6]^-$ ion. Chlorocomplexes such as $CsPaCl_6$ may be made by interaction of the component halides in thionyl chloride solution, but this method fails for complexes of protactinium (IV) owing to oxidation by the solvent, and oxygen-free concentrated hydrochloric acid must be used as solvent in the preparation of Cs_2PaCl_6. By the action of liquid N_2O_5 on $CsPaCl_6$ the complex nitrate $Cs[Pa(NO_3)_6]$

is obtained; niobium and tantalum yield oxytetranitrato complexes $CsM^{V}O(NO_3)_4$, whilst uranium does not form nitrato complexes at all under similar conditions (uranium does, of course, form them under other conditions).

In summary, it is seen that there is very little chemical similarity between protactinium and actinium or thorium, but a close resemblance to niobium and tantalum; only a few special points provide evidence that the element is a member of an inner transition series.

23.7 Uranium[1, 2, 6]

Uranium and the three following elements form a group in which, for the first time in the actinide series, close resemblances in properties are found. We shall describe the chemistry of uranium first and then discuss that of neptunium, plutonium and americium, with special emphasis on comparisons with uranium.

Several aspects of the nuclear chemistry of uranium were mentioned in Chapter 1. The naturally occurring isotopes, ^{238}U (99·3 per cent), ^{235}U (0·7 per cent) and ^{234}U (0·005 per cent), have half-lives of $4{\cdot}5 \times 10^9$, 7×10^8 and $1{\cdot}6 \times 10^5$ years respectively; all are α-emitters. Although uranium compounds are highly toxic and all unnecessary exposure to radioactive materials should be avoided, the activity of natural uranium is not such as to introduce complications into the study of its chemistry.

The richest source of uranium is pitchblende, an oxide of approximate composition U_3O_8. Uranium may be taken into solution by the action of either sulphuric acid and an oxidizing agent (air at 130° will serve) or a carbonate/bicarbonate mixture and compressed air at 100°; the products of these operations are uranyl sulphate (UO_2SO_4) or the anion $[UO_2(SO_4)_2]^{2-}$ and the complex carbonate ion $[UO_2(CO_3)_3]^{4-}$ respectively. Subsequent stages may involve the use of anion-exchange resins or solvent extraction of uranyl nitrate tetrahydrate from nitric acid solution with diethyl ether or methyl isobutyl ketone, or of uranium (IV) in some ill-defined form with tributyl phosphate in kerosene. Washing the organic extract with water results in the return of uranium to the aqueous layer, from which it may be precipitated as ammonium diuranate ($(NH_4)_2U_2O_7$) or the peroxide, $UO_2(O_2).2H_2O$; both of these substances yield successively yellow UO_3 and then greenish-black U_3O_8 when they are heated. The third oxide, black UO_2, is made by reduction of UO_3 with hydrogen at 700° or carbon monoxide at 350°. The trioxide gives rise to both uranyl salts, containing the linear UO_2^{2+} cation, and a range of 'uranates', some (but only some) of which contain discrete oxy-anions; the dioxide is exclusively basic. The dioxide has the fluorite structure; UO_3 is dimorphic, one form having the ReO_3 structure (based on ReO_6 octahedra sharing all corners) whilst the other is effectively $(UO_2)O$. The uranyl group is also discernible in the structure of U_3O_8, a mixed UO_2^{2+}/U^{4+} oxide. Other oxide phases, some non-stoichiometric, are also known, but the detailed discussion of the uranium–oxygen system lies beyond the scope of this book.

Metallic uranium (m.p. 1132°) is obtained by reduction of the tetrafluoride (made by the action of hydrogen fluoride on the dioxide) with calcium or magnesium. It is trimorphic; the α form, stable at ordinary temperatures, has a unique corrugated sheet structure in which each atom has only four nearest neighbours. Uranium combines directly with most other elements; the finely divided metal is pyrophoric in air. Hydrogen reacts at 250° to form the hydride UH_3 (described in Chapter 8), which is readily converted into other compounds (e.g. into UCl_3 by HCl at 250°, UF_4 by HF at 400° and UCl_4 by chlorine at 400°). Fluorine and the metal give UF_6, chlorine UCl_4, UCl_5 or UCl_6 according to conditions, bromine UBr_4, and iodine UI_3 or UI_4. The finely divided metal decomposes boiling water with formation of uranium dioxide and hydrogen; the hydrogen in turn forms some uranium hydride. Steam at 150° reacts vigorously to give the dioxide and the hydride.

The fluorides are among the most important compounds of uranium, the colourless volatile hexafluoride (which has a vapour pressure of 1 atm at 56°) because of its use in the separation of isotopes, and the green tetrafluoride as an intermediate in the preparation of the metal. The hexafluoride is commonly made by the action of fluorine or chlorine trifluoride on the tetrafluoride, though it is also obtained by the action of oxygen at 800°:

$$2UF_4 + O_2 \rightarrow UF_6 + UO_2F_2$$

Since UO_2F_2 can be reduced to UO_2 by the action of hydrogen, and the UO_2 reconverted into UF_4 by hydrogen fluoride, this reaction provides the basis for a preparation of UF_6 without the use of elemental fluorine. Uranium hexafluoride, which behaves as an ideal gas, forms a molecular lattice of perfectly symmetrical octahedral molecules. It is rapidly hydrolysed, and acts as a vigorous fluorinating agent towards organic compounds; it combines with alkali metal fluorides under ill-defined conditions to yield complexes of formula M^IUF_7 or $M_2{}^IUF_8$. Uranium tetrafluoride is a non-volatile inert solid, very sparingly soluble in water, from which it may be precipitated as a dihydrate. Steam hydrolyses it to UO_2 and HF, a reversal of the usual method of preparation. A number of fluorides intermediate between UF_4 and UF_6 are also known, and a lower fluoride, UF_3, is obtained by heating the tetrafluoride with the metal at 900°. Although UF_5, made from UF_4 and UF_6, is unstable with respect to disproportionation, the $[UF_6]^-$ ion is stable in concentrated aqueous or anhydrous hydrofluoric acid, and salts such as $Cs[UF_6]$ may be isolated from either medium.

The chlorides UCl_3, UCl_4, UCl_5 and UCl_6 are all known, their syntheses being summarized by the following scheme:

$$UO_2 \xrightarrow[500^\circ]{CCl_4} UCl_4$$

$$UCl_3 \xleftarrow[300^\circ]{Al} UCl_4 \xrightarrow[500^\circ]{Cl_2} UCl_5 \xrightarrow[150^\circ]{heat} UCl_4 + UCl_6$$

The trichloride, like the other trihalides, is essentially salt-like in character; the tetrahalides are somewhat more easily fusible, and the hexachloride is readily volatile.

The uranyl ion in aqueous solution is hydrolysed to some extent, with formation of species such as $U_2O_5{}^{2+}$ and $U_3O_8{}^{2+}$. The reduction chemistry of the ion is summarized conveniently in the form of the following potential diagram (see Section 6.5 for a discussion of such diagrams). The values here, as usual in the absence of any indication to the contrary, relate to unit hydrogen ion concentration.

$$UO_2{}^{2+} \xrightarrow{+0{\cdot}06} UO_2{}^{+} \xrightarrow{+0{\cdot}58} U^{4+} \xrightarrow{-0{\cdot}63} U^{3+} \xrightarrow{-1{\cdot}80} U$$

$$UO_2{}^{2+} \xrightarrow{+0{\cdot}32} U^{4+}$$

This diagram expresses the fact that the $UO_2{}^+$ ion is unstable with respect to the disproportionation

$$2UO_2{}^+ + 4H^+ \rightleftharpoons UO_2{}^{2+} + U^{4+} + 2H_2O$$

for which K is approximately 10^6 (compare $NpO_2{}^+$ and $PuO_2{}^+$). It also provides a background against which the descriptive chemistry of uranium in aqueous media may be interpreted. Reduction of solutions of $UO_2{}^{2+}$ with zinc, for example, gives U^{4+} and some U^{3+}, but the latter may be converted into the former by blowing air through the solution; oxidation by a great variety of oxidizing agents (e.g. Ce^{IV}, MnO_2, H_2O_2, PbO_2 and $Cr_2O_7^{2-}$) then regenerates UO_2^{2+}.

The mechanisms of some of these redox reactions have been studied with interesting results. When UO_2^{2+} is reduced by Cr^{2+} the bright-green ion $[(H_2O)_5CrOUO(H_2O)_n]^{4+}$ appears to be formed as an intermediate on the way to Cr^{3+} and U^{4+}. In the oxidation of U^{4+} by PbO_2, H_2O_2 or MnO_2, the oxygen in the UO_2^{2+} produced has been shown by ^{18}O tracer studies to come from the oxidant rather than from the water (with which UO_2^{2+} exchanges ^{18}O only very slowly). Further, although the U^{4+}/U^{3+} and UO_2^{2+}/UO_2^{+} systems are reversible, UO_2^{+}/U^{4+} and UO_2^{2+}/U^{4+} are not; electron transfer here is a fast process, but U—O bond fission or formation a relatively slow one.

We have already mentioned the uranyl complex sulphate and carbonate anions; stable complexes are also formed with acetate and, especially, fluoride in aqueous media. As in the case of other actinides, little is known about complexing of either UO_2^{2+} or U^{4+} with nitrogen-, phosphorus- or sulphur-containing ligands, but this may well be because few studies in non-aqueous media have yet been described. Within the framework of the class 'a' and class 'b' division of cations, however, those derived from uranium (and from the other actinides) are clearly in class 'a'.

23.8 Neptunium, plutonium and americium[1, 2, 6]

These three elements show a close relationship, with the stability of high oxidation states decreasing as the atomic number increases. This factor is so important in the separation of the elements that it is desirable to begin this section by giving the potential diagrams for all three elements before describing individual compounds. Mention is made later of compounds containing neptunium (VII) and plutonium (VII) for which accurate data are not yet available.

We shall discuss disproportionation later; for the present it will suffice to point out that all the MO_2^{2+} ions are much less stable to reduction than UO_2^{2+} (though there is only a small difference between NpO_2^{2+} and PuO_2^{2+}), that Pu^{4+} is much more easily reduced than Np^{4+}, and that all higher oxidation states of americium are powerful oxidizing agents. Solvent-extraction and precipitation may then be used to effect separations from solutions of different oxidation potentials. The nitrates $MO_2(NO_3)_2$ may be extracted

$$NpO_2^{2+} \xrightarrow{+1{\cdot}14} NpO_2^{+} \xrightarrow{+0{\cdot}74} Np^{4+} \xrightarrow{+0{\cdot}15} Np^{3+} \xrightarrow{-1{\cdot}83} Np$$

$$NpO_2^{2+} \xrightarrow{+0{\cdot}94} Np^{4+}$$

$$PuO_2^{2+} \xrightarrow{+0{\cdot}91} PuO_2^{+} \xrightarrow{+1{\cdot}17} Pu^{4+} \xrightarrow{+0{\cdot}98} Pu^{3+} \xrightarrow{-2{\cdot}03} Pu$$

$$PuO_2^{2+} \xrightarrow{+1{\cdot}04} Pu^{4+} \qquad PuO_2^{2+} \xrightarrow{+1{\cdot}03} Pu^{3+}$$

$$AmO_2^{2+} \xrightarrow{+1{\cdot}60} AmO_2^{+} \xrightarrow{+1{\cdot}04} Am^{4+} \xrightarrow{+2{\cdot}44} Am^{3+} \xrightarrow{-2{\cdot}32} Am$$

$$AmO_2^{+} \xrightarrow{+1{\cdot}74} Am^{3+} \qquad AmO_2^{2+} \xrightarrow{+1{\cdot}69} Am^{3+}$$

from nitrate solutions by diethyl ether or methyl isobutyl ketone; or the elements in the tetrapositive state may be extracted from 6M nitric acid solution with tributyl phosphate in kerosene. Only the tetra- and tripositive states can be precipitated (or coprecipitated) as fluorides; and coprecipitation with zirconium phosphate (or, rather surprisingly, bismuth phosphate) in acidic solution is almost specific for the tetrapositive states.

In the preparation of ^{239}Pu (of half-life $2{\cdot}4 \times 10^4$ years) in the pile by the sequence of reactions

$$^{238}U(n,\gamma)^{239}U \xrightarrow[23\ \text{min}]{\beta} {}^{239}Np \xrightarrow[2{\cdot}3\ \text{days}]{\beta} {}^{239}Pu$$

the short half-life of ^{239}Np means that it is never present in large amounts, and its concentration decays rapidly after irradiation ceases; however, ^{237}Np (of half-life $2{\cdot}2 \times 10^6$ years) is formed in small amounts by two successive (n,γ) reactions of ^{235}U and subsequent β-decay of ^{237}U. Dissolution of the irradiated uranium rods in nitric acid thus gives a solution containing UO_2^{2+}, NpO_2^{2+} and PuO_2^{2+} (as nitrates) and fission products. Solvent-extraction at this stage removes the actinides into the organic phase, from which neptunium and plutonium may be re-extracted by washing with an aqueous solution of a mild reducing agent, usually sulphur dioxide. Separation of neptunium and plutonium may be effected by bromate oxidation (which converts neptunium to NpO_2^{+}, but does not oxidize plutonium beyond Pu^{4+}) or hydrazine reduction (which converts Pu^{4+} to Pu^{3+}, but does not affect Np^{4+}), followed by precipitation of Pu^{4+} or Np^{4+}. Several other separation methods have been used, but all depend upon similar principles.

There are two isotopes of americium of half-life long enough for conventional studies, ^{241}Am (458 years) and ^{243}Am (7600 years). Both are obtained from ^{239}Pu by successive neutron captures to give ^{241}Pu and ^{243}Pu, which then undergo β-decay to give the americium isotopes. Separation of americium is readily achieved by oxidation of plutonium to PuO_2^{2+} and precipitation of AmF_3 or coprecipitation with LaF_3.

With these elements the effects of α-emission on the study of their chemistry become appreciable; there is, for example, considerable radiation-induced decomposition of water into hydrogen atoms and hydroxyl radicals, whose formation may lead to reduction of high oxidation states and production of hydrogen peroxide respectively. Thermal effects and radiation damage to solids become serious as half-lives decrease. Furthermore, the health hazards associated with working with strong α-emitters are very considerable unless rigid safety precautions are adopted; all of the work described in this and the following sections therefore has to be carried out in isolated systems.

The metals neptunium, plutonium and americium are all obtained by reduction of their fluorides with lithium or other electropositive metals at 1000–1200°. In general reactivity they resemble uranium, reacting readily with acids and with hydrogen. The stable oxides of all three elements are the dioxides, which are obtained by heating the nitrates or hydroxides of any oxidation states in air. Hydrated trioxides of neptunium and plutonium have been obtained by the action of ozone on the tetrahydroxides, but anhydrous trioxides are not known; Np_3O_8 is obtained by thermal decomposition of $(NH_4)_2Np_2O_7$, but Pu_3O_8 is less stable and has not been isolated.

A similar variation is found in the stabilities of halides. Neptunium and plutonium both form volatile hexafluorides closely resembling UF_6, but their preparation requires the use of fluorine at higher temperatures and the products are less stable with respect to decomposition into tetrafluoride and fluorine. Neptunium forms a tetrachloride and a tetrabromide, but neither compound can be isolated in the case of plutonium or americium, though chloro complexes of plutonium (IV), such as Cs_2PuCl_6, can be made from metal chlorides, plutonium trichloride and chlorine at 50°. The only halides, binary or complex, of americium (IV) are the tetrafluoride, obtained from the trifluoride and fluorine at 500°, and its complexes. A further interesting illustration of the decreasing stability of fluorine compounds of elements after uranium in high oxidation states is the action of NpF_6 and PuF_6 on sodium fluoride; elemental fluorine is liberated and complexes of neptunium (V) and plutonium (V) are formed. In anhydrous hydrogen fluoride, however, $CsNpF_6$ disproportionates, forming Cs_2NpF_6 and NpF_6. In aqueous solutions, on the other hand, neptunium (V), present in NpO_2^+, is stable at pH 0 with respect to disproportionation, though the margin of stability is not large, K for the reaction

$$2NpO_2^+ + 4H^+ \rightleftharpoons NpO_2^{2+} + Np^{4+} + 2H_2O$$

being 10^{-7}. At higher acidities, NpO_2^+ becomes unstable, and in 8M acid K is about 200.

In the case of plutonium, successive potentials for the reduction of PuO_2^{2+} are so similar that PuO_2^{2+}, PuO_2^+, Pu^{4+} and Pu^{3+} are all present in M $HClO_4$ solution. Hydrolysis and complex formation also occur, and since kinetic factors are often involved too (the PuO_2^+/Pu^{4+} system, for example, is not reversible), the interpretation of equilibria and kinetics in plutonium-containing solutions is a matter of great difficulty.

The highest oxidation states of all three elements, neptunium (VII), plutonium (VII) and americium (VI), are attained in oxygen-containing ions; NpO_5^{3-} and PuO_5^{3-} are obtained by oxidation of neptunium (VI) and plutonium (VI) in alkaline solution with ozone, hypochlorite or perdisulphate, or electrolytically, and may be isolated as Ba^{2+} or $[Co(NH_3)_6]^{3+}$ salts.[9] Alternatively, Li_5NpO_6 and Li_5PuO_6 may be made by heating Li_2O and the actinide dioxide in oxygen at 400°. For the system Np^{VII}/Np^{VI} E^0 has been estimated as approximately +2·0 V, which would make neptunium (VII) as powerful an oxidizing agent as $S_2O_8^{2-}$ or $Co^{3+}(aq)$. The conversion of americium (III) into AmO_2^+ or AmO_2^{2+} requires the use of hypochlorite, ozone or perdisulphate; characteristic compounds are $Rb(Am^{V}O_2)(CO_3)$ and $Na(Am^{VI}O_2)(CH_3COO)_3$. The use of alkaline media for the preparation of compounds of metals in high oxidation states follows naturally from consideration of the fundamental equilibria involved: conversion of M^{n+} into an oxygen-coordinated species $[M^{(n+2)+}(O^{2-})_m]^{(n+2-2m)+}$ may be represented by the general equation

$$M^{n+} + mH_2O \rightleftharpoons [M^{(n+2)}(O^{2-})_m]^{(n+2-2m)+} + 2mH^+ + 2e$$

whence the part played by removal of hydrogen ions in bringing about the increase in oxidation state is immediately apparent.

23.9 Curium and later elements[1, 2, 6]

The only remaining elements for which solid compounds have been prepared up to the present time are curium, berkelium, californium and einsteinium. The separation of these elements, however, involves the same problems as that of the later ones, so we shall consider curium first in some detail and then refer only briefly to the heavier elements.

The curium isotope ^{242}Cm (an α-emitter of half-life 162 days) may be obtained by an (α,*n*) reaction on ^{239}Pu or an (*n*,γ) reaction on ^{241}Am, followed by β-decay. As a consequence of this short half-life, the energy evolved by radioactive decay is as high as 122 J (25 cal) sec^{-1} g^{-1}, and observation of the existence of curium (IV) compounds is impossible. For chemical work, ^{244}Cm, with a half-life of 19 years, is much more satisfactory; it is obtained from ^{239}Pu by multiple neutron capture followed by decay of the ^{244}Am so produced. Curium exhibits only the tripositive oxidation state in solution, and hence orthodox chemical separation of it from americium and fission products (in which lanthanides are prominent) is impossible, and ion-exchange methods must be used. (Solvent-extraction methods may provide an alternative but have so far been much less employed.)

Two problems have to be solved: the separation of the actinides from the lanthanides and their subsequent separation from one another. The former process is effected by adsorbing the mixture of tripositive ions on a cation-exchange resin and selectively eluting the actinides, which form more stable chloride complexes, with 13 M hydrochloric acid. This also gives partial separation of the actinides from one another, but better results are obtained by dilution, readsorption and elution with citrate or a similar anion; α-hydroxy-isobutyrate has been used with conspicuous success, the elution order being as shown in Fig. 23.1. This was drawn before the isolation of nobelium and lawrencium, and it is now known that the predictions for these elements have been fulfilled. (With other complexing agents less regular elution orders are sometimes found.) After the ion-exchange separation, elements present in sufficient concentration may be precipitated as trifluorides or oxalates and thence converted into a few other compounds; in the case of curium, for

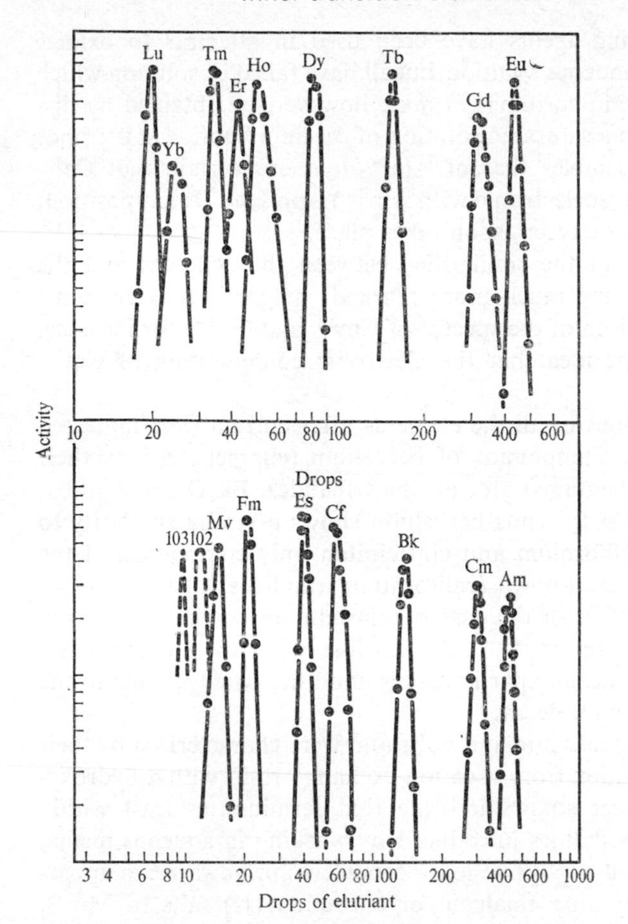

Figure 23.1 Elution of tripositive lanthanide ions (upper curves) and homologous tripositive actinide ions (lower curve) from Dowex-50 cation exchange resin with ammonium α-hydroxy-isobutyrate eluant. The predicted elution positions of elements 102 and 103 are indicated in the lower portion of the figure by broken lines

example, this may be done in the ways indicated in the following schemes:

$$Cm \xleftarrow[1250°]{Ba} CmF_3 \xrightarrow[400°]{F_2} CmF_4$$

$$Cm_2(C_2O_4)_3 \xrightarrow[O_2/O_3]{\text{heat in}} CmO_2 \xrightarrow{\text{heat}} Cm_2O_3 \xrightarrow{HCl} CmCl_3$$

Many oxidizing agents have been used in attempts to oxidize curium (III) in aqueous solution, but all have failed. A solution which appears to contain curium (IV) may, however, be obtained by dissolving CmF_4 in a saturated solution of caesium fluoride at 0°; since its spectrum resembles that of Am^{3+} it seems certain that Ce^{4+} (which would be isoelectronic with Am^{3+}) is present. Decomposition, with liberation of oxygen, soon takes place.

From this point the similarities between the actinides and the lanthanides become much more marked; magnetic measurements and the comparison of the spectra of Cm^{3+} and Gd^{3+}, for instance, both support the idea that the electronic configuration of Cm^{3+} is $5f^7$.

The preparations of all the elements after curium were indicated in Section 23.1. Compounds of berkelium (characterized by their X-ray powder diagrams) are: all the trihalides, Bk_2O_3, and BkF_4, BkO_2 and Cs_2BkCl_6. Thus berkelium shows a strong similarity to terbium. For californium and einsteinium only compounds of the tripositive state are known (californium trihalides, Cf_2O_3, CfOBr; and $EsCl_3$, EsOCl). In the case of einsteinium compounds, X-ray powder photographs have to be taken just below the melting point, when thermal annealing partly makes good the damage done to the lattice by radioactive decay.

Fermium, mendelevium and nobelium were characterized by their behaviour on elution from a cation-exchange resin with α-hydroxyisobutyrate. Tracer studies indicate that fermium, at least within the range of possibilities introduced by working in aqueous media, behaves only as a tripositive ion. Mendelevium, however, is apparently reduced by zinc amalgam or chromium (II) salts to Md^{2+}, since after reduction it is carried by $BaSO_4$ and is not extractable under the conditions used for tripositive actinides with 2-ethylhexylphosphoric acid as extractant; no evidence has been obtained for the existence of mendelevium (IV). Nobelium behaves as a dipositive ion so far as solubility relationships are concerned, unless it is oxidized by cerium (IV); after oxidation is is carried by lanthanum compounds.[10] This suggests that No^{2+} is the stable oxidation state in solution, and chemical evidence indicates E^0 for the No^{III}/No^{II} system is perhaps as high as $+1{\cdot}4$ V. (The value suggested for Md^{III}/Md^{II} is about zero.) If these observations are to be reconciled with the ion-exchange evidence for the identification of nobelium, the solutions used in the latter experiments must have contained the element in its higher oxidation state. Lawrencium, on the actinide hypothesis the analogue of lutecium, cannot be reduced in aqueous media.

The design, execution and interpretation of tracer experiments become ever more difficult with increase in atomic number, but strong evidence has recently been put forward to show that rutherfordium is not an actinide. The elution of the activity attributed to this element from a cation exchange resin by α-hydroxyisobutyrate is quite different from that of the actinides, and closely resembles the behaviour of zirconium and hafnium under comparable conditions.[11] The general features of the classification of the post-radon elements are therefore now clearly established.

References for Chapter 23

1 Katz, J. J., and Seaborg, G. T., *The Actinide Elements* (Methuen, 1957).
2 Cotton, F. A., and Wilkinson, G., *Advanced Inorganic Chemistry*, 2nd edn. (Interscience, 1966).
3 Asprey, L. B., and Cunningham, B. B., *Prog. Inorg. Chem.*, 1960, **2**, 267.
4 Keller, C., *Angew. Chem. (International)*, 1965, **4**, 903.
5 Wells, A. F., *Structural Inorganic Chemistry*, 3rd edn. (Oxford University Press, 1962).
6 Bagnall, K. W., in *Halogen Chemistry*, ed. V. Gutmann, Vol. III (Academic Press, 1967).
7 Brown, D., *Adv. Inorg. Chem. Radiochem.*, 1969, **12,** 1.
8 Brown, D., and Maddock, A. G., *Quart. Rev. Chem. Soc.*, 1963, **17,** 289.
9 Spitzyn, V. I., Gelman, A. D., Krot, N. N., Mefodiyeva, M. P., Zakharova, F. A., Komkov, Y. A., Shilov, V. P., and Smirnova, I. V., *J. Nuclear Inorg. Chem.*, 1969, **31,** 2733.
10 Maly, J., Sikkeland, T., Silva, R., and Ghiorso, A., *Science*, 1968, **160,** 1114.
11 Silva, R., Harris, J., Nurmia, M., Eskola, K., and Ghiorso, A., *Inorg. Chem. Letters*, 1970, **6,** 871.

Index